Tunneling Systems
in Amorphous
and Crystalline Solids

Springer
*Berlin
Heidelberg
New York
Barcelona
Budapest
Hong Kong
London
Milan
Paris
Singapore
Tokyo*

Pablo Esquinazi (Ed.)

Tunneling Systems in Amorphous and Crystalline Solids

With 197 Figures

 Springer

Editor

Professor Dr. Pablo Esquinazi
Universität Leipzig
Fakultät für Physik und Geowissenschaften
Abteilung Supraleitung und Magnetismus
Linnéstrasse 5
D-04103 Leipzig, Germany

Library of Congress Cataloging-in-Publication Data

Tunneling systems in amorphous and crystalline solids / Pablo Esquinazi, (ed.) p. cm.
ISBN 3-540-63960-8 (alk. paper)
1. Solid state physics. 2. Tunneling (Physics) 3. Amorphous substances. 4. Polycrystals.
I. Esquinazi, Pablo.
QC176.8.T8T86 1998
530.4'16--dc21 98-6386 CIP

ISBN 3-540-63960-8 Springer-Verlag Berlin Heidelberg New York

This work is subject to copyright. All rights are reserved, whether the whole or part of the material is concerned, specifically the rights of translation, reprinting, reuse of illustrations, recitation, broadcasting, reproduction on microfilm or in any other way, and storage in data banks. Duplication of this publication or parts thereof is permitted only under the provisions of the German Copyright Law of September 9, 1965, in its current version, and permission for use must always be obtained from Springer-Verlag. Violations are liable for prosecution under the German Copyright Law.

© Springer-Verlag Berlin Heidelberg 1998
Printed in Germany

The use of general descriptive names, registered names, trademarks, etc. in this publication does not imply, even in the absence of a specific statement, that such names are exempt from the relevant protective laws and regulations and therefore free for general use.

Typesetting: Camera-ready copy by editor
Cover design: *design & production* GmbH, Heidelberg
SPIN 10567452   57/3144 – 5 4 3 2 1 0 – Printed on acid-free paper

Preface

The aim of this book is to provide a wealth of information on the influence of quantum-tunneling systems on the low-temperature properties of solids. The nineteen contributors of this book review thermodynamic, acoustic, dielectric and optical properties of amorphous and crystalline solids in eight chapters. The newest experiments on long-time spectral diffusion and heat release, electric field and strain effects, as well as nonequilibrium phenomena are thoroughly discussed. Several theoretical contributions provide an overview of the most original ideas and theoretical methods used nowadays to understand the nature and dynamics of tunneling systems in solids and their interactions. The book provides full description on the experimental and theoretical details of the relevant experiments and theories.

Within the broad spectrum of properties that are related to the contribution of tunneling systems, I have selected special topics which will be of value for experienced as well as graduate students working on low-temperature research. Most of the chapters serve as useful introductory surveys, since they are not too specialized.

I wish to express my thanks to the following colleagues for their important comments on several chapters of this book: Ansel C. Anderson, Guy Bellessa, Dimitri Parshin, Roman Personov, Robert Silbey, James Skinner, Ulrich Weiss, and Helmut Wipf. I would like to thank Mrs. Annette Setzer for her help with the reference list, as well as to the co-workers at the Department of Superconductivity and Magnetism, Universität Leipzig, for their support.

Finally, I would like to thank the eighteen co-authors for their valuable contributions, their just-in-time finishing of the review chapters, and for several comments and ideas which influenced the final character of this book.

Leipzig P. Esquinazi
April 1998

Table of Contents

1. **Introduction** (P. Esquinazi) 1
   1.1 Tunneling Systems 1
   1.2 Content and Organization of the Book 3

2. **Heat Release in Solids**
   (A. Nittke, S. Sahling, and P. Esquinazi) 9
   2.1 A Simple System with Two Levels of Energy 9
   2.2 Phenomenological Theory for the Heat Release 12
       2.2.1 Generalities 12
       2.2.2 The Standard Tunneling Model 14
       2.2.3 The Time and Temperature Dependence
             of the Specific Heat 18
       2.2.4 Influence of a Finite Number of Tunneling Systems .. 23
       2.2.5 Influence of High-Order Tunneling Processes
             and a Finite Cooling Rate 24
   2.3 The Heat Release Within the Soft-Potential Model 27
       2.3.1 The Heat Release and Specific Heat 30
       2.3.2 Influence of Thermal Activation 31
   2.4 Experimental Details 35
       2.4.1 Quasi-static Measurements 35
       2.4.2 Calorimetric Measurements 37
   2.5 Experimental Results 38
       2.5.1 The Time Dependence of the Heat Release 38
       2.5.2 The Temperature Dependence of the Heat Release ... 44
       2.5.3 Influence of Thermal Activation 45
       2.5.4 Correlation Between the Heat Release
             and Other Low-Temperature Properties 54
   2.6 Conclusion and Outlook 55

3. **Crossover to Phonon-Assisted Tunneling
   in Insulators and Metals** (A. Würger) 57
   3.1 Introduction 57

3.2	The Spin–Boson Model		60
3.3	Polaron Transformation and Phonon Dressing		66
	3.3.1	Break-Down of Perturbation Theory	67
	3.3.2	Canonical Transformation	68
	3.3.3	Phonon Dressing	69
	3.3.4	Time Evolution	71
3.4	Crossover to Incoherent Tunneling		71
	3.4.1	Noninteracting Blip Approximation	72
	3.4.2	Nearest-Neighbor Blip Interactions	74
	3.4.3	Time Evolution in NIBA	77
	3.4.4	Two-State Dynamics Beyond NIBA	83
	3.4.5	The Undressing Effect	86
	3.4.6	Discussion	87
3.5	Phonon-Assisted Tunneling in Metals		91
	3.5.1	Blip Expansion for Zero Asymmetry	91
	3.5.2	Coherent Motion	93
	3.5.3	The Incoherent Rate	94
	3.5.4	Quantum Diffusion of Trapped Hydrogen in Niobium	98
	3.5.5	Rate Equations for Large Asymmetry Energy	100
	3.5.6	Resistance Fluctuations of Mesoscopic Wires	103
	3.5.7	Discussion	105
3.6	Phonon Dressing in Real Systems		108
3.7	Asymmetric Tunneling Systems		110
	3.7.1	Projection Method	111
	3.7.2	Approximations	112
	3.7.3	The Damping Kernel	114
	3.7.4	Crossover to Relaxation	117
	3.7.5	Low Temperatures: $T \ll T^*$	117
	3.7.6	High Temperatures: $T \gg T^*$	118
	3.7.7	How Large is the Maximum Tunnel Energy in Glasses?	119
	3.7.8	Sound Propagation in Amorphous Solids Above 5 K	122
3.8	Two-State Dynamics for Weak Phonon Coupling		125
	3.8.1	Perturbation Series	126
	3.8.2	Phase Relaxation: $W(z)$	129
	3.8.3	Energy Relaxation: $V(z)$	131
	3.8.4	Discussion	132
	3.8.5	Mode-Coupling Approximation (MCA)	133
	3.8.6	Comparison of Perturbation Theory and MCA	136
3.9	Summary		139

4. **Influence of Tunneling Systems on the Acoustic Properties of Disordered Solids** (P. Esquinazi and R. König) 145
   4.1 Acoustic Properties and Tunneling Systems 145
   4.2 Theoretical Remarks 147
       4.2.1 Resonant and Relaxation Processes 147

	4.2.2	The Standard Tunneling Model. Relaxation due to Phonons	150
	4.2.3	Relaxation due to Conduction Electrons	154
	4.2.4	Influence of the Acoustic Intensity	161
	4.2.5	Coherent Coupling Below 100 mK	165
	4.2.6	Acoustic Properties Above 1 K: Thermal Activation and Incoherent Tunneling	166
4.3	Experimental Details		168
	4.3.1	Experimental Methods for Low and High Frequencies	168
	4.3.2	The Vibrating Reed and Vibrating Wire Techniques	170
	4.3.3	The Influence of the Clamping	174
	4.3.4	Acoustic Experiments at Very Low Temperatures: Cryogenics and Sample Thermalization	176
4.4	Acoustic Properties of Amorphous Solids		178
	4.4.1	Dielectrics	178
	4.4.2	Normal-Conducting Amorphous Metals	185
	4.4.3	Superconductors	191
	4.4.4	Influence of Thermal Treatment on the Acoustic Properties of Amorphous Metals	194
	4.4.5	Amorphous Thin Films	197
4.5	Acoustic Properties of Polycrystalline Metals		199
	4.5.1	General Remarks	199
	4.5.2	Polycrystalline Superconductors	200
	4.5.3	Normal Metals. The Absence of Electron-Assisted Relaxation in Polycrystals	207
	4.5.4	The Influence of Thermal Treatment	215
	4.5.5	Acoustic Properties of Polycrystals at $T > 1K$	217
4.6	On the Origin of Tunneling Systems in Disordered Solids: Conclusion and Perspective		219

5. Interactions Between Tunneling Defects in Amorphous Solids (A. L. Burin, D. Natelson, D. D. Osheroff, and Yu. Kagan) 223

	5.0.1	Dielectric and Acoustic Properties	223
	5.0.2	Interaction Effects: Spectral Diffusion and Dephasing	225
5.1	Interactions and Equilibrium Properties		227
	5.1.1	Standard Tunneling Model Predictions	228
	5.1.2	Interactions Between Tunneling Systems: Spectral Diffusion	234
	5.1.3	Theoretical Approaches to the Relaxation of Tunneling Systems	241
	5.1.4	Many-Body Effects and Collective Excitations	242
	5.1.5	Interaction-Stimulated Relaxation of Tunneling Systems	250

		5.1.6	Equilibrium Acoustic and Dielectric Measurement Techniques 255

        5.1.6 Equilibrium Acoustic
             and Dielectric Measurement Techniques 255
        5.1.7 Equilibrium Acoustic and Dielectric Loss Data 258
        5.1.8 Equilibrium Dielectric Saturation
             at Very Low Temperatures 261
   5.2 Nonequilibrium Effects: Long-Time Relaxations
       and the Dipole Gap 263
       5.2.1 Nonequilibrium Experimental Techniques 263
       5.2.2 Experimental Results 267
       5.2.3 Nonequilibrium Behavior: General Remarks 277
       5.2.4 Nonequilibrium Behavior
            Without Interactions Between Tunneling Systems 278
       5.2.5 Weak Interactions: The Dipole Gap 279
       5.2.6 Discussion of the Experiments 288
       5.2.7 Anomalous Hysteretic Behavior
            and Ultralow Temperatures 292
   5.3 On the Universality of the Low-Temperature Properties 295
       5.3.1 Basic Facts 296
       5.3.2 Significance of $1/R^3$ Interactions 297
       5.3.3 The Renormalization Group Model 299
       5.3.4 A Key Identity 301
       5.3.5 General Model 304
       5.3.6 Tunneling Motion 309
       5.3.7 Discussion of the Results 311
   5.4 Conclusion and Remarks 315

6. Investigation of Tunneling Dynamics by Optical Hole-Burning Spectroscopy
(H. Maier, B. M. Kharlamov, and D. Haarer) 317

   6.1 Introduction 317
   6.2 Optical Spectra of Impurities in Solids..................... 318
       6.2.1 Crystals.. 318
       6.2.2 Amorphous Solids 322
   6.3 Basic Methods of Hole-Burning Spectroscopy 327
       6.3.1 Introduction 327
       6.3.2 Experimental Techniques 328
       6.3.3 Technical Limitations 333
   6.4 High-Barrier Versus Low-Barrier Tunneling 338
       6.4.1 Photochemical Hole Burning 338
       6.4.2 Nonphotochemical Hole Burning 344
       6.4.3 Hole Burning in a Model System: Benzoic Acid 347
       6.4.4 Conclusion 351
   6.5 Spectral Diffusion: Low-Barrier Tunneling 352
       6.5.1 Spectral Diffusion 352
       6.5.2 Theoretical Description of Spectral Diffusion 355

		6.5.3	Equilibrium Glass Dynamics 358

		6.5.3	Equilibrium Glass Dynamics 358
		6.5.4	Long-Time Equilibrium Dynamics: Nonclassical Distribution of Tunneling States 360
		6.5.5	Nonequilibrium Glass Dynamics 370
	6.6	Conclusion 386	
7.	**Tunneling of H and D in Metals and Semiconductors** (G. Cannelli, R. Cantelli, F. Cordero, and F. Trequattrini) 389		
	7.1	Introduction 389	
	7.2	Solid Solutions of Hydrogen............................. 390	
		7.2.1	The bcc Metals V, Nb and Ta 392
		7.2.2	The Rare Earths Sc, Y and Lu 393
		7.2.3	Trapping of Hydrogen by Impurities 394
	7.3	Experimental Techniques Revealing the Tunneling of Hydrogen 395	
		7.3.1	Specific Heat 395
		7.3.2	Acoustic Measurements 396
		7.3.3	Neutron Spectroscopy............................ 406
		7.3.4	Nuclear Magnetic Resonance...................... 409
	7.4	Long-Range Diffusion and Incoherent Hopping of Hydrogen in bcc Metals 411	
		7.4.1	Theories of Quantum Diffusion 411
		7.4.2	The Gorsky Effect: Long-Range Diffusion 413
		7.4.3	Hopping of Hydrogen near Interstitial Impurities 416
		7.4.4	Hopping of Hydrogen near Substitutional Impurities .. 418
	7.5	Coherent Tunneling and Fast Local Motion of Hydrogen 418	
		7.5.1	Hydrogen Trapped by Interstitial O,N and C in Nb and Ta: A Two-Level System 418
		7.5.2	Hydrogen Trapped by Substitutional Ti and Zr in Nb: Two- and Four-Level Systems 434
		7.5.3	Tunneling of H in hcp Rare Earths 449
		7.5.4	Motion and Delocalization of Untrapped Hydrogen in Nb, Ta and V 452
	7.6	Nonclassical Motion of Hydrogen in Doped Semiconductors................................ 455	
	7.7	Conclusion 457	
8.	**Microscopic View of the Low-Temperature Anomalies in Glasses** (A. Heuer) 459		
	8.1	Introduction 459	
	8.2	Phenomenological Description of the Low-Temperature Anomalies 461	
		8.2.1	The Tunneling Model 461
		8.2.2	Determination of Tunneling Parameters from Experiments 463

         8.2.3 Soft-Potential Model 464
   8.3 Double-Well Potentials in Computer Simulations 465
         8.3.1 The Scope of Computer Simulations
               in the Present Context 465
         8.3.2 Summary of Earlier Simulations 470
         8.3.3 Systematic Search of Double-Well Potentials
               for a Model Glass 471
         8.3.4 Application of Different Search Strategies......... 481
         8.3.5 Tunneling Systems in the Presence of Impurities ... 484
         8.3.6 Total Energy Landscape of a Glass-Forming System .. 487
   8.4 Coupling Between Tunneling Systems and Heat Bath 493
         8.4.1 Microscopic Origin of the Deformation Potential
               and the Velocity of Sound 494
         8.4.2 Numerical Evaluation of the Deformation Potential ... 497
         8.4.3 Relation Between the Deformation Potential
               and the Structure of DWP's 498
   8.5 Nature of Tunneling Systems Beyond Computer Simulations . 503
         8.5.1 1D Model Glass 504
         8.5.2 Spin Glass Like Model Glass 505
         8.5.3 Simple Models of Soft Modes 507
   8.6 Universality of the Low-Temperature Parameters 508
         8.6.1 Corresponding States 508
         8.6.2 Universal Relations for LJ Glasses 509
         8.6.3 Application for Different Types of Glasses 512
         8.6.4 Quantitative Universality: What Does it Express? ... 517
   8.7 Experimental Hints about the Microscopic Nature
       of the Soft Modes 519
         8.7.1 Relation to Strong and Fragile Glasses 519
         8.7.2 Cooling Rate Dependence of TS's 520
         8.7.3 The Microscopic Nature of Soft Modes in SiO_2 521
         8.7.4 The Properties of Defects 521
         8.7.5 Pressure Dependence 522
         8.7.6 Length-Scale Dependence........................... 523
   8.8 Summary and Outlook 523

9. **Beyond the Standard Tunneling Model:
   The Soft-Potential Model**
   (M. A. Ramos and U. Buchenau) 527
   9.1 Introduction 527
   9.2 Tunneling States and Soft Modes in Glasses 530
         9.2.1 Specific Heat 530
         9.2.2 Thermal Conductivity 531
         9.2.3 Coherent Neutron Scattering 532
         9.2.4 Temperature Dependence
               of Raman and Neutron Scattering 535

	9.2.5	Comparison Between Neutron and Specific-Heat Data	537
	9.2.6	More Recent Neutron Data	538
9.3	The Soft-Potential Model and its Parameters	541
	9.3.1	The Anharmonic Quartic Potential	541
	9.3.2	Assumptions	543
	9.3.3	Level Splittings and Matrix Elements	544
	9.3.4	The Distribution-Limiting Thermal Strain "Ansatz"	548
	9.3.5	Other Approaches	550
9.4	Predictions of the Soft-Potential Model	551
	9.4.1	Tunneling Density of States in Double-Well Potentials	551
	9.4.2	Vibrational Density of States	552
	9.4.3	Specific Heat	555
	9.4.4	Thermal Conductivity	559
	9.4.5	Acoustic Attenuation	564
9.5	Conclusion and Outlook	566

References 571

Index 592

List of Symbols

The symbols listed below are mostly of common use in the literature within the subject of the book. Note, however, that sometimes the same symbol is used for different quantities, or when necessary different symbols for the same quantity. Usual abbreviations are: TS, tunneling system; TLS, two-level system; DWP, double-well potential; SPM, soft-potential model.

α	Acoustic attenuation.
	$= \Delta_0/\Delta$: A useful substitutional parameter.
	Phonon-coupling parameter proportional to $\gamma^2/\varrho v^5 \hbar$ with dimension s^2.
$\alpha(\lambda)$	Optical density $= -\log(I(\lambda)/I_0)$.
$\tilde{\alpha}$	$= k_B^2 \alpha/\hbar^2$, dimension K^{-2}.
β	$1/(k_B T)$.
γ	Averaged effective coupling constant or deformation potential between TS's and phonons.
	Rate of phase memory loss.
$\gamma_{l,t}$	$\simeq \partial \Delta / 2 \partial u_{ik}$: Effective deformation potential for longitudinal or transversal phonons.
Γ	Effective gyromagnetic ratio, when modeling TLS dynamics in analogy to NMR.
	Relaxation rate of a TLS.
	Optical line width.
$\delta \epsilon'$	TS contribution to the real part of dielectric response.
$\delta \epsilon''$	TS contribution to the imaginary part of dielectric response.
$\tan \delta$	$= G/(\omega C) = \epsilon''/\epsilon'$: Ratio of dissipative part of material's impedance to capacitive part.
Δ	The asymmetry of a DWP.
$\overline{\Delta}$	Width of the distribution function for the TLS asymmetries.
Δ_0	The tunneling matrix element or energy (dressed, zero-temperature value, as observed experimentally).
$\tilde{\Delta}_0$	Dressed or renormalized tunneling matrix element for $T > 0$ K.
Δ_b	Bare tunneling energy.
Δ_{el}	The tunneling energy dressed by electron-hole excitations.

List of Symbols

Δ_*	The tunneling energy renormalized by the electron-polaron effect.
Δ_{0c}	Critical value of tunneling parameter, from Landau–Zener criterion.
Δ_{0k}	Effective tunneling parameter of an excitation involving k defects.
Δ_{0*}	Characteristic scale of tunneling parameter of primary defects.
$\Delta_{0,\min}$	A minimum value of tunneling amplitude corresponding to very long relaxation times.
$\Delta_{0,p}$	Effective tunneling parameter of a pair excitation.
Δ_p	Effective asymmetry energy of a pair excitation.
$\Delta_s(T)$	Superconducting energy gap.
$\Delta\epsilon'/\epsilon'$	$= \epsilon'(T,\omega,E_{ac}) - \epsilon'(T_0,\omega,E_{ac})/\epsilon'(T_0,\omega,E_{ac})$.
$\Delta v/v$	$= v(T,B,\epsilon) - v(T_0,B_0,\epsilon_0)/v(T_0,B_0,\epsilon_0)$.
$\epsilon_{ik}(t)$	Elastic strain tensor ($= \frac{1}{2}u_{ik}(t)$).
ϵ_i	Elastic strain tensor in Voigt notation: $1 = xx$, $2 = yy$, $3 = zz$, $4 = yz$, $5 = xz$, $6 = xy$.
ϵ_0	Reference applied strain.
$\epsilon'(\omega)$	Real part of dielectric response at frequency $\omega = 2\pi f$; the capacitive contribution.
$\epsilon''(\omega)$	Imaginary part of dielectric response; the dissipative contribution.
κ	The thermal conductivity.
λ	The Gamow factor or barrier strength.
	The wavelength of light.
$\lambda(T)$	Wavelength of thermal phonon.
λ_{ij}	Elastic dipole $\partial \epsilon_{ij}/\partial c$, ϵ_{ij}: strain, c: concentration of atoms or defects per mole.
λ_{ik}^α	Elastic dipole tensor of a defect in state α.
Λ	Width of Lorentzian distribution of energy shifts from spectral diffusion.
ν	Hopping rate between two sites, $2\nu = \nu_{12} + \nu_{21} = \tau^{-1}$.
	Phonon frequency.
$\xi:$	$= \ln(L/r_{\min})$, a rescaled length parameter used in renormalization group analysis of interacting defects.
ρ	Electron density of states.
ϱ	Mass density
σ	Cross-section for neutron scattering.
$\sigma_{ik}(t)$	Elastic stress tensor.
σ	or σ_i, the Pauli spin matrices.
σ_{ij}	External stress applied to a crystal.
τ, τ_1	Longitudinal relaxation time of tunneling systems (TS's) determined by the emission or absorption of phonons ($\tau_{1,p} = \tau_p$) and/or conduction electrons ($\tau_{1,e} = \tau_e$).

τ_2	Transverse (dephasing) relaxation time of TS's determined by the interaction between them.
τ_{12}	$= (\tau_1^{-1} + \tau_2^{-1})^{-1}$
τ_0	Longitudinal interaction-driven relaxation time for "thermal" TS.
	Prefactor of the thermally activated relaxation time.
$\tau_{1,\min}$	Minimum longitudinal relaxation time of TS's, due to phonons ($\tau_{p,\min}$) or electrons ($\tau_{e,\min}$).
τ_c	The value of τ_1 for TS's with energy splittings of $\sim k_B T$ and tunneling matrix elements Δ_{0c}.
τ_e	Longitudinal relaxation time of TS's due to electron-assisted interaction.
τ_p	Longitudinal relaxation time of TS's due to phonon-assisted interaction.
$\tau_{p,\min}$	Minimum phonon-assisted relaxation time.
τ_P	Relaxation time for a coupled pair of TLS's.
τ_{R1}	Relaxation time of TS's due to first-order Raman process.
τ_s	Switching time of external perturbation.
τ_{TA}	Thermally activated relaxation time of TS's.
τ_{1T}	Longitudinal phonon relaxation time for "thermal" TS's, those with $E \approx \Delta_0 \approx k_B T$.
$\tau_{\max}$	Maximum relaxation time of TS's.
$\tau_{\min}$	Minimum relaxation time of TS's.
τ_η^{-1}	Resonant phonon scattering rate for a specific polarization η.
τ_*	Characteristic time for thermal transport from pair to pair of TS's.
ϕ_0	Energy of applied perturbation.
ϕ_i	Energy shift of primary centers due to random field used in renormalization group calculations.
Φ_i	The phase of the wave function of the TLS i.
χ	$= P_0 U_0$, dimensionless parameter characterizing the size of the TS's interactions.
$\Psi_{L,R}$	"Left well" and "right well" basis states for a TLS.
$\Psi_\pm$	Energy eigenstate basis states for a TLS.
ω	Angular frequency of a measurement.
ω_0	Characteristic frequency of the tunneling "particle" for under-barrier motion.
Ω	$= 2\pi \times$ an attempt frequency for tunneling between wells of a TLS.
	$= (K/\alpha)^{1/2}$.
a	Size of an effective "unit cell" in the amorphous host.
A	A parameter proportional to $\gamma^2/\varrho v^5 \hbar^4$ with dimensions $J^{-3}s^{-1}$.
B_0	Reference magnetic field in Tesla.

XVIII List of Symbols

B	Effective magnetic field, when modeling TLS dynamics in analogy to NMR.
c	Atomic (molar) ratio of the number of H atoms or defects to the number of metal atoms (moles if there is more than an atom per formula unit) n_H/n_M.
	Atomic ratio of the substitutional atom.
C	Equilibrium capacitance of a sample [Farads].
	The specific heat.
	A constant with the value $P_0\gamma^2/\varrho v^2$.
C_0	Reference capacitance.
C_p	The specific heat at constant pressure.
C_{ph}	Heat capacity of phonon subsystem.
C_{ijhk}	Elastic stiffness tensor.
C_{ik}	Elastic stiffness tensor in Voigt notation: $1 = xx, 2 = yy, 3 = zz, 4 = yz, 5 = xz, 6 = xy$.
d	Separation distance in the configurational coordinate of the two wells of a TLS.
	Dimensionality of the system.
D	Diffusion coefficient.
D_{ik}	$= \partial E/\partial u_{ik} = (\Delta/E)(\partial\Delta/\partial u_{ik}) + (\Delta_0/E)(\partial\Delta_0/\partial u_{ik})$: The deformation potential or modulation of E under the influence of a strain. It is usually simplified as $D = D_{ii} = 2\gamma\Delta/E$.
E	The interlevel spacing.
$\boldsymbol{E}$	The electric field. To avoid confusion with the interlevel spacing E, $\vec{F}$ is also used.
E_{ac}	Magnitude of ac electric field used for capacitance measurement.
E_{dc}	Magnitude of the applied dc electric field to perturb the system.
E_B	Binding energy.
E_0	Ground state energy of the potential wells.
$\mathcal{E}$	Binding energy of the atoms.
E_p	Energy splitting of a pair excitation.
E_α	Elastic energy of a defect in state α.
g	Distribution function of random fields seen by primary defect centers in renormalization group calculations.
G	Effective equilibrium conductance of a sample [Siemens].
I	Acoustic intensity.
$I(\lambda)$	Light intensity transmitted by the sample.
I_0	Incident light intensity.
I_{c1}	Critical acoustic intensity for the resonant process.
I_{c2}	Critical acoustic intensity for the relaxation process.
$J(\omega)$	Spectral function.
k	Number of defects participating in a collective excitation.

List of Symbols XIX

k_*	Number of defects involved in the dominant collective excitation at a particular point in the renormalization group analysis of interacting defects.
K	Coupling strength between a particle (TS, muon, etc.) and conduction electrons.
$l_{\text{res,rel}}$	Mean free path of phonons for resonant ('res') or relaxational ('rel') mechanism.
L	Maximum effective radius of interactions; at zero temperature, equal to sample size.
L_k	$= \ln(\Delta_{0k}/W)$, a rescaled tunneling parameter of excitations involving k defects.
m	Mass of a sample.
M_{ik}	$= (1/2)[(\Delta/E)(\partial \Delta_0/\partial u_{ik}) - (\Delta_0/E)(\partial \Delta/\partial u_{ik})]$: The deformation potential of the off-diagonal elements describing the strength of the coupling between two-level systems and the phonons for the resonant interaction. It is usually simplified as $M \simeq -\gamma \Delta_0/E$.
$\tilde{N}_0$	Difference in the population of the two-levels at a given temperature and at thermal equilibrium.
N	Number of two-level systems with a given energy.
n	Spatial density of primary defect centers.
n_T	Spatial density of "thermal" TS's.
n_r	Spatial density of pairs of resonant TS's.
n_α	Fraction of defects in the state α, c_α/c.
p	Average magnitude of the dipole moment of a TLS.
	The effective "mass" parameter used in computer simulations as a measure of the number of particles involved in the transfer between DWP's.
p_{ik}^α	Double force tensor of a defect in state α.
$P(..,..,..)$	The density of states of TS's.
P_0	Constant density of states of TS's.
P_2	Density of states of pair excitations.
P_n	Density of states of n-center excitations.
P_s	Constant density of states of TS's within the soft-potential model.
Q^{-1}	Elastic energy loss coefficient or internal friction.
$r_{\min}$	Some minimum size, at which the R^{-3} interaction energy equals the high energy cutoff, W.
$r(E, \Delta_0, T)$	Tunneling system relaxation rate.
R	Distance from a particular TLS.
	Cooling rate in heat release experiments.
R_*	Average distance between resonant pairs of TS.
R_c	Cutoff radius for coherent coupling between a TLS and its neighbors.

XX   List of Symbols

$R(T)$	Anelastic relaxation strength.
R_T	Thermally limited cutoff radius for coherent coupling between a TLS and its neighbors.
$\mathbf{S}$	$= \boldsymbol{\sigma}/2$: The spin operator, a vector of Pauli matrices.
$S_{T,\epsilon}$	$= \partial(\Delta\epsilon'/\epsilon')/\partial\log(T)$: Logarithmic temperature 'slope' of TLS dielectric response.
$S_{T,v}$	$= \partial(\Delta v/v)/\partial\log(T)$: Logarithmic temperature 'slope' of TLS acoustic response.
$S_{t,\epsilon}$	$= \partial(\Delta\epsilon'/\epsilon')/\partial\log(t)$: Logarithmic time 'slope' of nonequilibrium TLS dielectric response.
$S_{t,v}$	$= \partial(\Delta v/v)/\partial\log(t)$: Logarithmic time 'slope' of nonequilibrium TLS acoustic response.
S_{ijhk}	Elastic compliance tensor.
t	Time since application of electric or strain perturbing field, or after reaching the measuring temperature T_0 after cooling the sample from the charging temperature T_1.
t_A	Time needed to cool the sample from T_1 to T_0.
t_d	Time interval between sample cooling and hole-burning.
t_m	$= \chi\xi$, a scaling parameter used in renormalization group analysis of interacting defects.
t_0	Time required to determine the initial hole width after burning.
T_1	"Charging" temperature in heat release experiments.
T_0	Measuring temperature in heat release experiments.
	Reference temperature in Kelvin (sometimes used as T_{ref}).
	$= (2\pi^2\tilde{\alpha})^{-1/2} \propto 1/\gamma$, crossover temperature to overdamped, or incoherent tunneling (of the order of T^*).
T_c	Superconducting critical temperature.
	Crossover temperature where the thermally activated relaxation time equals the tunneling relaxation time.
T_{eff}	Effective "charging" temperature in hole-burning and heat release experiments.
T_{el}	$= \Delta_0/\pi K k_B$, the temperature at which conduction electrons drive a crossover to incoherent motion (Δ_0 is the dressed tunnel energy).
T_f	Final measuring temperature in hole-burning experiments.
T_g	The glass transition temperature.
T_i	"Charging" temperature in hole-burning experiments.
T_m	The same as $T_{\max}$ but used for dielectric data.
$T_{\min}$	Temperature where a minimum in C/T^3 is observed (C: specific heat); it indicates the crossover frequency between two-level states and higher-frequency excitations.

List of Symbols

$T_{\max}$	The crossover temperature between resonant and relaxational processes. In acoustic experiments it indicates the temperature of the maximum in the sound velocity.
	Temperature of the maximum in C/T^3 (C: specific heat).
T_s	Temperature of the sound attenuation maximum in glasses.
T^*	Freezing temperature; below it and for typical cooling rates the TS's remain in a nonequilibrium state and contribute to the heat release.
	Crossover temperature to incoherent tunneling motion.
T_0^*	Temperature below which one expects to see delocalized collective excitations.
T'	Temperature below which interaction-driven relaxation is expected to be more effective than phonon-driven relaxation.
T_ϕ	Effective temperature of TS's after application of perturbation.
u_{ik}	$= \partial u_i/\partial x_k + \partial u_k/\partial x_i$: The strain tensor.
u	The displacement of the medium.
	$= \Delta_0/E$ a useful substitutional parameter.
	Internal energy per unit volume.
$u_{\min}$	The minimum possible value of $\Delta_{0,\min}/E_{\max}$.
U_{km}	Interaction energy between collective excitations consisting of k and m defects.
U_0	Mean magnitude of coefficient of interaction strength, i.e. interaction energy $\propto U_0/R^3$.
v	The sound velocity.
V	Potential barrier or barrier height.
v_0	Unit cell volume.
$V_0(x,y,...)$	Equilibrium potential energy (without strain).
$\mathcal{V}$	Sample volume.
w	A probability.
w_b	The probability that a TLS that had left the infinite cluster of pair excitations has returned.
W	Fundamental energy parameter of the soft-potential model, of the order of the energy difference between ground state and first excited state.

List of Contributors

Ulrich Buchenau
Institut für Festkörperforschung
Forschungszentrum Jülich GmbH
D-52425 Jülich, Germany

Alexander L. Burin
Department of Chemistry
Northwestern University
2145 Sheridan Road
Evanston, IL 60208-3113, USA

Gaetano Cannelli
Dipartimento di Fisica
Università della Calabria
I-87036 Arcavacata di Rende (CS)
and INFM, Italy

Rosario Cantelli
Dipartimento di Fisica
Università di Roma "La Sapienza"
Piazzale A. Moro 2
I-00185 Roma and INFM, Italy

Francesco Cordero
Istituto di Acustica "O. M. Corbino"
CNR
Area di Ricerca di Tor Vergata
Via del Fosso del Cavaliere
I-00133 Roma and INFM, Italy

Pablo Esquinazi
Abt. Supraleitung und Magnetismus
Fakultät für Physik und
Geowissenschaften, Universität Leipzig
Linnéstr. 5
D-04103 Leipzig, Germany

Dietrich Haarer
Experimentalphysik IV
Universität Bayreuth
D-95440 Bayreuth, Germany

Andreas Heuer
MPI für Polymerforschung
Ackermannweg 10
D-55128 Mainz, Germany

Yuri Kagan
Russian Scientific Center
"Kurchatov Institute"
123182 Moscow, Russia

Boris M. Kharlamov
Institute of Spectroscopy
Russian Academy of Sciences
142092 Troitsk, Moscow Region
Russia

Reinhard König
Experimentalphysik V
Universität Bayreuth
D-95440 Bayreuth, Germany

Hans Maier
MPI für Plasmaphysik
Boltzmannstr. 2
D-85748 Garching, Germany

Douglas Natelson
Department of Physics
Varian Laboratory
Stanford University
Stanford, CA 94305-4060, USA

Andreas Nittke
Abt. Supraleitung und Magnetismus
Fakultät für Physik und
Geowissenschaften, Universität Leipzig
Linnéstr. 5
D-04103 Leipzig, Germany

Douglas D. Osheroff
Department of Physics
Varian Laboratory
Stanford University
Stanford, CA 94305-4060, USA

Miguel Angel Ramos
Departamento de Física
de la Materia Condensada, C-III
Universidad Autónoma de Madrid
Cantoblanco
E-28049 Madrid, Spain

Sven Sahling
Institut für Tieftemperaturphysik
Technische Universität Dresden
Zellescher Weg 16
D-01217 Dresden, Germany

Francesco Trequattrini
Dipartimento di Fisica
Università di Roma "La Sapienza"
Piazzale A. Moro 2
I-00185 Roma and INFM, Italy

Alois Würger
Institut Max von Laue–Paul Langevin
Avenue des Martyrs
B. P. 156
F-38042 Grenoble Cedex 9, France

1. Introduction

Pablo Esquinazi

The low-temperature properties of amorphous and disordered solids depend to a large extent on the contribution of – or on the interaction with – tunneling systems (TS's). The extensive experimental work in the last nearly 30 years provided a rather universal picture of the behavior of the low-temperature properties. This amazing universality captivated researchers since the work of Zeller and Pohl (1971) on the thermal properties of amorphous solids clearly showed remarkable differences from their crystalline counterparts. The interest has been enhanced continuously, due to the fact that some of the observed "anomalies" – in comparison with the well understood low-temperature properties in "Debye-like" crystalline solids – are not only measured in glasses but also in a large number of solids with some kind of disorder. This book deals with tunneling entities, i.e. tunneling defects with two energy levels, their nature, their dynamics and their influence on different solid state properties at low temperatures. After more than 15 years of the publication of review volumes on the low-temperature properties of amorphous solids (see for example Phillips (1981a)) the authors of this book are convinced that a review volume on tunneling systems in amorphous and crystalline solids and their influence on some specific properties is needed. The next eight chapters of this book contain the contribution of scientists that investigated low-temperature properties of solids affected by quantum tunneling processes. We are rather confident that the broad spectrum of new theoretical ideas and experimental results described in this book will help to understand a step further one of the most fascinating aspects of the physics of glasses and disordered solids.

1.1 Tunneling Systems

A novice reader would certainly like to know as soon as possible the kind of tunneling and tunneling entity we are referring to. When we talk about tunneling we mean quantum tunneling. If a given particle can be in one of two (or more) possible energy levels separated by a potential barrier, classical physics predicts a nonzero probability for the transport of the particle across the barrier only if the particle energy is larger than the barrier height. This process is known as thermally activated. However, quantum transport makes the jump of the particle "under" the barrier feasible even for particle energies much lower than the barrier height.

In this book we do not describe the quantum-mechanical reasons for this process. Our interest is focused on the energy and temperature dependence of the quantum-tunneling process that, via the coupling of the tunneling entities with phonons and/or conduction electrons in the solid and via the interaction between them, influences the behavior of several dynamic and transport properties. Generally, the tunneling processes we deal with become evident at temperature $T < 20$ K approximately, though we also discuss examples where quantum tunneling or diffusion of particles at higher temperatures is important (Chaps. 3 and 7).

Tunneling Systems in Amorphous Materials. Simply put, the tunneling systems in amorphous solids are believed to be atoms, a small group of them, or more complicated clusters that have the possibility to 'move' between *two* similar energy states separated by a barrier. Therefore, we call them "two-level systems" (TLS's). Due to the existence of a wide range of local environment in the disordered atomic lattice, the distribution of the intrinsic parameters of the TLS's – like their potential barrier and, therefore, their relaxation times and energies – is assumed to be very broad. The other ingredient of the phenomenological tunneling model (Anderson et al. (1972) and Phillips (1972)) used to understand the low-temperature properties of amorphous solids and discussed thoroughly in this book, is the assumption of a coupling of these entities to phonons and conduction electrons. If the dynamics of these systems is characterized by a tunneling rate, then we call them "tunneling systems" (TS's). Both names, TLS and TS, are used in this book. Usually, if one wants to emphasize their 'two'-level nature, "TLS" is used to denote them.

The tunneling model does not specify the nature of the tunneling entities. Actually, their unspecified nature is an advantage for the phenomenology, though the researcher not working in this field may be sometimes unsatisfied with several of the assumptions of the phenomenology. However and whatever the nature of the TS's in amorphous and disordered solids, scientists were able to experimentally show that specific parameters of the model that depend on the density of states of TS's and their interaction with phonons are surprisingly similar for nearly all dielectric and metallic glasses (in as-quenched state). Several ideas on this issue are discussed in this book. In particular, a microscopic view of the tunneling two-level systems in amorphous solids is given in Chap. 8.

Tunneling of Light Atoms. We also discuss properties which are influenced by well-characterized tunneling entities. Cannelli, Cantelli, Cordero and Trequattrini review in Chap. 7 the properties of metals and semiconductors with specific tunneling systems formed by the light atoms H or D. The absence of a broad distribution of tunneling parameters facilitates the characterization of the quantum entity and its dynamics.

Beyond the Standard Tunneling Model. Several chapters of this book go beyond the "standard" tunneling regime (below 1 K) and provide an overview of the main experimental facts as well as the newest, most original ideas discussed nowadays in the literature to understand properties at $T > 1$K. Clearly, one would ask if the same tunneling entities responsible for the low-temperature behavior ($T < 1$ K) are also responsible for the "high"-temperature ($T > 1$ K) anomalies, and if their density of states and/or the energy and temperature-dependent relaxation rate remain the same or not. It is now well known that the "standard" one-phonon relaxation rate of "weakly damped" TS's used to describe properties below 1 K does not provide the correct answer to the observed "high"-temperature properties. On this issue, Würger (Chap. 3) reviews the crossover to phonon-assisted tunneling in insulators and metals. He discusses the crossover to incoherent tunneling with rising temperature, the implications of the phonon dressing on the tunneling energy, as well as the effect of conduction electrons. In particular the interaction of the TS's with conduction electrons attracts special attention in two other chapters, Chaps. 4 and 7. This is in part due to the influence of conduction electrons to the quantum motion of light atoms like H or D in crystalline Nb, and also because of the apparent nonuniversal behavior of the properties of amorphous and disordered metals as compared to that of amorphous dielectrics (Chap. 4).

As pointed out by Ramos and Buchenau in Chap. 9, there are strong reasons to extend the standard tunneling model to understand the properties of glasses above a few Kelvin. Following numerical simulations of model glasses reviewed by Heuer in Chap. 8, low-energy barriers appear to be necessary to have the required tunneling splitting of the order of 1 K. Therefore, it seems reasonable to expect a potential distribution with vanishing energy barriers and small restoring force, i.e. a set of single-wells "soft" potentials.

1.2 Content and Organization of the Book

This book reviews thermodynamic (Chaps. 2, 7, and 9), acoustic (Chaps. 4, 5, 7, and 9), thermal transport (Chap. 9), dielectric (Chap. 5) and spectroscopic (Chap. 6) properties, at temperatures $T < 100$ K to the mK region (Chaps. 2, 4, 5, and 6) or even below (Chap. 4), as well as neutron spectroscopy data (Chaps. 7 and 9). In the low-temperature region (~ 100 mK $\leq T < 5$ K) the assumption of independent tunneling entities and the so-called one-phonon process for the tunneling rate are usually sufficient to understand the observed anomalies in amorphous insulators. At even lower temperatures, however, experimental results obtained in the last years indicate that the interaction between the tunneling entities may become important and new approaches are needed. The experimental evidence and the last published ideas on this subject are described by Burin, Natelson, Osheroff and Kagan in Chap. 5, whereas in Chap. 4, König and I review some acoustic data at very low temperatures related to this subject.

The chapters of this book are independently written reviews and an advanced reader can read them, in principle, without a predefined order. However, we have taken care to present the rather extensive spectrum of the physics involved, in an accessible way for beginners, giving also special attention to the cross-references. On the one hand, we tried to reduce repetitions as much as possible, without degrading a fluent and pedagogical reading of a given subject. On the other hand, the cross-references will help the reader to gain more information on a specific issue that may not be described thoroughly in a chapter or is interpreted in a different way. We note that we do not try to show only "one side of the medal" by giving preference to some particular interpretation. The broad spectrum of reviews provides partially complementary or even different approaches that try to explain old and new experimental evidence discussed nowadays in the literature.

All chapters describe the basic concepts and equations needed to understand and fit experimental data. Several well-known equations used in the literature are in part derived in detail. The experimental chapters (Chaps. 2, 4, 5, 6, and 7) provide full details on the used experimental techniques. The brief description of the chapters of the book written below gives a rough idea of their contents and distribution.

Basic Phenomenology and its Extension. The beginner will be comfortable to start with the phenomenology, i.e. with the "standard tunneling model" in the following chapter of the book. Together with Nittke and Sahling, we provide in Chap. 2 a relatively easy view of it and its usefulness in the understanding of thermodynamic properties like the heat release and specific heat below a few Kelvin. Chapter 2 reviews the basic equations within the standard tunneling model and the soft-potential model for the heat release. Because the relaxation time of the tunneling systems can be as long as weeks or months, it is experimentally possible to measure the heat released by the ensemble of those entities in an amorphous or disordered sample after a relatively rapid cooling of the sample. Thus, the time dependence of the heat release provides information on the dynamics of the tunneling systems. This chapter is also, to our knowledge, the first extensive review on the experimental data of the time and temperature dependence of the heat release in amorphous and disordered solids. The time-dependent heat release has a special importance within the phenomenology, since it provides directly the density of states of tunneling systems.

The book starts and ends with the phenomenology of independent tunneling "defects" and soft modes. In the last chapter, Chap. 9, Ramos and Buchenau go beyond the standard tunneling model and provide a useful review ("for experimentalists") of the main reasons for the introduction of the soft-potential model, its main predictions, the comparison with experimental data like specific heat, thermal conductivity and acoustic attenuation, and the description of some experimental results from neutron scattering. From the reviews included in these two chapters, the beginner will recognize the special features shown by the low-temperature properties of glasses. Whatever the personal view of the reader to the phenomenology, the successful description of the low-temperature anomalies of glasses in such a broad range of temperature and time or frequency in terms of the tunneling model and its extension, will not remain overlooked and will enhance the interest to look inside the other chapters.

Acoustic and Dielectric Properties. Chapters 4 and 5 review the low-temperature acoustic properties of amorphous and disordered solids. In Chap. 4, König and I summarize the main predictions of the tunneling model for the acoustic properties, within the "standard" assumptions. Further, we review the last published ideas and experimental data related to the TS-electron interaction and its influence on the acoustic properties, an issue that is not yet completely understood. Special emphasis is given to the description of the experimental methods used for the measurement of the acoustic properties in the kHz range and at very low temperatures (down in the sub-millikelvin region) and the acoustic properties of polycrystalline metals. We review the deviations of the experimental data from the standard predictions of the tunneling model and discuss them briefly in terms of the interaction between TS's as well as incoherent tunneling.

After discussing the predictions of the standard model of noninteracting tunneling defects to the equilibrium dielectric properties of amorphous solids, Burin, Natelson, Osheroff and Kagan present in Chap. 5 a detailed analysis of the consequences of the interaction between tunneling defects. We will learn that through the dipolar interaction between tunneling systems and at low enough temperatures, other tunneling entities like "pairs" or even more complicated clusters may exist, and they appear to be necessary to understand the low-temperature properties of amorphous solids. They review and discuss the newest experimental results of nonequilibrium dielectric relaxation after the application of electric and strain fields, as well as nonequilibrium acoustic behavior after the application of dc electric fields. Chapter 5 is not only an "experimental" chapter, but also provides a review of the main ideas and equations of the interacting model. The authors show that the interaction between the tunneling defects, within a renormalization group model, can lead to the observed universality of the low-temperature properties of amorphous and disordered solids. They demonstrate how the coupling of tunneling systems to phonons leads to a long-range interaction resulting in a dephasing rate caused by the spectral diffusion.

Acoustic properties at low (kHz) and higher (MHz) frequencies of crystalline solids which are influenced by the quantum diffusion of H and D are reviewed by Cannelli and co-workers in Chap. 7. These authors consider and derive in detail the response of a tunneling system under the application of a stress and its influence on the acoustic properties taking into account the symmetry and geometry of the tunneling entity in a crystalline atomic lattice. They show that the coherent tunneling motion below 10 K of those light atoms can be accounted for satisfactorily by the existing models, taking into account their interactions with phonons and conduction electrons. This chapter also discusses results from specific heat, neutron spectroscopy and nuclear magnetic resonance measurements.

Optical Hole-Burning Spectroscopy. This book has a single chapter, Chap. 6, that deals with optical spectroscopy. This highly developed and nowadays standard experimental technique and its findings related to tunneling dynamics in amorphous solids were often not given the necessary attention in the discussion of low-temperature properties of amorphous solids in the literature, other than the specific one on their optical properties. This was perhaps partially due to the difficulty of researchers to understand the optical hole-burning technique and the spectral diffusion mechanism. The review in Chap. 6 written by Maier, Kharlamov and Haarer will change this situation. The authors introduce the reader with the necessary background to understand the optical techniques and hole-burning spectroscopy. They show that optical spectroscopy methods provide a broad time scale – from subpicoseconds to months – to study the dynamics of tunneling defects. The authors review some of the recent achievements to understand the spectral diffusion mechanism, the time-dependent experiments on the equilibrium and nonequilibrium dynamic properties of tunneling systems in amorphous samples, and the relation to heat release and dielectric experiments described in the other chapters.

New Theoretical Approaches. As emphasized above, in Chap. 5, Burin and co-workers describe in detail the main ideas and predictions of the interacting model, as well as provide a comparison with recently published experimental results. In addition, this book has another two theoretical chapters, Chaps. 3 and 8. In Chap. 3, Würger develops new theoretical ideas to deal with the effects of phonons and conduction electrons that couple to two-level systems. He uses and describes in detail weak-coupling perturbation theory, mode-coupling as well as strong-coupling approaches and discusses their results on the tunneling rate in dielectric and metallic glasses. His treatment covers also H-diffusion in crystalline Nb and mesoscopic wires, as well as the crossover to incoherent tunneling at high temperatures. For the high-temperature regime or "dilute" limit, i.e. independent tunneling defects, Würger disregards the elastic interaction between adjacent two-level systems via phonon exchange.

Heuer, on the other hand, provides in Chap. 8 a microscopic view of the tunneling entities based on computer simulation studies. He describes carefully the scope, the search algorithms, as well as the main results obtained from computer simulations related to the microscopic nature of the tunneling entities in amorphous materials. He also discusses the results on the deformation potential (or coupling constant) of the tunneling entities to phonons obtained by the simulations, as well as possible reasons for the observed quantitative universality of the low-temperature properties in glasses, using different concepts and arguments as those presented in Chap. 5.

At the end of the book we provide a single alphabetically ordered reference list.

2. Heat Release in Solids

Andreas Nittke, Sven Sahling and Pablo Esquinazi

In this chapter we will discuss two consequences of the tunneling model that is used to understand the low-temperature properties of amorphous and disordered solids: a *time-dependent* specific heat and the corresponding *time-dependent* heat release. This "heat release" (often referred to as "energy relaxation") is the warming-up effect observed in a sample that is cooled from a temperature T_1 to a measuring temperature T_0 without an external heat input. To our knowledge, it appears that the first comment in literature on heat release is given by Greywall (1978). He found a time-dependent heat leak in a calorimeter made from Plexiglas (mass $m=15$g) that he correctly brought in connection with the amorphous nature of this material, but he did not extend his considerations. This task has been done later by Zimmermann and Weber (1981a), who actually introduced heat-release measurements as a new method to investigate the properties of amorphous solids. Together with the first systematic measurements of the heat release in Suprasil W, Zimmermann and Weber (1981a) were the first to propose a quantitative analysis of the data. After this publication, numerous measurements of the heat release on various materials have been done. Nowadays this method can be considered as established to characterize the anomalous low-temperature properties of disordered solids due to tunneling processes.

2.1 A Simple System with Two Levels of Energy

The purpose of this and the next section is to review the phenomenology necessary to understand the low-temperature heat release and the specific heat measured in amorphous solids. We will consider mainly the developments since 1972, when Anderson et al. (1972) and Phillips (1972) – independently from each other – suggested a model based on two level systems (TLS's) now called "tunneling model", which has proved to be a useful model to describe the properties of amorphous solids at temperature below 10K.

In the early 70's it became apparent that the Debye model is not adequate to describe the low temperature ($T \leq 1K$) specific heat and thermal conductivity of amorphous solids (Zeller and Pohl (1971), for further references see Stephens (1976)). Zeller and Pohl (1971) found that the specific heat of different glasses at low temperature varies nearly linearly ($\sim T^{1.2}$)

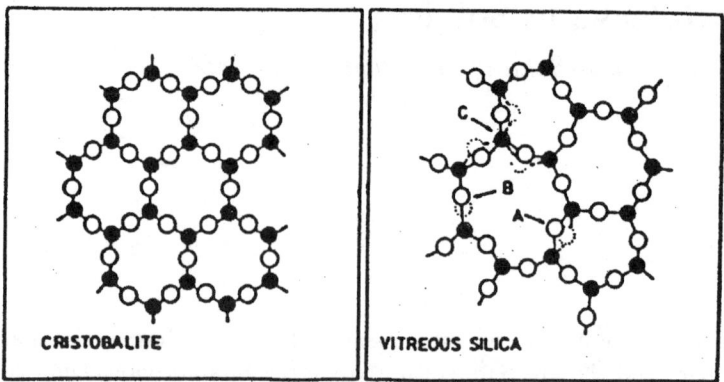

Fig. 2.1. Comparison between a crystalline state (cristobalite) and the amorphous state (vitreous silica) of SiO_2 (two-dimensional representation). In vitreous silica, the arrows indicate three possible metastable states (Hunklinger and Arnold (1976)).

with temperature. Further, thermal conductivity of these materials varies nearly as the square of the temperature ($\kappa \sim T^{1.8}$), in contrast to crystalline insulators where both properties vary with temperature as T^3. The need for a new model to describe the data was met by the tunneling model (Anderson et al. (1972), Phillips (1972)). This model started without detailed microscopic assumptions. The basic idea is that the random character of the atomic positions in amorphous solids enable some atoms or groups of atoms to move between two preferable sites. A picture often used to illustrate this situation is shown in Fig. 2.1, where a crystalline state of SiO_2 (cristobalite) is compared with an amorphous state (vitreous silica). In the tunneling model, these states are described as particles in double-well potentials as shown in Fig. 2.2 and are therefore called "two-level systems" (TLS's). Although the suggestion from Fig. 2.1 is apparent, in most of the samples studied, it is not yet well defined what the particles in the double-well potential really are. In fact, this remains one of the important open questions concerning the tunneling model.

For simplicity, it is assumed that the ground states of each well have the same energy $E_0 = \hbar\omega_0$. The energy difference between the minima is named the asymmetry Δ. If we use the quantum states "particle left" (Ψ_L) and "particle right" (Ψ_R) as basis, the Hamiltonian of such a system is

$$H = \frac{1}{2}\begin{pmatrix} \Delta & -\Delta_0 \\ -\Delta_0 & -\Delta \end{pmatrix}. \tag{2.1}$$

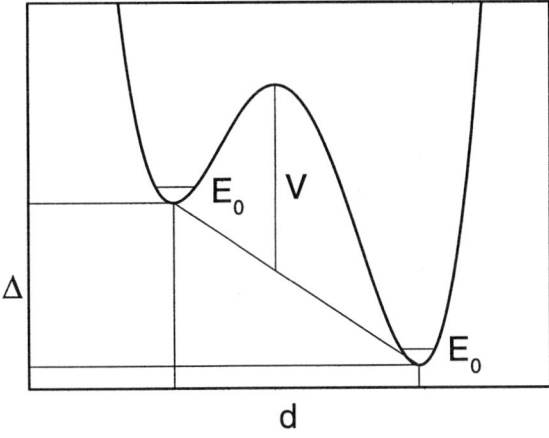

Fig. 2.2. The potential of a two-level system (TLS) with barrier height V, asymmetry energy Δ, well separation d and ground state energy E_0. Note that the magnitude of the parameters in the picture are arbitrary.

Δ_0 is the tunnel splitting which describes the coupling between the two states.[1] Using the Wentzel–Kramers–Brillouin (WKB) approximation, it is possible to express Δ_0 in terms of a tunneling parameter λ (the Gamow parameter)

$$\Delta_0 = \hbar\omega_0 e^{-\lambda}. \tag{2.2}$$

λ can be written in terms of the other characteristic potential parameters as

$$\lambda = \sqrt{\frac{2mV}{\hbar^2}} d. \tag{2.3}$$

Here, m denotes the mass of the particle, V is the barrier height and d is the separation between the two minima along the configurational coordinate. The separation is not necessarily spatial but could also be, e.g. the rotational separation of two states where a group of molecules can rest in two energetically favored orientations.

The (Ψ_L, Ψ_R) states are not the energy eigenstates which we denote as (Ψ_+, Ψ_-). The energy difference between the two energy eigenstates is[2]

$$E = \sqrt{\Delta^2 + \Delta_0^2}. \tag{2.4}$$

---

[1] Rigorously speaking, this Δ_0 is the "experimentally" determined dressed tunneling energy. The difference between bare Δ_b and dressed tunneling energies will become clear in Chap. 3, see Sect. 3.3.
[2] A simple approach to diagonalize the Hamiltonian (2.1) is given in a footnote in page 426.

2.2 Phenomenological Theory for the Heat Release

2.2.1 Generalities

In this section we will develop the basic concepts of the theory for the heat release due to the relaxation of TLS's in glasses. We consider a glass or a disordered material containing an ensemble of N-independent noninteracting and *identical* TLS's in contact to a thermal bath at a temperature T_1. $N_\uparrow$ and $N_\downarrow$ are the numbers of TLS's in the upper ($\uparrow$) and the lower ($\downarrow$) state with $N = N_\uparrow + N_\downarrow$. The difference in the population $\tilde{N}$ is $\tilde{N} = N_\downarrow - N_\uparrow$.

Despite the lack of long-range order and translational invariance in amorphous materials, phonons are still a useful concept in such a media. In particular, at intermediate and low temperatures, when the wavelengths of collective vibrations are long compared to the intermolecular spacing, the vibrational heat capacity is well approximated by the Debye model.[3]

A TLS can only change its state if the energy conservation law is fulfilled which is done in insulators through the coupling of the TLS's to thermal phonons: The phonons induce a local strain which modulates Δ and Δ_0 and enables the tunneling entity (e.g. the atoms or groups of atoms in the simple standard model) to move to the other energy state. If the final state of this process has lower energy than the initial one, the energy difference is given to the phonon system, otherwise it is taken from the phonon system. The coupling between the TLS's and the phonon system will be taken into account by the coupling constant γ. The values for γ that are usually found in amorphous solids and some polycrystalline metals are of the order of 1 eV, which is surprisingly large.

If we regard N *identical* two-level systems at a temperature T_1, we can calculate the difference in the population in thermal equilibrium with standard thermodynamics. Be $w_\downarrow$ the probability for one TLS to be in the lower state and $w_\uparrow$ the probability to be in the upper state, the ratio of these probabilities for a large enough ensemble of TLS's is given by

$$\frac{w_\uparrow}{w_\downarrow} = e^{-\frac{E}{k_B T_1}}. \tag{2.5}$$

With $w_\downarrow + w_\uparrow = 1$ we can easily find for the difference in the population $\tilde{N}_0$ *in thermal equilibrium* at a temperature T_1

$$\tilde{N}_0(T_1) = N \tanh\left(\frac{E}{2k_B T_1}\right). \tag{2.6}$$

Suppose that the sample is now cooled rapidly to a temperature $T_0 < T_1$. In thermal equilibrium, the difference in the population at temperature T_0 is $\tilde{N}_0(T_0)$. This means that $(\tilde{N}_0(T_1) - \tilde{N}_0(T_0))$ TLS's have to change from the

---

[3] Note that the low-temperature specific heat of amorphous solids does not follow the Debye prediction due to the contribution of the TLS's and soft modes, see further in this chapter and also Chap. 9.

2.2 Phenomenological Theory for the Heat Release

upper (↑) to the lower (↓) state and give the energy difference to the phonon system. If this relaxation from one state to the other were infinitely fast, this would be the end of the story: The TLS's give an additional contribution to the specific heat and that is all. But the relaxation process is all but fast for *all* TLS's: Experimental results show relaxation times up to some weeks! In other words: The ensemble of TLS's will reach its thermal equilibrium after some time and will release the energy stored in the $(\tilde{N}_0(T_1) - \tilde{N}_0(T_0))$ two-level systems during the process to equilibrium, see Fig. 2.3. This energy is given to the phonon system, released in the form of heat. Thus, if we would leave the sample in an adiabatic situation the temperature of the sample would increase.

Let us now calculate the time dependence of the heat released per unit time $\dot{Q}$ from the ensemble of relaxing TLS's. First, the heat dQ released during the relaxation process is just the energy difference E times the number of TLS's which switch from one state to the other within the time period dt

$$\dot{Q}(t) = \dot{N}_\uparrow(t)E = \frac{1}{2}\dot{\tilde{N}}(t)E, \qquad (2.7)$$

where the dot denotes the time derivative.

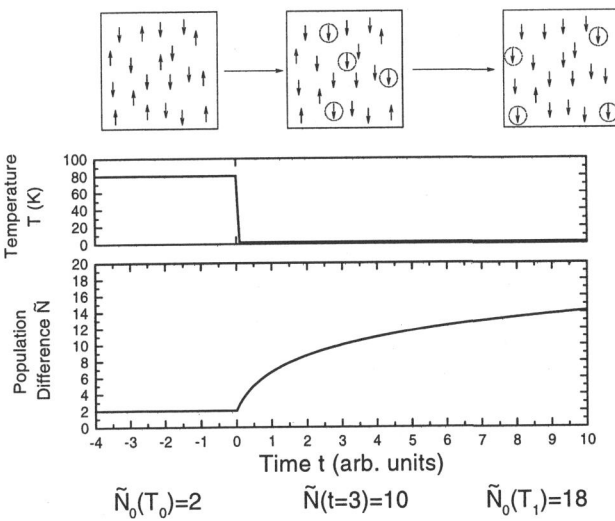

Fig. 2.3. Two-level systems in a glass sample after a temperature change from T_1 to $T_o < T_1$ at $t=0$. In the upper picture, each vector represents a TS in the upper (↑) or in the lower (↓) state. Those TS's which have changed their states are marked by a circle. The situations at $t \leq 0$, at $t = 3$ (arb. units) and for $t \to \infty$ are sketched. In the middle the temperature as a function of time is plotted and the lower plot shows the time evolution of the corresponding population difference $\tilde{N}$. Remember that $\tilde{N}_0 = \tilde{N}(t \to \infty)$.

The *dynamical* behavior of $\tilde{N}(t)$ (where $\tilde{N}$ is now supposed to be time dependent and not necessarily the thermal equilibrium value) can be described assuming the relaxation time approximation

$$\frac{\mathrm{d}(\tilde{N}(t) - \tilde{N}_0(T))}{\mathrm{d}t} = -\frac{\tilde{N}(t) - \tilde{N}_0(T)}{\tau(T)}, \tag{2.8}$$

where $\tau(T)$ is the relaxation time of a TLS with energy difference E at a temperature T. This can be rewritten as

$$\dot{\tilde{N}}(t) = \frac{\partial \tilde{N}_0(T)}{\partial T} \frac{\mathrm{d}T}{\mathrm{d}t} - \frac{\tilde{N}(t) - \tilde{N}_0(T)}{\tau(T)}. \tag{2.9}$$

Equation (2.9) can only be solved analytically for $T = \mathrm{const.}$, because then $\mathrm{d}T/\mathrm{d}t = 0$ and it simplifies to

$$\dot{\tilde{N}}(t) = -\frac{\tilde{N}(t) - \tilde{N}_0(T)}{\tau(T)}. \tag{2.10}$$

For the solution of this differential equation, we find

$$\tilde{N}(t) = \left(\tilde{N}(0) - \tilde{N}_0(T_0)\right) \mathrm{e}^{\frac{-t}{\tau(T_0)}} + \tilde{N}_0(T_0), \tag{2.11}$$

where $\tilde{N}(0)$ is the difference in the population at $t = 0$: $\tilde{N}(0) = \tilde{N}_0(T_1)$. If we insert this into (2.11) and take the time derivative, we obtain

$$\dot{\tilde{N}}(t) = N\left(\tanh\frac{E}{2k_\mathrm{B}T_0} - \tanh\frac{E}{2k_\mathrm{B}T_1}\right)\frac{\mathrm{e}^{-\frac{t}{\tau(T_0)}}}{\tau(T_0)}. \tag{2.12}$$

2.2.2 The Standard Tunneling Model

Now we need to have a thorough look at the relaxation rate τ^{-1} because this relaxation rate contains all the information about the relaxation process. Following Jäckle (1972) we consider the effect of a strain induced by an elastic wave. Such an elastic wave adds a perturbation H_1 to the Hamiltonian H (2.1) which reads in $(\Psi_\mathrm{L}, \Psi_\mathrm{R})$ basis as

$$H_1 = \frac{1}{2}\begin{pmatrix} \delta\Delta & 0 \\ 0 & -\delta\Delta \end{pmatrix}, \tag{2.13}$$

if we neglect a possible variation of the off-diagonal matrix element Δ_0, see (3.11) and the discussion in that section. The change of the asymmetry $\delta\Delta$ is related to the local elastic strain tensor δu_{ik} via the coupling constant γ,

$$\delta\Delta = \gamma 2\delta u_{ik}. \tag{2.14}$$

Usually the tensorial character of u_{ik} is neglected and the averaged magnitude u is used. For a more general approach and the general definition of the coupling constants, considering also the change in the tunneling energy Δ_0, see Sect. 4.2.2, and also Sect. 7.3.2 for the case of single crystals where the symmetry of the atomic lattice should be taken into account.

To calculate the relaxation rate between the TLS and the phonon bath it is easiest to work in the energy eigenstates basis, (Ψ_+, Ψ_-). The complete Hamiltonian there reads:

$$H = \frac{1}{2}\begin{pmatrix} E & 0 \\ 0 & -E \end{pmatrix} + \frac{1}{2}\begin{pmatrix} D & 2M \\ 2M & -D \end{pmatrix}u = H_0 + H_1, \qquad (2.15)$$

where

$$D \simeq 2\gamma\Delta/E \quad \text{and} \quad M \simeq -\gamma\Delta_0/E. \qquad (2.16)$$

Further, the elastic strain can be expressed in terms of phonon creation and annihilation operators. Considering single-phonon transition processes, the relevant matrix element connects an initial state $|\Psi_+; \emptyset\rangle$ (TLS in upper state, no phonon) with the final state $|\Psi_-; \boldsymbol{k}, j\rangle$ (TLS in lower state, phonon of wave vector $\boldsymbol{k}$ and polarization j) via the elastic interaction part of (2.15), H_1. The matrix element works out (Jäckle (1972)) to be

$$\langle \Psi_-; \boldsymbol{k}, j | H_1 | \Psi_+; \emptyset \rangle = \sqrt{\frac{k}{2\varrho v_j}}\gamma_j \frac{\Delta_0}{E}. \qquad (2.17)$$

Here ϱ is the mass density of the sample and v_j denotes the respective sound velocity. With this matrix element and applying the golden rule, one finds, for the one-phonon relaxation rate τ_p^{-1} (sometimes written as the longitudinal relaxation rate τ_1^{-1}) of a particular TLS with energy difference E and tunnel splitting Δ_0,

$$\tau_p^{-1} = \left(\frac{\gamma_l^2}{v_l^5} + \frac{2\gamma_t^2}{v_t^5}\right)\frac{\Delta_0^2 E}{2\pi\varrho\hbar^4}\coth\left(\frac{E}{2k_BT}\right) = A\Delta_0^2 E \coth\left(\frac{E}{2k_BT}\right), (2.18)$$

where the indices l and t refer to the longitudinal and transversal phonon branches. For further calculations it will be useful to introduce $\tau_{p,\min}^{-1}$ which is the relaxation rate of the symmetrical double-well potential ($\Delta = 0$ and $E = \Delta_0$)

$$\tau_p^{-1} = \tau_{p,\min}^{-1}\left(\frac{\Delta_0}{E}\right)^2. \qquad (2.19)$$

The standard tunneling model assumes that at low temperature ($T < 2K$) the one-phonon process is the dominating mechanism; other relaxation mechanisms will be discussed in Sect. 2.2.5. We remark that if one considers the low-temperature situation where the tunneling transition of the TLS's is the most important relaxation mechanism, one usually denotes a two-level system as a "tunneling system" (TS).

Until now we have always considered an ensemble of *identical* TS's. One can easily imagine that for a glass with its random character this is not true. Moreover, the parameters of the TS's will be widely distributed, and this leads us to the second assumption of the tunneling model. Anderson et al. (1972) and Phillips (1972) proposed that the density of states P for the TS's should be constant in terms of the parameters Δ and λ:

$$P(\Delta, \lambda)\, d\Delta\, d\lambda = P_0\, d\Delta\, d\lambda. \tag{2.20}$$

Sometimes it is convenient to use as independent variables Δ and Δ_0. The density of states of TS can then be written as

$$P(\Delta, \Delta_0)\, d\Delta\, d\Delta_0 = \frac{P_0}{\Delta_0}\, d\Delta\, d\Delta_0. \tag{2.21}$$

Ultrasonic absorption experiments (Hunklinger et al. (1973), Golding et al. (1973)) showed that it is possible to saturate the acoustic absorption at any particular frequency with ultrasound of high intensity. The results support the idea that in glasses and at low temperature a large number of excitations exist, each of them with a small number of accessible states. Through phonon echo measurements, Golding and Graebner (1976) and Golding et al. (1977) demonstrated that the underlying excitations do indeed behave as if they obey the proposed two-level system Hamiltonian, thus showing that the excitations are two-state in nature.

For further calculation of the heat release we need $P(E, \tau_p)$. In practice, the time to cool down the sample from T_1 to T_0 is longer than 1s. Tunneling systems with shorter relaxation times (than 1s) reach thermal equilibrium during the cooling process and therefore heat-release measurements are only sensitive to TS's with long relaxation times ($\tau_p > 1$s). Thus, we find from (2.19) with $\tau_{p,\min} < 10^{-6}$s,

$$\frac{\Delta_0}{E} = \left(\frac{\tau_{p,\min}}{\tau_p}\right)^{\frac{1}{2}} < 10^{-3}. \tag{2.22}$$

This means that $E \approx \Delta$ (2.4) and we can set $dE = d\Delta$. From (2.2) and (2.19), we obtain

$$d\lambda = \frac{1}{2\tau_p}\, d\tau_p, \tag{2.23}$$

and finally

$$P(E, \tau_p)\, dE\, d\tau_p = \frac{P_0}{2\tau_p}\, dE\, d\tau_p. \tag{2.24}$$

In some cases it is useful to have this distribution function in terms of the asymmetry and the barrier height V between the potential wells. Following the work of Tielbürger et al. (1992) and assuming two well-defined harmonic potentials, it can be shown that in a first approximation $\lambda = V/E_0$. In this case the distribution function is

$$P(\Delta, V)\, d\Delta\, dV = \frac{P_0}{E_0}\, d\Delta\, dV. \tag{2.25}$$

If we introduce a new variable u (do not be confused with the local strain) with

$$u = \frac{\Delta_0}{E}, \tag{2.26}$$

2.2 Phenomenological Theory for the Heat Release

the distribution function P can be expressed depending on E and u:

$$P(E,u)\,dE\,du = \frac{P_0}{u\sqrt{1-u^2}}\,dE\,du. \tag{2.27}$$

Figure 2.4 shows a plot of the distribution function $P(E,u)$ at constant E.

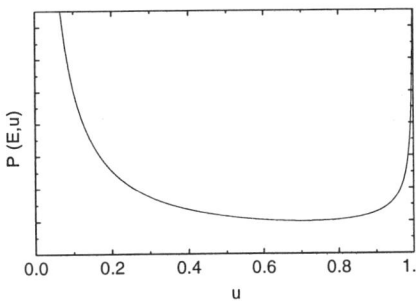

Fig. 2.4. Distribution function $P(E,u)$ for $E = $ const. plotted against $u = \Delta_0/E$.

For the calculation of the heat release, the number of TLS's N will be replaced with the double integral over the distribution function P. Physically it is not acceptable to choose infinite limits of integration (as it would follow from the assumptions of the tunneling model) because this would lead to an infinite number of TLS's per unit volume of the sample.

Thus, finite limits of integration have to be applied. For the original distribution function $P(\Delta,\lambda)$ this would be $\Delta_{\max}$ and $\lambda_{\max}$. These limits are transformed together with the distribution function. For the distribution function $P(E,u)$, this leads to $E_{\max}$ and $u_{\min}$, where u runs from $u_{\min}$ to 1. Although $P(E,u)$ has a singularity at $u=1$, this point is allowed as a limit because the singularity is integrable there. The absolute value of $u_{\min}$ and its possible determination from experiments will be discussed in Sect. 2.2.4.

We can now write down an equation for the time-dependent heat release $\dot{Q}$; starting with (2.7) and inserting (2.12), we obtain

$$\dot{Q}(t) = \frac{1}{2}EN\left(\tanh\frac{E}{2k_BT_0} - \tanh\frac{E}{2k_BT_1}\right)\frac{1}{\tau(T_0)}e^{-\frac{t}{\tau(T_0)}}. \tag{2.28}$$

Replacing the total number N of the TS's with the integrals in E and u, the heat release of a solid with an ensemble of TS's can be written as

$$\dot{Q}(t) = \frac{P_0 V}{2}\int_0^{E_{\max}} dE\, E\left(\tanh\frac{E}{2k_BT_0} - \tanh\frac{E}{2k_BT_1}\right)$$

$$\times \int_{u_{\min}}^{1}\frac{du}{u\sqrt{1-u^2}}\frac{1}{\tau(T_0)}e^{-t/\tau(T_0)}, \tag{2.29}$$

where $\mathcal{V}$ denotes the volume of the sample. Equation (2.29) can only be solved numerically because $\tau(T_0)$ depends on E and u. With this equation, it is possible to investigate the influence of other relaxation processes other than the direct one.

If we regard only the direct process (which is reasonable because heat-release measurements are done at low temperatures), we can start with (2.28) and replace the number of TS's with the integrals in E and τ_p. In this case, we obtain

$$\dot{Q}(t) = \frac{P_0 \mathcal{V}}{2} \int_0^{E_{\max}} dE\, E \left(\tanh\frac{E}{2k_B T_1} - \tanh\frac{E}{2k_B T_0} \right)$$

$$\times \int_{\tau_{\min}}^{\tau_{\max}} d\tau_p \frac{P(E,\tau_p)}{\tau_p(T_0)} e^{-t/\tau_p(T_0)}. \qquad (2.30)$$

For usual experimental conditions and most glasses, integral (2.30) is insensitive to the limits of integration. One can show that the main contribution to the heat release comes from TS's with $E \approx 2k_B T_1$ and $\tau \approx t$. Indeed, the calculation of the heat release according to (2.30) with limits $0.05t \leq \tau \leq 5t$ yields 95% of the total value. For the case $E \ll E_{\max}$, (2.30) leads to

$$\dot{Q}(t) = \frac{\pi^2 k_B^2}{24} P_0 \mathcal{V} (T_1^2 - T_0^2) \frac{1}{t} e^{-t/\tau_{\max}}. \qquad (2.31)$$

For $t < 0.1\tau_{\max}$ the result is independent of $\tau_{\max}$ and we obtain

$$\dot{Q}(t) = \frac{\pi^2 k_B^2}{24} P_0 \mathcal{V} (T_1^2 - T_0^2) \frac{1}{t}. \qquad (2.32)$$

Usually, heat-release experiments are analyzed with (2.32). This is the most accurate method to determine P_0 since the heat release is not influenced by the coupling constant γ (like in scattering experiments).

2.2.3 The Time and Temperature Dependence of the Specific Heat

We consider now the specific heat of a solid which contains TS's. Additionally to the phonon specific heat, there will be a contribution of the TS's because via their coupling to the phonons, part of the energy applied to the solid will be absorbed by the TS's. As stated above, the temperature dependence of the specific heat of amorphous materials lead to the development of the tunneling model. How can we calculate the excess specific heat? First, we know from thermodynamics that the specific heat $C_1(E,T)$ of a single TLS with energy splitting E is

$$C_1(E,T) = \frac{E^2}{k_B T^2} \frac{e^{-E/k_B T}}{(1+e^{-E/k_B T})^2}. \qquad (2.33)$$

2.2 Phenomenological Theory for the Heat Release

To obtain the specific heat of an *ensemble* of TS's, we have to weigh this specific heat $C_1(E,T)$ with the density of states of TS's and integrate over E:

$$C_{\text{TS}}(T) = \int_0^\infty P(E) C_1(E,T)\, dE = \frac{\pi^2}{6} k_B^2 P_0 T \qquad (2.34)$$

with $P(E) = P_0$.[4] With this linear term it is possible to understand semi-quantitatively the specific heat of some amorphous solids.

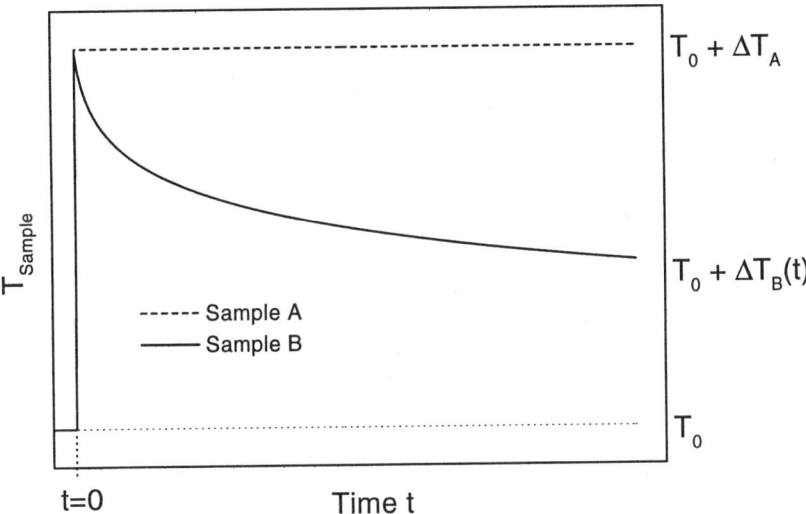

Fig. 2.5. Heat-pulse experiment to measure the specific heat of a solid without (A) and with (B) TS's. In both cases, the same amount of energy has been applied.

But there is another effect which should occur if the tunneling model were valid: the measured excess specific heat due to TS's should be time dependent. To illustrate this, we show in Fig. 2.5 the time evolution of a typical specific heat experiment on a crystal (without TS, sample A) and on a solid containing TS's (sample B). Let us assume that both samples are isolated from the surrounding and have no intrinsic relaxation time, i.e. the phonon heat conductivity is infinite and both samples have the same specific heat due to phonons. At $t = 0$, an infinitely short heat pulse ΔQ is applied which changes the temperature of A from T_0 to $T_0 + \Delta T_A$ with no further time dependence, see Fig. 2.5. The same heat pulse would cause in B the same initial temperature change $\Delta T_B(t = 0) = \Delta T_A$, however, immediately after

---

[4] Due to the exponential function in (2.33) for high enough E_{max}, the integral is independent of E_{max} and for calculation E_{max} can be set as infinite. This will be used also in the rest of this section.

reaching $T_0 + \Delta T_{\rm B}(t=0)$, the temperature begins to fall. The reason for this behavior can easily be understood: Shortly after the heat pulse, the TLS's are out of thermal equilibrium and they relax according to (2.8). In this case, the TS's absorb energy from the phonon system and as a consequence, the temperature falls. In other words: We find $\Delta T_{\rm B}$ to be time dependent, and because the specific heat is defined as $C = \Delta Q/\Delta T$, the specific heat itself can be regarded as a time-dependent property.

To calculate the time dependence of the specific heat, we introduce a distribution function $P(E,T,t)$ which depends also on the measuring time scale t. Following Black (1978) this distribution function can be written as

$$P(E,T,t) = \int_{\sqrt{\tau_{\min}/t}}^{1} du \frac{P_0}{u\sqrt{1-u^2}} = \frac{1}{2} P_0 \ln \frac{4t}{\tau(E)}, \qquad (2.35)$$

where $\tau_{\min}$ is the minimum relaxation time due to the interaction with phonons or conduction electrons. Equation (2.35) is only valid in the time range $\tau_{\min} \ll t \leq \tau_{\min}/u_{\min}^2$; for longer times, the distribution function is time independent due to the cutoff in $P(E,u)$ according to

$$P(E,T,t \to \infty) = \int_{u_{\min}}^{1} P(E,u) du = P_0 \ln\left(\frac{2}{u_{\min}}\right). \qquad (2.36)$$

The time- and temperature-dependent specific heat of a solid with TS's and according to the tunneling model can now be written as

$$C_{\rm TS}(T,t) = \int_0^\infty C_1(E,T) P(E,T,t) dE. \qquad (2.37)$$

We can now insert (2.35) into (2.37) and obtain

$$C_{\rm TS}(T,t) = \int_0^\infty \frac{E^2}{k_{\rm B} T^2} \frac{e^{-E/k_{\rm B}T}}{(1+e^{-E/k_{\rm B}T})^2} \frac{1}{2} P_0 \ln\left(\frac{4t}{\tau(E)}\right) dE. \qquad (2.38)$$

Here, the only time dependence is in the logarithm which is still valid after the integration over E.

An approximation for $C_{\rm TS}(T,t)$ is obtained taking into account only the dominant phonons ($E \approx 2.4 k_{\rm B} T$, Black (1978)) and only the one-phonon process (2.18):

$$C_{\rm TS}(T,t) \approx \frac{\pi^2}{12} k_{\rm B}^2 T P_0 \ln\left(66.24 A T^3 t\right). \qquad (2.39)$$

with A given by (2.18).

Using this equation, Zimmermann and Weber (1981b) have been able to explain the nonlinear temperature dependence observed in the specific heat of vitreous silica below 1K. Figure 2.6 shows the specific heat of Suprasil W and Suprasil I and the fits to the data according to (2.39) using a measuring time of $t = 10$ s.

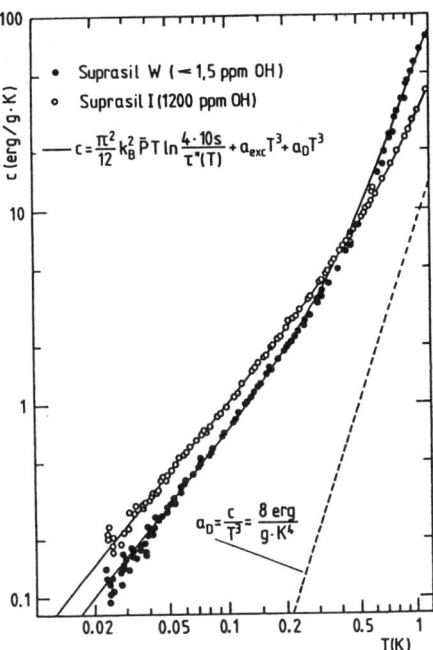

Fig. 2.6. Specific heat as a function of temperature for two vitreous silica or suprasil (a-SiO$_2$) samples with different OH concentrations. The continous lines are obtained with (2.39) including the Debye contribution due to phonons. Taken from Zimmermann and Weber (1981b).

Before, there had been some discussions on how to explain these deviations from the linear temperature dependence (2.34) by modeling an appropriate distribution function P by introducing additional parameters like cutoffs or additional states.

Because we have used (2.35) to derive (2.39), the latter also is limited to the time range $\tau_{\min} \ll t \leq \tau_{\min}/u_{\min}^2$. We have implicitly assumed that all TLS's with $\tau < t$ ($\tau > t$) contribute to the specific heat with probability 1 (0). This makes the use of (2.39) rather limited; it breaks down at very low temperature and/or in the μs time scale as well as for very long times where C_{TS} should become time independent.

Following Deye and Esquinazi (1989) these restrictions can be overcome if one assumes that at a given time t, the probability p that a TS with relaxation time $\tau(E, T, u)$ couples to the perturbation is

$$p(E, T, t, u) = 1 - \exp\left(-\frac{t}{\tau(E, T, u)}\right). \tag{2.40}$$

The effective distribution function can then be expressed as

$$P(E,T,t,u_{\min}) = \int_{u_{\min}}^{1} p(E,T,t,u) \frac{P_0 \, du}{u\sqrt{1-u^2}} \tag{2.41}$$

without limitation in the time scale. Inserting this new time-dependent distribution function (2.41) into (2.37), we obtain

$$C_{\mathrm{TS}}(T,t) = \int_0^\infty C_1(E,T) \int_{u_{\min}}^1 \left[1 - \exp\left(-\frac{t}{\tau(E,T,u)}\right)\right]$$

$$\times \frac{P_0}{u\sqrt{1-u^2}} \, du \, dE. \tag{2.42}$$

This equation can be solved only numerically, but one can determine the time and temperature dependence of the specific heat for the following temperature and time limits:

$$t = 0 \quad , \quad C_{\mathrm{TS}} = 0, \tag{2.43}$$
$$t/\tau \ll 1 \quad , \quad C_{\mathrm{TS}} \propto t \cdot T^n, \tag{2.44}$$
$$t/\tau \geq 1 \quad , \quad C_{\mathrm{TS}} \propto T \cdot \ln\left(t \cdot T^{n-1}\right), \tag{2.45}$$
$$t \to \infty \quad , \quad C_{\mathrm{TS}} = \frac{\pi^2}{6} k_B^2 T P_0 \ln\left(\frac{2}{u_{\min}}\right), \tag{2.46}$$

where $n = 4, 8, 2$ correspond to a relaxation due to one-phonon process (2.18), two-phonon process (first-order Raman process, Doussineau et al. (1980)), and Korringa-like relaxation due to the interaction of TS's with conduction electrons (Golding et al. (1978)). For a short description of the above-mentioned processes, see Sect. 2.2.5.

In this approach, no new parameter has been introduced. According to Deye and Esquinazi (1989), for a more general ansatz with two new parameters,

$$p = 1 - \exp\left(-a\left(\frac{t}{\tau}\right)^m\right), \tag{2.47}$$

from the vitreous silica data and with reasonable values for the phonon-TS coupling constant γ, one finds $a = 1 \pm 0.5$ and $m = 1 \pm 0.4$. For a better determination of a and m, lower temperature ($T < 0.1$K) and shorter time scale (μs) data would be necessary.

The validity of the time dependence of the specific heat is demonstrated in Fig. 2.7, where experimental data on vitreous silica are plotted together with the numerical results from (2.42). We will now study the connection between the time-dependent specific heat $C_{\mathrm{TS}}(T,t)$ and the heat release of an amorphous solid. The energy needed to heat the solid from a temperature T_0 to a temperature T_1 (or which has to be removed if the solid is cooled down) can be expressed as the integral of the specific heat over the temperature

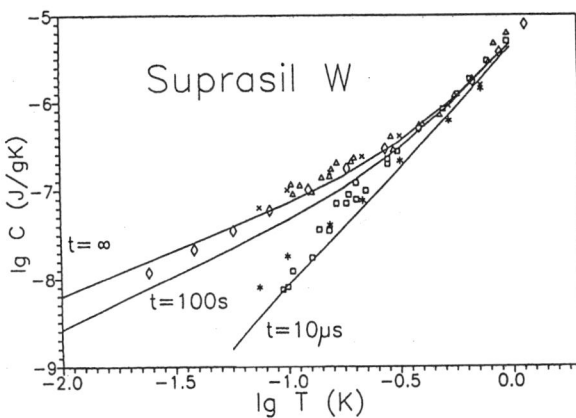

Fig. 2.7. Specific heat of Suprasil W as a function of temperature at different measuring times. The lines are results of numerical calculations using (2.42) with parameters $P_0 = 1.6 \times 10^{38} (\text{Jg})^{-1}$, $Ak_B^3 = 0.4 \times 10^7 \text{s}^{-1}\text{K}^{-3}$, $u_{\min} = 10^{-6}$. Taken from Deye and Esquinazi (1989), where also the original references are supplied.

$$Q(t) = \int_{T_0}^{T_1} C(T,t) \mathrm{d}T. \tag{2.48}$$

The time derivation leads to the heat the sample absorbs (or releases) per unit time at a time t:

$$\dot{Q}(t) = \int_{T_0}^{T_1} \dot{C}(T,t) \mathrm{d}T. \tag{2.49}$$

We can now insert (2.39) and we immediately find (2.32) which holds for $t < \tau_{\max}$, as stated there.

2.2.4 Influence of a Finite Number of Tunneling Systems

In Sect. 2.2.2 it was necessary to introduce finite integration limits to avoid a nonphysical infinite number of TS's. What is the consequence of such cutoffs for the heat release? Are they physically important or only mathematical parameters? If we look at (2.32), which is obtained disregarding the integration limits, obviously this equation is only valid in a limited time interval because for $t = 0$ the heat release would be infinite as well as the time integral over the heat release for $t \to \infty$.

Deye and Esquinazi (1989) have carried out numerical calculations of the heat release to check the significance of the value of $u_{\min}$. They found essentially two possible effects of $u_{\min}$ on the heat release of amorphous materials. The first effect is that the parameter $u_{\min}$ determines a crossover time t_c

where the time dependence of the heat release changes from t^{-1} to $t^{-1.67}$. Experimental time scales do not exceed 10^6s significantly, and with typical parameters for glasses (e.g. a-SiO$_2$), Deye and Esquinazi (1990) found that with $u_{\min} \sim 10^{-6}$, a direct effect of $u_{\min}$ on the heat release should be measurable. Long time heat-release measurements were not able to unambiguously detect an influence of $u_{\min}$ in glasses. Currently, an upper value $u_{\min} \sim 10^{-10}$ seems possible there, though there are hints from ultrasound experiments (Tielbürger et al. (1992)) that $u_{\min}$ might be smaller. However, in polycrystalline materials, a change in the time dependence of the heat release towards an exponential time dependence has been found, see Sect. 2.5.1.

A second consequence of $u_{\min}$ on the heat release is that the temperature dependence for higher T_1 should be weaker than the $(T_1^2 - T_0^2)$ dependence predicted by the tunneling model and should eventually become independent of T_1. This saturation indeed has been extensively measured (see for example Fig. 2.21 and Sect. 2.5.2), but again a value of $u_{\min} \approx 10^{-6}$ would be necessary to explain the experimental results. The saturation of the heat release with high T_1 could also be explained with a cutoff in E (Schwark et al. (1985), Koláč et al. (1986)). An alternative explanation of this effect is given in Sect. 2.3.2 based on thermally activated processes.

2.2.5 Influence of High-Order Tunneling Processes and a Finite Cooling Rate

In the calculations presented above, we have taken into account only the one-phonon process for the relaxation of the TLS's. However, in principle there could be other additional processes which could influence the relaxation. Possible relaxation mechanisms are:

1. First-order Raman process: A two-phonon process where one phonon resonantly excites the tunneling entity, then the tunneling process occurs and finally a phonon is emitted which carries the resulting energy difference. The relaxation rate for the process is given by (Doussineau et al. (1980))

$$\tau_{R1}^{-1} = u^2 R T^7 F_7\left(\frac{E}{2k_B T}\right), \qquad (2.50)$$

with

$$F_7(x) = \frac{x}{70}(x^2 + \pi^2)\left(x^4 - \pi^2 x^2 + \frac{10}{3}\pi^4\right)\coth(x). \qquad (2.51)$$

R is a constant proportional to $(\partial^2 \Delta/\partial u^2)^2$.

2. Electron-assisted tunneling in normal-conducting metals: A process where a conduction electron in a metal is inelastically scattered by a TS and carries the energy difference from the relaxation process. In a simple approach the electron-TS relaxation rate follows a Korringa-like law and can be written as (Golding et al. (1978))

2.2 Phenomenological Theory for the Heat Release

$$\tau_e^{-1} = u^2 KE \coth\left(\frac{E}{2k_\mathrm{B}T}\right). \tag{2.52}$$

The constant K describes the coupling between TS's and conduction electrons.[5]

3. Thermally activated relaxation: If we consider an ensemble of particles in thermal equilibrium, we know from thermodynamics that the energy of the particles is distributed. The probability that one particular particle with temperature T has an energy E is proportional to $\mathrm{e}^{-E/k_\mathrm{B}T}$. If E is larger than the potential height V of the TS, the particle can overcome the barrier *without tunneling*. As in the one-phonon process, the energy difference is given to the phonon system. However, because of the different relaxation mechanism, the relaxation rate is different and follows an Arrhenius law

$$\tau_\mathrm{TA}^{-1} = \tau_0^{-1} \mathrm{e}^{-V/k_\mathrm{B}T}, \tag{2.53}$$

τ_0 is the so-called attempt frequency, i.e. the frequency at which the particle touches the barrier.

4. Incoherent tunneling: Usually, dissipation in the quantum-mechanical system is not taken into account. If one does so, one finds that dissipation destroys the quantum coherence between the two wells. The corresponding relaxation mechanism is described in detail in Sect. 3.324. For heat-release measurements, this effect seems to be not relevant because mainly symmetrical TS's are affected which contribute little to the heat release [see (2.22)], but more research is necessary to confirm this.

The effective relaxation rate τ^{-1} is the sum of all relaxation rates

$$\tau^{-1} = \tau_\mathrm{R1}^{-1} + \tau_e^{-1} + \tau_\mathrm{TA}^{-1}. \tag{2.54}$$

Numerical simulations show that at temperature below $\approx 2K$, all the above-mentioned relaxation mechanisms can be neglected compared to the direct process. However, in normal-conducting disordered metals, the electron-assisted tunneling might be important and should be taken into account, see Sects. 4.2.3 and 4.4.2.

The above-mentioned relaxation mechanisms are relevant at higher temperatures, i.e. $T > 2$ K. In this case, the cooling process would have to be taken into account. As stated before, it is not possible to find an analytical solution of (2.29) for a finite cooling rate. However, one can simulate the cooling process numerically by dividing it into discrete cooling steps with *infinite* cooling rate but *finite* time duration. This method is illustrated in Fig. 2.8: The cooling function $T(t)$ is divided into several steps i of width Δt_i. $\tilde{N}_i$ denotes the difference in the population at the beginning of the ith time

---

[5] The constant K used in (2.52) has dimension $\mathrm{s}^{-1}/\mathrm{J}$. In some cases a unitless coupling constant is used within a similar equation. The relationship between these constants is given by the factor $2\hbar/\pi$, see also (3.8).

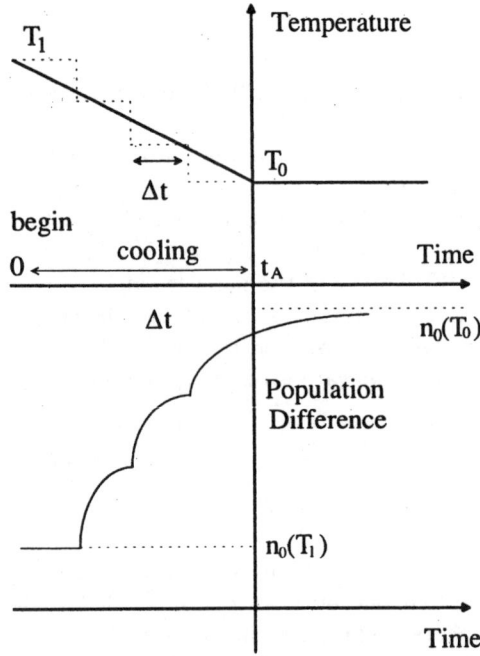

Fig. 2.8. Illustration of the numerical method to simulate the cooling process. The cooling procedure from T_1 to T_0 is devided into discrete steps with infinite cooling rate but of finite size, and the population difference is calculated iteratively. Taken from Nittke et al. (1995).

step, $\tilde{N}_{i+1}$ at the end of it, T_i is the corresponding temperature. The population difference at the measuring temperature is then calculated iteratively according to the formula

$$\tilde{N}_{i+1} = \tilde{N}_0(T_i) - \left(\tilde{N}_0(T_i) - \tilde{N}_i\right) \exp(-\Delta t_i/\tau(T_i)) . \qquad (2.55)$$

Because τ depends on both integration variables (e.g. E and u), this iteration has to be passed through for every value of the integrand. A possible influence of high-order processes in this picture would change the difference in the population number and thus also have an effect on the heat release.

Such numerical simulations have been carried out by Nittke et al. (1995) for the first-order Raman process and the thermally activated process. It has been found that the first-order Raman process leads to heat release which is the same as with only the direct process but with a lower "effective" charging temperature. According to these simulations, the thermally activated process additionally can change slightly the time dependence to $t^{-\alpha}$ with $\alpha < 1$. A finite cooling rate has a negligible effect on the heat release.

Figure 2.9 shows the heat release at $t = 1.5 \times 10^5$s calculated with and without thermally activated processes and a finite cooling rate. The saturation with charging temperature T_1 is an experimental fact observed in all amorphous materials, see Sect. 2.5.2; however, the crossover region from $\dot{Q} \propto (T_1^2 - T_0^2)$ to $\dot{Q} =$ const. is much broader in the experiments than in the simulation.

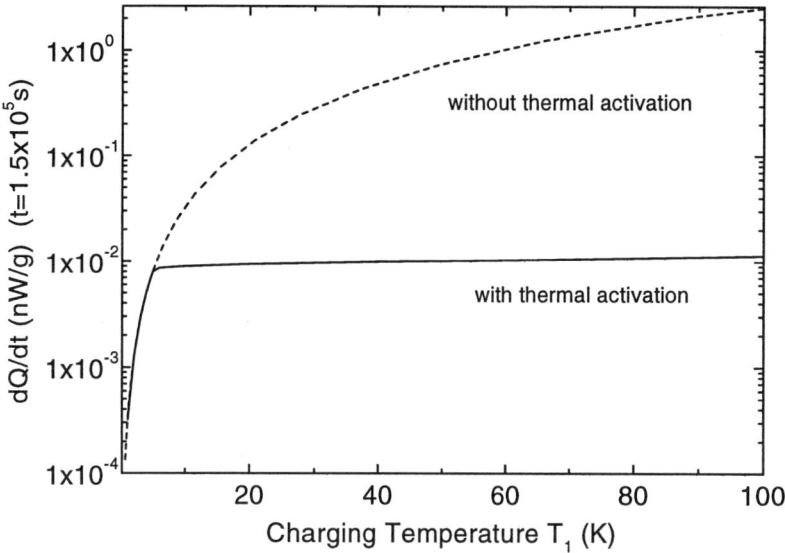

Fig. 2.9. Heat release at $t = 1.5 \times 10^5$ s as a function of charging temperature T_1 calculated with and without thermally activated processes and a finite cooling rate. Taken from Nittke et al. (1995).

A detailed analysis of the influence of thermal activation will be presented in Sect. 2.3.2, where it will be discussed in the framework of the soft-potential model. There, also the broadening of the transition region will be analyzed.

2.3 The Heat Release Within the Soft-Potential Model

In this section we describe the main results of the soft-potential model (SPM) for the heat release without going into conceptual details and reasons for the introduction of this model. This will be done in Chap. 9. We shortly note that the physical picture is essentially the same as in the tunneling model, but the shape of the potential is not that from two harmonic minima but a quartic potential. As we will see below and in Chap. 9, the success of the SPM lies in its predictions of the low-temperature and high-temperature ($T > 1$ K) properties of amorphous solids. In this chapter we will follow the original notation used by Karpov et al. (1983).

The soft-potential model (SPM) assumes that in disordered solids localized modes exist with weak and even negative restoring force constants. Such behavior can be described by the following soft anharmonic-oscillator potentials

$$V(x) = \mathcal{E}\left\{\eta\left(\frac{x}{a}\right)^2 + \xi\left(\frac{x}{a}\right)^3 + \left(\frac{x}{a}\right)^4\right\}, \tag{2.56}$$

where $\mathcal{E}$ is the binding energy of the atoms, x is their displacement and a is a distance of the order of the interatomic spacing. The values of the dimensionless parameters η and ξ are distributed due to the disorder of a glass (Karpov et al. (1983)).

We obtain an energy scale of (2.56) from the purely quartic potential ($\eta = \xi = 0$), where the spacing between the ground state and the first excited state is

$$W = \mathcal{E}\eta_L^2, \qquad (2.57)$$

with $\eta_L = (\hbar^2/(2Ma^2\mathcal{E}))^{\frac{1}{3}}$.

The analysis of experimental data within the soft-potential model shows that the effective mass M of the soft mode is much larger than the average mass m of the atoms constituting the glass (see Buchenau et al. (1991)), i.e. the soft mode is a collective motion of a group of atoms. Therefore, and also due to some fluctuations of $\mathcal{E}$, one expects that W varies from mode to mode. However, it seems that most low-temperature properties are not sensitive to a possible distribution of W and the calculation with a single average value of W yields a good agreement with the experimental data. It will be shown later (see Sect. 2.3.1) that the heat release is more sensitive to a distribution of W.

Whether (2.56) describes a single- or a double-well potential depends on the parameters η and ξ. For $\eta < \eta_L^2/\xi^2$, we have double-well potentials with the splitting energy E of the lowest two levels given by (2.4) but with

$$\Delta \approx \frac{W}{\sqrt{2}} \frac{|\xi|}{\sqrt{\eta_L}} \left(\frac{|\eta|}{\eta_L}\right)^{\frac{3}{2}}, \qquad (2.58)$$

$$\Delta_0 \approx W \exp\left[-\frac{\sqrt{2}}{3}\left(\frac{|\eta|}{\eta_L}\right)^{\frac{3}{2}}\right], \qquad (2.59)$$

and a barrier height between the two minima of

$$V \simeq \frac{W}{4}\left(\frac{\eta}{\eta_L}\right)^2. \qquad (2.60)$$

In the low-temperature range ($|\eta|, |\xi| \ll 1$), the soft-potential model reproduces the picture of the tunneling model and after the determination of the distribution function $P(\eta, \xi)$ in the vicinity of $\eta = \xi = 0$, we have to carry out the same calculations as in Sect. 2.2.

Following Il'in et al. (1987) the distribution function $P(\eta, \xi)$ equals $P(\eta, -\xi)$ because of the absence of a preferred direction in the glass. Thus, the function $P(\xi)$ has a maximum or a minimum at $\xi = 0$ and will be constant in the vicinity of zero. If we assume that most of the atoms in glasses move in standard (nonsoft) atomic potentials, the distribution $P(\eta, \xi)$ as a function of η has a maximum value for $\eta = 1$ and falls down slowly with decreasing η,

2.3 The Heat Release Within the Soft-Potential Model

so that $P(0,\xi) > 0$. However, the states with $\eta = 0$ (zero restoring force constant) are suppressed due to the interaction between soft- and high-frequency modes (see also Buchenau et al. (1991)). This leads to a linear dependence of the distribution function on $|\eta|$:

$$P(\eta,\xi) = \frac{P_0}{2}|\eta|, \qquad (2.61)$$

where P_0 is a constant *not identical* to that from the standard tunneling model, see also Chap. 9. In comparison with the tunneling model, the soft-potential model has essentially one more free parameter given by the energy W, or in some cases as for the heat release, a distribution of W is necessary to explain the data within the model.

Let us compare now the distribution function $P(\Delta, \lambda)$ of the soft-potential model and the tunneling model, starting from

$$P(\eta,\xi)\mathrm{d}\eta\mathrm{d}\xi = \frac{P_0}{2}|\eta|\mathrm{d}\eta\mathrm{d}\xi. \qquad (2.62)$$

With (2.58), (2.59) and

$$\lambda \equiv \ln\frac{\hbar\omega_c}{\Delta_0} = \ln\frac{\hbar\omega_c}{W} + \frac{\sqrt{2}}{3}\left(\frac{|\eta|}{\eta_L}\right)^{\frac{3}{2}}, \qquad (2.63)$$

we obtain

$$P(\Delta,\lambda) = 2\frac{P_0\eta_L^{7/2}}{W|\eta|}, \qquad (2.64)$$

where the factor 2 in (2.64) comes from taking positive and negative values of the asymmetry, see Buchenau et al. (1992). In contrast to the tunneling model this distribution function is not constant but decreases with increasing η or barrier height V [see (2.60)]. Usually, different experiments cover only a very small range of η and it is difficult to check the differences in the time or frequency dependence. Moreover, in scattering processes no difference will be observed for the value $P_0\gamma^2$, since the coupling constant γ is proportional to $\sqrt{|\eta|}$ within the soft-potential model and $P_0\gamma^2$ is constant in agreement with the experimental data and the tunneling model (see Buchenau et al. (1992)). Nevertheless, we will show in Sect. 2.3.1 that the different distribution functions of the two models lead to significant differences in the heat release.

In addition to the tunneling modes the soft-potential model contains quasi-harmonic states with one-well potentials ($9\xi^2/32 < \eta$). These modes influence the low-temperature features of glasses essentially; in particular they cause an additional contribution to the specific heat (at $T > 1.8\ W/k_B$) and the "plateau" of the thermal conductivity, for which the tunneling model provides no reasonable explanation, see Chap. 9. However, the relaxation times of these states are short and they will not influence the long-time heat release.

2.3.1 The Heat Release and Specific Heat

A detailed analysis of the heat release within the soft-potential model has been performed by Parshin and Sahling (1993). We start with (2.28), replacing $P(E, u)$ by the corresponding distribution function $P(E, u)$ from the soft-potential model. Equation (2.64) with (2.4), (2.58) and (2.59) leads to

$$P(E, u) \simeq \left(\frac{2}{9}\right)^{\frac{1}{3}} \frac{P_0 \eta_L^{5/2}}{W} \frac{1}{u\sqrt{1-u^2}} \frac{1}{L^{\frac{2}{3}}}, \qquad (2.65)$$

where $L = \ln(W/Eu)$, see also (9.30) and details below. The relaxation time τ can be expressed through the same variables too [see (2.18)]:

$$\frac{1}{\tau_p} = \frac{u^2}{\tau_{p,\min}(E)}, \qquad (2.66)$$

where

$$\frac{1}{\tau_{p,\min}(E)} = \frac{K_3 E^3}{8 k_B^3} \qquad (2.67)$$

and

$$K_3 = \frac{4 k_B^3 \gamma^2}{\pi \varrho \hbar^4 v^5}. \qquad (2.68)$$

In our further calculations, we will neglect the weak dependence of L on E and assume a constant deformation potential γ. The integration over E and u gives for $t \gg \tau_{p,\min}$

$$\dot{Q} \simeq \frac{\pi^2 k_B^2}{24} \left(\frac{2}{9}\right)^{\frac{1}{3}} \frac{P_0 \eta_L^{5/2}}{W} v \left(T_1^2 f(T_1, t) - T_0^2 f(T_0, t)\right) \frac{1}{t}, \qquad (2.69)$$

where

$$f(T, t) = \ln^{-\frac{2}{3}} \left(\frac{W}{k_B T} \left(\frac{t}{\tau_{p,\min}(k_B T)}\right)^{\frac{1}{2}}\right). \qquad (2.70)$$

The corresponding calculation of the heat capacity gives

$$C(T) \simeq \frac{\pi^2 k_B^2}{2} \left(\frac{2}{9}\right)^{\frac{1}{3}} v \frac{P_0 \eta_L^{5/2}}{W} T \ln^{\frac{1}{3}} \left(\frac{W}{k_B T} \left(\frac{t}{\tau_{p,\min}(k_B T)}\right)^{\frac{1}{2}}\right). \qquad (2.71)$$

Equation (2.71) can be obtained with (2.69) using (2.49). An improved expression for $C(T)$, taking into account the numerical estimate of the tunneling splitting, is given in Chap. 9, see (9.44).

With respect to the time and temperature dependence of the heat release and specific heat, both models provide similar results. Precise heat-release measurements over 3–4 decades in time would be necessary to distinguish between the time dependencies predicted by the different models. But, as a consequence of the $|\eta|$-dependence in (2.61), one expects that the absolute value of the heat release in comparison to the value calculated within the tunneling model will be smaller for the same corresponding values of the specific heat – if P_0 is assumed to be the same for both models – since the main contribution to the heat release comes from TS's with larger $|\eta|$-values (longer relaxation times).

2.3.2 Influence of Thermal Activation

At higher temperature T_1 or T_0 thermally activated relaxation of the TLS's will influence the heat release. To estimate the temperature where this occurs we first calculate the crossover temperature T_c at which the following relation is fulfilled:

$$\frac{1}{\tau_p} = \frac{1}{\tau_{\text{TA}}} = \frac{1}{t}. \tag{2.72}$$

τ_{TA}^{-1} is the relaxation rate due to thermal activation

$$\frac{1}{\tau_{\text{TA}}} = \frac{1}{\tau_0} \exp\left(-\frac{V}{k_{\text{B}}T}\right), \tag{2.73}$$

with $\tau_0 \simeq 10^{-12}$s. τ_p^{-1} is given by (2.66) with (2.59) and (2.60):

$$\frac{1}{\tau_p} = \frac{1}{\tau(E)} \exp\left(-\frac{8}{3}\left(\frac{V}{W}\right)^{\frac{3}{4}}\right), \tag{2.74}$$

where

$$\tau(E) = \tau_{p,\min}(E)\left(\frac{E}{W}\right)^2, \tag{2.75}$$

and t is certain experimental time where (2.72) is fulfilled. Using (2.72) for the elimination of V, we have

$$k_{\text{B}}T_c = W \frac{l(t)}{\ln\left(\frac{t}{\tau_0}\right)}, \tag{2.76}$$

where

$$l(t) = \left(\frac{3}{8}\ln\left(\frac{t}{\tau(E)}\right)\right)^{\frac{4}{3}}. \tag{2.77}$$

Typical values $t = 1$h, $\tau_0 = 10^{-12}$s, $\tau(E) = 10^{-10}$s lead to $k_{\text{B}}T_c = 0.74\,W$, i.e. the crossover will occur at low temperatures, since W typically has a value of a few Kelvin, e.g. $T_c \sim 3$ K, with $W \simeq 4$ K for a-SiO$_2$.

For $T_0 < T_1 < T_c$ the thermal activation can be neglected and (2.69) describes the heat release in the tunneling regime.

For $T_0 < T_c < T_1$ (intermediate case), the cooling starts from a temperature where the thermal activation process dominates, but the measurement is performed at a temperature where the thermal activation is negligible. In this case we have to consider the influence of thermal activation during the cooling procedure of the sample at the beginning of the measurement only. For $T_c < T_0 < T_1$ the thermal activation influences the relaxation of the TLS's during cooling and during the measurement too.

Intermediate Case. The main contribution to the heat release at $T_0 < T_c$ and time t comes from TLS's with $\tau = t$ and the corresponding barrier height V. Let us consider now the behavior of all TLS's with this barrier height V during cooling with a high but finite cooling rate. At high temperature, according to (2.73), the relaxation time is so short that all TLS's are in thermal equilibrium with the phonon system during cooling and will not contribute to the heat release after cooling is finished. That means, at high enough temperature T_1, the heat release is independent of T_1. However, with decreasing temperature, the relaxation time increases rapidly and below a certain freezing temperature T^* the cooling rate will be large enough that all TLS's remain in the nonequilibrium state and contribute to the heat release at T_0. This will occur at least at $T = T_c$, since below T_c the relaxation time of the two level systems is nearly temperature independent ($\tau(T_c) \simeq \tau(T_0)$) due to (2.74) (heat-release measurements have sense only for $t_A \ll t$, where t_A is the time necessary for cooling the sample from T_1 to T_0). Thus, (2.69) is correct for the intermediate case also, if we reduce T_1 to T^*. The calculation of T^* gives, see Parshin and Würger (1992),

$$k_B T^* \approx \frac{V}{\ln\left(k_B T^{*2}/\tau_0 V |R|\right)}, \tag{2.78}$$

where R is the cooling rate. If the cooling rate is not constant, we can use the cooling rate at T^* as a first approximation. The calculation of T^* shows that $T_c < T^* < 1.4 T_c$ and we can replace T^* by T_c within the logarithmical function. From (2.74), we find

$$V = W l(t), \tag{2.79}$$

where l(t) is given by (2.77). With (2.76) we finally arrive at

$$k_B T^* = W \frac{l(t)}{\ln\left(\frac{W l(t)}{k_B \tau_0 |R| \ln^2(t/\tau_0)}\right)}. \tag{2.80}$$

With typical values $t = 1$ h, $\tau_0 = 10^{-12}$ s, $\tau(E) = 10^{-10}$ s and $W = 4$ K, we obtain $k_B T^* = 0.85\, W$ and $T^*/T_c = 1.15$. Due to the very strong temperature dependence of the thermal activation relaxation time (2.73), the freezing temperature is near the crossover temperature. Thus, at $T^* \approx T_c \approx W/k_B$ the T_1 dependence of the heat release changes $\dot{Q}$ is constant for $T_1 > T^*$ and proportional to $T_1^2 - T_0^2$ for $T_1 < T^*$. We will show now that the temperature range ΔT^* where the temperature dependence changes is small ($\Delta T^*/T^* \simeq 0.2$). Using the distribution function $P(\tau, E)$ of the tunneling model (2.24) for $\tau \gg \tau_{p,\min}$ the heat release in the tunneling range is given by

2.3 The Heat Release Within the Soft-Potential Model

Table 2.1. Calculated values for V (potential barrier height), T_c (crossover temperature), $T^\star$ (freezing temperature) and $\Delta T^\star/T^\star$ (relative temperature range, where the T_1-dependence of $\dot Q$ changes) for a-SiO$_2$ and LiCl·7H$_2$O. The characteristic energy W was deduced from the temperature $T_{\min}$ where a minimum of $C(T)/T^3$ was observed: $W \simeq 2k_B T_{\min}$. The coefficient K_3 [see (2.68)] was determined from ultrasonic experiments. $\tau(E)$ was calculated with (2.75) and $E = 2.4 k_B T^\star$.

Material →	a-SiO$_2$			LiCl·7H$_2$O		
W/k_B (K)	4			14		
K_3 (s^{-1}K^{-3})	4×10^8			4.4×10^9		
$\tau(E)$ (s)	1.6×10^{-9}			2.8×10^{-13}		
τ (s)	180	3600	18000	180	3600	18000
V/k_B (K)	90.7	104.0	111.3	418	468	495
T_c (K)	2.76	2.90	2.97	12.7	13.1	13.2
$T^\star$ (K)	3.05	3.50	3.75	13.4	15.0	15.9
$k_B T_c/W$	0.69	0.73	0.74	0.91	0.94	0.94
$k_B T^\star/W$	0.76	0.88	0.93	0.96	1.07	1.14
$\Delta T^\star/T^\star$ for t=1h	0.21			0.17		
$T^\star$ (K) for t=1h	3.4			14.7		

$$\dot Q = \frac{\pi^2 k_B^2}{24} \mathcal{V} P_0 (T_1^2 - T_0^2) \int_{\tau_1}^{\infty} \frac{1}{\tau^2} e^{-t/\tau} d\tau, \qquad (2.81)$$

where $\tau_{p,\min} \ll \tau_1 \ll t$. Two-level systems with very long or short relaxation times compared to the experimental time scale will not contribute to the heat release. Indeed, the integration of (2.81) from $0.05t$ to $5t$ yields 95% of the exact value and $\Delta T^\star$ can be calculated from

$$\Delta T^\star = T^\star(5t) - T^\star(0.05t). \qquad (2.82)$$

The corresponding values for a-SiO$_2$ with $t = 1$h, $W = 4$K, $\tau_0 = 10^{-12}$s, $|R| = 10^{-2}$Ks^{-1} and $\tau(E) = 3 \times 10^{-10}$s [calculated for $E = 2.4 k_B T^\star$ with (2.67) and (2.75) and $K_3 = 4 \times 10^9$s^{-1}K^{-3} deduced from ultrasonic experiments] are given in Table 2.1. Indeed, the ratio $\Delta T^\star/T^\star \simeq 0.2$, which enables us not only to determine the important (average) value of W from the T_1-dependence of the heat release but also to answer the question as to whether a distribution of W exists. Such a distribution of W would lead to a broadening of $\Delta T^\star$, see (2.80).

Thermal Activation Range. In this range ($T_c < T_0 < T_1$) both the calculation and the measurement of the heat release is much more complicated. The heat release is caused by TLS's with much larger barrier heights than in the tunneling range. For example, the main contribution to the heat release in a-SiO$_2$ at $T_0 = 6$ K comes from TLS's with $V/k_B \simeq 215$ K [see (2.73)]. *These* TLS's give a remarkable contribution to the heat release in the *tunneling range* ($T_0 < 3$K) after only 1000 years. With (2.78) we can calculate the corresponding freezing temperature $T^\star(V/k_B = 215K) \simeq 7.1$K ($\tau_0 = 10^{-12}$s,

$|R| = 10^{-2}$K/s, T^{*2} in the logarithmical function can be replaced by T_0^2). Above T^* the occupation number equals the equilibrium occupation. For high enough T_1 ($T_1 > 1.3T_0$) the heat release will be independent on T_1 and we have to use the frozen occupation $\tilde{N}_0(T^*)$:

$$\dot{Q} = \sum_{\text{TLS}} \frac{E}{\tau} \left(\frac{1}{e^{E/k_B T^*} + 1} - \frac{1}{e^{E/k_B T_0} + 1} \right) e^{-t/\tau} \Theta(T^* - T_0). \tag{2.83}$$

With the Heaviside step function $\Theta(x)$ ($\Theta(x) = 1$ for $x > 0$, $\Theta(x) = 0$ for $x < 0$), we take into account the fact that TLS's with $T^* \leq T_0$ do not contribute to the heat release. Transforming the distribution function $P(\xi, \eta)$ to $P(V, E)$ and expressing V in terms of τ according to (2.73), we obtain after integration over E,

$$\dot{Q} \simeq \frac{\pi^2 k_B^2}{24} \frac{P_0 \eta_L^{\frac{5}{2}}}{W^{\frac{5}{4}}} V \int_{t_0}^{\infty} \frac{1}{\tau^2} e^{-t/\tau} \frac{k_B T_0}{(k_B T_0 \ln \frac{\tau}{\tau_0})^{\frac{3}{4}}} (T^{*2} - T_0^2) \, d\tau. \tag{2.84}$$

$\Theta(x)$ was eliminated by the introduction of the lower limit of integration t_0, determined by the condition $T^*(t_0) = T_0$. From (2.78) with (2.73) we obtain, replacing T^* in the logarithmic function by T_0,

$$t_0 = \frac{T_0}{|R| \ln \frac{t_0}{\tau_0}} \approx \frac{T_0}{|R| \ln \left(\frac{1K}{|R|\tau_0} \right)}. \tag{2.85}$$

For typical values of heat-release experiments ($T_0 = 10$K, $\tau_0 = 10^{-12}$s, $|R| = 10^{-2}$K/s), we obtain $t_0 \approx 30$s. Thus, our further calculations will provide a good approximation for $t \gg t_0 \simeq 30$s. Using (2.78) and (2.73) we find for the τ-dependence of T^*:

$$k_B T^* \approx \frac{V}{\ln \left(\frac{k_B T^{*2}}{\tau_0 |R| V} \right)} \approx \frac{k_B T_0 \ln \frac{\tau}{\tau_0}}{\ln \frac{t_0}{\tau_0}}. \tag{2.86}$$

With $y = 1/\tau$, (2.84) can be transformed into

$$\dot{Q} \simeq \frac{\pi^2}{24} \frac{P_0 \eta_L^{5/2}}{W} \frac{W^2}{\ln^2 \frac{t_0}{\tau_0}} \left(\frac{k_B T_0}{W} \right)^{\frac{9}{4}} \frac{V}{t} \int_0^{\frac{t}{t_0}} e^{-y} \frac{\ln^2 \left(\frac{t}{\tau_0 y} \right) - \ln^2 \left(\frac{t_0}{\tau_0} \right)}{\ln^{\frac{3}{4}} \left(\frac{t}{\tau_0 y} \right)} \, dy. \tag{2.87}$$

Since $t \gg t_0$, we finally obtain

$$\dot{Q} \simeq \frac{\pi^2}{12} \frac{P_0 \eta_L^{5/2}}{W} \frac{W^2}{\ln^2 \frac{t_0}{\tau_0}} \left(\frac{k_B T_0}{W} \right)^{\frac{9}{4}} f(t) \frac{1}{t}, \tag{2.88}$$

where

$$f(t) = \frac{\left(\ln \left(\frac{tt_0}{\tau_0^2} \right) \right)^{\frac{1}{2}} \ln \left(\frac{t}{t_0} \right)}{\left(\ln \left(\frac{t}{\tau_0} \right) \right)^{\frac{3}{4}}}. \tag{2.89}$$

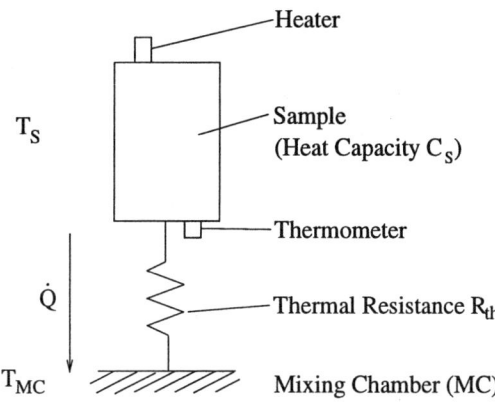

Fig. 2.10. Experimental setup for the quasi-static measurements. The sample is connected to the cold bath through a weak link (thermal resistance R_{th}).

Numerical calculations show that $(\frac{1}{t})f(t)$ is proportional to $t^{-0.76}$ for $20 < t/t_0 < 200$.

From (2.86) we can see that $(T^\star - T_0)/T_0$ is a small value for all TLS's contributing to the heat release. Since we can neglect the contribution of TLS's with $\tau \geq t$ to the heat release, we find for typical values ($t_0 = 30$ s, $t = 1$ h, $\tau_0 = 10^{-12}$s) that $(T^\star(36000\text{s}) - T_0)/T_0 = 0.2$.

Thus, in the thermal activation range, all dependencies change in comparison to the tunneling regime: the heat release is independent of T_1 (for $T_1 > 1.2T_0$), it is proportional to $T_0^{9/4}$ and roughly proportional to $t^{-0.76}$ (instead of t^{-1} in the tunneling regime).

2.4 Experimental Details

There are two different methods which are used to measure the heat release of solids. We will present here both experimental setups and discuss their advantages and difficulties.

2.4.1 Quasi-static Measurements

This is the method which has been used by Zimmermann and Weber (1981a) for the first experiments of the heat release in amorphous solids. The idea is to measure the heat release via the temperature difference caused by the heat flow through a thermal resistance. The setup used for this kind of measurement is shown in Fig. 2.10: The sample with a heat capacity C_S is mounted on a sample holder and is connected to the bath through a thermal resistance R_{th}. Additionally, there are a heater H and a thermometer T_S mounted on the sample.

To understand the working principle of this method, we regard a state where the sample releases *no* heat. The bath temperature is T_0, which is ideally (assuming zero heat leak) equal to the temperature of the sample T_S. Now, we turn on the heater, it applies a certain, well tunable electrical power P_{el} to the sample. The sample temperature increases and reaches a static state with a time constant $\tau_i(T_S) \simeq R_{th}(T_S)C_S(T_S)$, assuming that $(T_s - T_0)/T_0 \ll 1$. In this static state, the following equation describes the relation between the applied electrical power P_{el} and the sample temperature T_S

$$P_{el} = \frac{T_S - T_0}{R_{th}(T_S)} = f(T_S). \tag{2.90}$$

If the sample releases heat due to the relaxation of TLS's, the same equation applies, but we have to keep in mind that it is valid only in the static state. This means that we can use this method without further refinement only if the time scale of heat release is much larger than the internal relaxation time τ_i (this is the reason for the name "quasi-static").

Decreasing the temperature of the bath, the temperature of the sample follows it with an (temperature-dependent) internal relaxation time $\tau_i(T) = R_{th}(T)C_S(T)$. After the final bath temperature has been achieved and after waiting an appropriate time for the steady state to establish, the heat released by the sample can be found directly from the sample temperature.

For the design of such an experiment the value of the thermal resistance R_{th} should be chosen carefully because it determines two important quantities, namely the sensitivity and the accessible time range. Because the time scale of the measurements has to be much larger than the internal relaxation time τ_i, for a given sample with heat capacity C_S, the thermal resistance should be as *small* as possible. On the other hand, to get a large temperature change for a given change in the heat release, the thermal resistance should be as *large* as possible. Thus, it depends strongly on the experiment which value for the thermal resistance is the optimum.

In practice, it is possible to produce a relatively wide range of values for R_{th} from rather high values (only the sample holder and the wires to the thermometer and to the heater serve as a weak link to the bath) to very small values which can be achieved simply with a copper wire.

Because of the importance of the right choice of the thermal resistance, we give an example: We want to measure the heat release of a 10g Suprasil sample at $T = 200$ mK. The heat capacity for this sample is $C_S \simeq 2$nJ/mK. We assume that the error in the measurement of the temperature (stability of the bath temperature plus sensitivity of both thermometers) is 0.1 mK. We want to measure the long time relaxation ($t \geq 1$ h) after cooling it down from nitrogen temperature ($T = 77$ K), so a measuring time scale in the 100 s range is appropriate. Then, the thermal resistance has to be $R_{th} = \tau_i/C_S = 50$ mK/nW.

Is this value for R_{th} reasonable? No, because we know from the literature (or we will learn in the course of the experiment) that most amorphous materials show a heat release of 0.1–1 nW/g shortly (≈ 3 h) after cool down

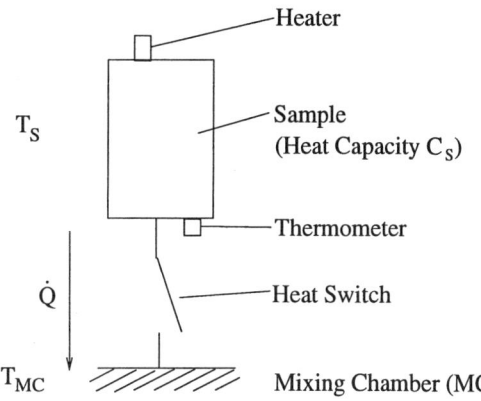

Fig. 2.11. Experimental setup for a calorimeter. Apart from measuring the specific heat, it is possible to measure the heat release of a sample with this setup too.

which would lead to a temperature rise of the sample of at least 50 mK. This is not acceptable at $T_0 = 200$ mK, so we better use a thermal resistance of 5 mK/nW which provides a temperature difference of ≤ 50 mK for rather short times. If the temperature difference exceeds about 20% of T_0 then the thermal resistance has to be decreased. In this case, $\tau_i = 10$ s and the sensitivity is 0.02 nW which usually is also the limit of the stability of the background signal.

The advantage of this setup is its great simplicity and its high accuracy for small absolute values of the heat release. The most disturbing disadvantage of this method is that if the thermal resistance turns out to be not well chosen, it may be necessary to do the experiment once again.

2.4.2 Calorimetric Measurements

The experimental setup used here is similar to a calorimeter (Fig. 2.11): The sample is isolated from the surroundings and can be connected to the cold bath by a heat switch. For cooling down, the heat switch is closed. When the sample has reached the bath temperature, the heat switch is opened and the sample starts to warm up due to the internal heat release from the relaxing TLS's. With a known specific heat of the sample (which can be determined easily with this setup) the heat release can be calculated from the warm-up curve. For every data point of the heat release, one warm-up curve has to be recorded.

This setup has been used extensively in the 1 K temperature range. An advantage of this method is the very short cool-down time which makes measurements on short time scales (some minutes) possible.

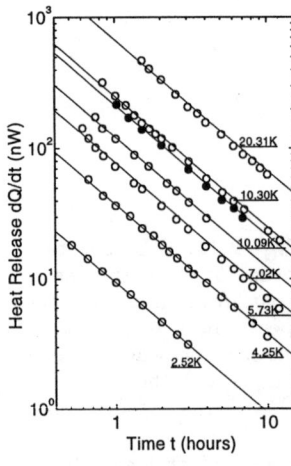

Fig. 2.12. Heat release of 110 cm^3 LiCl·7H$_2$O after rapid cooling of the sample from different charging temperature T_1 to 1.5 K (o) or 4.2 K (•). The straight lines are proportional to t^{-1}. Data taken from Sahling (1989).

2.5 Experimental Results

In this section, we review some results of the heat-release measurements. In Sect. 2.5.1 we discuss the observed time dependence; the data are classified according to the material class. In Sect. 2.5.2 we present experimental data on the temperature dependence of the heat release. One peculiar deviation from the temperature dependence predicted by the tunneling model will be discussed in Sect. 2.5.3 in terms of thermal activation within the framework of the soft-potential model.

2.5.1 The Time Dependence of the Heat Release

In Table 2.2 we have compiled different data published in the literature taking into account the time dependence of the heat release. A short discussion of the peculiarities observed in the different classes of materials is given below.

Dielectric Anorganic Materials. This is the group of materials which follow the predictions of the tunneling model very well. We present here the two materials which were examined thoroughly in the past: LiCl·H$_2$O (Fig. 2.12) and a-SiO$_2$ (Fig. 2.13). The data are plotted in double logarithmic plots, where a t^{-1} dependence appears as a straight line with slope -1. The figures show that both materials follow the t^{-1} time dependence predicted by the tunneling and soft-potential models. This is the case even for high charging temperature as can be seen in the a-SiO$_2$ sample. We will see later that at higher temperature additional processes are important and it can not be taken for granted that this time dependence holds after the cooling procedure.

Table 2.2. Comprehensive compilation of the time dependence of heat-release measurements. τ_{max} is the upper limit of the relaxation time spectrum and a is the exponent in the time dependence t^{-a} of the heat release. If a range is given for a, the time dependence of the heat release follows no power law.

Amorphous Dielectrics

Material	a	τ_{max} (h)	Reference
a-SiO$_2$	1.0	> 2	Zimmermann and Weber (1981a)
a-SiO$_2$	1.0	> 200	Schwark et al. (1985)
LiCl·7H$_2$O	1.0	> 48	Sahling (1990)

Amorphous Metals

Material	a	τ_{max} (h)	Reference
Fe$_{80}$B$_{14}$Si$_6$	1.0	> 48	Koláč et al. (1986)
Co$_{69}$Fe$_{4.5}$Cr$_2$Si$_{2.5}$B$_{22}$	1.0	> 48	Sahling et al. (1986a)

Amorphous Organic Materials

Material	a	τ_{max} (h)	Reference
Stycast	0.7	> 200	Schwark et al. (1985)
Epoxy resin	0.76	> 1000	Koláč et al. (1987)
PMMA	0.6–1.1	> 2	Zimmermann (1984)
PMMA	0.9	> 170	Nittke et al. (1995)
PS	0.7–3.2	200	Nittke et al. (1995)
Pentanol-2	0.69	> 10	Bunyatova et al. (1990)
3-methylpentane/ 2,3-dimethylbutane	0.6–7	30–50	Sahling (1992)

Glass–like Crystalline Materials

Material	a	τ_{max} (h)	Reference
Pb$_{0.915}$La$_{0.085}$(Zr$_{0.65}$Ti$_{0.35}$)O$_3$	1.2–2.0	0.3–3	Sahling et al. (1988)
(Sr$_{0.55}$Ba$_{0.45}$)Nb$_2$O$_3$	2.0	≈ 0.5	Mattausch et al. (1996)
KH$_4$PO$_4$	-	< 0.2	Sahling et al. (1986b)
YBa$_2$Cu$_3$O$_7$	1.0	> 50	Sahling and Sahling (1988)
YBa$_2$Cu$_3$O$_6$	1.0	> 20	Sahling and Sahling (1989)
Bi$_2$CaSr$_2$O$_8$	1.2–2.0	0.5–1.5	Sahling and Sievert (1990)
(ZrO$_2$)$_{0.87}$(CaO)$_{0.13}$	1.0	> 15	Abens et al. (1996)
NbTi-H	1.0	> 100	Schwark et al. (1985)
NbTi-D	1.0	> 100	Schwark et al. (1985)
NbTi	1.0	> 10	Abens and Sahling (1996)
Al	0.2–1.0	≥ 330	Nittke and Esquinazi (1996)

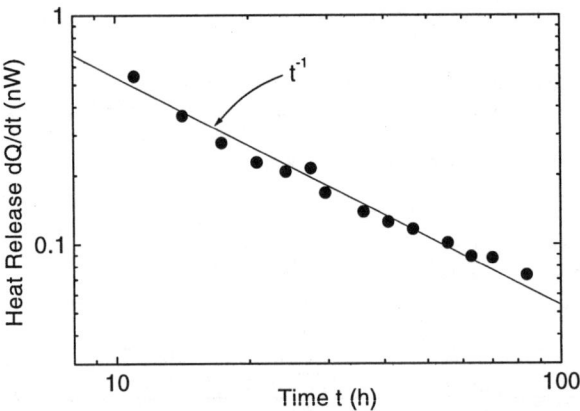

Fig. 2.13. Heat release of a-SiO$_2$ after cool down from nitrogen temperature to a measuring temperature of $T_0 = 20$ mK. Data taken from Schwark et al. (1985).

Amorphous Metals. In amorphous metals, the t^{-1} time dependence predicted by the tunneling model is found, too. As an example the heat release of Fe$_{80}$B$_{14}$Si$_6$ is shown in Fig. 2.14. This measurement demonstrates the universality of heat-release measurements: It is very difficult and partially impossible to determine the distribution parameter P_0 of amorphous metals, since the main contribution to the low temperature heat capacity is given by the electrons (and magnons in Fe$_{80}$B$_{14}$Si$_6$), and in scattering experiments the scattering on electrons (and magnons) dominates. In heat-release measurements electrons in principle may influence the relaxation process of TS's. But this influence is important at lower temperatures ($T_0 \ll 1$ K) only (see Chaps. 3 and 4). For $T_1 < 6$ K the temperature dependence of the tunneling model was found and the distribution parameter $P_0 = 2.3 \times 10^{37}$J^{-1}g^{-1} or $P_0 = 1.5 \times 10^{44}$J^{-1}m^{-3} (with mass density $\rho = 6.5$ g/cm^3) was obtained. Thus, heat-release measurements on amorphous metals yield nearly the same distribution parameter as for a-SiO$_2$ (see Table 2.5).

Organic Materials. Organic materials often show deviations from the t^{-1} time dependence. In Fig. 2.15 we present heat-release data for some organic compounds, mainly epoxy resins, after cool down from nitrogen temperature measured by Schwark et al. (1985). It is typical for organic materials that the time dependence is weaker than t^{-1}. Within the usual time window of the experiments and within the error, the time dependence of the heat release can be mostly fitted by a power law. However, there are also organic materials which show a t^{-1} time dependence of the heat release, like PMMA (Fig. 2.16), Vespel [see (d) in Fig. 2.15] and polystyrene (PS) for $T_1 \leq 1$K (Fig. 2.17).

The above-mentioned deviations have not yet been fully investigated and it is not clear up to now what causes them. Possible explanations are: (a) the complicated nature of these materials leads to a nonuniform distribution

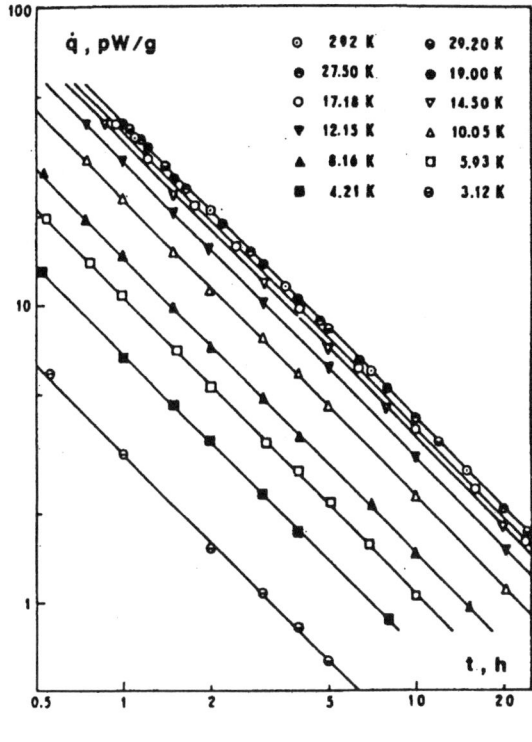

Fig. 2.14. Heat release of $Fe_{80}B_{14}Si_6$ at $T_0 = 1.3$ K. Taken from Koláč et al. (1986).

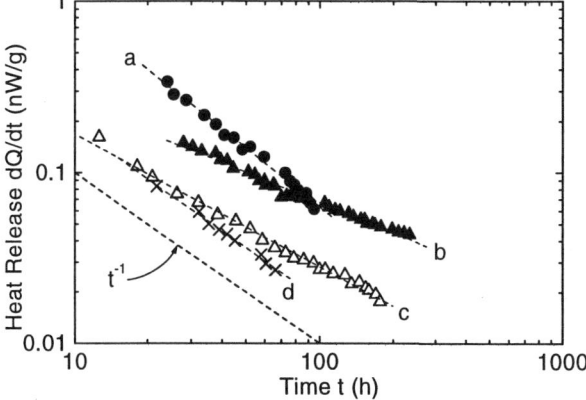

Fig. 2.15. Heat release of (a) Dimethyl-Siloxan, (b) Stycast 2850 F, (c) Stycast 1266, (d) Vespel. $T_1 = 80$ K, $T_0 = 20$ mK; for further details, see Schwark et al. (1985).

function of TS or (b) during cool down the TS's are nonuniformly relaxed. In this context it is interesting to note that PMMA shows a much larger heat release (by a factor of 10) than PS after cooling them from nitrogen temperature (Nittke et al. (1995)). This is surprising because at low temperature they exhibit similar properties like the specific heat, thermal conductivity and the sound velocity and attenuation. However, the heat release for low charging temperatures is different only by a factor of 3, see Table 2.4.

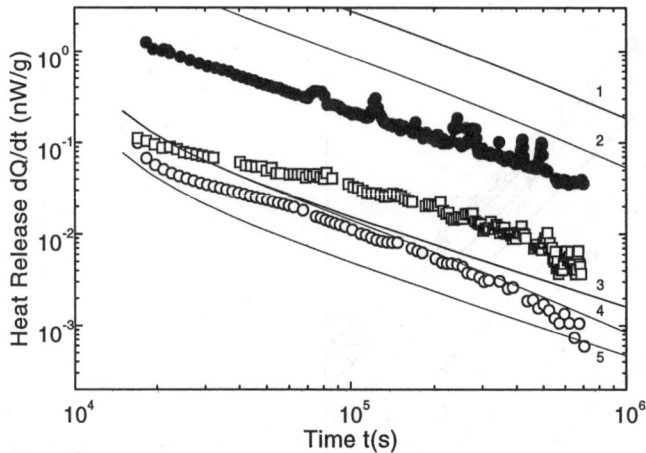

Fig. 2.16. Heat release of PMMA (•, $T_0 = 90$ mK) and of polystyrene (PS) (o: $T_0 = 90$ mK, □: $T_0 = 300$ mK) after cooling the sample from $T_1 = 80$ K. The continuous lines 1–5 are calculated according to the standard tunneling model. A detailed explanation of these lines is given by Nittke et al. (1995).

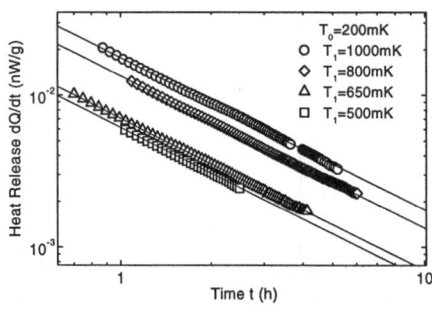

Fig. 2.17. Heat release of PS at $T_0 = 200$ mK with various charging temperatures as indicated in the plot. The straight lines have a t^{-1} dependence. Data are taken from Nittke et al. (1995).

Crystalline Materials. Considering the experimental facts in amorphous materials which stimulated the tunneling model one would not expect to find a measurable heat release in polycrystalline materials. However, the acoustic results of polycrystalline metals (see Sect. 4.5) reveal a surprising similarity to amorphous materials. Therefore, we would expect to find a measurable heat release if the density of states and distribution of relaxation time are broad enough. The heat release of a few polycrystalline samples was measured, see Table 2.2. For example, in $Pb_{0.915}La_{0.085} \cdot Zr_{0.65}Ti_{0.35}$ (PLZT, Sahling et al. (1986b)), a heat release comparable to that found in usual glasses has been measured, see Fig. 2.18. In contrast to other heat-release measurements on glasses, a peculiar deviation from the t^{-1} time dependence of the heat release has been found. Obviously, the data obey no power law but can be described by the function $\dot{Q} \propto t^{-1}e^{-t/\tau_{max}}$, see (2.31). This can be interpreted as τ_{max} – the upper limit in the relaxation time spectrum – is so short in these materials that influences the t dependence of $\dot{Q}$ in the relatively short time scale.

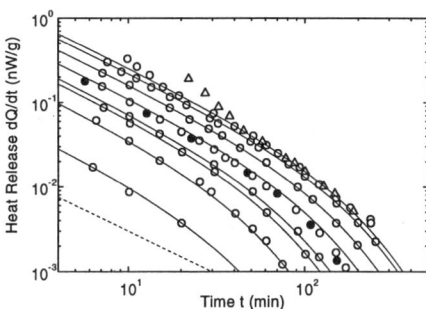

Fig. 2.18. Heat release of polycrystalline $Pb_{0.915}La_{0.085} \cdot Zr_{0.65}Ti_{0.35}$ (PLZT) after cooling from T_1 to $T_0 = 1.3$ K. Taken from Sahling et al. (1986b).

A measurable heat release has been found in polycrystalline aluminum. The heat release of a highly pure (6N) aluminum sample ($m = 61.8$ g) has been measured after cooling it from $T_1 = 80$ K to a measuring temperature $T_0 = 200$ mK (Nittke and Esquinazi (1996)). The experimental data are shown in Fig. 2.19. In contrast to the results on PLZT, the heat release in aluminum shows more an evidence for a *minimum* relaxation time than for a maximum. In agreement with vibrating reed measurements (see Sect. 4.5), no influence of conduction electrons could be found in measurements in a magnetic field higher than the critical field. Within the resolution of the

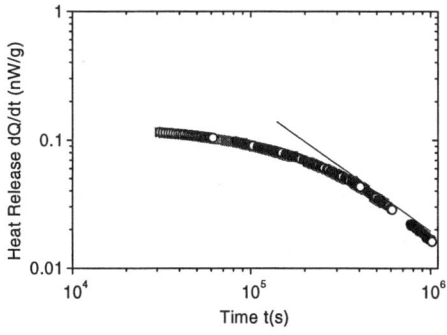

Fig. 2.19. Heat release of a polycrystalline aluminium sample ($m = 61.8$ g) at $T = 200$ mK. The solid line represents the corresponding heat release of a-SiO_2. Data are taken from Nittke and Esquinazi (1996).

experiment no heat release could be found in less pure aluminum at higher temperature ($T_0 > 1$ K, Sahling (1989)). In comparison with the heat-release behavior of amorphous insulators, the heat release of Al also shows a striking difference. Nittke and Esquinazi (1996) tried to find the density of states of TS's in Al "charging" the sample at $T_1 = 2$ K for a given period of time and interrupting at different times its relaxation produced by systems charged at 80 K. Usually one would expect an increase of the heat release after cooling the sample from 2 K to the measuring temperature of $T_0 = 0.2$ K. However, the opposite effect (although very weak) has been found. This would indicate that there are less TS's which can be excited at 2 K as assumed by the standard tunneling model.

In summary, based on the heat-release results in polycrystalline materials, one finds that polycrystalline materials show partially glassy behavior, containing TS's like glasses but with probably nontypical distribution functions.

2.5.2 The Temperature Dependence of the Heat Release

According to (2.32) the heat release should depend on the temperature as $\dot{Q} \propto (T_1^2 - T_0^2)$, i.e. it depends on the charging temperature T_1 as well as on the measuring temperature T_0.

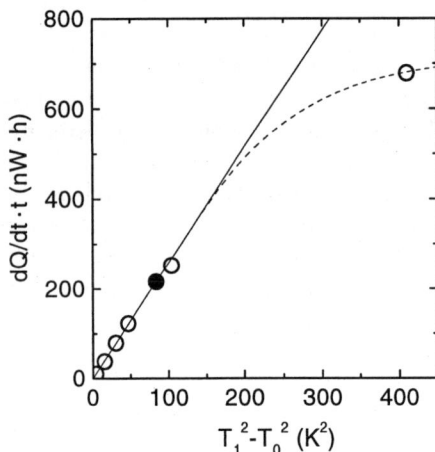

Fig. 2.20. The product $\dot{Q} \cdot t$ plotted versus $T_1^2 - T_0^2$ for a 110 cm^3 LiCl·7H$_2$O. o: $T_0 = 1.3$ K, •: $T_1 = 4.2$ K. The dotted line is a guide to the eye. Data are taken from Sahling (1989).

For practically all investigated materials, (2.32) is fulfilled in the temperature region $T_1 \leq 3$ K; for LiCl·7H$_2$O even up to $T_1 \leq 10$ K. This can be seen in Figs. 2.20 and 2.21, where $\dot{Q} \cdot t$ is plotted versus $T_1^2 - T_0^2$, which, according to (2.32), should show a linear dependence. Even for the organic materials which show non-t^{-1} time dependence, the temperature dependence predicted by (2.32) holds if one regards the heat release at a certain time,

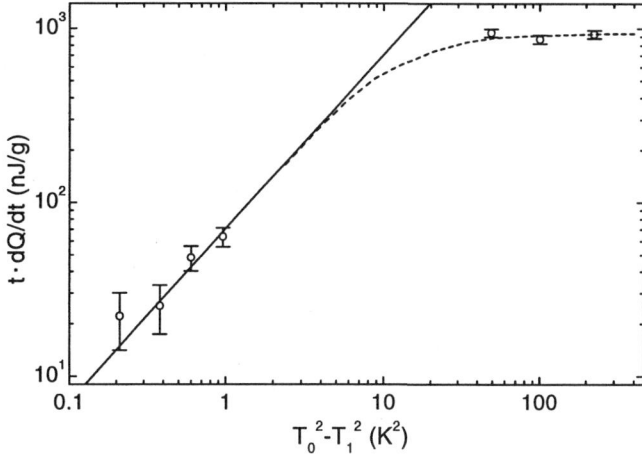

Fig. 2.21. $\dot{Q} \cdot t$ plotted versus $T_1^2 - T_0^2$ for PS. The solid line represents a fit according to (2.32), the dashed line is a guide to the eye. Data taken from Nittke et al. (1995).

e.g. at $t = 1$ h. This is a clear evidence that in this temperature region the tunneling model works very well.

However, for higher charging temperature all materials studied show clear deviations from this law. The temperature dependence of the heat release becomes weaker and eventually saturates, i.e. it becomes independent of T_1 as can be observed in Figs. 2.20 and 2.21. Possible explanations for this effect will be discussed in the following section.

2.5.3 Influence of Thermal Activation

A first attempt to explain the saturation of the heat release $\dot{Q}$ with the charging temperature T_1 within the tunneling model was made by Schwark et al. (1985) assuming the existence of a cutoff energy E_f in the density of states $P(E)$ for the TLS's:

$$P(E) = P_0 \frac{1}{1 + \exp\left((E - E_f)/k_B T_b\right)}. \tag{2.91}$$

Indeed, with (2.91) an excellent fit for all experimental heat release data with $T_b \simeq 0$ (sharp cutoff) can be obtained, see also Koláč et al. (1986). However, these calculations lead to unexpected low values for E_f and there are no arguments from other low temperature experiments or theories that confirm the existence of such a cutoff energy. Another explanation has been given by Deye and Esquinazi (1990) analyzing the consequence of a minimum value $u_{\min} = \Delta_{0\min}/E_{\max}$ in the distribution function $P(E, u)$ [see (2.27) and Sect. 2.2.4]. In this case $E_{\max}$ can be larger ($E_{\max} \simeq k_B T_g$, where T_g is the glass transition temperature) and the new fit parameter $\Delta_{0\min}$ determines a

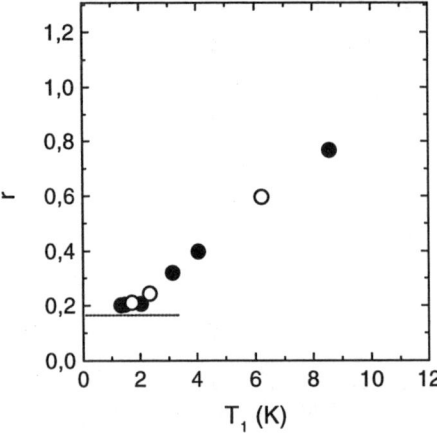

Fig. 2.22. Ratio r of power released 1h after cooling from T_1 with short waiting time ($t_M = 0.3$ h) to that after cooling from equilibrium $T_1 (t_M = 20$ h) at the measuring temperatures: (•) $T_0 = 1.3$ K, (○) $T_0 = 1.15$ K. The horizontal line is calculated with the standard tunneling model, see text. Data taken from Koláč et al. (1987).

cutoff $V_{\max}$ in the distribution of the barrier heights. Again, a good fit to the experimental data (at least for a-SiO$_2$) can be obtained.

While both attempts to explain the saturation at high T_1 start with a modification of the distribution function (cutoff in E or V), "short time heating" experiments carried out by Sahling et al. (1986a) indicated that there is an additional relaxation process which reduces at higher temperature the relaxation times of the TLS's contributing to the heat release for a give time and low temperature T_0 ($T_0 \ll T_f \equiv E_f/k_B$), so that the relaxation time τ of the TLS becomes very short at $T_0 \simeq T_f$. The results of such an experiment are given in Fig. 2.22. First, the heat release of epoxy resin $\dot{Q}(T_1, T_0, t)$ was measured with waiting time $t_M = 20$ h to reach the equilibrium state at T_1. In the next step, the measurement was performed with a very short waiting time $t_M = 0.2$ h at T_1. Calculations within the tunneling model indicate that the ratio of the heat release for $t_M = 0.2$h and for $t_M = 20$ h is small ($\dot{Q}(t_M = 0.2$ h$, t = 1$ h$)/\dot{Q}(t_M = 20$ h$, t = 1$ h$) \simeq 0.2$) and that it is nearly independent of T_1 *if there were no other relaxation process* (dotted line in Fig. 2.22). However, Fig. 2.22 shows clearly that this ratio increases rapidly with increasing T_1. At $T_1 \approx T_f$ the waiting time 0.2 h and 20 h give the same value of the heat release measured 1h after cooling to T_0, i.e. in 0.2 h the occupation of the TLS's contributing to the heat release after cooling reaches the equilibrium occupation at $T_1 \approx T_f$. This is only possible if the relaxation time of these TLS's is much smaller than the one expected from the tunneling model given by (2.18). Although high-order tunneling processes may contribute to the shorter relaxation of the TLS, the heat-release data indicate that thermally activated relaxation dominates the heat release at higher temperature. The influence of incoherent tunneling has not yet been calculated. Note that the thermal activation of TLS's leads to a "quasi-cutoff" in the energy dependence similar to (2.91). After cooling from $T_1 > T^\star$, where $T^\star$ is the freezing temperature, using the dominant phonon approximation, one

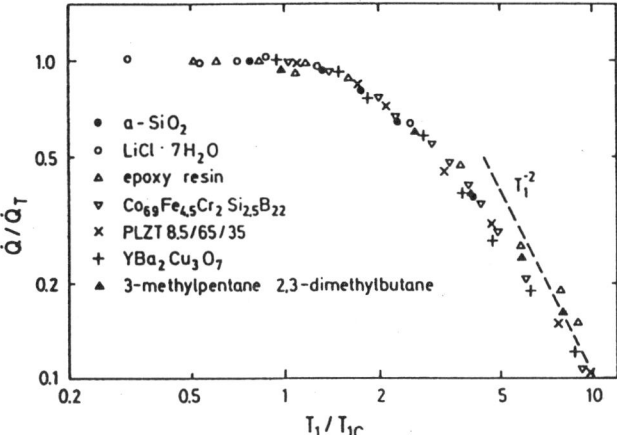

Fig. 2.23. Heat release $\dot{Q}$ of various amorphous and glasslike crystalline materials normalized to the heat-release value extrapolated from the tunneling regime $\dot{Q}_T$ (2.32) as a function of T_1/T_{1c} at $T_0 < T_c$. T_{1c} is the upper limit of T_1, where $\dot{Q}$ is proportional to $T_1^2 - T_0^2$. Taken from Parshin and Sahling (1993).

finds that the main contribution to the heat release comes from TLS's with energy $E^\star \approx 2.4 k_B T^\star \approx E_f$ (see Table 2.1). However, $E^\star$ depends weakly on the cooling rate in this case.

According to the calculation in Sect. 2.3.2 we will distinguish between the cases $T_0 < T_c < T_1$ (intermediate case) and $T_c < T_0 < T_1$ (thermal activation range).

Intermediate Case ($T_0 < T_c < T_1$). In Sect. 2.3.2 two important results were obtained: the change in the T_1-dependence from $\dot{Q} \propto T_1^2 - T_0^2$ to $\dot{Q} = $ const. occurs at $T_1 \approx W$, and the range where the T_1-dependence shifts from one behavior to the other is small, i.e. $\Delta T^\star/T^\star \simeq 0.2$ if there were no distribution of W, and with the assumption that only the thermally activated rate is active at $T > W/k_B$.

First, we will compare the second conclusion with experimental data. Figure 2.23 shows the ratio of the heat release $\dot{Q}$, measured at T_1, to the heat release $\dot{Q}_T$, measured in the tunneling range and extrapolated according to $T_1^2 - T_0^2$ to higher T_1, as a function of T_1/T_{1c}. T_{1c} is the upper limit of T_1, where $\dot{Q}$ is proportional to $T_1^2 - T_0^2$ corresponding to the lowest value of the freezing temperature $T_1^\star$. The values of T_{1c} are given in Table 2.1. Let $T_2^\star$ be the upper value of the freezing temperature. Then, $\dot{Q}(T_1)$ is constant for $T_1 > T_1^\star$ and $\dot{Q}(T_1)/\dot{Q}_T$ is proportional to T_1^{-2}. From Fig. 2.23 we see that this occurs at $T_2^\star \approx 7 T_{1c}$, i.e. in the range $(T_2^\star - T_1^\star)/\bar{T}^\star \approx 6 T_{1c}/4 T_{1c} = 1.5$, where $\bar{T}^\star = (T_1^\star + T_2^\star)/2$.

For further analysis we introduce some distribution function $G(T^\star) \approx G(W)$ and assume that $G(W)$ (or $G(T^\star)$) is a Gaussian distribution with the maximum at $\bar{W}$ (or $\bar{T}^\star$):

$$G(T^\star) = \frac{1}{\sqrt{2\pi}\sigma} \exp\left((T^\star - \bar{T}^\star)^2/(2\sigma^2)\right). \tag{2.92}$$

Within less than 5% deviation, this can be replaced by an approximation which is much more convenient for further calculations:

$$G(T^\star) = 4T^\star a \frac{\tanh aT^{\star 2}}{\cosh^2 aT^{\star 2}}, \tag{2.93}$$

where a is a constant.

It can be shown that the distribution function $G(T^{\star 2})$ coincides with the absolute value of the second derivative of the function $\dot{Q}(T_1^2)$ on its argument T_1^2 and we obtain $G(T^\star)$ from $G(T^{\star 2})$:

$$G(T^\star) = 2T^\star G(T^{\star 2}). \tag{2.94}$$

Thus, if we fit the experimental data $\dot{Q}(T_1^2)$ with

$$\dot{Q}(T_1^2) = A \tanh aT_1^2, \tag{2.95}$$

where A and a are constants, the second derivative on T_1^2 gives $G(T^{\star 2})$ and then (2.94) leads to (2.93) which is nearly identical with the Gaussian distribution (2.92).

The distribution function obtained in this way from the experimental data $\dot{Q}(T_1^2, 1\text{ h})$ of LiCl·7H$_2$O is shown in Fig. 2.24b (full line). The dots represent a Gaussian distribution (2.92) with the same position of the maximum at $\bar{T}^\star = 16$ K and the dispersion $\sigma = 5.35$ K. Since we know that for LiCl·7H$_2$O we have $k_B T^\star/W = 1$ at $t = 1$ h (see Table 2.1) the average value for $\bar{W}$ is also $\bar{W}/k_B = 16$ K.

For all other materials investigated, it was not possible to find a good fit with (2.95). Instead, it was necessary to add one or two more terms with different values of $\bar{T}^\star$ (or $\bar{W}$):

$$\dot{Q}(T_1^2) = A \tanh aT_1^2 + B \tanh bT_1^2 + C \tanh cT_1^2. \tag{2.96}$$

Figure 2.24a shows the distribution function for a-SiO$_2$ with three peaks, the first one with a maximum at $T_{m1}^\star = 3.5$K, the second one with a maximum at higher temperature $T_{m2}^\star = 20.9$ K (and probably the third one with $T_{m3}^\star = 24.2$ K). This result can be interpreted in such a way that three kinds of TLS's exist in a-SiO$_2$ with different average values $W_1 = 4.2$K ($T^\star/W \simeq 0.83$ for $t = 1$ h), $W_2 = 25$ K and $W_3 = 29$ K. 78% of all TLS's belong to the peak at the lowest temperature. Nevertheless, the TLS's corresponding to the other two peaks give a remarkable contribution to the heat release for $T_1 > W_3/k_B$ due to the large W_2 and W_3: $(\dot{Q}_2 + \dot{Q}_3)/\dot{Q}_1 \approx 11$, where $\dot{Q}_1$, $\dot{Q}_2$ and $\dot{Q}_3$ are the saturation values for the respective group of TLS's. We see that the heat release is very sensitive to TLS's with high freezing temperature (and W), even if their concentration is small. In Figs. 2.24c,d the distribution function of an amorphous metal and polycrystalline YBa$_2$Cu$_3$O$_7$ with remarkable structures are given.

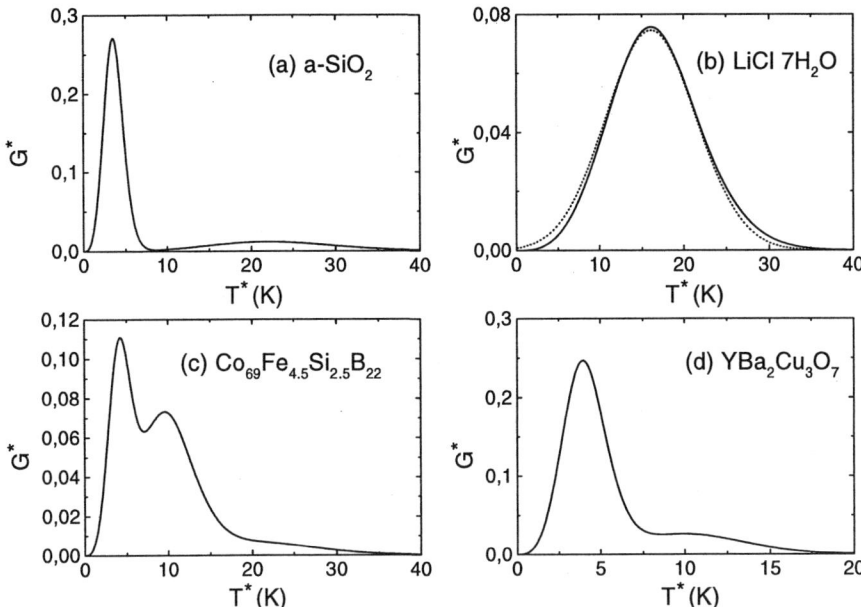

Fig. 2.24. Distribution functions $G(T^*)$ for (a) a-SiO$_2$, (b) LiCl·7H$_2$O, (c) amorphous metal Co$_{69}$Fe$_{4.5}$Cr$_2$Si$_{2.5}$B$_{22}$ and (d) polycrystalline YBa$_2$Cu$_3$O$_7$ deduced from heat-release data. T^* is a freezing temperature (see text). Data are taken from Parshin and Sahling (1993).

From Figs. 2.24a–d it is apparent that the distribution functions are different for different materials. There is no more "universal behavior" of glasses. That means that this curve is sensitive to the microscopic origin of the TLS's. From this point of view it seems reasonable to investigate the distribution function of different glasses in detail and to try to understand from where the different peaks originate.

Before we begin with this task, we will try to answer the question: Does this distribution function $G(T^*)$ represent a real physical situation or is it just a fit within a complicated theory?

To answer this, we will first compare the value W_1 deduced from the position of the first maximum of $G(T^*)$ with W obtained from specific heat data. According to Il'in et al. (1987) W can be calculated from the temperature $T_{\min}$ where a minimum of the function C/T^3 is observed:

$$W \approx 1.8 k_\mathrm{B} T_{\min} \, . \tag{2.97}$$

The most precise data of the specific heat we have for a-SiO$_2$ have been presented by Buchenau et al. (1991), where $T_{\min} = (2.1 \pm 0.4)$ K, see Chap. 9. This leads to $W/k_\mathrm{B} = (3.8 \pm 0.7)$ K which is in excellent agreement with $W_1/k_\mathrm{B} = 4.2$ K deduced from the position of the first maximum. The con-

Table 2.3. Model parameters for the calculation of the heat release and the specific heat data: E_f – cutoff energy obtained from the best fit of heat-release data with (2.91), T_{1c} – the lowest temperature T_1 where $\dot{Q}$ is still proportional to $T_1^2 - T_0^2$, T_{m1}^*, T_{m2}^*, T_{m3}^* – the temperature of the first, second and third maximum in the distribution function $G(T^*)$ [see Eqs. (2.93) and (2.96)], T_{av}^* – the average freezing temperature, W_1 – the characteristic energy, obtained from the first peak of the distribution function $G(T^*)$, $T_{\min}$ – temperature where C/T^3 has a minimum, C - specific heat, W – characteristic energy, calculated from $T_{\min}$ with (2.97).

material	E_f/k_B (K)	$2.4\,T_{av}^*$ (K)	T_{1c} (K)	T_{m1} (K)	T_{m2} (K)	T_{m3} (K)	$T_{\min}$ (K)	W/k_B (K)	W_1/k_B (K)
a-SiO$_2$	13	19	2.0	3.5 (78%)	20.9 (13%)	24.2 (9%)	2.0	4	4.2
LiCl·7H$_2$O	48	40	8.0	16.1 (100%)			$\geq$7.0	$\geq$ 13	16.1
Fe$_{80}$B$_{14}$Si$_6$	20	18	3.3	4.8 (68%)	11.9 (23%)				5.8
Co$_{69}$Fe$_{4.5}$Cr$_2$Si$_{25}$B$_{22}$	24	22	4.0	4.1 (33%)	9.6 (50%)	20.5 (11%)			4.9
Epoxy resin	16.5	17.5	2.5	4.6 (79%)	15.9 (21%)		1.4	25	5.5
Pentanol 2	23	23	3.8	5.6 (55%)	13.6 (45%)				6.7
3MP/-2.3DB	16	15	2.5	4.1 (79%)	13.9 (21%)				4.9
YBa$_2$Cu$_3$O$_7$	17	13	2.3	3.9 (79%)	10 (21%)				4.7
PLZT	7.5	8	1.3	2.1 (67%)	4.1 (24%)	10.0 (9%)	1.9	3.4	2.6
YBa$_2$Cu$_3$O$_6$	16	15	2.5						
Bi$_2$CaSr$_2$Cu$_2$O$_8$	8	8	1.5						

tribution of the second and third peak to the specific heat is too small to be observed.

Specific heat data of LiCl·7H$_2$O are available up to 7 K only. These data give $T_{\min} \geq 7$ K and $W/k_B \geq 13$ K which again is in agreement with the maximum of the distribution function $W/k_B \simeq T^* = 16$ K.

Such an agreement between W and W_1 could not be obtained for epoxy resin (see Table 2.1). This indicates (together with the unusual time dependence of the heat release, see Sect. 2.5.1) that there are some additional TLS's in the long-time range of relaxation times. Those TLS's determine mainly the heat release but much less the specific heat and experiments for which scattering processes are important.

Thus, in agreement with the soft-potential model, the thermal activation leads to a maximal change of the T_1-dependence of $\dot{Q}$ just at $T_1 \approx W/k_B$. However, the range ΔT^* where the T_1-dependence changes, is much larger than expected and might indicate a distribution of W. The Gauss-like distribution for a given average value $\bar{W}$ could be caused by the distribution of the effective mass of the modes, which is usually much larger than the mass of a single atom. The observation of two (or more) peaks in the distribution function indicates that there might be different kinds of motion of atoms (different binding energy and different effective mass) contributing to the heat release. It should be noted that these results (Fig. 2.24) depend on the assumption of thermally activated mechanism for the TLS's above a certain temperature. If other mechanisms for the relaxation of the TLS's were important, different functions G would be obtained.

Thermal Activation Range ($T_c < T_0 < T_1$). Since $T_c \approx W/k_B$, measurements at rather high temperature $T_0 \geq 3$ K are necessary to prove (2.98). Such experiments are difficult because the sensitivity of heat-release measure-

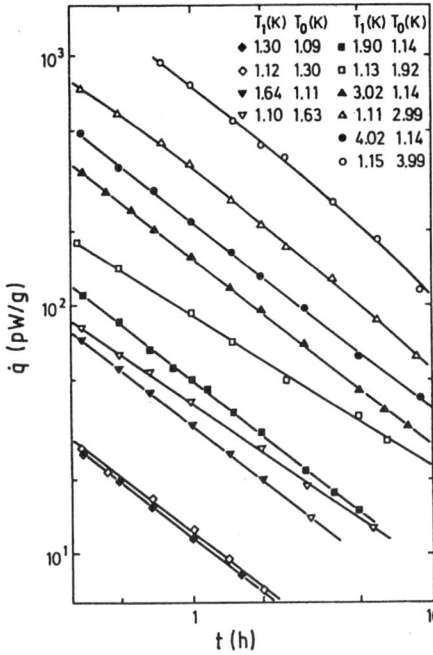

Fig. 2.25. Heat release (*solid symbols*) after cooling and the absolute value of the heat absorption after heating (*open symbols*) of an epoxy resin sample from T_1 to T_0 as a function of time t. Taken from Parshin and Sahling (1993).

ments decreases rapidly with increasing T_0. Systematical measurements of the heat release with $T_1 > T_c$ were performed on epoxy resin only (Koláč et al. (1986)).

The first maximum of the distribution function $G(T^\star)$ for epoxy resin is at $T^\star_{m1} \simeq 4.6$K, but as a consequence of the broad distribution of $T^\star$, the thermal activation process influences the heat release at a much lower temperature. Figure 2.25 shows the heat release of epoxy resin after cooling from T_1 to T_0 and the absolute value of the heat absorption after rapid heating from T_0 to T_1. Without the influence of thermal activation, we have

$$\dot{Q}(T_1, T_0, t) \simeq |\dot{Q}(T_0, T_1, t)|. \tag{2.98}$$

This was observed for the lowest values 1.3 K and 1.0 K only. Since the thermal activation gives an additional contribution to the heat absorption at $T_0 \geq 1.63$ K, the absolute value of the heat absorption is larger than the corresponding heat release. It can also be observed that the time dependence of the heat absorption shifts from $t^{-0.76}$ at $T_0 = 1.3$K to $t^{-0.62}$ at $T_0 = 1.92$K, in qualitative agreement with the prediction, see (2.88) and (2.89), where the time dependence shifts from t^{-1} in the tunneling range to $t^{-0.76}$ in the thermal activation range.

At $T_0 \geq 3.0$ K and for long times, the time dependence becomes stronger, which is probably the consequence of a cutoff $V_{\max}$. This can be seen much better in Fig. 2.26, where the heat absorption was measured for different

Table 2.4. Parameters of the standard tunneling model obtained from different properties for polystyrene (PS) and polymethyl methacrylate (PMMA). The specific heat can be written in terms of two contributions $C(T) = c_1(T) + c_3 T^3$.

Property	Parameter	Polystyrene	PMMA
Specific heat	c_1 (μJ/(gK^2))	4.6±0.5 [a] 5.1 [b]	3.0±0.3 [a] 4.6 [b]
	c_3 (μJ/(gK^4)) P_0(10^{38} J^{-1}g^{-1})	93±18 [a] 7.5 [a] 6.9 [b]	77±23 [a] 4.0 [a] 6.2 [b]
Thermal conductivity	κ(10^{-3} W/mK) ($T<0.7$ K)	$19(T/K)^{1.93}$ [a] $20(T/K)^{1.87}$ [b] $15.6(T/K)^{1.78}$ [c]	$28(T/K)^{1.84}$ [a] $33(T/K)^{1.81}$ [b] $29(T/K)^{1.77}$ [c]
	$P_0\gamma^2$ (10^6 J/m^3)	1.2 [a] 1.3 [b] 1 [c]	0.92 [a]
	γ (eV)	0.25 [a]	0.28 [a]
Internal friction	C (10^{-4})	8.3 [a] 5.3 [d]	2.6 [a] 3.2 [e] C_l=3.7 [f] C_t=5.7 [f]
	$P_0\gamma^2$ (10^6 J/m^3) γ (eV)	2.4 [a] 0.35 [a]	0.92 [a] 0.30 [a] 0.27 [f]
	E_0(K)	13±2	20±5 [a]
Sound velocity	C (10^{-4})	11±2 [a] 3.6-4.1 [d]	4.6±0.8 [a] C_l=2.0 [f] C_t=3.7 [f]
	$P_0\gamma^2$ (10^6 J/m^3)	3.2 [a] 3-3.4 [d]	1.7 [a]
	γ (eV)	0.41 [a]	0.40 [a] 0.27 [f]
Heat release	P_0(10^{38} J^{-1}g^{-1})	9.0 [a]	30 [g]

[a] Nittke et al. (1995)
[b] Stephens (1973)
[c] Stephens et al. (1972)
[d] Duquesne and Bellesa (1979)
[e] Crissman et al. (1964)
[f] Federle and Hunklinger (1982)
[g] Zimmermann (1984)

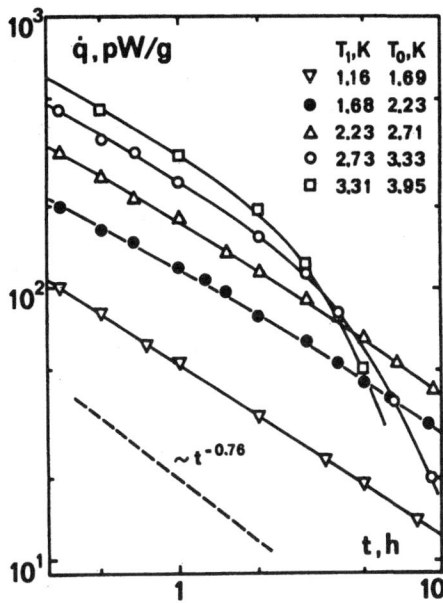

Fig. 2.26. Heat absorption after subsequent small steps between 1.16 K and 3.96 K. $\dot{q}$ is the heat absorbed in 1 g of EPILOX T-20-20. The solid lines are guides to the eye, the broken line is $\dot{q} \propto t^{-0.76}$ for comparison. Taken from Koláč et al. (1987).

T_0 and relatively small $\Delta T = T_1 - T_0$. In the short time range ($t \leq 1$ h), where the influence of the cutoff $V_{\max}$ is small, the temperature dependence of $\dot{Q}/\Delta T$ is roughly proportional to $T^{\frac{9}{4}}$ as predicted by (2.88), see Fig. 2.27.

Thus, these first investigations are in qualitative agreement with the theory. Since the heat release in epoxy resin shows an unusual time dependence

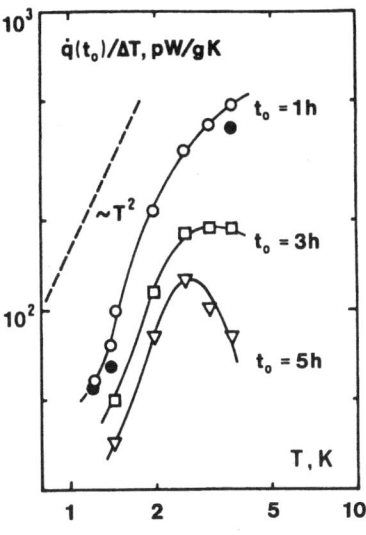

Fig. 2.27. Temperature dependence of the heat release (absorption) of EPILOX T-20-20 at fixed time t_0 after cooling (heating). Here $\dot{q}(t_0)/\Delta T$ is the power absorbed (released) at the time t_0 after transition from T_1 to T_0 divided by $\Delta T = |T_1 - T_0|$; $T = (T_1+T_0)/2$; (o) absorption; (•) release. Taken from Koláč et al. (1987).

even in the tunneling range ($\dot{Q}$ is proportional to $t^{-0.76}$ instead of t^{-1}), for a closer examination of the theory, similar experiments with other glasses (e.g. a-SiO$_2$) are necessary.

Table 2.5. Comparison between the calculated values for P_0 deduced from specific heat ($P_{0,C}$) and from heat-release ($P_{0,Q}$) measurements using the soft-potential model equations.

Material	a-SiO$_2$	LiCl·7H$_2$O	PMMA	PS	Epoxy Resin
$P_{0,C}$ (10^{44} J^{-1}m^{-3})	5.3[a] 7.5[b]	31 ± 6[c]	4.6[d] 7.1[e]	7.9[d] 7.2[e]	10[f]
$P_{0,Q}$ (10^{44} J^{-1}m^{-3})	1.0[g] 3.3[h]	10.1 ± 0.3[c]	34[i]	9.4[d]	7.8[j]
$P_{0,C}/P_{0,Q}$	1.6–7.5	3.1 ± 0.7	0.14–0.21	0.8–0.88	1.28

[a] Zimmermann and Weber (1981b) (Suprasil W)
[b] Zimmermann and Weber (1981b) (Suprasil I)
[c] Sahling (1989)
[d] Nittke et al. (1995)
[e] Stephens (1973)
[f] Scheibner and Jäckel (1985)
[g] Schwark et al. (1985)
[h] Zimmermann and Weber (1981a)
[i] Zimmermann (1984)
[j] Koláč et al. (1987)

2.5.4 Correlation Between the Heat Release and Other Low-Temperature Properties

As mentioned in the introduction, the success of the tunneling model is based on the fact that it describes a wide range of low-temperature properties with a consistent and relatively small set of parameters. Table 2.4 shows a compilation of parameters needed to describe some low-temperature properties of polystyrene (PS) and polymethyl methacrylate (PMMA). Those parameters are essentially the density of states P_0 and the coupling constant γ. Within the accuracy one expects from a phenomenological model, the values agree quite well.

Table 2.4 gives us a hint to a peculiar feature of the heat release which makes it special among the low-temperature properties of amorphous and disordered solids: the measurement of the heat release provides directly the density of states of the TS's, i.e. only experimental parameters are needed for its determination. All the other low-temperature measurements, the coupling between TS's and phonons or conduction electrons, have to be taken into account (or in the case of the specific heat the parameter $u_{\min}$) to obtain P_0.

Despite the overall agreement of the different values, there are some differences which are usually ascribed to the complicated nature of the investigated materials or to some deviations from the simple phenomenological theory.

In Table 2.5 we show the values of the density of states of TS's from the measurements of the heat release and specific heat using the equations of the soft-potential model. Depending on the material, we note that the density of states obtained from these two properties can differ considerably. This difference is not yet clarified.

2.6 Conclusion and Outlook

In this chapter it has been shown that the heat release after a relatively rapid cooling of the sample is a feature typical of disordered, amorphous and some polycrystalline solids. From measurements of the heat release the density of states of TS's can be obtained *directly*, and in some cases (amorphous metals) even unique. It has also been shown that in the tunneling regime ($T \leq 4$ K) experiments are well described by the tunneling model. Furthermore, we have presented a detailed analysis of the temperature dependence of the heat release. At higher temperature, the tunneling model does not adequately describe the observations. Within the soft-potential model it is possible to understand the effect of thermal activation which appears to play an important role in the temperature range above the tunneling range.

From heat-release data it is possible to derive information about the density of states of TS's with long relaxation times. It still remains open as to what extent incoherent tunneling, which appears to describe acoustic data at $T > 1$ K well (see Sect. 4.5.5 and Chap. 3), influences the heat release at $T_1 > 1$ K.

The direct measurement of the density of states of TS's through the heat release provides the possibility to test currently discussed theories on the origin of the universal behavior of amorphous solids below ~ 1 K. One of the proposed models is based on the interaction between TS's, see Sect. 5.1.4 and references therein. The interaction model predicts not only a different dynamical behavior of the TS's at low enough temperatures, but also an *increase* of the density of states of TS's *decreasing* the size of the sample. Since, according to theory, a logarithmic dependence on the sample size is expected, to test this prediction, one should measure samples of nm size to measure a clear increase of the density of states of TS's with respect to the bulk sample. Heat-release experiments are currently performed in small size samples. They may provide information on a probable influence of the sample size on the density of states of TS's and also on their relaxation processes.

Acknowledgment

Part of the research presented in this review has been possible with the support of the Deutsche Forschungsgemeinschaft.

3. Crossover to Phonon-Assisted Tunneling in Insulators and Metals

Alois Würger

3.1 Introduction

Tunneling defects in various materials are well described in terms of quantum motion of a bistable system. The thermal behavior at helium temperature first indicated the existence of such defects; as examples we note the Schottky anomaly of weakly doped alkali halides reported by Narayanamurti and Pohl (1970), and the linear specific heat of oxide glasses below 5 K observed by Zeller and Pohl (1971).

Yet the study of dynamic and transport properties permits a much deeper insight into the physics of the host solid, since they are strongly affected by the interaction with collective degrees of freedom, such as elastic waves and, in metals, electron-hole excitations. Hunklinger and Arnold (1976) have reviewed the particular low-temperature relaxation properties of amorphous solids that arise from tunneling between distinct local configurations. Interstitial hydrogen trapped by an oxygen impurity in niobium is confined to two quantum states on tetragonal sites; the strongly damped tunneling motion has been observed by neutron scattering (Wipf (1997)). Proton transfer in molecular crystals has been investigated by NMR, neutron and light scattering, and analyzed in terms of a two-state model by Skinner and Trommsdorff (1988a). More recently, Golding et al. (1992) and Chun and Birge (1993) have shown that the conductance of mesoscopic wires may be significantly affected by two-state defects.

Because of these many experimental realizations, much theoretical effort has been devoted to the dissipative two-state system. Applying lowest-order perturbation theory on the defect-phonon coupling, Jäckle (1972) derived a relaxation contribution to the motional spectrum of the TS's in oxide glasses and thus resolved the anomalous temperature variation of the sound velocity below about 1 K [cf. Piché et al. (1974)]. Using a diagram technique, Maleev (1981) obtained an incoherent background and discussed thermal properties of glasses (Maleev (1983)).

When retaining next-order corrections to the one-phonon rate one finds that the defect-phonon interaction effectively increases with rising T, thus prohibiting a weak-coupling approach at high temperatures; in terms of the perturbation series, multiphonon contributions exceed that of the direct process (Würger, 1997c). The strong-coupling case was tackled with diagrammatic perturbation theory (Pirc and Gosar (1969), Becker and Keller (1986)), path integral methods (Leggett et al. (1987), Niu (1991)), and mode-coupling theory (Beck et al. (1979), Neu and Würger (1994a)). An approach based on Bloch equations was used by Silbey and Harris (1989). More recently, Kehrein and Mielke (1997) treated the defect-phonon coupling by means of flow equations developed originally for strongly correlated electrons. There seems to be general agreement on the behavior at low temperature where weakly damped tunneling oscillations prevail and where the damping rate involves the direct or one-phonon process only. As to multiphonon contributions that are relevant above a few Kelvin, however, contradictory results have been obtained in different approaches.

The controversial point may be cast in the question of whether or not the thermal motion destroys the coherent tunneling oscillations. On the one hand, it was claimed that damping of a two-level system by phonons is always weak, resulting in coherent tunneling motion at all temperatures. On the other hand, several works indicated a crossover to incoherent motion with rising temperature; each of them, however, found a different high-temperature relaxation rate.

In metallic hosts, coupling to conduction electrons provides a second damping mechanism. The constant density of states at the Fermi surface gives rise to an infrared singularity, similar to that derived earlier by Nozières and de Dominicis (1969) for X-ray absorption in metals and by Anderson and Yuval (1969) for the Kondo problem. This singularity governs most features of the two-state dynamics and may lead to spontaneous symmetry breaking at zero temperature (Bray and Moore (1982)). The resulting damping function for a dissipative two-state system has been derived by Grabert and Weiss (1985) and Fisher and Dorsey (1985). The relevance for macroscopic quantum coherence has been discussed by Leggett and Garg (1985). The influence of a composite bath, involving electrons and phonons, has been studied by Grabert (1992).

Dissipative tunneling of a bistable defect may be described in terms of a two-state polaron which, in turn, is closely related to quantum diffusion on a lattice, as considered first by Holstein (1959). Later on, the role of processes involving one or two phonons has been pointed out by Flynn and Stoneham (1970), Kagan and Klinger (1974), and Teichler and Seeger (1981). Muon relaxation in alkali halides and diffusion of ^{3}He in a ^{4}He crystal have been discussed by Kagan and Prokof'ev (1992). A comprehensive review of theoretical developments and their application to interstitial hydrogen in metals has been given recently by Grabert and Schober (1997). As a by-product, these latter works deal with the damped two-state system. We will see in Sect. 3.4, however, that the two-state polaron presents a few subtle features which are missed when simply using the theory for diffusion on a lattice.

Finally, we briefly discuss the relevance of the elastic interaction that arises from phonon exchange of adjacent two-state systems. For N defects in a volume $\mathcal{V}$, the average nearest-neighbor distance is given by $n^{-1/3}$, where $n = N/\mathcal{V}$ is the density. As soon as the distance $n^{-1/3}$ is smaller than the phonon mean free path or, equivalently, smaller than the coherence length of resonant acoustic waves, the defect-phonon coupling gives rise to an effective interaction of adjacent defects in glasses. When investigating the consequences of this interaction on the low-temperature behavior of glasses, Joffrin and Levelut (1975) found that spin-flip processes strongly affect the phase coherence of the TS's. With well-known values for the parameters, they derived a transverse rate $1/T_2 \approx 10^9$ sec^{-1}, corresponding to a few hundred millikelvin. On the contrary, there is little effect on the energy relaxation, or longitudinal relaxation rate $1/T_1$. Later on, Maleev (1988) refined the perturbation theory, and pointed out the relevance of interaction for identical defects. This case is realized by off-center impurities in alkali halides, with dipolar or elastic interaction. In recent years, the dynamics of such defects has been studied in detail by Enss et al. (1997); a review has been given by the present author (Würger, 1997a).

Regarding the low-temperature behavior of TS's in glasses, the ideas of Joffrin and Levelut (1975) and Maleev (1988) have been taken up by Burin and Kagan (1994a); cf. Chap. 5 of this book. More recently, Enss and Hunklinger (1997) applied a theory for interacting impurities (Würger, 1997a) on TS's in oxide glasses, and they showed that the anomalies of sound propagation observed at very low temperatures ($T < 100$ mK) can be explained by incoherent tunneling of interacting TS (see, however, the discussion on this topic on p. 181 and the following). The important point is that the elastic interaction does not depend on the energy splitting, but only on the distance. Since the most relevant TS's have an energy splitting of the order of $k_\mathrm{B}T$, the elastic coupling is irrelevant at not too low temperatures. Below 100 mK, however, the elastic interaction is comparable to the energy of thermal TS's, thus destroying the phase coherence of the two-state dynamics.

In the present contribution, we will disregard the interaction of adjacent defects. This assumption is justified at sufficiently high temperature, or in the dilute limit. In this chapter we are mainly concerned with the effects of phonons and, in the case of a metallic host, the interplay of phonons and conduction electrons. Because of their larger density of states at small frequencies, electron-hole excitations are more efficient at low temperatures, whereas phonons prevail at higher T. In Sect. 3.2 we introduce the spin-boson model and the bath spectral density for acoustic lattice modes and for electron-hole excitations.

Sections 3.3–3.7 contain our main results. After transforming to a small-polaron representation and discussing the resulting phonon-dressing effect, we develop in Sect. 3.4 a strong-coupling approach for such a two-state polaron. In terms of a blip expansion, we evaluate the first and second order of the resulting series for the self-energy, corresponding to the *noninteracting blip approximation* (NIBA) and to lowest-order corrections to NIBA, respectively. Regarding the two-state dynamics, we find a crossover to incoherent tunneling and an exponentially increasing high-temperature relaxation rate. It turns out that, for phonon coupling, NIBA does not constitute a controlled approximation, contrary to the situation encountered for conduction electrons with small Kondo parameter.

In Sect. 3.5 we study two-state defects in metals, where dissipation arises from both phonons and conduction electrons. While symmetric defects, like trapped hydrogen in Nb, are best discussed in terms of a blip expansion, the case of large asymmetry, as realized, e.g. by defects in mesoscopic wires, is more conveniently described by coupled rate equations for the two states. In Sect. 3.6 we discuss the dressing effect and relate our results to experimental findings for defects in insulating and metallic materials.

Section 3.7 deals with asymmetric TS in insulating glasses. We address a few implications of phonon dressing on the distribution function for the tunnel energy and discuss the sound velocity at higher temperature. In Sect. 3.8 we derive a weak-coupling perturbation theory, and we calculate the first two terms of the corresponding series for the rate. The lowest-order contribution is identical to the well-known one-phonon process; the next-order correction provides a precise criterion for the strong-coupling approaches.

3.2 The Spin-Boson Model

Two-state systems are best described in terms of a double-well potential for some collective coordinate q. At low temperatures only the ground states in the two wells, $|R\rangle$ and $|L\rangle$, are relevant. Then all quantum features are accounted for by a bare tunneling amplitude Δ_b which depends exponentially on the particle mass m, the potential barrier V, and the distance of the wells d, according to (2.2) and (2.3).

Because of the lack of crystal symmetry in amorphous solids, the minima of the double-well potential are not degenerate in general, but separated by an asymmetry energy Δ.

The quantum states $|R\rangle$ and $|L\rangle$ give rise to a two-level system whose operators are most conveniently expressed in terms of Pauli matrices,

$$\sigma_z \equiv |L\rangle\langle L| - |R\rangle\langle R|, \qquad \sigma_x \equiv |R\rangle\langle L| + |L\rangle\langle R|, \tag{3.1}$$

where the discrete coordinate $q = \frac{1}{2}d\sigma_z$ takes the values $\pm\frac{1}{2}d$. The simplest dynamic model is given by the two-state pseudospin system, whose reduced coordinate σ_z is linearly coupled to a heat bath,

$$H = \frac{1}{2}\Delta_b\sigma_x + \frac{1}{2}\Delta\sigma_z + \frac{1}{2}\sigma_z\sum_k \hbar\lambda_k\left(b_k + b_k^\dagger\right) + \sum_k \hbar\omega_k b_k^\dagger b_k, \tag{3.2}$$

where the bath operators obey Bose commutation relations $[b_k, b_{k'}^\dagger] = \delta_{kk'}$. As for the pseudospin operators, we adopt the notation of Leggett et al. (1987), which is more convenient for a strong-coupling approach. In the literature, σ_z is sometimes chosen to be diagonal in the energy eigenstates of the uncoupled spin. For elastic waves, the coupling constants vary as $\lambda_k \propto k\omega_k^{-1/2}$; cf. (3.13). In terms of the bath spectral density (3.4), they are defined through (3.7) for electron-hole excitations and in (3.19) for phonons.

Here, a word concerning notation is in order. The tunnel matrix element in (3.2) reads Δ_b instead the usual Δ_0. As we will see below, the coupling term in (3.2) results in a dressed tunnel matrix element which is smaller than the bare value Δ_b. Explicit expressions are given in Eqs. (3.39)-(3.43). We stress that only the dressed quantities Δ_0 and $\tilde{\Delta}_0$ can be observed experimentally. In order to meet the common notation used in this book, we denote by Δ_0 the dressed value that is actually measured at low temperatures.

For the case of dilute tunneling centers, the time evolution of the bath operators is independent of the coupling to the TS's, i.e.,

$$b_k^\dagger(t) = e^{i\omega_k t} b_k^\dagger, \qquad b_k(t) = e^{-i\omega_k t} b_k. \tag{3.3}$$

Accordingly the heat bath is entirely characterized by the spectral function

$$J(\omega) = \frac{\pi}{2} \sum_k \lambda_k^2 \delta(\omega - \omega_k). \tag{3.4}$$

Equations (3.2) and (3.4) state the so-called spin-boson problem which, with an appropriate choice for the spectral function $J(\omega)$, accounts for various situations in solid-state physics and chemistry (for a review see Leggett et al. (1987)). Two particularly interesting cases are defined by the linear and cubic spectral functions, arising from coupling to electron-hole excitations and elastic waves, respectively. (We discuss atomic tunneling only, where the energy Δ_b is much smaller than the cut-off of the spectral function. Then $J(\omega)$ shows a simple power law behavior in the relevant frequency range.)

Electron-Hole Excitations. When studying muon diffusion in a metal, Kondo (1976) pointed out how screening by conduction electrons affects the dynamics of the charged particle. In terms of quasi-particles with wave vector k and spin σ, the interaction potential at the muon position r reads as

$$\frac{1}{N} \sum_{k,k',\sigma} U_{kk'} e^{i(k-k')\cdot r} c_{k,\sigma}^\dagger c_{k',\sigma}. \tag{3.5}$$

When reducing the coordinate to two values, $r = \frac{1}{2}\sigma_z d$, and replacing the coupling potential by a constant, $U_{kk'} \equiv U_0$, the relevant part of (3.5) simplifies to

$$\sigma_z \frac{U_0}{N} \sum_{k,k',\sigma} i \left[1 - \cos\left(\tfrac{1}{4}(k-k')\cdot d\right)^2\right] c_{k,\sigma}^\dagger c_{k',\sigma}. \tag{3.6}$$

The resulting damping function is equivalent to that of a bosonic heat bath, as in (3.2), with linear spectral density,

$$J_{\text{el}}(\omega) = \pi K \omega, \tag{3.7}$$

and coupling strength K. Yamada et al. (1985) derived the general expression

$$K = 2 \left[\frac{1}{\pi} \arctan \frac{\sqrt{1-x} \tan \delta}{\sqrt{1 + x \tan^2 \delta}} \right]^2, \tag{3.8}$$

where the phase shift reads as $\tan \delta = -\pi \rho U_0$, ρ is the electron density of states, and $x = [\sin(k_F d)/k_F d]^2$, with the Fermi wave vector k_F. From this result one easily obtains the relation $0 \leq K \leq \frac{1}{2}$ and, for $\rho U_0 \ll 1$, one recovers Kondo's weak-coupling result $K = 2U_0^2 \rho^2 (1 - x)$. The expression (3.8) is identical to the parameter α of Leggett et al. (1987) and, for weak coupling, to the quantity b of Kagan and Prokof'ev (1986).

The linear spectral density J_{el} leads to a frequency independent damping function at low frequency and finite temperature; for this reason it is often referred to as the case of Ohmic dissipation. As a most striking feature, a logarithmic infrared singularity arises in any order of perturbation theory from the linear frequency dependence. Summing up these logarithmic terms, Kondo (1984) derived both the dressed tunnel matrix element and the muon diffusion rate $\Gamma \propto T^{2K-1}$. Kagan and Prokof'ev (1986) found similar results when studying the electronic polaron effect for a heavy particle in a metal. The underlying infrared singularity had been considered earlier, albeit in a different context, by Anderson and Yuval (1969) and Nozières and de Dominicis (1969).

Dissipative effects on the two-state system have been investigated by many authors, mostly in terms of functional integral methods and the *noninteracting blip approximation* (NIBA). Chakravarty and Leggett (1984) obtained the rate at finite temperatures. Grabert and Weiss (1985) and Fisher and Dorsey (1985) calculated the frequency-dependent damping function. Leggett et al. (1987) studied the crossover from coherent oscillations at low temperature to incoherent motion at higher T; for small K, the incoherent rate turned out to be identical to Kondo's hopping rate. Weiss and Wollensak (1989) have shown that the proper treatment of a finite asymmetry energy requires to go beyond NIBA. The functional integral approach to the Ohmic damping model has been reviewed by Weiss (1993). Aslangul et al. (1986) pointed out how certain results could be derived in second-order Born approximation in terms of polaron operators. Self-consistent mode-coupling schemes were used by Zwerger (1983) and Götze and Vujicic (1988). The relaxation behavior of the two-state model was studied by Kagan and Maksimov (1980); for a recent review see Kagan and Prokof'ev (1992).

Experimental evidence for screening by conduction electrons has been obtained from sound propagation in metallic glasses by Golding et al. (1978) and Weiss et al. (1980) (see Chap. 4 for more details), and from neutron scattering on interstitial hydrogen by Wipf et al. (1987) (see Chap. 7 for a detailed discussion). More recently, Golding et al. (1992) and Chun and Birge (1993) studied the dynamics of a single two-level system by means of the conductance fluctuations in submicrometer metallic wires, confirming quantitatively the validity of the Ohmic damping model below 1 K; as to the elastic coupling at higher temperatures, cf. Würger (1997d).

Phonon Damping. The other case of physical relevance is realized in insulating materials where acoustic phonons provide the most efficient damping mechanism. Despite the lack of translational symmetry, disordered solids exhibit low-frequency vibrations which are similar to those of their crystalline counterparts and thus may be described in terms of a harmonic crystal. Following Kittel (1963), we write the vibrational amplitude at position x as

$$R(x) = \mathcal{V}^{-1/2} \sum_{ks} e(ks)(2\varrho\omega_{ks})^{-1/2} \left[e^{i\mathbf{k}\cdot\mathbf{x}} b_{ks} + e^{-i\mathbf{k}\cdot\mathbf{x}} b_{ks}^\dagger \right], \tag{3.9}$$

where $\mathcal{V}$ is the volume of the sample and ϱ its mass density. ks are wave vector and branch index of the normal modes, and $e(ks)$ the corresponding polarization vector. The bath operators fulfil the commutations relation $[b_{ks}, b_{k's'}^\dagger] = \delta_{k,k'}\delta_{s,s'}$.

The linear coupling to the tunneling defect is written in terms of the elastic strain tensor $\varepsilon_{\nu\mu} = \partial_\mu R_\nu$,

$$\varepsilon_{\nu\mu}(x) = \mathcal{V}^{-1/2} \sum_{ks} e_\nu(ks) k_\mu (2\varrho\omega_{ks})^{-1/2} \left[i e^{i\mathbf{k}\cdot\mathbf{x}} b_{ks} - i e^{-i\mathbf{k}\cdot\mathbf{x}} b_{ks}^\dagger \right]. \tag{3.10}$$

In principle both the tunnel frequency Δ_b and the bias Δ are functionals of the elastic strain ε. Yet most works resort to two basic approximations with respect to the coupling: First, the tunnel frequency Δ_b is assumed to be independent of the strain and, second, the bias is linearized as

$$\Delta[\varepsilon_{\nu\mu}] = \Delta + 2 \sum_{\nu\mu} \gamma_{\nu\mu} \varepsilon_{\nu\mu}, \tag{3.11}$$

where Δ is the usual asymmetry energy and $\gamma_{\nu\mu}$ the deformation potential.

The first assumption may be justified by working out the effect of elastic strain on Δ_b in terms of some simple atomic potential; in fact the tunnel frequency turns out to be hardly affected by the strain. The second one, i.e. the linear strain dependence of (3.11), has been discussed by Leggett et al. (1987); in view of experimental data for various systems, the quadratic and higher-order terms in ε would seem to be irrelevant.

The elastic energy $\gamma_{\nu\mu}$ depends on the configuration of the tunnel defect with respect to propagation and polarization directions of the normal modes. For impurity atoms in crystals, the tensor character of the strain field may give rise to significant selection rules for transitions mediated by elastic waves. [Cf. Grabert and Schober (1997) for interstitial hydrogen, and Chap. 2 of Würger (1997a) for lithium impurities in alkali halides.] Yet no such effect is expected for amorphous solids.

In order to simplify the coupling potential (3.11), we evaluate the strain at $x = 0$ and, following Jäckle (1972), replace the tensor product by

$$\sum_s \gamma_s \varepsilon_s \equiv \frac{1}{2} \sum_{ks} \lambda_{ks} (ib_{ks} - ib^\dagger_{ks}), \tag{3.12}$$

where the sum runs over one longitudinal and two transverse acoustic phonon branches and where we have defined the quantities

$$\frac{1}{2}\lambda_{ks} = \frac{1}{\sqrt{\mathcal{V}}} \frac{\gamma_s k}{\sqrt{2\varrho \hbar \omega_{ks}}}. \tag{3.13}$$

The spectral function defined in (3.4) is given by the Fourier transform of the elastic response function,

$$J_{\mathrm{ph}}(\omega) = \sum_s \gamma_s^2 \int_{-\infty}^{\infty} dt\, e^{i\omega t} \langle [\varepsilon_s(t), \varepsilon_s] \rangle = \frac{\pi}{2} \sum_{ks} \lambda_{ks}^2 \delta(\omega - \omega_{ks}). \tag{3.14}$$

Evaluation of the spectral density requires to specify the dispersion law ω_{ks}. In Debye's approximation it reads as

$$\omega_{ks} = v_s |\boldsymbol{k}|, \tag{3.15}$$

with longitudinal and transverse sound velocities v_l and v_t. Writing the discrete sum over $\boldsymbol{k}$ as an integral and replacing the Brillouin zone by a sphere with radius k_D, we have

$$\frac{1}{\mathcal{V}} \sum_{ks} \ldots \to \sum_s \int \frac{d\boldsymbol{k}}{(2\pi)^3} \ldots \to \frac{1}{2\pi^2} \sum_s \int_0^{k_D} dk\, k^2 \ldots. \tag{3.16}$$

Normalization of both sides requires

$$k_D = (6\pi^2 N/\mathcal{V})^{1/3}, \tag{3.17}$$

where N is the number of atoms in the sample and $\mathcal{V}$ its volume.

Inserting (3.13)–(3.16) in the spectral density (3.14), we recover the well-known expression

$$J_{\mathrm{ph}}(\omega) = \left(\frac{\gamma_l^2}{v_l^5} + \frac{2\gamma_t^2}{v_t^5} \right) \frac{\omega^3}{2\pi \varrho \hbar}. \tag{3.18}$$

The cubic frequency dependence arises from the coupling constant, $\lambda_k \propto \sqrt{\omega_k}$, and the Debye density of states, $\sum_k \delta(\omega - \omega_k) = \mathrm{const.} \times \omega^2$. For notational convenience we write

$$J_{\mathrm{ph}}(\omega) = \pi \alpha \omega^3 \equiv \pi \tilde{\alpha} (\hbar^2/k_B^2) \omega^3, \tag{3.19}$$

where we have defined two parameters: α has dimension (frequency)$^{-2}$, and $\tilde{\alpha}$ has (temperature)$^{-2}$. (When setting $\hbar = k_B$, one finds $\alpha = \tilde{\alpha}$.) The former is related to the material constants through

$$\alpha = \frac{1}{2\pi^2 \hbar \varrho} \left(\frac{\gamma_l^2}{v_l^5} + \frac{2\gamma_t^2}{v_t^5} \right) \equiv \frac{3\gamma^2}{2\pi^2 \hbar \varrho v^5}, \tag{3.20}$$

where v and γ are appropriate average values.

3.2 The Spin-Boson Model

In the isotropic Debye model, a frequency cut-off for each phonon branch s is given by $v_s k_D$. For later use we define the Debye temperatures Θ_s,

$$k_B \Theta_s = \hbar v_s (6\pi^2 N/\mathcal{V})^{1/3}, \tag{3.21}$$

the appropriate average value Θ (Kittel (1963)),

$$3/\Theta^3 = 1/\Theta_l^3 + 2/\Theta_t^3, \tag{3.22}$$

and the corresponding Debye frequency ω_D,

$$\hbar \omega_D = k_B \Theta. \tag{3.23}$$

A heat bath with cubic spectral density leads to damping phenomena which are basically different from Ohmic dissipation. Whereas the latter case is determined by an infrared singularity, the cubic spectral function does not cause any anomaly at low frequency; yet as to the weight of thermal phonons, the frequency variation causes a strong enhancement with temperature. Contrary to Ohmic damping, there is no well-defined dimensionless small parameter for the phonon heat bath. It turns out, however, that the temperature

$$T_0 = (2\pi^2 \tilde{\alpha})^{-1/2} \tag{3.24}$$

provides a most useful parameter and permits the specification the ranges of weak and strong coupling. Note that T_0 is *inverse* proportional to the elastic coupling potential γ, i.e. large elastic coupling results in small T_0 and vice versa. For real systems, T_0 takes values between 1 and 100 K; cf. Table 3.1 on p. 109.

In insulators, there are four relevant parameters, Δ_b, T, T_0, and Θ. For tunneling of atoms in solids, the tunnel energy Δ_b is small as compared to the Debye energy $k_B \Theta$,

$$\Delta_b \ll k_B \Theta. \tag{3.25}$$

With respect to T, three cases are to be considered: The ranges of low and intermediate temperatures, $T < \Delta_b/k_B$ and $\Delta_b/k_B < T < T_0$, are well understood; except for a constant reduction of the bare tunnel frequency, they are reasonably well described by the lowest-order perturbation theory of Jäckle (1972). The main concern of the present work is the change which occurs when temperature approaches T_0, and the dynamics at high temperatures $T > T_0$. Accordingly, a relevant dimensionless parameter is provided by T^2/T_0^2, or $2\pi^2 \tilde{\alpha} T^2$.

In order to simplify certain integrals over phonon modes, we mainly consider temperatures well below the Debye temperature,

$$T \ll \Theta. \tag{3.26}$$

Yet our results are easily generalized to the case $\Theta < T$. We will see that the rate for strong coupling, i.e. $T_0 < T$, looks quite different for $T \ll \Theta$ and $\Theta < T$.

Deviations from the Cubic Bath Spectrum. The cubic spectral density has been derived assuming linear coupling to the bath operators and the absence of selection rules for elastic transitions. We conclude this section with a brief discussion of the phonon model.

Regarding the assumption concerning the linear strain dependence in (3.11), Leggett et al. (1987) have shown that, in most cases, linear coupling to the oscillator heat bath provides a proper description of dissipative tunneling. As a counterexample we note the hindered rotation of a methyl group in a molecular crystal: In optical hole-burning experiments, Gradl et al. (1992) observed, at intermediate temperatures, a T^7 power law for the nuclear spin conversion rate of methyl groups, characteristic of two-phonon or Raman scattering. This result confirmed previous theoretical evidence for the relevance of the quadratic coupling term (Würger (1990)). For TS in amorphous solids, two-phonon scattering has been discussed by Doussineau et al. (1980) and, with respect to the asymmetry energy, by Silbey and Trommsdorff (1990); yet it would seem that so far there is no unambiguous experimental confirmation for the Raman process being relevant in glasses. Regelmann et al. (1994) have shown that a quadratic coupling term would result in an Ohmic contribution to the bath spectrum and hence considerably modify the damping function.

Finally we turn to the frequency dependence of the scattering potential, $\lambda_k \propto k\omega_k^{-1/2}$, that gives $J_{\rm ph}(\omega) = \pi\alpha\omega^3$ for acoustic phonons. At an impurity site with inversion symmetry, the couplings vary as $\lambda_k \propto k^2\omega_k^{-1/2}$, i.e. as $\lambda_k \propto \omega_k^{3/2}$ in the limit of large wavelength, leading to a spectral density $J_{\rm ph}(\omega) \propto \omega^5$ instead of the cubic law. Such a modification occurs, e.g. for certain transitions of [111] off-center impurities in alkali halides Würger (1997a). Yet TS's in amorphous solids and impurities on sites with low symmetry, as H in $Nb(OH)_x$, are well described by the cubic law. Finally we note that, qualitatively, the findings of this chapter hold true even for the bath spectrum involving the fifth power of frequency.

The cubic spectral density relies on the dispersion law $\omega_{ks} = v_s|\boldsymbol{k}|$ which ceases to be valid at the boundary of the Brillouin zone, i.e. for frequencies of the order of ω_D. Hence, even for a monoatomic crystal, the sharp cutoff of the cubic law at the Debye frequency is unphysical. In crystals with basis there are, moreover, optical branches; the law $\omega_{ks} = v_s|\boldsymbol{k}|$ holds true in a restricted part of $\boldsymbol{k}$-space only. For a more detailed discussion see p. 110.

3.3 Polaron Transformation and Phonon Dressing

The spin-phonon model as stated in (3.2) has been investigated extensively in perturbative and self-consistent approaches. In both cases the spin–phonon interaction $\frac{1}{2}\hbar f\sigma_z$ is treated as a perturbation; the resulting series expansion is either truncated at finite order or approximated by a partial summation as, e.g. in the noncrossing approximation of Sect. 3.8.

Before transforming the Hamiltonian to a strong-coupling representation, we point out the limitation of a power-series expansion in terms the spin-phonon coupling. Here and in the next section we consider the case of zero asymmetry, $\Delta = 0$.

3.3.1 Break-Down of Perturbation Theory

The heat bath affects the two-state dynamics in two ways: First, it gives rise to dissipation with a finite relaxation rate that reads, to quadratic order in the phonon coupling, as

$$\Gamma = \pi \sum_k \lambda_k^2 (1 + 2n_k) \delta(\omega_k - \Delta_b/\hbar). \tag{3.27}$$

Second, it results in a shift of the tunnel energy

$$\delta\Delta_b = -\Delta_b \sum_k (\lambda_k/\omega_k)^2 (1 + 2n_k). \tag{3.28}$$

By inserting the phonon spectral density (3.19), we recover the rate obtained by Jäckle (1972),

$$\Gamma = \pi\alpha(\Delta_b/\hbar)^3 \coth(\Delta_b/2k_B T). \tag{3.29}$$

The shift of the tunnel energy is negative and reduces the bare value Δ_b according to

$$\Delta_b + \delta\Delta_b = \Delta_b \left[1 - \tfrac{1}{2}\tilde{\alpha}\Theta^2 - \tfrac{1}{3}\pi^2\tilde{\alpha}T^2\right], \tag{3.30}$$

resulting in a slowing down of the tunneling motion.

Perturbation theory is appropriate as long as higher-order corrections are small. A proper criterion for such a weak-coupling condition requires a dimensionless coupling parameter. Yet (3.29) and (3.30) provide not only one but *three* such parameters. From the rate Γ we may split off the factor $\alpha(\Delta_b/\hbar)^2$ or, equivalently,

$$\tilde{\alpha}(\Delta_b/k_B)^2. \tag{3.31}$$

The shift of the tunnel energy involves two more dimensionless parameters; the first one is constant,

$$\tilde{\alpha}\Theta^2, \tag{3.32}$$

whereas the second one depends on temperature,

$$\tilde{\alpha}T^2. \tag{3.33}$$

It is by no means clear from the outset, which of these quantities is the most significant one.

There has been some confusion in the literature, arising when only one of the above dimensionless parameters is taken into account. When discussing the influence of the phonon heat bath, Leggett et al. (1987) expand (on pages 54–58 of their article) the self-energy in powers of a parameter b which is closely related to (3.31) and (3.32). Thereby, these authors miss temperature-dependent corrections in terms of (3.33) and are led to conclusions which cease to be valid at temperatures fulfilling $\tilde{\alpha}T^2 \geq 1$.

From (3.30) one may conjecture that finite-order perturbation theory breaks down as soon as $\tilde{\alpha}T^2$ or $\tilde{\alpha}\Theta^2$ approaches unity. In terms of (3.24) this means that T or Θ are of the order of T_0. Considering the numbers for real physical systems gathered in Table 3.1 on p. 109, we find that (3.32) is rarely small; for configurational defects it is even much larger than unity, thus prohibiting the truncation of the perturbation series at finite order.

As we will see below, the parameter (3.33) is the most relevant one, in view of the temperature dependence of tunnel energy and damping rate. Both the tunnel frequency and the damping rate change significantly in the crossover region $2\pi^2\tilde{\alpha}T^2 \equiv T^2/T_0^2 \approx 1$. The high-temperature range $T > T_0$ proves to be of particular interest.

3.3.2 Canonical Transformation

A proper treatment of the two-state dynamics thus requires a strong-coupling approach, including terms of any order of the parameters $\tilde{\alpha}T^2$ and $\tilde{\alpha}\Theta^2$. The canonical transformation

$$S = \exp\left[-\tfrac{1}{2}\sigma_z \sum_k u_k (b_k - b_k^\dagger)\right] \tag{3.34}$$

provides a representation for the spin-phonon model that turns out to be an appropriate starting point. Here we have defined for each vibrational mode a dimensionless coupling constant

$$u_k = \lambda_k/\omega_k. \tag{3.35}$$

In physical terms, S is the translation operator which shifts the coordinate of each phonon mode, x_k, by a distance $\pm a_k$. This is most obvious when using the momentum operator $\hat{p}_k$ of mode k; then (3.34) reads as

$$S = \exp[-\sigma_z \sum_k \hat{p}_k a_k]. \tag{3.36}$$

For $\gamma_l = \gamma = \gamma_t$ and $v_t = v = v_l$, we have $a_k = (\sqrt{3}\gamma/4Mv\hbar\omega_k)$. Applying (3.34) on the Hamiltonian (3.2) and discarding the asymmetry energy Δ, we find with $H \to e^S H e^{-S}$,

$$H = \frac{1}{2}\Delta_b(\sigma_+ B_- + \sigma_- B_+) + \sum_k \hbar\omega_k b_k^\dagger b_k, \tag{3.37}$$

where we have dropped a constant and used

$$B_\pm = \exp\left[\pm \sum_k u_k(b_k - b_k^\dagger)\right] \tag{3.38}$$

and the ladder operators $\sigma_+ = |L\rangle\langle R|$ and $\sigma_- = |R\rangle\langle L|$. These operators are well known from studies on polaron motion by Holstein (1959), and dissipative two-state dynamics (see, e.g. Leggett et al. (1987)).

3.3.3 Phonon Dressing

For the original Hamiltonian (3.2), the off-diagonal matrix element between the two quantum states $|L\rangle$ and $|R\rangle$ is given by the bare tunnel energy, $\langle L|H|R\rangle = \frac{1}{2}\Delta_b$. This is no longer true for the transformed Hamiltonian (3.37), since its off-diagonal matrix elements involve the bath operators $B_\pm$.

When performing the thermal average $\langle\ldots\rangle_B$ of $2\langle L|H|R\rangle = \Delta_b B_-$ with respect to the bath variables, we find the dressed tunnel energy

$$\tilde{\Delta}_0 \equiv 2\langle\langle L|H|R\rangle\rangle_B = \Delta_b \langle B_-\rangle_B. \tag{3.39}$$

The reduction factor may be written as an exponential,

$$\langle B_\pm\rangle_B = e^{-\frac{1}{2}W}, \tag{3.40}$$

whose argument involves (3.35) and Bose factors $n_k = [e^{\beta\hbar\omega_k} - 1]^{-1}$,

$$W = \sum_k u_k^2(1 + 2n_k), \tag{3.41}$$

and defines the well-known Debye–Waller factor e^{-W}. Inserting the continuous spectral density (3.19) in (3.41) gives

$$W = \tilde{\alpha}\Theta^2 + \tfrac{2}{3}\pi^2\tilde{\alpha}T^2. \tag{3.42}$$

When expanding the exponential in (3.39) in powers of W, we find $\tilde{\Delta}_0 = \Delta_b[1 - \frac{1}{2}W + \ldots]$ which agrees with the perturbation expansion for the tunnel energy, (3.28) and (3.30).

For notational convenience we define the zero-temperature value of the reduced tunnel energy,

$$\Delta_0 = \Delta_b \exp(-\tfrac{1}{2}W_0), \tag{3.43}$$

with the argument of the exponential evaluated at $T = 0$,

$$W_0 = \sum_k u_k^2 = \tilde{\alpha}\Theta^2. \tag{3.44}$$

The constant Δ_0 contains the zero-temperature Debye–Waller factor, while the thermal motion reduces the tunnel energy further to the value $\tilde{\Delta}_0$. Note $\tilde{\Delta}_0 \leq \Delta_0 < \Delta_b$. Together with the slow-tunneling condition (3.25), (3.43) implies the inequality

$$\Delta_0 \ll k_B T_0, \tag{3.45}$$

that will be useful for the discussion of the significance of two-phonon corrections to the rate. Except for a numerical factor, it is identical to the self-consistency condition of Leggett et al. (1987).

The reduction of the off-diagonal matrix element (3.39) is commonly referred to as *phonon-dressing effect*. It arises from the fact that the phonon states corresponding to the two positions of the defect are not the same; more precisely, the quantum state $|L\rangle$ is correlated with a phonon state $|\Psi_L\rangle$, and $|R\rangle$ with its counterpart $|\Psi_R\rangle$. Thus the energy overlap matrix element between the left and right configurations comprises two factors; besides the bare tunnel energy Δ_b, the phonon states give rise to a reduction factor $\langle \Psi_L | \Psi_R \rangle = \mathrm{e}^{-\frac{1}{2}W_0}$ at zero temperature.

Interaction with conduction electrons in a metallic host gives rise to a formally similar *screening effect*. The temperature dependence of the reduced tunnel energy is different from (3.42), however, due to the much weaker frequency dependence of the bath spectral density.

In Fig. 3.1 we sketch the potential energy as a function of the defect position q and one collective coordinate x_k labeled by a wave number k. Black dots indicate the potential minima; in two-state approximation the variable q is restricted to the values $\pm \frac{1}{2} d\sigma_z$. For zero phonon coupling, $\lambda_k = 0$, the tunneling motion is independent of the vibrational coordinate x_k, as shown in Fig. 3.1a, and its frequency is given by the bare tunnel matrix element Δ_b.

In real solids, however, the defect atom polarizes its surroundings and drags the adjacent atoms when tunneling from one well to the other. This is schematically shown in Fig. 3.1b, where the two potential minima occur at different values $\pm a_k$ of the vibrational coordinate. As a consequence, the way of the tunneling particle through the barrier becomes longer, and the tunnel matrix element is reduced accordingly.

So far we have discussed the case of zero temperature where, except for the zero-point motion, the particles are confined to the potential minima. At finite T, the defect coordinate q is still restricted to the discrete values $\pm \frac{1}{2} d\sigma_z$, whereas the vibrational coordinate fluctuates in a range given by the number of quanta in mode k, resulting in a temperature-dependent term in the reduction factor, $\tilde{\Delta}_0 = B \Delta_\mathrm{b}$, which is nothing else but a Debye–Waller factor for phonon scattering from the two-level system.

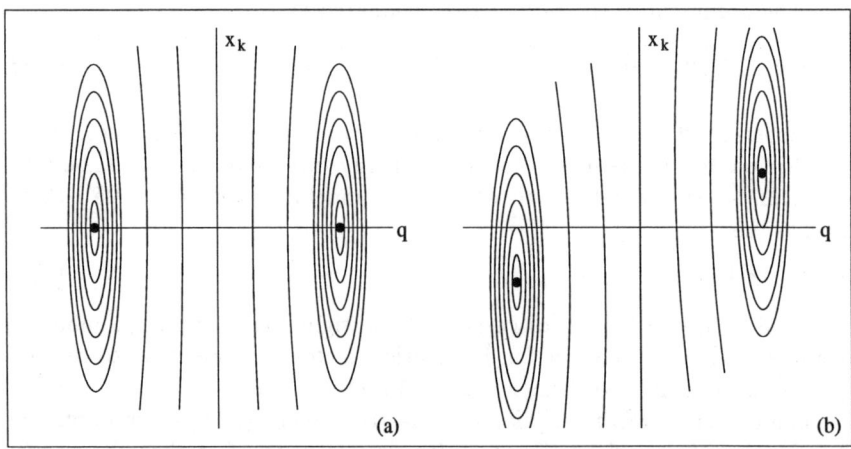

Fig. 3.1. Potential energy landscape for a defect coordinate q and one phonon mode x_k for (**a**) zero defect-phonon coupling and (**b**) finite coupling.

3.3.4 Time Evolution

In the Heisenberg picture, spin operators evolve in time according to

$$\sigma_\alpha(t) = e^{iHt/\hbar}\sigma_\alpha e^{-iHt/\hbar} \equiv e^{i\mathcal{L}t}\sigma_\alpha, \quad (3.46)$$

with the quantum Liouville operator $\mathcal{L}$. When passing from (3.2) to (3.37) through the canonical transformation S, we have not transformed the Pauli matrices; the Hamiltonian (3.37) is still written in terms of the original operators σ_z and $\sigma_\pm = \frac{1}{2}(\sigma_x \pm i\sigma_y)$. As a consequence, the forms (3.2) and (3.37) do not lead to the same time dependence. This is not surprising, since in general σ_i and $S\sigma_i S^\dagger$ are different operators; for finite phonon coupling, i.e. $u_k \neq 0$, one finds, e.g. $[S, \sigma_\pm] \neq 0$.

The case of the operator σ_z, however, is particular: Since it commutes with the transformation (3.34), $[S, \sigma_z] = 0$, time evolution of the expectation value $\langle \sigma_z(t) \rangle$ may be performed with either representation of the Hamiltonian. In the following sections, we investigate the two-state dynamics in terms of (3.37), whereas Sect. 3.8 deals with the time evolution arising from (3.2).

3.4 Crossover to Incoherent Tunneling

Starting from the equation of motion for the pseudospin operators, we first present a simple derivation for the noninteracting blip approximation. Then we show that for a phonon heat bath, certain blip–blip interactions contribute significantly to the rate, and we evaluate explicitly the resulting corrections to NIBA.

Following Leggett et al. (1987), we consider a particle which dwells in the left well at $t = 0$ and whose position evolves in time according to (3.37),

$$P(t) = \langle \sigma_z(t) \rangle, \quad (3.47)$$

with the initial condition $P(t = 0) = 1$. This condition is met by the definition for the average

$$\langle \ldots \rangle = \mathrm{tr}(\rho \ldots), \quad (3.48)$$

with a statistical operator that factorizes at $t = 0$,

$$\rho = \rho_S \rho_B, \quad (3.49)$$

where the pseudospin part

$$\rho_S = \tfrac{1}{2}(1 + \sigma_z) = |L\rangle\langle L| \quad (3.50)$$

projects on the quantum state $|L\rangle$. The remaining factor,

$$\rho_B = e^{-\beta H_B}/\mathrm{tr}(e^{-\beta H_B}), \quad (3.51)$$

describes the heat bath in thermal equilibrium, with $H_B = \sum_k \hbar\omega_k b_k^\dagger b_k$.

A blip expansion for the two-state dynamics is most readily derived from the equation of motion $\hbar\dot\sigma_\alpha = [H, \sigma_\alpha]$ for the spin operators σ_z and $\sigma_\pm$,

$$\hbar\dot\sigma_z = i\Delta_b(B_+\sigma_- - B_-\sigma_+), \qquad \hbar\dot\sigma_\pm = \mp\tfrac{i}{2}\Delta_b B_\pm \sigma_z. \tag{3.52}$$

When we insert the formal integral of the second equation,

$$\sigma_\pm(t) = \sigma_\pm(t') \mp \frac{i}{2\hbar}\Delta_b \int_{t'}^{t} d\tau B_\pm(\tau)\sigma_z(\tau), \tag{3.53}$$

in the first one, we find an integro-differential equation for $\sigma_z(t)$. When integrating over t, inserting the r.h.s. repeatedly in the l.h.s., and taking the thermal average according to (3.48), we obtain the infinite series for (3.257),

$$\begin{aligned}P(t) &= 1 - \int_0^t d\tau_1 \int_0^{\tau_1} d\tau_2 \langle K(\tau_1, \tau_2)\rangle \\ &+ \int_0^t d\tau_1 \int_0^{\tau_1} d\tau_2 \int_0^{\tau_2} d\tau_3 \int_0^{\tau_3} d\tau_4 \langle K(\tau_1, \tau_2)K(\tau_3, \tau_4)\rangle + \ldots,\end{aligned} \tag{3.54}$$

where a 'blip' is described by the kernel

$$K(t, t') = \frac{1}{2}\frac{\Delta_b^2}{\hbar^2}(B_+(t)B_-(t') + B_-(t)B_+(t')). \tag{3.55}$$

The operators $B_\pm$ shift the phonons between the potential minima corresponding to $\sigma_z = \pm 1$. In view of Fig. 3.1 they account for the fact that tunneling occurs between different bath configurations.

In this section we propose two different approximations for $P(t)$, which are both based on a partial resummation of the series (3.54). In a first step, we merely retain the average of the kernel $K(t, t')$, corresponding to the it noninteracting blip approximation. In a second step, we retain certain blip–blip interactions that result in an additional self-energy contribution.

3.4.1 Noninteracting Blip Approximation

Much work on the spin-boson model relies on the noninteracting blip approximation (NIBA); for the Ohmic damping model with spectral density (3.7), it has been reviewed by Leggett et al. (1987) and Weiss (1993). In a field-theoretical language, a blip consists of an instanton pair which describes the motion along the classical trajectory in an inverted double-well potential. In terms of the Hamiltonian (3.37), a blip corresponds to the scattering from one well to the other and back to the original well, while the 'dressed' particle drags its phonon cloud.

Since the kernel (3.55) accounts for a single blip, the noninteracting blip approximation P_1 is obtained by replacing $K(t, t')$ in $P(t)$ with the average $\langle K(t, t')\rangle$,

3.4 Crossover to Incoherent Tunneling

$$P_1(t) = 1 - \int_0^t d\tau_1 \int_0^{\tau_1} d\tau_2 \langle K(\tau_1, \tau_2) \rangle$$
$$+ \int_0^t d\tau_1 \int_0^{\tau_1} d\tau_2 \int_0^{\tau_2} d\tau_3 \int_0^{\tau_3} d\tau_4 \langle K(\tau_1, \tau_2) \rangle \langle K(\tau_3, \tau_4) \rangle + \ldots \quad (3.56)$$

The infinite series on the r.h.s. gives rise to the integral equation

$$P_1(t) = 1 - \int_0^t d\tau \int_0^\tau d\tau' \Sigma_1(\tau - \tau') P_1(\tau'), \quad (3.57)$$

where we have defined the self-energy

$$\Sigma_1(t - t') = \langle K(t, t') \rangle. \quad (3.58)$$

In turns out that it is convenient to separate the static part of Σ_1 from its time-dependent fluctuations,

$$\Sigma_1(t) = (\tilde{\Delta}_0/\hbar)^2 + \Gamma(t). \quad (3.59)$$

Because of $\lim_{t\to\infty} \langle K(t,0) \rangle = (\tilde{\Delta}_0/\hbar)^2$, the second term $\Gamma(t)$ vanishes for long times.

To second order in the operators $B_\pm$, there are two different bath correlation functions. Following Mahan (1981), we express both

$$\langle B_\pm(t) B_\mp(t') \rangle = B^2 e^{\varphi(t-t')}, \quad (3.60)$$

$$\langle B_\pm(t) B_\pm(t') \rangle = B^2 e^{-\varphi(t-t')}, \quad (3.61)$$

in terms of a phase that is given by the coupled phonon propagator

$$\varphi(t) = \sum_k u_k^2 \left[n_k e^{i\omega_k t} + (1 + n_k) e^{-i\omega_k t} \right], \quad (3.62)$$

with the Bose occupation numbers

$$n_k \equiv n(\omega_k) = [e^{\beta \hbar \omega_k} - 1]^{-1}. \quad (3.63)$$

Since (3.55) involves products of bath operators with opposite labels only, the self-energy is given by (3.60), and its time-dependent part reads as

$$\Gamma(t) = (\tilde{\Delta}_0/\hbar)^2 [e^{\varphi(t)} - 1]. \quad (3.64)$$

When taking the Laplace transform of (3.57) and (3.58), we have

$$P_1(z) = -[z + \Sigma_1(z)]^{-1} = -\left[z + \Gamma(z) - z^{-1}(\tilde{\Delta}_0/\hbar)^2\right]^{-1}. \quad (3.65)$$

This formal result has been derived by Leggett et al. (1987) in a functional integral approach as the NIBA; it has been widely used for the two-state model with Ohmic damping.

From (3.64) it is clear that NIBA is identical to lowest Born approximation, or second-order perturbation theory, with respect to the fluctuating parts of the operators $B_\pm$ (Aslangul et al. (1986), Würger (1997b), Würger (1998a)).

3.4.2 Nearest-Neighbor Blip Interactions

When taking the average of each factor K in (3.56), we have neglected correlations of subsequent 'blips' in the exact series (3.54). Now we take into account certain blip–blip interactions which arise from correlations of different factors K in the series for $P(t)$.

There is no unique scheme for improving the NIBA result (3.65). An infinite series for the corrections in terms of

$$\delta K(\tau, \tau') = K(\tau, \tau') - \langle K(\tau - \tau') \rangle \tag{3.66}$$

has been derived in Würger (1997b). Yet it turns out that truncating at finite order does not conserve the analyticity of $P(z)$ in the upper complex half plane. In Würger (1997b) this problem has been avoided by considering the spectral function $P''(\omega) = \Im P(\omega + i\epsilon)$, which is well behaved in any case.

Here we proceed in a different way and derive an expression for $P(z)$ that is analytic in the upper half plane. In order to do so, we need to retain a self-energy correction Σ_2 that involves an infinite number of terms, corresponding itself to a partial resummation of the series (3.54).

Since the kernel K comprises two operators $B_\pm$ at different times, each term of the series (3.54) involves $2n$ factors $B_\pm(\tau_i)$ whose time arguments are ordered according to $t \geq \tau_1 \geq ... \geq \tau_{2n} \geq 0$. There is a subtle point concerning the labels of subsequent factors $B_\pm$. Those operators $B_\pm$ that belong to one blip, i.e. that form a factor K, have correlated signs as is obvious from (3.55). As a consequence, the sign of the phase φ in (3.64) is positive. On the other hand, the signs of subsequent operators $B_\pm$ from different blips are not correlated, thus giving rise to positive and negative phases, according to (3.60) and (3.61).

In Fig. 3.2 we schematically show the irreducible part arising from two subsequent blips,

$$\langle \delta K(\tau_1, \tau_2) \delta K(\tau_3, \tau_4) \rangle = (\tilde{\Delta}_0/\hbar)^4 e^{\varphi(\tau_1 - \tau_2) + \varphi(\tau_3 - \tau_4)} [\cosh(\Phi) - 1], \tag{3.67}$$

with the function

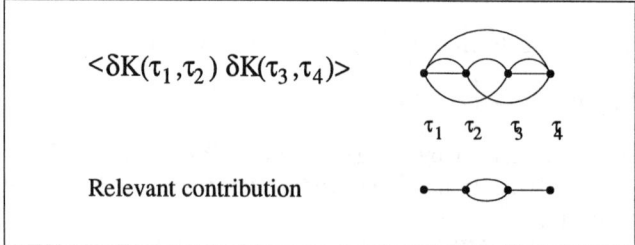

Fig. 3.2. Diagrammatic illustration of lowest-order correction to NIBA. Black points represent the vertices of strength $\tilde{\Delta}_0/\hbar$; horizontal lines are equal to unity. Half circles account for phonon correlation functions $[e^{\pm \varphi} - 1]$.

$$\Phi(\tau_1,\tau_2,\tau_3,\tau_4) = \varphi(\tau_1-\tau_4) + \varphi(\tau_2-\tau_3) - \varphi(\tau_1-\tau_3) - \varphi(\tau_2-\tau_4).$$

Black points in Fig. 3.2 denote the vertex strength $\tilde{\Delta}_0/\hbar$. Regarding time evolution, horizontal lines indicate unity, and the half circles the phonon correlations $[e^{\pm\varphi} - 1]$. A power series expansion in terms of φ shows that significant corrections arise from the correlation of the adjacent operators at times τ_2 and τ_3 only,

$$\langle \delta K(\tau_1,\tau_2)\delta K(\tau_3,\tau_4)\rangle \to (\tilde{\Delta}_0/\hbar)^2 \Gamma_{\text{g}}(\tau_2-\tau_3). \tag{3.68}$$

This expression does not depend on the times τ_1 and τ_4, since in our approximation there are no time correlations of $\langle B_\alpha(\tau_1)\rangle$ and $\langle B_\beta(\tau_4)\rangle$.

In general, the most important corrections to NIBA arise from adjacent pairs of operators $B_\pm$ that belong to *different* factors K in (3.54),

$$\frac{1}{4}\sum_{\alpha\beta}\langle B_\alpha(\tau_{2n}) B_\beta(\tau_{2n+1})\rangle. \tag{3.69}$$

As a consequence, the labels α and β are uncorrelated, and the resulting two-time correlation function involves terms with both e^φ and $e^{-\varphi}$; it reads as

$$\Gamma_{\text{g}}(t) = (\tilde{\Delta}_0/\hbar)^2 \left[\cosh\bigl(\varphi(t)\bigr) - 1\right]. \tag{3.70}$$

The term '1' in brackets has to be subtracted, since the uncorrelated part of (3.69), i.e. $\langle B_\alpha(\tau_{2n})\rangle\langle B_\beta(\tau_{2n+1})\rangle = B^2$, has already been counted in (3.58).

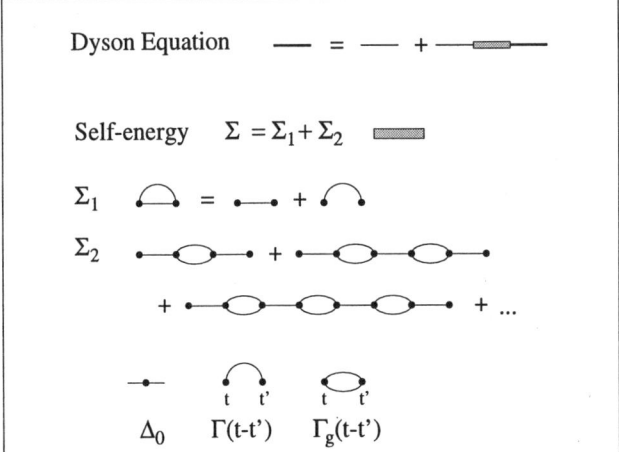

Fig. 3.3. Diagrammatic illustration of the blip expansion. The Dyson equation is given by the integral equation (3.73). Black points represent the vertices of strength $\tilde{\Delta}_0/\hbar$; the thin lines are equal to unity. Phonon correlation functions $\Gamma(t-t')$ and $\Gamma_{\text{g}}(t-t')$ are defined in (3.64) and (3.70). All intermediate times in $\Sigma_2(t-t')$ are to be integrated over, as in (3.72).

The second-order correction has been given explicitly above. Proceeding in a similar fashion with the third-order term, we neglect all contributions but

$$\langle \delta K(\tau_1, \tau_2) \delta K(\tau_3, \tau_4) \delta K(\tau_5, \tau_6) \rangle \to (\tilde{\Delta}_0/\hbar)^2 \Gamma_g(\tau_2 - \tau_3) \Gamma_g(\tau_4 - \tau_5), \quad (3.71)$$

which is illustrated in Fig. 3.3. Collecting terms arising from any number of factors δK and integrating over inner time arguments, we find the self-energy correction

$$\Sigma_2(t - t') = -\frac{\tilde{\Delta}_0^2}{\hbar^2} \int_{t'}^{t} d\tau_1 \int_{t'}^{\tau_1} d\tau_2 \Gamma_g(\tau_1 - \tau_2)$$
$$\times \left[1 - \int_{t'}^{\tau_2} d\tau_3 \int_{t'}^{\tau_3} d\tau_4 \Gamma_g(\tau_3 - \tau_4) + \ldots \right]. \quad (3.72)$$

Since $\Gamma_g(t)$ is an even function of $\varphi(t)$, the self-energy Σ_2 contains even powers of φ only; in particular, there is no correction linear in φ.

Rearranging the series (3.54) in powers of $\Sigma(t) = \Sigma_1(t) + \Sigma_2(t)$ and repeating the summation, we obtain an infinite series that reads schematically as

$$P_2 = 1 - \int\!\!\int \Sigma + \int\!\!\int\!\!\int\!\!\int \Sigma\Sigma + \ldots$$

and that leads to an integral equation for P_2,

$$P_2(t) = 1 - \int_0^t d\tau \int_0^\tau d\tau' [\Sigma_1(\tau - \tau') + (\Sigma_2(\tau - \tau')] P_2(\tau'). \quad (3.73)$$

Upon Laplace transformation we find

$$P_2(z) = -[z + \Sigma_1(z) + \Sigma_2(z)]^{-1}. \quad (3.74)$$

The above choice for the correlations retained in Σ_2 was partly motivated by the fact that the form (3.72) is given as a sum of multiple convolution integrals. Each term of this series may be viewed as the $2n$-fold convolution of n factors Γ_g with $n+1$ factors 1. Hence the Laplace transform is given as a geometric series,

$$\Sigma_2(z) = -(\tilde{\Delta}_0/\hbar)^2 \frac{1}{z} \sum_{n=1}^{\infty} \left(-\frac{\Gamma_g(z)}{z} \right)^n, \quad (3.75)$$

which leads to the resolvent form

$$\Sigma_2(z) = -(\tilde{\Delta}_0/\hbar)^2 \left[(z + \Gamma_g(z))^{-1} - z^{-1} \right]. \quad (3.76)$$

After inserting this expression in (3.74) we find

$$P_2(z) = -\left[z + \Gamma(z) - \frac{(\tilde{\Delta}_0/\hbar)^2}{z + \Gamma_g(z)} \right]^{-1}. \quad (3.77)$$

Comparison with the NIBA result (3.65) shows that the blip–blip interactions do not simply add a term to the smooth part of the self-energy, $\Gamma(z)$, but rather replace the singular part $(\tilde{\Delta}_0/\hbar)^2 z^{-1}$ with $(\tilde{\Delta}_0/\hbar)^2 [z + \Gamma_g(z)]^{-1}$.

When linearizing $\Sigma_2(z)$ with respect to Γ_g, i.e. when retaining only the first term $z^{-2}(\tilde{\Delta}_0/\hbar)^2 \Gamma_g$ of the series (3.75), we recover the corrections to NIBA derived previously (Würger (1997b)); the factor z^{-2} caused the analyticity problem mentioned below (3.66). We will see below that the full expression (3.77) is analytic in the upper half plane, i.e. the analytic behavior has been regularized by the partial summation in (3.75).

The above expression for P_2 can be formally improved by retaining bath correlations between the times τ_{2n-1} and τ_{2n} in the self-energy Σ_2. It turns out, however, that this modification is of little consequence for the propagator. We finally remark that (3.77) can be derived equally well in a different approach based on a strong-coupling perturbation theory in terms of fluctuation operators $\xi_\pm = B_\pm - B$; cf. Würger (1998a).

3.4.3 Time Evolution in NIBA

Equations (3.65) and (3.77) contain the formal results for the spin dynamics. In order to obtain a more explicit form for $P(t)$, we apply a Markov approximation to the damping spectra and evaluate the rates. We start with the NIBA expressions given in (3.64) and (3.65).

The spectrum $\Gamma''(\omega) = \Im \Gamma(\omega + i0)$ is regular at zero frequency, because $\lim_{t \to \infty} \Gamma(t) = 0$ vanishes sufficiently fast in the long-time limit. Since the real part is roughly linear in frequency, and the imaginary part almost constant, we may put

$$\Gamma(z) = z\Xi + i\Gamma_1, \tag{3.78}$$

where Ξ is given by the derivative $\partial_z \Gamma'(z)$, evaluated at the poles $\pm \tilde{\Delta}_0/\hbar$; cf. (A7). The rate is determined by the spectral function

$$\Gamma''(\omega) = \frac{1}{2} \int_{-\infty}^{\infty} dt e^{i\omega t} \left[e^{\varphi(t)} - 1 \right], \tag{3.79}$$

again taken at the poles of $P_1(z)$,

$$\Gamma_1 = \frac{1}{2}[\Gamma''(\tilde{\Delta}_0/\hbar) + \Gamma''(-\tilde{\Delta}_0/\hbar)]. \tag{3.80}$$

Inserting $\Gamma(z)$ in (3.65),

$$P_1(z) = -z \left[z^2(1+\Xi) - (\tilde{\Delta}_0\hbar)^2 + iz\Gamma_1 \right]^{-1} \tag{3.81}$$

gives rise to a $\mathcal{Z}$-factor $[1+\Xi]^{-1}$, where Ξ is given by the Kramers–Kronig integral $\Gamma'(\omega)$.

As shown in the Appendix, the frequency derivative Ξ is much smaller than unity. Accordingly we neglect this term in (3.81). Then

$$P_1(z) = -z[z^2 - (\tilde{\Delta}_0\hbar)^2 + iz\Gamma_1]^{-1} \tag{3.82}$$

exhibits two complex poles at $-i\Gamma_\pm$, with

$$\Gamma_\pm = \tfrac{1}{2}\Gamma_1 \pm \sqrt{\tfrac{1}{4}\Gamma_1^2 - (\tilde{\Delta}_0/\hbar)^2}. \tag{3.83}$$

Taking the inverse Laplace transform we find

$$P_1(t) = \frac{\Gamma_+}{\Gamma_+ - \Gamma_-} e^{-\Gamma_+ t} - \frac{\Gamma_-}{\Gamma_+ - \Gamma_-} e^{-\Gamma_- t}. \tag{3.84}$$

This equation constitutes the solution of the two-state dynamics in NIBA; we still have to evaluate the roots $\Gamma_\pm$ or, more precisely, the rate Γ_1.

The Damping Rate Γ_1

As we will see, the two-state dynamics is quite different for temperatures below and well above T_0, as defined in (3.24). We start by considering the low-temperature case $T \ll T_0$.

Low Temperatures: $T \ll T_0$. In this range we may expand the factor $(e^\varphi - 1)$ in (3.64) in powers of φ and truncate after a few terms. We start by retaining the linear term only; when evaluating the spectrum $\varphi''(\omega)$ at the bare poles in (3.81),

$$\Gamma_1 = \frac{1}{2}\tilde{\Delta}_0^2 [\varphi''(\tilde{\Delta}_0) + \varphi''(-\tilde{\Delta}_0)], \tag{3.85}$$

and inserting the cubic spectral function (3.19) in the coupled phonon propagator,

$$\varphi''(\omega) = \pi \sum_k u_k^2 [n_k \delta(\omega + \omega_k) + (1 + n_k)\delta(\omega - \omega_k)], \tag{3.86}$$

we find

$$\Gamma_1 = \pi\alpha(\tilde{\Delta}_0/\hbar)^3 \coth(\tilde{\Delta}_0/2k_B T). \tag{3.87}$$

As temperature approaches T_0, terms of second and higher order in φ become important when calculating the damping rate. The quadratic term involves a convolution of two factors φ''. Since the convolution integral depends little on frequency in the relevant range $\hbar|\omega| < k_B T$, it may be evaluated at $\omega = 0$,

$$\frac{1}{\pi}\int_{-\infty}^{\infty} d\omega\, \varphi''(\omega)\varphi''(-\omega) = \frac{8}{3}\pi^3(\hbar/k_B)\tilde{\alpha}^2 T^3. \tag{3.88}$$

More rigorously, the condition (3.45) assures that the two-phonon term is small for temperatures $k_B T \approx \Delta_0$, where φ'' significantly varies with frequency in the range $\hbar|\omega| \approx \Delta_0$.

When replacing the coth function in (3.89) by its inverse argument, we find the series

$$\Gamma_1 = 2\pi\alpha\tilde{\Delta}_0^2 k_B T \hbar^{-3} \left(1 + \tfrac{2}{3}\pi^2\tilde{\alpha}T^2 + \ldots\right). \tag{3.89}$$

The quadratic correction clearly shows the relevance of the parameter $\tilde{\alpha}T^2$. For high temperatures the whole series in terms of φ has to be retained.

3.4 Crossover to Incoherent Tunneling

An Exact Expression for $\Gamma''(\omega)$. The coupled phonon propagator $\varphi(t)$ is not invariant under time reversal but rather satisfies the relation $\varphi(-t) = \varphi(t - i\beta\hbar)$. It turns out convenient to define the function

$$\bar{\varphi}(t) = \varphi(t - \tfrac{1}{2}i\beta\hbar), \tag{3.90}$$

which is symmetric under time reversal and reads explicitly as

$$\bar{\varphi}(t) = \sum_k u_k^2 \frac{\cos(\omega_k t)}{\sinh(\beta\hbar\omega_k/2)}. \tag{3.91}$$

Since $\Gamma(t)$ is an analytic function of time, we may shift the time integration from the real axis into the complex t-plane. As the phonon density of states vanishes in the limit of zero frequency, the phase varies rapidly with time for $t \to \infty$. Thus the Fourier integral (3.79) is determined by thermal frequencies; the integration contour for large t is immaterial, and, with $t \to t - \tfrac{1}{2}i\beta\hbar$, we have instead of the spectrum (3.79)

$$\Gamma''(\omega) = \frac{1}{2}(\tilde{\Delta}_0/\hbar)^2 e^{\beta\hbar\omega/2} \int_{-\infty}^{\infty} dt\, e^{i\omega t} \left[e^{\bar{\varphi}(t)} - 1 \right]. \tag{3.92}$$

When replacing the sum over k in $\bar{\varphi}(t)$ by an integral and increasing the cutoff frequency ω_D to infinity, Grabert (1992) obtained the phase

$$\bar{\varphi}(t) = 2\alpha \left(\frac{\pi}{\hbar\beta}\right)^2 \frac{1}{\cosh(\pi t/\beta\hbar)^2}. \tag{3.93}$$

Since the Fourier transformation of $[e^{\bar{\varphi}(t)} - 1]$ cannot be performed in closed form, we expand the exponential in a power series, and transform each term separately,

$$\Gamma''(\omega) = (\tilde{\Delta}_0/\hbar)^2 e^{\beta\hbar\omega/2} \frac{\beta\hbar}{\pi} \sum_{n=1}^{\infty} \frac{\phi^n}{n!} A_n(\beta\hbar\omega). \tag{3.94}$$

For notational convenience we have split off a prefactor in powers of

$$\phi = 2\alpha(\pi/\hbar\beta)^2 = \frac{T^2}{T_0^2}; \tag{3.95}$$

the remaining Fourier integral arising from the exponential series is given by

$$A_n(\beta\hbar\omega) = \frac{\pi}{\hbar\beta} \int_{-\infty}^{\infty} dt\, e^{i\omega t} \frac{1}{\cosh(\pi t/\beta\hbar)^{2n}}. \tag{3.96}$$

The parameter ϕ plays the role of a dimensionless coupling constant. At low temperatures, $T \ll T_0$, it is much smaller than unity, and the series (3.94) may be restricted to its first term. We stress that (3.93)–(3.95) are valid only for a Debye temperature much larger than temperature, $T \ll \Theta$.

When noting the value of the Fourier integral $A_1(x) = (x/2)\sinh(x/2)^{-1}$, we easily recover the above first-order result (3.87). This linear term has been derived previously by Leggett et al. (1987), where the series (3.94) has been truncated at lowest order. In physical terms this means that multiphonon processes have been discarded. In view of the expansion parameter $\phi = T^2/T_0^2$ it is clear that this approximation breaks down as temperature approaches T_0.

The Series at Zero Frequency. In view of (3.85) we need to evaluate $\Gamma'''(\omega)$ at $\omega = \tilde{\Delta}_0/\hbar$. For temperatures well above $\tilde{\Delta}_0/k_B$, the argument of the function (3.96) is much smaller than unity. Since $A_n(x)$ depends weakly on x for $|x| \ll 1$, we may put $\omega = 0$ in (3.94),

$$\Gamma'''(0) = \tilde{\Delta}_0^2 \frac{\beta\hbar}{\pi} \phi F(\phi), \qquad (3.97)$$

where the infinite series has been absorbed in the factor

$$F(\phi) = \sum_{n=1}^{\infty} \frac{\phi^{n-1}}{n!} A_n, \qquad (3.98)$$

with the coefficients $A_n \equiv A_n(0)$. The integral (3.96) at $\omega = 0$ can be found in Gradstein and Ryshik (1981),

$$A_n = 4^{n-1} \frac{(n-1)!^2}{(2n-1)!}. \qquad (3.99)$$

One easily verifies $F(\phi) \to 1$ for $\phi \to 0$. With the definition of ϕ, the first few terms of the series read as

$$\Gamma'''(0) = 2\pi\alpha \tilde{\Delta}_0^2 k_B T \hbar^{-3} \left[1 + \frac{1}{3}\phi + \frac{4}{45}\phi^2 + \frac{2}{105}\phi^3 + ... \right]. \qquad (3.100)$$

The linear correction $[1 + \frac{1}{3}\phi]$ is identical to that of (3.89). When expanding the coth function in powers of its inverse argument, we find that (3.87) and (3.100) agree with respect to the leading contribution.

Saddle-point integration. At high temperatures, $T \gg T_0$, the spectrum at zero frequency, $\Gamma'''(0)$, may be evaluated by saddle-point integration. The expansion of the phase $\bar{\varphi}(t)$ in powers of t reads as

$$\bar{\varphi}(t) = \varphi_0 - \tfrac{1}{2}\varphi_2 t^2 + \tfrac{1}{4!}\varphi_4 t^4 + ..., \qquad (3.101)$$

with coefficients

$$\varphi_{2n} = \sum_k \frac{\lambda_k^2}{\omega_k^2} \frac{\omega^{2n}}{\sinh(\hbar\omega_k/2k_B T)}. \qquad (3.102)$$

Truncating this series after the quadratic term, evaluating the Gaussian integral, and calculating lowest-order corrections with respect to φ_4, we find the rate

$$\Gamma_{\text{SPI}} = \frac{\tilde{\Delta}_0^2}{\hbar^2} \sqrt{\frac{\pi}{2\varphi_2}} e^{\varphi_0} \left(1 + \frac{\varphi_4}{8\varphi_2^2} + ... \right). \qquad (3.103)$$

This expression is valid in the whole temperature range $T \gg T_0$. Here we restrict the evaluation of the rate to T well below the Debye temperature, $T \ll \Theta$. In this case the first expansion coefficient,

$$\varphi_0 = \sum_k (\lambda_k/\omega_k)^2 \sinh(\hbar\omega_k/2k_B T)^{-1}, \qquad (3.104)$$

turns out to be identical to the parameter ϕ defined above,

$$\varphi_0 = T^2/T_0^2 \equiv \phi \qquad (T \ll \Theta). \qquad (3.105)$$

The next terms fulfil the relations $\varphi_2 = \phi^2/\alpha$ and $\varphi_4 = 4\phi^3/\alpha^2$. This permits us to rewrite (3.103) as a series in powers of $1/\phi$,

$$\Gamma_{\text{SPI}} = \frac{\tilde{\Delta}_0^2}{\hbar^2}\sqrt{\frac{\pi\alpha}{2}}\frac{1}{\phi}e^\phi\left(1+\frac{1}{2\phi}+\ldots\right). \qquad (3.106)$$

Inserting the coupling parameter (3.95) in the rate formula and using $\tilde{\Delta}_0 = \Delta_0 \exp(-\frac{1}{6}\phi)$, we recover the diffusion rate of Holstein (1959):

$$\Gamma_{\text{SPI}} = \frac{\Delta_0^2 T_0}{2\sqrt{\pi}\hbar k_B T^2}e^{\frac{2}{3}T^2/T_0^2} \qquad (T_0 \ll T \ll \Theta). \qquad (3.107)$$

We have dropped corrections of the order ϕ^{-1}; the saddle-point integration provides the correct result for $\phi \gg 1$, or $T \gg T_0$.

Interpolation Formula. For practical purposes, the series expansion (3.98) is not convenient, since it converges slowly. From Stirling's formula, $n! \approx (n/e)^n\sqrt{2\pi n}$ for large n, we find that the main contributions to the series stem from terms with n of the order of ϕ. Thus for temperatures about ten times larger than T_0, several hundred terms have to be retained in order to assure convergence of the series (3.98).

For this reason we propose a simple interpolation formula for the correction factor $F(\phi)$,

$$F(\phi) = \frac{e^\phi - 1}{\phi}(1 + 4\phi/\pi)^{-1/2}, \qquad (3.108)$$

that correctly describes the limits $F(\phi) \to 1$ for $\phi \to 0$ and $\phi F(\phi) \to \sqrt{\pi/4\phi}e^\phi$ for $\phi \gg 1$. The error in the intermediate range about $\phi \approx 1$ does not exceed 15 per cent.

As to the spectrum $\Gamma''(\omega)$, the factor $F(\phi)$ assures the correct high-temperature behavior, where $\Gamma''(\omega)$ depends little on frequency. At low temperature, multiphonon corrections are immaterial, and $\Gamma''(\omega)$ tends towards the one-phonon spectrum $\varphi''(\omega)$. In order to meet these conditions we write

$$\Gamma''(\omega) = 2\pi(\tilde{\Delta}_0/\hbar)^2 \alpha\omega[1 + n(\omega)]F(\phi), \qquad (3.109)$$

which holds true for frequencies $\hbar|\omega| < \max(k_B T_0, k_B T)$. In terms of the interpolation formula (3.108), the rate Γ_1 reads as

$$\Gamma_1 = \pi\alpha(\tilde{\Delta}_0/\hbar)^3 \coth(\tfrac{1}{2}\beta\tilde{\Delta}_0)F(\phi). \qquad (3.110)$$

At low temperatures, $T \ll T_0$, this rate tends towards the one-phonon expression (3.87), whereas in the opposite limit it approaches Γ_{SPI}.

Beyond Debye Temperature. Finally we consider the case where T is not small compared to the Debye temperature. Following Holstein (1959) and Mahan (1981), we may write $\Gamma(t) = (\Delta_{\rm b}/\hbar)^2 \exp(-W + \varphi_0 - \tfrac{1}{2}\varphi_2 t^2)$. As to the time-independent part of the exponential, it turns out convenient to evaluate

$$-W + \varphi_0 = \sum_k u_k^2 [-(1 + 2n_k) + \sinh(\hbar\omega_k/2k_{\rm B}T)^{-1}]. \qquad (3.111)$$

For $T > \Theta$ we may expand the temperature factors in powers of $x_k = \tfrac{1}{2}\beta\hbar\omega_k$. Retaining the leading terms of $(1 + 2n_k) = x_k^{-1} + \tfrac{1}{2}x_k + ...$ and $\sinh(x_k)^{-1} = x_k^{-1} + O(x_k^2)$ only, we obtain

$$-W + \varphi_0 = -\frac{V}{k_{\rm B}T}, \qquad (3.112)$$

with the activation energy

$$V = \frac{1}{4}\sum_k u_k^2 \hbar\omega_k = \frac{1}{6}k_{\rm B}\tilde{\alpha}\Theta^3. \qquad (3.113)$$

The coefficient of the quadratic term reads as (Niu (1991), Würger (1997a))

$$\varphi_2 = \frac{4}{3}(\hbar/k_{\rm B})^2 \tilde{\alpha}\Theta^3 T = 8V k_{\rm B}T/\hbar^2 \qquad \text{for } T \geq \Theta. \qquad (3.114)$$

The Fourier transform of the Gaussian $\Gamma(t)$ is easily calculated,

$$\Gamma''(\omega) = e^{-V/k_{\rm B}T}\sqrt{\frac{\pi}{2\varphi_2}} e^{-\omega^2/2\varphi_2}. \qquad (3.115)$$

After inserting the above expression for φ_2 and evaluating the spectrum at zero frequency, we recover the expression derived first by Holstein (1959),

$$\Gamma_{\rm SPI} = \frac{\Delta_{\rm b}^2}{\hbar}\sqrt{\frac{\pi}{16V k_{\rm B}T}} e^{-V/k_{\rm B}T} \qquad (T \geq \Theta), \qquad (3.116)$$

which shows Arrhenius behavior with activation energy V.

Time Evolution

The dressed tunnel energy $\tilde{\Delta}_0$ decreases exponentially with temperature, whereas the rate Γ_1 increases with rising T. The relation $\tfrac{1}{2}\hbar\Gamma_1 = \tilde{\Delta}_0$ defines the implicit equation for the temperature T^*,

$$T^* = T_0\sqrt{\tfrac{6}{5}\log\left(2\sqrt{\pi}k_{\rm B}T^{*2}/\Delta_0 T_0\right)}, \qquad (3.117)$$

which separates two qualitatively different regimes for the time evolution of the TS. We discuss these two cases arising from (3.84) in more detail.

Coherent Oscillations at low T. Low temperatures imply $\frac{1}{2}\hbar\Gamma_1 < \tilde{\Delta}_0$ and, in view of (3.83), a pair of complex conjugate frequencies,

$$\Gamma_\pm = \tfrac{1}{2}\Gamma_1 \pm i\sqrt{(\tilde{\Delta}_0/\hbar)^2 - \tfrac{1}{4}\Gamma_1^2} \equiv \Gamma_t \pm i\omega_t. \tag{3.118}$$

Accordingly, $P(t)$ shows weakly damped oscillations,

$$P_1(t) = \cos(\omega_t t)\exp(-\Gamma_t t), \tag{3.119}$$

with frequency ω_t and the transverse rate

$$\Gamma_t = \frac{\pi}{2}\alpha(\tilde{\Delta}_0/\hbar)^3 \coth(\tilde{\Delta}_0/2k_B T)\left[1 + \frac{1}{3}\phi + \frac{4}{45}\phi^2 + \ldots\right]. \tag{3.120}$$

We have neglected a small phase shift in $P_1(t)$.

Here the reader is referred to the discussion of the dressed tunnel energy (3.39). At low temperatures, $T \ll T_0$, the temperature-dependent dressing is insignificant, i.e. $\tilde{\Delta}_0 = \Delta_0$. Then (3.119) reduces to weakly damped tunnel oscillations with frequency $\Delta_0/\hbar$, and Γ_t is the usual one-phonon rate.

With rising temperature the expansion parameter $\phi = T^2/T_0^2$ approaches unity, and the corrections result in a strong increase of the rate Γ_t and drive the system towards the aperiodic case $\omega_t = 0$ at $T = T^*$.

Overdamped Motion at High T. For $T \geq T^*$, both roots $\Gamma_\pm$ are real, resulting in two distinct relaxation features in (3.84). The first one, whose amplitude is larger than unity, decays rapidly with the larger rate Γ_+. In the long-time limit, however, the second term prevails, since its rate Γ_- is small; because of the negative amplitude, $P_1(t)$ is negative for $t \gg 1/\Gamma_+$.

The latter feature is of rather academic interest, since the amplitude $\Gamma_-/(\Gamma_+ - \Gamma_-)$ becomes negligibly small as T significantly exceeds T^*. This is seen easily when noting $\Gamma_- \approx (\tilde{\Delta}_0/\hbar)^2/\Gamma_1$ for $\Gamma_1 \gg \tilde{\Delta}_0/\hbar$. In this case the second root Γ_- may be discarded; with $\Gamma_1 \approx \Gamma_+$, we obtain for (3.84)

$$P_1(t) = \exp(-\Gamma_1 t). \tag{3.121}$$

In summary, with rising temperature the two-state dynamics (3.84) shows a crossover from damped oscillations to incoherent tunneling, as is obvious from the limiting cases (3.119) and (3.121).

3.4.4 Two-State Dynamics Beyond NIBA

Now we turn to the improved propagator $P_2(z)$ given in (3.77), which accounts for corrections to NIBA. As in (3.82), we drop the real parts of the self-energy, Γ' and Γ'_g, and resort to a Markov approximation for the spectra,

$$\gamma_g = \frac{1}{2}\bigl(\Gamma''_g(\tilde{\Delta}_0/\hbar) + \Gamma''_g(-\tilde{\Delta}_0/\hbar)\bigr). \tag{3.122}$$

It turns out convenient to define odd and even contributions of (3.97),

$$\gamma_{\rm g} = \pi\alpha(\tilde{\Delta}_0/\hbar)^3 \coth(\tilde{\Delta}_0/2k_{\rm B}T) \sum_{n=1}^{\infty} \frac{1}{(2n)!} \phi^{2n-1} A_{2n}\left(\tfrac{1}{2}\beta\tilde{\Delta}_0\right), \qquad (3.123)$$

$$\gamma_{\rm u} = \pi\alpha(\tilde{\Delta}_0/\hbar)^3 \coth(\tilde{\Delta}_0/2k_{\rm B}T) \sum_{n=0}^{\infty} \frac{1}{(2n+1)!} \phi^{2n} A_{2n+1}\left(\tfrac{1}{2}\beta\tilde{\Delta}_0\right). \qquad (3.124)$$

Accordingly, the rate Γ_1 may be written as the sum of both parts,

$$\Gamma_1 = \gamma_{\rm u} + \gamma_{\rm g}. \qquad (3.125)$$

Inserting these rates in (3.77), we have

$$P_2(z) = -\left[z + i\Gamma_1 - \frac{(\tilde{\Delta}_0/\hbar)^2}{z + i\gamma_{\rm g}}\right]^{-1}, \qquad (3.126)$$

whose poles are given by the roots of the quadratic form in the denominator,

$$\Gamma_\pm = \gamma_{\rm g} + \tfrac{1}{2}\gamma_{\rm u} \pm \sqrt{\tfrac{1}{4}\gamma_{\rm u}^2 - \tilde{\Delta}_0^2/\hbar^2}. \qquad (3.127)$$

Taking the inverse Laplace transformation, we obtain

$$P_2(t) = \sum_\pm \frac{\Gamma_\pm - \gamma_{\rm g}}{\Gamma_\pm - \Gamma_\mp} \exp\left(-\Gamma_\pm t\right), \qquad (3.128)$$

whose complex frequencies are given by the roots (3.127). Equation (3.128) with the rates (3.110) and (3.123) constitutes our solution for the spin-phonon problem.

The qualitative behavior is very similar to that of the NIBA solution. The argument of the square root in (3.127) changes sign as a function of temperature. At low temperature, the relation $\tilde{\Delta}_0 > \tfrac{1}{2}\hbar\gamma_{\rm u}$ gives rise to two complex poles in (3.126) and to underdamped oscillations of $P_2(t)$. With increasing temperature, the system passes through the aperiodic case $\tilde{\Delta}_0 = \tfrac{1}{2}\hbar\gamma_{\rm u}$ and finally reaches the range of incoherent motion, where both roots (3.83) are purely imaginary, i.e. $\Gamma_\pm$ are real. We deal separately with these two cases.

Coherent Motion at Low T. In view of the temperature dependence of both rate and tunnel energy, this range may be labeled as weak-coupling or low-temperature case.

The complex roots (3.127) read as $\Gamma_\pm = \Gamma_t \pm i\omega_t$, where the effective tunnel frequency ω_t and the transverse damping rate are given by

$$\omega_t = \sqrt{(\tilde{\Delta}_0/\hbar)^2 - \tfrac{1}{4}\gamma_{\rm u}^2}, \qquad \Gamma_t = \tfrac{1}{2}\gamma_{\rm u} + \gamma_{\rm g}. \qquad (3.129)$$

Accordingly we find damped oscillations,

$$P_2(t) = \frac{\cos(\omega_t t + \delta)}{\cos(\delta)} \exp\left(-\Gamma_t t\right), \qquad (3.130)$$

with a phase shift defined by $\tan\delta = (\gamma_u/2\omega_t)$. At zero temperature, the tunnel frequency ω_t is almost identical to $\Delta_0/\hbar$. In the range $T \approx T_0$ it decreases exponentially according to (3.42), and finally vanishes at the crossover temperature T^* as defined in (3.135) below.

At very low temperatures the rate is dominated by the one-phonon contribution to γ_u,

$$\Gamma_t = \frac{\pi}{2}\alpha(\tilde{\Delta}_0/\hbar)^3 \coth(\tilde{\Delta}_0/2k_BT) + O(\alpha^2) \quad (T \ll T_0). \tag{3.131}$$

With rising temperature, multiphonon terms become more important; in order to permit a comparison with the results of NIBA and those of the perturbation theory of Sect. 3.8, we give the exact expression for the lowest-order term of γ_g, resulting from (3.122),

$$\gamma_g = \pi\alpha(\tilde{\Delta}_0/\hbar)^3 \coth(\tilde{\Delta}_0/2k_BT)\left[\tfrac{1}{3}\phi + \tfrac{2}{105}\phi^3 + \ldots\right]. \tag{3.132}$$

These corrections are identical to the even-order terms of the NIBA result (3.120). From (3.129) we obtain the transverse damping rate

$$\Gamma_t = \frac{\pi}{2}\alpha(\tilde{\Delta}_0/\hbar)^3 \coth(\tilde{\Delta}_0/2k_BT)\left[1 + \frac{2}{3}\phi + \frac{4}{45}\phi^2 + \frac{4}{105}\phi^3 + \ldots\right]. \tag{3.133}$$

Note that Γ_t involves both γ_u and γ_g, whereas the phase shift depends on the odd-order terms γ_u only.

Comparison with (3.309) shows that the rate Γ_t agrees with the exact result from perturbation theory. Thus in the range $T \approx T_0$, the corrections (3.132) arising from blip–blip interactions are of the same order of magnitude as the rate obtained in NIBA. Strictly speaking, NIBA is not a controlled approximation, although it gives qualitatively correct results.

Incoherent Tunneling: $\tfrac{1}{2}\hbar\gamma_u > \tilde{\Delta}_0$. For the aperiodic case $\tfrac{1}{2}\gamma_u = \tilde{\Delta}_0/\hbar$, the two poles $-i\Gamma_\pm$ merge on the imaginary axis, $\Gamma_+ = \Gamma_-$. When increasing the temperature further, the rate $\tfrac{1}{2}\gamma_u$ exceeds $\tilde{\Delta}_0/\hbar$. Then both poles of (3.126) are purely imaginary and increase with rising temperature; accordingly the motion is best described as incoherent tunneling between the two states $\sigma_z = \pm 1$ with two different relaxation rates $\Gamma_\pm$.

Even for very high temperatures, the 'small' rate Γ_- differs from Γ_+ merely by a factor of two, $\Gamma_- \approx \gamma_g \approx \tfrac{1}{2}(\gamma_g + \gamma_u) \approx \tfrac{1}{2}\Gamma_+$. Due to the exponential increase of the rates, however, the amplitude of the term involving Γ_- in (3.128) vanishes rapidly. As a consequence, in the incoherent regime, $P(t)$ is well described by the simpler function

$$P(t) = e^{-\Gamma t} \quad \text{with } \Gamma = \Re\Gamma_+, \tag{3.134}$$

whose relaxation rate is given by the dissipative (real) part of the root Γ_+. Yet it turns out that this rate differs little from the NIBA result (3.110), i.e. at higher temperature we have $\Gamma = \Gamma_1 = \gamma_u + \gamma_g$. This expression confirms the statement, made previously by Leggett et al. (1987), that blip–blip interactions are irrelevant in the incoherent regime.

The Crossover Temperature. The crossover to incoherent motion described above occurs at $2\tilde{\Delta}_0 = \hbar\gamma_u$, requiring a value for $\phi = T^2/T_0^2$ well beyond unity. In this range we may use $\gamma_u \approx \frac{1}{2}\Gamma_1$. Inserting the tunnel energy (3.43) and the definition (3.24) in the condition $2\tilde{\Delta}_0 = \hbar\gamma_u$, we find an implicit equation for the crossover temperature T^*,

$$T^* = T_0\sqrt{\tfrac{6}{5}\log\left(4\sqrt{\pi}k_B T^{*2}/T_0\Delta_0\right)}. \tag{3.135}$$

With the slow-tunneling condition (3.25), one easily proves the inequality $\Delta_0 \ll k_B T_0$. Therefore the crossover temperature T^*, which is defined by $\omega_t = 0$ in (3.129), is larger than T_0. Comparing T^* with the corresponding result from NIBA, (3.117), we find that both expressions differ merely by a factor of 2 in the argument of the logarithm.

As a caveat, we recall that the present discussion is subject to the inequality (3.26). In view of this restriction, it is clear that the crossover temperature is a meaningful quantity only when $T^* \ll \Theta$. In this case, incoherent tunneling sets in at T^*, and our formal results are valid for all temperatures. In the range $T \geq \Theta$, we simply have to use the corresponding rate (3.116).

If, on the other hand, the solution of (3.135) is of the order of the Debye temperature or even larger, this indicates that the system never satisfies the crossover condition $\frac{1}{2}\gamma_u = \tilde{\Delta}_0/\hbar$. Still, we must use (3.116) as temperature approaches Θ, but this rate never exceeds the tunnel frequency. Hence we find weakly damped oscillations at all temperatures, with the dissipation rate (3.116).

Since T_0 and T^* do not differ very much, such a large value for T^* implies, via (3.24), a small coupling parameter $\tilde{\alpha}$ and a Debye–Waller factor that is close to unity. Hence the case where the motion does not become overdamped is closely related to weak phonon dressing; vice versa, the occurrence of the incoherent limit requires a strong dressing effect.

3.4.5 The Undressing Effect

Phonon coupling affects the tunneling motion in two respects. First, the polaron effect leads to a reduction of the tunnel frequency, while, second, thermal fluctuations give rise to dissipation and, at higher temperature, to incoherent tunneling between the two wells. The two-state polaron has been discussed on p. 69, illustrating the physical meaning of the polaron transformation. Here we focus on the phonon-induced enhancement of the rate with rising temperature, which may be considered as arising from an undressing effect.

At low T, the rate Γ_t is smaller than the tunnel frequency $\tilde{\Delta}_0/\hbar$, and the TS shows weakly damped oscillatory motion. With rising temperature, however, multiphonon processes become important, and at some point the rate exceeds $\tilde{\Delta}_0$. Then a particle starting in the left well has lost its phase memory before reaching the right well for the first time. Such a motion we call incoherent tunneling.

In view of Fig. 3.1 on p. 70, for $k_B T \gg \hbar\omega_k$ thermal fluctuations permit the coupled system to explore a wide range along the x_k-axis, and thus to penetrate the barrier horizontally at any point. With rising temperature, an increasing number of vibrational modes relax from their frozen-in zero-temperature state. In terms of Fig. 3.1 this means that the distance of the two minima along the x_k-axis becomes irrelevant, since the thermal fluctuations are larger anyhow. This effect strongly enhances the damping rate. Note that in the incoherent regime, the barrier crossing time, $1/\Gamma$, is shorter than the inverse tunnel matrix element, $\hbar/\tilde{\Delta}_0$.

This picture is supported by a comparison of the expressions for the rate at low and high temperatures. For $T \ll T_0$, the one-phonon rate (3.87) involves the dressed tunnel energy $\tilde{\Delta}_0$; in this range the bare value Δ_b has no physical meaning, since almost all vibrational modes are frozen in the minima shown in Fig. 3.1. In the opposite limit, however, as T approaches the Debye temperature, the rigid phonon cloud has molten and, accordingly, the rate (3.116) depends on the bare tunnel energy.

Two more remarks are in order. *First*, the undressing effect involves the rate only, yet not the tunnel matrix element. The latter, in fact, becomes exponentially small. This is not surprising and can, again, be understood in terms of Fig. 3.1. The dressed tunnel energy (3.39) involves the thermal average over all phonon states. Yet the phonon overlap matrix element decreases exponentially with increasing average distance, i.e. with rising temperature. *Second*, although the rate varies with the bare tunnel energy at high temperatures, (3.116) assures that it is smaller than the corresponding frequency $\Delta_b/\hbar$; cf. the discussion on p. 89. In the incoherent regime we have $\tilde{\Delta}_0 \ll \hbar\Gamma \ll \Delta_b$.

3.4.6 Discussion

The blip expansion provides a strong-coupling approach for the spin-phonon model; our solution is given by (3.128). The corresponding rates (3.110) and (3.122) comprise terms of any order in the coupling parameter $\tilde{\alpha}$ or, more precisely, in the dimensionless quantity ϕ. Here we discuss the main results and some special features of the blip expansion.

Which are the Relevant Parameters? We have seen on p. 67 that the coupling constant $\tilde{\alpha}$ gives rise to three dimensionless parameters (3.31)–(3.33). Roughly speaking, these parameters express the bare tunnel energy Δ_b, the Debye temperature Θ, and temperature T in units of the crossover temperature T_0. If one of these parameters is smaller than unity, the corresponding effects on the two-state dynamics may be treated in second-order perturbation theory in terms of the spin-phonon couplings λ_k.

As to the first one, $\tilde{\alpha}(\Delta_b/k_B)^2$, we assume it to be small throughout this contribution. As a consequence, all rates depend quadratically on the tunnel energy $\Delta_{\rm b}$, or on its dressed value $\tilde{\Delta}_0$. This choice was motivated by the fact that most real systems fulfil this inequality $\tilde{\alpha}(\Delta_{\rm b}/k_{\rm B})^2 \ll 1$. A thorough investigation of the opposite case, $\tilde{\alpha}(\Delta_{\rm b}/k_{\rm B})^2 > 1$, is still lacking; we expect the more interesting features to arise from the frequency renormalization, or $\mathcal{Z}$-factor. Finally we note that (3.25) is a much stronger condition than the 'self-consistency condition' of Leggett et al. (1987), which involves the reduced tunnel energy and reads in our notation $\tilde{\alpha}(\Delta_0/k_{\rm B})^2 \ll 1$.

The present theory is concerned with the remaining two parameters, (3.32) and (3.33). We have seen on p. 69 that both $2\pi^2\tilde{\alpha}T^2$ and $\tilde{\alpha}\Theta^2$ contribute to the reduction of the tunnel energy, as expressed by the dressed values $\tilde{\Delta}_0$ and Δ_0, respectively. This means that all vibrational modes contribute to the frequency renormalization. On the contrary, dissipation arises mainly from thermal bath frequencies; as a consequence, all corrections to the one-phonon rate involve the temperature-dependent parameter $\phi = 2\pi^2\tilde{\alpha}T^2$, as is most obvious from the series (3.98).

NIBA and Beyond. The two-state dynamics obtained from NIBA shows a crossover from weakly damped tunnel oscillations at low temperature to incoherent motion at higher T. This general picture is not changed by corrections arising from blip–blip interactions. It turns out that these additional terms double the even-order contributions to the rate in the coherent regime. Equations (3.120) and (3.133) show explicitly that NIBA misses a factor of 2 in that respect.

In the incoherent regime, however, the difference between NIBA and the improved approach disappears, since the respective roots Γ_+ tend both towards the NIBA rate Γ_1. This confirms a previous statement by Leggett et al. (1987), who claimed on formal grounds that blip–blip interactions are irrelevant in the case of overdamped motion.

The Motional Spectrum at Zero Frequency. The imaginary part of $P(z)$ provides the motional spectrum according to $P''(\omega) = \Im P(\omega+i0)$. When considering the NIBA spectrum $P_1''(\omega)$, we find that it vanishes at zero frequency, $P_1''(0) = 0$, and gives rise to a surprising feature of the relaxation behavior. Evaluating (3.84) for the overdamped case,

$$P_1(t) = [1+\xi]e^{-\Gamma_1 t} - \xi e^{-\xi\Gamma_1 t}, \tag{3.136}$$

with the small parameter $\xi = (\tilde{\Delta}_0/\hbar\Gamma_1)^2 \ll 1$, one finds that $P_1(t)$ is not a smoothly damped function. At short times, $\Gamma_1 t < 1$, it is governed by the first term, whereas in the long-time limit, the second term prevails, resulting in a negative $P_1(t)$. Thus, even far in the incoherent regime, $P_1(t)$ shows half an oscillation and then approaches zero from negative values.

This unphysical behavior is removed when taking blip–blip interactions into account. The additional self-energy contribution leads to a finite value of the motional spectrum at zero frequency, $P_2''(0) = \gamma_{\rm g}(\hbar/\tilde{\Delta}_0)^2$; the resulting relaxation function $P_2(t)$ is positive for all times.

Beyond the Debye Temperature. The strong increase of the rate for temperatures beyond T_0 constitutes the most significant finding of this chapter. At temperatures well beyond T_0, multiphonon corrections enhance the rate and lead to an exponential increase, proportional to $\exp(\tfrac{2}{3}T^2/T_0^2)$.

This exponential law ceases to be valid as T approaches the Debye temperature Θ. According to (3.116), the rate shows activated behavior in the range about Θ, and decreases at temperatures well above. Yet the precise form of these laws should not be taken too seriously, because for real materials the phonon spectrum differs significantly from that of the Debye model; cf. the discussion on p. 110. Moreover, one should be aware of the limited validity of the two-state model; at such high temperatures the truncation to the ground states in the double-well potential cannot be justified in general.

The high-temperature expression (3.116) provides, however, an upper bound for the rate. When considering the most relevant case $T_0 < \Theta$ and discarding numerical factors, we find that the relaxation rate Γ_1 obeys the inequality $\hbar\Gamma_1 < (\Delta_b^2/k_B\Theta)$. Since, for atomic tunneling, the Debye energy $k_B\Theta$ in general exceeds by far the matrix element Δ_b, we may conclude that the rate Γ_1 is much smaller than the bare tunnel frequency, $\Gamma_1 \ll \Delta_b/\hbar$.

Z Factor. In previous work on the two-state dynamics, only the dissipative part of the self-energy has been retained. The reactive (or real) part is negligible as long as its derivative with respect to frequency is small compared to unity. The quantity Ξ in (3.78) is determined by the derivative $\partial_\omega \Gamma'(\omega)$. For the case of a small bare tunnel energy, $\Delta_b \ll \hbar\omega_D$, we have shown that $\partial_\omega \Gamma'(\omega)$ is small in fact; see Appendix and (A23).

Yet we have not addressed the general case where the ratio $\Delta_b/\hbar\omega_D$ is not small. Although most real systems satisfy the slow-tunneling condition $\Delta_b \ll \hbar\omega_D$, a thorough study of the opposite limit would seem desirable. We will see on pp. 109 and 121, that in oxide glasses the maximum tunnel energy $\Delta_0^{\max}$ is hardly one order of magnitude smaller than the effective Debye energy $\hbar\omega_D$; as a result, the bare value Δ_b may be comparable to the effective value $\hbar\omega_D$.

Comparison with Strong-Coupling Perturbation Theory. The results of this section can be derived equally well from a strong-coupling perturbation theory that is based on a series expansion in terms of the fluctuation operators $\xi_\pm = B_\pm - \langle B_\pm \rangle$. In such an approach, the pseudospin motion is given by a four-dimensional propagator matrix, whose self-energy reduces to a 4×4 rate matrix Γ_{ij}. When evaluating the latter to second order in the fluctuation operators $\xi_\pm$, it turns out that the NIBA result is identical to the diagonal element Γ_{zz}, whereas the corrections arising from blip–blip interactions form the next diagonal element Γ_{yy}; cf. Würger (1998a).

This correspondence links the present blip expansion, which is based on a one-dimensional integro-differential equation for σ_z, to the matrix approach. In order to obtain from the present theory the lowest-order terms of the 4×4

rate matrix, we had to perform an infinite partial summation of the self-energy diagrams. The final result for the propagator, as given in (3.128), is identical to that of Würger (1998a).

Relation to Previous Work. Contrary to the Ohmic case, the phonon heat bath with cubic spectral density has attracted little attention in the past. There are but a few works on this problem, whose results, however, disagree with respect to the high-temperature behavior. We briefly discuss the relation to the present theory.

In a functional integral approach, Leggett et al. (1987) derived the NIBA and studied the case of ohmic dissipation in detail. When extending their results to the cubic bath spectrum, they find a damping rate that is identical to our first-order expression (3.87), i.e. these authors retain only the first term of the series (3.98) or (3.100). Since both multiphonon contributions to NIBA and corrections arising from blip–blip interactions are small for $\phi \ll 1$, the phonon-driven rate of Leggett et al. (1987) is valid at low temperatures $T \ll T_0$.

More recently, Niu (1991) investigated the two-state dynamics in limiting cases which, in our notation, correspond to $T \ll T_0$ and $T \gg \Theta$. As to the former, he confirmed the one-phonon rate, whereas above the Debye temperature Θ, he recovered Holstein's diffusion rate (3.116). Similar results have been reported earlier by Pirc and Gosar (1969) and Kondo (1984).

The works mentioned so far rely on the small-polaron representation (3.37). For completeness we mention the rate obtained by Jäckle (1972) in a Boltzmann equation approach, which, in fact, is similar to our one-phonon rate, but with the bare tunnel energy Δ_b instead of the renormalized value in (3.87). We end with a remark on the mode-coupling approach which will be discussed in detail on p. 137. According to (3.327) and (3.328), the resulting high-temperature rate decreases with rising temperature as T^{-2}. Comparing (3.328) and the result from saddle-point integration (3.107), we find that these rates are very similar, except for the exponential temperature factor. Of course, this factor missing in (3.328) is a most serious drawback of the mode-coupling approach; it is linked to neglecting the real part of the memory function in Sect. 3.8.5.

In summary, in this section we have presented an expansion scheme which permitted us to evaluate the rate in terms of the noninteracting blip approximation, and to calculate correction terms beyond NIBA. As to the NIBA result, we confirm, at low temperatures $T \ll T_0$, the one-phonon rate of Leggett et al. (1987). As soon as temperature approaches T_0, multiphonon terms give rise to a strong enhancement of the rate and drive the crossover to incoherent motion.

In the range $T \approx T_0$, however, significant corrections to NIBA arise from blip–blip interactions. Although it still gives qualitatively correct results, NIBA is not a controlled approximation in this intermediate range. For typical values of T_0, see Table 3.1 on p. 109.

3.5 Phonon-Assisted Tunneling in Metals

In the preceding section we have studied a two-state defect in an insulating solid, whose low-frequency vibrational modes give rise to a cubic bath spectral density given by Eq. (3.19). Here, we complete this treatment by considering a defect in a metallic host whose spectral density comprises a term linear in frequency, due to electron-hole excitations in the conduction band. Such a two-state system with a linear, or 'Ohmic', bath spectrum (3.7) has been investigated in detail by various authors. For a survey, we refer to the book of Weiss (1993) and to the review article by Leggett et al. (1987).

Yet in real metals, coupling to both conduction electrons and elastic waves has to be taken into account, and the actual bath spectrum comprises both (3.7) and (3.19). We thus consider the composite spectral density

$$J(\omega) = \pi K \omega + \pi \alpha \omega^3. \tag{3.137}$$

The change from the linear to the cubic law occurs at a frequency $\Omega = \sqrt{K/\alpha}$ which, in terms of (3.24), may be written as $\Omega = \sqrt{2K\pi k_B T_0}/\hbar$. Incoherent tunneling with such a composite spectral density has been studied in detail by Grabert (1992); more recently, Grabert and Schober (1997) have given a comprehensive review.

Phonon coupling results in a strong temperature dependence of the damping rate. Accordingly, it provides an effective dissipation mechanism at high T, whereas conduction electrons are most relevant at low temperatures. Accordingly there is a crossover from the electron driven rate at low T to phonon-assisted jumps at higher temperature.

Such a behavior has been observed for interstitial hydrogen trapped by oxygen or nitrogen impurities in niobium, and for configurational defects in submicron wires. Since, in the former case, the crystal provides a symmetric two-state potential, i.e. $\Delta = 0$, we first generalize the blip expansion of the last section to the metallic case, by taking into account coupling to both phonons and conduction electrons.

In the second part of this section, when considering tunneling between two unlike quantum wells, we need to add an asymmetry energy to the Hamiltonian (3.37). Restricting our treatment to the case where the asymmetry is much larger than the tunnel energy, we may use a simple master equation approach, where the rate is given by a golden rule expression.

For both examples, we express the theoretical results in terms of microscopic quantities and we present a detailed comparison with experimental data, based on known values for the elastic deformation potential.

3.5.1 Blip Expansion for Zero Asymmetry

Formally, we proceed as in Sect. 3.4. We assume that the particle starts from the left well at $t = 0$, and thus consider $P(t) = \langle \sigma_z(t) \rangle$, with the initial value

$P(t=0) = 1$. The essential difference arises from the fact that the bath comprises, according to (3.137), both electron-hole excitations and phonons.

For later convenience, we adopt a notation which slightly differs from that of Sect. 3.4, and write for the self-energy contribution arising from NIBA

$$\Sigma_1(t) = (\Delta_b/\hbar)^2 e^{\psi(t)}. \tag{3.138}$$

The phase used here contains a time-independent constant,

$$\psi(t) = \sum_k \frac{\lambda_k^2}{\omega_k^2}\Big[(1 + 2n_k)[\cos(\omega_k t) - 1] - i\sin(\omega_k t)\Big], \tag{3.139}$$

contrary to the notation in (3.62). In Sect. 3.4, this constant has been absorbed in the effective tunnel energy $\tilde{\Delta}_0$, by means of the factor (3.40). We repeat that the sum in the phase ψ involves both phonons and conduction electrons.

According to previous work on the Ohmic model (cf. e.g. Leggett et al. (1987)), the conduction electrons drive a crossover to incoherent motion at a temperature T_{el} which is roughly given by $\Delta_{el} = \pi K k_B T_{el}$, where Δ_{el} is the dressed tunnel energy. In the preceding section, we have found that the phonon-assisted rate, too, would lead to incoherent motion, just above the temperature T_0.

Moreover, we have seen that the corrections due to blip interactions are significant for phonon coupling in the coherent regime only. Saying it the other way round, NIBA is a valid approximation (i) in the incoherent range and (ii) where dissipation by conduction electrons prevails. From this we conclude that NIBA describes the whole range of temperatures, if we have $T_{el} < T_0$, in terms of the temperatures defined above. Throughout this section, we assume this inequality holds, i.e. we assume that the crossover to incoherent motion is driven by the Ohmic bath and that phonon damping becomes relevant only at higher T.

Accordingly, we retain the self-energy contribution from NIBA only, and find after Laplace transformation

$$P_1(z) = -[z + \Sigma_1(z)]^{-1}. \tag{3.140}$$

The self-energy is determined by the function e^ψ which may be written as a product of two time-dependent factors,

$$e^{\psi(t)} = \Gamma_{el}(t)\Gamma_{ph}(t), \tag{3.141}$$

where $\Gamma_{el}(t)$ and $\Gamma_{ph}(t)$ comprise the effects of electron-hole excitations and phonons, respectively. As to the former, we quote the well-known result for positive times

$$\Gamma_{el}(t) = e^{i\pi K}[\pi k_B T/\hbar\Delta_b \sinh(\pi t k_B T/\hbar)]^{2K}. \tag{3.142}$$

(Cf. e.g. Leggett et al. (1987)). The phonon factor reads $\Gamma_{ph}(t) = e^{\psi_{ph}(t)}$, where $\psi_{ph}(t)$ is given by (3.139) with the sum involving phonon modes only.

The damping spectrum $\Gamma''_{\rm el}(\omega) = \Im \Gamma_{\rm el}(\omega + i0)$ in NIBA is obtained from (Grabert and Weiss (1985), Fisher and Dorsey (1985))

$$\Delta_{\rm b}^2 \Gamma_{\rm el}(z) = \frac{\Delta_{\rm el}}{\hbar} \left(\frac{2\pi k_B T}{\Delta_{\rm el}}\right)^{2K-1} \frac{\Gamma(K + z\hbar/2\pi k_B T)}{\Gamma(1 - K + z\hbar/2\pi k_B T)}, \tag{3.143}$$

where Γ is the complex gamma function and $\Delta_{\rm el}$ the tunnel energy dressed by electron-hole excitations at $T = 0$. Here we will use an approximate expression that is valid for small K,

$$\Delta_{\rm b}^2 \Gamma''_{\rm el}(\omega) = \frac{1}{\hbar} \Delta_{\rm el}^2 \left(\frac{2\pi k_B T}{\Delta_{\rm el}}\right)^{2K} \frac{2\pi K \omega[1 + n(\omega)]}{(2\pi K k_B T/\hbar)^2 + \omega^2}, \tag{3.144}$$

as given by Dattagupta et al. (1989).

An approximate expression for the phonon counterpart has been derived in Sect. 3.4. In terms of the expansion (3.94), the spectrum $\Gamma''_{\rm ph}$ reads as

$$\Gamma''_{\rm ph}(\omega) = e^{-W} \left[\pi \delta(\omega) + e^{\beta \hbar \omega/2} \frac{\beta \hbar}{\pi} \sum_{n=1}^{\infty} \frac{\phi^n}{n!} A''_n(\beta \hbar \omega)\right]. \tag{3.145}$$

This expression for the phonon spectrum is valid in the whole temperature range.

For T well below the Debye temperature, the Debye–Waller factor is given by (3.42), and the coefficient ϕ by (3.105). We repeat $W = W_0 + \frac{1}{3}\phi = \alpha \omega_{\rm D}^2 + \frac{1}{3}T^2/T_0^2$, where the zero-temperature value W_0 is defined in (3.44); the temperature $T_0 = (2\pi^2 \tilde{\alpha})^{-1/2}$ accounts for the onset of multiphonon processes.

In (3.109) we have given an approximate expression which is valid at thermal frequencies, $\hbar|\omega| \lesssim k_B T$, and temperatures well below Θ,

$$\Gamma''_{\rm ph}(\omega) = \pi e^{-W} \left[\delta(\omega) + 2\alpha \omega[1 + n(\omega)]F(\phi)\right]. \tag{3.146}$$

Here, $n(\omega) = [e^{\beta \hbar \omega} - 1]^{-1}$; the multiphonon correction factor $F(\phi)$ is given by (3.98) or (3.108).

In any order, the self-energy can be written as a convolution of the spectra $\Gamma''_{\rm el}(\omega)$ and $\Gamma''_{\rm ph}(\omega)$. Here, we note the result in NIBA,

$$\Sigma''_1(\omega) = \frac{1}{\pi}(\Delta_{\rm b}/\hbar)^2 \int d\omega' \, \Gamma''_{\rm ph}(\omega - \omega') \Gamma''_{\rm el}(\omega'). \tag{3.147}$$

All spectra fulfil the detailed balance condition, e.g. $\Gamma''_{\rm el}(-\omega) = e^{-\hbar \beta \omega} \Gamma''_{\rm el}(\omega)$.

3.5.2 Coherent Motion

At low temperatures, one finds a pair of complex poles and a branch point on the imaginary axis (cf. Leggett et al. (1987)). For $K \ll 1$, the latter is of little relevance, and $P_1(z)$ shows a two-pole structure,

$$P_1(z) = -\frac{z}{z^2 + iz\Gamma - (\tilde{\Delta}_0/\hbar)^2}. \tag{3.148}$$

The effective tunnel energy involves dressing by conduction electrons and phonons. The effect of the former is absorbed in $\Delta_{\rm el}$. At zero temperature, the dressed value reads (Chakravarty and Leggett (1984)) as

$$\Delta_{\rm el} = [\cos(\pi K)\Gamma(1-2K)]^{1/(2-2K)}\Delta_{\rm b}(\Delta_{\rm b}/D)^K, \tag{3.149}$$

where D is the band width of conduction electrons. For small K, this reduces to the simple result $\Delta_{\rm el} = \Delta_{\rm b}(\Delta_{\rm b}/D)^K$, given by Kondo (1976).

Phonon coupling gives rise to the Debye–Waller factor e^{-W}. For later convenience we split off the zero-temperature value e^{-W_0} and define

$$\Delta_0^{2(1-K)} = e^{-W_0}\Delta_{\rm el}^{2(1-K)}. \tag{3.150}$$

For $k_{\rm B}T > \Delta_{\rm el}$ and small K, the temperature-dependent part is accounted for by

$$\tilde{\Delta}_0^2 = e^{-\frac{1}{3}\phi}(k_{\rm B}T/\Delta_{\rm el})^K \Delta_0^2. \tag{3.151}$$

Note that, with rising temperature, screening by conduction electrons *increases* the tunnel energy, whereas phonon dressing leads to a *decrease*. In this section, we will use mainly the zero-temperature value Δ_0.

For weak coupling, or low temperature, (3.148) exhibits a pair of complex poles. Taking the inverse Laplace transform we find with $\hbar\Gamma \ll \tilde{\Delta}_0$

$$P(t) = \cos(\tilde{\Delta}_0 t/\hbar)\exp(-\tfrac{1}{2}\Gamma t) \qquad (\text{low } T), \tag{3.152}$$

where the rate comprises the one-phonon term and the corresponding contribution arising from the scattering of conduction electrons,

$$\Gamma = \pi K(\tilde{\Delta}_0/\hbar)\coth(\tfrac{1}{2}\beta\tilde{\Delta}_0) + \pi\alpha(\tilde{\Delta}_0/\hbar)^3 \coth(\tfrac{1}{2}\beta\tilde{\Delta}_0). \tag{3.153}$$

Since the coupled phonon density of states is small at low frequency, the one-phonon term in the damping rate is negligible in general, i.e. at low temperatures electron-hole excitations provide a much more efficient dissipation mechanism.

Yet the argument invoking the phonon density of states does not hold for the frequency renormalization. Even for $T \to 0$, phonon coupling leads to a significant reduction of the tunnel energy, by means of the zero-temperature Debye–Waller factor e^{-W_0}.

3.5.3 The Incoherent Rate

With rising temperature, both conduction electrons and phonons may drive a crossover to incoherent tunneling. When replacing the coth function in (3.153) by its inverse argument, one finds that the electronic part of the rate defines a crossover temperature $T_{\rm el}$ through $\Delta_0 = \pi K k_{\rm B} T_{\rm el}$. Calculation of the overdamped rate, however, requires a more careful analysis; cf. Kondo (1984) and Chakravarty and Leggett (1984). The phonon part of (3.153) is always smaller than $\tilde{\Delta}_0$; hence it does not lead to incoherent motion, as shown by Leggett et al. (1987). Yet at temperature T_0, multiphonon processes become important and cause a substantial change of the two-state dynamics.

Phonon-Driven Crossover. For $T_0 \ll T_{\text{el}}$, damping by conduction electrons is relevant at temperatures $T \ll T_0$ only. Overdamping occurs at T_0 and is due to phonon coupling, whereas the temperature T_{el} is of no physical significance. This case may be treated in terms of the theory of Sect. 3.4, by adding $\pi K(\tilde{\Delta}_0/\hbar)\coth(\tfrac{1}{2}\beta\tilde{\Delta}_0)$ to the low-temperature rate, and replacing the reduced tunnel frequency by (3.150). The results do not differ from those reported in the last section.

Electron-Driven Crossover. The opposite case $T_{\text{el}} \ll T_0$ is more interesting, both from a theoretical point of view and since it describes real systems. At temperature T_{el}, conduction electrons drive a crossover to incoherent motion, as obtained, e.g. by Leggett et al. (1987) in terms of the electronic damping function (3.142). Yet above T_0, phonon coupling results in a significant change of the temperature dependence of the damping rate. As discussed above (3.140), NIBA is valid in the whole range of temperatures.

Here we are interested in this second crossover at T_0. Accordingly, we consider the two-state dynamics well above T_{el}, where $P(z)$ involves a single relaxation pole on the imaginary axis,

$$P(z) = \frac{-1}{z + i\Gamma} \qquad (T \gg T_{\text{el}}), \tag{3.154}$$

with the relaxation rate $\Gamma = \Sigma_1''(0)$.

When inserting the spectra in the convolution integral of (3.147), the first term of (3.146) gives the rate $e^{-\tfrac{1}{3}\phi}\tilde{\Gamma}_{\text{el}}$, where

$$\tilde{\Gamma}_{\text{el}} = \frac{\Delta_0}{\hbar}\frac{1}{K}\left(\frac{2\pi k_B T}{\Delta_0}\right)^{2K-1}. \tag{3.155}$$

Note that the tunnel energy Δ_0 accounts for the zero-temperature phonon Debye–Waller factor only.

The remaining second part of the phonon spectrum Γ_{ph}'' is significant for sufficiently high temperatures only, $T \geq T_0$. Evaluating the convolution integral and inserting the definition (3.24), we obtain

$$\Gamma = e^{-\tfrac{1}{3}\phi}\tilde{\Gamma}_{\text{el}}[1 + 2KI_K\phi F(\phi)]. \tag{3.156}$$

Here, by substituting $x \equiv \beta\hbar\omega'$, the convolution has been cast in the form

$$I_K = \frac{1}{\pi}\int dx \frac{2\pi K}{(2\pi K)^2 + x^2}\frac{(x/2)^2}{\sinh(x/2)^2}. \tag{3.157}$$

Thereby, we have reduced calculation of the rate Γ to the integral I_K that can easily be done numerically.

We briefly discuss an approximate expression for this integral. In the limiting cases with respect to $2\pi K$, we easily find $I_K = 1$ for $2\pi K \ll 1$ and $I_K = (3K)^{-1}$ for $2\pi K \gg 1$. Most real systems satisfy $2\pi K \leq 1$. In order to account for the intermediate case, we simplify (3.157) by replacing the last factor with an appropriate step function, $[x/2\sinh(x/2)]^2 \to \Theta(\tfrac{1}{3}\pi^2 - |x|)$. Performing the remaining integral,

$$I_K = \frac{2}{\pi} \arctan\left(\frac{\pi}{6K}\right), \tag{3.158}$$

we find that it yields the exact results for the above given limiting cases.

The incoherent rate (3.156) constitutes a main result of this section. Defects in real metals are described by a finite Kondo parameter K and a temperature T_0. For $T \ll T_0$, the second term in brackets is negligible, and we find $\tilde{\Gamma}_{el} e^{-\frac{1}{3}\phi}$, i.e. the rate for damping by conduction electrons reduced by a Debye–Waller factor. As temperature approaches T_0, the factors involving $\phi = T^2/T_0^2$ dominate the rate, resulting in an exponential increase, $\Gamma \propto \exp(\frac{2}{3}T^2/T_0^2)$.

Beyond Debye Temperature. The above result is valid for T well below the Debye temperature Θ. At temperatures above Θ, both coefficients φ_0 and φ_2 may be expressed in terms of the energy

$$V = \frac{1}{12\pi^2} \frac{k_B \Theta^3}{T_0^2}; \tag{3.159}$$

cf. (3.112) and (3.114), resulting in the rate

$$\Gamma_{act} = \frac{\Delta_{el}^2}{\hbar} \sqrt{\frac{\pi}{16 V k_B T}} e^{-V/k_B T} I_K \left(\frac{2\pi k_B T}{\Delta_{el}}\right)^{2K}, \tag{3.160}$$

whose temperature dependence is characterized by the Arrhenius behavior with activation energy V and an additional factor $T^{-1/2}$. Since we have dropped the 'direct process' (Holstein (1959)), the limit $K \to 0$ gives with $\Delta_{el} \to \Delta_b$ and $I_K \to 1$ the phonon-driven rate (3.116).

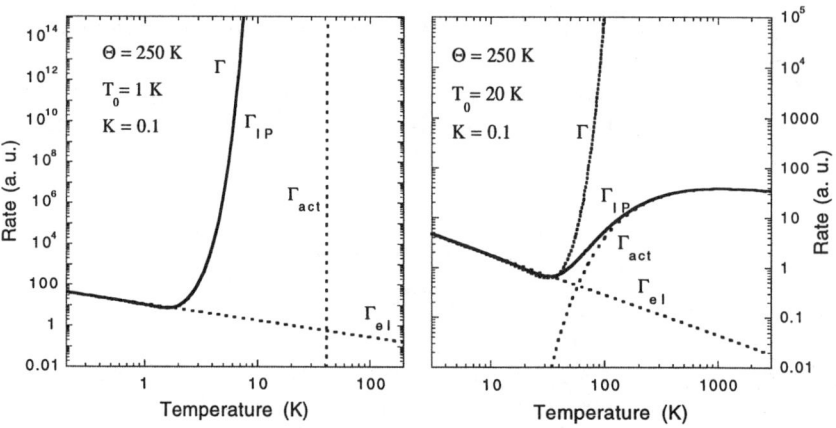

Fig. 3.4. Comparison of various expressions for the damping rate in metals. Γ_{el} as given in (3.155), arises from dissipation by conduction electrons only. The rate Γ, cf. (3.161), takes phonon coupling into account; it is valid for $T \ll \Theta$ only. Γ_{act} denotes the thermally activated high-temperature result (3.160), whereas Γ_{IP} provides a smooth interpolation between these limiting cases.

3.5 Phonon-Assisted Tunneling in Metals

An Interpolation Formula. Equation (3.156) holds true for T much smaller than the Debye temperature, say $T < \frac{1}{10}\Theta$, only. Hence, for typical values of Θ of a few hundred Kelvin, this expression cannot be used beyond about 30 K. On the other hand, the activated high-temperature rate Γ_{act} is valid for $T \geq \Theta$ only. For various real systems, however, the most interesting temperature range covers the intermediate regime.

For this reason we consider more closely the case where T is not small compared to the Debye temperature, $T \approx \Theta$. The correction factor $F(\phi)$ has been derived assuming $\varphi_2 = \phi^2/\alpha$ which, however, holds true for $T \ll \Theta$ only. As temperature approaches Θ, we have to replace the factor $(1 + 4\phi/\pi)^{-1/2}$ by $[1 + (2/\pi)\varphi_2(\hbar/\pi k_B T)^2]^{-1/2}$. Then the rate is given by the interpolation formula

$$\Gamma_{\text{IP}} = \frac{\Delta_{\text{el}}^2}{\hbar^2}\left(\frac{2\pi k_B T}{\Delta_{\text{el}}}\right)^{2K} e^{-W}\left[\frac{\hbar}{2\pi K k_B T} + I_K(e^{\varphi_0} - 1)\sqrt{\frac{\pi}{2\varphi_2}}\right], \quad (3.161)$$

where the quantities W, φ_0, and φ_2 have to be calculated *numerically* from (3.41) and (3.102). One easily verifies that this expression reduces to (3.156) for $T \ll \Theta$, whereas in the opposite case, $T > \Theta$, we recover the activated rate Γ_{act}.

In Fig. 3.4 we compare the above approximations for two values of the crossover temperature, $T_0 = 1$ K and $T_0 = 20$ K. According to Table 3.1, the chosen values for T_0, Θ, and K are relevant for the real systems to be discussed below. In both cases, screening by conduction electrons with $K = 0.1$ governs the rate in the limit of low temperatures.

The first set of parameters, $T_0 = 1$ K, $\Theta = 250$ K, and $K = 0.1$ describes strong phonon coupling, as realized, e.g. by configurational defects in Bi wires. Since, in the crossover range, the condition $T \ll \Theta$ is well satisfied, we may use the expression (3.156), which differs little from the numerical evaluation of the interpolation formula (3.161); in the plot the curves labelled Γ and Γ_{IP} cannot be distinguished. As to the activated rate Γ_{act}, it is too small by many orders of magnitude because of the large activation energy $V/k_B = 132000$ K; it becomes correct only at temperatures well above Θ.

The situation is quite different for the second value $T_0 = 20$ K. Since the crossover to phonon-dominated damping occurs at about 40 K, the condition $T \ll \Theta$ does not hold true in the relevant temperature range. On the other hand, the activated rate is valid only as temperature approaches Θ. As a consequence, neither of the analytical expressions (3.156) and (3.160) properly describes the crossover regime. These formal considerations are confirmed in Fig. 3.4. At the crossover, the activated rate Γ_{act} is too small by almost one order of magnitude, whereas Γ exhibits too strong an exponential increase, as compared to Γ_{IP}. Above 100 K, the rate shows activated behavior, with $V/k_B = 330$ K, before levelling off towards the $T^{-1/2}$ law, at temperatures well beyond V/k_B.

Two conclusions may be drawn from Fig. 3.4. First, phonon damping exceeds the electron-driven rate above T_0; for $K = 0.1$, the crossover occurs at about $2T_0$. Second, the temperature dependence of the rate above T_0 is determined by the ratio T_0/Θ. For sufficiently small T_0, say $T_0 \ll \frac{1}{10}\Theta$, we find a strong exponential increase according to $\exp(\frac{2}{3}T^2/T_0^2)$, whereas in the opposite case the rate shows Arrhenius behavior.

3.5.4 Quantum Diffusion of Trapped Hydrogen in Niobium

As a first example we discuss the quantum motion of interstitial hydrogen trapped by an oxygen or nitrogen impurity atom in niobium. The hydrogen motion between two tetrahedral sites is well described as a two-state TS[1]. Inelastic neutron scattering at low temperature ($T < 10$ K) revealed coherent motion; for oxygen traps, the tunnel energy takes a value $\Delta_0/k_B = 2.4$ K. According to Wipf et al. (1987), the damping by conduction electrons leads to a Kondo parameter $K = 0.055$.

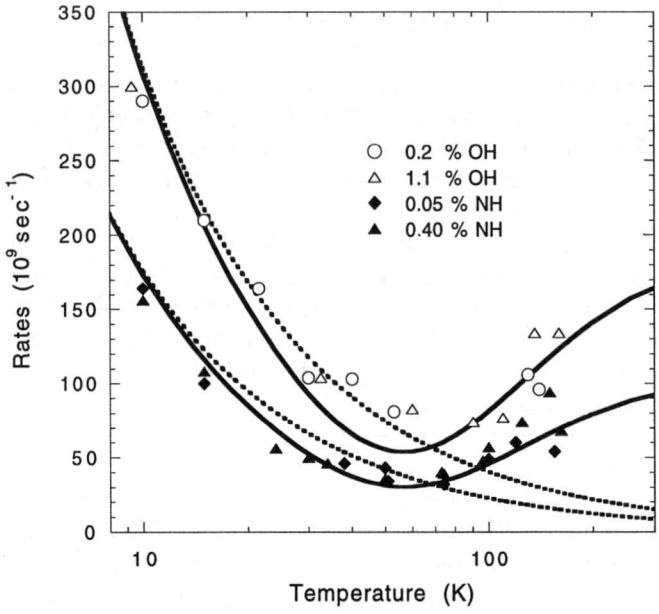

Fig. 3.5. Diffusion rate for hydrogen in Nb(OH)$_x$. Circles are data points for $x = 0.002$, and triangles for $x = 0.011$. For details see Steinbinder et al. (1988). The dashed line has been calculated for damping by conduction electrons, whereas the solid line accounts for both electrons and phonons, according to Γ_{IP} as given in (3.161). The data are taken from Steinbinder et al. (1991).

---

[1] A detailed discussion of such systems is given in Chap. 7.

Here we are interested in the incoherent motion well above 10 K. In Fig. 3.5 we plot the jump rates observed for hydrogen concentrations between $x = 0.0005$ and $x = 0.011$ by Steinbinder et al. (1988) and Steinbinder et al. (1991) through quasi-elastic neutron scattering. At such high densities, the elastic interaction of adjacent impurities gives rise to an asymmetry energy Δ which is different on each impurity site. According to (3.175) below, a finite asymmetry energy reduces the jump rate significantly. The data of Fig. 3.5 have been corrected for this effect by assuming a Lorentzian distribution for Δ (Steinbinder et al. (1991), Wipf (1997)), and the corresponding rates for zero asymmetry have been extracted. As expected, the resulting rates do not depend on concentration x; thus the data for a given trap system at different x fall on a single curve.

Accordingly, we have plotted the theoretical jump rates for zero asymmetry. The dashed line is given by (3.155), which is essentially identical to previous results of Kondo (1984), Grabert and Weiss (1985), and Leggett et al. (1987). We have used $\Delta_0/k_B = 2.4$ K for oxygen traps and $\Delta_0/k_B = 1.8$ K for hydrogen trapped by nitrogen, which differ little from values given by Wipf (1997), 2.4 and 1.9 K, respectively.

Temperature Dependence. The ohmic damping model provides a good description up to about 60 K, whereas it fails in view of the increase of the rate at higher temperatures, which requires us to take phonon coupling into account. Yet it turns out that (3.156) is not valid in the relevant temperature range; with $\Theta = 250$ K, the condition $T \ll \Theta$ is not satisfied at temperatures of about 100 K. The solid line is calculated from (3.161) with $T_0 = 23$ K; the Debye–Waller factor and the coefficients φ_0 and φ_2 have been evaluated numerically using (3.41) and (3.102) with a sharp cutoff at ω_D. Below 20 K, the solid line differs little from the dashed one, i.e. there are no temperature-dependent phonon effects. Up to about 70 K, the Debye–Waller factor results in a slight reduction of the rate, whereas above 70 K, multiphonon processes lead to an enhancement of the rate. Note, however, that the increase is much weaker than the exponential law expected for $T \ll \Theta$; beyond 100 K it is well described as activated behavior. By chance, the activation energy $V/k_B = 246$ K is almost identical to the Debye temperature.

Given that there is a single free parameter T_0 for the two solid lines describing hydrogen tunneling at O and N traps, the fits agree well with the data. Their quality could be improved by using a more realistic high-frequency cutoff for the Debye model. This would be of little significance, however, because the additional parameters are not easily related to observable quantities. Furthermore, T_0 and K need not be identical at different trap atoms.

Influence of the Trap Atom. Although the data scatter significantly above 100 K, they clearly indicate that the dependence of the rate on the properties of the trap does not change in the temperature range considered. The tunnel energy is most sensitive to the mass of the surrounding atoms; it changes by 30% when substituting the nitrogen trap by an oxygen impurity. A much weaker dependence is expected for the librational frequency in each minimum of the the double-well potential and for the barrier height, i.e. the rates for classical barrier crossing should be very much the same for both trap systems.

From the ratio of the rates persisting to almost 200 K, we conclude that even at such high temperatures, phonon-assisted tunneling governs the motion of trapped hydrogen. For a more detailed discussion of the high-temperature diffusion law, the reader is referred to Chap. 7.

3.5.5 Rate Equations for Large Asymmetry Energy

Up to now we have considered a symmetric double-well potential. Now we turn to the case of large asymmetry energy, $\Delta \gg \Delta_0$. Formally, the theory of subsection 3.5.1 is easily generalized to the case where the Hamiltonian contains an asymmetry term,

$$H = \frac{1}{2}\Delta_b \left(\sigma_+ B_- + \sigma_- B_+\right) + \frac{1}{2}\Delta\sigma_z + H_B. \tag{3.162}$$

As a consequence, the first-order self-energy (3.138) acquires a factor $\cos(\Delta t/\hbar)$, i.e. $\Sigma_1(t) = (\Delta_b/\hbar)^2 e^{\psi(t)} \cos(\Delta t/\hbar)$.

A more serious problem, however, arises from the finite long-time limit of $P(t)$. In the symmetric case, the poles of $P(z)$ occur in the lower half of the complex plane, i.e. $P(t)$ is exponentially damped. Yet for finite asymmetry, $P(z)$ displays an undamped pole at $z = 0$, resulting in a finite value for $P(t \to \infty)$. For Ohmic damping, this problem has been tackled by Leggett et al. (1987) and Weiss and Wollensak (1989), by splitting off the $1/z$ pole. It turns out, however, that the resulting stationary state is correct for large asymmetry only.

Here we propose a different approach which consists in deriving a set of coupled rate equations for the quantum states $|L\rangle$ and $|R\rangle$. It is comparatively simple and yields correct results for the case where the asymmetry energy is much larger than the tunnel matrix element, $\Delta_0 \ll \Delta$. In terms of the resolvent $P(z)$, our rate equations do not account for the inelastic resonances at $z = \pm\Delta/\hbar$. In the limit of large asymmetry, the corresponding residues are negligible anyhow.

We start from the projections $P_L = |L\rangle\langle L|$ and $P_R = |R\rangle\langle R|$ and the statistical operator $\rho = \rho_S \rho_B$ with $\rho = cP_L + (1-c)P_R$; ρ_B describes phonons and conduction electrons in thermal equilibrium. Time evolution of the occupation numbers is given by

$$p_L(t) = \langle P_L(t)\rangle = \mathrm{tr}\left(\rho e^{i\mathcal{L}t} P_L\right) \tag{3.163}$$

and $p_R = 1 - p_L$.

In order to set up a perturbation theory, we write $H = H_0 + H_1$, with $H_0 = \frac{1}{2}\Delta\sigma_z + H_B$ and $H_1 = \frac{1}{2}\Delta_b(\sigma_+ B_- + \sigma_- B_+)$. After separating the

3.5 Phonon-Assisted Tunneling in Metals

Liouville operator accordingly, $\mathcal{L} = \mathcal{L}_0 + \mathcal{L}_1$, with $\hbar\mathcal{L}_0 A = [H_0, A]$ etc., the derivative of the time-evolution operator in (3.163) reads as

$$\frac{\partial}{\partial t}e^{i\mathcal{L}t} = ie^{i\mathcal{L}_0 t}\mathcal{L} + i^2\int_0^t d\tau e^{i\mathcal{L}(t-\tau)}\mathcal{L}_1 e^{i\mathcal{L}_0\tau}\mathcal{L}. \tag{3.164}$$

When inserting this in (3.163) and using $P_R P_L = 0$, $\langle\sigma_\pm\rangle = 0$, etc., the first term on the r.h.s. vanishes, and we find the exact equation

$$\dot{p}_L(t) = \frac{1}{4}\int_0^t d\tau\left\langle e^{i\mathcal{L}(t-\tau)}P_R \Xi_+(\tau) - e^{i\mathcal{L}(t-\tau)}P_L \Xi_-(\tau)\right\rangle, \tag{3.165}$$

where time evolution of the bath correlations

$$\Xi_\pm(t) = \tfrac{1}{4}(\Delta_{\mathrm{b}}/\hbar)^2\left[e^{\pm i\Delta\tau/\hbar}B_\pm(0)B_\mp(\tau) + e^{\mp i\Delta\tau/\hbar}B_\pm(\tau)B_\mp(0)\right] \tag{3.166}$$

is given by $\mathcal{L}_0$. A similar equation for $\dot{p}_R$ is derived by exchanging L and R, and the labels of $\Xi_\pm$.

In order to solve this set of equations, we resort to two standard approximations: *First*, in order to decouple spin and bath parts of (3.165), we replace the functions $\Xi_\pm(t)$ by their thermal averages $\langle\Xi_\pm(t)\rangle$, and thus obtain coupled integro-differential equations for $\langle p_L\rangle$ and $\langle p_R\rangle$,

$$\dot{p}_L(t) = \int_0^t d\tau\left[p_R(t-\tau)\langle\Xi_+(\tau)\rangle - p_L(t-\tau)\langle\Xi_-(\tau)\rangle\right]. \tag{3.167}$$

Second, since the bath correlations decay much faster than the pseudospin state, we resort to a Markov approximation,

$$\langle\Xi_\pm(t)\rangle \to \Gamma_{\uparrow\downarrow}\delta(t),$$

which results in the coupled rate equations

$$\dot{p}_L(t) = -\Gamma_\uparrow p_L(t) + \Gamma_\downarrow p_R(t), \tag{3.168}$$

$$\dot{p}_R(t) = -\Gamma_\downarrow p_R(t) + \Gamma_\uparrow p_L(t). \tag{3.169}$$

Equations (3.168) and (3.169) constitute our formal solution for the two-state dynamics. The rates $\Gamma_{\uparrow\downarrow}$ remain to be calculated. When noting

$$\langle\Xi_\pm(t)\rangle = \frac{1}{4}(\Delta_{\mathrm{b}}/\hbar)^2[e^{\psi(t)}e^{\mp i\Delta t/\hbar} + e^{\psi(-t)}e^{\pm i\Delta t/\hbar}],$$

we may write the rates in terms of the Fourier transforms at zero frequency,

$$\Gamma_{\uparrow\downarrow} = \frac{1}{2}\int_{-\infty}^\infty dt\langle\Xi_\pm(t)\rangle = \frac{1}{4}(\Delta_{\mathrm{b}}/\hbar)^2\int_{-\infty}^\infty dt e^{\psi(t)}e^{\mp i\Delta t/\hbar}. \tag{3.170}$$

The latter result, in turn, is expressed in terms of the spectra (3.146) and (3.144),

$$\Gamma_{\uparrow\downarrow} = \frac{1}{4\pi}(\Delta_{\mathrm{b}}/\hbar)^2\int_{-\infty}^\infty d\omega\, \Gamma_{\mathrm{ph}}''(-\omega)\Gamma_{\mathrm{el}}''(\omega \pm \Delta/\hbar), \tag{3.171}$$

where $\Gamma_\uparrow$ is given by the plus sign.

One easily confirms the detailed balance condition $\Gamma_\downarrow = e^{\beta\Delta}\Gamma_\uparrow$ for the rates. The stationary state is obtained by setting $\dot{p}_L = 0$; Eqs. (3.168) and (3.169) yield $p_L = e^{-\beta\Delta}p_R$ in thermal equilibrium.

Now we are ready to consider the quantity $P(t) = \langle\sigma_z(t)\rangle$. Noting $\sigma_z = P_L - P_R$ and inserting the above results, we find

$$P(t) = P(\infty) + [P(0) - P(\infty)]e^{-\Gamma t}. \tag{3.172}$$

Here, the initial state $P(0) = 2c$ is arbitrary, see (3.163), whereas $P(\infty) = -\tanh(\frac{1}{2}\beta\Delta)$ gives the stationary solution. Moreover we have defined the relaxation rate

$$\Gamma = \Gamma_\uparrow + \Gamma_\downarrow. \tag{3.173}$$

In proves to be convenient to evaluate the rate Γ, and to deduce the upward and downward scattering rates $\Gamma_\uparrow$ and $\Gamma_\downarrow$ afterwards.

When inserting the spectra in the convolution integral (3.171), the first term of (3.146) gives the rate derived by Weiss and Wollensak (1989):

$$\Gamma_{el} = \tfrac{1}{2}(\Delta_b/\hbar)^2[\Gamma''_{el}(\Delta/\hbar) + \Gamma''_{el}(-\Delta/\hbar)], \tag{3.174}$$

times e^{-W}. Absorbing the zero-temperature Debye–Waller factor in the reduced tunnel energy Δ_0 and inserting the spectrum (3.144), we find

$$\tilde{\Gamma}_{el} = \frac{1}{\hbar}\Delta_0^2\left(\frac{2\pi k_B T}{\Delta_0}\right)^{2K}\frac{\pi K\Delta}{(2\pi K k_B T)^2 + \Delta^2}\coth(\Delta/2k_B T). \tag{3.175}$$

Comparison with (3.155) shows that, at low temperature, a finite asymmetry reduces the relaxation rate.

As in (3.156), the second term of the phonon spectrum contributes at high temperatures $T \geq T_0$ only. We will restrict the subsequent evaluation of the convolution integral to the case $\Delta \ll k_B T_0$, which is most relevant for the real systems to be discussed below. Accordingly, we neglect Δ in the second term and, after inserting (3.6) and the integral (3.157), we obtain

$$\Gamma = e^{-\frac{1}{3}\phi}\left[\tilde{\Gamma}_{el} + 2\frac{\Delta_0}{\hbar}\left(\frac{2\pi k_B T}{\Delta_0}\right)^{2K-1}I_K\phi F(\phi)\right]. \tag{3.176}$$

This result has been derived for $\Delta \ll k_B T_0$. It may be simplified significantly for not too small a Kondo parameter, satisfying $\Delta \ll 2\pi K k_B T_0$. Considering (3.144) and (3.157), one finds

$$\Gamma = \tilde{\Gamma}_{el}e^{-\frac{1}{3}\phi}\left[1 + 2K\,I_K\phi F(\phi)\right], \tag{3.177}$$

for $\Delta \ll 2\pi K k_B T_0$. (Since $K < \frac{1}{2}$, this inequality implies $\Delta \ll k_B T_0$.)

Comparing the relaxation rate Γ with that obained for the symmetric case, (3.156) reveals that both are equal for $\Delta \ll k_B T$. This is not a surprising result, since the relaxation behavior is governed by thermal frequencies.

As T approaches the Debye temperature, the factor $F(\phi)$ has to be corrected according to the discussion on p. 97. We refrain from repeating the above argument, since the system to be studied here is well described by (3.177).

3.5.6 Resistance Fluctuations of Mesoscopic Wires

The conductance of submicron wires may be significantly affected by the presence of bistable atomic configurations. Each local arrangement of the defect constitutes a different obstacle for conducting electrons, and thus results in a different value for the resistance. Such fluctuations have been observed by Zimmermann et al. (1991), Golding et al. (1992), Chun and Birge (1993), and others. Measurements of the temperature dependence of the jump rates between the two configurations permitted a detailed study of the defect motion. In this sense, the resistance fluctuations of mesoscopic wires provide a unique probe for the dynamics of a *single* two-state defect in a metal.

So far, little is known about the microscopic nature of these defects; they might be located at the grain boundaries of the polycrystalline material, or arise from lattice imperfections or impurity atoms. In terms of a phenomenological two-state model, such a defect is characterized by the tunnel frequency $\Delta_0/\hbar$, the asymmetry energy Δ, the coupling constant for conduction electrons, K, and the elastic deformation potential, γ. The former three parameters, Δ_0, Δ, and K, have been derived previously by Chun and Birge

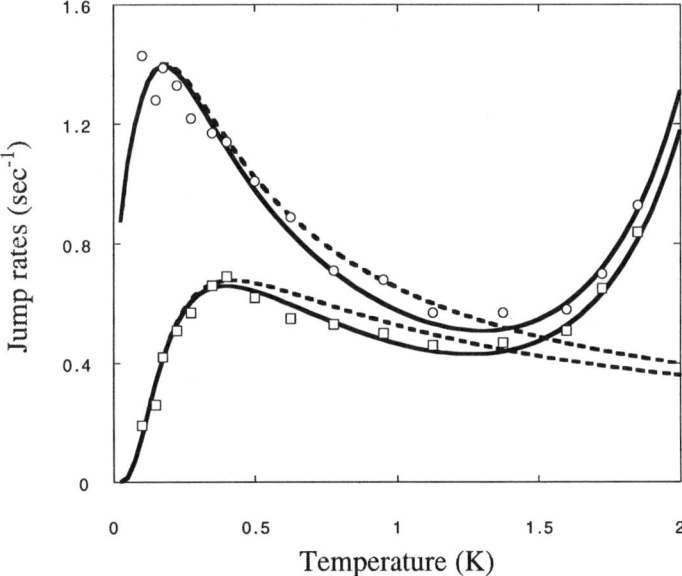

Fig. 3.6. Upward and downward jump rates for a configurational defect in a mesoscopic Bi wire. The dashed lines arise from damping by conduction electrons; the solid lines account for both electron and phonon coupling. The data points are from Chun and Birge (1993).

(1993) from the temperature dependence of the resistance jumps below 1 K. More recently, Birge et al. (1996) have measured γ by applying static strain on the sample.

In Fig. 3.6 we show data for the upward and downward jump rates of a single two-state defect. The solid lines are calculated from (3.177); the upward and downward rates $\Gamma_\uparrow$ and $\Gamma_\downarrow$ are obtained from $\Gamma = \Gamma_\uparrow + \Gamma_\downarrow$ and $\Gamma_\uparrow = e^{-\beta\Delta}\Gamma_\downarrow$. The dashed lines give the corresponding rates driven by conduction electrons only.

Comparison of the solid and dashed lines shows that phonon effects are immaterial at temperatures below 250 mK. Up to about 1 K, phonon coupling reduces the rate, due to the temperature-dependent Debye–Waller factor $e^{-\frac{1}{3}\phi}$ in (3.177). Well above 1 K, multiphonon scattering results in an exponential increase. The rate shows a characteristic minimum at a temperature slightly above T_0. Since T_0 is by more than two orders of magnitude smaller than the Debye temperature, we may use the analytic result (3.177); cf. the discussion of p. 97 and Fig. 3.4.

The rate Γ depends on the quantities Δ, Δ_0, K, and T_0. By adjusting $\tilde{\Gamma}_{el}$ to the data below 1 K, Chun and Birge (1993) derived the values for the asymmetry $\Delta/k_B = 0.213$ K, the effective tunnel frequency $\Delta_0/\hbar = 2.5 \times 10^5$ s^{-1}, and the Kondo parameter $K = 0.16$. The corresponding rates resulting from $\tilde{\Gamma}_{el}$ are given as dashed lines.

With a crossover temperature T_0 of about 1 K, and noting $2\pi K \approx 1$, the parameters K and Δ satisfy the relation $\Delta \ll 2\pi K k_B T_0$. Hence we may use the simple expression (3.177). The second term of the rate involves the quantities I_K and T_0. The solid lines of Fig. 3.6 have been calculated with $I_K = 0.71$ and $T_0 = 0.85$ K. As for I_K, the integral (3.157) gives $I_K = 0.71$, whereas from the approximate expression (3.158) we would obtain $I_K = 0.82$.

The relation between the crossover temperature and the elastic deformation energy,

$$k_B T_0 = \gamma^{-1}(\hbar^3 \varrho v^5/3)^{1/2}, \qquad (3.178)$$

is easily established from (3.20) and (3.24). By inserting known values for the mass density $\varrho = 9.8$ g/cm^3 and the average sound velocity $v = 1230$ m/s, we derive $T_0 = 0.85$ K the elastic deformation potential, $\gamma = 1.6$ eV. Through the application of isotropic strain, Birge et al. (1996) obtained the value $\gamma = 2.3 \pm 0.4$ eV. This reasonably good agreement permits us to conclude that the increasing rate above 1 K is due to phonon-assisted hopping of the two-state system accounted for in (3.177), rather than to activated relaxation as suggested by Cukier et al. (1995).

Concerning the phonon effects, there is no free parameter in (3.177). The adjusted value, $\gamma = 1.6$ eV, agrees well with that observed by Birge et al. (1996), $\gamma = 2.3 \pm 0.4$ eV. The discrepancy may well be due to the mesoscopic nature of the Bi wire, since its vibrational spectrum for wavelengths longer than its diameter is different from that of a three-dimensional body. For transverse modes whose half wavelength is equal to the sample height, $\frac{1}{2}\lambda =$

0.02 μm, or to the sample width, $\frac{1}{2}\lambda = 0.05$ μm, we find frequencies of 27 GHz and 11 GHz, corresponding to 1.2 K and 0.5 K, respectively. Accordingly, at temperatures below 1 K, the number of thermal phonons would show significant deviations from the Debye law (3.19).

3.5.7 Discussion

Most of the comments on tunneling in insulators in Sect. 3.4.6 apply equally well to metallic hosts. There are, though, a few features that seem worthy of a more detailed discussion.

Validity of NIBA. For weak coupling, $K \ll 1$, NIBA provides a controlled approximation for the Ohmic damping model. In the symmetric case, the corrections arising from blip–blip interactions are negligible. First note that the only dimensionless coupling parameter is given by the Kondo parameter K. When pushing the perturbation expansion beyond the first-order rate (3.153), one finds a term proportional to K^2, which is negligible for weak coupling. For this reason, there are no significant corrections to NIBA for the Ohmic bath with coupling $K \ll 1$. Regarding phonon coupling, NIBA is valid according to the argument above Eq. (3.154), since the motion is already overdamped when multiphonon processes set in at temperature T_0.

Imaginary-Time Approach. In a different approach based on an imaginary-time functional integral technique and NIBA, Grabert (1992 and 1997) has evaluated the partition function, which, like $P(t) = p_L(t) + p_R(t)$ above, consists of two parts, $Z = Z_+ + Z_-$. Due to dissipation, the corresponding free energy $F_\pm$ acquires an imaginary part, which reads in our notation

$$\Gamma_{\uparrow\downarrow} = \frac{1}{4}(\Delta_b/\hbar)^2 e^{\mp\beta\Delta/2} \int_{-\infty}^{\infty} dt\, e^{-\Lambda(t)\mp i\Delta t/\hbar}, \qquad (3.179)$$

where the phase is related to (3.139) through $-\Lambda(t) = \psi(t - \frac{1}{2}i\hbar\beta)$.

Like $\bar{\varphi}(t)$ in Sect. 3.4, $\Lambda(t)$ is real and a symmetric function of time, whereas $\psi(t)$ satisfies $\psi(-t) = \psi(t - i\hbar\beta)$. Now we see that (3.179) can be obtained from (3.170) by shifting the time argument according t by an imaginary constant to $(t - \frac{1}{2}i\hbar\beta)$, i.e. by distorting the Fourier integral from the real t-axis into the complex plane.

By inserting the composite spectral density in the phase $\Lambda(t)$ and performing the frequency integral, Grabert (1992) obtained the result

$$\Lambda(t) = 2K \ln\left[\frac{\beta\hbar\omega_c}{\pi}\cosh\left(\frac{\pi t}{\hbar\beta}\right)\right] - \frac{T^2}{T_0^2}\left[\frac{1}{\cosh(\pi t/\hbar\beta)^2} - \frac{1}{3}\right], \qquad (3.180)$$

where ω_c is the cutoff frequency of the bath spectrum. Since terms of the order $(k_B T/\hbar\omega_c)$ have been dropped, this expression for the phase is valid for temperatures well below the Debye temperature only.

Now the rate formula can be evaluated exactly. The relaxation rate $\Gamma = \Gamma_\uparrow + \Gamma_\downarrow$ is given by the expression arising from electron damping, $\tilde{\Gamma}_{\text{el}}$, times a phonon reduction factor,

$$\Gamma = \tilde{\Gamma}_{\text{el}} e^{-\frac{1}{3}\phi} {}_2F_2(K + i\beta\Delta/2\pi, K - i\beta\Delta/2\pi; K, K + \tfrac{1}{2}; \phi), \tag{3.181}$$

where the latter involves, besides the Debye–Waller term $e^{-\frac{1}{3}\phi}$, the generalized hypergeometric series ${}_2F_2$ (cf. Gradstein and Ryshik (1981)). This expression simplifies significantly for zero asymmetry; approximating moreover the gamma function in $\tilde{\Gamma}_{\text{el}}$ as $\Gamma(K) \approx 1/K$ and $\Gamma(1-K) \approx 1$, we find

$$\Gamma = \frac{\Delta_0}{\hbar} \frac{1}{K} \left(\frac{2\pi k_B T}{\Delta_0}\right)^{2K-1} e^{-\frac{1}{3}\phi} M(K; K + \tfrac{1}{2}; \phi), \tag{3.182}$$

where the Kummer function reads as

$$M(K; K + \tfrac{1}{2}; \phi) = \sum_{n=0}^{\infty} \frac{(K)_n}{(K + 1/2)_n} \frac{1}{n!} \phi^n, \tag{3.183}$$

with $(a)_n = a(a+1)(a+2)\cdots(a+n-1)$. It turns out to be interesting to discuss a few limiting cases with respect to the coupling parameters K and ϕ.

The case of zero phonon coupling or, equivalently, zero temperature, is achieved by putting $\phi \to 0$. Since all terms but that with $n = 0$ vanish in (3.183), we find $M(K; K + \tfrac{1}{2}; 0) = 1$ and therefore $\Gamma = \Gamma_{\text{el}}$. As expected, switching off the interaction with elastic waves removes the phonon correction factor, and the rate is given by Γ_{el}. In (3.156), this case is accounted for the '1' in brackets.

In the opposite case of strong coupling, or high temperature, the parameter in Eq. (3.33) by far exceeds unity, $\phi \gg 1$. Then the series in (3.183) is dominated by contributions with large n. Each product $(K)_n$ starts with a factor K; accordingly, the factors $(K)_n/(K + 1/2)_n$ are linear in K for $K \to 0$. This corresponds to the factor K of the second term in brackets in (3.156). Together with the factor $1/K$ in (3.182), it assures that we recover the phonon-assisted rate. [This point is not discussed correctly in Würger (1997a), p. 182.]

For $K \to 0$, the term $n = 0$ of the series (3.183) gives rise to a singular contribution, which corresponds to the 'diagonal transition' (Holstein (1959)) or to the constant of the self-energy (3.59). In this limit, the equation

$$\lim_{K \to 0} \frac{1}{K}\left[M(K; K + \tfrac{1}{2}; \phi) - 1\right] = \phi F(\phi) \tag{3.184}$$

relates the remaining part to our correction factor (3.98).

Finally we repeat that the expression for the phase (3.180) is valid for the case of large Debye temperature, $T \ll \Theta$, only. As discussed on p. 99, this condition is not satisfied by hydrogen in Nb in the relevant temperature range above T_0.

Phonon-Assisted Tunneling vs. Jumps over the Barrier. In Figs. 3.5 and 3.6 we have fitted experimental data for hydrogen in Nb and defects in Bi wires with the rate obtained for phonon-assisted tunneling. It should be kept in mind, however, that thermally activated jumps over the barrier would lead to a qualitatively similar, strongly increasing rate. Thermal activation to excited levels in the double-well potential would imply the break-down of the two-state approximation for the tunneling system.

In principle, it should be possible to distinguish whether the data obey the law for phonon-assisted tunneling, $\Gamma \propto T^{-2}\exp(T^2/T_0^2)$, or that of thermal activation, $\Gamma \propto \exp(-E/k_B T)$. In the usual picture, E is the harmonic excitation energy in the two wells. Yet the data available at present do not reach sufficiently high temperatures, in order to provide such a criterion. Nonetheless, there are good reasons to consider phonon-assisted tunneling as the relevant process.

First, the only parameter of the phonon-assisted rate, T_0 or, equivalently, the elastic deformation potential γ, agrees quite well with those obtained from independent measurements. The value used for defects in Bi wires, $\gamma = 1.6$ eV, is close to that obtained by Birge et al. (1996), $\gamma = 2.3 \pm 0.4$ eV. Therefore, the theoretical results, a minimum at about 1 K and a strongly increasing rate above, are definite and do not admit much deviation. The quality of the fits is remarkable.

Second, when fitting the data of Fig. 3.6 by adding an activated rate $\Gamma_{\text{act}} = \omega_0 \exp(-E/k_B T)$ to that driven by conduction electrons, Birge et al. (1996) derived an activation energy $E/k_B = 19.1$ K. In view of the very small tunnel energy, $\Delta_0/k_B = 1.9 \times 10^{-7}$ K, one would expect a high barrier and, accordingly, a much larger libration energy E; then the onset of activated relaxation would occur at a much higher temperature.

Third, in the case of trapped hydrogen, the phonon-assisted tunneling rate itself shows Arrhenius behavior, with an activation energy given by (3.159), which can hardly be distinguished from that arising from jumps over the barrier. The ratio of the rates for different trap atoms, however, rather supports the relevance of phonon-assisted tunneling. In the whole temperature range, the data shown in Fig. 3.5 exhibit a ratio of about 2 for hydrogen at oxygen and nitrogen traps. This value agrees well with the tunneling rate $\Gamma \propto \Delta_0^{2(1-K)}$ and values for Δ_0 measured at low temperature in the coherent regime (Wipf (1997)). The over-barrier jump rate does not depend on the tunnel energy; it would be a strange coincidence, if replacing an oxygen trap by a nitrogen would have the same effect on the rates arising from phonon-assisted tunneling and from activated barrier crossing.

In summary, for TS in metallic hosts, coupling to both conduction electrons and phonons needs to be taken into account. At low temperatures, the conduction electrons provide the more efficient damping mechanism, where the incoherent rate $\Gamma \propto T^{2K-1}$ decreases with rising T. The strong increase of the phonon contribution leads to a minimum of the relaxation rate at the temperature T_0 and an exponential increase above. The crossover temperature is inversely proportional to the elastic deformation potential γ; for hydrogen impurities in Nb we find $T_0 \approx 23$ K, and $T_0 \approx 0.85$ K for configurational defects in Bi wires.

3.6 Phonon Dressing in Real Systems

In Sects. 3.4 and 3.5 we have studied how coupling to elastic waves affects two-state defects in insulators and metals. In the latter section, we have discussed in some detail two experiments on TS's in metallic hosts, namely hydrogen impurities in niobium and configurational defects in Bi wires, and we have derived the elastic deformation potential by relating the observed data to Eq. (3.178). Here, we extend this discussion to TS's in insulating materials.

According to (3.20) and (3.24), the elastic deformation potential γ and the temperature T_0 are related through

$$T_0 = \gamma^{-1}\sqrt{\hbar^3 \varrho v^5 / 3k_B^2}. \tag{3.185}$$

The average sound velocity v and the mass density ϱ are well known; thus (3.185) constitutes a useful relation between γ and T_0.

In Table 3.1 we compare these parameters for defects in a few crystalline and amorphous materials. From the elastic deformation potential γ measured by Hunklinger and Arnold (1976) for a-SiO$_2$ and by Enss (1996) for KCl:Li, we calculate the crossover temperature T_0. For Nb(OH)$_x$ and amorphous Bi, we proceed in the opposite way: The values for T_0 have been taken from Figs. 3.6 and 3.5; then we calculate the deformation potential according to (3.185). The resulting values for γ agree well with those measured independently; cf. the discussion on p. 104 for the defects in Bi wires.

According to (3.43), the dressed tunnel energy at zero temperature, Δ_0, is related to the bare value through T_0 and Θ. Clearly, only the dressed matrix element can be observed by experiment, whereas Δ_b cannot be measured independently. Equation (3.43) is useful nonetheless, since it permits us to derive an upper bound for the tunnel energy.

Impurity Atoms in Crystals. 'Tunneling' is a meaningful concept for a sufficiently high barrier only, i.e. if the bare energy Δ_b is significantly smaller than the energy difference of the lowest two levels for zero barrier. As a simple estimate, we calculate for a particle with mass m in a box with length d the energy of the ground state, $\epsilon_0 = (\pi^2 \hbar^2 / 2md^2)$, and of the first excited level, $\epsilon_1 = 4\epsilon_0$. For a ^{7}Li atom in a rectangular box of length 2 Å, the energy difference $\epsilon \equiv \epsilon_1 - \epsilon_0$ takes the value $\epsilon = 2$ meV, which corresponds to 23 K. For the lighter hydrogen atom we find $\epsilon = 14$ meV, or 161 K. (The actual distance of Li off-center positions in KCl is 1.4 Å, that of the interstitial states in niobium 1.17 Å.)

Since the bare tunnel energy Δ_b of a lithium impurity must be significantly smaller than ϵ, it can hardly exceed 10 Kelvin. This estimate agrees qualitatively with the reduction factor $e^{-\frac{1}{2}W_0} = 0.85$, calculated from Table 3.1, and the measured value $\Delta_0/k_B = 1.1$ K. From the Debye–Waller factor being close to unity, we may conclude that Δ_0 differs little from Δ_b, and that phonon dressing is of little significance for tunneling of Li in KCl.

This discussion relies on the two-state approximation. A detailed study of the actual eight-state problem shows that, for symmetry reasons, the dressing effect of lithium tunneling along a crystal axis is still weaker. As another consequence of the symmetry properties of [111]-defects, there is no one-phonon rate for these transitions; cf. Würger (1997a). A thorough discussion of the dressing effect for off-center impurities in alkali halides has been given by Shore and Sanders (1975).

As to interstitial hydrogen in Nb, we obtain from (3.43) with $W_0 = 5.93$ the reduction factor $e^{-\frac{1}{2}W_0} = 0.05$, which is still compatible with the measured values for Δ_0/k_B of about 2 K.

Configurational Defects. The situation is different for TS's that arise from local configurational disorder. Their large elastic coupling energy leads to a crossover temperature T_0 which, according to Table 3.1, is significantly smaller than Θ, thus resulting in a strong phonon-dressing effect.

The tunnel energy measured for defects in Bi wires, $\Delta_0/k_B \approx 2\times10^{-7}$ K, is by about seven orders of magnitudes smaller than those observed for impurity atoms. The calculated reduction factor $e^{-\frac{1}{2}W_0}$ is very small indeed and – at least qualitatively – confirms the law (3.43). Note, however, that the values for the effective Debye temperature given in Table 3.1 are subject to various uncertainties, due to our simplistic phonon model.

Regarding oxide glasses, the above argument imposes quite severe restrictions on the parameters. As to the deformation potential γ, values between

Table 3.1. Parameters for tunneling defects in various materials. For a-Bi and Nb(OH)$_x$, the crossover temperature T_0 is taken from the fits in Figs. 3.6 and 3.5; the elastic deformation γ is calculated according to (3.185). For the insulating systems, γ is derived from measured values for the low-temperature rate Γ_1, and T_0 is obtained from (3.185). The Debye temperature Θ is given by (3.22).

	ϱ (g/cm^3)	$v_{l,t}$ (km/s)	γ (eV)	T_0 (K)	Θ (K)	Δ_0/k_B (K)
Nb(OH)$_x$	8.4	5.1/2.1	0.2	23	249	2.4
KCl:Li	2.0	3.9/2.4	0.06	85	234	1.65/1.1 [a]
Bi	9.8	2.3/1.1	1.6	0.85	108	1.9×10^{-7}
a-SiO$_2$	2.2	5.8/3.8	2.6	4.8	461	≤ 5

[a] 1.65 for the lighter isotope ^{6}Li, and 1.1 for ^{7}Li impurities

2 and 3 eV have been reported. In a two-pulse echo experiment, Hunklinger and Arnold (1976) observed symmetric TS's with a dressed tunnel energy $\Delta_0/k_B \approx 35$ mK. From the logarithmic variation of the sound velocity measured up to 3 K, it follows that the distribution of the tunneling model, $P(\Delta_0) = P_0/\Delta_0$, is valid up to tunnel energies of 5 K. Assuming an upper bound for the bare tunnel energy Δ_b of about 25 K and, accordingly, a maximum reduction factor $e^{-\frac{1}{2}W_0}$ of $\frac{1}{5}$, we may conclude from $T_0 \approx 5$ K that the effective Debye temperature Θ cannot exceed 50 K, contrary to the value of Table 3.1 We will come back to this issue on p. 119.

We conclude by pointing out a few flaws concerning the phonon model introduced in Sect. 3.1 First, the dispersion law of Eq. (3.15) is valid for acoustic modes only, far from the Brillouin zone boundary. Second, the frequency dependence of the coupling energies λ_k in (3.43) holds only for vibrational modes with wavelengths larger than the size of the two-state system, d. For impurities in crystals, the two-state distance d is significantly smaller than the lattice constant, which, in turn is comparable to the shortest wavelength. Configurational defects, however, are supposed to involve reorientation of several atoms. As a consequence, the phase factors in (3.10) may vary rapidly over the size of the defect, and thus suppress the coupling constants λ_k for short wavelengths. The effective Debye temperature in (3.42) should be replaced by a smaller value accordingly.

A third proviso concerns the basic assumption of a harmonic solid, which assures the existence of $3N$ extended vibrational modes, i.e. three modes per atom. Its validity is not obvious for glasses, where local oscillators seem to constitute a significant number of degrees of freedom (Buchenau et al. (1992)). The corresponding reduction of the number of phonon modes leads to a smaller value for the effective Debye temperature. Hence the presence of such local oscillators could explain the discrepancy between the value for Θ given in Table 3.1 and that discussed above and on p. 119.

3.7 Asymmetric Tunneling Systems

In the preceding sections we have studied the motion of a two-level systems whose asymmetry energy Δ is either zero or, on the contrary, much larger than the tunnel energy. In amorphous solids, however, one finds TS's with a broad distribution of Δ_0 and Δ which, in general, satisfies neither of the mentioned limits. Here we propose a perturbative approach which relies on a matrix representation for the reduced time-evolution operator.

We start from the Hamiltonian in the polaron representation. When setting up a perturbation theory in terms of the fluctuation operators $\xi_\pm = B_\pm - \langle B_\pm \rangle$, Hermitian combinations prove to be a convenient choice,

$$\xi_g = \frac{1}{2}(B_+ + B_- - 2), \qquad \xi_u = \frac{1}{2i}(B_+ - B_-). \tag{3.186}$$

Accordingly we separate the Hamiltonian into two parts,

$$H = H_0 + H_1, \qquad (3.187)$$

where the first one describes the uncoupled system,

$$H_0 = \frac{1}{2}\tilde{\Delta}_0 \sigma_x + \frac{1}{2}\Delta \sigma_z + \sum_k \hbar \omega_k b_k^\dagger b_k, \qquad (3.188)$$

with the reduced tunnel energy (3.39), and the second term contains the interaction,

$$H_1 = \frac{1}{2}\Delta_b \sigma_x \xi_g + \tfrac{1}{2}\Delta_b \sigma_y \xi_u. \qquad (3.189)$$

The treatment of the two-state dynamics in this section relies on a perturbation expansion in terms of the spin-phonon coupling H_1, which is derived from a projection technique and a usual decoupling approximation. Since we project on the three pseudospin operators, asymmetry and tunnel energies are treated on an equal footing. Starting from the equations of motion for the spin operators,

$$\hbar \dot{\sigma}_x = -\Delta \sigma_y + \Delta_b \xi_u \sigma_z, \qquad (3.190)$$

$$\hbar \dot{\sigma}_y = \Delta \sigma_x - \tilde{\Delta}_0 \sigma_z - \Delta_b \xi_g \sigma_z, \qquad (3.191)$$

$$\hbar \dot{\sigma}_z = \tilde{\Delta}_0 \sigma_y + \Delta_b \xi_g \sigma_y - \Delta_b \xi_u \sigma_x, \qquad (3.192)$$

we proceed in the standard way by projecting on spin fluctuations.

3.7.1 Projection Method

The relevant dynamic space is spanned by the fluctuating part of the three spin operators σ_α. Thus we subtract the thermal average,

$$\delta \sigma_\alpha = \sigma_\alpha - \langle \sigma_\alpha \rangle, \qquad (3.193)$$

and define the correlation matrix

$$\delta G_{\alpha\beta}(t - t') = (\delta\sigma_\alpha(t)|\delta\sigma_\beta(t')), \qquad (3.194)$$

where we have introduced the short-hand notation for the symmetrized correlation function

$$(A(t)|A'(t')) = \tfrac{1}{2}\langle A(t)A'(t') + A'(t')A(t)\rangle \qquad (3.195)$$

for Hermitian operators A and A'.

The elements $|A)$ form a linear space of infinite dimension with the inner product $(A|A')$. Mori's method relies on the projection on an appropriate low-dimensional subspace; with the static correlation matrix

$$\eta_{\alpha\beta} = (\delta\sigma_\alpha|\delta\sigma_\beta), \qquad (3.196)$$

we define operators $\mathcal{P}$ and $\mathcal{Q}$,

$$\mathcal{P} = \sum_{\alpha\beta} |\delta\sigma_\alpha)(\eta^{-1})_{\alpha\beta}(\delta\sigma_\beta| = 1 - \mathcal{Q}, \qquad (3.197)$$

where $\mathcal{P}$ projects on the three-dimensional subspace spanned by (3.193), and $\mathcal{Q}$ on the complement.

After inserting the time-evolution operator $e^{i\mathcal{L}t}$ in the correlation function and taking the Laplace transform, one finds the somewhat formal result

$$\delta G_{\alpha\beta}(z) = (\delta\sigma_\alpha|[\mathcal{L}-z]^{-1}\delta\sigma_\beta). \qquad (3.198)$$

When making use of the resolvent identity, one finally obtains the matrix equation

$$\delta G(z) = -\eta \frac{1}{z\eta + \hat{\Omega} + \hat{M}(z)} \eta, \qquad (3.199)$$

with the frequency matrix

$$\hat{\Omega}_{\alpha\beta} = (\mathcal{P}\mathcal{L}\delta\sigma_\alpha|\delta\sigma_\beta) \qquad (3.200)$$

and the time-dependent memory function

$$\hat{M}_{\alpha\beta}(t) = (\mathcal{Q}\mathcal{L}\delta\sigma_\alpha|e^{-i\mathcal{Q}\mathcal{L}\mathcal{Q}t}|\mathcal{Q}\mathcal{L}\delta\sigma_\beta). \qquad (3.201)$$

With the definitions

$$\Omega = \eta^{-1}\hat{\Omega}, \qquad M = \eta^{-1}\hat{M}, \qquad (3.202)$$

Equation (3.199) takes the simpler form

$$\delta G(z) = -\eta \frac{1}{z + \Omega + M(z)}. \qquad (3.203)$$

3.7.2 Approximations

Although the resolvent matrix (3.203) provides an exact formal expression for the two-state dynamics, it cannot be evaluated as it stands. Here we resort to a standard approximation scheme, which is similar to the perturbation theory applied by Würger (1998a). We are going to expand both the statistical operator and the time evolution in terms of the bath fluctuations ξ_g and ξ_u.

The statistical operator. — When expanding the statistical operator in terms of the spin-phonon coupling H_1, we obtain the series

$$\rho(\beta) = \rho_0(\beta) - \int_0^\beta d\beta' \rho_0(\beta - \beta') H_1 \rho_0(\beta') + \ldots . \qquad (3.204)$$

We keep only the first term which describes the thermal equilibrium state with respect to H_0,

$$\rho = \rho_S \rho_B, \qquad (3.205)$$

where the bath part is given by $\rho_B = e^{-\beta H_B}/\text{tr}(e^{-\beta H_B})$, and the spin part by

3.7 Asymmetric Tunneling Systems   113

$$\rho_S = \tfrac{1}{2}[-(u\sigma_x + \bar{u}\sigma_z)\tanh(E/2k_BT)]. \tag{3.206}$$

Here we have used the definitions

$$E = \sqrt{\tilde{\Delta}_0^2 + \Delta^2}, \qquad u^2 = \tilde{\Delta}_0^2/E^2, \qquad \bar{u}^2 = \Delta^2/E^2. \tag{3.207}$$

The latter constants fulfil $u^2 + \bar{u}^2 = 1$. All thermal mean values are calculated with respect to the uncoupled density operator ρ, i.e. $\langle\ldots\rangle = \text{tr}(\rho\ldots)$.

Both the static correlations $\eta_{\alpha\beta}$ and the frequency matrix $\Omega_{\alpha\beta}$ involve the spin part ρ_S only. A straightforward calculation yields

$$\eta = \begin{pmatrix} 1-u^2s^2 & 0 & -u\bar{u}s^2 \\ 0 & 1 & 0 \\ -u\bar{u}s^2 & 0 & 1-\bar{u}^2s^2 \end{pmatrix}, \quad \Omega = \frac{i}{\hbar}\begin{pmatrix} 0 & \Delta & 0 \\ -\Delta & 0 & \tilde{\Delta}_0 \\ 0 & -\tilde{\Delta}_0 & 0 \end{pmatrix}, \tag{3.208}$$

with the spin polarization $s = \tanh(E/2k_BT)$.

The Memory Matrix. The actual difficulty in evaluating (3.203) arises from the frequency dependent memory kernel. First we consider the vertices $\mathcal{QL}\delta\sigma_\alpha$, which contain the component of the derivate $\mathcal{L}\delta\sigma_\alpha$ that lies outside the relevant space spanned by $\delta\sigma_\alpha$. When taking into account (3.197), the statistical operator ρ, and the relations $\langle\xi_g\rangle = 0 = \langle\xi_u\rangle$, one finds

$$\begin{aligned} i\hbar|\mathcal{QL}\delta\sigma_x) &= \Delta_b|\xi_u\sigma_z), \\ i\hbar|\mathcal{QL}\delta\sigma_y) &= -\Delta_b|\xi_g\sigma_z), \\ i\hbar|\mathcal{QL}\delta\sigma_z) &= \Delta_b|\xi_g\sigma_y - \xi_u\sigma_x). \end{aligned} \tag{3.209}$$

Comparison with (3.190)–(3.192) reveals that only the coupling terms, i.e. those involving ξ_g and ξ_u, survive the action of the projection $\mathcal{Q}$.

According to our approximation scheme, we keep only terms quadratic in ξ_α. Since both vertex factors in (3.201) are already linear, we drop the coupling part in the time evolution and replace in the exponential the Liouville operator $\mathcal{QLQ}$ by $\mathcal{QL}_0\mathcal{Q}$, the latter being defined through $\mathcal{L}_0 A = (1/\hbar)[H_0, A]$.

The 'vectors' (3.209) are elements of $\mathcal{Q}$ space. Since $\mathcal{L}_0$ does not scatter between the $\mathcal{P}$ and $\mathcal{Q}$ subspaces, the factors $\mathcal{Q}$ in the exponential of (3.201) are meaningless, and we may replace $\mathcal{QL}_0\mathcal{Q}$ with $\mathcal{L}_0$. Thus the memory function reads in quadratic order in ξ_α as

$$\hat{M}_{\alpha\beta}(t) = (\mathcal{QL}\delta\sigma_\alpha|e^{-i\mathcal{L}_0 t}|\mathcal{QL}\delta\sigma_\beta), \tag{3.210}$$

where both the thermal average and time evolution are performed with respect to H_0. Inserting (3.209) and noting that the Hermitian conjugate of $i|A)$ is $-i(A|$, we obtain explicit expressions in terms of correlation functions of σ_α and ξ_α.

Considering that ξ_g contains only even powers of phonon operators, and ξ_u only odd powers, we find $\langle\xi_g(t)\xi_u(t')\rangle = 0$. Taking this selection rule into account, we find the finite matrix elements of the memory kernel

$$\hat{M}_{xx}(t) = (\Delta_b/\hbar)^2(\xi_u(t)\sigma_z(t)|\xi_u\sigma_z), \tag{3.211}$$

$$\hat{M}_{yy}(t) = (\Delta_{\rm b}/\hbar)^2 (\xi_{\rm g}(t)\sigma_z(t)|\xi_{\rm g}\sigma_z), \tag{3.212}$$

$$\hat{M}_{zz}(t) = (\Delta_{\rm b}/\hbar)^2 (\xi_{\rm g}(t)\sigma_y(t) - \xi_{\rm u}(t)\sigma_x(t)|\xi_{\rm g}\sigma_y - \xi_{\rm u}\sigma_x), \tag{3.213}$$

$$\hat{M}_{yz}(t) = (\Delta_{\rm b}/\hbar)^2 (\xi_{\rm g}(t)\sigma_z(t)|\xi_{\rm g}\sigma_y) = \hat{M}_{zy}(-t). \tag{3.214}$$

The remaining entries $\hat{M}_{xz}$, $\hat{M}_{zx}$, $\hat{M}_{xy}$, and $\hat{M}_{yx}$ vanish. Thus the correlation matrix $\delta G(z)$ is block-diagonal.

3.7.3 The Damping Kernel

With (3.195) the entries $M_{\alpha\beta}(t)$ are given as a sum of products containing each a spin and a bath factor, e.g.

$$\hat{M}_{xx}(t) = \frac{1}{2}(\Delta_{\rm b}/\hbar)^2 \left[\langle \xi_{\rm u}(t)\xi_{\rm u}\rangle\langle \sigma_z(t)\sigma_z\rangle_0 + \langle \xi_{\rm u}\xi_{\rm u}(t)\rangle\langle \sigma_z\sigma_z(t)\rangle_0 \right]. \tag{3.215}$$

For further use we note the symmetrized bath correlation functions

$$\Gamma_\alpha(t) = \frac{1}{2}\langle \xi_\alpha(t)\xi_\alpha + \xi_\alpha\xi_\alpha(t)\rangle, \tag{3.216}$$

with $\alpha = {\rm g, u}$, and the uncoupled spin correlations

$$G^0_{\alpha\beta}(t-t') = \frac{1}{2}\langle \sigma_\alpha(t)\sigma_\beta(t') + \sigma_\beta(t')\sigma_\alpha(t)\rangle_0, \tag{3.217}$$

with $\alpha, \beta = x, y, z$.

The link between ordinary correlations in (3.215) and the symmetric ones (3.216), (3.217) is established most easily in frequency space through the relations (3.223); with (3.211)–(3.214), we find

$$\hat{M}''_{xx}(\omega) = (\Delta_{\rm b}/\hbar)^2 (G^{0''}_{zz} * \Gamma''_{\rm u})(\omega), \tag{3.218}$$

$$\hat{M}''_{yy}(\omega) = (\Delta_{\rm b}/\hbar)^2 (G^{0''}_{zz} * \Gamma''_{\rm g})(\omega), \tag{3.219}$$

$$\hat{M}''_{zz}(\omega) = (\Delta_{\rm b}/\hbar)^2 (G^{0''}_{yy} * \Gamma''_{\rm g})(\omega) + (\Delta_{\rm b}/\hbar)^2 (G^{0''}_{xx} * \Gamma''_{\rm u})(\omega), \tag{3.220}$$

$$\hat{M}''_{yz}(\omega) = (\Delta_{\rm b}/\hbar)^2 (G^{0''}_{zy} * \Gamma''_{\rm g})(\omega) = \hat{M}''_{zy}(-\omega), \tag{3.221}$$

$$\hat{M}''_{xz}(\omega) = (\Delta_{\rm b}/\hbar)^2 (G^{0''}_{zx} * \Gamma''_{\rm u})(\omega) = -\hat{M}''_{zx}(\omega). \tag{3.222}$$

Thus each entry of the memory kernel is given as a convolution of spin and bath correlation spectra, where the weighted convolution integral reads as

$$(f''(\omega) * g''(\omega)) \equiv \frac{1}{\pi}\int_{-\infty}^{\infty} d\omega' f''(\omega')g''(\omega-\omega')\frac{w(\omega)}{w(\omega')w(\omega-\omega')}, \tag{3.223}$$

with $w(\omega) = \cosh(\frac{1}{2}\beta\hbar\omega)$. The undamped spin correlations are easily calculated from (3.217),

$$G^{0''}(\omega) = \pi\delta(\omega)\begin{pmatrix} u^2 & 0 & u\bar{u} \\ 0 & 0 & 0 \\ u\bar{u} & 0 & \bar{u}^2 \end{pmatrix} + \frac{\pi}{2}\sum_\pm \delta(\omega \pm E/\hbar)\begin{pmatrix} \bar{u}^2 & \pm i\bar{u} & -u\bar{u} \\ \mp i\bar{u} & 1 & \pm iu \\ -u\bar{u} & \mp iu & u^2 \end{pmatrix}. \tag{3.224}$$

3.7 Asymmetric Tunneling Systems

The matrices η and $G^{0\prime\prime}(\omega)$ permit us to calculate the memory kernel $M = \eta^{-1}\hat{M}$. This result is, however, cumbersome and of little practical interest. In order to simplify the analysis, we consider the case of small tunnel energy, $\tilde{\Delta}_0 \ll k_BT$, and we use expressions which are correct in the cases $\Delta \ll \tilde{\Delta}_0$ and $\tilde{\Delta}_0 \ll \Delta$ and which provide a good approximation in the intermediate range.

First we show that the off-diagonal part of M is insignificant. The entries M''_{xy} and M''_{yx} vanish because of $\langle \xi_g(t)\xi_u(t')\rangle = 0$. As may be seen from the undamped spectra $G''_{ij}(\omega)$, the entries M''_{xz} and M''_{zx} are proportional to $u\bar{u}$. Therefore they are negligible for both limiting cases $\Delta \ll \tilde{\Delta}_0$ and $\tilde{\Delta}_0 \ll \Delta$; in the intermediate case they do not change the damping behavior significantly.

As for the remaining matrix elements M''_{yz} and M''_{zy}, they are small for the following reason. The spectra G''_{zy} and G''_{yz} are odd functions of frequency and hence vanish at $\omega = 0$. Since in Markov approximation, the memory matrix is to be evaluated at small frequency, the entries M''_{yz} and M''_{zy} are small as compared to the diagonal ones and may be neglected (Würger (1998a)).

Now we calculate the diagonal elements. In the case of large asymmetry, $\tilde{\Delta}_0 \ll \Delta$, we drop u in η and $G^{0\prime\prime}$, and replace $\bar{u}$ by unity; thus we obtain

$$M''_{xx}(\omega) = (\Delta_b/\hbar)^2 \Gamma''_u(\omega), \qquad M''_{yy}(\omega) = (\Delta_b/\hbar)^2 \Gamma''_g(\omega),$$

$$M''_{zz}(\omega) = \frac{1}{2}(\Delta_b/\hbar)^2 \sum_{\pm} \Gamma''(\omega \pm E/\hbar) w(\omega) w(E/\hbar)^{-1} w(\omega \pm E/\hbar)^{-1}, \quad (3.225)$$

with $w(\omega) = \cosh(\beta\hbar\omega/2)$. Applying a Markov or pole approximation, we evaluate M''_{xx} and M''_{yy} at $E \approx \Delta$, and M''_{zz} at zero frequency, and find

$$M''_{xx}(E/\hbar) = \gamma_u,$$

$$M''_{yy}(E/\hbar) = \gamma_g,$$

$$M''_{zz}(0) = \gamma_u + \gamma_g \equiv \Gamma \qquad (3.226)$$

for $\tilde{\Delta}_0 \ll \Delta$. Here we have used the rates of odd and even order, (3.123) and (3.124).

In the opposite case of small asymmetry energy $\Delta \to 0$, one finds very similar rates which differ merely in the frequency argument of M''_{yy} (Würger (1998a)). Therefore we may use (3.226) for all values of Δ.

Now we are prepared to calculate the correlation matrix (3.203). In order to invert $[z + \Omega + iM'']$, we need its determinant $D = |z + \Omega + iM''|$,

$$D = (z + i\Gamma)(z^2 + iz\Gamma - \gamma_u\gamma_g - (E/\hbar)^2) + i\gamma_g(\tilde{\Delta}_0/\hbar)^2. \qquad (3.227)$$

When multiplying the matrices η and $[z + \Omega + iM'']^{-1}$ according to (3.203), we find that the relevant matrix element δG_{zz} comprises two terms,

$$\delta G_{zz}(z) = -D^{-1}\left[\eta_{zz}\bigl((z+i\gamma_g)(z+i\gamma_u) - (\Delta/\hbar)^2\bigr) \right.$$
$$\left. + \eta_{zx}\tilde{\Delta}_0\Delta/\hbar^2\right]. \tag{3.228}$$

To proceed further we have to calculate the roots of (3.227) and the corresponding residues in (3.228).

We start by looking for approximate roots of the determinant. For further convenience we define the reduction factor

$$\tilde{u}^2 = \tilde{\Delta}_0^2/(E^2 + \hbar^2\gamma_u\gamma_g). \tag{3.229}$$

The rates are given in terms of $\tilde{u}^2$ and (3.123), (3.124),

$$\Gamma_l = \gamma_u + (1-\tilde{u}^2)\gamma_g, \qquad \Gamma_t = \tfrac{1}{2}\left[\gamma_u + (1+\tilde{u}^2)\gamma_g\right]. \tag{3.230}$$

After regrouping the terms appropriately, we rewrite (3.227) as

$$D = (z+i\Gamma_l)\left[(z+i\Gamma_t)^2 - (\tilde{E}/\hbar)^2\right] + R, \tag{3.231}$$

where the remaining constant reads as $R = (\tilde{u}^2\gamma_g)^2(\Gamma + \tilde{u}^2\gamma_g)$, and the effective oscillation frequency is given by

$$\tilde{E}^2 = E^2 - \tfrac{1}{4}\hbar^2(\gamma_u-\gamma_g)^2 - \tilde{u}^2\hbar^2\gamma_g(\gamma_u+\gamma_g) + \tfrac{3}{4}\hbar^2(\tilde{u}^2\gamma_g)^2. \tag{3.232}$$

The complex roots of (3.231) as obtained by means of Cardani's formula are too cumbersome to be of any practical use. Yet it turns out that the constant R is negligible in the limits of both low and high temperatures, i.e. for both small and large damping rates. Thus we set $R=0$; then the determinant factorizes in a relaxation pole with width Γ_l and a pair of complex poles $-i\Gamma_t \pm \tilde{E}/\hbar$.

When inserting the determinant in (3.226) and evaluating the pole contributions, we obtain lengthy expressions involving the various rates defined above. The following argument permits a significant simplification. For temperatures well below T_0 as defined in (3.24), all rates are much smaller than the two-level splitting. In the opposite case, $T \gg T_0$, the rates are large, but the tunnel energy $\tilde{\Delta}_0$ is exponentially small, and so is the factor $\tilde{u}$ as defined in (3.229).

Moreover we may set $\gamma_u = \gamma_g = \tfrac{1}{2}\Gamma$ for $T \gg T_0$. The energy splitting $\tilde{E}$ is well approximated by E, i.e. it is given by Δ for $T \gg T_0$. Evaluating both terms in (3.228) accordingly, we find

$$\delta G_{zz}(z) = \left[\bar{u}^2(1-s^2) + u^2 - \tilde{u}^2\right]\frac{-1}{z+i\Gamma_l}$$
$$+ \frac{\tilde{u}^2}{2}\sum_\pm \frac{\exp(\pm i\delta)}{\cos\delta}\frac{-1}{z+i\Gamma_t \pm \tilde{E}/\hbar}. \tag{3.233}$$

[There is no simple expression for the phase shift δ; yet for zero asymmetry one recovers $\tan\delta = (\hbar\gamma_u/2\tilde{E})$.]

3.7.4 Crossover to Relaxation

Together with the rates (3.230), the resolvent function G_{zz} constitutes the main result of this section. For further convenience we note its inverse Laplace transform

$$\delta G_{zz}(t) = \left[\bar{u}^2(1-s^2) + u^2 - \tilde{u}^2\right] e^{-\Gamma_l t} + \tilde{u}^2 \frac{\cos(\tilde{E}t/\hbar + \delta)}{\cos\delta} e^{-\Gamma_t t}. \quad (3.234)$$

This correlation function contains the dissipative two-state dynamics for all relevant cases, i.e. for all values of the parameter $\tilde{\alpha}T^2$, and of the ratio of tunnel and asymmetry energies, $\tilde{\Delta}_0/\Delta$. It seems worth noting that at $t = 0$ (3.234) fulfils the relation

$$\delta G_{zz}(t=0) = 1 - \bar{u}^2 s^2 = 1 - \langle \sigma_z \rangle^2, \quad (3.235)$$

which follows from the definition of the fluctuation operators (3.193).

In order to simplify the prefactor of the relaxation term in (3.234), we define a crossover temperature T^* through the condition $E^2 = \hbar^2 \gamma_u \gamma_g$. In this range we may use $\gamma_u = \gamma_g = \frac{1}{2}\Gamma$ and the high temperature limit (3.103), and thus find

$$T^* = T_0 \sqrt{\tfrac{3}{2} \log(4\sqrt{\pi} k_B T^* E / T_0 \Delta_0^2)}. \quad (3.236)$$

We note the general relation $T^* > T_0$. Setting $\Delta = 0$, we recover the result for the symmetric case, (3.135). An insignificant numerical difference arises from the fact that, in deriving (3.236), we have assumed E independent of temperature; yet this assumption holds true for $\tilde{\Delta}_0 \ll \Delta$ only.

The definition of the crossover temperature T^* is motivated by the behavior of the reduction factor $\tilde{u}^2$ defined in (3.229). With (3.207) we find $\tilde{u}^2 = u^2$ for $T \ll T^*$, whereas in the opposite case we may put $\tilde{u}^2 = 0$. In the sequel of this section we discuss the main features of (3.234) by considering separately low and high temperatures with respect to T^*.

3.7.5 Low Temperatures: $T \ll T^*$

In this range the rates are small as compared to the two-level spacing, resulting in underdamped oscillations with frequency $E/\hbar$. As a consequence, the phase shift is small, $\delta \ll 1$, and we may put $\tilde{u}^2 - u^2 = 0$ in (3.234); inserting the spin polarization $s = \tanh(\beta E/2)$, we obtain

$$\delta G_{zz}(t) = \frac{\Delta^2}{E^2} \cosh(E/2k_B T)^{-2} e^{-\Gamma_l t} + \frac{\tilde{\Delta}_0^2}{E^2} \frac{\cos(Et/\hbar + \delta)}{\cos\delta} e^{-\Gamma_t t}. \quad (3.237)$$

At first sight, this result is similar to that of Hunklinger and Arnold (1976), obtained by solving the Bloch equations for $\langle\sigma_\alpha\rangle$. There are, however, two essential differences. First, our rates are not restricted to the one-phonon process, but contain terms of any order. Second, the prefactor of the oscillating part involves the renormalized tunnel frequency $\tilde{\Delta}_0$, instead of the bare value Δ_b.

This second issue gives rise to a strong temperature dependence of the oscillation amplitude, $(\tilde{\Delta}_0^2/E^2) = (\Delta_0^2/E^2)\exp(-\frac{1}{3}T^2/T_0^2)$, whose possible impact on observable quantities is discussed below.

Before turning to the high-temperature case, we briefly discuss the rates in the limit of zero asymmetry; from (3.230) we find

$$\Gamma_l = \gamma_u, \quad \Gamma_t = \gamma_g + \tfrac{1}{2}\gamma_u \quad (\Delta \to 0). \tag{3.238}$$

The transverse rate comprises both γ_g and γ_u, in agreement with the result (3.129) of the blip expansion in Sect. 3.4. The longitudinal rate Γ_l, however, contains terms of odd order in the coupling parameter α only, i.e. energy relaxation involves processes with one, three, five, etc. phonons. This result has been confirmed in two complementary perturbative approaches (Würger (1997c), Würger (1998a)).

The physical origin of this symmetry property has been discussed in terms of phonon scattering between energy eigenstates by Egorov and Skinner (1995). The one-phonon term, or direct process, involves a single scattering event from the upper state to the lower, or vice versa; the corresponding rate is quadratic in the elastic deformation potential γ or, equivalently, linear in the constant α. The two-phonon term in the perturbation series requires one up-scattering and one down-scattering process; it does not contribute to energy relaxation, since the corresponding initial and final states are the same. This argument is easily generalized to n-phonon term in the perturbation series; one finds that relaxation is due to contributions with an odd number of phonons.

The perturbation expansion given by Egorov and Skinner (1995) yields the correct rate at $T = 0$; its n-phonon term is based on the emission of n quanta. At finite temperature, however, additional processes, involving both emission and absorption, have to be taken into account, enhancing significantly the temperature dependence of the rate. In a diagrammatic approach Pirc and Gosar (1969) obtained rates that are, at first sight, similar to (3.230). For zero asymmetry, however, longitudinal and transverse rates are interchanged, as compared with (3.238), i.e. this approach does not satisfy the above symmetry property of the relaxation rate Γ_l.

3.7.6 High Temperatures: $T \gg T^*$

According to the above discussion, the factor $\tilde{u}^2$ is negligible now, and so is the oscillating part in the correlation function,

$$\delta G_{zz}(t) = \left[\frac{\Delta^2}{E^2}\cosh(E/2k_BT)^{-2} + \frac{\tilde{\Delta}_0^2}{E^2}\right]e^{-\Gamma_l t}. \tag{3.239}$$

As a consequence, the dynamic susceptibility has no 'resonant' part. As a result of the crossover to overdamped motion, the corresponding amplitude in (3.237) shifted to the relaxation part in (3.239). Thus the response of the two-level system is purely relaxational, with a rate Γ_1 that increases exponentially with temperature.

So far no restriction has been placed on the ratio of tunnel and asymmetry energies, the above results being valid for any value of $(\tilde{\Delta}_0/\Delta)$. Considering the limit $\Delta \to 0$ we find that the crossover temperature (3.236) almost coincides with the corresponding expression (3.135). Moreover, a glance at (3.130) and (3.134) convinces one immediately that the results for the limiting cases (3.237) and (3.239) merge, for $\Delta \to 0$, into those obtained for a symmetric two-level system.

There is, however, a discrepancy concerning the onset of relaxation, which is smooth in (3.234) but abrupt in (3.128). The resolvent function (3.233) always comprises a relaxation pole, even at low temperature and zero asymmetry; its amplitude $u^2 - \tilde{u}^2 = (\hbar^2 \gamma_u \gamma_g / E^2)$ is finite for $\Delta = 0$ and $T \ll T^*$, albeit much smaller than that of the oscillating part. On the other hand, in the incoherent or high-temperature case, (3.234) still contains an oscillatory term with frequency $(\Delta/\hbar)$; its amplitude $\tilde{u}^2$, however, is negligible for $T \gg T^*$.

In summary, the three-pole structure of (3.233) provides a smooth crossover from weakly damped oscillations to relaxational motion. This crossover is driven by the relaxation amplitude which increases continuously from zero at $T = 0$ to unity for $T \gg T^*$. This picture is different from that drawn in Sect. 3.4 on p. 85. There the change in the two-state dynamics occurs sharply at temperature T^*, where the two complex poles meet on the imaginary axis.

3.7.7 How Large is the Maximum Tunnel Energy in Glasses?

Starting from a bare tunnel energy of a few Kelvin, the discussion of p. 108 provides an estimate for an upper bound for the dressed value Δ_0. While we found at least qualitative agreement for defects in Bi and impurities in crystals, a more serious discrepancy arises for tunneling in oxide glasses. Here we briefly discuss two relevant experiments in view of phonon dressing.

Two-Pulse Echo Decay in Vitreous Silica. Nonlinear response functions probe a well-defined subensemble of TS's and thus allow a thorough comparison with theory. As an example we show data of Hunklinger and Arnold (1976) for the decay of two-pulse echoes. The resonance condition singles out defects with an energy splitting $E = \hbar\omega$, where $\omega/2\pi = 760$ MHz is the oscillation frequency of the applied elastic waves. Since such an experiment probes mainly systems with small asymmetry energy, we have $\omega \approx \Delta_0$, resulting in $\Delta_0/k_B \approx 35$ mK. In Fig. 3.7 we plot the observed relaxation rate Γ_1 as a function of temperature. The solid line is calculated from Γ_1 in (3.230). The coupling strength $\tilde{\alpha}$ involves the elastic deformation potential as the only free parameter; for the fit we have used $\gamma = 2.6$ eV. [Hunklinger and Arnold (1976) derived a slightly different value $\gamma = 3$ eV.]

Unfortunately, the data cover only the temperature range where the rate is dominated by the direct, or one-phonon, process. They do not extend to temperatures where, according to (3.230), we expect an exponential increase of the rate. Nonetheless, this experiment permits us to draw a conclusion with respect to the phonon-dressing effect in glasses.

As mentioned above, (3.43) with T_0 and Δ_0 from Table 3.1 indicates an effective Debye temperature of at most 60 K. On the other hand, the Debye model for a harmonic crystal gives, according to (3.21), a value of 461 K. An even worse discrepancy is encountered when analyzing low-temperature sound-velocity data in view of phonon dressing.

Sound Velocity. Most dynamic experiments on glasses involve *linear* response functions with respect to a time-dependent elastic or electric field. Since they require an average over the broad distribution for the parameters of the two-state systems, the observed acoustic and dielectric properties are less specific with respect to the dynamics of a single defect.

The interaction of elastic waves with TS's results in a relative change of the sound velocity,

$$\frac{\delta v}{v} = -\frac{1}{V}\sum_{TS}\frac{\gamma^2}{2\varrho v^2 \hbar}\chi'(\omega), \qquad (3.240)$$

where $\chi(\omega) = \chi'(\omega) + i\chi''(\omega)$ is the complex elastic susceptibility of a single defect and the sum runs over all TS's; see Sect. 4.2. At low temperatures the relaxation rates γ_1 of thermal two-level systems are much smaller than the

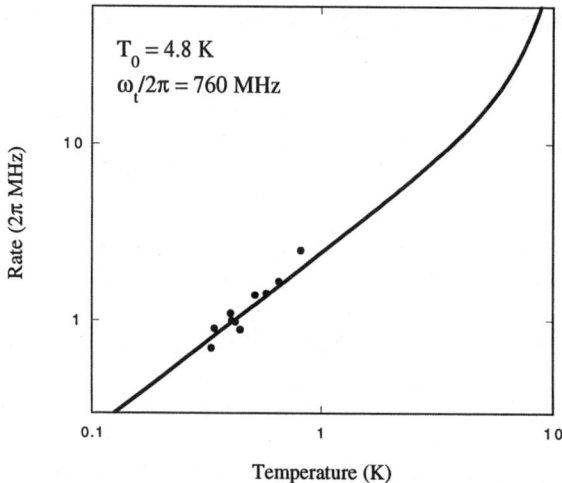

Fig. 3.7. Relaxation rate of two-state systems in vitreous silica, measured by two-pulse echoes. The main contribution stems from symmetric double-well potentials. Data are taken from Hunklinger and Arnold (1976); the solid line is given by Γ_1 in (3.230) with $T_0 = 4.8$ K.

3.7 Asymmetric Tunneling Systems

external frequency ω, i.e. $\omega\tau_1 \gg 1$; then only the 'resonant' part of the susceptibility, $\chi' = (\tilde{\Delta}_0^2/E^3)\tanh(E/2k_BT)$, is relevant. With the distribution function of the parameters, $P(\Delta, \Delta_0) = P_0/\Delta_0$, the sum in (3.240) gives

$$\frac{\delta v}{v} = \frac{\gamma^2 P_0}{\varrho v^2}\log(T/T_{\text{ref}}) \qquad (\Gamma_1 \ll \omega \ll k_BT/\hbar), \tag{3.241}$$

where P_0 is a constant, and T_{ref} a reference temperature. This logarithmic law has been first derived by Piché et al. (1974) and since then applied to the low-temperature sound velocity of various glasses.

We recall a few assumptions which are made implicitly when deriving (3.241).

(i) Weak Damping. The logarithmic law ceases to be valid as soon as relaxation sets in, i.e. when $\Gamma_1 \approx \omega$. This requires sufficiently low temperatures, $T < T_0$, since for $T > T_0$ the strong increase of the relaxation rate leads to $\Gamma_1 \gg \omega$ and thus invalidates (3.241). Depending on the applied frequency, the logarithmic variation (3.241) has been reported up to several K (cf. Phillips (1981a)), which is consistent with the requirement $T < T_0$. According to our theory, the law (3.241) ceases to be valid as temperature approaches T_0.

(ii) Large Tunnel Energy. The logarithmic law arises from thermal two-level systems with small bias Δ, i.e. with a tunnel energy close to temperature. Since it has been observed up to several K, the maximum tunnel energy $\Delta_0^{\max}/k_B$ can hardly be smaller than 5 K.

Yet the argument on p. 108 provides an upper bound of about 20 K for the bare tunnel frequency Δ_b. Hence we find that the reduction factor due to phonon dressing, $B_0 = \Delta_0/\Delta_b$, cannot be significantly smaller than unity, say $B_0 \geq \frac{1}{4}$. Inserting $T_0 = 4.8$ K in (3.43), we conclude that the effective Debye temperature Θ can hardly exceed 40 K, which is by one order of magnitude smaller than the value of Table 3.1.

According to (3.21), the value of Θ is a measure for the number of extended vibrational modes $\sum_{k,s} = 3N$. If we try to explain the above discrepancy in such a way, we are led to conclude (i) that only a small fraction of the degrees of freedom are harmonic vibrations and (ii) that the remaining modes do not contribute to the polaron effect. Indeed, it has been postulated previously that there are no high-frequency phonons in glasses. Yu and Leggett (1988) suggested the existence of diffusive modes, whereas Buchenau et al. (1992) put forward arguments in favor of anharmonic local oscillators, see Chap. 9.

It turns out that the question about the maximum tunnel energy in glasses is intimately connected to the value of Θ or, in more physical terms, to the number of phonon modes. In order to settle this question it would seem most desirable to determine the range of validity of the tunneling model, i.e. to have a more precise bound for the maximum tunnel frequency $\Delta_0^{\max}$.

3.7.8 Sound Propagation in Amorphous Solids Above 5 K

Acoustic waves in crystals travel at the same velocity, independent of frequency. This is different in amorphous solids, where relaxation of two-state systems leads to an intricate dependence on frequency and temperature. Jäckle (1972) and Piché et al. (1974) have shown that the tunneling model and weak coupling to phonons provide a good description for the logarithmic laws observed for the sound velocity and for the plateau occurring in the attenuation below 3 K. See Sect. 4.4.

Temperature Dependence. Sound propagation at higher temperatures, say above 5 K, however, is poorly understood. So far, neither the absorption peak at about 50 K nor the linear sound velocity have been given a satisfying explanation. Tielbürger et al. (1992) have taken into account thermally activated relaxation, besides the one-phonon damping rate; with an appropriate distribution of activation energies, these authors have been able to fit both absorption and sound velocity data.

Applying a mode-coupling approximation, Neu and Würger (1994a) and Rau et al. (1995a) derived a crossover to incoherent tunneling at about 5 K. Besides the formal drawbacks discussed in Sect. 3.8, this theory suffers from the uncertainties concerning high-frequency phonons, discussed on p. 110. Any modification in the phonon spectrum at frequencies corresponding to 20 or 30 K would significantly change the damping at temperatures of about 10 K.

Both approaches, thermal activation and incoherent tunneling, give a strong increase of sound attenuation above 5 K and a linear variation of the sound velocity,

$$\frac{\delta v}{v} = C \frac{T}{T_*} \log(\omega \tau_0), \qquad (3.242)$$

where $C = P_0 \gamma^2 / \rho v^2$, and T_* and τ_0 depend on the respective model parameters. There is no doubt that thermal activated relaxation will prevail at higher temperature. On the other hand, incoherent tunneling accounts, without adjustable parameters, for the qualitative change that occurs at 5 K.

Frequency Dependence. The linear law in T for the sound velocity has already been discussed by Bellessa (1978) in view of experimental data on vitreous silica and amorphous Pd-Si; cf. Sects. 4.4.1 and 4.4.2. Both compounds show a linear temperature dependence according to (3.242), like other oxide glasses (cf. Rau et al. (1995a)) and polycrystalline metals (cf. Gaganidze et al. (1995), see Sect. 4.5). Yet regarding the variation with frequency, Bellessa's data do not agree with (3.242). Although the sound velocity shows a logarithmic frequency dependence, the prefactor is not proportional to T. It seems rather to be constant for amorphous Pd-Si, and to depend very weakly on T for vitreous silica. Bellessa (1978) concludes that the data cannot be explained by thermal activation only.

3.7 Asymmetric Tunneling Systems

We do not pretend to solve this puzzle here, but we intend to show that phonon-assisted tunneling is likely to be responsible for the discrepancy between (3.242) and experimental data, with respect to the dependence on ω. (On this point, cf. the discussion in Sect. 4.4.2.) We start with the observation that for temperatures well above 5 K, 'resonant' processes are not significant. Moreover, sound propagation is determined by asymmetric defects, and we may simplify the spectral function of (3.239) as $\delta G''_{zz}(\omega) = \cosh(\Delta/2k_\mathrm{B}T)^{-2}\Gamma_1/(\omega^2 + \Gamma_1^2)$. Applying the fluctuation-dissipation theorem $\chi''(\omega) = 2\tanh(\frac{1}{2}\beta\hbar\omega)\delta G''_{zz}(\omega)$ and the Kramers–Kronig-relation, we obtain the complex susceptibility

$$\chi(\omega) = \frac{\hbar}{k_\mathrm{B}T}\frac{1}{\cosh(\Delta/2k_\mathrm{B}T)^2}\frac{i\Gamma_1}{\omega + i\Gamma_1}, \qquad (3.243)$$

where the relaxation rate for thermal systems is given by (3.230) or (3.87).

Calculation of the sound velocity requires us to perform the average with respect to $P(\Delta_0) = P_0/\Delta_0$, with P_0 constant. The distribution for Δ_0 results in a broad relaxation spectrum whose bounds are supposed to fulfil $\Gamma_\mathrm{min} \ll \omega \ll \Gamma_\mathrm{max}$. Since χ factorizes, the integrals over Δ and Δ_0 may be done separately. After substituting $r = (\Delta_0/\Delta_0^\mathrm{max})^2$ and defining

$$I_\Delta = \frac{1}{2k_\mathrm{B}T}\int_0^{\Delta_\mathrm{max}} d\Delta \cosh(\Delta/2k_\mathrm{B}T)^{-2} \qquad (3.244)$$

with the maximum bias Δ_max, one obtains the relaxation contribution to the sound velocity,

$$\frac{\delta v}{v} = \frac{1}{2}CI_\Delta \log(\omega/\Gamma_\mathrm{max}). \qquad (3.245)$$

The logarithmic dependence on frequency is characteristic of a broad relaxation spectrum.

Random Bias. A more remarkable feature, however, arises when performing the integral (3.244) for a broad bias distribution, $\Delta_\mathrm{max} \gg k_\mathrm{B}T$, as it is usually assumed for glasses. In this case, $I_\Delta = 1$, and we find a variation of the sound velocity,

$$\frac{\delta v}{v} = \frac{1}{2}C\log(\omega/\Gamma_\mathrm{max}), \qquad (3.246)$$

which depends logarithmically on frequency and whose prefactor is constant with respect to temperature. Thus phonon-assisted tunneling provides exactly the additional term postulated by Bellessa (1978), in order to describe the data for a-SiO$_2$ and a-Pd-Si.

The maximum rate occurring in (3.246) is not a constant, but strongly depends on T. Well above T^*, it is given by the result from saddle-point integration,

$$\Gamma_\mathrm{max} = \Gamma_\mathrm{SPI}, \qquad (3.247)$$

for large Debye temperature, $T \ll \Theta$. According to the above discussion, this condition is not always well justified in glasses. As temperature approaches the value Θ, the exponential increase of (3.247) turns over to activated behavior, $\Gamma_{\max} \propto e^{-V/T}$, with $V = \Theta^3/12\pi^2 T_0^2$. [For a more detailed discussion, cf. pp. 166–168 of Würger (1997a); yet note that a factor of $\frac{1}{2}$ is missing in the definition for V.] It is important that this change in $\Gamma_{\max}$ does not affect the law (3.246). As to the internal friction, we merely mention that the imaginary part of (3.243) yields the usual result with the plateau value $\pi C/2$.

Small Bias. Now we turn to the opposite case of a narrow bias distribution, $\Delta_{\max} \ll k_B T$, as it could be realized for tunneling defects in a crystalline environment. Assuming $\Gamma_{\min} \ll \omega \ll \Gamma_{\max}$ and $T_0 < T \ll \Theta$, and using the integral $I_\Delta = (\Delta_{\max}/2k_B T)$ and the expression (3.247), we obtain

$$\frac{\delta v}{v} = C \frac{\Delta_{\max}}{4 k_B T} \log(\hbar\omega/\pi k_B T) - C \frac{1}{3} \frac{\Delta_{\max}}{k_B T_0^2} T. \tag{3.248}$$

Thus for a narrow bias distribution, phonon-assisted tunneling results in quite a complicated dependence on T and ω. The first term varying with $\log \omega$ is typical for relaxation; its prefactor depends inversely on T.

More remarkably, the second term results in a linear decrease of the sound velocity with temperature. It arises from the prefactor $1/T$ carried by the relaxational susceptibility, multiplied by $\log \Gamma_{\max}$ which is quadratic in T.

In summary, there are several open questions concerning sound propagation in glasses above 5 K. A thermally activated relaxation mechanism correctly describes the T dependence of sound velocity and attenuation as observed for various oxide glasses, albeit with a number of free parameters.

On the other hand, both the mode-coupling approach and the present strong-coupling theory indicate that the crossover to incoherent tunneling is responsible for the qualitative change in the sound propagation at about 5 K. The crossover temperature depends mainly on the elastic deformation potential, without further adjustable parameters. This agreement supports the relevance of incoherent motion, despite the flaws concerning the phonon spectrum.

Because of the different temperature dependence of the rate, the present approach gives results that are different from those obtained from the mode-coupling approximation. Two surprising features arising from phonon-assisted incoherent tunneling are to be mentioned in particular.
(i) For the distribution of the tunneling model, we find that the sound velocity varies logarithmically with frequency, with a temperature independent prefactor, as observed by Bellessa (1978). To our knowledge, (3.246) provides the only theoretical explanation for Bellessa's data. In order to account for the linear temperature dependence, one would be led to assume that some TS relax according to the phonon-assisted rate, whereas activated barrier crossing provides the most efficient mechanism for others.
(ii) A narrow bias distribution, on the other hand, results in a linear decrease

with temperature; such a behavior has been found for various compounds. Equation (3.248) does *not* apply to TS's in oxide glasses, since there is strong evidence for a broad distribution of asymmetry energies, and since probably T_0 is not much smaller than the effective Debye temperature. Yet that expression might well be relevant for TS's in mixed crystals or alloys.

3.8 Two-State Dynamics for Weak Phonon Coupling

In this section we consider the symmetric two-state system, i.e. the Hamiltonian (3.2) with zero asymmetry energy Δ,

$$H = \frac{1}{2}\Delta_b \sigma_x + \sum_k \hbar\omega_k b_k^\dagger b_k + \frac{1}{2}\hbar f \sigma_z \equiv H_S + H_B + H_1, \qquad (3.249)$$

where the elastic coupling energy $\hbar f$ is given by

$$f = \sum_k \lambda_k (b_k + b_k^\dagger). \qquad (3.250)$$

For sufficiently weak phonon coupling, a perturbation expansion in terms of f provides a well-defined approximation scheme; cf. Würger (1997c).

Time evolution is written in terms of the quantum Liouville operator $\mathcal{L}$, whose action on a pseudospin σ_α is given by the von Neumann equation

$$\dot{\sigma}_\alpha = (\mathrm{i}/\hbar)[H, \sigma_\alpha] \equiv \mathrm{i}\mathcal{L}\sigma_\alpha. \qquad (3.251)$$

Formal integration yields

$$\sigma_\alpha(t) = \mathrm{e}^{\mathrm{i}\mathcal{L}(t-t')}\sigma_\alpha(t'). \qquad (3.252)$$

For later convenience we note the equation of motion explicitly

$$\dot{\sigma}_x = -f\sigma_y, \quad \dot{\sigma}_y = f\sigma_x - (\Delta_b/\hbar)\sigma_z, \quad \dot{\sigma}_z = (\Delta_b/\hbar)\sigma_y. \qquad (3.253)$$

The dynamics of the two-state system is described most conveniently in terms of a reduced time evolution operator

$$\mathcal{U}(t) = \langle \mathrm{e}^{\mathrm{i}\mathcal{L}t}\rangle_B. \qquad (3.254)$$

The subscript B indicates a partial trace over the bath,

$$\langle \ldots \rangle_B = \mathrm{tr}_B(\rho_B \ldots). \qquad (3.255)$$

Accordingly, the time-dependent spin polarization reads as

$$\langle \sigma_\alpha(t)\rangle_S = \langle \mathcal{U}(t)\sigma_\alpha\rangle_S, \qquad (3.256)$$

with $\langle \ldots \rangle_S = \mathrm{tr}_S(\ldots)$ indicating the partial trace over the two-state variables.

The dissipative two-state dynamics is entirely determined by the time evolution of σ_z and σ_x. Following Leggett et al. (1987), we define

$$P(t) = \langle \sigma_z(t)\rangle \qquad (3.257)$$

as the time-dependent expectation value of the reduced two-state coordinate σ_z, with the initial value $P(t=0) = 1$. For zero phonon coupling, $\lambda_k = 0$, it shows coherent oscillations with the tunnel frequency $\Delta_b/\hbar$, $P(t) = \cos(\Delta_b t/\hbar)$. Two effects are expected to arise from phonon coupling: (i) phonon dressing reduces the tunnel energy to an effective value $\tilde{\Delta}_0$, and (ii) thermal bath fluctuations result in a loss of phase coherence in terms of an exponential damping factor, $P(t) = \cos(\tilde{\Delta}_0 t/\hbar)e^{-\Gamma_t t}$.

The second quantity of interest is given by

$$R(t) = \langle \sigma_x(t) \rangle. \tag{3.258}$$

In the case of zero phonon coupling, we have $R(t) = 0$ for all times, since σ_x is diagonal in the energy eigenstates of the uncoupled system. Taking into account the interaction with phonons results in a finite lifetime of the spin states. Then the average $R(t)$ provides two relevant quantities: In the long-time limit, it tends towards the equilibrium spin polarization, $R(t \to \infty) = \langle \sigma_x \rangle_{\text{eq}}$. The corresponding relaxation time determines the lifetime of the spin states.

3.8.1 Perturbation Series

The operator $\mathcal{U}(t)$ can be represented as a 4×4 matrix, which acts on the space spanned by the identity operator σ_0 and the Pauli matrices σ_x, σ_y, σ_z,

$$\mathcal{U}_{ij}(t) = \tfrac{1}{2}\text{tr}(\sigma_i \mathcal{U}(t)\sigma_j), \tag{3.259}$$

with $i, j = 0, x, y, z$. The element σ_0 is needed in order to obtain a closed algebra with respect to multiplication, $\sigma_i \sigma_0 = \sigma_i$ and $\sigma_i^2 = \sigma_0$; note $\text{tr}(\sigma_0) = 2$.

According to (3.249) the Liouville operator consists of three parts $\mathcal{L} = \mathcal{L}_S + \mathcal{L}_B + \mathcal{L}_1$, which are defined by $\hbar \mathcal{L}_S A = [H_S, A]$, etc. Here A is an arbitrary composite operator which in general involves both spin and bath degrees of freedom. The perturbation series for $\mathcal{U}(t)$ is set up by putting $\mathcal{L}_0 = \mathcal{L}_S + \mathcal{L}_B$, expanding the time-evolution operator in terms of $\mathcal{L}_1$, and performing the partial trace over bath coordinates.

In view of (3.259), we need to represent all quantities appearing in the perturbation series as matrices. The unperturbed time evolution of spin operators is given by

$$\check{\mathcal{U}}_{ij}(t) = \tfrac{1}{2}\text{tr}(\sigma_i e^{i\mathcal{L}_S t}\sigma_j); \tag{3.260}$$

integrating the equation of motion (3.253) for $f = 0$, and noting $\dot{\sigma}_0 = 0$, one finds in a straightforward fashion

$$\check{\mathcal{U}}(t) = \begin{pmatrix} 1 & 0 & 0 & 0 \\ 0 & 1 & 0 & 0 \\ 0 & 0 & \cos(\Delta_b t/\hbar) & \sin(\Delta_b t/\hbar) \\ 0 & 0 & -\sin(\Delta_b t/\hbar) & \cos(\Delta_b t/\hbar) \end{pmatrix}. \tag{3.261}$$

The time evolution of the bath operators is given by (3.3).

3.8 Two-State Dynamics for Weak Phonon Coupling

The coupling part of the Liouville operator, $\mathcal{L}_1$, involves composite operators whose spin parts develop according to (3.261) and bath parts according to (3.3). Yet each factor $\mathcal{L}_1$ acting as a commutator with the whole object to its right, gives rise to a subtlety with respect to the time ordering of the bath operators; cf. Würger (1998a).

Here we merely quote the result of a partial resummation in terms of bath correlation functions, leading to an integral equation for the time evolution operator (3.259),

$$\mathcal{U}_{ij}(t) = \check{\mathcal{U}}_{ij}(t) - \int_0^t d\tau \int_0^\tau d\tau' \check{\mathcal{U}}_{ik}(t-\tau)\Sigma_{kl}(\tau-\tau')\mathcal{U}_{lj}(\tau'). \tag{3.262}$$

The self-energy is given as a series

$$\Sigma_{ij}(t) = \Sigma_{ij}^{(1)}(t) + \Sigma_{ij}^{(2)}(t) + \ldots \tag{3.263}$$

whose first two terms read as

$$\Sigma_{ij}^{(1)}(t-t') = \Lambda_{ik}\check{\mathcal{U}}_{kl}(t-t')\Lambda_{lj}\phi_{kj}(t,t'), \tag{3.264}$$

$$\begin{aligned}\Sigma_{ij}^{(2)}(t-t') &= -\int_{t'}^t d\tau \int_{t'}^\tau d\tau' \Lambda_{ik}\check{\mathcal{U}}_{kl}(t-\tau)\Lambda_{lm}\check{\mathcal{U}}_{mn}(\tau-\tau') \\ &\quad \times \Lambda_{np}\check{\mathcal{U}}_{pq}(\tau'-t')\Lambda_{qj}\phi_{kmpj}(t,\tau,\tau',t').\end{aligned} \tag{3.265}$$

Besides the free spin propagator $\check{\mathcal{U}}$, the self-energy involves the matrix

$$\Lambda = \begin{pmatrix} 0 & 0 & 0 & 1 \\ 0 & 0 & -i & 0 \\ 0 & i & 0 & 0 \\ 1 & 0 & 0 & 0 \end{pmatrix}. \tag{3.266}$$

The $2n$-point bath correlation functions $\phi_{kj}(t,t')$, $\phi_{kmpj}(t,\tau,\tau',t')$, etc. can be expressed in terms of response and correlation function with respect to the strain field,

$$\chi(t-t') = \tfrac{1}{2}\langle [f(t), f(t')]\rangle, \tag{3.267}$$

$$\psi(t-t') = \tfrac{1}{2}\langle \{f(t), f(t')\}\rangle. \tag{3.268}$$

There are eight finite functions of second order,

$$\phi_{0x}(t,t') = \phi_{0y}(t,t') = \phi_{zx}(t,t') = \phi_{zy}(t,t') \equiv \chi(t-t'), \tag{3.269}$$

$$\phi_{xx}(t,t') = \phi_{xy}(t,t') = \phi_{yy}(t,t') = \phi_{yx}(t,t') \equiv \psi(t-t'). \tag{3.270}$$

As to the fourth-order correlations, we note explicitly those which will be needed later,

$$\begin{aligned}\phi_{xyxy}(t,\tau,\tau',t') &= \phi_{yxyx}(t,\tau,\tau',t') \\ &= \psi(t-t')\psi(\tau-\tau') + \psi(t-\tau')\psi(\tau-t'),\end{aligned} \tag{3.271}$$

$$\phi_{zxyx}(t,\tau,\tau',t') = \chi(t-t')\psi(\tau-\tau') + \chi(t-\tau')\psi(\tau-t'), \tag{3.272}$$

$$\phi_{0zxy}(t,\tau,\tau',t') = \chi(t-t')\chi(\tau-\tau') + \chi(t-\tau')\chi(\tau-t'). \tag{3.273}$$

For further use we write down the phonon propagators

$$\psi(t) = \sum_k \lambda_k^2 (1 + 2n_k) \cos(\omega_k t), \tag{3.274}$$

$$\chi(t) = -i \sum_k \lambda_k^2 \sin(\omega_k t), \tag{3.275}$$

where the Bose occupation numbers read as $n_k = [e^{\beta \hbar \omega_k} - 1]^{-1}$. We mention two general properties of the self-energy matrix Σ, which will considerably simplify the evaluation of the perturbation series (3.263).

First, from the structure of the bath correlations it follows that, in any order of the perturbation expansion, the first and fourth columns of the matrix $\Sigma_{ij}(t)$ vanish,

$$\Sigma_{i0}(t) = 0 = \Sigma_{iz}(t) \quad \text{for } i = 0, x, y, z. \tag{3.276}$$

[Thus Σ is not symmetric; when choosing $\text{tr}_B(\cdots \rho_B)$ instead of (3.255), we would find the adjoint matrix $\Sigma^\dagger$.]

Second, for zero bias the Hamiltonian (3.2) is invariant under the canonical transformation

$$\sigma_z \to -\sigma_z, \quad \sigma_y \to -\sigma_y, \quad b_k \to -b_k. \tag{3.277}$$

As a consequence any correlation function involving an odd number of these operators vanishes; in terms of the self-energy this condition requires

$$\Sigma_{ij}(t) = 0 = \Sigma_{ji}(t) \quad \text{for } i = 0, z \text{ and } j = x, y. \tag{3.278}$$

Accordingly, the self-energy matrix splits in two 2×2 blocks; taking into account both (3.276) and (3.278), we have

$$\Sigma(t) = \begin{pmatrix} 0 & \Sigma_{0x}(t) & 0 & 0 \\ 0 & \Sigma_{xx}(t) & 0 & 0 \\ 0 & 0 & \Sigma_{yy}(t) & 0 \\ 0 & 0 & \Sigma_{zy}(t) & 0 \end{pmatrix}. \tag{3.279}$$

Thus solving the two-state dynamics has been reduced to calculating the four entries of (3.279). The first two terms of the perturbation series, (3.264) and (3.265), have been evaluated in Würger (1998a); here we merely quote the main results.

In order to evaluate the propagator $\mathcal{U}$, we need to take the Laplace transform and apply a Markov approximation. Applying (A3) to (3.262), using the convolution theorem, and solving for $\mathcal{U}(z)$, we obtain

$$\mathcal{U}(z) = \frac{-1}{-\check{\mathcal{U}}(z)^{-1} + \Sigma(z)}, \tag{3.280}$$

where the uncoupled spin dynamics is given by

$$-\check{\mathcal{U}}(z)^{-1} = \begin{pmatrix} z & 0 & 0 & 0 \\ 0 & z & 0 & 0 \\ 0 & 0 & z & -i\Delta_b/\hbar \\ 0 & 0 & i\Delta_b/\hbar & z \end{pmatrix}. \tag{3.281}$$

From the determinant

$$\det(\breve{\mathcal{U}}(z)^{-1}) = -z^2(z^2 - (\Delta_b/\hbar)^2),\qquad(3.282)$$

one finds that the uncoupled propagator exhibits a double pole at $z = 0$ and a pair of poles at $z = \pm\Delta_b$.

The entries of the self-energy matrix are smooth functions of frequency. In order to calculate the resonance energies of $\mathcal{U}(z)$, we expand the self-energy $\Sigma(z)$ in a power series about the real poles z_0,

$$\Sigma(z) = \Sigma(z_0) + (z - z_0)\left[\frac{d\Sigma(z)}{dz}\right]_{z=z_0} + \cdots,\qquad(3.283)$$

and we truncate after the linear term. After inserting this approximate expression in (3.280), one finds that the determinant of $\mathcal{U}(z)^{-1}$ is given by a fourth-order polynomial in z. Each root of this polynomial provides a resonance, which acquires an imaginary part after evaluating the dissipative terms of Σ at the corresponding frequency.

Since both $\breve{\mathcal{U}}$ and Σ are block-diagonal, the propagator may be written as

$$\mathcal{U}(z) = \begin{pmatrix} \mathcal{V}(z) & 0 \\ 0 & \mathcal{W}(z) \end{pmatrix},\qquad(3.284)$$

with 2×2 matrices $\mathcal{V}$ and $\mathcal{W}$, where $\mathcal{V}$ acts on the subspace labelled by 0 and x, and $\mathcal{W}$ on that spanned by y and z. This particular form of the propagator follows directly from (3.260) and the invariance under the canonical transformation (3.277). Owing to the fact that $\mathcal{U}(z)$ is block-diagonal, the submatrices $\mathcal{V}$ and $\mathcal{W}$ may be dealt with separately. We start with $\mathcal{W}$.

3.8.2 Phase Relaxation: $\mathcal{W}(z)$

The submatrix $\mathcal{W}$ describes tunneling oscillations with the bare frequency Δ_b. Phonon coupling affects this motion in two respects. First, it results in a renormalization of the tunnel energy and, second, it destroys the phase coherence. These two effects will be accounted for by a reduced tunnel energy $\tilde{\Delta}_0$ and a damping rate Γ_t.

After inserting the bath correlations (3.269)–(3.270), one easily finds that there is a single first-order term,

$$\Sigma^{(1)}_{yy}(t) = \breve{\mathcal{U}}_{xx}(t)\psi(t);\qquad(3.285)$$

in second order, both $\Sigma^{(2)}_{yy}$ and $\Sigma^{(2)}_{zy}$ contribute (Würger (1998a)).

The reduced tunnel energy is calculated by expanding the reactive part of the self-energy in powers of z, and truncating after the linear term. The derivatives

$$\partial_z \Sigma^{(1)}_{yy} = \sum_k \lambda_k^2 (1 + 2n_k)\omega_k^{-2} \equiv \delta_1\qquad(3.286)$$

and $\partial_z \Sigma^{(2)}_{yy}(z) \equiv \delta_2$ define the $\mathcal{Z}$-factor $\mathcal{Z} = [1 + \delta_1 + \delta_2]^{-1}$. Since we are going to truncate the perturbation series for the rate at second order in the coupling constant $\tilde{\alpha}$, we need to consider only the first-order correction in the $\mathcal{Z}$-factor,

$$\mathcal{Z} = [1 + \delta_1]^{-1}. \tag{3.287}$$

When inserting the spectral density in (3.286), we find a constant and a temperature-dependent part,

$$\delta_1 = \tilde{\alpha}\Theta^2 + \tfrac{2}{3}\pi^2\tilde{\alpha}T^2. \tag{3.288}$$

This expression for δ_1 is valid for $T \ll \Theta$ only.

In order to account for damping we evaluate the dissipative part of Σ at the effective tunnel frequency $\tilde{\Delta}_0/\hbar$. The characteristic equation

$$\det\left(\mathcal{W}(z)^{-1}\right) = \mathcal{Z}^{-1}\left[z^2 - (\tilde{\Delta}_0/\hbar)^2 + iz\mathcal{Z}\Gamma(z)\right] \tag{3.289}$$

involves the renormalized tunnel energy

$$\tilde{\Delta}_0^2 = \mathcal{Z}\Delta_b^2 \tag{3.290}$$

and the dissipation kernel

$$\Gamma(z) = \Im\Sigma_{yy}(z) + (\Delta_b/\hbar z)\Re\Sigma_{zy}(z), \tag{3.291}$$

which is a symmetric function of z. Solving the quadratic equation (3.289) and taking $\Gamma(z)$ at $z = \tilde{\Delta}_0/\hbar$ leads to the transverse rate

$$\Gamma_t = \tfrac{1}{2}\mathcal{Z}\Gamma(\tilde{\Delta}_0/\hbar), \tag{3.292}$$

and to the effective resonance frequency

$$\omega_t^2 = (\tilde{\Delta}_0/\hbar)^2 - \tfrac{1}{4}\mathcal{Z}^2\left[\Im\Sigma_{yy}(\tilde{\Delta}_0/\hbar)\right]^2. \tag{3.293}$$

After matrix inversion and dropping an insignificant term Σ_{zy} in the residue, we find

$$\mathcal{W}(z) = -\frac{1}{(z+i\Gamma_t)^2 - \omega_t^2}\begin{pmatrix} \mathcal{Z}z & -i\mathcal{Z}\Delta_b/\hbar \\ i\mathcal{Z}\Delta_b/\hbar & z + 2i\Gamma_t \end{pmatrix}. \tag{3.294}$$

In the limit of very small coupling the factor $\mathcal{Z}$ is unity; the rate is given by the first-order term

$$\Gamma_t = \tfrac{1}{2}\pi\alpha(\Delta_b/\hbar)^3 \coth(\Delta_b/2k_BT) \quad \text{for } \delta_1 \ll 1. \tag{3.295}$$

The terms of higher order are relevant at higher temperatures $k_BT \gg \Delta_b$, where the coth function may be replaced by its inverse argument, $\coth(x) \approx x^{-1}$ for $x \ll 1$. From (3.292) we obtain

$$\Gamma_t = \frac{\pi}{2}\alpha(\tilde{\Delta}_0/\hbar)^3 \coth(\tilde{\Delta}_0/2k_BT)\left[1 + \tfrac{4}{3}\pi^2\tilde{\alpha}T^2 + O(\tilde{\alpha}^2T^4)\right]. \tag{3.296}$$

There are two significant features. First, the rate Γ_t involves the reduced tunnel energy $\tilde{\Delta}_0$ and thus carries a reduction factor $\mathcal{Z}$. Second, the linear and quadratic terms in brackets differ by a factor $\tfrac{4}{3}\pi^2\tilde{\alpha}T^2$. This indicates that the perturbation series (3.263) corresponds to a power series in terms of the dimensionless parameter $\tilde{\alpha}T^2$, besides the temperature dependence of the reduced tunnel energy.

The propagator $\mathcal{W}(t)$ is obtained as the inverse Laplace transform of (3.294). When using (3.256) and (3.260), we find the thermal average of the reduced coordinate (3.257) to be given by the lower diagonal element of $\mathcal{U}(t)$; assuming $\Gamma_t < \tilde{\Delta}_0$, we have

$$P(t) = e^{-\Gamma_t t} \cos(\omega_t t + \delta)/\cos(\delta), \qquad (3.297)$$

with $\tan(\delta) = \Gamma_t/\omega_t$.

3.8.3 Energy Relaxation: $\mathcal{V}(z)$

For zero phonon coupling, $\mathcal{V}$ has two undamped poles at zero frequency. Already to first order in $\tilde{\alpha}$, both relevant entries of Σ give finite contributions,

$$\Sigma^{(1)}_{0x}(t) = i\breve{\mathcal{U}}_{zy}(t)\chi(t), \qquad \Sigma^{(1)}_{xx}(t) = \breve{\mathcal{U}}_{yy}(t)\psi(t). \qquad (3.298)$$

When expanding $\Sigma_{xx}(z)$ and $\Sigma_{0x}(z)$ about $z_0 = 0$ and inverting the 2×2 matrix $\mathcal{V}(z)$, one of the poles requires a finite damping rate,

$$\Gamma_1 = \mathcal{Z}\Sigma_{xx}(z=0), \qquad (3.299)$$

where the $\mathcal{Z}$-factor obtained from

$$1 + \partial_z \Sigma_{xx} = \mathcal{Z}^{-1} \qquad (3.300)$$

is essentially identical to (3.287).

The discussion of the relaxation part of the propagator, $\mathcal{V}$, will be restricted to the first order with respect to the coupling parameter $\tilde{\alpha}$. Separating the two poles and defining $s_0 = \tanh(\tfrac{1}{2}\beta\Delta_b)$, we find

$$\mathcal{V}(z) = -\frac{1}{z}\begin{pmatrix} 1 & -s_0 \\ 0 & 0 \end{pmatrix} - \frac{1}{z + i\Gamma_1}\begin{pmatrix} 0 & s_0 \\ 0 & \mathcal{Z} \end{pmatrix}. \qquad (3.301)$$

Inserting the phonon spectral density, we find the longitudinal rate

$$\Gamma_1 = \mathcal{Z}\pi\alpha(\Delta_b/\hbar)^3 \coth(\Delta_b/2k_B T). \qquad (3.302)$$

Inverse Laplace transformation of (3.301) finally gives

$$\mathcal{V}(t) = \begin{pmatrix} 1 & -s_0 \\ 0 & 0 \end{pmatrix} + e^{-\Gamma_1 t}\begin{pmatrix} 0 & s_0 \\ 0 & \mathcal{Z} \end{pmatrix}. \qquad (3.303)$$

The first term accounts for the long-time limit, whereas the second one, which disappears for $t \to \infty$, describes the relaxation behavior of the two-state system.

Note that, for finite δ_1, we have $\mathcal{Z} < 1$ and the initial value $\mathcal{V}(t=0)$ is not equal to identity. As a consequence, the spectrum of the correlation function $C(t-t') = \langle \sigma_x(t)\sigma_x(t')\rangle$ does not satisfy a general sum rule, since its integral gives $\int d\omega C''(\omega) = \mathcal{Z}\pi$ instead of π. The missing spectral weight, $1 - \mathcal{Z}$, has been lost in the above pole approximation.

The equilibrium value of the spin polarization $\langle \sigma_i \rangle$ is given by the long-time limit of $\mathcal{U}(t)$ according to

$$\langle \sigma_i \rangle = \lim_{t \to \infty} \langle \sigma_0 \sigma_i(t) \rangle = \lim_{t \to \infty} \mathcal{U}_{0i}(t). \tag{3.304}$$

From (3.303) and (3.302) we find $\langle \sigma_0 \rangle = 1$, $\langle \sigma_x \rangle = -s_0$, and $\langle \sigma_y \rangle = 0 = \langle \sigma_z \rangle$, and thus the statistical operator in thermal equilibrium

$$\rho_S^{eq} = \tfrac{1}{2}[\sigma_0 - \sigma_x \tanh(\Delta_b/2k_B T)]. \tag{3.305}$$

3.8.4 Discussion

Validity of the Perturbative Approach. From the above results it is clear that truncation of the perturbation series at first or second order yields proper results as long as the quantity δ_1 is much smaller than unity; in terms of temperature T, Debye temperature Θ, and coupling constant $\tilde{\alpha}$, this condition reads

$$\tilde{\alpha} T^2 \ll 1 \tag{3.306}$$

$$\tilde{\alpha} \Theta^2 \ll 1. \tag{3.307}$$

Thus the perturbative regime is characterized by weak coupling and low temperature. The opposite case requires a strong-coupling approach as discussed in Sect. 3.4.

Phonon Dressing. According to (3.290), the reactive part of the self-energy reduces the tunnel energy from its bare value Δ_b by a factor $\sqrt{\mathcal{Z}}$. In physical terms, this corresponds to the phonon-dressing effect: The particle drags a phonon cloud when tunneling from one well to the other, which leads to a slowing-down of its motion.

The reduction factor $\mathcal{Z}$ comprises a constant and a temperature dependent part. At $T = 0$ only the former is relevant; when retaining the second-order term δ_1 only, the reduced tunnel energy reads as

$$\Delta_0^2 = \Delta_b^2 \left[1 + \sum_k (\lambda_k/\omega_k)^2\right]^{-1} \qquad (T = 0). \tag{3.308}$$

After inserting (3.288), we find $\Delta_0^2 = \Delta_b^2[1+\tilde{\alpha}\Theta^2]^{-1}$. At finite T, the thermal lattice motion results in a further reduction of the effective tunnel energy, $\tilde{\Delta}_0^2 = \Delta_b^2[1 + \delta_1]^{-1}$, with δ_1 as in (3.288).

As already mentioned in the introductory section below (3.3), we have identified the effective tunnel energy at $T = 0$ with the parameter Δ_0, used in the other chapters of this book and, quite generally, in the literature on TS's. This choice is motivated by the fact that the quantity observed in experiments at temperatures well below T_0 corresponds to our Δ_0. This is most obvious by noting that the linear term of (3.310) below, corresponds to the usual one-phonon rate.

3.8 Two-State Dynamics for Weak Phonon Coupling 133

Temperature Dependence of the Rate. Two competing effects give rise to the temperature dependence of the damping rate (3.296) which, for $\tilde{\Delta}_0 \ll k_B T$, reads as

$$\Gamma_t = \pi \alpha \tilde{\Delta}_0^2 k_B T / \hbar^3 \left[1 + \tfrac{4}{3}\pi^2 \tilde{\alpha} T^2 + O(\tilde{\alpha}^2 T^4)\right]. \tag{3.309}$$

The factor $\tilde{\Delta}_0$ decreases with rising T, whereas the higher-order terms in brackets enhance the rate. When expanding the $\mathcal{Z}$-factor in powers of $\tilde{\alpha}T^2$ and retaining the linear term only, we find $\tilde{\Delta}_0^2 = \Delta_0^2[1 - \tfrac{2}{3}\pi^2 \tilde{\alpha} T^2]$ and, after inserting this in (3.309),

$$\Gamma_t = \pi \alpha \Delta_0^2 k_B T / \hbar^3 \left[1 + \tfrac{2}{3}\pi^2 \tilde{\alpha} T^2 + O(\tilde{\alpha}^2 T^4)\right]. \tag{3.310}$$

Since Δ_0 is constant with respect to T, (3.310) proves that the second term of the perturbation series enhances the rate Γ_t.

The correction term in brackets depends on whether the first-order rate is calculated with the temperature-dependent tunnel energy $\tilde{\Delta}_0$ or with the constant Δ_0. In this chapter we have used $\tilde{\Delta}_0$ rather than the Δ_0; in particular, the discussion of multiphonon correction in Sect. 3.4 referred to (3.296) and (3.309).

3.8.5 Mode-Coupling Approximation (MCA)

Here we review the main features of a mode-coupling theory for the two-state system and compare the damping rate with the result of the above perturbative approach. For more details on the mode-coupling approximation we refer to the original work by Neu and Würger (1994a); the case of a finite asymmetry energy is treated in Rau et al. (1995a).

The mode-coupling approximation permits us to derive self-consistent equations for the spectra of the symmetrized correlation functions

$$C_i(t - t') = \tfrac{1}{2}\langle \delta\sigma_i(t)\delta\sigma_i(t') + \delta\sigma_i(t')\delta\sigma_i(t) \rangle \equiv (\delta\sigma_i(t)|\delta\sigma_i(t')) \tag{3.311}$$

of spin-fluctuation operators $\delta\sigma_i = \sigma_i - \langle \sigma_i \rangle$. The spectral function $C_i''(\omega)$ is defined as the Fourier transform of (3.311), according to (A1).

The symmetric correlation function $(A|B) := \tfrac{1}{2}\langle AB + BA \rangle$, cf. (3.311), provides an inner product on the space of quantum operators. Formally, the Laplace transform of (3.311) can be expressed as a resolvent matrix element

$$C_i(z) = (\delta\sigma_i | [\mathcal{L} - z]^{-1} | \delta\sigma_i). \tag{3.312}$$

According to Mori (1965), it can be reduced to a simple complex function by applying the so-called resolvent identity. For the longitudinal correlation function one obtains

$$C_x(z) = \frac{-\eta}{z + \Gamma_1(z)}, \tag{3.313}$$

where the memory kernel $\Gamma_1(z)$ depends in an intricate fashion on the dynamics of the pseudospins. For $\Delta_b \ll k_B T$, the factor $\eta = 1 - \langle \sigma_x \rangle^2$ may be replaced by unity.

For the transverse counterpart $C_z(z)$, Mori's reduction scheme has to be applied once more, resulting in

$$C_z(z) = \frac{-1}{z - \dfrac{(\Delta_b/\hbar)^2}{z + 2\Gamma_2(z)}}. \tag{3.314}$$

Both memory functions $\Gamma_1(z)$ and $\Gamma_2(z)$ are given by resolvent matrix elements of composite operators $f\sigma_i$, with the elastic strain field (3.250).

The mode-coupling approximation relies on decoupling two-time correlation functions of such composite operators into products of single-operator correlations, and on simplifying the projected time evolution, according to

$$\langle f\sigma_i e^{-i\mathcal{L}_{QQ} t} f\sigma_i \rangle \to \langle f(t)f \rangle \langle \sigma_i(t)\sigma_i \rangle. \tag{3.315}$$

Here, $\langle ... \rangle$ denotes the thermal average, and the dynamics on the left-hand side is governed by a projected Liouville operator $\mathcal{L}_{QQ}$. In the weak-coupling limit this approximation reduces to the first Born approximation; cf. Rau et al. (1995a).

After applying the factorization (3.315) and inserting the coupled phonon spectrum,

$$\psi''(\omega) = \frac{\pi}{2}\sum_k \lambda_k^2 (1 + 2n_k)[\delta(\omega - \omega_k) + \delta(\omega + \omega_k)], \tag{3.316}$$

one derives the longitudinal damping function

$$\Gamma_1''(\omega) = (\hbar/\Delta_b)^2 \left(\omega^2 C_z''(\omega) * \psi''(\omega)\right). \tag{3.317}$$

Putting $\eta = 1$, we obtain for the transverse kernel

$$2\Gamma_2''(\omega) = \left(C_x''(\omega) * \psi''(\omega)\right). \tag{3.318}$$

Equations (3.317) and (3.318) constitute a dynamic mean-field approximation to the correlated motion of spin and bath operators. Together with (3.313) and (3.314), they form a closed set of coupled nonlinear integral equations which – at least in principle – can be solved self-consistently. Neu and Würger (1994a) have proposed an approximate analytic solution that relies on the following simplifications of the mode-coupling integrals:
(i) The damping spectra are parameterized by an appropriate ansatz. A glance at the results from perturbation theory turns out to be instructive. For frequencies much smaller than $k_B T/\hbar$, the first-order term of the self-energy, increases with the square of frequency, whereas the second-order contribution does not depend significantly on frequency. Thus one is led to the ansatz

$$\Gamma_1''(\omega) = \gamma_1 + \kappa_1 \omega^2, \qquad \Gamma_2''(\omega) = \gamma_2 + \kappa_2 \omega^2, \tag{3.319}$$

whose frequency dependence turns out to be essential for consistent solutions.
(ii) The real parts of the memory kernels are neglected, i.e. we drop Γ' in the continued fractions for $C_i(z)$. When retaining the dissipative parts Γ'' only, and taking the imaginary part according to (A5), we obtain

3.8 Two-State Dynamics for Weak Phonon Coupling

$$C_x''(\omega) = \frac{\eta \Gamma_1''(\omega)}{\omega^2 + \Gamma_1''(\omega)^2}, \tag{3.320}$$

$$C_z''(\omega) = \frac{2\Gamma_2''(\omega)(\Delta_b/\hbar)^2}{(\omega^2 - (\Delta_b/\hbar)^2)^2 + 4\omega^2 \Gamma_2''(\omega)^2}. \tag{3.321}$$

These correlation spectra have been evaluated numerically by Neu and Würger (1994a).

(iii) In order to obtain an analytic solution and an explicit expression for the correlation function, we need to apply moreover a Markov approximation and replace the damping spectra by constants.

The analytic solution shows quite a different behavior in two ranges that may be classified as weak- and strong-coupling limits. Here we quote the results for the resolvent function

$$C_z(z) = -\frac{z + i2\Gamma_t}{(z + i\Gamma_t)^2 - \omega_t^2}, \tag{3.322}$$

in terms of the transverse damping rate Γ_t and the effective tunnel frequency

$$\omega_t = \sqrt{(\Delta_b/\hbar)^2 - \tfrac{1}{4}\Gamma_t^2}. \tag{3.323}$$

We discuss two limiting cases with respect to the parameter $\tilde{\alpha}T^2$, which correspond to weak and strong damping,

$$\begin{aligned}\tilde{\alpha}T^2 \ll 1 &: \quad \Gamma_t \ll \Delta_b/\hbar, \\ \tilde{\alpha}T^2 \gg 1 &: \quad \Gamma_t \gg \Delta_b/\hbar.\end{aligned} \tag{3.324}$$

(More precisely, there is a small range of parameters with $\tilde{\alpha}T^2 < 1$ and $\Gamma_t > \Delta_b/\hbar$, as discussed by Neu and Würger (1994a).)

Damped Oscillations. At sufficiently low temperature we find from the first inequality of (3.324) the transverse damping rate

$$\Gamma_t = \pi \alpha \Delta_b^2 (k_B T/\hbar) + \tfrac{1}{2}\pi^4 (k_B/\hbar)\tilde{\alpha}^2 T^5 \qquad (\tilde{\alpha}T^2 \ll 1). \tag{3.325}$$

The tunneling frequency ω_t is close to $\Delta_b/\hbar$; hence the time-dependent correlation function reads as

$$C_z(t) = \cos(\omega_t t)\exp(-\Gamma_t t) \qquad (\tilde{\alpha}T^2 \ll 1), \tag{3.326}$$

where we have dropped a small phase shift.

Incoherent Tunneling. In the opposite case of high temperature, the transverse damping rate reads as

$$\Gamma_t = \gamma_2 = \pi^2 k_B \hbar^{-1} \sqrt{\tilde{\alpha}/8}\, T^2 \qquad (\tilde{\alpha}T^2 \gg 1). \tag{3.327}$$

According to (3.323) the frequency ω_t is imaginary; thus the resolvent function (3.322) exhibits two imaginary poles $\tfrac{1}{2}\Gamma_t \pm (\tfrac{1}{4}\Gamma_t^2 - (\Delta_b/\hbar)^2)^{1/2}$, and the tunneling motion is purely relaxational. In the limit $\hbar\Gamma_t \gg \Delta_b$ the amplitude of the fast decaying contribution vanishes, and we find

$$C_z(t) = \exp(-\Gamma t), \qquad \Gamma = 2(\Delta_b/\hbar)^2/\gamma_2 \tag{3.328}$$

for high temperatures, $\tilde{\alpha}T^2 \gg 1$.

3.8.6 Comparison of Perturbation Theory and MCA

In deriving the above results we have resorted to two different approximations. *First*, the mode-coupling approximation involves the decoupling of spin-bath correlation functions according to (3.315), which is a basic feature of any mode-coupling theory; once such an approach has been chosen, the outcome can differ only little from our Eqs. (3.317) and (3.318) for the memory spectra. *Second*, in order to obtain an analytic solution, we have discarded the real parts of the memory functions, Γ'_i, and chosen the quadratic law (3.319) for the spectra.

We first address the renormalization of the tunnel energy which would arise from the real part of the memory functions. Next we discuss the temperature dependence of the rate, in comparison with the results of the above perturbation theory.

Renormalization of the Tunnel Energy. Since the Debye temperature Θ is much larger than temperature T and tunnel energy Δ_b, the Kramers–Kronig integral (A6) for $\Gamma'_2(\omega)$ is dominated by high frequencies, well beyond $k_B T/\hbar$. In this range, the damping spectrum is identical to the bath spectral function, $\Gamma''_2(\omega') = \frac{1}{2}\psi''(\omega')$, according to (3.317). Thus at small frequency ω, the leading term of the real part reads as

$$\Gamma'_2(\omega) = \tfrac{1}{2}\omega\tilde{\alpha}\Theta^2 \equiv \tfrac{1}{2}\omega\delta; \tag{3.329}$$

the temperature dependent contributions are at most of the order of magnitude $\omega\tilde{\alpha}T^2$ [cf. the corresponding results from perturbation theory, Eqs. (3.286)–(3.288)].

When inserting $2\Gamma'_2(z) = z\delta$ in the resolvent function (3.314), defining the factor $\mathcal{Z} = [1+\delta]^{-1}$ and the renormalized tunnel frequency and damping rate

$$\tilde{\Delta}_0^2 = \mathcal{Z}\Delta_b^2, \qquad \tilde{\Gamma}_t = \mathcal{Z}\Gamma''_2(\tilde{\Delta}_0/\hbar), \tag{3.330}$$

we find the expression

$$C_z(z) = -\frac{z + i2\tilde{\Gamma}_t}{(z + i\tilde{\Gamma}_t)^2 - \tilde{\Delta}_0^2/\hbar^2}. \tag{3.331}$$

This form is identical to (3.322), except for the renormalization factors in (3.330). The reduction factor $\mathcal{Z}$ agrees with the first-order term of the result from perturbation theory, (3.288) and (3.287). Given the fact that δ is not small in general, i.e. we may have $\mathcal{Z} \ll 1$, it would seem much desirable to use (3.331) rather than (3.322).

Yet in order to treat both spectra C''_x and C''_z on an equal footing, the same procedure has to be applied on (3.313). Noting $\Gamma'_1(z) = z\delta$ and defining the rate $\tilde{\Gamma}_1 = \mathcal{Z}\Gamma''_1(0)$, we find for the longitudinal spectrum

$$C''_x(\omega) = \mathcal{Z}\frac{\tilde{\Gamma}_1}{\omega^2 + \tilde{\Gamma}_1^2} + (1-\mathcal{Z})R''(\omega). \tag{3.332}$$

Whereas the integrated weight of the transverse spectrum C''_z remained unchanged, i.e. $\int d\omega C''_z(\omega) = \pi$, the Lorentz function arising from (3.313) carries a reduction factor $\mathcal{Z}$. Hence we have added the function R'', in order to fulfil the sum rule for $C''_x(\omega)$. Except for the normalization condition $\int d\omega R''(\omega) = \pi$, little is known about R'', thus rendering arbitrary the calculation of the convolution integral (3.318).

In the above treatment, we have refrained from this renormalization procedure, in order to keep the set of mode-coupling equations closed. Yet it is clear from (3.330) that its effects are by no means insignificant. The small-polaron approach of Sect. 3.4 accounts for the frequency renormalization through the reduction factors $\langle B_\pm \rangle$; cf. the discussion on p. 69.

Noncrossing Approximation. According to (3.315) the mode-coupling approximation consists in decoupling the memory functions and inserting the full propagator in the spin part. Since the remaining bath-spectral function (3.316) defines the self-energy in first Born approximation, MCA amounts to a partial resummation of all noncrossing diagrams in the perturbation series.

Here we make this argument more quantitative by comparing the rates as they are obtained from second-order perturbation theory and the weak-coupling result of the present mode-coupling approach. The damping function (3.291) comprises terms of first and second order. In a diagrammatic expansion, the latter may be separated in contributions arising from crossing diagrams, Y''_C, and noncrossing or 'rainbow' diagrams, Y''_R. With the first-order term Y''_1, the damping function may be written as

$$\Gamma(\tilde{\Delta}_0/\hbar) = Y''_1 + Y''_R + Y''_C, \tag{3.333}$$

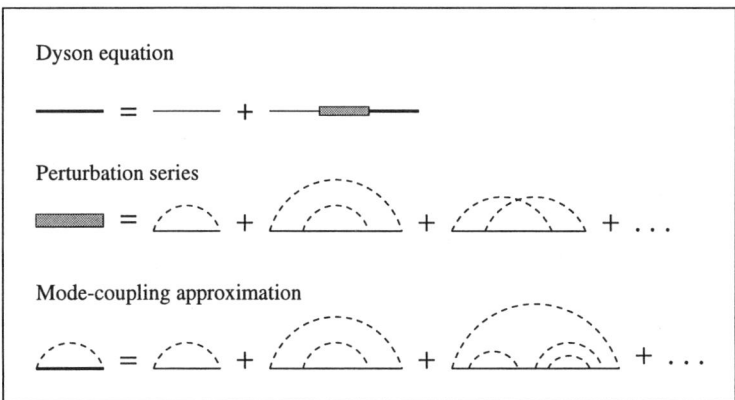

Fig. 3.8. Diagrammatic representation of the perturbation series and the mode-coupling approximation. The Dyson equation illustrates the matrix equation (3.262), where the thick solid line represents the full spin propagator (3.254) and the thin line the uncoupled propagator (3.260). Phonon propagators (3.267), (3.268) are drawn as dashed lines.

corresponding to the schematic representation in Fig. 3.8.

The noncrossing approximation amounts to retaining only diagrams of the rainbow type in the perturbation series. In the second-order term, this corresponds to Y_R'', or to the second diagram of the perturbation series for the self-energy in Fig. 3.8. Discarding small corrections proportional to $\tilde{\Delta}_0^4$, we find the contribution from the rainbow or noncrossing diagram

$$Y_R'' = \tfrac{32}{15}\pi^5(k_B/\hbar)\tilde{\alpha}^2 T^5 + \tfrac{8}{3}\pi^3(\hbar/k_B)\tilde{\alpha}^2 \Delta_b^2 T^3. \tag{3.334}$$

[The present notation slightly differs from that in Würger (1997c). Here, Y_R'' and Y_C'' comprise both Σ_{yy}'' and Σ_{zy}''; thus there are no constant terms that, in the final result, would cancel each other anyhow.]

In the perturbation series, the crossing diagram contributes the same term proportional to T^5 but with a minus sign,

$$Y_C'' = -\tfrac{32}{15}\pi^5(k_B/\hbar)\tilde{\alpha}^2 T^5;$$

as a consequence, the resulting rate (3.296) contains only the remaining term proportional to T^3.

Yet when keeping only the noncrossing diagrams of perturbation theory (cf. Fig. 3.8) and dropping small terms, the rate to second order in $\tilde{\alpha}$ reads as $\gamma_{NC} = \tfrac{1}{2}\mathcal{Z}(Y_1'' + Y_R'')$; with (3.334) we have

$$\gamma_{NC} = \pi\hbar^{-3}\tilde{\alpha}\tilde{\Delta}_0^2 k_B T + \tfrac{16}{15}\pi^5(k_B/\hbar)\tilde{\alpha}^2 T^5. \tag{3.335}$$

Comparison with the mode-coupling result (3.325) reveals that both rates agree with respect to the essential features. There are two minor differences: In the mode-coupling result, the $\mathcal{Z}$-factor is missing, since we have neglected the real part of the memory kernel (cf. the above discussion). The respective second-order terms differ by a numerical factor of $\tfrac{32}{15}\pi$.

We conclude this discussion with a few statements on the validity of the mode-coupling approximation for the spin-phonon model.
(i) The agreement of (3.325) and (3.335) with respect to the temperature dependence confirms that, at least for the spin-boson problem, the mode-coupling approximation corresponds to a resummation of noncrossing diagrams, i.e. to the *noncrossing approximation*.
(ii) Yet it is clear from the expressions for Y_R'' and Y_C'' that the noncrossing approximation is not well justified for the dissipative two-state dynamics. It largely overestimates the second-order contribution, since it misses the cancellation of the leading terms of Y_R'' and Y_C''.
(iii) In our treatment, we have resorted to an additional approximation, when neglecting the real parts of the memory functions, Γ_i'. In view of the significant reduction factors derived for real systems, our dropping $\mathcal{Z}$ in (3.331) is by no means an innocuous approximation.
(iv) According to (3.324) and (3.326), the MCA provides a crossover to incoherent motion where $\tilde{\alpha}T^2 \approx 1$. This qualitative result is strongly supported by the perturbation series (3.296).

In summary, although it shows the proper weak-coupling dynamics and accounts for the crossover to incoherent tunneling at higher temperature, the noncrossing approximation does not provide a proper description of the dissipative two-state dynamics.

3.9 Summary

In this chapter, we have studied phonon coupling of two-state systems in both insulating and metallic materials, using weak-coupling perturbation theory, a mode-coupling approximation, and several strong-coupling approaches based on a small-polaron representation. Here, we briefly summarize the main results, and discuss their experimental relevance.

(i) Damping Rate. In lowest-order perturbation theory, we confirm the one-phonon rate (3.295), as obtained by Jäckle (1972). Higher-order corrections lead to a renormalization of the tunnel frequency, $\Delta_0^2 = \mathcal{Z}\Delta_b^2$, and an infinite series in terms of multiphonon contributions,

$$\Gamma_t = \pi\alpha\tilde{\Delta}_0^2(k_BT/\hbar)\left[1 + \tfrac{4}{3}\pi^2\tilde{\alpha}T^2 + O(\tilde{\alpha}^2T^4)\right]. \tag{3.336}$$

The two-phonon correction factor $\tfrac{4}{3}\pi^2\tilde{\alpha}T^2$ is an exact result, and provides a thorough criterion for any strong-coupling approximation; cf. (3.133).

(ii) Crossover to Incoherent Tunneling. Above $T_0 = (2\pi^2\tilde{\alpha})^{-1/2}$, the rate strongly increases with temperature, resulting in the break-down of finite-order perturbation theory and driving a crossover to overdamped, or incoherent, dynamics for the two-state system. For $T \ll \Theta$, the corresponding relaxation rate,

$$\Gamma_1 = 2\pi\alpha\Delta_0^2(k_BT/\hbar)\,\mathrm{e}^{-\tfrac{1}{3}\phi}\,\frac{\mathrm{e}^\phi - 1}{\phi(1 + 4\phi/\pi)^{1/2}}, \tag{3.337}$$

is given by the one-phonon rate multiplied by a Debye–Waller factor $\mathrm{e}^{-\tfrac{1}{3}\phi}$, and a correction factor accounting for multiphonon processes, with $\phi = T^2/T_0^2$. Accordingly, the rate increases exponentially with temperature for $T > T_0$; cf. p. 81. As T approaches the Debye temperature Θ, the correction factor tends towards an activated law, proportional to $\exp(-V/k_BT)$.

For oxide glasses, T_0 takes values of about 5 K. As discussed on p. 123, incoherent tunneling could well be at the origin of the puzzling variation of sound velocity observed for several materials between 5 and 20 K.

(iii) Phonon-Assisted Tunneling in Metals. At temperatures well below T_0, dissipation in metals is governed by coupling to conduction electrons, with a relaxation rate $\Gamma \propto T^{2K-1}$ that decreases with rising temperature. Above T_0, however, phonons strongly enhance the rate and give rise to a minimum as a function of T. The rate is given by a convolution of electron and phonon damping spectra,

$$\Gamma = \Delta_0 \left(\frac{2\pi K k_B T}{\Delta_0}\right)^{2K-1} e^{-\frac{1}{3}\phi} \left[1 + 2K I_K \frac{e^\phi - 1}{(1 + 4\phi/\pi)^{1/2}}\right], \qquad (3.338)$$

where the first factor is Kondo's rate, while the second one accounts for phonon dressing, and the correction term in the third factor describes phonon-assisted tunneling. We recall $\phi = T^2/T_0^2$.

Both hydrogen impurities in Nb and tunneling defects in mesoscopic Bi wires exhibit such a minimum for the rate. As shown in the figures on pp. 98 and 103, phonon-assisted tunneling accounts quantitatively for the relaxation dynamics observed for these systems.

Acknowledgment

Many people have contributed to this article, through discussions, helpful comments, good questions, and otherwise; I should like to express my thanks to N.O. Birge, D. Bruß, C. Enss, F. Gebhard, H. Grabert, S. Hunklinger, R. Kühn, P. Neu, F. Pistolesi, O. Terzidis, B. Thimmel, G. Weiss, and P. Wölfle. I particularly thank H. Horner for critical discussions and his valuable advice.

Appendix

Spectral Representation

Fourier Transformation. We consider an arbitrary two-time correlation function $f(t)$. Its spectrum is obtained by Fourier transformation,

$$f''(\omega) = \frac{1}{2} \int_{-\infty}^{\infty} dt\, e^{i\omega t} f(t), \qquad (A1)$$

$$f(t) = \frac{1}{\pi} \int_{-\infty}^{\infty} d\omega\, e^{-i\omega t} f''(\omega). \qquad (A2)$$

For evaluating continued fractions, the Laplace transform is more advantageous; note the factor i in our definition,

$$f(z) = i \int_0^\infty dt\, e^{izt} f(t) \qquad (\Im z > 0). \qquad (A3)$$

Inserting (A2) and exchanging time and frequency integrations yields the spectral representation

$$f(z) = \frac{1}{\pi} \int_{-\infty}^{\infty} d\omega\, \frac{f''(\omega)}{\omega - z}. \qquad (A4)$$

In the limit of real z we obtain for reactive and dissipative parts of f

$$\lim_{\eta \to 0} f(\omega + i\eta) = f'(\omega) + i f''(\omega); \qquad (A5)$$

the latter gives the correlation spectrum $f''(\omega)$, and is connected to the reactive part through a Kramers–Kronig relation

$$f'(\omega) = \frac{1}{\pi} \int_{-\infty}^{\infty} d\omega' \frac{f''(\omega')}{\omega' - \omega}. \tag{A6}$$

In most cases one deals with real functions $f(t)$. Then both $f'(\omega)$ and $f''(\omega)$ are real, i.e. (A5) separates real and imaginary parts of $f(z)$.

Reactive Part Γ'

Here we consider the reactive part of the self-energy in NIBA. When expanding the function $\Gamma(z)$ about the poles $\pm\tilde{\Delta}_0/\hbar$, we find that the damping rate is given by the spectrum $\Gamma''(\omega)$, according to (3.80). From the real part we obtain both a constant and the term $z\Xi$, with the derivative

$$\Xi = \frac{1}{2}[\partial_\omega \Gamma'(\omega) - \partial_\omega \Gamma'(-\omega)]_{\omega=\tilde{\Delta}_0/\hbar}. \tag{A7}$$

The constant will be neglected for two reasons. First, it is much smaller than the resonance frequency $\tilde{\Delta}_0/\hbar$; for the term linear in α one finds $2\alpha(\tilde{\Delta}_0/\hbar)^2\omega_D$, which is negligible in view of (3.25). Second, this constant part of Γ' is unphysical; it vanishes when considering symmetrized bath correlations $\Gamma(t) + \Gamma(-t)$; cf. Würger (1998a).

In this appendix we intend to show that the derivative Ξ is much smaller than unity. As a consequence, neither the resonance energy nor the spectral function of the propagator (3.140) are significantly modified by the real part Γ'.

Since the Kramers–Kronig integral (A6) is determined by high-frequency contributions, we need to use the exact expression rather than the approximate (3.109), when evaluating $\Gamma'(\omega)$. We expand the exponential in (3.79) in a power series in $\varphi(t)$ and take the Fourier transform of each term,

$$\Gamma''(\omega) = \tilde{\Delta}_0^2 \sum_{n=1}^{\infty} \frac{1}{n!} \kappa_n''(\omega), \tag{A8}$$

where we use the shorthand notation

$$\kappa_n''(\omega) = \frac{1}{2}\int_{-\infty}^{\infty} dt\, e^{i\omega t}\varphi(t)^n. \tag{A9}$$

The derivative Ξ is given by a corresponding series,

$$\Xi = \frac{d}{d\omega}\tilde{\Delta}_0^2 \sum_{n=1}^{\infty} \frac{1}{n!}\gamma_n(\omega)\bigg|_{\omega=\tilde{\Delta}_0/\hbar}, \tag{A10}$$

with

$$\gamma_n(\omega) = \tfrac{1}{2}[\kappa_n'(\omega) - \kappa_n'(-\omega)]. \tag{A11}$$

Due to the frequency dependence of $\varphi''(\omega)$, the Kramers–Kronig integral is determined by high frequencies. At moderate frequency and temperature, i.e. $T \ll \Theta$ and $|\omega| \ll \omega_D$, we may replace $\Gamma'(\omega)$ by its value at $T=0$ and $\tilde{\Delta}_0$ by Δ_0; finite-temperature corrections are small.

Hence we consider the zero-temperature limit for the coupled phonon spectrum,

$$\varphi''(\omega) = \begin{cases} 2\pi\alpha\omega & \text{for } 0 \leq \omega \leq \omega_D \\ 0 & \text{else} \end{cases} \quad (T=0). \tag{A12}$$

According to (A9), the Kramers–Kronig integral $\kappa_n'(\omega)$ may be written in terms of an n-fold convolution of $\varphi''(\omega)$. After inserting the zero-temperature expression (A12), we obtain

$$\gamma_n'(\omega) = \omega(2\alpha)^n \int_0^{\omega_D} d\omega_1 \cdots \int_0^{\omega_D} d\omega_n \frac{\omega_1 \cdots \omega_n}{(\omega_1 + \ldots + \omega_n)^2 - \omega^2}. \tag{A13}$$

The first two terms are easily integrated,

$$\gamma_1'(\omega) = 2\omega\alpha \log(\omega_D/\omega), \tag{A14}$$

$$\gamma_2'(\omega) = \omega(2\alpha)^2 \omega_D^2 \left[\log(2) - \frac{1}{2}\right], \tag{A15}$$

where we have neglected corrections of the order ω/ω_D. The presence of the logarithmic factor renders the first term a bit particular; because of the factors $\omega_1 \cdots \omega_n$ in (A13), there is no such factor in the higher orders, as shown explicitly in (A15) for the quadratic term.

For this reason, we may expand the integrand of (A13) in powers of ω^2, and integrate the term of order zero,

$$\kappa_n(\omega) = \omega(2\alpha)^n \omega_D^{2(n-1)} I_n [1 + O(\omega/\omega_D)]. \tag{A16}$$

Here we have substituted $x_i \equiv \omega/\omega_D$ and defined the integral

$$I_n = \int_0^1 dx_1 \cdots \int_0^1 dx_n \frac{x_1 \cdots x_n}{(x_1 + \ldots + x_n)^2}. \tag{A17}$$

With increasing n, the coefficients I_n tend towards zero; we give those for $n = 2, 3, 4$,

$$I_2 = \log(2) - \frac{1}{2}, \tag{A18}$$

$$I_3 = \frac{3}{8}[\log(3) - 1], \tag{A19}$$

$$I_4 = \frac{9}{4}\log(3) - \frac{28}{9}\log(2) - \frac{11}{36}, \tag{A20}$$

which indicate a rapid convergence of the series (A10).

In order to obtain an upper bound for that series, we resort to the following approximations for the terms of order $n \geq 2$. Since the integrand is positive,

3.9 Summary

discarding the term $x_2 + \cdots + x_n$ in the denominator provides an upper limit for I_n. Then the n integrals factorize, resulting in the inequality

$$I_n < 2^{1-n} \quad \text{for } n \geq 2. \tag{A21}$$

Inserting (A14) and (A21) in the expression for $\Gamma'(\omega)$ and using $W_0 = \alpha \omega_D^2$, we obtain

$$\Xi < 2\alpha(\Delta_0/\hbar)^2 \left[(\log(\hbar\omega_D/\Delta_0) - 2) + \frac{e^{W_0} - 1}{W_0} \right]. \tag{A22}$$

In physical terms, $\gamma'_n(\omega)$ describes a frequency shift due to n-phonon processes. The sum of these terms is negligible if $\Xi \ll 1$. In this case, the '$\mathcal{Z}$-factor' $[1 + \Xi]^{-1}$ is close to unity.

Noting the relation $\Delta_0^2 = \Delta_b^2 e^{-W_0}$ and the slow-tunneling condition (3.25), one finds that the second term in brackets is immaterial for both limits $W_0 \ll 1$ and $W_0 \gg 1$. As for the first term, it is small for a large Debye–Waller factor, $W_0 \gg 1$. The opposite case, $W_0 \leq 1$, requires the slightly more restrictive condition $\alpha(\Delta_b/k_B)^2 \log(\hbar\omega_D/\Delta_b) \ll 1$.

In summary we have shown that the derivative is much smaller than unity,

$$\Xi \ll 1, \tag{A23}$$

which justifies our neglecting the reactive part of the self-energy in Sect. 3.4.

Our derivation of (A23) relied on the condition (3.25). In view of the parameters of oxide glasses, a more thorough investigation of the real part Γ' would seem most interesting, considering the case where (3.25) is not satisfied.

4. Influence of Tunneling Systems on the Acoustic Properties of Disordered Solids

Pablo Esquinazi and Reinhard König

4.1 Acoustic Properties and Tunneling Systems

In this chapter we review the acoustic properties of disordered solids at low temperatures, i.e. we discuss the propagation of sound in amorphous materials (dielectrics and metals) as well as in polycrystalline metals in terms of sound attenuation (or internal friction) and sound velocity. Along with the study of thermal (see Chap. 2), dielectric (see Chap. 5), and some optical (see Chap. 6) properties of amorphous or disordered solids at $T \leq 10$ K, the results of the investigation of the low-temperature acoustic properties indicate the existence of tunneling entities in the disordered atomic structure with a broad distribution of relaxation times and an almost energy-independent density of states (at least for $T < 5$ K). There is a general consensus that these low-energy excitations associated with quantum-tunneling processes are responsible for the so-called "glasslike" behavior of the low-temperature properties of disordered solids.

In amorphous dielectrics and superconductors for $T \ll T_c$ (T_c: superconducting transition temperature), the influence of tunneling systems (TS's) can be observed in the specific heat and thermal conductivity. In normal metals (amorphous and polycrystalline), the contribution of conduction electrons to the specific heat usually overwhelms that of the TS's. The acoustic properties *in the kHz range*, however, provide an excellent tool to study the interaction between electrons and/or phonons and TS's in almost every solid because the conduction electrons–phonon interaction is usually negligible. Moreover, due to the relatively large phonon wavelength, the interaction of the acoustic wave with grain boundaries in polycrystalline solids at kHz frequencies does not play any role either. Measurements of the internal friction and sound velocity provide information on the density of states (P) of TS's and their coupling to phonons (γ) or conduction electrons (K). An additional advantage of acoustic measurements in the kHz range is the possibility of measuring the acoustic properties at ultralow temperatures ($T < 5$ mK) as the energy dissipation in the sample (proportional to the third power of the frequency) can be kept at extremely small levels ($< 10^{-15}$ W), thus avoiding self-heating effects.

Although important and interesting results were obtained using ultrasonic phonon frequencies (in the MHz range), the emphasis in this chapter

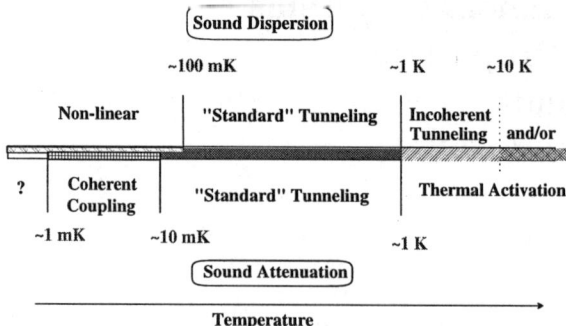

Fig. 4.1. Different tunneling regimes described in this chapter and observed in the acoustic properties of amorphous and polycrystalline samples. The temperatures that separate different tunneling regimes from each other are only approximate temperatures and do not mean sharp crossovers.

will be given to the experimental methods and the results in the kHz range obtained during the last ten years of research in this field. We will discuss the behavior of the internal friction and sound dispersion of disordered solids below ≈ 10 K only where the interaction of phonons with TS's is, although not completely understood, rather well-characterized and a systematic behavior can be observed. The temperature dependence of the acoustic properties of a large amount of amorphous solids and disordered crystals above and below ~ 5 K has been recently reviewed by White and Pohl (1996) and Topp and Cahill (1996). Although the experimental data suggest a connection between the low- and high-temperature regimes (White and Pohl (1996)), it becomes clear (Topp and Cahill (1996)) that no simple universality in the acoustic properties can be observed at relatively high temperatures. A distribution of barrier heights V added to nonsimple and sample-dependent atomic reorientations may play a major role at high temperatures.

Figure 4.1 shows schematically the possible mechanisms that appear to be responsible for the behavior of the acoustic properties of very different disordered solids at low temperatures. The expression "Standard Tunneling" stands for the standard tunneling model, i.e. the "standard" resonant and relaxation mechanisms proposed in the 70's (for a review see Hunklinger and Arnold (1976)) summarized in some detail in Sects. 4.2.1–4.2.3, while other mechanisms are described in the rest of the theoretical part of this and other Chap. of the book. The temperatures that separate the different tunneling regimes described in Fig. 4.1 are only estimates and do not imply that strict separation of the mechanisms needs to exist. These "crossover" temperatures may depend on the phonon frequency as well as on intrinsic sample parameters. It is one of the aims of this chapter to review some of the experimental data that provide some support for the different tunneling regimes described in Fig. 4.1.

4.2 Theoretical Remarks

The propagation of an acoustic perturbation in the atomic lattice is characterized by changes in some physical properties like temperature, density or pressure. If an acoustic wave (phonon) propagates in a solid, the amplitude of this wave decreases with time, i.e. the energy of this wave is absorbed and converted into heat. Although the change in temperature due to this energy absorption is usually too small to be measured in the experiments described in this chapter, the acoustic methods are yet sensitive enough to measure the attenuation of the wave. In an homogeneous medium the energy absorption is due to a phase lag between the changes in density and pressure. One possibility to describe this kind of absorption is given by the relaxation mechanism that we will describe briefly in Sect. 4.2.2.

Assuming the solid responds linearly to the stress imposed by an acoustic wave propagating in the positive direction of the x-axis, we can write for the variation of the local density ϱ, for example, that undergoes a sinusoidal change

$$\varrho = \varrho_0 e^{-\alpha x/2} \mathrm{Re}\left(e^{i\omega(t-x/v)}\right), \tag{4.1}$$

with α the coefficient of intensity absorption or attenuation per unit length (sometimes also referred to as inverse phonon mean free path l^{-1}), ϱ_0 the amplitude of the density variation, $\omega = 2\pi\nu$ the angular frequency of the acoustic wave, t the time and v the sound velocity in the elastic medium without perturbation.[1]

In a cycle, the energy loss per unit volume is given by $\Delta E = \alpha \lambda E_0$, where E_0 is the energy stored per unit volume and λ the wavelength of the oscillation. For experiments in the kHz range and for the characterization of the attenuation, it is usual to define a unitless quantity, the internal friction $Q^{-1}(T)$. The internal friction or inverse of the quality factor is related to the attenuation coefficient by

$$Q^{-1} = \alpha v/\omega. \tag{4.2}$$

We will see below that not only the attenuation but also the relative change of sound velocity (the so-called sound dispersion) provides important information for the characterization of the interaction between phonons and TS's.

4.2.1 Resonant and Relaxation Processes

Before we review the theoretical treatments of the interaction of phonons (Sect. 4.2.2) and electrons (Sect. 4.2.3) with TS's, we would like to describe here an early experiment performed by Hunklinger et al. (1972) that provided strong support for the remarkable fact that systems with *two energy states*

---
[1] Note that the absolute change of the sound velocity due to the interaction of phonons with tunneling entities is extremely small but can be very well measured.

are resonant scatterers of phonons and that the sound attenuation depends nonlinearly on the acoustic intensity.

As described in the phenomenology of two-level TS's (see Chap. 2) we assume that in the (disordered) sample a system of N two-level systems with energy difference E is in thermal equilibrium at a given temperature T. The coupling of an acoustic wave to this system is described by a coupling constant γ that enters in the relaxation time $\tau(E,T)$ of the TS's. If the energy of the acoustic wave is of the order of the energy difference of the TS's, phonons can be resonantly absorbed. As pointed out by Jäckle (1972), this resonant absorption of phonons corresponds to the transition of the tunneling entity from the ground state to an excited state and will depend on the population of the upper level and therefore on temperature. If the acoustic wave has enough energy to change the population of the upper levels, a saturation of this resonant absorption can be achieved. In other words, the attenuation of the wave should be inversely proportional to the acoustic intensity but proportional to the population difference in thermal equilibrium $\tilde{N}_0 \propto \tanh(E/2k_\mathrm{B}T)$.

A detailed calculation of the resonant absorption is not trivial because not only the longitudinal but also the transverse relaxation time $\tau_2(T)$ has to be taken into account. This means that the resonant absorption depends on the recombination time τ_1 in the process of absorption and emission of a phonon by a two-level tunneling system as well as on the effective number of TS's taking part in the resonant absorption and determined by the linewidth τ_2^{-1}. A detailed description of the calculation for the resonant absorption is given by Hunklinger and Arnold (1976) using the formal equivalence between the Hamiltonian of a two-level system perturbed by an elastic field and the Hamiltonian for a spin with $\mathbf{S} = 1/2$ in an oscillating magnetic field. A more detailed discussion will be given below. Here, we only consider the main expression for the ultrasonic attenuation due to the resonant interaction as given by Hunklinger and Arnold (1976):

$$\alpha_\mathrm{res} \simeq \frac{\pi P_0 \gamma^2}{\varrho v^3} \frac{\omega \tanh(\hbar\omega/2k_\mathrm{B}T)}{\sqrt{1+\frac{I}{I_{c1}}}}, \tag{4.3}$$

where $I_{c1} \propto 1/M^2 \tau_1 \tau_2$ is a critical intensity that is typically of the order of $10^{-7} \mathrm{W/cm^2}$ and M is a deformation potential, see Sect. 4.2.2. Figure 4.2 shows the power dependence of the ultrasonic attenuation of an amorphous dielectric sample for longitudinal waves at three different frequencies.

The main experimental features of the nonlinear acoustic intensity dependence of the ultrasonic attenuation can be derived from (4.3). For $\hbar\omega \ll k_\mathrm{B}T$, $\omega\tau_2 \gg 1$, one obtains $\alpha_\mathrm{res} \propto \omega^2/T$ for $I/I_{c1} \ll 1$ and $\alpha_\mathrm{res} \propto (\omega^2/T)(I_{c1}/I)^{1/2}$ at $I/I_{c1} \gg 1$. Both dependencies are nicely observed in experiments (Arnold et al. (1974)), see Fig. 4.2 and the results of Golding et al. (1973) and Bachellerie (1974) too.

In general, the intensity dependence of the ultrasonic attenuation is difficult to measure because the acoustic intensities used in the experiments are

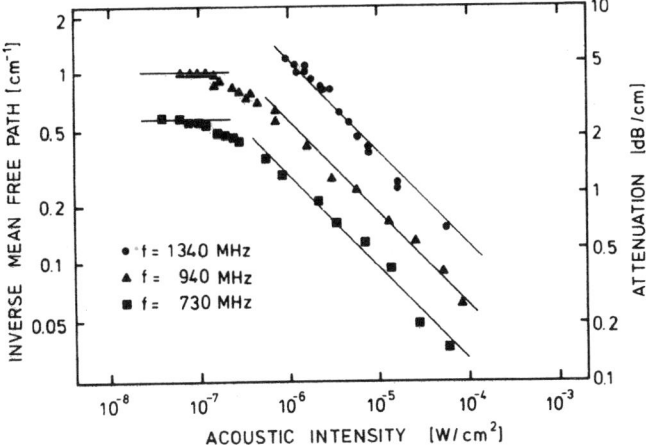

Fig. 4.2. Attenuation of longitudinal ultrasonic waves at three different frequencies as a function of the acoustic intensity in borosilicate glass at 0.48 K. At low intensities the attenuation depends quadratically on frequency and decreases as $I^{-1/2}$ at larger intensities (straight lines) (from Arnold et al. (1974)).

usually larger than the critical one. Due to the frequency dependence of the resonant absorption, the internal friction measured in acoustic experiments in the kHz range does not depend on the resonant interaction. However, the resonant interaction provides a clear and well measurable temperature dependence for the sound velocity at low enough temperatures.

Another mechanism, the relaxation absorption, contributes to both the sound attenuation and to the sound dispersion. This relaxation mechanism is related to an effective "viscosity" of the medium. It is observed in very different media as, for example, in liquids (in this case one introduces a shear viscosity) and is explained as due to the phase delay between stress and imposed strain rate. The relaxing system, in our case the TS's with the characteristic (longitudinal) relaxation time τ_1, is exposed to a sinusoidal perturbation or strain given by

$$\epsilon_{ik}(t) = \frac{1}{2}u_{ik}(t) = \epsilon^0_{ik}\sin(\omega t), \tag{4.4}$$

where u_{ik} refers to the local strain tensor ($i, k = 1, 2, 3$). The acoustic wave couples to the TS's through a coupling constant (see next section) and changes their thermal equilibrium. The TS's try then to relax to a new equilibrium state emitting and absorbing thermal phonons. The general expression for the sound attenuation and sound dispersion is given by relations of the type

$$\alpha = \frac{\omega}{\varrho v^3}\mathrm{Im}(\chi(\omega)) \propto f(T)\frac{\omega\tau_1}{1+\omega^2\tau_1^2}, \tag{4.5}$$

$$\frac{\Delta v}{v} = -\frac{1}{2\varrho v^2}\mathrm{Re}(\chi(\omega)) \propto f(T)\frac{1}{1+\omega^2\tau_1^2}, \qquad (4.6)$$

where $\chi(\omega)$ is an effective acoustic susceptibility that, like the function $f(T)$, depends on the details of the model and the medium where the wave propagates.[2] In the case of the TS's, $f(T)$ depends on their coupling to phonons or conduction electrons, their density of states and the energy derivate of the population difference between the two levels.

The relaxation absorption has the following interesting characteristics: for a given relaxation time, the attenuation vanishes at low frequencies, it has a maximum at $\omega\tau_1 \sim 1$ and tends to a constant nonzero value at frequencies higher than the relaxation frequency $2\pi\tau_1^{-1}$. At a given frequency and, for simplicity, taking only the temperature dependence of τ_1 [see (3.295) and (3.131)] the sound velocity saturates at high and low temperatures having a maximum change with temperature where the attenuation has a maximum at $\omega\tau_1 \sim 1$. Obviously the measurement of both acoustic properties is important if quantitative deductions are to be made.

As one might expect, the observed behavior in most of the materials studied indicates that a single relaxation time cannot describe adequately the experimental data. The behavior of the acoustic data in amorphous and disordered solids suggests the existence of a rather broad distribution of relaxation times. This can be taken into account by integrating (4.5) and (4.6) with an appropriate distribution function. This assumption, as well as the bounds of the distribution and the coupling between TS's and phonons, are the main ingredients of the tunneling model that are summarized in the following section. For a detailed calculation of the results see the analogous treatment of the dielectric properties in Sect. 5.1.1 [see also Hunklinger and Arnold (1976); see Chap. 2 for an introduction of the phenomenology].

4.2.2 The Standard Tunneling Model. Relaxation due to Phonons

The standard tunneling model assumes the existence of entities with an energy splitting E (2.4) that can interact with phonons. This interaction is given by the Hamiltonian matrix of the form

$$H_1(t) = -\epsilon(t)\left(\frac{1}{2}D\sigma_3 + M\sigma_1\right), \qquad (4.7)$$

where $\sigma_i (i = 1, 2, 3)$ are the Pauli matrices. The acoustic wave is represented by the applied strain $\epsilon = \epsilon_0 \sin(\omega t)$ where ω is the angular frequency and t the time. This wave produces a local strain given by the strain tensor $u_{ik} = \partial u_i/\partial x_k + \partial u_k/\partial x_i$, where $u_{i,k}$ is the local displacement of the "particle". The model assumes two deformation potentials or coupling constants

---

[2] For a derivation of the general relations for the relaxation interaction, see Sect. 7.3.2.

$$D_{ik} = \frac{\partial E}{\partial u_{ik}} = \frac{\Delta}{E}\frac{\partial \Delta}{\partial u_{ik}} + \frac{\Delta_0}{E}\frac{\partial \Delta_0}{\partial u_{ik}}, \qquad (4.8)$$

$$M_{ik} = \frac{1}{2}\left(-\frac{\Delta_0}{E}\frac{\partial \Delta}{\partial u_{ik}} + \frac{\Delta}{E}\frac{\partial \Delta_0}{\partial u_{ik}}\right). \qquad (4.9)$$

D_{ik} describes the shift of the energy splitting E and M_{ik} the strength of the coupling between TS's and phonons for the resonant interaction. For simplicity only the isotropic case will be considered, i.e. $D_{ik} = D\delta_{ik}$ and the same applies for M (for the anisotropic case, see Sect. 7.3.2). Each of the equations described above is valid for both the longitudinal and transverse phonon polarizations. An important simplification in the model is the relationship between these coupling constants and the derivatives of the asymmetry and tunneling energy: it is assumed that the variation of the tunneling energy with the local strain is negligible in comparison with the variation of the asymmetry Δ [see (3.11) and the discussion in that section]. Defining a coupling constant γ_j for the longitudinal $(j = l)$ or transverse $(j = t)$ polarization, the simplified constants D and M are then

$$D_j \simeq 2\gamma_j \frac{\Delta}{E}, \quad M_j \simeq -\gamma_j \frac{\Delta_0}{E}, \quad \gamma_j = \frac{1}{2}\frac{\partial \Delta}{\partial u_j}. \qquad (4.10)$$

In what follows and to simplify the notation we will not write explicitly the subindex j. The polarization depends on the experimental method used to study the acoustic properties.

Referring to the similarity of this problem to that of a spin $S = \frac{1}{2}$ system in an oscillating field, Golding et al. (1973) and Hunklinger and Arnold (1976) solved the Bloch-like equations and ended up with two pairs of equations for sound attenuation and dispersion that, due to their structure, may be identified as resonant and relaxation interactions. The standard tunneling model assumes a constant distribution of TS's in terms of the asymmetry and tunneling parameter (see Sect. 2.2.2). Sometimes it is convenient to express this distribution in terms of the energy splitting E and the parameter $u = \Delta_0/E$, i.e. $P(E, u) = P_0/u(1 - u^2)^{1/2}$. In this case, the equations for the relaxation interaction are

$$Q_{\mathrm{rel}}^{-1} = \frac{P_0 \gamma^2}{\varrho v^2} \int_{E=0}^{E=E_{\max}} dE \int_{u=u_{\min}}^{u=1} du \frac{\sqrt{1-u^2}}{uk_{\mathrm{B}}T}\mathrm{sech}^2\left(\frac{E}{2k_{\mathrm{B}}T}\right)\frac{\omega\tau_1}{1+\omega^2\tau_1^2} \qquad (4.11)$$

$$\left.\frac{v(T) - v(T_0)}{v(T_0)}\right|_{\mathrm{rel}} = -\frac{P_0 \gamma^2}{2\varrho v^2} \int_{E=0}^{E=E_{\max}} dE \int_{u=u_{\min}}^{u=1} du \frac{\sqrt{1-u^2}}{uk_{\mathrm{B}}T}$$
$$\times \mathrm{sech}^2(E/2k_{\mathrm{B}}T)[1/(1+\omega^2\tau_1^2)]. \qquad (4.12)$$

At low enough temperatures and small relaxation time τ_2, the intensity-dependent resonant contribution to the sound dispersion is given by (Cordié and Bellessa (1981))

$$\left.\frac{v(T)-v(T_0)}{v(T_0)}\right|_{\text{res}} = -\frac{1}{2\varrho v^2}\int_{E=0}^{E=E_{\max}} dE \int_{u=u_{\min}}^{u=1} du P(E,u) M^2 \tau_2^2$$

$$\times \frac{\tanh(E/2k_B T)}{\hbar^2}\left(\frac{E_0-E}{1+(I/I_{c1})+(E_0-E)^2\hbar^{-2}\tau_2^2}\right.$$

$$\left. -\frac{E_0+E}{1+(I/I_{c1})+(E_0+E)^2\hbar^{-2}\tau_2^2}\right), \qquad (4.13)$$

where the critical acoustic intensity $I_{c1} = \Gamma\hbar^2\varrho v^3/2M^2\tau_1$ (Γ is the linewidth) and E_0 is the energy of the applied acoustic wave. T_0 is an arbitrary reference temperature. Taking into consideration the results of typical experiments at acoustic frequencies ($\nu < 50$ kHz $\ll k_B T/h$), we can approximate (4.13), taking into account that the main contribution arises from "thermal" TS's with energy $E \sim k_B T$ (as for the relaxation process). With $\omega_0 = E_0/\hbar \ll E/\hbar$ and $I \leq I_{c1}$, (4.13) has the simple form

$$\left.\frac{v(T)-v(T_0)}{v(T_0)}\right|_{\text{res}} = -\frac{1}{2\varrho v^2}\int\int P(E,u) 2\frac{M^2}{E}\tanh(E/2k_B T) dE du. \quad (4.14)$$

It is possible to integrate (4.14) using the distribution function $P(\Delta, \Delta_0)$ $d\Delta d\Delta_0 = (P_0/\Delta_0)d\Delta d\Delta_0$. This case is particularly useful when we include the shift of the asymmetry due to the acoustic wave (see Sect. 4.2.4). For the case the resonant part of the absorption is not saturated (the acoustic intensity $I \ll I_{c1}$, see Sect. 4.2.1), the standard model predicts

$$Q_{\text{res}}^{-1} = \frac{\pi P_0 \gamma^2}{\varrho v^2}\frac{\hbar\omega}{2k_B T}, \quad \hbar\omega \ll k_B T, \omega\tau_2 \gg 1; \qquad (4.15)$$

$$Q_{\text{res}}^{-1} = \frac{\pi P_0 \gamma^2}{\varrho v^2}, \quad \hbar\omega > k_B T. \qquad (4.16)$$

Usually, in our acoustic experiments we have $\hbar\omega \ll k_B T$; therefore, the resonant contribution to the internal friction is negligible in comparison with the relaxation contribution, i.e. $Q_{\text{rel}}^{-1}(\omega\tau_1 < 1) = \pi C/2 \gg Q_{\text{res}}^{-1}(k_B T \gg \hbar\omega)$, with $C = P_0\gamma^2/\varrho v^2$.

Figure 4.3A shows the resonant (1) and relaxation (2) contribution to the sound dispersion according to the standard model and assuming that the TS's relax through the one-phonon process with $\tau_1(E,T) = \tau_p(E,T)$, see (2.18).

The total contribution to the sound dispersion is given by a simple addition of the two contributions $\Delta v/v|_{\text{rel}} + \Delta v/v|_{\text{res}}$ and is depicted by curve (3) in Fig. 4.3A. From this addition we obtain a maximum in the sound velocity at the crossover temperature $T_{\max} \propto (\omega/Ak_B^3)^{1/3}$ that occurs when $\omega\tau_{1,\min} \sim 1$, with $\tau_{1,\min} \sim AE^3$ the minimum longitudinal relaxation time, that in this case is due to the one-phonon process [$A = \pi\alpha/\hbar^3$ where α is given by (3.20)]. The temperature dependence of the sound dispersion below and above the maximum is logarithmic with slopes $+C$ and $-C/2$, respectively.

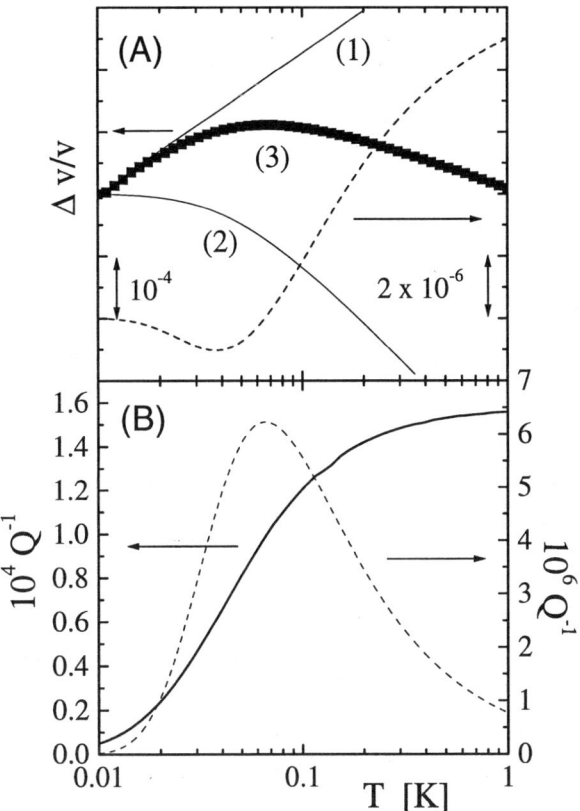

Fig. 4.3. Predictions of the standard tunneling model for the sound dispersion and internal friction taking into account the relaxation of TS's with phonons only through the one-phonon process. (A) Curves (1) and (2) show the temperature dependence of the relative change of sound velocity due to the resonant and relaxation processes, calculated with (4.11), (4.12), (4.14) using the parameters $C = 10^{-4}$, $\nu = 10^3\,\mathrm{Hz}$, and $Ak_B^3 = 8.5 \times 10^7\,\mathrm{s}^{-1}\mathrm{K}^{-3}$ that are appropriate for amorphous isolators. Curve (3) is obtained adding the curves (1) and (2). (B) The continuous line represents Q^{-1} as a function of temperature taking into account the relaxation interaction only. The parameters are the same as in (A). The dashed lines in both figures are obtained assuming only tunneling systems with energy 0.09 K $\leq E \leq$ 0.1 K and $0 \leq \Delta_0 \leq 0.1$ K.

The internal friction due to the relaxation process is shown in Fig. 4.3B. In the same figure, we plot the relaxation absorption calculated assuming a narrow distribution in energy for P. The standard model predicts an internal friction $\propto T^3$ at temperatures far below $T_{\max}$ and a temperature independent region above $T_{\max}$. This "plateau" is given by $Q^{-1}(T \gg T_{\max}) = \pi C/2$ and

Table 4.1. Predictions of the standard tunneling model for the acoustic properties of dielectric glasses. The parameters are $C = P_0\gamma^2/\varrho v^2$, $a = (\pi^3/24)(\gamma_l^2/v_l^5 + 2\gamma_t^2/v_t^5)(k_B^3/\varrho\hbar^4)$.

Interaction	Property	$\omega\tau_{1,\min} \gg 1$ $(T < T_{\max})$	$\omega\tau_{1,\min} \ll 1$ $(T > T_{\max})$
Resonant	Q^{-1}	negligible at acoustic frequencies	
Resonant	$\Delta v/v$	$+C\ln(T/T_0)$	
Relaxation	Q^{-1}	aCT^3/ω	$(\pi/2)C$
Relaxation	$\Delta v/v$	negligible	$-(3/2)C\ln(T/T_0)$

is independent of the relaxation mechanism of the TS's. Table 4.1 summarizes the results obtained from the standard tunneling model at temperatures $T \ll T_{\max}$ and $T \gg T_{\max}$, assuming that only the one-phonon process is responsible for the relaxation of the TS's.

In several cases the experimental results cannot be accounted for by the standard model in a quantitative manner. Therefore, scientists often tried to modify slightly its standard assumptions, as for example: a different dependence for the density of states, the addition of a coupling that changes the tunneling energy Δ_0 with strain, the addition of a third energy level, etc. Perhaps the most appealing assumption to be modified is the constancy of the density of states P as a function of Δ and λ. To our knowledge, none of these modifications gained consent among scientists to be treated as something inherent and important. However, we would like to draw some attention to a modification of the density of states $P(E, u)$ introduced *ad hoc* by Hunklinger (1984), mainly because of pedagogical reasons rather than because of its significance. Hunklinger (1984) suggested the introduction of a parameter W that changes the relative contribution of the two branches of the distribution function: the slowly relaxing systems with $u \ll 1$ and the fast ones with $u \approx 1$. Within this approach the distribution function is given by $P(E, u) = P_0[(u/\sqrt{1-u^2}) + W(\sqrt{1-u^2}/u)]$. The original distribution is obtained for $W = 1$. Other values of W quantitatively change the theoretical results; it is interesting to recognize the influence of the branches of P on the acoustic properties. For example, taking only the fast relaxing TS's (i.e. with $W = 0$) the logarithmic slope for the sound velocity below $T_{\max}$ remains but with a slope that is 50% smaller than the original with $W = 1$.

4.2.3 Relaxation due to Conduction Electrons

In the last subsection we summarized the main results of the "standard" tunneling model necessary to understand, at least qualitatively, the anomalies of the acoustic properties of amorphous and some crystalline solids, assuming that the relaxation of the TS's is due to phonons only. However, for metallic samples, another channel for the relaxation due to conduction electrons is possible. The experimental results of the acoustic properties of metallic samples (amorphous and crystalline) obtained during the last ≈ 13 years – some of which we will review in this chapter – reveal that there is no universal behavior as is the case for insulating glasses. There are at least three important theoretical approaches that provide possible answers to this difficult problem and that deserve our attention. Due to historical reasons, rather than to its universality, the first approach described below might be included inside the tunneling model as the "standard" one. For a theoretical introduction to this problem and the influence of conduction electrons to the coherent and incoherent tunneling rates, see Sect. 3.5 and references therein.

"Standard" Korringa-Like Relaxation. The rather strong interaction between TS's and conduction electrons in amorphous metals was proven experimentally by Golding et al. (1978) and Doussineau et al. (1978). In metallic glasses the need of a "giant" ultrasonic acoustic amplitude to achieve a saturation of the attenuation clearly indicates that the TS's can relax by several orders of magnitude faster in comparison with insulating glasses. It was proposed that the relaxation rate of the TS's due to conduction electrons follows a Korringa-like interaction similar to the relaxation of nuclear spins through their interaction with conduction electrons (Golding et al. (1978)). Assuming a single and constant coupling K (see the footnote on p. 25), the relaxation rate is given by (Black (1981))

$$\tau_{1,e}^{-1} = K E u^2 \coth\left(\frac{E}{2k_\mathrm{B}T}\right) . \tag{4.17}$$

In the standard approach, the coupling between conduction electrons and TS's is weak enough and, therefore, its influence on the acoustic properties can be accounted for considering time-dependent perturbation theory on a second-order level on the eigenstates of the isolated TS's (Black (1981)).

If both phonons and conduction electrons contribute independently to the relaxation of TS's, the total relaxation rate is given by $\tau_1^{-1} = \tau_{1,p}^{-1} + \tau_{1,e}^{-1}$. The predictions of the standard tunneling model for the case that only electrons contribute to the relaxation of the TS's are given in Table 4.2 for the limiting cases of very low and high temperatures. Note that the results for the resonant interaction do not depend on the relaxation channel (phonons or electrons) in the temperature range of interest of the experiments described here. The value of the plateau ($\omega\tau \ll 1$) in the internal friction is also independent of the relaxation mechanism.

Figure 4.4 shows the theoretical predictions for the temperature dependence of the acoustic properties of a normal conducting metallic glass in the regime $\hbar\omega \ll k_\mathrm{B}T$. At kHz-frequencies this condition is fulfilled in the whole temperature range plotted in Fig.4.4; for much larger frequencies (i.e. for $\hbar\omega > k_\mathrm{B}T$), the sound velocity becomes temperature independent, see Bellessa (1978). At low enough temperatures, i.e. $\omega\tau_{1,e,\min} \gg 1$, the sound velocity increases logarithmically with temperature with a slope C due to

Table 4.2. Predictions of the standard tunneling model for the acoustic properties of metallic glasses. The parameters are $C = P_0\gamma^2/\varrho v^2, b = (\pi^3/24)K^2(k_B/\hbar)$. T_0 is an arbitrary reference temperature.

Interaction	Property	$\omega\tau_{1,\min} \gg 1$ ($T < T_{\max}$)	$\omega\tau_{1,\min} \ll 1$ ($T > T_{\max}$)
Resonant	Q^{-1}	negligible at acoustic frequencies	
Resonant	$\Delta v/v$	$+C\ln(T/T_0)$	
Relaxation	Q^{-1}	bCT/ω	$(\pi/2)C$
Relaxation	$\Delta v/v$	negligible	$-(1/2)C\ln(T/T_0)$

the resonant interaction. At higher temperatures the relaxation due to electrons sets in at $\omega\tau_{1,e,\min} \sim 1$ and the sound velocity depends also logarithmically on temperature but with a slope $+C/2$. At still higher temperatures $\omega\tau_{1,p,\min} \ll 1, \tau_{1,e} \gg \tau_{1,p}$ the sound velocity decreases logarithmically with a slope $-C/2$ due to the relaxation of the TS's to phonons. According to the standard theory the internal friction should show a temperature-independent behavior to lower temperatures in comparison with the dielectric case (see Fig.4.4B) down to $\omega\tau_{1,e,\min} \sim 1$. At even lower temperatures Q^{-1} is expected to decrease linearly with temperature.

The effect of the condensation of quasiparticles into the superconducting state on the electron-assisted relaxation mechanism of the TS's in amorphous metals was taken into account theoretically by Black and Fulde (1979) for the simplified case when $E \ll \Delta_s(T)$, where $\Delta_s(T)$ is the superconducting energy gap. Based on arguments from the BCS-theory, one finds for the relaxation rate of TS's due to the interaction with electrons (Black and Fulde (1979))

$$\tau_{1,e,s}^{-1} = 4Ku^2k_BT/(1 + \exp(\Delta_s(T)/k_BT)). \tag{4.18}$$

The acoustic properties of superconducting glasses depend strongly on the superconducting transition temperature T_c. If T_c is in the temperature region where electrons dominate the relaxation of TS's, a kink at T_c and an *increase* of the sound velocity at $T < T_c$ is expected because of the rapidly increasing number of quasiparticles. If T_c lies in the region where phonons determine the relaxation of TS's, one expects the same behavior as in dielectric glasses and no special anomaly should be observed at T_c. If T_c lies in an intermediate region then one expects though not pronounced but clear features around T_c. For the internal friction at acoustic frequencies, the standard theory predicts a *decrease* at $T \leq T_c$ if T_c lies in the electron dominated temperature region. If T_c lies at higher temperatures where phonons dominate, i.e. $\omega\tau_{1,p,\min} \ll 1$, then no differences in the acoustic properties between insulating and superconducting glasses should be observed. These features can be recognized easily by the reader using the curves plotted in Fig. 4.4.

The measured acoustic properties of superconducting metallic glasses in the superconducting and normal states and at GHz-frequencies (Weiss et al.

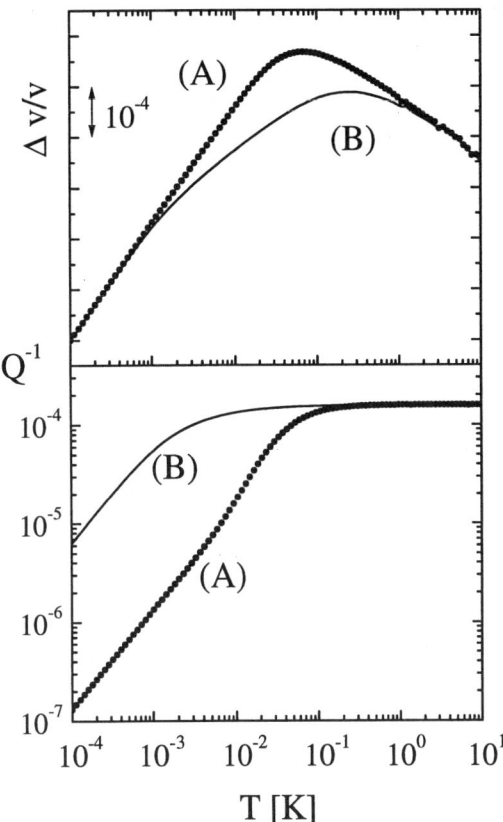

Fig. 4.4. Predictions of the standard tunneling model for the acoustic properties taking into account the relaxation of TS's with phonons and conduction electrons. The parameters are: $C = 10^{-4}, \nu = 10^3$ Hz, $Ak_B^3 = 8.5 \times 10^7 \mathrm{s}^{-1}\mathrm{K}^{-3}$ and a coupling constant to electrons $Kk_B = 10^5$ $\mathrm{s}^{-1}\mathrm{K}^{-1}$ [curve (A)] and $5 \times 10^6 \mathrm{s}^{-1}\mathrm{K}^{-1}$ [curve (B)] is used.

(1982), Arnold et al. (1982a)) appeared to be consistent to these theoretical considerations. However, experiments at high and low frequencies (Esquinazi et al. (1986)) showed that the situation is actually more complicated.

Strong Coupling Theory. The assumption that the coupling between conduction electrons and TS's can be treated as a weak perturbation is generally not valid at low temperatures. Vladár and Zawadowski (1983) pointed out that the motions of the tunneling entity and of the conduction electrons screening cloud are strongly correlated. This strong coupling between conduction electrons and TS's might give rise to a "bound state" that would affect the intrinsic parameters of the tunneling entity. The strong coupling regime is formed at temperatures below a so-called Kondo temperature T_K that depends on the dimensionless coupling constants v_x and v_z between TS's and conduction electrons for off-diagonal and diagonal interaction, respectively. Usually, the standard tunneling model assumes $v_x \ll v_z$. However, Vladár and Zawadowski (1983) showed that v_x might increase towards low temperatures, thus being more important for the relaxation of the TS's. Therefore, this model provides an additional channel for the relaxation of the tunneling systems to electrons.

One of the interesting results of this model that may at least partly contribute to a comprehension of the behavior observed in metallic glasses, is the reduction of the energy splitting E. This reduction depends on the symmetry of the tunneling entity. At low temperatures, the reduction of E due to the electron screening could be more than two orders of magnitude if $\Delta_0 > \Delta$. But for TS's with $\Delta_0 < \Delta$ the reduction of E is smaller. This indicates that $\sim 15\%$ of the TS's may enter in a strong coupling regime with a screened E. To our knowledge there are no published calculations based on this model that unambiguously show to be compatible with the experimentally determined acoustic data for metallic glasses and polycrystals reviewed below. The main obstacle for such calculations is the unknown temperature and energy dependence of the coupling constants as well as the behavior in the region $T < T_K$. Nevertheless, as we will see below the experimental data appear to support the idea of a strong coupling regime. We should mention that recent experiments of the differential resistance in point contacts with structural disorder (Ralph and Buhrman (1992)) seem to provide evidence supporting the strong coupling between TS's and conduction electrons described by the Kondo-like model of Vladár and Zawadowski (1983).

The Electron-Polaron Effect. The interaction of TS's in an amorphous metal has been considered by Kagan and Prokof'ev (1988, 1990) taking into account the role of the electron-polaron effect. This effect is due to the slow electronic excitations that do not follow the tunneling entity. Therefore, the corresponding part of the electron wave function, i.e. the nonadiabatic part, orientates towards the center of the double-well potential. The main results of this theory for the acoustic properties depend on the parameter b that is proportional to the density of states at the Fermi surface and the averaged value at the Fermi surface of the difference between the matrix element of the tunneling entity interacting with electrons in the two sides of the two-well potential. The parameter b can be identified as the electron-TS's coupling constant K introduced above. According to the authors, it is expected that in most cases $b \lesssim 0.5$.[3]

An important result that follows from the theoretical work of Kagan and Prokof'ev (1986) is that the energy splitting E is now normalized to $E_* = \sqrt{\Delta^2 + \Delta_*^2}$ because the electron-polaron effect renormalizes the tunneling amplitude (see also Sect. 3.5.2)

$$\Delta_* = \Delta_0 (\Delta_0/\omega_0)^{b/(1-b)}, \qquad (4.19)$$

where ω_0 is a characteristic frequency of the tunneling entity underbarrier motion. In the superconducting state the renormalization given above re-

---

[3] In this section we use the notation given by Kagan and Prokof'ev (1988, 1990), see also Sect. 3.5. In particular, see Sect. 3.5.4 and Chap. 7 for the influence of conduction electrons on the quantum-tunneling motion of H.

mains if $E_* > 2\Delta_s(T)$. However, the other limit is of importance and the TS's characterized by $E_* < 2\Delta_s(T)$ will have a transition from a strong to weak coupling regime entering in the superconducting state. In the low-temperature limit $T \ll T_c$, the renormalized tunneling energy is given by $\Delta_{*s} = \Delta_0(2\Delta_s(0)/\omega_0)^b$. The theoretical results for the acoustic properties taking into account the electron-polaron effect are more complicated than those from the standard theory. Following Kagan and Prokof'ev (1988, 1990) the results can be summarized as follows. The sound dispersion due to the resonant interaction in the normal state is given by

$$\left.\frac{\Delta v}{v}\right|_{\text{res}} = (1-b)C\ln(T/T_0), \quad E_{\max} > \Delta_{*,\max}. \tag{4.20}$$

It is important to note that according to (4.20) the slope of the logarithmic dependence of the sound dispersion due to the resonant interaction is not universal, as in the standard tunneling model, but depends on the characteristics of the TS's-electron interaction as well as on an intrinsic property of the metal through the coefficient b. In this regime and due to the renormalization of Δ_0, the value of P_0 obtained from acoustic data in the normal state is smaller by a factor $1/2 \leq (1-b) \leq 1$.

For the sound dispersion due to the relaxation interaction it is necessary to introduce the existence of a low energy cutoff $\Delta_{0,\min}$ in the distribution function $P(\ln(\Delta_0))$. The authors obtain

$$\left.\frac{\Delta v}{v}\right|_{\text{rel}} \simeq -\frac{C}{2}\ln(2\pi bT/\omega), \quad T(\omega/2\pi bT)^{1/2(1-b)} > \Delta_{*,\min}, \tag{4.21}$$

$$\left.\frac{\Delta v}{v}\right|_{\text{rel}} \simeq -(1-b)C\ln(T/T_0), \quad T(\omega/2\pi bT)^{1/2(1-b)} < \Delta_{*,\min}. \tag{4.22}$$

The first case describes the expected dependence of the sound dispersion at high frequencies and/or high temperatures and/or small b-values, and the second case applies at low frequencies and/or low temperatures. The overall renormalization of sound velocity is given as in the standard model by the sum of the two contributions

$$\frac{\Delta v}{v} \simeq C(1/2-b)\ln(T/T_0), \quad T(\omega/2\pi bT)^{1/2(1-b)} > \Delta_{*,\min}, \tag{4.23}$$

$$\frac{\Delta v}{v} \simeq \text{const.}, \quad T(\omega/2\pi bT)^{1/2(1-b)} < \Delta_{*,\min}. \tag{4.24}$$

This results in a nonuniversal logarithmic slope $C(1/2-b)$ for the case of high frequencies and/or high temperatures and a temperature-*independent* sound velocity in the other case, in clear contrast to the results of the standard tunneling model, see Table 4.2.

The renormalization of the sound velocity in the superconducting state gives the following results:

$$\left.\frac{\Delta v}{v}\right|_{\text{res}} \simeq C\ln(T/T_0) - Cb\ln(2\Delta_s(T)/k_BT_0^*), \quad \Delta_* > 2\Delta_s(T) > 0, \tag{4.25}$$

$$\frac{\Delta v}{v} \simeq \frac{C}{2} \ln(T/T_0') - Cb\ln(\Delta_s(T)/T_0''), \qquad (4.26)$$

$$[T\omega(1+\exp(\Delta_s(T)/k_\mathrm{B}T))/(4\pi b)]^{1/2}(2\omega_0/e\Delta_s(T))^b > \Delta_{0,\min},$$

$$\frac{\Delta v}{v} \simeq \mathrm{const.}, \qquad (4.27)$$

$$[T\omega(1+\exp(\Delta_s(T)/k_\mathrm{B}T))/(4\pi b)]^{1/2}(2\omega_0/e\Delta_s(T))^b < \Delta_{0,\min},$$

where T_0, T_0^*, T_0' and T_0'' are different constants. As in the standard tunneling model (4.18), the relaxation contribution becomes negligible at temperatures $k_\mathrm{B}T < \hbar\omega(1+\exp(\Delta_s(T)/k_\mathrm{B}T))/4\pi b$. The striking theoretical results listed above indicate that at low enough frequencies, there should be practically no change of the sound velocity with temperature in the normal-conducting state, as well as at the crossing to the superconducting state, until the superconducting gap is large enough. At $T < \Delta_{0,\min}^2(e\Delta_s(T)/2\omega_0)^{2b}2\pi b/\omega$ and if both processes contribute to the relaxation of TS's, the sound velocity should show no temperature dependence. Therefore, a "latent" temperature interval may exist near T_c where no significance difference in the temperature dependences of the sound velocity in the normal and superconducting states can be observed.

If the internal friction is determined only by the relaxation contribution, as is the case at acoustic frequencies, with this theoretical approach one obtains

$$Q^{-1} = (C\pi/2)P(\ln(\Delta_0'(T))/P_0. \qquad (4.28)$$

In contrast to the standard model, the internal friction may depend on temperature even in the region $\omega\tau_{1,e} < 1$. This is due to the assumption that the distribution function $P(\ln(\Delta_0))$ must decrease for $\Delta_0 \to 0$ because $P(\lambda) \to 0$ for $\lambda \to \infty$. In the normal state, $\Delta_0' \propto T^{0.5-b}$ and in the superconducting state at $T \ll T_c$, $\Delta_0' \propto T^{1/2}\exp(\Delta_s(T)/2k_\mathrm{B}T)$. Therefore, a slow increase of Q^{-1} with temperature in the normal phase is expected. For $T < T_c$, Δ_0' increases exponentially with decreasing temperature and, therefore, the distribution function increases too. As a result the internal friction might be larger in the superconducting state than in the normal state, in clear contrast with the expectation from the standard theory. The results of Kagan and Prokof'ev (1988, 1990) indicate that different temperature dependences can be observed depending on the value of the cutoff $\Delta_{*,\min}$ or Δ_0' that, as well as the distribution function $P(\ln(\Delta_0))$, are unknown parameters. Therefore, a quantitative comparison between theory and experiment is not possible without particular assumptions.

Within the same ideas described above, Coppersmith and Golding (1993) discussed in detail the influence of the strong coupling between conduction electrons and TS's within a linear response approximation in the incoherent ($k_\mathrm{B}T > \Delta_*$) and coherent or low-temperature region ($k_\mathrm{B}T < \Delta_*$), where the standard tunneling model should be valid. They compared the predictions of the theory in the two above-mentioned regions with previously published data of PdSiCu amorphous metal in the GHz range (Golding et al. (1978)). Although a better agreement between strong coupling theory and experiment is obtained (in comparison with the standard tunneling model), several discrepancies still remain. A comparison of the experimental acoustic data at kHz-frequencies with theory is presented in Sect. 4.4.2.

Defect Interactions in Metallic Glasses. In Chap. 5 (see Sect. 5.1.2) we discuss in detail how a strain-mediated dipolar-like interaction between defects may explain experimental data of dielectrics at very low temperatures. Support for the dipolar-like interaction model is also provided by the anomalous temperature dependence of the internal friction at very low temperatures discussed below in this chapter. In relation with the electron-TS's problem, Coppersmith (1993) proposed that if the density of states P_0 is determined by the interaction coupling constant $g \sim \gamma^2/\varrho v^2$, as one can deduce from the interaction model (Yu and Leggett (1988)), then P_0 should be larger in the superconducting state than in the normal state. This conclusion is obtained if one assumes that: (a) in the normal state, in addition to the strain-mediated interaction $\propto g_{\mathrm{ph}}/r^3$, the RKKY-like interaction between TS's and electrons would also decay as g_{RKKY}/r^3 where r is the distance between defects; (b) $P_0 \propto 1/g_{\mathrm{eff}}$ where $g_{\mathrm{eff}} \sim (g_{\mathrm{ph}}^2 + g_{\mathrm{RKKY}}^2)^{1/2}$; (c) the significant length scale of the interaction at a given temperature T is of the order of $(g/k_{\mathrm{B}}T)^{1/3} \gg l_e, \xi$ – the electron mean free path and the superconducting coherence length, respectively – at $T \sim 0.1$ K. In this case one might expect that $P_0 \propto 1/g_{\mathrm{eff}}$ is larger in the superconducting state than in the normal state because $g_{\mathrm{ph}} < g_{\mathrm{eff}}$. It is interesting to note that, within this idea, the electron mean free path can play a role in the electron-TS's interaction problem. If the electron mean free path is much larger than the characteristic length of the defect or RKKY interaction, one might expect to observe probably no electronic influence on the acoustic properties. An experimental test to these ideas was performed recently and will be described in Sect. 4.5.3, see also Sect. 4.4.4.

4.2.4 Influence of the Acoustic Intensity

Relaxation due to Phonons. In Sect. 4.2.1 we have pointed out that if the acoustic wave has enough energy to change the population of the upper energy levels, a nonlinear absorption is observed. This resonant absorption can be saturated at large enough acoustic intensities. In this section, we discuss another kind of nonlinear behavior for the acoustic properties, in particular for the sound dispersion in the temperature region where the resonant interaction provides the main contribution.

This nonlinear effect occurs when the probing sound wave produces a modulation of the characteristic energy of the TS's of the order of the thermal energy. Suppose we perform a typical acoustic experiment at $T = 1$ mK and at a frequency $\nu = 1$ kHz with a maximum strain of the probing wave

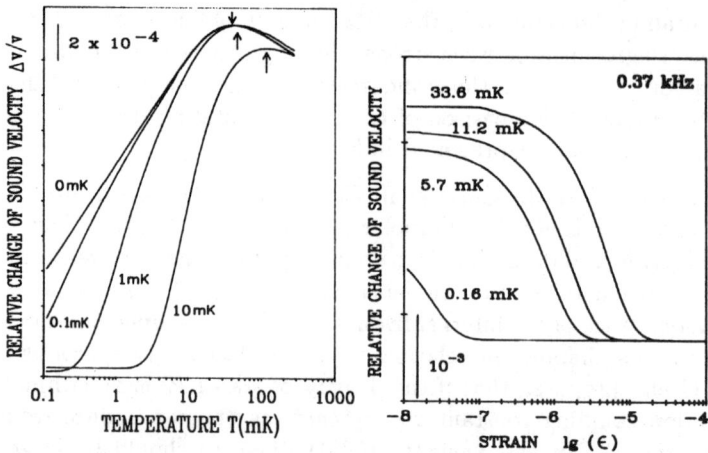

Fig. 4.5. Left figure: Theoretical curves for the relative change of sound velocity of a dielectric glass vs. temperature at 400 Hz following the standard tunneling model but with a population difference $\propto \tanh((E+d_0)/2k_BT)$. The numbers represent the strain field d_0 in mK. The arrows indicate the position of $T_{\max}$ (taken from Esquinazi et al. (1992)). Right figure: Strain dependence of the sound velocity at different temperatures following the standard tunneling model and assuming a homogeneously distributed and time-independent strain d_0 with a coupling constant between phonons and TS's $\gamma = 1$ eV (after König et al. (1993)).

$\epsilon_0 = 10^{-7}$. Taking a coupling constant between phonons and TS's of $\gamma = 1$ eV (as is the case for amorphous SiO_2, see Sect. 4.4.1), the modulation energy $d_0 = \gamma\epsilon_0 = 1$ mK becomes comparable to the thermal energy. The minimum relaxation time of TS's at $T = 1$ mK can be estimated at $\tau_{1,\min} \sim 10$ s. This value is much larger than the period of the acoustic wave $2\pi\omega^{-1} \sim 10^{-3}$ s, i.e. the TS's have no time to relax within one period of the wave. One can estimate the change produced by the acoustic wave at this low temperature, assuming an effective change in the population of the energy levels in equilibrium, i.e. $n_0 \propto \tanh(E/2k_BT) \to \tanh((E+d_0)/2k_BT)$. Numerical calculations using (4.14) with the shifted E in the population difference and including the relaxation contribution provide the curves shown in Fig. 4.5 for the sound velocity as a function of temperature at constant strains (left figure) or as a function of strain at constant temperatures (right figure).

From the numerical results we identify three main features: (1) the position of the maximum of the sound velocity at $T_{\max}$ depends on the applied strain; (2) the sound velocity increases at $T < T_{\max}$ stronger than the predicted logarithmic temperature dependence, and (3) it saturates at low temperatures. The saturation value is approximately independent of the applied strain. The integration of (4.14) over one period of time of the probing wave results in a time dependence of the tunneling splitting, assuming that the strain shifts the asymmetry as

$$E(t) = \sqrt{\Delta_0^2 + (\Delta + 2\gamma d_0 \sin(\omega t))^2}. \tag{4.29}$$

This triple numerical integration has been carried out by Esquinazi et al. (1994) and the results resemble qualitatively those presented in Fig. 4.5.

Within the same idea described above, Parshin (1993) and later Stockburger et al. (1995), based on an earlier work of Galperin et al. (1984), published a rigorous treatment of this nonlinear problem within the adiabatic approximation, i.e. $\hbar\omega \ll E, \hbar \mid dE/dt \mid \ll E^2$. As pointed out by Parshin (1993), the differentiation between resonant and relaxation processes is artificial since at the working acoustic frequencies both contributions come from thermally activated TS's with energy $E \sim k_B T$. Therefore, a different theoretical approach is necessary for a thorough treatment of this nonlinear effect.

In the limit of slow relaxation of the TS's in the saturation regime, i.e. $T \ll T_{max}$ and $\gamma\epsilon_0 T^2/2 \ll k_B T_{max}^3$, the sound dispersion can be approximated by (Stockburger et al. (1995))

$$\frac{\Delta v}{v}(T) - \frac{\Delta v}{v}(T = 0 \text{ K}) \simeq 8.8\, C \left(\frac{k_B}{\gamma\epsilon_0}\right)^2 T^2 \ln\left(\frac{2k_B T}{\Delta_{0,min}}\right). \tag{4.30}$$

In the temperature region around T_{max} the theoretical analysis is more difficult and an analytical approximation for the limits $\gamma\epsilon_0/2 \ll 2k_B T$ and $\gamma\epsilon_0/2 \ll 2k_B T(T_{max}/T)^{3/2}$ has been obtained by Stockburger et al. (1995). Their results resemble those plotted in Fig. 4.5 (left figure) and, as we will see below, agree with the experimental data obtained in dielectric and some polycrystalline metals. Note that in this treatment, TS's with very small tunneling energy Δ_0 contribute significantly to the acoustic response (the nonlinear regime occurs in the region $\omega\tau \gg 1$) leading to a divergence unless a lower energy limit $\Delta_{0,min}$ is assumed for the distribution of TS's.

In the slow relaxation limit ($\omega\tau_1 \gg 1$) Stockburger et al. (1995) computed also the internal friction. The nonlinear internal friction is reduced from the linear-response result $Q \propto T^3$ (see Table 4.1) according to

$$Q_{nlr}^{-1}(T) \simeq \frac{2k_B T}{\gamma\epsilon_0} Q^{-1}(T). \tag{4.31}$$

Relaxation due to Conduction Electrons. As discussed in Sects. 4.2.1 and 4.2.2, the acoustic properties due to the resonant process depend on the acoustic intensity I [see (4.3) and (4.13)]. We have pointed out above that one of the main differences between a normal-conducting amorphous metal and an insulating glass is that the acoustic intensity necessary to observe the saturation of attenuation is several orders of magnitude larger. This experimental fact has been ascribed to the significantly shorter relaxation time due to the coupling of the TS's to the conduction electrons. However, various experimental results (see Sects. 4.4.2 and 4.4.3) indicate that for a comprehension of the acoustic properties in normal and superconducting amorphous metals, the existence of a shorter relaxation time only is not sufficient to explain the experimental results; it seems that a new approach to the problem is required.

A particular approach to the problem of the electron-tunneling system interaction applied to the acoustic properties of amorphous metals (including nonlinear response) was studied theoretically by Stockburger et al. (1994). They assume that the conduction electrons behave collectively as if they were bosons with a spectral density of the Ohmic form [Weiss (1993), see (3.7) and (3.137)]. The authors avoid using the description of the nonlinear behavior in terms of occupation number and of energy levels as in the case of phonon relaxation. This is due to the very short relaxation time of the TS's arising from the electron-assisted relaxation; the charge fluctuation of conduction electrons destroys the quantum coherence of the tunneling process. For typical experiments with mechanical oscillators and assuming the standard tunneling model (Korringa-like relaxation time, see Sect. 4.2.3), the nonlinear regime occurs in the region $\omega \ll \tau_{1,e}^{-1} \propto \Delta_0^2$. Therefore, TS's with very small Δ_0 do not significantly affect the acoustic properties in the experimental temperature range and the final result for the relative change of sound velocity or resonance frequency of the mechanical oscillator (vibrating reed) does not depend on $\Delta_{0,\min}$ in contrast to the case for dielectric glasses (Stockburger et al. (1995)).

Following Stockburger et al. (1994), in the saturation regime ($\gamma\epsilon_0 \gg k_B T$) the change in the resonance frequency of the reed (sample) is given by

$$\frac{\Delta\omega}{\omega} = \frac{C}{2}\left[\ln\left(\frac{\hbar\bar{\epsilon}}{k_B T_0}\right) - f(\ln(\hbar\bar{\epsilon}/2k_B T)) + 0.59\right], \tag{4.32}$$

where $T_0 = \hbar\pi\alpha\Delta_{0,\max}^2/k_B\omega$, $\alpha = \eta d^2/2\pi\hbar$, $\bar{\epsilon} = \gamma u_{\max} d\kappa^2/\hbar\sqrt{3}$, $u_{\max}$ is the maximum reed amplitude, η is a viscosity coefficient for the motion of the tunneling entity in the electron gas, d is the thickness of the reed, $\kappa = 1.875/l$ for the fundamental bending mode, l is the length of the reed, and $f(x) = (x/(x-1))\ln(x) + 0.19$.

The corresponding equation for the linear regime (compare with Table 4.2) ($\gamma\epsilon_0 \ll k_B T$) is (Stockburger et al. (1994))

$$\frac{\Delta\omega}{\omega} = \frac{C}{2}\left[\ln\left(\frac{T}{T_0}\right) - 0.33\right]. \tag{4.33}$$

In the nonlinear regime, the internal friction is not temperature independent as expected from the linear standard tunneling model but decreases with temperature according to (Stockburger et al. (1994))

$$Q^{-1} = \frac{\pi}{2}Cg(\ln(\hbar\bar{\epsilon}/2k_B T)), \tag{4.34}$$

where $g(x) = (x - 1 - \ln(x))/(x - 1)^2$.

4.2.5 Coherent Coupling Below 100 mK

As noted in Chap. 3 and as will be discussed later in this chapter, the assumption of independent tunneling entities with a distribution of asymmetry and tunneling energies given by (2.20) that interact with phonons provides a reasonably good description of the low-temperature properties below 1 K. To some extent a good description is also obtained if conduction electrons determine the relaxation of the TS's.

The assumption of independent TS's, however, can be valid only if the interaction of the medium (phonons and/or conduction electrons) with a tunneling entity ($\sim \hbar/\tau_1(T)$) is stronger than the interaction energy between TS's. We have pointed out already in the discussion of the conduction electron-assisted relaxation (Sect. 4.2.3) that Yu and Leggett (1988) stressed the importance of the interaction with a $1/R^3$-dependence between defects (with R the distance between them) that may be responsible for the anomalous properties of amorphous systems at low temperatures. Based on this idea and stimulated by the internal friction results on vitreous silica (Sect. 4.4.1), Burin and Kagan (1993) (see also Burin and Kagan (1994b)) proposed that at low enough temperatures, the interaction between tunneling or two-level systems leads to a subsystem of collective excitations and a coherent dynamic coupling, i.e. transitions involving two or more two-level systems. This idea will be discussed in more detail in Chap. 5. We would like to point out here some key concepts that may explain the observed temperature dependence of the internal friction in vitreous silica at very low temperatures.

Following Burin and Kagan (1994b), the coherent coupling between TS's leads to a new tunneling entity with collective excitations. Considering only a pair of TS's and *up–down* transitions, i.e. from a state $\uparrow\downarrow \longrightarrow \downarrow\uparrow$, the tunneling energy and asymmetry of the pair is given by $\Delta_{0,p} = U_{12}\Delta_{0,1}\Delta_{0,2}/2E_1E_2$ and $\Delta_p = E_1 - E_2$, where the indices 1,2 indicate the tunneling system 1 or 2, and U_{12} is the coupling between them that is proportional to $1/R_{12}^3$ (for more details see Sect. 5.1.4). The coherent coupling is destroyed if $\Delta_{0,p}\tau_1 < \hbar$, where τ_1 is the relaxation time of an independent tunneling system. For the formation of a coherently coupled pair, there should exist a distance smaller than a radius R_c that is estimated as follows:

$$\tau_{1,\min}^{-1} \simeq \frac{\gamma^2}{\varrho\hbar^4 v^5} k_B^3 T^3, \quad \text{for } \Delta, \Delta_0 \sim k_B T. \tag{4.35}$$

The interaction between TS's is of the order (in Jm^3) $U_0 \sim \gamma^2/\varrho v^2$, then

$$\tau_{1,\min}^{-1} \simeq \frac{U_0}{\hbar R_c^3}, \text{ with } R_c = \frac{\hbar v}{k_B T}. \tag{4.36}$$

For tunneling amplitudes of the pair $\Delta_{0,p} > U_0/R_c^3$ the coherent coupling will be important. For typical parameters of dielectric glasses one estimates $R_c \sim 1$ μm at 100 mK. Furthermore, Burin and Kagan (1994b) estimate the probability for an individual tunneling system to find a resonance up–down partner in the coherence region determined by R_c^3. This probability is $w \sim C\ln(\Theta_D/T)\pi^2/2$, that is of the order of 0.01 at $T = 1$ K and $\Theta_D = 1000$ K. The density of states of *up–down* excitations formed by the pairs is given by $P_2(\Delta_p, \Delta_{0,p}) = P_0 C k_B T \pi^3/(12\Delta_{0,p}^2)$ that applies for $\Delta_{0,p} > U_0/R_c^3$, see (5.76). Due to the tunneling amplitude dependence of the density of states of pair excitations, this density of states becomes larger than the one for isolated TS's for $\Delta_{0,p} < k_B T C$. Burin and Kagan (1994b) estimate the relaxation rate for the system of thermal TS's (i.e. $\Delta_0, \Delta \sim k_B T$) that are forming pairs; this rate is given by $\tau_0^{-1} \sim 10 k_B T C^3/\hbar$, see (5.97). The crossover between the individual phonon-assisted relaxation (one-phonon process) and the collective excitations is obtained by equating $\tau_{p,\min}$ to $\tau_{1,\min}$. A crossover would be observable at temperatures $T < T'$, where T' is given by (5.102). Replacing in this equation values appropriate for SiO_2, one obtains a crossover temperature $T' \sim 30$ mK, assuming that the interaction with transversal phonons is significant, see the discussion below (5.112).

An estimate of the internal friction for $T < T'$ can be easily done for the limit $\omega\tau_{\min} \ll 1$. In this case the internal friction due to the relaxation process can be approximated as $Q^{-1} \sim C/\omega\tau_{\min}$. Replacing the rate $\tau_{1,\min}^{-1}$ by $\tau_{p,\min}^{-1}$, one obtains a linear temperature dependence (Burin and Kagan (1994b)) due to the collective excitations of the ensemble of TS's, see Sect. 5.1.5.

4.2.6 Acoustic Properties Above 1 K: Thermal Activation and Incoherent Tunneling

Several authors tried to correlate the high-temperature behavior of the acoustic properties ($T > 1$ K) with properties at low temperatures (Hunklinger and Arnold (1976), Phillips (1990)). In particular, it has been proposed that the maximum in the attenuation of phonons observed in amorphous materials at $T > 10$ K can be interpreted assuming a thermally activated relaxation of the TS's. As for the heat release (see Chap. 2), in order to compute the influence of thermal activation, a thermally activated relaxation time $\tau_{TA} = \tau_0 \exp(+V/k_B T)$ is introduced. Furthermore, an important underlying assumption is that both processes, thermal activation and quantum tunneling, are independent and the total rate is simply given by the sum of both.

Tielbürger et al. (1992) estimated the temperature dependence of the acoustic properties assuming two well-defined harmonic potentials. As discussed in Chap. 2 (see Sect. 2.2.2), the authors obtained $\lambda = V/E_0$, where E_0 is the zero point energy. Just above the temperature-independent region due to tunneling, the internal friction is given by

$$Q^{-1} = \frac{\pi C k_B T}{E_0}. \tag{4.37}$$

Sound attenuation proportional to the temperature has been observed by Duquesne and Bellesa (1979) in polystyrene at ultrasonic frequencies. In the same temperature region ($\omega\tau_0 \ll 1$), the sound dispersion can be written as (Tielbürger et al. (1992))

$$\frac{\Delta v}{v} = \frac{Ck_BT}{E_0}\ln(\omega\tau_0). \tag{4.38}$$

The linear temperature dependence as well as the logarithmic dependence in $\omega\tau_0$ of the sound velocity (4.38) has been first published by Bellessa (1978). He observed a linear temperature dependence above a few Kelvin in vitreous silica as well as in an amorphous metal, but not the predicted $\ln(\omega)$ change for the slope of the sound velocity as a function of temperature (4.38) (Bellessa (1978)). By comparing different experimental data for SiO_2, however, Tielbürger et al. (1992) claimed that a reasonable agreement can be found, although the physical meaning of some prefactors, like τ_0, remains unclear. A possible explanation for the striking behavior of the frequency and temperature dependence of the sound velocity in amorphous PdSiCu, based on phonon-assisted tunneling, is given by Würger on p. 122 (Chap. 3). A comparison of the thermally activated model to data for amorphous polymers is given in Sect. 4.4.1.

Strictly speaking, the crossover from quantum to thermal activation might be more complicated than the simple addition of the two rates. Theoretical work on the crossover from tunneling to thermally activated motion of a string across a barrier has been done by Affleck (1981), Ivlev and Mel'nikov (1987) and recently by Gorokhov and Blatter (1996). To our knowledge, no attempt has been made so far to apply these theoretical concepts to the problem of TS's in disordered media.

A new explanation for the acoustic properties above 1 K has been proposed by Neu and Würger (1994b) based on a relaxation mechanism arising from overdamped TS's, see the discussion in Sect. 3.7.8. Neu and Würger (1994b) proposed that above a few Kelvin incoherent tunneling rather than a coherent oscillation of the TS's affects substantially the dynamics of the TS's. In this regime, the relaxation interaction between phonons and TS's is enhanced, see (3.337) and details in Chap. 3.

At temperatures $T > T^*$ all thermal TS's ($E \leq k_BT$) are overdamped and their motion is no longer coherent. Following Neu and Würger (1994b), (see 3.135), the crossover temperature is given by

$$T^* \approx \frac{2\hbar}{\pi^2 k_B}\frac{1}{\tilde{\gamma}}, \tag{4.39}$$

where $\tilde{\gamma} = \gamma/\pi(\varrho v^5\hbar)^{1/2}$. In the incoherent tunneling regime the relative change of the sound velocity is given by

$$\frac{\Delta v}{v} = \beta(T)T, \qquad \beta(T) = -0.446\frac{C}{T^*}\ln\left(\frac{k_BT^2}{\hbar\omega T^*}\right). \tag{4.40}$$

Note that there is a direct proportionality between the parameters C and β. Since C can be obtained from the logarithmic slope of the sound velocity or from the plateau in the internal friction below 1 K, this proportionality can be experimentally verified. Experimental evidence that supports the model of incoherent tunneling above a few Kelvin is discussed in Sect. 4.5.5.

One should not confuse the incoherent tunneling referred to in this section with another kind of "incoherent tunneling" used by Enss and Hunklinger (1997) to interpret the low-temperature acoustic properties of oxide glasses. In the work of Enss and Hunklinger (1997) it is assumed that elastic interactions below 100 mK can be comparable to the thermal energy. Then the interaction may decrease the tunneling amplitude for the elastically coupled TS's and change or destroy the phase coherence (see also the discussion on p. 181).

4.3 Experimental Details

The investigation of the sound velocity and internal friction of materials at low temperatures provides information about the density of the TS's and their coupling to phonons and/or electrons. According to the frequency dependence of the internal friction and of the sound velocity (see Sect. 4.2.1), the choice of the experimental method (and therefore of the frequency range) plays a crucial role for the study of the acoustic properties of a given material. In Sect. 4.3.1, experimental methods for acoustic measurements at low and high frequencies are briefly summarized, although we mainly concentrate our discussion on the results obtained with two methods suitable for the kHz-frequency range: the vibrating reed and the vibrating wire method both described in Sect. 4.3.2 in more detail. As far as the theoretical treatment of the two methods is concerned, they are both based on the same mathematical description of a driven harmonic oscillator. From the experimental point of view, both have in common that the sample to be investigated has to be clamped on one (reed) or on two (wire) ends. This clamping of the sample onto the sample holder may cause significant contributions to the overall damping. Therefore, Sect. 4.3.3 contains a brief discussion of the influence of the clamping on the experimental results. Finally, as we report on measurements of the acoustic properties of materials in a temperature range of five orders of magnitude ($\approx 10^{-4}$K $< T <$ 10K), this section will be closed with a short description of the experimental environment used to perform these experiments down to such very low temperatures (Sect. 4.3.4).

4.3.1 Experimental Methods for Low and High Frequencies

The early results on the investigation of the acoustic properties of insulating glasses in the MHz- and GHz-frequency range provided important support

for the interpretation of the "anomalous" specific heat and thermal conductivity in terms of two-level TS's:
- In the ultrasonic frequency range the attenuation of a sound wave in amorphous insulators shows a strong dependence on the applied acoustic intensity with a saturation at high acoustic intensity and power-independent attenuation at low intensity (Hunklinger et al. (1972), Golding et al. (1973), Hunklinger et al. (1973), see Sect. 4.2.1).
- The study of the relative change of the sound velocity in different glasses (see e.g. Piché et al. (1974), Hunklinger and Piché (1975)) at frequencies 30MHz$\leq \nu \leq$150MHz and in the temperature range $\sim$0.3K$\leq T \leq$4.2K performed at acoustic intensities where the ultrasonic attenuation is completely saturated, revealed the characteristic temperature dependence with a frequency-dependent maximum $T_{\max}$ at about 2K. The coupling of the two-level systems to the phonons (sound wave) due to the resonant interaction leads to a logarithmic increase in the sound velocity at low temperatures, but as the relaxation process becomes more and more important at higher temperature the sound velocity decreases at $T >$2K.

These studies of the acoustic properties of glasses at ultrasonic frequencies showed the intensity dependence of the resonant attenuation ($\sim I^{-1/2}$, Hunklinger et al. (1972)) as well as the logarithmic temperature dependence of the sound velocity (see e.g. Hunklinger and Piché (1975)) as expected from the tunneling model. However, the position of the maximum in the sound velocity at about 2K is close to the temperature where in addition to the one-phonon process, other relaxation processes, e.g. phonon processes of higher order, incoherent tunneling or relaxation due to thermal activation, become increasingly important, thus making the analysis in terms of resonant and relaxation contributions to the sound velocity difficult. Ultrasonic experiments are usually performed with transducers made of a piezoelectric bulk crystal, ceramic or of a thin film of e.g. ZnO deposited onto either one or two faces of the sample. Before sputtering ZnO onto the faces, they are polished and coated with metal films (usually Cr and Au or Au) for a ground plane. The transducer generates a sound wave in the sample and the variation of the amplitude of the echo of the acoustic signal provides information on the attenuation of the sound wave in the sample (see e.g. Golding and Graebner (1976)). In case of a single transducer, it simultaneously serves as transmitter and receiver. Apart from the investigation of velocity and damping of the sound wave, experiments using different sequences of high frequency pulses provide information on the relaxation times τ_1 (longitudinal relaxation time) and τ_2 (dephasing time) similar to the equivalent magnetic experiment of saturation recovery and spin echoes. For a detailed discussion of pulse echo experiments, we refer to Phillips (1987).

According to the standard tunneling model, with decreasing frequency the crossover temperature $T_{\max}$ is shifted towards lower temperatures away from the above-mentioned additional relaxation processes at Kelvin temperatures.

Therefore, in order to obtain independent information on resonant and relaxation processes, investigations of the acoustic properties of glasses at kHz-frequencies are necessary. With the maximum of the sound velocity located at about 0.1K it is possible to distinguish between resonant (at $T < T_{\max}$) and resonant plus relaxation (at $T > T_{\max}$) processes in the sound velocity. Moreover, the internal friction of insulating glasses is then only determined by the relaxation contributions (Raychaudhuri and Hunklinger (1984)) leading to an almost constant internal friction ("plateau") at $T > T_{\max}$ (for details see Sect. 4.2.2). A further advantage of the study of the acoustic properties in the kHz-range is given by the possibility of extending the measurements to very low temperatures because the energy dissipation in the sample decreases with decreasing frequency as ν^3. On the other hand, it is no longer possible at these low frequencies to study separately the longitudinal and shear phonons and their different coupling to the TS's.

Low frequency sound experiments are usually performed with the sample itself acting as a driven mechanical resonator vibrating at its natural frequency that is mainly determined by the geometrical dimension of the sample as well as its density and Young modulus. The most common techniques used in this frequency range are vibrating reed and vibrating wire techniques that are exclusively used in the experiments described in this chapter and are discussed in more detail in the following section.

4.3.2 The Vibrating Reed and Vibrating Wire Techniques

Whereas the vibrating reed technique is a well-known tool for the investigation of the acoustic properties of dielectric and metallic samples (Berry and Pritchet (1975), Raychaudhuri and Hunklinger (1984)), the main field for the experimental application of the vibrating wire technique was until recently essentially restricted to the investigation of the low-temperature dynamic properties of quantum liquids such as the viscosity of normal- and superfluid ^{3}He and liquid ^{3}He-^{4}He mixtures (Guénault and Pickett (1990), König et al. (1994a)). Here, the applicability of a vibrating wire is based on the sensitivity of the damping (of its motion in the liquid) to the scattering of ^{3}He quasiparticles. As a result of the strong temperature dependence of the mean free path of the ^{3}He quasiparticles, a vibrating wire in liquid ^{3}He provides a very sensitive thermometer down to temperatures of almost 100 μK. In experiments with quantum liquids, the damping due to the liquid is usually several orders of magnitudes larger than the intrinsic damping of the wire itself. Therefore, although there are some indications in experiments of other groups about an "anomalous" temperature dependence of the intrinsic wire properties (Morishita et al. (1989), Guénault et al. (1983)), the internal friction of a vibrating wire was considered to be negligible and the technique was used for the study of the properties of the liquids only.

In the vibrating reed as well as in the vibrating wire technique, the sample to be investigated acts as a driven oscillator excited to resonant vibrations by external driving forces. For the *vibrating reed* technique (Fig. 4.6), the sample is clamped on one end between two blocks usually made of a material with good thermal conductivity as e.g. copper or silver, and it is excited to oscillations by applying an ac-voltage of typically 1 V to an electrode located near one side of its free end. A second electrode electrostatically detects the vibrations of the sample and should be placed as close as possible to the other side of the free end of the sample in order to gain maximum sensitivity of the signal.

The *vibrating wire* (Fig. 4.7) is usually bent into a semicircular loop with the plane of the loop parallel to a (small) static magnetic field that is provided by a superconducting solenoid and that is typically of order of a few tens of milliteslas. By passing an alternating current (1nA$< I <$1mA) through the wire, the Lorentz force acts on the wire which then behaves as a driven oscillator. Compared with the vibrating reed technique, there are different boundary conditions for the equation of motion for the vibrating wire technique since both ends of the sample are clamped to the sample holder.

In both techniques, information on the acoustic properties of the material is provided either from the analysis of the resonance curve obtained from a slow sweep of the frequency through the resonance (frequency step per second $<6\Gamma/\Delta t$; Γ is the damping in s^{-1}) of the resonator or from the free decay of the signal. Figure 4.8 shows a typical example of the phase-sensitive detection of the in-phase and out-of-phase signal of the resonator in the vicinity of its resonance frequency ν_0. In the linear response regime of the resonator, the x-signal is described by a Lorentzian curve; its half-height width $\Delta\nu$ (as well

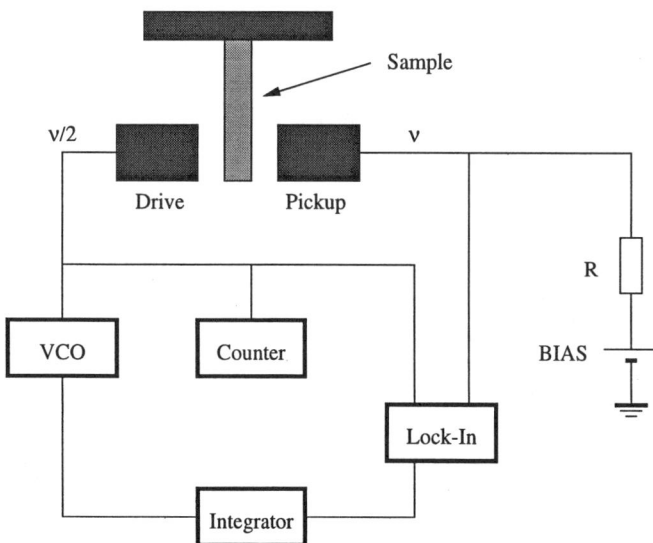

Fig. 4.6. Schematic drawing of a vibrating reed setup

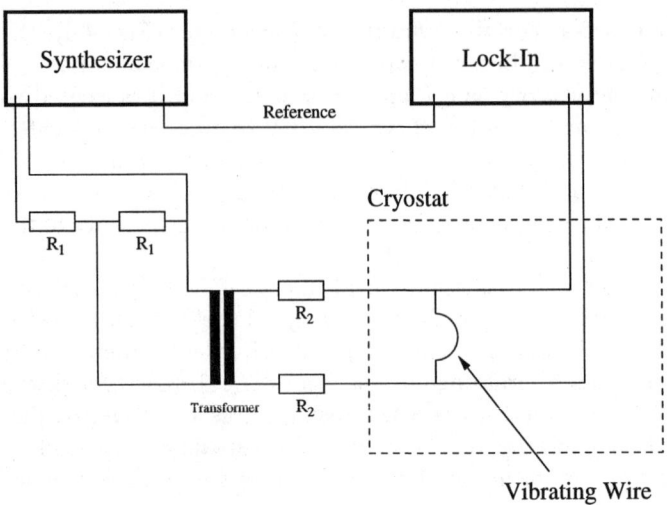

Fig. 4.7. Schematic drawing of a vibrating wire resonator.

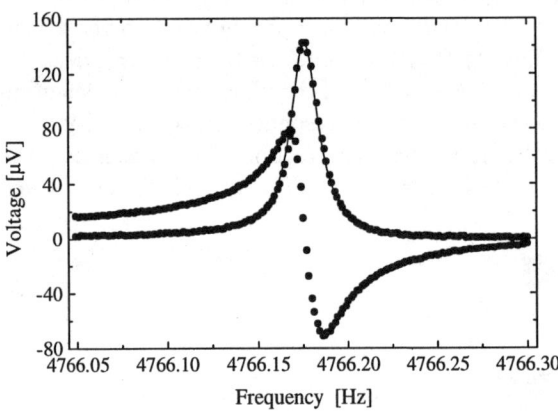

Fig. 4.8. In phase and out of phase signal of a resonance curve of a vibrating wire. The continuous line follows a Lorentzian function.

as the width between the two extrema of the y-signal) is a measure for the internal friction $Q^{-1} = \Delta\nu/\nu_0 = 2\Gamma/\nu_0$ of the material. From the free decay of the sample, the Q-value is obtained from $Q = \pi\Delta\nu t_0$ where t_0 is the decay time of the amplitude of vibration. The accuracy of the determination of the acoustic properties can be improved when the sample is locked to its resonance by a phase locked looped utilizing the large sensitivity of the y-signal between its two extrema to changes of the frequency.

The equations of motion for vibrating reed and vibrating wire provide the resonance frequencies

$$\nu_0 = \frac{1}{2\pi} \frac{s_k^2}{L^2} \sqrt{\frac{EI_a}{\varrho A}}, \tag{4.41}$$

4.3 Experimental Details

with the Young Modulus E and the geometrical moment of inertia I_a of the sample, L is the length of the reed or wire. Taking into account the different cross-section areas ($A = bd$ for the reed – b is the width and d the thickness – $A = a^2\pi$ for the wire), as well as the different geometrical moments of inertia ($I_a = bd/12$ for the reed, $I_a = \pi a^4/2$ for the wire), the fundamental frequency of the vibrating reed is

$$\nu_0 = \frac{d}{4\pi\sqrt{3}} \frac{s_{0R}^2}{L^2} \sqrt{\frac{E}{\varrho}}, \tag{4.42}$$

and of the vibrating wire

$$\nu_0 = \frac{a}{4\pi} \frac{s_{0W}^2}{L^2} \sqrt{\frac{E}{\varrho}}. \tag{4.43}$$

The different boundary conditions for the motion of the sample lead to different coefficients $s_{0R} = 1.875$ (reed) and $s_{0W} = 4.730$ (wire) for the fundamental mode. The Young modulus sound velocity $v_E = \sqrt{E/\varrho}$ is related to the longitudinal sound velocity v_l by

$$v_l = v_E \left(1 - \frac{2\sigma^2}{1-\sigma}\right)^{-1/2}, \tag{4.44}$$

with the Poisson ratio $2\sigma = (v_l^2 - 2v_t^2)/(v_l^2 - v_t^2)$. The subscripts l and t stand for longitudinal and transverse phonon polarization. As the dimensions of the mechanical resonator in an acoustic experiment are usually not known to the precision necessary to determine the *absolute* value of the sound velocity, the vibrating reed and vibrating wire technique are only applicable for measurements of *relative* changes in sound velocity $\Delta v_E/v_E = [v_E(T) - v_E(T_0)]/v_E(T_0)$ that is equivalent to relative changes of the resonance frequency $\Delta \nu/\nu_0 = [\nu(T) - \nu(T_0)]/\nu_0(T_0)$, where T_0 is an arbitrary reference temperature.

The oscillation of the resonator leads to a time-dependent strain ϵ that has its maximum value at the fixed end(s) of the reed (wire) and can be calculated from the amplitude u of the oscillation of the upper free end of the reed or in the middle of the loop of the wire, respectively, according to Esquinazi et al. (1992),

$$\epsilon \approx 1.8 \frac{du}{l^2} \quad \text{(reed)}, \tag{4.45}$$

$$\epsilon \approx 28.2 \frac{au}{l^2} \quad \text{(wire)}. \tag{4.46}$$

The numerical constants in (4.45) and (4.46) result from the equation of motion of the resonator.

We conclude this section in which we described the main properties of the vibrating reed and vibrating wire technique by summarizing the differences

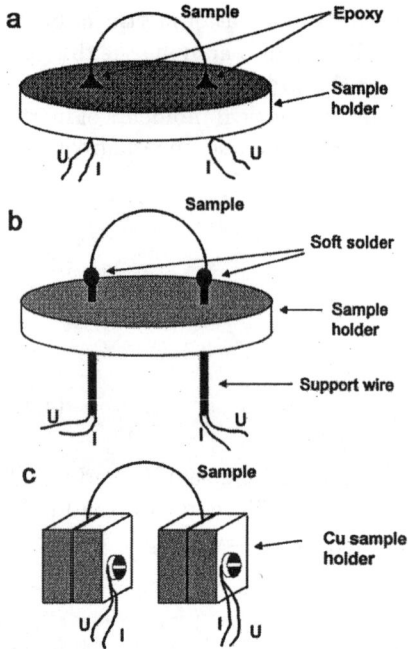

Fig. 4.9. Three different sample holders for vibrating wire experiments to study possible contributions from the clamping to the acoustic properties, after König et al. (1995).

between both techniques:
(i) Vibrating wire measurements require a steady magnetic field; i.e. in contrast to the vibrating reed technique, zero-field measurements are not possible.
(ii) Since the vibrating wire technique is based on the action of the Lorentz force, the sample to be investigated has to be a conducting material. This restriction is not valid for the vibrating reed technique.
(iii) For samples of comparable size, the typical amplitude of oscillation (that has to be small enough in order to avoid self-heating effects, see Sect. 4.3.4) is at least one order of magnitude smaller in the vibrating reed technique, thus resulting in smaller values for the strain too. In spite of these restrictions, the vibrating wire technique is a very attractive method because sample preparation and mounting of the sample are relatively easy, and several wires can be investigated simultaneously.

4.3.3 The Influence of the Clamping

In the experiments described in this work, the low-temperature values of the internal friction differs by more than two orders of magnitude among different materials (e.g. $\approx 10^{-4}$ for amorphous SiO_2 and $\approx 10^{-6}$ for annealed polycrystalline Ta). In order to be able to extract information of the bare damping of the sample as precisely as possible, a careful study of possible

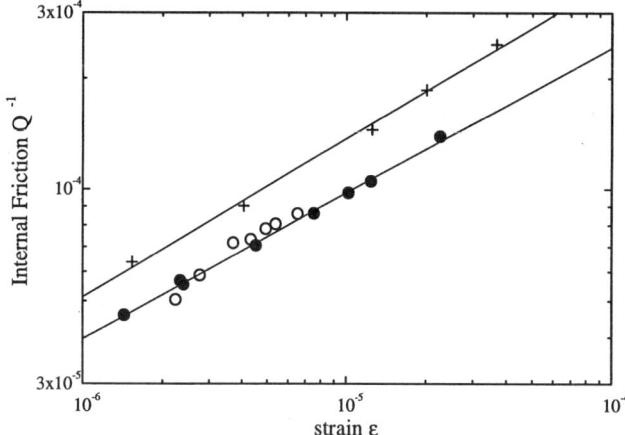

Fig. 4.10. Q^{-1} as a function of the applied strain for a Pt vibrating wire attached to different sample holders. (+) sample holder of Fig. 4.9a, $\nu = 84$ Hz. ($\bullet$, o) sample holder of Fig. 4.9b, $\nu = 1.5$ kHz.

contributions of the clamping of the sample to the measured overall damping is necessary. We have studied the contributions of various sample holders for vibrating wire measurements that differed significantly in the way the sample was attached to them (König et al. (1995)).

Figure 4.9 shows three typical sample holders: (a) The wire is fed through two tiny holes in a Ag or Cu sample holder and fixed with a tiny amount of varnish (either GE 7031 or Stycast 2850FT). (b) A wire of much larger diameter than the sample wire is fed through two holes in the sample holder and is electrically isolated from the main body of the sample holder. The wire to be investigated is soft soldered to the larger wire. (c) Each end of the sample is clamped between two small Cu blocks. Both pairs of blocks are electrically isolated from each other and from the main body of the sample holder. To avoid squeezing the sample between the Cu blocks, small slits are scratched into each Cu block where the sample ends are fixed. This last method is comparable with the mounting of a vibrating reed sample to its sample holder.

Contributions of the sample clamping to the damping should be observable, for example, by comparing the strain dependence of the internal friction for different sample holders (König et al. (1995)). Figure 4.10 shows the strain dependence of Q^{-1} for a platinum wire of diameter $\phi = 25$ μm taken at a constant temperature of ≈ 15 mK with the wire attached to two different sample holders as shown in Fig. 4.9a ($\nu = 84$ Hz) and Fig. 4.9b ($\nu = 1.5$ kHz). Although there is a small difference in the magnitude of the internal friction, the Pt wires show similar strain dependences for both clampings in the whole strain range investigated. Moreover, there is no observable influence from the clamping on the sound velocity.

4.3.4 Acoustic Experiments at Very Low Temperatures: Cryogenics and Sample Thermalization

Most of the measurements at temperatures below 50mK have been performed with the sample holder attached to the experimental flange of a Cu nuclear demagnetization refrigerator (Gloos et al. (1988)). At present, nuclear adiabatic demagnetization is the only technique to perform experiments at electronic temperatures of ~ 10 μK or even below (Pobell (1992)). In order to achieve thermal equilibrium along the sample holder at very low temperatures, the sample holder is usually annealed at about 850° to 900°C in vacuum for a few hours to heal lattice distortions, thus improving its thermal conductivity significantly. The samples are attached to this sample holder as described in the previous section. Electrical contact is usually made via superconducting NbTi wires that are heat sunk all the way along the sample holder, thus minimizing the heat leak into the sample. The larger sensitivity to stray capacitances requires the use of coaxes for both electrodes of a reed setup, whereas vibrating wire measurements may be performed as simple four terminal measurements using ordinary twisted pairs of superconducting wire for excitation current and induced voltage.

Thermometry at mK and μK temperatures is based on the superconducting transitions of five samples (tungsten, beryllium, iridium, $AuAl_2$, $AuIn_2$) of a NBS superconducting fixed point device (Soulen and Dove (1979)). The transition temperatures of these samples are between $\approx$ 15 mK and $\approx$ 210 mK. At higher temperatures germanium and carbon resistors are used for thermometry. Thermometry in the lower mK range uses the Curie susceptibility of a PdFe thermometer (Jutzler et al. (1986)). Finally, the nuclear susceptibility of platinum is used for NMR thermometry at μK temperatures. We have to emphasize that for all experiments performed in the μK-temperature range, the given temperature is always the electronic temperature of the Pt-NMR thermometer. The exact temperature of the wire or reed samples in vacuum in the μK regime is unknown.

Experiments at very low temperatures require particular conditions concerning the dissipation of energy that might cause self-heating of the sample. The dissipated energy of a mechanical resonator is given by (Esquinazi et al. (1992))

$$\dot{q} = Q^{-1} u^2 \omega^3 m/8, \qquad (4.47)$$

where m is the mass of the resonator. Equation (4.47) clearly indicates that a large amplitude of oscillation and a large resonance frequency strongly increase the energy dissipation leading to self-heating of the sample at the lowest temperatures. Self-heating effects are often revealed in a distinct strain-

and/or temperature dependence of either the internal friction or of the sound velocity of the sample; this matter will be discussed along with the experimental data in the respective sections. In this section, we restrict ourselves to the discussion of some general aspects concerning sample thermalization.

Self-heating effects and sample thermalization are of course strongly related to the properties of the material itself. The cooling of a sample to the lowest temperatures depends on whether the thermal conductivity of the sample is governed by phonons only as it is the case in dielectrics or in metals in the superconducting state far below the superconducting transition temperature, or whether electrons play the most important role in the heat transfer within the sample. Another important experimental parameter is the small sample size that is typically of order of some tens of microns and may therefore also contribute to a restriction in the thermalization of a sample particularly at very low temperatures. Furthermore, an additional damping mechanism is related to the oscillation of metallic samples in a magnetic field that causes eddy currents with an energy dissipation proportional to the square of the applied magnetic field (Voncken et al. (1997)). Obviously, this source of self-heating is an intrinsic problem of the vibrating wire technique as a magnetic field is a fundamental prerequisite for this method. However, due to the small magnetic fields used in the experiments described in this work, eddy current effects are usually negligible. An intensive study of the influence of high magnetic fields of up to 11.5 T on the behavior of vibrating wires has been performed by Voncken et al. (1997).

A comparison between the vibrating reed and the vibrating wire technique reveals that self-heating effects are more pronounced in the vibrating wire technique mainly due to the larger amplitude of oscillation. The estimated energy dissipation in a SiO_2 vibrating reed due to acoustic excitation at a strain of $\epsilon = 10^{-7}$ is $\dot{q} = 3.8 \times 10^{-15}$ W resulting in a temperature increase of at most 80 μK at a temperature of the nuclear stage of 600 μK. With the maximum strain $\epsilon = 10^{-8}$ applied in the measurements on the amorphous PdSiCu, the calculated $\dot{q} = 2 \times 10^{-17}$ W yields an even smaller influence on the sample thermalization. The estimated temperature difference between sample and sample holder amounts to less than 10 μK at $T = 100$ μK. In contrast, heating due to dissipation in the wires is usually of the order of 10^{-11} to 10^{-12} W (Esquinazi et al. (1992)), and the temperature increase in the middle of the wire is estimated to be of order of a few mK at the lowest temperatures. In normalconducting wires and at very low temperature, Joule heating has also to be considered as a serious source of energy dissipation: the use of an excitation current of 10 μA results in an energy dissipation due to Joule heating of ≈ 1 pW (with a typical electrical resistance of the wire of a few mOhms).

Self-heating effects reveal themselves in a (nonphysical) saturation of the acoustic properties at the lowest temperatures, and in particular the investigation of the strain dependences of the sound velocity and of the internal friction at a constant temperature clearly provide good indications for the upper limit of excitation allowed in order to avoid heating effects. As this work summarizes experimental results of very different types of materials [amorphous insulators, amorphous metals (normalconducting and superconducting), polycrystalline metals (normalconducting and superconducting)], we refer for a discussion of amplitude-dependent effects to the respective section in which the acoustic properties of a certain class of samples is presented.

4.4 Acoustic Properties of Amorphous Solids

4.4.1 Dielectrics

The acoustic properties of amorphous dielectrics have been studied by many groups, both experimentally and theoretically (for the latest short reviews, see Doussineau (1989) and Hunklinger (1989)). At ultrasonic frequencies the logarithmic increase in sound velocity with temperature due to the resonance interaction between phonons and TS's (see Fig. 4.3 and Table 4.1) can be easily measured above 100 mK, as for example in vitreous silica (Piché et al. (1974)). Usually, at frequencies of the order of 100 MHz, the maximum in the sound velocity occurs at $T_{\max} \sim 1$ K. Because other processes (incoherent tunneling and/or thermally activated relaxation) start to influence the acoustic properties at $T \geq 1$ K, the contribution of relaxation processes is not easily obtained from the experimental data, neither from the sound dispersion nor from the internal friction. This is not the case for measurements at acoustic frequencies, typically below 100 kHz. In this section, we will discuss the acoustic properties of SiO_2 and of two polymers: Polymethylmethacrylate (PMMA) and Polystyrene (PS). Figure 4.11 shows the relative change of the sound velocity for Suprasil I (SiO_2 with ~ 1200 ppm OH^-) at three different frequencies (Esquinazi et al. (1992)). The 400 Hz and 1.8 kHz data were obtained with the vibrating reed technique, whereas the data at 30 MHz result from an ultrasonic equipment on a sample from the same batch. At $T > 30$ mK similar results were obtained for Suprasil W (SiO_2 with negligible OH^- content, Raychaudhuri and Hunklinger (1984)). The results above 100 mK are typical for the glasslike behavior and can be explained with the standard tunneling model (see Fig. 4.3): the logarithmic decrease at $T > T_{\max}$ at frequencies of 400 Hz and 1.8 kHz (due to the resonant and relaxation interaction) is $\sim -1/2$ of the logarithmic increase measured at 1.8 kHz and 30 MHz below $T_{\max}$. From these logarithmic slopes, one obtains the parameter $C = (2.7 \pm 0.1) \times 10^{-4}$. It should be noted, however, that nonlinear effects below 100 mK (see Sect. 4.2.4) may influence the temperature dependence of the sound velocity in such a way that fortuitous agreements with the standard tunneling model cannot be ruled out. According to our experience with other materials with glasslike acoustic properties, the ratio of the quasi-logarithmic slopes of the sound velocity below and above its maximum

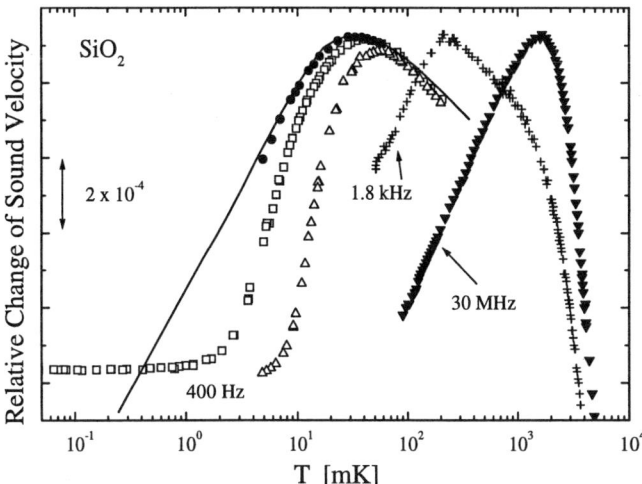

Fig. 4.11. Relative change of sound velocity for vitreous silica (SiO$_2$) as a function of temperature at three different frequencies. The data at 400 Hz were taken for the following strain amplitudes: (•) 2.5×10^{-8}, (□) 10^{-7}, (△) 10^{-6}. The continuous line is calculated from the standard tunneling model (4.12 and 4.14) with $C = 2.7 \times 10^{-4}$, $Ak_\mathrm{B}^3 = 8.5 \times 10^7$ s^{-1} K^{-3} (taken from Esquinazi et al. (1992)).

at T_max can be negligibly small but also much larger than 2, dependent on the frequency and the applied strain of the measurement in this temperature range ($T < 100$ mK), see p. 181.

The data at $T < 100$ mK reveal that the sound velocity is strongly dependent on the strain amplitude. We observe a shift of T_max with strain, a nonlogarithmic temperature dependence below T_max and a saturation at low temperatures (see Fig. 4.11). All these nonlinear effects are observed when the energy of the sound wave or strain field is of the order of the thermal energy and they can be explained by introducing an energy shift in the asymmetry [see (4.29)] or, qualitatively, by introducing an effective change of the population difference (see Sect. 4.2.4 and Fig. 4.5).

The saturation of the sound velocity observed at low temperatures is not related to self-heating of the sample; the shift of the whole curve of the sound velocity to higher temperature with increasing strain as well as the fact that the strain energy does not affect the internal friction at $T \geq 10$ mK speaks against self-heating effects, see Fig. 4.13. This strain dependence of the sound velocity – that was also observed in a silicon single crystal by Kleiman et al. (1987) – has been verified later in vitreous silica by Classen et al. (1994). A quantitative comparison of the sound velocity data with the theory in the saturation regime (i.e. $T \ll T_\mathrm{max}$ and $\gamma \epsilon_0 T^2/2 \ll k_\mathrm{B} T_\mathrm{max}^3$, see Sect. 4.2.4) can be done using the data at 400 Hz and strain $\epsilon_0 = 10^{-7}$ from Fig. 4.11. For this comparison an extrapolated value of the relative change

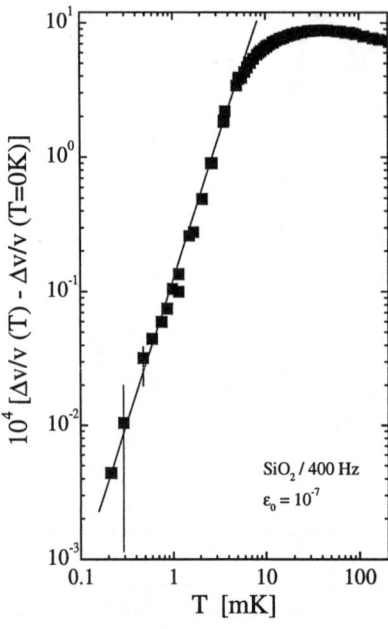

Fig. 4.12. Relative change of sound velocity corrected for its value extrapolated to $T = 0$ K as a function of temperature. The data were taken for vitreous silica at 400 Hz and $\epsilon_0 = 10^{-7}$ (from Esquinazi et al. (1992), see Fig. 4.11). The continuous line follows a law $T^2 \ln(T)$ and was obtained from (4.30) with the parameters: $C = 2.7 \times 10^{-4}, \gamma = 0.33$ eV, and $\Delta_{0,\min} = 1$ μK.

of sound velocity at $T = 0$ K has been subtracted. A fit to (4.30) provides the parameter $\gamma = 0.33$ eV and $\Delta_{0,\min} = 1$ μK. Good agreement is found between theory (Stockburger et al. (1995)) and experiment in the saturation regime, see Fig. 4.12. We should note that the fit is rather insensitive to the value $\Delta_{0,\min}$ since it enters logarithmically in (4.30).

Figure 4.13 shows the corresponding experimental data for the internal friction for SiO_2 from Esquinazi et al. (1992). With decreasing temperature we observe a decrease of the internal friction followed by a broad plateau that extends to ≈ 100 mK. Towards lower temperatures, Q^{-1} decreases again, and finally saturates. From the plateau in the internal friction (see Table 4.1) we obtain $C = (2 \pm 0.3) \times 10^{-4}$ for 400 Hz and $C = (2.7 \pm 0.3) \times 10^{-4}$ for 1.8 kHz, in good agreement with the value derived from the sound velocity. The continuous line in Fig. 4.13A and curve (a) in Fig. 4.13B are numerical calculations following the standard tunneling model, see (4.11), with the same parameters used to calculate the theoretical curve for the sound velocity in Fig. 4.11.

Figure 4.13A shows an apparent agreement of the data for the internal friction of SiO_2 with the standard tunneling model above 10 mK. The same data as in Fig. 4.13A but on a double logarithmic scale follow closely a linear temperature dependence especially in the region between 3 mK and 30 mK, in disagreement with the standard tunneling model. This result is in agreement with the predicted linear T-dependence due to collective excitations as proposed by Burin and Kagan (1994b) (see Sect. 4.2.5). The experimental crossover temperature $T' \sim 30$ mK is of the order of that estimated from (5.102).[4]

The subtraction of a background contribution estimated to be 2×10^{-5}, see Fig.4.13B, has a relatively large influence on the temperature dependence of the remaining data of the internal friction. Therefore, it is not clear if the quasi-linear T-dependence observed above 3 mK also holds at lower temperatures. It should be noted that the internal friction measured at 400 Hz shows negligible dependence with strain for $\epsilon_0 \lesssim 10^{-7}$, whereas it seems to *decrease* for larger strains below 20 mK (Esquinazi et al. (1992)). This result, along with a possible higher T-dependence than linear, is in qualitative agreement with the theoretical prediction in the slow relaxation limit (nonlinear regime) if we assume that in the linear regime $Q^{-1} \propto T$ due to the collective excitations, see (4.31).

As we have pointed out above, the apparent agreement between standard theory and the experimentally determined temperature dependence of the sound velocity at $T > 3$ mK for small applied strain (Fig. 4.11) is not always observed. Classen et al. (1994), for example, measured similar slopes for $\Delta v/v$ above and below $T_{\max}$ below 100 mK, in contradiction to the standard tunneling model that predicts a ratio 2/1 (Sect. 4.2.2). Comparable slopes above

---

[4] For this estimation it is assumed that the interaction constant is mainly determined by the transversal phonons, i.e. $U_0 \simeq (\gamma_l^2/v_l^2) + 2(\gamma_t^2/v_t^2) \sim 10(\gamma_t^2/v_t^2)$, see Burin and Kagan (1995).

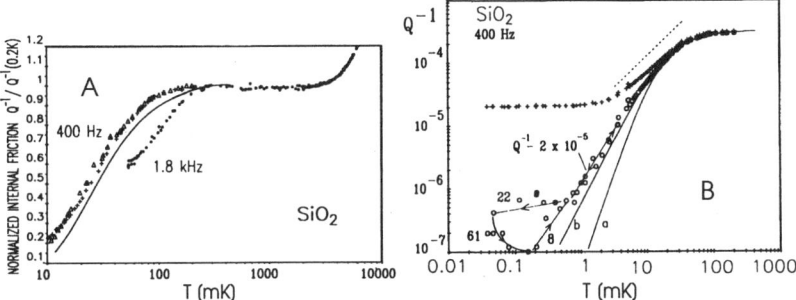

Fig. 4.13. (A): Normalized internal friction for Suprasil I vs. temperature at two different phonon frequencies (from Esquinazi et al. (1992)). The data at 400 Hz were taken at strains $(\Delta)10^{-6}$ and $(+)10^{-7}$. The absolute values are $Q^{-1}(0.2K) = 3.1 \times 10^{-4}$ (4.24×10^{-4}) at 400 Hz (1.8 kHz). The full line has been obtained numerically following the standard tunneling model, see (4.11). (B): Low-temperature data for $\nu = 400$ Hz. (o) after subtraction of a clamping contribution of 2×10^{-5}. The numbers indicate the elapsed time in hours at the indicated steps of demagnetization of the nuclear stage. The continuous curve (a) was obtained according to the standard tunneling model. Curve (b) was obtained assuming a quasi-two dimensional density of states for phonons (after Esquinazi et al. (1992)). The dashed line indicates a linear temperature dependence.

and below $T_{\max}$ have also been observed in SiO_2 thin films (Gaganidze et al. (1997)) as well as in some polycrystalline metals, see Figs. 4.28 and 4.30, though the ratio $\sim 2/1$ was observed too (Fig. 4.22). This difference between sometimes similar samples is not yet clarified. It currently appears that the quasi-linear temperature dependence of the internal friction at $T < 100$ mK as well as the ratio between the slopes in the temperature dependence of the sound velocity can be understood within two different theoretical pictures, both based on the interaction among TS's. The first is that introduced in Sect. 4.2.5 (see also Chap. 5): due to the relaxation of pairs of TS's a linear T-dependence below the crossover at T' is predicted. Concerning the sound velocity the following behavior is expected: if the crossover temperature (5.102) $T' < T_* = T_{\max}$, where T_* is defined at $\omega \tau_1(T_*) \sim 1$, then we expect to have a change in the logarithmic slope of v from C to $-C/2$ at $T_{\max}$ as observed in Fig. 4.11 for the smallest applied strain, i.e. $T_{\max} \sim 70$ mK and $T' \sim 30$ mK. However, if $T' > T_*$, the slope of v would change from $C/2$ to $-C/2$ at $T_{\max}$. It is important to note that in this case, $T_{\max}$ should be nearly independent of frequency (Burin (1997b)). If $T' > T_*$ and at $T \ll T_*$, a change of slope from $C/2$ to C with decreasing temperature should be observed when the relaxation to phonon mechanism becomes negligible. Within the theory of Burin and Kagan (1994b) it would be possible to have nonuniversal behavior depending on the sample parameters and the frequency of the measurements.

The second possible explanation for the deviations of the internal friction and sound velocity from the standard tunneling model is based on the incoherent tunneling of TS's coupled through elastic interactions at low enough temperatures. According to Enss and Hunklinger (1997), using this concept good agreement between theory and experiment for SiO_2 can be obtained. It is still unclear if the concept of incoherent tunneling can explain other experimental facts described in Chap. 5. Due to nonlinear effects, a clear experimental proof is, however, difficult.

At kHz-frequencies and at temperatures above $T_{\max}$ and below a few Kelvin, amorphous dielectrics show a temperature-independent region in the internal friction, the so-called "plateau" as predicted by the standard tunneling model.

At higher temperatures, the internal friction may increase linearly with temperature as is the case of SiO_2. Tielbürger et al. (1992) interpreted this increase due to a thermally activated relaxation of TS's, see Sect. 4.2.6. They applied (4.37) to various experimental data from literature and obtained a zero point energy $E_0 \sim 15$ K for SiO_2. A linear decrease with increasing temperature of the sound velocity above a few Kelvin is generally observed in amorphous dielectrics. As mentioned before, Tielbürger et al. (1992) showed that in a broad frequency range (10^2 Hz – 10^{10} Hz) and between 5 K and 25 K, the slope of the sound velocity $\partial(\ln v)/\partial T$ for SiO_2 follows well the predicted logarithmic dependence in the applied frequency, see (4.38). However, this result remains somehow controversial since it does not seem to be supported by results published by other groups (Bellessa (1978)). A possible explanation for this discrepancy is given by Würger in terms of phonon-assisted tunneling, see Sect. 3.7.8.

The interpretation of the data in the temperature range between a few Kelvin and 20 K within the tunneling model including thermally activated processes is also controversial because of the values of the free parameters E_0 and τ_0 necessary to fit the data. As discussed in Sect. 4.2.6, another explanation in terms of incoherent tunneling (Neu and Würger (1994b)) may also explain the linear increase with temperature of the internal friction as well as the linear decrease and frequency dependence of the slope of the T-dependence of the sound velocity above a few Kelvin up to ~ 20 K. Within this approach no free parameters are necessary to explain the experimental data, since they are obtained from the low temperature tunneling region ($0.1\,\mathrm{K} \leq T \leq 1\,\mathrm{K}$) where the standard tunneling model applies. A comparison of the data of SiO_2 and polycrystalline metals and their interpretation with the theory of incoherent tunneling is discussed in Sect. 4.5.5.

We would like to emphasize here that the universalities observed in the acoustic properties of amorphous dielectrics below a few Kelvin are not always observed at higher temperatures. In Chap. 2 the heat release of two polymers, PMMA and PS, and its relation with low-temperature thermal conductivity and specific heat has been discussed. The heat release shows large differences between both amorphous polymers. Here, we concentrate on the internal friction of these polymers that is presented in Figs. 4.14A and B. As expected, below a few Kelvin and because $\omega\tau_m \ll 1$ holds, the internal friction at kHz frequencies is temperature independent with a value equal to $\pi C/2$ (see table 4.1). From the value of C obtained from the plateau and using the coupling constant obtained from thermal conductivity measurements (4.48), we derive a density of states of TS's in PS $P_0 = 9 \times 10^{38}$ 1/Jg that is a factor ≈ 2.5 larger than for PMMA (Nittke et al. (1995)). Within the standard tunneling model, the thermal conductivity at temperatures below 1K may be calculated using the values for $P_0\gamma^2$ derived from the internal friction at the plateau; these data for the thermal conductivity are in very good agreement with experimental data for PMMA but a factor of two smaller for PS (Nittke et al. (1995)). This discrepancy may be partially caused by a background contribution to the measured internal friction that was not subtracted from the measurements used to compute the parameter C.

The internal friction results depicted in Fig. 4.14 show a qualitative difference above ~ 3 K for both polymers. PMMA shows a decrease in the internal friction reaching a minimum at ~ 30 K and increasing monotonously towards higher temperatures. On the contrary, the internal friction of PS shows the typical temperature dependence measured for other amorphous materials. It increases with temperature above 5 K, reaching a frequency-dependent maximum at ~ 40 K, although, the overall behavior of the internal friction (Fig.

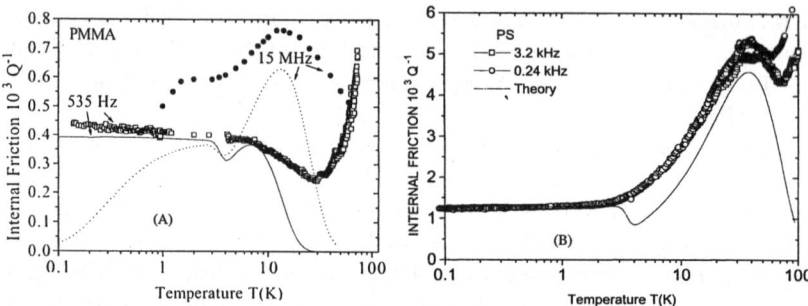

Fig. 4.14. Internal friction as a function of temperature for (A) PMMA and (B) PS. The full circles in (A) indicate the attenuation taken from ultrasonic measurements at 15 MHz (from Federle and Hunklinger (1982)). Solid and dotted lines represent the calculated values following the standard tunneling model including a thermally activated relaxation rate. For PS (B) the solid line was calculated at 0.24 kHz (Nittke et al. (1995)).

4.14) as well as that of the sound velocity observed in PS can be described assuming thermally activated relaxation including a Gaussian distribution $P(\Delta,V) = P_0 \exp(-V^2/2\sigma_0^2)/E_0$ of the barrier height V above a few Kelvin (Gilroy and Phillips (1981),Tielbürger et al. (1992),Nittke et al. (1995)). However, a too small relaxation time prefactor $\tau_0 \sim 10^{-17}$ s as well as a very broad distribution width ($\sigma \sim 1200$ K) are necessary to fit the data.

The predicted linear temperature increase of the internal friction (4.37) (Neu and Würger (1994b)) at $\nu = 535$Hz is not observed in PMMA, see Fig. 4.14A. It is tempting to interpret the decrease of the internal friction above 3 K as due to a cutoff in the density of states at ~ 15 K (Köhler et al. (1989)). However, ultrasonic attenuation results at 15 MHz (Fig. 4.14A) show a maximum at ~ 12 K that indicate the influence of thermally activated and/or incoherent tunneling. Nittke et al. (1995) have found a set of parameters that fit the low and high frequency data reasonably well with the assumption of thermally activated relaxation and without the need of an energy cutoff in the density of states of TS's. The continuous curves shown in Fig. 4.14A have been obtained using $E_0 = 10$ K, $\tau_0 = 10^{-13}$ s and a rather small distribution width of the barrier heights $\sigma_0 = 150$ K.

We should note that, as the absorption peak in SiO_2 occurs at ~ 30 K (at $\nu \sim 10$ kHz) (see, for example, Classen et al. (1994)), the reason for the attenuation peak at $T > 5$ K in several amorphous materials is not yet clear; both relaxation processes – thermally activated relaxation and incoherent tunneling – may contribute to the attenuation of phonons [see the fits to the attenuation peak in SiO_2 using thermally activated relaxation by Tielbürger et al. (1992) and incoherent tunneling by Neu and Würger (1994b)].

The acoustic properties found for the two polymers clearly indicate that, in spite of a similar behavior in specific heat and thermal conductivity, qualitative and quantitative differences in the acoustic properties are observed at

$T > 1$ K. The different behavior in particular of the internal friction between both polymers shows that the amorphous structure alone is not sufficient to guarantee universality in the acoustic properties at $T > 1$ K. The results obtained in PMMA indicate that the density of states of TS's might be restricted to much smaller energies in comparison with PS or SiO_2.

4.4.2 Normal-Conducting Amorphous Metals

In amorphous metals the conduction electrons provide another channel for the relaxation of the TS's (see Sect. 4.2.3). In contrast to amorphous insulators, only a few measurements of the acoustic properties of normal-conducting amorphous metals exist. Most of them were performed at ultrasonic frequencies on amorphous $Pd_{77.5}Si_{16.5}Cu_6$ (PdSiCu) (Bellessa (1977), Golding et al. (1978), Doussineau et al. (1978), Cordie and Bellessa (1981), Park et al. (1981)), CoP (Bellessa (1977)) and NiP (Bellessa (1977), Doussineau and Robin (1980)). Generally speaking, those experiments reveal a logarithmic increase of the sound velocity with temperature at 0.05 K $< T < 2$ K and a rather weak temperature dependence for the phonon attenuation (at 0.2 K$< T < 1$ K), in qualitative agreement with the standard tunneling model including a Korringa-like relaxation rate, see Sect. 4.2.3. Because at ultrasonic frequencies both the relaxation and the resonant process are expected to contribute to the phonon attenuation (Sects. 4.2.1 and 4.2.2), measurements of the dependence of the acoustic properties on the acoustic intensity are necessary in order to differentiate among the possible mechanisms.

However, experimental data on the acoustic intensity dependence of the sound velocity and attenuation of a normal-conducting amorphous metal at ultrasonic frequencies are rare. The results of Golding et al. (1978) show that the acoustic intensity necessary to saturate the phonon attenuation is several orders of magnitude larger than for insulating glasses. This remarkable observation is ascribed to the shorter relaxation time (Korringa-like) of the TS's due to their interaction with conduction electrons. Furthermore, the results of Golding et al. (1978) also indicate a nonsaturable attenuation that is much larger than the standard theory for the relaxation process predicts; a smaller saturable (resonant) contribution is observed, too, in disagreement with theoretical predictions. Some of these results can be accounted for taking into account the strong coupling between conduction electrons and TS's and the influence of incoherent and coherent tunneling (Coppersmith and Golding (1993)). However, as noted by Coppersmith and Golding (1993), the agreement between theory and experiment is still not satisfactory and further experiments are necessary. An acoustic intensity dependence of the sound velocity and attenuation was observed in PdSiCu at $T = 10$ mK and 500 MHz shear waves by Cordie and Bellessa (1981) . They interpreted this interesting result using the standard tunneling model (4.13) and showed that from these results a very short transversal relaxation time $\tau_2 = 2$ nsec can be obtained.

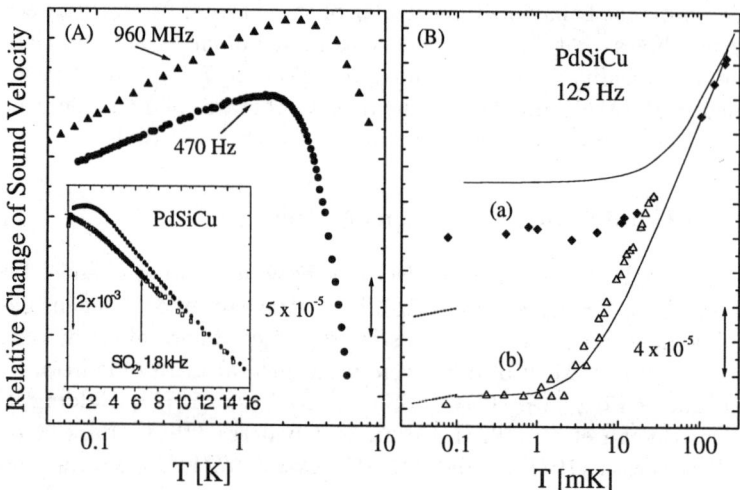

Fig. 4.15. Relative change of sound velocity of PdSiCu as a function of temperature. (A) The data at 470 Hz was obtained from Gaganidze and Esquinazi (1996b) with the vibrating reed technique and the data at 960 MHz is taken from Golding et al. (1978) (longitudinal waves). The inset shows data for SiO_2 (1.8 kHz, from Esquinazi et al. (1992)) with a linear temperature scale. The same vibrating reed data for PdSiCu as the main figure but with the y-axis multiplied by a factor 5.9 is also shown in the inset. (B) The same as in (A) but at two different acoustic intensities ($I_a \sim 5 \times 10^{-4}$ W/cm^2) with a ratio ~ 400. The full lines have been obtained taken into account the standard tunneling model and only the resonant process using (4.13) with $\tau_2 = 10$ ns and $I/I_{c1} = 0.5(200)$ for the lower (upper) curve (from Esquinazi et al. (1992)). The short dashed lines are obtained with (4.32).

Figure 4.15A shows the relative change of the sound velocity of amorphous PdSiCu in the temperature range 0.07 K $\leq T \leq 10$ K at two frequencies that differ by more than six orders of magnitude. At a first glance, the observed temperature dependence appears to be similar to that for dielectric glasses. But as pointed out by Raychaudhuri and Hunklinger (1984), two important differences have to be emphasized: (1) the maximum in the sound velocity at 470 Hz is observed at $T_{max} \sim 1.6$ K (in dielectric glasses $T_{max} \sim 0.1$ K ! at similar acoustic frequencies); (2) T_{max} shows a very weak dependence on the measuring frequency: T_{max} increases by a factor of $\simeq 1.5$ when the frequency is increased by a factor of 10^6.

According to the standard tunneling model and assuming a Korringa-like relaxation time for $\tau_1(E,T)$ (4.17), a logarithmic increase of the sound velocity is expected if the coupling constant between conduction electrons and TS's is large enough (from a comparison with the theoretical results from Fig. 4.4 one estimates $Kk_B \gg 10^6$ s^{-1}K^{-1}). In this temperature range $\omega\tau_{1,min} \ll 1$ holds and the sound velocity should increase logarithmically with a slope $C/2$ (see Table 4.2); at $T < T_{max}$ the phonon-mediated relaxation is negligible. From the logarithmic increase of the sound velocity below $T_{\max}$ we obtain $C = (4.4 \pm 0.1) \times 10^{-5}$ in comparison with the values $6.5 \times 10^{-5}, 2.8 \times 10^{-5}$ and 1.1×10^{-4} obtained from vibrating reed measurements at 1 kHz (Raychaudhuri and Hunklinger (1984)) and longitudinal and transverse sound wave ultrasonic measurements (Bellessa and Bethoux (1977)), respectively. The maximum in the sound velocity at $T_{\max} \simeq 1.6$ K does not occur at $\omega \tau_{1,\min}$ (this should occur at $T > 5$ K). As in dielectric glasses, a strong relaxation contribution leads to an approximately *linear* decrease of the sound velocity at $T > 4$ K, see inset in Fig. 4.15A. Raychaudhuri and Hunklinger (1984) interpreted the position of the maximum as due to a crossover from an electron-dominated relaxation at $T < T_{\max}$ to a phonon-dominated regime at higher temperatures, i.e. the condition $\tau_{1,e} \simeq \tau_{1,p}$ determines the position of the maximum. Using this condition and the one-phonon relaxation process, these authors could explain semiquantitatively the position of the maximum. However, recent results obtained for dielectric glasses (see Sect. 4.4.1 and Neu and Würger (1994b)) and polycrystalline metals (Gaganidze et al. (1995), Esquinazi (1996)) (see also Sect. 4.5) indicate that above 1 K the linear decrease of the sound velocity can be quantitatively explained in terms of incoherent tunneling. Therefore, a new analysis of the data in terms of the theoretical treatment of Neu and Würger (1994b) is necessary, see the discussion on p. 122. The maximum in the sound velocity is then expected to occur near the crossover from coherent to incoherent tunneling (Sect. 3.7.8).

Although no electron-assisted relaxation is included in the theoretical treatment of Neu and Würger (1994b), it is interesting to compare, at least qualitatively, some experimental facts with their predictions. We note also that at $T > 1$ K and if the crossover temperature $T_0 < T_{el}$ (see p. 95) one may expect that phonon-assisted tunneling overwhelms the conduction-electrons contribution to the relaxation, see Sect. 3.5.

The temperature dependence of the sound velocity of PdSiCu above 1 K is not exactly linear but shows a positive curvature, see inset in Fig. 4.15A. It tends to an approximately linear T-dependence above 10 K. The data of PdSiCu shown in the inset of Fig. 4.15A were multiplied by a factor 5.9 in order to match approximately the linear slope of the SiO_2 data at high temperatures. According to the theory of Neu and Würger (1994b), the slope of the linear temperature decrease of the sound velocity at high temperatures is proportional to the logarithmic slope below 1 K and therefore to the parameter C, see (4.40). From the comparison of the C-parameter obtained at low temperatures for PdSiCu and SiO_2, we obtain $C_{SiO_2}/C_{PdSiCu} = (2.7 \pm 0.3) \times 10^{-4}/(4.4 \pm 0.1) \times 10^{-5} = 6.1 \pm 0.7$. This ratio is in very good agreement with the ratio of the temperature coefficients β of the linear temperature decrease at high temperatures for the two samples $(\beta_{SiO_2}/\beta_{PdSiCu} \sim 6)$, in agreement with the theoretical prediction. Moreover, for a large number of amorphous and disordered solids the temperature coefficient $\beta \sim -3.5 \times 10^{-5}[\text{kg}/\text{Kms}^2]1/\varrho v^2$ (Nava (1994)) that indicates that $P_0\gamma^2 \sim 10^6$ kg/ms^2 independent of the material. This was also compiled by Berret and Meißner (1988), and can be accounted for by the theory of Neu and Würger (1994b) as shown by Gaganidze et al. (1995). Note that one should not expect that every disordered sample follows rigorously this relationship because β as well as C depend on the thermal treatment whereas the elastic modulus ϱv^2 remains practically unchanged, see Sect. 4.4.4. The experimental values for PdSiCu ($\beta \sim -0.6 \times 10^{-4}$ K^{-1}, $\varrho = 10.5$ g/cm^3, Young modulus sound velocity $v_E \simeq 3.6 \times 10^5$ cm/s), for SiO$_2$ and for polycrystalline metals are in agreement (within a factor two) with that prediction, see Fig. 4.37 in Sect. 4.5.3.

In the following we would like to discuss the sound velocity data of PdSiCu at lower temperatures. Figure 4.15B shows the temperature dependence of the sound velocity measured at 125 Hz with the vibrating reed technique at two acoustic powers. Note that the sound velocity in PdSiCu *increases* with acoustic intensity at a given temperature in contrast to SiO$_2$ (compare Figs. 4.15B and 4.11). Evidently the explanation used for SiO$_2$ is not sufficient to understand the nonlinear behavior observed in PdSiCu. From theory we

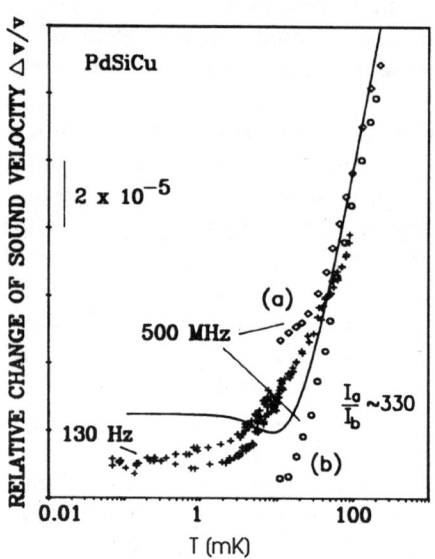

Fig. 4.16. Relative change of sound velocity as a function of temperature for PdSiCu at 130 Hz (vibrating reed data from Esquinazi et al. (1992)) and ultrasonic data at 500 MHz (taken from Cordie and Bellessa (1981)) at two acoustic intensities with a ratio of ~ 330. The full line was obtained from (4.13) with $I/I_{c1} = 1$, $\tau_{2,e} = 2$ ns and with 500 MHz as phonon frequency. After Esquinazi et al. (1992).

know that the standard tunneling model – taking into account the electron-assisted relaxation through a Korringa-like relaxation time only – might not be applicable due to the strong coupling between conduction electrons and TS's. The nonlinear response of a harmonically driven two-state system in a fermionic environment has been calculated by Stockburger et al. (1994) (Sect. 4.2.4). From this theory we can estimate the relative change of the resonance frequency at constant temperature and at two different acoustic intensities (Fig. 4.15B). The experiment (Esquinazi et al. (1992)) was performed using a PdSiCu reed ($l = 15$ mm, $d = 66\mu$m, $w = 3$mm) with a maximum reed amplitude (for $T < 10$ mK) of $u_{max} \simeq 2 \times 10^{-7}$ m and 10^{-8} m for the upper and lower curve, respectively. With an assumed coupling constant $\gamma = 1$ eV and the values of u_{max} given above, (4.32) results in an increase of $(5 \pm 1.5) \times 10^{-5}$. This result reasonably agrees with the experimental data, taking into account that the value of the coupling constant γ used in (4.32) is only an estimate for this compound. Note that (4.32) is only valid in the temperature range where $\hbar\epsilon > 2k_{\rm B}T$. No analytical expression is known for the intermediate range between saturation and the linear regime observed at high temperatures. Equation (4.32) also predicts an increase of $\sim 1.1 \times 10^{-5}$ in the sound velocity between 0.01 mK and 0.1 mK that approximately matches the weak temperature dependence observed in PdSiCu below 1 mK (Fig. 4.15B).

Although the theory of Stockburger et al. (1994) is apparently successfully applicable to the sound dispersion results of normal-conducting PdSiCu, some experimental facts still remain unexplained. Figure 4.16 shows the sound velocity data taken by Cordie and Bellessa (1981) at 500 MHz in PdSiCu, obtained at two acoustic intensities, together with the data of Esquinazi et al. (1992) at $\nu =130$ Hz. It appears that the sound velocity at MHz-frequencies saturates in the same temperature range as the vibrating reed experiment, and it shows similar dependence on the acoustic intensity at a phonon frequency that is 10^6 times larger. The nonlinear response developed by Stockburger et al. (1994) is not applicable at such high frequencies (it is valid for $\hbar\omega/k_{\rm B}T \ll 2\pi\alpha$, see Sect. 4.2.4). The similarity between the vibrating reed and ultrasonic velocity data is even more remarkable when we compare them with the predictions of the standard tunneling model neglecting the relaxation process, i.e. calculating the temperature dependence of the sound velocity due to resonant interaction. The curves in Fig. 4.15 are calculated with (4.13) using parameters obtained from ultrasonic and vibrating reed data. It is striking that using the resonant interaction of the standard tunneling model alone all the observed features of the sound velocity of PdSiCu at low temperatures can be reproduced.

The temperature and acoustic intensity dependence of the internal friction in PdSiCu differ strongly from those measured in dielectric glasses. Internal friction data at $T > 0.05$ K are shown in Fig. 4.17. The temperature dependence as well as the absolute value of the internal friction agree with previously published results (Raychaudhuri and Hunklinger (1984)). The internal friction of PdSiCu shows a temperature-independent region at $0.7 \text{ K} \leq T \leq 3$ K and starts to decrease at lower temperatures. The internal friction depends on the acoustic intensity; this dependence is already clearly resolved at temperatures as high as $T \sim 1$ K (Fig. 4.17). These results are in clear contrast to those obtained for SiO_2, see Sect. 4.4.1. From

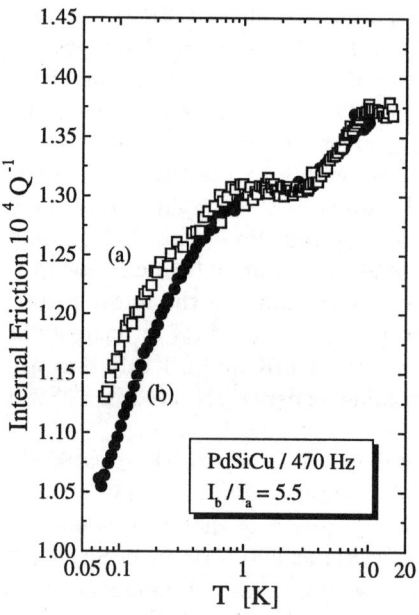

Fig. 4.17. Internal friction as a function of temperature for PdSiCu at 470 Hz and at two acoustic intensities differing by a ratio of 5.5 (from Gaganidze and Esquinazi (1996b)). The upper curve (a) was measured with a smaller acoustic power as curve (b).

the standard tunneling model we expect a plateau in the internal friction at temperatures where $\omega\tau_{1,e} \ll 1$ holds. Taking the coupling constant between electrons and TS's from literature we expect that $\omega\tau_{1,e} \sim 1$ should be reached at $T \ll 0.1$ mK. The deviation of the internal friction in PdSiCu from the expected temperature-independent behavior is at least partially related to the finite acoustic intensity of the measurements. This becomes apparent from the results presented in Fig. 4.17 as well as the acoustic power dependence of the internal friction at constant temperature measured at very low temperatures by Esquinazi et al. (1992). Due to the acoustic frequencies of our vibrating reed experiments, the internal friction is determined by relaxation processes only, therefore no acoustic intensity dependence is expected from the standard tunneling model. The experimental results indicate a *decrease* of the internal friction with acoustic power in agreement with previously reported results in amorphous metals at ultrasonic frequencies by Araki et al. (1979) and Arnold et al. (1982b). Its explanation is still not yet clear (Arnold et al. (1982b), Galperin et al. (1985a)). According to Stockburger et al. (1994), in the saturation regime ($\gamma\epsilon_0 \gg k_BT$) the internal friction *increases* with acoustic intensity [(4.34) can be approximated as $Q^{-1} \propto I^{0.2}$] in qualitative disagreement with the experimental results. We speculate that the large decrease of the relaxation absorption with acoustic power measured in PdSiCu overwhelms the rather weak increase predicted by Stockburger et al. (1994).

4.4.3 Superconductors

In Sect. 4.2.3 we have discussed the expected changes in the acoustic properties of an amorphous or disordered metal in its superconducting state below a critical temperature T_c. The expected behavior of the acoustic properties strongly depends on the region in which T_c lies. Within the standard tunneling model [i.e. the coupling between electrons and TS's is described only by the Korringa-like relaxation time $\tau_{1,e}$, (4.17) and (4.18)], and assuming that in the superconducting state the influence of the quasiparticles on the relaxation time decreases exponentially with temperature, it is not difficult to predict the expected variation of the acoustic properties. For example, from the numerical results shown in Fig. 4.4 and assuming $T_c = 0.02$ K, we conclude that with decreasing temperature, a decrease of the electronic relaxation rate in the superconducting state will *increase* the sound velocity and *decrease* the internal friction, compare curves (A) and (B). However, if $T_c > 1$ K, i.e. in the region where the relaxation of the TS's to phonons dominate, no special feature at T_c is expected within the standard model. For $T \ll T_c$ the number of thermally activated quasiparticles is negligibly small and the amorphous superconductor should behave like a dielectric glass.

One of the first acoustic measurements on amorphous superconductors ($Pd_{30}Zr_{70}$, $T_c = 2.6$ K and $Cu_{60}Zr_{40}$, $T_c = 0.3$ K) were published by Raychaudhuri and Hunklinger (1984). These authors showed that at $T \ll T_c$ the amorphous metal behaves indeed like a dielectric glass, as far as the acoustic properties are concerned. They also observed some features that appeared to be in contradiction with the expectations from the standard tunneling model. In particular, they observed for both amorphous superconductors a local minimum of the internal friction at T_c. Because the resolution of their measurements was not high enough and because no normal state data were available, it was not clear to what extent the standard treatment was not sufficient to explain the data.

The first measurement of the acoustic properties of the amorphous superconductor $Pd_{30}Zr_{70}$ in the superconducting *and* normal state was published by Neckel et al. (1986). Their results along with the subsequent studies on the same material at ultrasonic frequencies (Esquinazi et al. (1986)) and on the amorphous superconductor $Cu_{70}Zr_{30}$, and $(Mo_{1-x}Ru_x)_{0.8}P_{0.2}$ (Lichtenberg et al. (1990)) in both the normal and superconducting states clearly show that the standard approach for the conduction electrons–TS's interaction is not sufficient to explain the experimental data.

Figure 4.18A shows the temperature dependence of the internal friction of amorphous $Pd_{30}Zr_{70}$ (right axis) in the normal and superconducting state. In the superconducting state, the internal friction shows a kink at $T_c = 2.6$ K and increases with decreasing temperature below 1 K, reaching a nearly temperature independent region between 400 mK and 80 mK. From a comparison of the internal friction and the sound velocity data with the results obtained for dielectric glasses we conclude that below ~ 0.5 K the amorphous super-

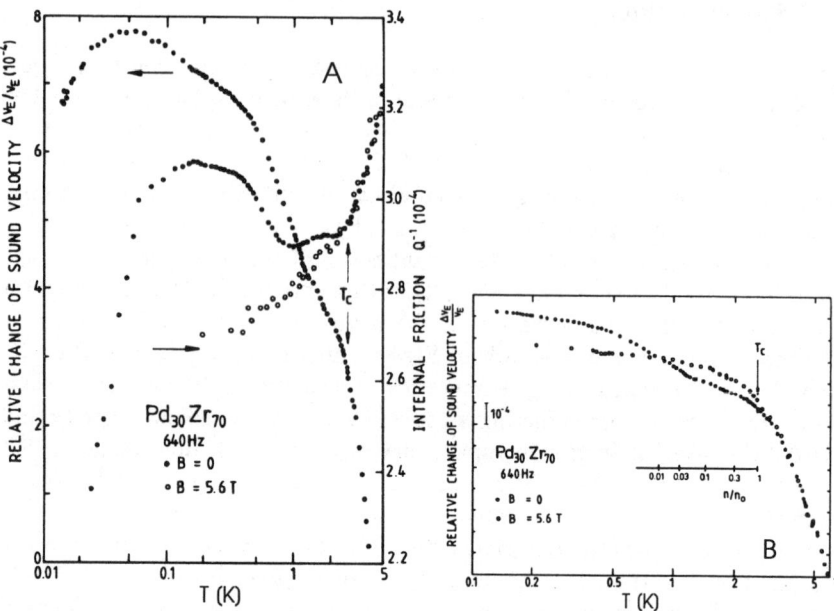

Fig. 4.18. (A) Internal friction (right axis) as a function of temperature for amorphous $Pd_{30}Zr_{70}$ in the normal (o) and superconducting states (•) at 640 Hz. Left axis: Temperature dependence of the sound velocity change in the superconducting state. (B) Temperature dependence of the sound velocity change in the normal and superconducting states. n/n_0 is the number of thermally activated quasiparticles in the superconducting states according to the BCS theory. After Esquinazi et al. (1986).

conductor behaves like a dielectric glass. From the "plateau" in the internal friction in this temperature range we deduce a parameter $C = 1.9 \times 10^{-4}$; from the slope in the change of the sound velocity, $C = 1.6 \times 10^{-4}$. These values are similar to that of SiO_2 ($C \simeq 2.7 \times 10^{-4}$). The results in the normal state, obtained by suppressing superconductivity with a field larger than the upper critical field $H_{c2}(0)$, clearly indicate that the internal friction is smaller than in the superconducting state (for $\omega\tau_1 < 1$) in contradiction with the expectations from the standard tunneling model. This result would agree with the strong coupling theory of Vladár and Zawadowski (1983) (Sect. 4.2.3) that predicts an effective decrease of the density of states of TS's in the normal state due to the screened tunneling splitting energy.

The change of the sound velocity with temperature shows several remarkable features related to the influence of conduction electrons at similar temperatures where the internal friction shows a characteristic behavior, see Figs. 4.18A and B. Below 0.5 K, the temperature dependence of the sound velocity resembles that of dielectric glasses. At 0.5 K < T < 1 K the sound velocity deviates from the logarithmic temperature dependence decreasing more steeply (compare with the internal friction in the same range). A further kink is observed at $T \sim 1.1$ K. In the normal state and at 0.8 K $< T < T_c$ the sound velocity is larger than in the superconducting state, and no maximum is observed down to 0.2 K. This result is in clear contradiction with the expected behavior from the standard theory, according to which one would expect a maximum in the sound velocity at $T \sim 2$ K. Furthermore, in the normal state the sound velocity should be smaller than in the superconducting state. To our knowledge, a semiquantitative comparison with the predictions of the strong coupling theory has not yet been published.

Qualitatively, some of the measured features are in agreement with the theory of Kagan and Prokof'ev (1988, 1990), see (4.21) and below. For example, according to (4.24), at low enough temperatures and frequencies and due to the renormalization of Δ_0, the sound velocity in the normal state is temperature independent. This is approximately what is observed experimentally below 1 K, see Fig. 4.18B. At higher frequencies, this theory predicts a logarithmic temperature dependence with a positive slope $C(1/2 - b)$, see (4.23); the results in amorphous $Pd_{30}Zr_{70}$ in the normal state and at 620 MHz would agree with this prediction. From the measurements at 620 MHz (Esquinazi et al. (1986)) we obtain that the ratio of the slopes of the logarithmic temperature increase of the sound velocity below 1 K between the normal and superconducting state is ~ 0.2. From theory, see (4.23) and Table 4.2, one expects this ratio to be $\sim 2(0.5 - b)$ for amorphous $Pd_{30}Zr_{70}$ then $b \sim 0.4$. This value for b is more than twice as large as for the normal-conducting amorphous metal PdSiCu obtained by Coppersmith and Golding (1993) from sound velocity ($b < 0.2$) and attenuation data ($b \sim 0.15$) at ultrasonic frequencies. We might conclude, therefore, that within the theory of Kagan and Prokof'ev (1988, 1990), the qualitative differences in the acoustic properties between the normal-conducting amorphous metal PdSiCu and amorphous $Pd_{30}Zr_{70}$ (compare Figs. 4.15, 4.16 and 4.17 with Fig. 4.18) are due to the difference in the coupling parameter b.

A comparison between the theoretical results of Kagan and Prokof'ev (1988, 1990) and the experimental data in the superconducting state is not an easy task. We give here a possible qualitative interpretation. According to this theory, the almost temperature independent behavior of the sound velocity in the normal state (at $T < 2$ K, see Fig. 4.18B) should remain below and near T_c. Just below T_c, a temperature dependence of the relative change of sound velocity given by (4.26) is expected to set in. However, with decreasing temperature and due to the increase of the superconducting gap, the inequality in (4.27) holds and again the sound velocity should show no temperature dependence. This is the "latent" temperature interval in the sound velocity discussed by Kagan and Prokof'ev (1988, 1990) and mentioned in Sect. 4.2.3, it would explain the similar temperature dependence of the sound velocity between the normal and superconducting states at 1 K $< T <$ T_c, see Fig. 4.18B. We note that, within this theory, the behavior of the sound

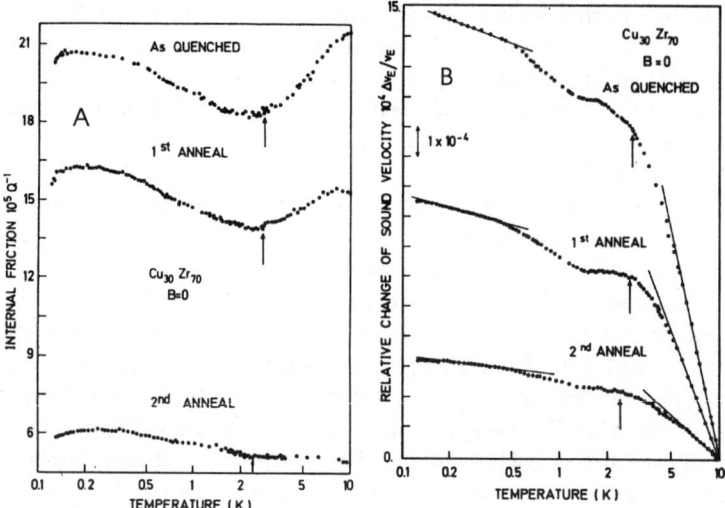

Fig. 4.19. (A) Internal friction and (B) the relative change of sound velocity as a function of temperature for amorphous $Cu_{30}Zr_{70}$ at 740 Hz and for three sample states. The first annealed state was achieved after annealing the sample for 15 minutes at 473 K, and for 20 hours at the same temperature for the second annealing state. The arrows indicate the corresponding critical temperatures $T_c = 2.806, 2.715$ and 2.436 K, respectively. The solid lines in (B) are only guides to the eye. After Esquinazi and Luzuriaga (1988).

velocity near T_c depends on the parameters $\omega_0, b, \Delta_s(T)$ and $\Delta_{0,\min}$ and on the temperature region where T_c is located.

At $T \ll T_c$ (below 0.5 K in Fig. 4.18) the logarithmic decrease of the sound velocity with temperature is due, as already mentioned, to the relaxation of TS's by phonons and its slope is given by the sum of the resonant $(+C)$ and relaxation $(-3C/2)$ processes. At $T > 0.5$ K the amount of quasiparticles is large enough to influence the dynamic of the TS's. A crude estimation of the decrease of the sound velocity observed between 0.5 K and ~ 1.3 K can be obtained from the second term at the right-hand site of (4.26) (the first is due to the sum of resonant and relaxation-to-electrons processes), $Cb\ln(\Delta_s(T=0)/k_B T_0) \sim 10^{-4}$.

4.4.4 Influence of Thermal Treatment on the Acoustic Properties of Amorphous Metals

It is well known that annealing of an amorphous material, i.e. its thermal treatment at a given temperature for a given time below the crystallization temperature, produces changes in the topological and possibly also chemical short-range order through local atomic rearrangements. This is particularly interesting because up to a certain degree of annealing, the amorphous nature of the structure remains (Chen (1980), Chen (1983), Lasocka and Matyja (1981)), and a more "relaxed" structure with a lower concentration of "extrinsic" defects is obtained. Therefore, heat treatment provides to a certain extent the possibility to study the relationship between intrinsic amorphous structure and the glasslike low-temperature properties. Properties such as low-temperature thermal conductivity (Ravex et al. (1981), Esquinazi et al. (1982), Grondey et al. (1983), Gronert et al. (1986)), specific heat (Grondey et al. (1983), Gronert et al. (1986), Ravex et al. (1984)), sound dispersion (Matey and Anderson (1978), Cibuzar et al. (1984), Esquinazi and Luzuriaga (1988)) and internal friction (Barmatz and Chen (1974),Esquinazi and Luzuriaga (1988)) have been reported to change upon annealing the sample.

Thermal conductivity measurements on amorphous Zr-based alloys indicate a change of the interaction between phonons and TS's on annealing, in particular, a reduction of the product $P_0 \gamma^2$ is observed (Ravex et al. (1981), Esquinazi et al. (1982), Grondey et al. (1983), Gronert et al. (1986)). On the other hand, specific heat measurements (Grondey et al. (1983),Gronert et al. (1986)) indicate that the density of states P_0 would remain fairly constant with annealing, at least in the early relaxation state of the amorphous metal. Without the need of assuming a change in the distribution function $P(E, u)$ with annealing, acoustic measurements on the same type of Zr-based amorphous alloys indicate that the coupling constant γ and not P_0 changes with annealing, in agreement with the thermal conductivity and specific heat results (Esquinazi and Luzuriaga (1988)).

Esquinazi and Luzuriaga (1988) studied the acoustic properties of the amorphous superconductor $Cu_{30}Zr_{70}$ in the normal and superconducting states and their changes with annealing. The heat treatments were performed at a constant temperature between 423 K and 523 K for a certain period of time below the glass transition temperature of 630 K of the amorphous alloy (Calvayrac et al. (1983), Altounian et al. (1982)). The variation with annealing of the electrical resistivity, superconducting critical temperature, upper critical field, penetration depth and of the Young modulus indicates that two well-defined regimes exist [see Esquinazi and Luzuriaga (1988) and references therein]. In the first regime (annealing up to 20 h at 473 K), a variation of the superconducting critical temperature of $\sim$ 15 % and a nearly constant electrical resistivity at 4 K is observed. In this regime, homogenization of the material occurs. In the second regime, devitrification or microcrystallization sets in along with a possible phase separation. The results of Esquinazi and Luzuriaga (1988) presented in this section were obtained in the first annealed state.

The general behavior of the acoustic properties in the superconducting and normal states observed in the as-quenched and annealing states of the amorphous superconductor $Cu_{30}Zr_{70}$ (Fig. 4.19) is similar to those observed in $Pd_{30}Zr_{70}$, see Sect. 4.4.3 and Esquinazi et al. (1986). Figure 4.19 clearly

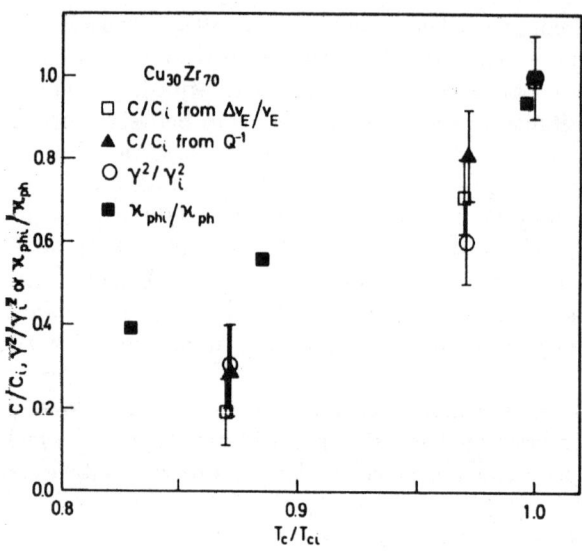

Fig. 4.20. Relative change upon annealing of the thermal conductivity κ_{ph} of phonons obtained at 0.5 K as a function of the superconducting critical temperature (i means initial or as-quenched state) for amorphous $Cu_{30}Zr_{70}$ (from Esquinazi et al. (1982)). In the same figure, the relative change of the parameter C obtained from the internal friction (Fig. 4.19A) and from the sound velocity (Fig. 4.19B) is included. The relative change of γ^2 obtained from the internal friction data (o) is plotted, too (Esquinazi and Luzuriaga (1988)).

shows that the annealing of an amorphous material decreases the magnitude of the glasslike anomalies. From the logarithmic slope of the sound velocity in the superconducting state ($T \ll T_c$) we obtain a decrease of the parameter $C = (1.6 \pm 0.08) \times 10^{-4}$ in the as-quenched state to $C = (0.30 \pm 0.035) \times 10^{-4}$ after the second annealing process. The decrease of the parameter C obtained from the "plateau" region in the internal friction is consistent, within experimental error, with that obtained from the sound velocity, see Fig. 4.20.

According to the standard tunneling model and within the resonant interaction between phonons and TS's, the thermal conductivity is given by

$$\kappa = \frac{\varrho k_B^3 v}{6\pi \hbar^2}(P_0\gamma^2)^{-1}T^2 \ . \tag{4.48}$$

Therefore, the slope of the nearly quadratic temperature dependence of κ measured at $T \ll T_c$ in amorphous $Cu_{30}Zr_{70}$ (Esquinazi et al. (1982)) provides the product $P_0\gamma^2$ and enables a comparison with the results from the acoustic measurements.

The relative change of C and $P_0\gamma^2$ with annealing obtained from the sound velocity, internal friction and thermal conductivity are compared in Fig. 4.20 as a function of the reduced superconducting critical temperature that indicate the annealed state of the amorphous sample. From the temperature dependence of the internal friction, Esquinazi and Luzuriaga (1988) estimated the change of the coupling constant γ with annealing and concluded that surprisingly the decrease of the parameter C with annealing can be ascribed to a very large extent to the decrease of the coupling constant and not to a change of the density of states P_0 in the early stage of annealing (Fig. 4.20). Similar results were obtained for polycrystalline Pt, see next section. Note that the decrease of γ (and C) and the constancy of P_0 with annealing disagrees with the predicted relationship $P_0 \propto 1/\gamma^2$ by Coppersmith (1993), see p. 161.

It is also interesting to note that the relative decrease of the internal friction and of the slope of the sound velocity below 0.5 K and above 5 K are almost equal (Fig. 4.19). This result indicates that the low- and high-temperature regimes (below and above 1 K) are correlated and could be explicable using the same assumptions. The model of incoherent tunneling allows a correlation between the high- and low-temperature behavior without introducing new free parameters. This is shown in Sect. 4.5.5 where a systematic study of this correlation in polycrystalline metals is reviewed. In agreement with this result, White and Pohl (1996) performed a survey of the available acoustic data in amorphous solids as well as quasicrystals above 5 K and found an empirical correlation between the value of the internal friction at the "plateau" with the slope of the linear variation of the sound velocity (see Sect. 4.5.5 and also Topp and Cahill (1996)).

4.4.5 Amorphous Thin Films

Although the theoretical treatment of the acoustic properties of amorphous solids in terms of the existence of TS's is rather successful, less is known about the nature of the tunneling entities, and the striking similarity of the glasslike behavior among this large number of sometimes very different materials cannot be explained either. As an important approach towards an explanation, in particular of the universality of the experimental data, the suggestion of Yu and Leggett (1988) has to be considered. They proposed that the reason for this universal glasslike behavior might be related to an *interaction* among defects mediated by phonons and dependent on the distance R between defects as $\propto R^3$ in three dimensions. Especially experiments performed at very low temperatures should be sensitive for this kind of strain-mediated indirect interaction that provides an additional relaxation channel along with the "normal" relaxation process due to thermal phonons.

Effects related to the interaction between tunneling systems might be studied in experiments with samples of reduced dimensionality. The properties of a sample should be affected by its dimensions when the interaction energy U_0 exceeds the thermal energy $k_\mathrm{B}T$. Thus, the existence of strain-mediated effects may be observable in a low temperature region defined by

$$T < \frac{U_0}{k_\mathrm{B} L^3}, \tag{4.49}$$

where the sample size L provides the maximum interaction distance.

Different experimental works that studied the acoustic properties or phonon transmission of amorphous thin films mostly at very high frequencies had been published [see, for example, Mebert et al. (1990), Härdle (1985), Rothenfusser et al. (1983)]. We would like to discuss in this section only recent studies of vitreous silica in the kHz range and at very low temperatures. White and Pohl (1995) have investigated the internal friction of amorphous SiO_2 films in the temperature range 50 mK$< T <$ 20 K and at film thicknesses between 0.75 nm and 1000 nm. The internal friction Q_{film}^{-1} of their films shows a plateau above $T = 100$ mK almost identical to the results of amorphous bulk SiO_2. They do not observe any change in the C-parameter of the standard tunneling model with the film thickness and in the temperature range of their investigations. Furthermore, they conclude that the interaction is limited to distances smaller than 0.75 nm which is the smallest thickness used in their experiment.

The study of the acoustic properties of amorphous SiO_2 films was extended to much lower temperatures (0.1 mK$< T <$1 K) by Gaganidze et al. (1997). Using the vibrating reed technique, Gaganidze et al. (1997) were able to investigate sound velocity *and* internal friction of film samples of thicknesses between 20 nm and 500 nm and at frequencies of $\approx$ 10 kHz. The parameter C extracted from the logarithmic temperature dependence of the sound velocity was found to be rather independent of the film thickness, thus confirming the results for the internal friction of White and Pohl (1995).

The sound velocity data, however, show clear deviations from bulk behavior: The crossover temperature T_{max} from the resonant to the resonant plus relaxation regime significantly decreases with decreasing film thickness. This is in clear contradiction to the behavior expected from the standard tunneling model according to which T_{max} only depends on the frequency of the measurement that is a constant in this experiment. An explanation of the observed behavior taking into account the bulk phonon mediated relaxation relation for tunneling systems according to the standard tunneling model fails, as this would lead to an increase of the coupling constant γ due to the decrease of T_{max} to unreasonably large values for γ of up to 40 eV for the $\approx$ 21 nm SiO_2 film.

A first approach to explain the observed shift of T_{max} with the film thickness assumes the TS's to consist of complex collective excitations of interacting defects that are expected to be very sensitive to the dimensionality of the phonon spectrum. The assumption that the glassy film is acoustically decoupled from the host reed leads to a change of the phonon spectrum of the film that in turn affects the relaxation rate of the TS's. The coupling describing the interaction of defects with the two-dimensional phonons now depends on the film thickness, and eventually the relaxation rate depends on the film thickness too. Although this "acoustic mismatch" and the reduced dimensionality of the interaction between TS's might explain the experimentally observed shift of T_{max} qualitatively (see Gaganidze et al. (1997)), more theoretical and experimental study is necessary to verify this interpretation.

4.5 Acoustic Properties of Polycrystalline Metals

4.5.1 General Remarks

The temperature and strain dependence of the acoustic properties of amorphous dielectrics and metals discussed so far in Sects. 4.4.1–4.4.3 were for a long time considered to be exclusively restricted to the class of amorphous materials. The surprising universality of the behavior of the low-temperature acoustic and thermodynamic properties of these materials led to the term of so-called "glasslike" properties of amorphous materials. Apart from amorphous dielectrics and metals, glasslike low-temperature properties were only studied – and found – in a rather large variety of special materials like Li-KCl (Narayanamurti and Pohl (1970), Wang et al. (1992)) or H-Nb(O) (Morr et al. (1989)), where it is well accepted that the tunneling entities are well-identified atoms or groups of atoms, see Chap. 7. In addition to these particular materials, materials that have been exposed to a special treatment, as neutron-irradiated quartz were investigated (Laermans and Esteves (1988), Keppens and Laermans (1996), Laermans and Keppens (1995), Peeters et al. (1997)). The growing interest in the acoustic properties of *amorphous materials* at low temperatures was stimulated by the unexpected results of the specific heat measurements on a number of non-crystalline solids published by Zeller and Pohl (1971); simple *polycrystalline metals* like silver or copper, however, were ignored from this research for a long time. It was only in the past few years that the experimental confirmation of the – sometimes even quantitative – similarity between the low-temperature acoustic properties of amorphous materials and polycrystalline metals led to experimental and theoretical activities in the search for an explanation of these surprising results.

In this section the results of our studies of the acoustic properties on polycrystalline metals performed at audio frequencies in the range $0.1\text{ kHz} \leq \nu \leq 20\text{ kHz}$ and at temperatures $10^{-4}\text{ K} \leq T \leq 10\text{ K}$ are reviewed. The measurements of the sound velocity and internal friction of the polycrystalline superconductors Nb, NbTi, and Ta in their superconducting state show the same characteristic features as observed in amorphous dielectrics such as SiO_2 (Esquinazi et al. (1992)): a maximum in the sound velocity at $T_{max} \sim 0.1\text{ K}$ with quasi-logarithmic temperature dependences above and below T_{max}, and a constant internal friction in the temperature range of $\sim 0.1\text{ K} \leq T \leq 1\text{ K}$ with a decrease of Q^{-1} towards lower temperatures. This resemblance is a clear indication of the existence of TS's in polycrystalline superconductors

Table 4.3. Investigated polycrystalline superconductors. T_c: superconducting transition temperature. B_c: critical magnetic field for superconductivity at $T = 0$ K.

Material	ϕ [μm]	Frequency [kHz]	T_c [K]	B_c [T]	Remarks
Ta	125	1.83	4.48	0.083	annealed
	125	5.45			
Nb	120	1.48	9.50	0.198	
Nb(Cu)	165	0.5			Nb filaments in Cu matrix
NbTi	20	0.6	9.5	12.2	surface treated with acid
	20	0.9			one filament
	40	3.7			surface mechan. treated
	50	0.37			
	50	1.47			with insulation

and of the interaction of these TS's with phonons due to the absence of quasiparticles in the superconducting state of the samples. A possible interaction of conduction electrons with TS's in polycrystalline metals is discussed in studies of the acoustic properties of Ta and Al in their normal- and superconducting states, and of normal-conducting metals like Ag, Cu, Pt, Pd and alloys like Manganin, PtRh and PtW. Measurements of the sound velocity of polycrystals at $T > 1$ K show a linear temperature of $\Delta v/v$ that confirms the observed similarity between the acoustic properties of polycrystals and amorphous materials at high temperatures too, see Sect. 4.5.5.

The results of these experiments clearly demonstrate that the "glasslike" behavior usually addressed solely to amorphous solids is of a much more general nature and is basically observed too in almost every polycrystalline metal investigated. Furthermore, as a consequence of the observed similarities in sound velocity and internal friction between amorphous insulators and polycrystalline normal-conducting metals, we conclude that the interaction between conduction electrons and TS's in polycrystals is negligible. This is in clear contrast to the results of the interaction of electrons and tunneling systems observed in amorphous metals. We would also like to note that the acoustic properties of polycrystalline metals with small amount of magnetic Fe-impurities (Esquinazi et al. (1996b)) show a different behavior to that presented in this section.

4.5.2 Polycrystalline Superconductors

The study of the acoustic properties of polycrystalline superconductors provides the possibility to compare internal friction and sound velocity of polycrystalline metals with the properties obtained from the investigation of amorphous dielectrics since at $T \ll T_c$ the number of quasiparticles is negligible, and the TS's interact with phonons only. The requirement of a high

4.5 Acoustic Properties of Polycrystalline Metals 201

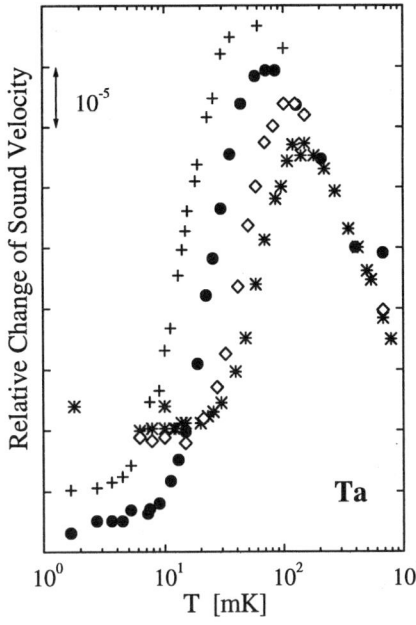

Fig. 4.21. Temperature dependence of the relative change of sound velocity of Ta in the superconducting state at four different strains (+) 4.1×10^{-6}, (•) 1.8×10^{-5}, (◊) 5.8×10^{-5}, (∗) 7.8×10^{-5} at a frequency $\nu = 5.45$ kHz.

transition temperature (and a large critical field in cases where the vibrating wire technique is applied) with respect to the temperature range of the measurements is fulfilled for all samples investigated, see Table 4.3.

Figure 4.21 shows the relative change of sound velocity of a Ta wire ($\phi = 125$ µm) at 1 mK$\leq T \leq$ 1 K at four different strains $4.1 \times 10^{-6} \leq \epsilon \leq 7.8 \times 10^{-5}$ (Esquinazi et al. (1992)) at a resonance frequency of 5.45 kHz. The measurements were performed with the vibrating wire technique in an applied magnetic field of $B = 20$ mT that is clearly below the critical field B_c, see Table 4.3.

With decreasing temperature we observe an increase of the sound velocity that is independent of the applied strain, and a maximum of the sound velocity at $T_{\max}$. Its position depends on the strain and is shifted to lower temperatures the lower the strain is. At $T < T_{\max}$ the sound velocity is strain dependent, decreases with decreasing temperature and approaches a constant value at the lowest temperatures. At the two largest strains this saturation starts at significantly higher temperatures indicating the influence of self-heating. A detailed discussion of amplitude-dependent (nonlinear) effects follows at the end of this section. These temperature and strain dependences observed for the relative change of sound velocity of superconducting Ta strongly resemble the behavior observed for the acoustic properties of amorphous dielectrics, see Sect. 4.4.1 and Fig. 4.11. Similar behavior is observed in *all* investigated polycrystals in their superconducting state. In Fig. 4.22

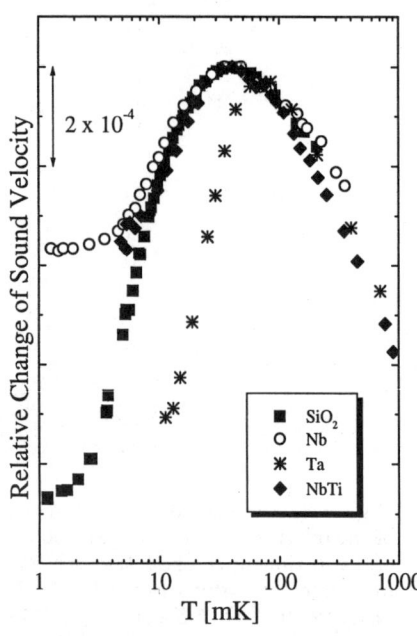

Fig. 4.22. Relative change of sound velocity for vitreous silica SiO$_2$ (0.4 kHz) compared with $\Delta v/v$ for Nb (1.5 kHz), Ta (5.5 kHz), and NbTi (0.37 kHz) in their superconducting state. The $\Delta v/v$ scale has to be divided by a factor of 4 (10) for Nb (Ta).

we compare the sound velocity of the polycrystalline superconductors Ta, NbTi and Nb at $T < T_c$ with the relative change of sound velocity of vitreous silica (SiO$_2$) at almost the same strain ($\epsilon \sim 10^{-6}$). The temperature dependence of the sound velocity above the maximum provides the parameter $C = (4 \pm 1) \times 10^{-5}$ for Ta, $(6 \pm 1) \times 10^{-5}$ for Nb, and $(4 \pm 1.5) \times 10^{-4}$ for NbTi. For NbTi the C value differs only by a factor of two from that obtained for vitreous silica [$C = (2.7 \pm 0.1) \times 10^{-4}$, Esquinazi et al. (1992)].

Studies of the internal friction of the superconducting polycrystals confirm the similarity of the acoustic properties between superconducting polycrystalline metals and amorphous insulators. Figures 4.23 and 4.24 show the temperature dependence of the internal friction of Ta and Nb in the superconducting state. Again, we observe features that are characteristic of the interaction of TS's with phonons: an almost constant internal friction at ~ 0.1 K $< T < 1$ K and a decrease of Q^{-1} towards lower temperatures. The saturation at the largest strains below $T \sim 20$ mK (see Fig. 4.23) is again caused by self-heating of the sample. The C values derived from the plateau in the internal friction are $C \simeq 2.6 \times 10^{-5}$ for Ta, $C \simeq 0.6 \times 10^{-4}$ for Nb and 3.5×10^{-4} for NbTi; they are in good agreement to those obtained from the sound velocity data. In contrast to the results obtained for the sound velocity there is within experimental error no strain dependence of the internal friction observable in the strain range of our investigations (apart from effects caused by self-heating).

4.5 Acoustic Properties of Polycrystalline Metals   203

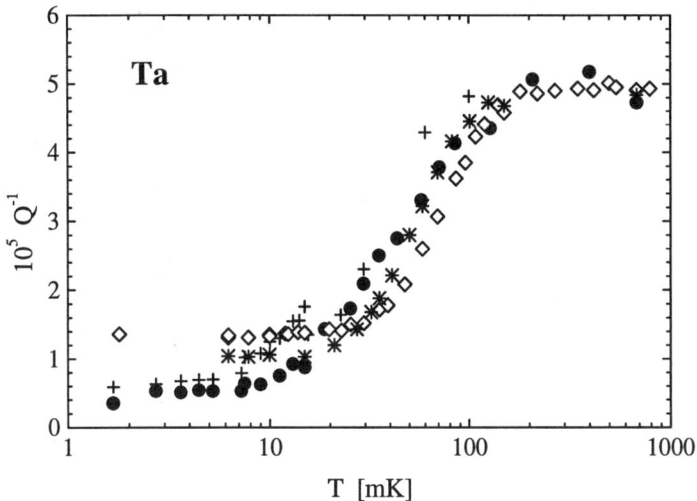

Fig. 4.23. Internal friction of Ta at four different strains (+) 4.1×10^{-6}, (•) 1.8×10^{-5}, (◊) 5.8×10^{-5}, (∗) 7.8×10^{-5} taken at a frequency of 5.45kHz.

It is known that in the polycrystalline metals Nb and Ta, both with a bcc crystal structure, hydrogen atoms may be associated with the tunneling entities (see Chap. 7). However, the tunneling of hydrogen in a polycrystal differs significantly from the properties of TS's in disordered solids as far

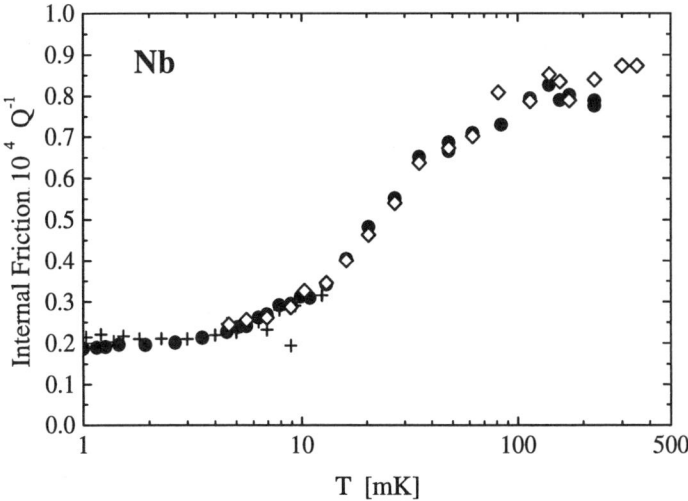

Fig. 4.24. Internal friction of Nb at four different strains ϵ: (+) 3.9×10^{-7}, (•) 2.0×10^{-6}, (◊) 3.6×10^{-6}, taken at a frequency of 1.5kHz.

as the distribution of the tunneling parameters is concerned. We certainly cannot rule out that the tunneling of interstitials in the investigated bcc-metals may contribute to the observed "glasslike" behavior. The striking similarity of the temperature dependences of $\Delta v/v$ and of Q^{-1} as well as of the strain dependence of the relative change of sound velocity of these metals with the results obtained for SiO_2 are nevertheless indications for the existence of a similar energy dependence and a broad distribution of relaxation times in the polycrystals as it is found in amorphous solids.

These low frequency measurements provide several experimental observations of the acoustic properties of polycrystalline metals in the superconducting state that the standard tunneling model fails to explain as is also the case for amorphous dielectrics, see Sect. 4.4.1:
(1) The strong dependence of the relative change of sound velocity on the applied strain below $T_{\max}$, and
(2) the apparent saturation of internal friction and sound velocity at the lowest temperatures. Whereas the saturation at large strains is caused by self-heating of the sample, the saturation at low strains has different origins for $\Delta v/v$ and Q^{-1}: the saturation in $\Delta v/v$ is attributed to the change of the population number of the TS's produced by the strain field, see Sects. 4.4.1 and 4.2.4. The saturation of Q^{-1} includes an additional contribution of the clamping of the sample. Subtraction of this constant background from Q^{-1}, however, still does not lead to the temperature dependence $Q^{-1} \propto T^3$ as expected from the standard (one-phonon) interaction of TS's with phonons.

The different strain dependences of the change of sound velocity (see Fig. 4.21) and of the internal friction (Figs. 4.23 and 4.24) at $T < 100$ mK are related to different relaxation mechanisms of the TS's in the frequency range of our measurements: the temperature dependence of Q^{-1} is governed only by relaxation processes in the whole temperature range, whereas the sound velocity probes mainly the resonant processes at temperatures below $T_{\max}$ leading to strain-dependent effects. As discussed in Sect. 4.4.1, this strain dependence of the sound velocity can be explained as a change of the population number of the TS's due to the strain wave (Esquinazi et al. (1992)). The similarity of the strain dependence with that observed in SiO_2 should be regarded as further evidence that in superconducting polycrystals the TS's mainly interact with phonons. In agreement with the standard tunneling model and as observed in amorphous dielectrics, the position of $T_{\max}$ of the sound velocity of polycrystals at comparable strain depends on the frequency of the measurements and scales according to $\omega T^3 \simeq$ constant as observed in NbTi (Esquinazi et al. (1993)).

It is interesting to note that the similarity of polycrystalline superconductors with amorphous materials is not only restricted to the acoustic properties as described above. Wässerbach (1978) (see also Wässerbach (1987)) reports a thermal conductivity κ of plastically deformed Nb and Ta single crystals with a temperature dependence of $\kappa \propto T^2$ for Ta at 0.3 K$< T <1$ K and

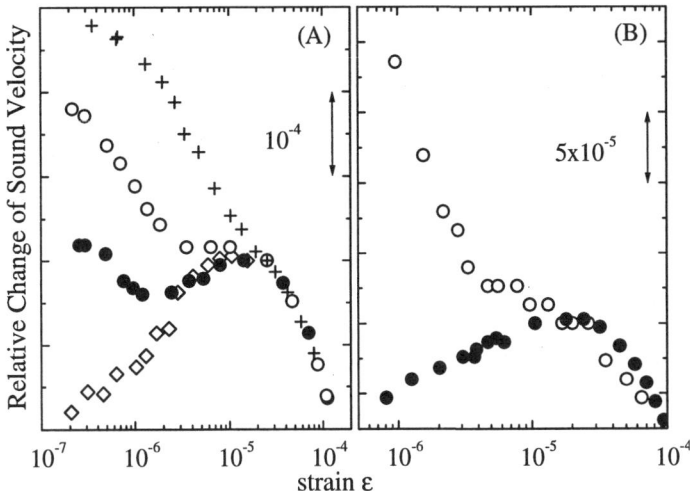

Fig. 4.25. Strain dependence of the change of sound velocity for two NbTi wires of the same diameter at two different frequencies: (A) $\nu = 0.37$ kHz at temperatures $T = 0.2$ mK ($\diamond$), 5.7 mK ($\bullet$), 11.2 mK ($\circ$), and 33.2 mK (+); (B) $\nu = 3.76$ kHz at $T = 6.4$ mK ($\bullet$), and 22.0 mK ($\circ$).

for Nb at 0.3 K< T <5 K. In the analysis, this temperature dependence is attributed to the scattering of phonons by static strain fields. A similar T-dependence of the thermal conductivity is also reported for NbTi in its superconducting state by Schmidt (1978), i.e. $\kappa \propto T^{1.85}$. Even in Al in its superconducting state at $T \ll 1$ K, a glasslike thermal conductivity $\kappa \propto T^2$ was observed (Gloos et al. (1990)), see Sect. 4.5.3.

Strain-Dependent Effects

Studies of the strain dependence of the sound velocity at constant temperatures are necessary in order to clarify whether the observed temperature dependence of the sound velocity at a given strain has a physical meaning or whether the (nonlinear) effects are spurious. The typical shape of the strain dependence of the relative change of sound velocity is discussed for two NbTi wires at resonance frequencies that differ by an order of magnitude ($\nu = 0.37$ kHz and $\nu = 3.76$ kHz, Fig. 4.25). The strain dependence of $\Delta v/v$ is the sum of three different contributions that can be identified as follows:
(1) The change in the population number causes a decrease of $\Delta v/v$ with increasing strain at *small* values of ϵ (e.g. at $\epsilon < 3 \times 10^{-6}$ at $T = 11.2$ mK for NbTi at $\nu = 0.37$ kHz, see Fig. 4.25A).
(2) Self-heating of the sample (that is of course increasingly pronounced the lower the temperature is) results in an increase of $\Delta v/v$ with increasing strain. This increase in $\Delta v/v$ may compensate the decrease due to the change of the population number resulting in an almost strain-independent

interval ($3 \times 10^{-6} < \epsilon < 10^{-5}$ for NbTi at $\nu = 0.37$ kHz at $T = 11.2$ mK). Self-heating is less pronounced at higher temperatures and has almost disappeared at $T = 33.2$ mK (at $\nu = 0.37$ kHz). The different strain dependence of the sound velocity of both NbTi samples at almost the same temperature ($T = 5.7$ mK at $\nu = 0.37$ kHz and $T = 6.4$ mK at $\nu = 3.76$ kHz) is related to the frequency dependence of the energy dissipation of the resonator [see (4.47)].

(3) At high enough strains a nonlinear restoring force (i.e. $\omega_0^2 x + \beta x^3$ with $\beta < 0$) leads again to a now temperature-independent decrease of the resonance frequency (at about $\epsilon > 10^{-5}$ for both wires).

Strain-dependent effects are strongly dependent on material and size of the sample. Although the resonance frequency of the Ta sample is larger than that of the NbTi wires, we observe a decrease in the sound velocity (with increasing strain at $\epsilon \leq 10^{-5}$) even at $T = 4.45$ mK (Fig. 4.26). This behavior is attributed to the larger thermal conductivity of the Ta wire in addition to its larger radius ($a_{Ta} \approx 3 a_{NbTi}$). The decrease in the sound velocity of Ta due to the nonlinear restoring forces at high strains is smaller than in NbTi and was not observed in Ta at $T < 35$ mK. At higher temperatures ($T > 200$ mK) where self-heating effects and the strain dependence of $\Delta v/v$ due to the phonon–TS's interactions are negligible, the influence of nonlinear restoring forces can still be recognized (see Fig. 4.27, König et al. (1993)).

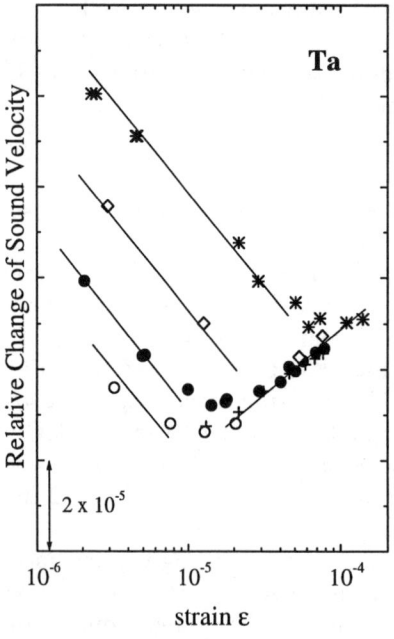

Fig. 4.26. Strain dependence of the change of sound velocity of Ta at a resonance frequency $\nu = 5.45$ kHz and at $T = 4.45$ mK (○), 6.20 mK (+), 7.50 mK (●), 14.9 mK (◇), and 35.4 mK (∗). The lines are only guides for the eye.

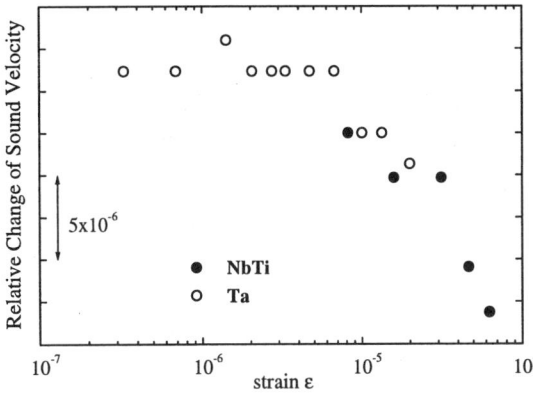

Fig. 4.27. Strain dependence of the relative change of sound velocity of Ta (o, $\nu = 5.45$ kHz) and NbTi (•, $\nu = 3.7$ kHz) measured at $T = 400$ mK (Ta) and $T = 500$ mK (NbTi).

As a consequence of the amplitude dependence of the resonance frequency, the resonance curves show non-Lorentzian lineshapes (König et al. (1993)). This effect is particularly pronounced for the Ta wire due to its very small internal friction. The observed asymmetric lineshapes can be approximated by a numerical simulation using a logarithmic amplitude dependence for the measured resonance frequency $\omega' \sim \omega(1 \pm f \mid \ln u \mid)$, where u is the reduced amplitude, ω is the resonance frequency at $u = 1$, and f is a numerical constant. The sign in brackets depends on the decrease $(-)$ or increase of the sound velocity $(+)$ with the applied strain (Esquinazi et al. (1992)).

4.5.3 Normal Metals. The Absence of Electron-Assisted Relaxation in Polycrystals

Information on the interaction of *electrons* with TS's in polycrystalline metals can be obtained, for example, from a comparison of the sound velocity and internal friction in the superconducting and normal state of the materials. This can be easily achieved in an experimental setup with the sample situated in a field coil that enables us to switch between the normal $(B > B_{c2}, B_c)$ and superconducting state $(B < B_{c2}, B_c)$. Results obtained on Al in both states as discussed in the following indicate that the electron-assisted relaxation in polycrystals is negligible. Studies of the acoustic properties of nonsuperconducting materials like Pt, Ag and Cu strongly support this result. A summary of important parameters of the normal metals investigated in this work is given in Table 4.4.

Figure 4.28 shows the temperature dependence of $\Delta v/v$ and Q^{-1} of an aluminum sample (Al wire, $\phi = 25$ μm, König et al. (1995)) in its normal and superconducting state. Within experimental error, we cannot distinguish between the acoustic properties of both phases of this Al sample. This striking result indicates that the phonon–TS interaction is not altered in the presence of electrons in normal-conducting polycrystalline aluminum. Additional

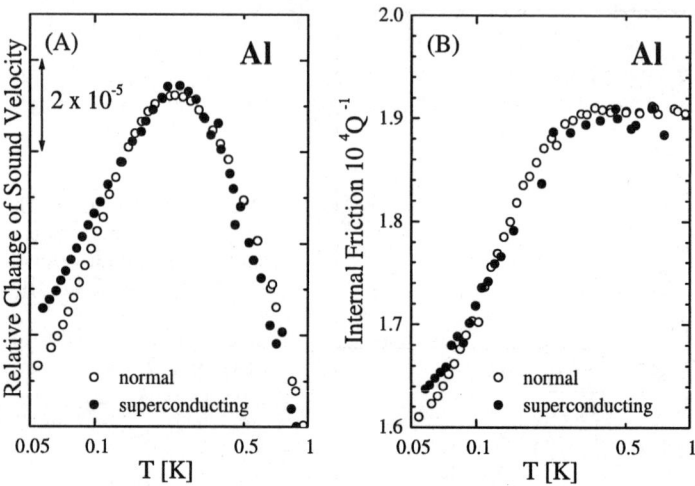

Fig. 4.28. Temperature dependence of sound velocity (A) and internal friction (B) of Al at $\nu = 7.7$ kHz in the normal (○, $B = 10.7$ mT) and superconducting (●, $B = 4.7$ mT) state. The applied strains were $\epsilon \simeq 7 \times 10^{-5}$ (superconducting state) and $\epsilon \simeq 9 \times 10^{-5}$ (normal state). This small difference in the applied strain accounts for the deviation among the data at the lowest temperatures. After König et al. (1995).

support for the absence of the electron–TS interaction in Al is provided by measurements of the sound velocity at different strains (Fig. 4.29) where similar behavior as in amorphous SiO_2 and in amorphous superconductors is observed: a maximum in the sound velocity that shifts to lower temperatures with decreasing strain and a strain dependence of the sound velocity below this maximum. The value derived for the parameter C either from the slope of $\Delta v/v$ above $T_{\max}$ or from the plateau of Q^{-1} is 6×10^{-5} and 1.2×10^{-4}, respectively. This is only a factor of 4 (2) smaller as obtained for SiO_2.

Further evidence for the absence of electron-assisted relaxation in polycrystalline metals is provided by the studies of the acoustic properties of platinum. In a series of measurements, we have investigated the temperature and strain dependence of the sound velocity and internal friction of Pt wires at various resonance frequencies in the range 0.08 kHz $\leq \nu \leq 1.5$ kHz with the result of a striking similarity to the acoustic properties of amorphous *insulators*. Figure 4.30 shows the temperature dependence of a Pt wire ($\phi = 25$ μm) at $\nu = 1.5$ kHz at three different strains that is again similar to the characteristic temperature and strain dependences of amorphous insulators.

From Fig. 4.30 we conclude that the acoustic properties of platinum are governed by the interaction of *phonons* with TS's. However, the acoustic properties of Pt also show some differences compared to the acoustic properties of amorphous insulators or polycrystalline superconductors. In particular, we note that $T_{\max} \propto \nu^{1/3}$ was not observed for Pt. This deviation from the behavior expected from the standard tunneling model, however, might either be attributed to a different strain distribution in the wires (that were taken from different batches) or it might be related to an additional contribution to the strain from the clamping and/or bending of the wire (König et al. (1995)).

Particular emphasis has to be taken to the saturation of the sound velocity at temperatures below $\approx$ 5 mK that results from the change of the population number of the TS's. This saturation in Pt cannot be caused by self-heating of the sample as the respective $T_{\max}$ shifts to higher temperatures with increasing strain; furthermore, the corresponding data of the internal friction do not show any strain dependence. From the logarithmic slope of the sound velocity above $T_{\max}$ we obtain $C = (7 \pm 1.5) \times 10^{-5}$ for Pt that is only about a factor of four smaller than for vitreous silica.

Among all polycrystalline metals investigated, copper and silver are the only materials that clearly show a different behavior in their acoustic properties compared with the above discussed results for Pt, Al and the super-

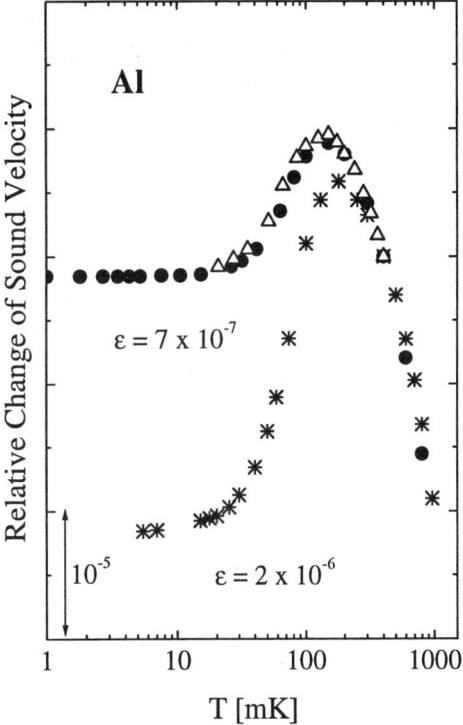

Fig. 4.29. Temperature dependence of the relative change of sound velocity of an Al wire ($\phi = 5$ μm, $\nu = 0.32$ kHz) in the superconducting (•, $B = 3$ mT; △, $B = 0.8$ mT) and normal (∗, $B = 33$ mT) state. Magnetic field and excitation current for both measurements in the superconducting state were chosen to provide the same strain $\epsilon = 7 \times 10^{-7}$ that confirms the reproducibility of the results. The applied strain at $B = 33$ mT was $\epsilon = 2 \times 10^{-6}$ that explains the shift of $T_{\max}$ and the different slope at $T < T_{\max}$.

Table 4.4. Properties of some of the investigated normal-conducting polycrystals.

Material	Dimensions [μm]	Frequency [kHz]	Remarks
Ag	7x2x0.03 [a]	0.25	Vibrating Reed[b]
	25	2.70	
	125	2.47	as-received
	125	2.31	annealed
Al	5	0.3	$T_c = 1.14$K, $B_c = 10.5$ mT
	125	0.6	
Cu	25	0.3	
	125	0.37	
Pd	7×2×0.2 [a]	0.20	
	5×2×0.1 [a]	20	
Pt	25	0.084	RRR = 460[c]
	25	0.76	
	25	1.5	annealed
PtW	25	1.44	($Pt_{92}W_8$) RRR = 1
PtRh	5×2×0.1 [a]	12	($Pt_{70}Rh_{30}$)

[a] Dimensions of the numbers in mm
[b] Two samples with different impurity content of 1ppm and 100ppm
[c] The RRR (Residual Resistivity Ratio) of 460 applies to all nonannealed Pt samples

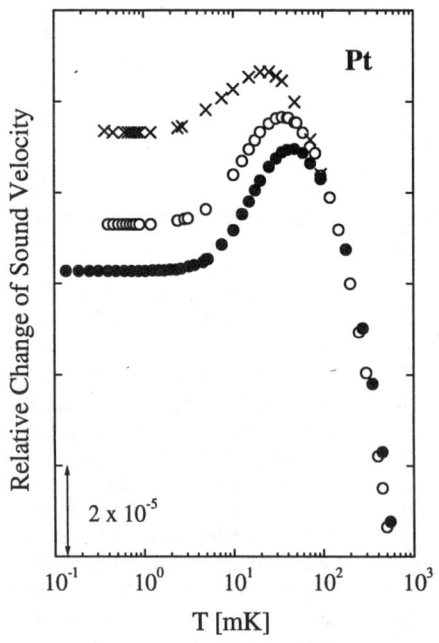

Fig. 4.30. Temperature dependence of the relative change of sound velocity for Pt at $\nu = 1.5$ kHz with $\epsilon = 1 \times 10^{-6}$ (×), $\epsilon = 7 \times 10^{-6}$ (o), and $\epsilon = 20 \times 10^{-6}$ (•).

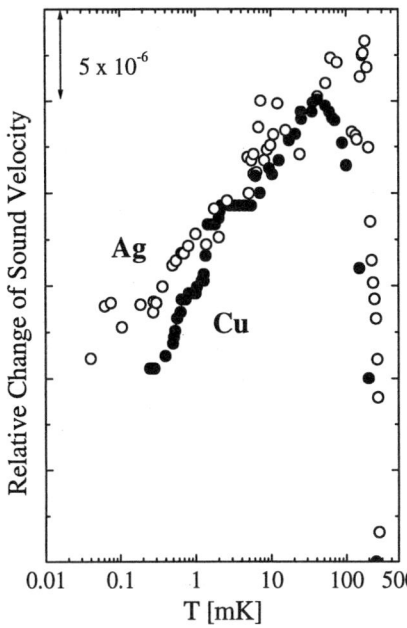

Fig. 4.31. Relative change of sound velocity of a Cu vibrating wire (●, $\nu = 0.25$ kHz) and of a Ag vibrating reed (○, $\nu = 0.30$ kHz).

conductors Ta, Nb and NbTi, although at least a qualitative similarity in the acoustic properties is also found in these two metals. However, there do exist some significant deviations in the behavior of $\Delta v/v$ and Q^{-1}:

(1) The relative change of sound velocity of Cu and Ag also shows a maximum at about 0.1K (Esquinazi et al. (1992), Esquinazi et al. (1996a)) but the temperature dependence of $\Delta v/v$ at $T < T_{\max}$ is much weaker than at $T > T_{\max}$ (Fig. 4.31).

(2) The logarithmic temperature dependence of $\Delta v/v$ at $T < T_{\max}$ extends into the μK temperature range without any sign of saturation, and the slope of this temperature dependence of $\Delta v/v$ is at least one order of magnitude smaller as found for the other polycrystalline metals investigated (Esquinazi et al. (1996a), see Fig. 4.31).

(3) The internal friction shows an anomalously weak temperature dependence too. For a cold-rolled Ag sample, we have found $Q^{-1} \propto T^{1/3}$ (Esquinazi et al. (1992)).

(4) Neither sound velocity nor internal friction shows a strain dependence in the investigated strain range (Esquinazi et al. (1992)) that is in clear contrast to the observation of the amplitude-dependent acoustic properties of the amorphous metal PdSiCu (Sect. 4.4.2).

The extension of the logarithmic temperature dependence of the sound velocity down to $T \approx 100$ μK is a strong indication that self-heating effects can be excluded in these acoustic measurements on Ag and Cu. A similar observation of a temperature dependence ($\propto T^{1/3}$) extended over almost three decades down to ≈ 100 μK is made for the internal friction of Ag (after the subtraction of a constant background attributed to the clamping of the sample, see Esquinazi et al. (1992)). For Cu, however, the internal friction shows a significantly different behavior as observed in measurements with two Cu wires with different diameters ($\phi = 25$ μm and 125 μm) in the temperature range 0.06 mK$\leq T \leq$ 900 mK (Esquinazi et al. (1994)) and at strains $\approx 10^{-7} \leq \epsilon \leq 10^{-3}$. For $\epsilon > 10^{-6}$ and at $T < 300$ mK, the strain dependence of the internal friction can be described by $Q^{-1} \propto \epsilon^n$ with $n = (0.41\pm0.02)$. For $\epsilon < 10^{-6}$, Q^{-1} deviates from this dependence; the deviation is larger the higher the temperature. A possible explanation for this behavior might be given in terms of tunneling of pinned dislocations assuming that the combination of sufficiently low temperatures and high enough stress is more favorable for dislocations to leave their pinning centers as it could be due to thermal activation.

The observation of a negligible influence of the electrons on the relaxation mechanism of the TS's in the polycrystalline normal-conducting metals is in clear contrast to the results obtained for amorphous metals.[5] In amorphous superconductors in the normal state, no maximum in the sound velocity was observed down to 0.1K (Esquinazi and Luzuriaga (1988), see Sect. 4.4.3). The measurements of the acoustic properties of superconducting glasses in their normal and superconducting state show that the presence of conduction electrons could lead to a renormalization of the density of states and/or the coupling constant between phonons and tunneling systems.

Comparing the differing results for the acoustic properties of amorphous and polycrystalline metals, one may think about a possible influence of the electron mean free path l_e on the relaxation mechanism of the TS's. The mean free path is the property with the largest difference between both classes of materials, and in polycrystals the value of l_e varies by several orders of magnitude. Therefore, it was speculated that a simultaneous study of the dependence of the parameter C (that is proportional to the density of states of TS's) and of the coupling constant γ on the electron mean free path could provide indications for a possible interaction of electrons with TS's in polycrystals with a very small mean free path. Figure 4.32 shows the dependence of C and Fig. 4.33 of the coupling constant γ on the mean free path for a number of polycrystalline metals (Gaganidze and Esquinazi (1996a)). The values for C presented in Fig. 4.32 were obtained from the slope of the T-dependence of $\Delta v/v$ at $T > T_{\max}$. The coupling constant is derived from the position of the maximum taking into account that $\tau_m^{-1} \propto \gamma_t^2 T_{\max}^3$ (assuming $v_t < v_l$)

---

[5] As we have not studied the acoustic properties of the superconducting metals Nb, NbTi and Ta in their normal-conducting state, we restrict our discussion to the investigated normal-conducting metals, e.g. Ag, Al or Pt. We should note that in the two-phase system Zr-Nb the contribution of an electronic relaxation process is observed in the normal-conducting state of the sample (Thomas et al. (1980), Weiss et al. (1981)).

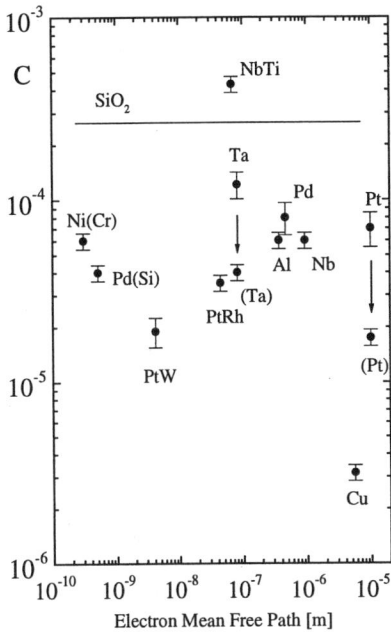

Fig. 4.32. The parameter $C = P_0\gamma^2/\varrho v^2$ as a function of the electron mean free path l_e for various polycrystalline metals. The straight line indicates the value $C = 2.7 \times 10^{-4}$ for SiO_2. The arrows indicate the decrease of C after annealing.

according to the standard tunneling model. Clearly, there is no observable correlation between neither C nor γ_t and l_e.

The decrease of the parameter C for Ta and Pt after the samples have been exposed to a thermal heat treatment (Fig. 4.32) has been observed in amorphous metals too (Esquinazi and Luzuriaga (1988), see Sect. 4.4.3). This observation, however, is of no significant importance concerning the striking conclusion that the acoustic properties of the polycrystalline metals resemble those of amorphous insulators much more than those of amorphous metals. The interaction between electrons and TS's in polycrystals seems to be of no particular importance.

The discussion of the origin of low-energy excitations in polycrystalline metals is guided by very similar arguments as is the case with amorphous materials: there do exist several explanations that may be applied to interpret the results of some particular materials but there is no common comprehension, neither for the universality of the acoustic properties of polycrystals nor for their similarity to those of amorphous materials. The existence of TS's of some kind in polycrystals may not be extraordinarily surprising, but excitations associated with defects in polycrystals are expected to have discrete energies, in sharp contrast to the excitation spectrum characteristic of glasses. One might for example speculate that light impurities like hydrogen are associated with the tunneling entities. This could be the case for e.g. Nb as has been discussed previously (Morr et al. (1989), Cannelli et al. (1986))

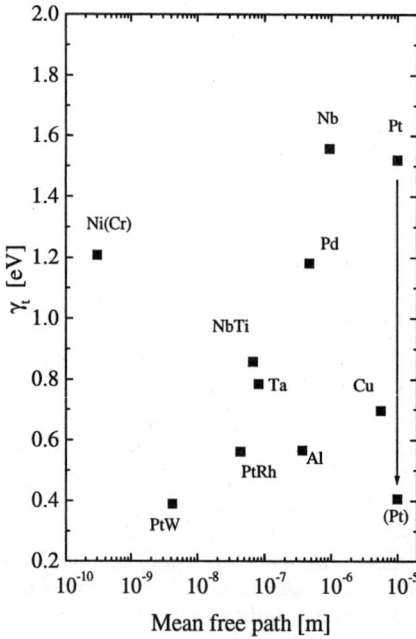

Fig. 4.33. The coupling constant γ_t between phonons and TS's as a function of the electron mean free path l_e. The arrow indicates the decrease of γ_t for Pt after the sample was annealed.

but this does not apply for metals such as platinum. Furthermore, the study of the acoustic properties of polycrystals clearly indicates that the universality of the observed acoustic properties and thus the origin of the tunneling entities have no correlation to the impurity concentration as the impurity content of the samples investigated varied from 1 ppm to almost 1000 ppm (according to the analysis of manufacturers of the samples). In order to verify quantitatively that the content of impurities does not account for a change in the acoustic properties, we have performed measurements on two almost identical Ag samples that only differed in their impurity content by two orders of magnitude (1 ppm and 100 pm). Within experimental error, we could not detect any difference in their acoustic properties (Thauer et al. (1990), Esquinazi et al. (1992)).

The annealing experiments discussed in Sect. 4.5.4 clearly reveal a strong influence of a thermal treatment of the sample on its acoustic properties. Annealing of a polycrystal leads to an increase of the grain size and to a reduction of the disordered zones. Therefore, one may speculate whether grain boundaries in polycrystals may be the origin for the TS's. However, it is difficult to understand how the very different microstructure of the polycrystals and the very small volume of their grain boundaries can lead to the observed universality in the acoustic properties among all polycrystals investigated – from our experimental observations we can conclude that grain boundaries *only* cannot be ascribed as the origin of the tunneling entities. It is interesting to note in this context that an amorphous material that was exposed to thermal treatment and thus contains the maximum possible volume of grain boundaries shows no or only weak anomalies in its acoustic behavior (Esquinazi and Luzuriaga (1988), see Sect. 4.4.4).

4.5.4 The Influence of Thermal Treatment

The results of the investigations of sound velocity and internal friction of *amorphous* materials exposed to thermal treatment as discussed in Sect. 4.4.4 revealed a decrease of the magnitude of glasslike anomalies after annealing. The acoustic properties of *polycrystalline* metals are strongly history dependent too, i.e. annealing or cold-rolling of the sample are leading to significant changes in the temperature dependence of the sound velocity as well as in the magnitude of the internal friction. For polycrystalline metals, this was first observed in measurements of the acoustic properties of an Ag vibrating reed (Thauer et al. (1990), Esquinazi et al. (1992)).[6] The maximum in the sound velocity of the as-received sample disappeared after the sample was annealed in vacuum at 1100 K for 2 h, and subsequent cold-rolling made the maximum reappear again. It is important to note that these studies of the acoustic properties of polycrystalline Ag samples containing different concentrations of impurities and the exposure to various mechanical and thermal treatments provided the first strong indications that some kind of "internal strain" (lattice defects and possibly the interaction among them) rather than the influence of impurities are likely to be the origin for the "glasslike" anomalies in polycrystals.

The annealing of the Ag samples described above was performed at a temperature close to the melting temperature of the metal ($T_{\text{melt}} = 1235$ K for Ag), and the study of the influence of annealing on amorphous metals was performed at a temperature close to the glass-transition temperature (see Sect. 4.4.4). However, for the observation of an effect of annealing on the acoustic properties, the exposure of the sample to a temperature close to a critical transition temperature does not seem to be a necessary prerequisite. The effect of thermal treatment on the acoustic properties of Pt and Ta, for example, is clearly observable although the melting temperatures of Pt ($T_{\text{melt}} = 2045$ K) and Ta ($T_{\text{melt}} = 3293$ K) are a factor of up to 2.5 larger than the highest annealing temperature available in our laboratory (≈ 1300 K).

We have thoroughly investigated the effect of annealing on the sound velocity and the internal friction of Pt wires ($\phi = 25$ μm). The wires investigated were taken from the same batch; the as-received and annealed samples were mounted onto the sample holder in a semicircular shape with almost the same diameter in order to obtain comparable resonance frequencies for the samples. One wire was annealed at $T = 1300$ K for 5 hours and particular

---

[6] An annealing effect was also observed in specific heat measurements on Zr-Nb, see Lou (1976).

care was taken that the heat treatment was performed with the wire already bent into its final shape of a semicircular loop. This annealing condition is important to ensure that no additional strain had to be applied to the wire while it was mounted to the sample holder. The annealing of the platinum

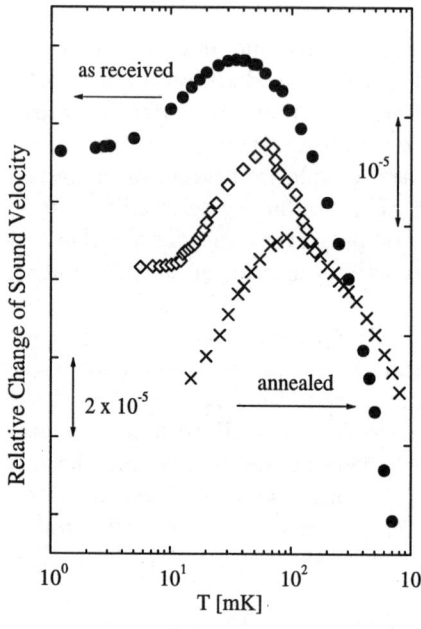

Fig. 4.34. Relative change of sound velocity as a function of temperature for Pt measured with the vibrating wire technique: as received ($\nu = 1.5$ kHz): (•) $\epsilon = 7 \times 10^{-6}$; annealed ($\nu = 1.4$ kHz): (◊) $\epsilon = 2.6 \times 10^{-6}$, (×) $\epsilon = 4 \times 10^{-6}$.

sample leads to a decrease in the logarithmic slope of the temperature dependence of the sound velocity at $T > T_{\max}$ by a factor of ≈ 4 and to a shift of $T_{\max}$ towards higher temperatures (Fig. 4.34).

Similar changes in the sound velocity were obtained in amorphous $Zr_{70}Cu_{30}$ after annealing (Esquinazi and Luzuriaga (1988), see Sect. 4.4.4). Within the framework of the standard tunneling model for equal strains and phonon frequencies, the shift of $T_{\max}$ can only be explained by a change in the coupling constant according to $\gamma_t^2 T_{\max}^3 \approx$ const. This leads to a decrease in the coupling constant after annealing to a value of about 30% of γ_t before annealing (König et al. (1995)). The decrease in the coupling constant may also be responsible for the change of the slope of the sound velocity above the maximum as $\Delta v/v \propto P_0 \gamma^2$.

The internal friction of the annealed Pt sample shows a decrease in its magnitude of about a factor 4 compared with the value obtained for the as-received sample at a similar strain (at $T = 10$ mK) of $\epsilon = (6 \pm 2) \times 10^{-6}$ (Fig. 4.35). This is in good agreement with the decrease of $\Delta v/v$ after annealing (Fig. 4.34). In contrast to measurements on different Pt samples (see e.g. König et al. (1994b)) the as-received sample in Fig. 4.35 does not show a plateau but a maximum in the internal friction that slightly shifts to higher temperatures after annealing. This might be explained by a deviation from the assumption of the standard tunneling model of an energy-independent distribution of TS's (or density of states of TS's). The observation of a shallow maximum in Q^{-1} rather than a plateau has also been seen in amorphous materials (Esquinazi et al. (1986), Esquinazi and Luzuriaga (1988)). The decrease of the coupling constant γ_t after the sample was exposed to a thermal treatment leads to a decrease of the parameter C that is shown for Ta and Pt in Fig. 4.32. It is important to note that after the annealing process, the sample sremain polycrystalline as seen by X-ray studies on Pt wires from the same batch and under the same annealing conditions as for the wires used for the acoustic experiments (Semmelhack et al.(1996)).

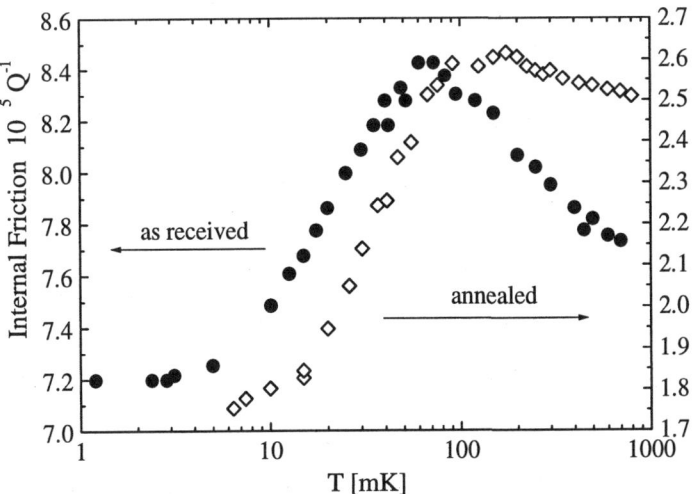

Fig. 4.35. Temperature dependence of the internal friction of Pt: *left scale*:(•) as received, $\nu = 1.5$ kHz, $\epsilon = 7 \times 10^{-6}$; *right scale*: (◊) annealed, $\nu = 1.4$ kHz, $\epsilon = 4 \times 10^{-6}$. The small differences in the applied strains cannot cause the decrease of Q^{-1} after annealing due to the very weak strain dependence of Q^{-1}, see Fig. 4.10.

4.5.5 Acoustic Properties of Polycrystals at $T > 1$K

The similarities of the acoustic properties between polycrystals and amorphous solids are not only limited to the temperature regime below 1K but also hold for higher temperatures. This is confirmed by studies of the temperature dependence of the sound velocity that show a rather universal linear dependence in amorphous and disordered solids. At temperatures above ≈1K,

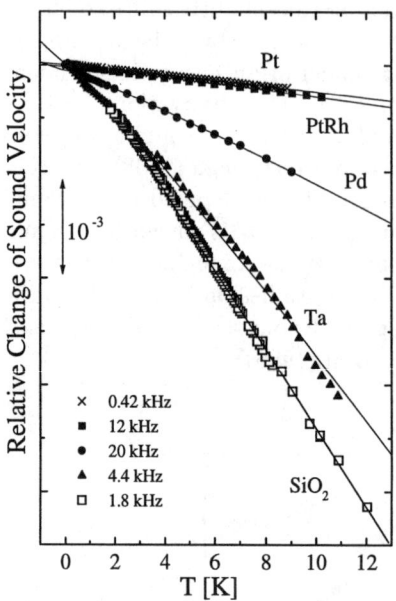

Fig. 4.36. Temperature dependence of the relative change of sound velocity for different polycrystalline samples and for SiO_2 for temperatures up to ~ 10 K. The continuous lines are obtained from $\Delta v/v = \beta(T) \cdot T$ with $\beta(T)$ given by (4.40) and C values from the data at $T < 1$ K.

the change of sound velocity can be described within a model proposed by Neu and Würger (1994b) (see Chap. 3) as an extension of the tunneling model without introducing new parameters. According to this model the relaxation interaction between phonons and TS's is enhanced due to the contribution of all TS's, even those with vanishing asymmetry. Taking into account incoherent tunneling of the TS's, the temperature dependence of the change of sound velocity can be described as $\Delta v/v = \beta(T) \cdot T$, see (4.40).

The parameter $\beta(T)$ (4.40) depends only weakly on temperature and is essentially determined by the parameter $C = P_0 \gamma^2 / \varrho v^2$ that can be obtained from the temperature dependence of the acoustic properties below 1K (logarithmic slope of the sound velocity or plateau of the internal friction). T^* is an effective crossover temperature from the logarithmic to the linear temperature dependence. The proportionality of $\beta(T)$ and C directly provides a correlation between the low temperature ($T < 1$ K, expressed by C) to the high temperature ($T > 1$ K, expressed by β) regime and is a crucial test for a further universality of the acoustic properties extended to the temperature range above 1 K.

Figure 4.36 shows the temperature dependence of the sound velocity of different polycrystals up to 10 K (Gaganidze et al. (1995)). This figure indeed confirms that at $T > 1$ K the relative change of sound velocity in polycrystals does not follow a logarithmic but a linear temperature dependence in agreement with the observations in amorphous solids. The derived values for

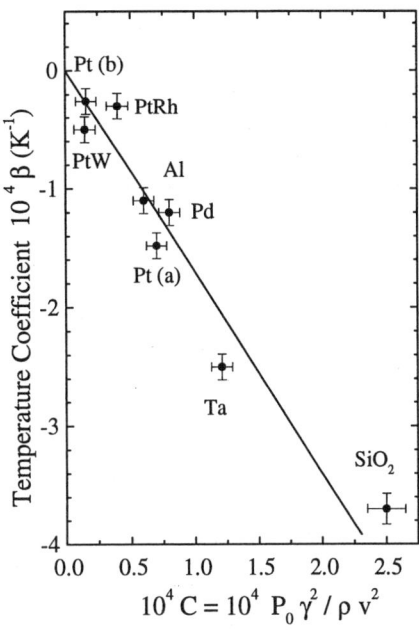

Fig. 4.37. Temperature coefficient β as a function of the parameter C. The straight line is a linear fit to the data given by $\beta = -1.9[\text{K}^{-1}]C$. The conditions for the Pt samples are: (a) as received; (b) annealed at 1300 K for 5 hours.

T^* are 10 K (5.7 K, 5.6 K, 3.7 K) for PtRh (Pt, Pd, Ta), very similar to $T^* \simeq 6$ K for vitreous silica. The coefficient β derived from the values for T^* and C as a function of C is shown in Fig. 4.37 for a frequency $\nu = 10$ kHz (Esquinazi (1996)). The dashed line is given by $\beta = -(1.9 \pm 0.3)[\text{K}^{-1}] \times C$ and confirms the linear relation between β and C. Furthermore, the coupling constant γ_t for the transversal phonon branch can be derived from T^* (Neu and Würger (1994b)). We obtain coupling constants $\gamma_t \simeq 1$ eV similar to those found in amorphous materials.

4.6 On the Origin of Tunneling Systems in Disordered Solids: Conclusion and Perspective

It is often stated that a tunneling system is an atom or a "small" group of atoms that tunnels between two almost equivalent positions in the amorphous structure. Theoretical models have been proposed and have tried to correlate the tunneling entities with "voids" (Cohen and Grest (1980)), "polymorphism" (Mon and Ashcroft (1978)), "disclination loops" (Rivier (1979)), special features obtained in computer-generated structural models (Banville and Harris (1980), Egami et al. (1980), Brandt and Kronmüller (1987)), or interstitials (Granato (1992)). It is appealing that these randomly distributed entities (or lattice defects) may have a uniform distribution of asymmetry levels Δ, however, these ideas lack the necessary inputs to understand the uniform distribution of $\ln(\Delta_0)$.

Although the question of the origin of the TS's and their broad distribution of energies and relaxation times as well as the remarkable universal behavior of the low-temperature properties found in disordered solids is, in general, not yet clearly answered, there is enough experimental evidence and theoretical work that point out the importance of the strain interactions between the entities, whatever they are as noted by Klein et al. (1978). Taking into account dipolar strain interactions between tunneling systems, it is in principle possible to understand the striking constancy of the parameter C in amorphous solids (Klein et al. (1978)), though the factor ϱv^2 is strongly material dependent (Freeman and Anderson (1986)). Going beyond the assumption of independent tunneling entities, Yu and Leggett (1988) argued that the elastic interaction between defects and not the defect itself dominates the energy scale of the low-energy excitations. Coppersmith (1993) noted that weakly interacting, localized TS's might be the result of strongly interacting defects (each of these a "two-level-system"); tunneling would be able only for those two-level systems with completely frustrated interactions. A further theoretical development in this direction was carried out recently by Burin and Kagan (1996a), see Sect. 5.3.

Many of the features of low-temperature glasslike anomalies can be found in crystals after irradiation at doses much lower than those needed for amorphization (Laermans and Esteves (1988)). In some special crystals like Li-KCl (Narayanamurti and Pohl (1970)), $(NaCl)_{1-x}(NaCN)_x$ (Watson and Pohl (1995)), or H-Nb(O) (Morr et al. (1989)), it is rather well accepted that the tunneling entities are well-identified atoms. It has been argued that random internal strains lead to low-energy excitations by decreasing the potential barrier heights increasing x in $(NaCl)_{1-x}(NaCN)_x$ (Watson and Pohl (1995)). Within the ideas described above, we should not be very surprised to observe low-temperature glasslike properties in simple polycrystalline metals, if the number of the defects is large enough. One would tend to argue that the tunneling entities and/or defects that produce the glasslike properties in polycrystalline metals are localized only at the grain boundaries. However, the very different microstructure of the polycrystalline samples, their similar C values compared with bulk amorphous materials, taking into account the very small volume of the grain boundaries, indicate that other atomic lattice defects and their elastic interactions contribute to the glasslike properties. As a matter of fact, acoustic measurements (Classen et al. (1997)) on a Al single crystal showed glasslike acoustic properties at low temperatures. As expected the magnitude of the glasslike acoustic properties is nearly two orders of magnitude smaller than in polycrystalline Al, but of the same order of those found in previously annealed and then cold-rolled polycrystalline Ag (Esquinazi et al. (1992)). The contribution of the dislocation lines and their pinning to the acoustic properties is not yet clear too. Internal friction results on pure Cu at very low temperatures show indications of tunneling and thermally activated relaxation of dislocations (Esquinazi et al. (1994)). Their contribution to the acoustic properties in well annealed crystals and polycrystals might overwhelm that from the TS's. A careful study should be carried out before conclusions on the (non)existence of glasslike acoustic properties are drawn.

From the acoustic properties of an amorphous metal and their change with thermal treatment, we can conclude that a relaxed amorphous state would show vanishing, small glasslike properties. We should conclude, therefore, that amorphousness or polycrystallinity are neither necessary nor sufficient conditions for the existence of glasslike anomalies at low temperatures.

We have reviewed recent experimental results of the acoustic properties (relative change of sound velocity and internal friction) of a large number of materials of different classes:
- amorphous dielectrics, in particular bulk SiO_2 and SiO_2 films;
- amorphous superconductors, e.g. $Pd_{30}Zr_{70}$;
- amorphous normal-conducting metals (PdSiCu);
- polycrystalline superconductors, e.g. tantalum, niobium;
- normal-conducting polycrystals, e.g. Ag and Cu.

Although there do exist deviations of some kind or other among the acoustic properties of the investigated materials, the major common feature of these studies is that the temperature dependences of the internal friction and the sound velocity reveal at least qualitatively a striking universality that was originally believed to be valid for amorphous materials (and here in particular for amorphous dielectrics) only. It is the unexpected results of the investigations of the acoustic properties of *polycrystals* with their very similar behavior to the amorphous materials – over a temperature range covering five orders of magnitude (0.1 mK$< T <10$ K) – that question the longstanding phrase of a "glasslike" behavior when we, for example, talk about a plateau in the internal friction or a maximum in the relative change of the sound velocity.

The standard tunneling model still acts as the framework for the interpretation of the properties observed. However, it does not seem to be too unexpected – basically because it is built on rather simple assumptions – that modifications have to be introduced in order to be able to explain the results of certain experiments testing parameters that go beyond the applicability of this model. We just want to mention for example the importance of amplitude-dependent effects when the energy of the sound wave (or strain field) is of the order of the thermal energy leading to a change of the population number of the TS's. As another example one may consider the very new studies of the influence of dimensionality of the sample on the experimental findings. Perhaps the strongest modification, however, has to be applied not to any particular parameter of the model itself but to our current thinking to accept that the presence of TS's and the description of at least part of their properties have to be extended to the class of polycrystalline metals too. And once, getting used to the fact that the polycrystalline metals obviously contain TS's that strongly influence the propagation of phonons, we are confronted with another astonishing result, as with polycrystals the electrons do not seem to contribute to the relaxation of the tunneling systems as is the case with amorphous metals. This might be comprehensible for polycrystals in their superconducting state (Nb, Ta) due to the formation of Cooper pairs; the surprising fact of the absence of an electron-assisted relaxation in normal-conducting metals, however, is not yet understood.

Apart from the investigation of a number of detailed problems that are not yet clarified, one might expect that future experiments do mainly concentrate on an answer to the question about the nature of the TS's. There is an important theoretical and experimental progress towards this direction mainly based on the works emphasizing the role of the interactions among TS's.

Acknowledgment

This research has been possible with the support of the Deutsche Forschungsgemeinschaft. The authors thank Frank Pobell for his guidance that triggered most of the work presented here. We would like to thank Ermile Gaganidze, Alexander Burin, Peter Smeibidl, Piotr Sekowski, Siegfried Hunklinger, and Georg Weiss for their support, comments and criticism of the work done in the last years.

5. Interactions Between Tunneling Defects in Amorphous Solids

Alexander L. Burin, Douglas Natelson, Douglas D. Osheroff, and Yuri Kagan

This review covers a wide range of experimental and theoretical studies of low-temperature properties of glasses. The emphasis is on the various effects of interactions between two-level tunneling systems, TS's. We present experimental evidence, both direct and indirect, for the existence of interactions.

Recent measurements of equilibrium internal friction and dielectric dissipation which deviate from the predictions of the noninteracting tunneling model are presented. The theoretical interpretation of this behavior in dielectric glasses at ultralow temperature is discussed in the context of recent theories of the interaction-induced TS relaxation.

In addition, recent nonequilibrium dielectric and acoustic measurements are reviewed and a theoretical explanation for their results based on the formation of a dipole gap is discussed.

The ability to explain the universality of low-temperature glassy behavior in terms of the interaction of some mobile defects is discussed in the light of recent experimental, numerical, and analytical results.

5.0.1 Dielectric and Acoustic Properties

Through their coupling to elastic fields, TS's modify the acoustic properties of amorphous materials; see Chap. 4 for a detailed description of this phenomenon. In brief, the perturbation of a given TS potential by the periodic local strain field from propagating sound can cause the TS to change states at some phase with respect to the driving strain. The dispersion of phonons that results is reflected in a modified sound speed and internal friction, first calculated by Jäckle (1972). Note that the situation is more complicated in disordered metals, where the TS model has been extended to include interactions between TS's and conduction electrons, as discussed in Sects. 4.2.3 and 4.4.2.

Similarly, those TS's that possess permanent electric dipoles can couple to oscillating electric fields. Because of the perturbations to their energy splittings by electric fields, TS's can be driven by an external driving field to switch states. These transitions occur at the driving frequency with some phase shift with respect to the driving field. Transitions can also be assisted by phonons. The oscillating polarization that results manifests itself as a

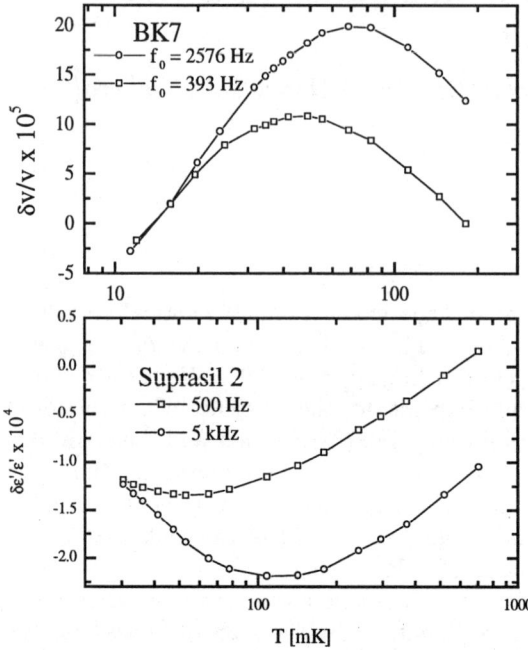

Fig. 5.1. Relative change of sound velocity $\delta v/v$ and real part of dielectric susceptibility $\delta\epsilon'/\epsilon'$ as a function of temperature for BK7 and Suprasil 2, respectively. Data from Natelson et al., unpublished.

modified dielectric susceptibility, $\tilde{\epsilon} = \epsilon' + \mathcal{I}\epsilon''$. The real part of the susceptibility corresponds to reactive response, whereas the imaginary component is proportional to dissipative behavior.

Typical behavior at acoustic frequencies and low drive amplitudes is shown in Fig. 5.1. The sound speed is nonmonotonic, rising logarithmically in temperature from very low T, independent of frequency; passing through a peak whose location is frequency dependent; and then decreasing logarithmically with increasing T. Note that the size of the high-temperature response is frequency dependent. The capacitive response, ϵ', is analogous to within a sign, showing a minimum rather than a maximum. The higher temperature, or "relaxational", regime is characterized by TS dynamics being dominated by phonon-driven transitions between states. The lower temperature, or "resonant", regime is considered to have TS transitions dominated by quantum tunneling.

Figure 5.2 shows the typical behavior of the dissipative part of the TS response (internal friction Q^{-1} for acoustic response, loss tangent $\tan\delta$ for dielectric response). In both cases the loss increases with increasing temperature until the crossover to the relaxational regime, at which point the loss approaches a frequency-independent plateau.

Section 5.1 contains a more detailed discussion of the standard tunneling model description of both the reactive and dissipative behaviors.

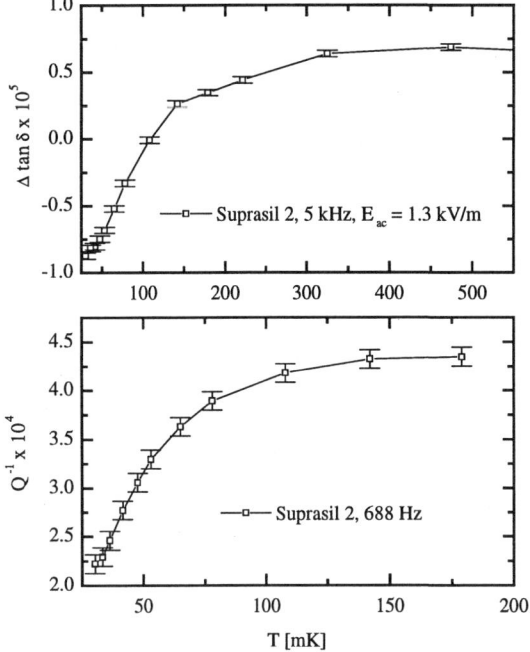

Fig. 5.2. Q^{-1} and $\Delta(\tan\delta)$ as a function of temperature for Suprasil 2, a form of SiO_2. Data from Natelson et al., unpublished.

5.0.2 Interaction Effects: Spectral Diffusion and Dephasing

Each TS couples to local strain fields, implying that each TS produces a strain field in the surrounding bulk material. In the standard tunneling model, this field is dipolar. The exact field configuration depends on the state of the TS. It is clear that two such TS's in close proximity must influence each other's behavior via their associated elastic fields. In field-theoretic language one considers this sort of interaction as resulting from the exchange of virtual phonons. A qualitatively identical effect, though electric in origin, occurs between TS's with sufficiently large electric dipole moments. A detailed discussion of this interaction mechanism is presented below, in Sect. 5.1.2.

Typical TS elastic coupling strengths are around $\gamma = 1$ eV. That is, the asymmetry energy Δ of a particular two-level system is shifted by an amount that is 1 eV times the local strain. The interaction between two such TS's separated by a distance r in a material with density ϱ and sound velocity v is given by

$$E_{\text{int}} = \frac{\gamma^2}{\varrho v^2} \times \frac{1}{r^3}. \tag{5.1}$$

For typical values of these parameters, TS's separated by 10 nm alter each other's asymmetry energies by an amount of order 20 mK.

Typical TS electric dipole moments are of the order of 1 Debye (= 10^{-18} esu cm = 3.33×10^{-30} C m). Such a TS in an electric field of magnitude 1 MV/m has its asymmetry shifted by 0.24 K. The dipole–dipole interaction at a distance of 10 nm is then approximately 7 mK.

The pulsed acoustic absorption saturation experiments described below show some of the first evidence that TS's interact with one another, rather than being essentially independent. This evidence is two-fold. First, consider the "hole-burning" experiment illustrated in Fig. 5.3 and performed by Arnold and Hunklinger (1975) on BK7, a common optical glass. In this case a high intensity ultrasound pulse was used to equalize the state populations of TS's with an energy splitting equal to the energy of the incoming acoustic radiation. Once their state populations were equalized, on average, these TS's could no longer absorb incident radiation at that frequency, since such an absorption is balanced by stimulated emission. Subsequent low-intensity pulses at fixed frequency while varying the frequency of the "hole-burning pulse" were used to probe this "hole" now in the acoustic absorption spectrum. This hole's width grew with increasing T and with increasing time (Arnold et al. (1978)).

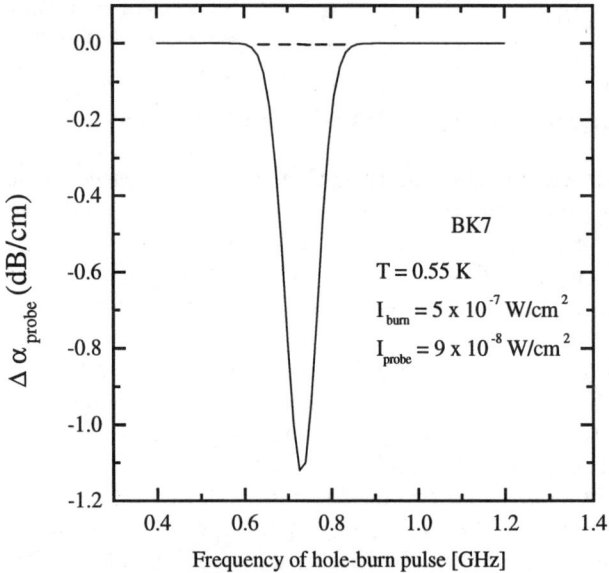

Fig. 5.3. Acoustic hole burning: resonant absorption of a weak probing pulse as function of the frequency of a strong saturating pulse. After Arnold and Hunklinger (1975).

The mechanism responsible for the growth in time of the saturation hole's width, spectral diffusion, was first suggested by Hunklinger et al. (1973) and was quantified by Black and Halperin (1977) and others (Hu and Walker (1977)). Consider a group of TS's of energy splitting $\hbar\omega_0$ whose absorption has been saturated as described above. As TS's near those saturated undergo transitions, either by tunneling or thermal activation, elastic interaction effects will cause the energy splittings of the saturated TS's to change from their original values, leading to a broadening of the hole in the absorption spectrum. Again, this is detailed below. Work on optical hole burning in amorphous materials is described in Chap. 6.

A second spectral diffusion effect was seen while studying spontaneous phonon echoes (Golding and Graebner (1976),Golding et al. (1977), Graebner and Golding (1979)). Two processes lead to the decay in amplitude of spontaneous echoes. In τ_1 processes, the TS's in question rethermalize with the bulk sample, and hence will not participate in any subsequent echo. In τ_2 processes, spectral diffusion leads to a loss of phase coherence between 'tipped' TS's, and incoherent TS's do not contribute to spontaneous echoes. Through stimulated echo experiments it is possible to measure τ_1 independently, and thus one can extract a value for the dephasing rate due to spectral diffusion between TS's.

While these spectral diffusion effects rapidly gained acceptance in the community as evidence that TS's interact, interactions were still not thought to play a significant role within the TS formalism. Black and Halperin (1977) state this explicitly. However, over time there has been a growing effort both to incorporate TS interactions into the theoretical framework and to establish the experimental importance of such interactions. Two main questions driving the investigation of TS interactions are: why does the standard TS model systematically deviate from experiment in thermal properties (e.g. κ is not quadratic in T, but varies as $T^{1.8}$); and why is the TS model applicable to so many systems with such a variety of microscopic structures (e.g. polymers, oxide glasses, and polycrystalline metals)?

5.1 Interactions and Equilibrium Properties

In this section we will consider the effects of TS interactions on the equilibrium properties of amorphous insulators at low temperatures. We begin by reviewing the basic predictions of the standard tunneling model whose ingredients have already been presented in Sects. 2.1 and 2.2.2. The calculation of the equilibrium dielectric response from the TS equations of motion is shown in detail.

We then discuss the general tensorial elastic interaction between TS's, making several approximations to allow tractable calculations. From this starting point, we show how interactions lead to the effects described above, spectral diffusion and TS dephasing. These calculations proceed by considering a particular TS and the effects it feels due to interactions with an environment composed of TS's that are treated as independent.

To address the possibility of collective TS effects, we examine a pair of interacting TS's. We find that, for particular TS parameters, this pair can be considered "resonant"; that is, its constituent TS's can couple resonantly together via their interaction, allowing the pair to behave as a single effective TS. Using energetic considerations we compute the distribution of such effective TS's as a function of their energies and tunneling parameters.

We move on to examine interactions between these effective TS's. Because the pair excitations have a different distribution than the standard tunneling model's TS's, the pairs are able to form large delocalized collective excitations at very low temperatures. Re-examining dephasing, now considering how a particular TS interacts with an environment composed of a cluster of correlated pair excitations, we find an enhanced dephasing rate, that we compare with experimental results.

Because of spectral diffusion, pair excitations wander in and out of the delocalized collective modes. We show that this itinerant behavior leads to an interaction-stimulated relaxation mechanism for TS's, with detectable effects in the low-temperature limits of the internal friction and dielectric loss tangent.

After pausing to review standard experimental techniques for dielectric and acoustic measurements, we present data that are quantitatively consistent with interaction-stimulated relaxation. Finally, we discuss intrinsic dielectric saturation at the lowest experimental temperatures, an effect that is currently unexplained.

5.1.1 Standard Tunneling Model Predictions

To understand the calculations presented in subsequent sections, it is useful to present the effects of an ac measuring field on the TS Hamiltonian. In what follows, we consider the TS's in question to have an electric dipole moment p and label the electric field used for measurements $E_{ac}(t)$. The perturbation to the Hamiltonian due to the measurement is then [in the basis of (2.1)]

$$H_{ac} = \begin{pmatrix} 1 & 0 \\ 0 & -1 \end{pmatrix} p \cdot E_{ac}(t). \tag{5.2}$$

When we change bases as in (2.15), this contribution to the Hamiltonian becomes

$$H_{ac} = \begin{pmatrix} \frac{\Delta}{E} & \frac{\Delta_0}{E} \\ \frac{\Delta_0}{E} & -\frac{\Delta}{E} \end{pmatrix} p \cdot E_{ac}(t). \tag{5.3}$$

The strain field u associated with thermal phonons may be written as a sum over modes, using raising and lowering operators. Considering single phonon transition processes, the relevant matrix element is that connecting an initial state $|\Psi_+; \emptyset\rangle$ (TS in upper state, no phonon) with the final state $|\Psi_-; \boldsymbol{k}, \eta\rangle$ (TS in lower state, phonon of wave vector $\boldsymbol{k}$ and polarization η) via the elastic interaction part of (2.15), H_1.

From Fermi's golden rule, it then follows that the tunneling relaxation rate τ_1^{-1} for a particular TS coupled to phonons as described above is given by (2.18). It is useful to absorb the material-dependent constants into a parameter A defined by the relation

$$\tau_1 = \frac{1}{A\Delta_0^2 E \coth(E/2k_{\rm B}T)}. \tag{5.4}$$

Note that the above rate expression leads to a minimum relaxation time $\tau_{1,\min}$ when $\Delta = 0$ and thus $\Delta_0 = E$.

From this point, there are two common paths of calculation. Consider the interaction of a particular TS with a small oscillatory driving field [elastic and electric fields both act on the asymmetry elements of the Hamiltonian, (2.13), but with different prefactors]. In the approach used by Jäckle (1972) and Anthony and Anderson (1979), one can calculate the absorption of energy from the driving wave by the TS, integrating over the TS parameter distribution to obtain the total absorption. This absorption has two regimes, resonant and relaxational (see Sect. 4.2.1). Once the absorption is known, the Kramers–Kronig relation is then used to obtain the nondissipative acoustic or capacitive response. Since superposition is implicit in the Kramers–Kronig relations, this method restricts its applicability to the limit of linear response. The other approach (for example, Hunklinger and Raychaudhuri (1986)) to calculate the real part of the acoustic and dielectric reponse is to directly integrate the equations of motion of the system under the influence of a driving external field. Again, once the solutions are obtained for a particular TS, the distribution (2.20) [or (2.27)] is integrated to get the total system response. This is the approach we follow here, specifically the method of Carruzzo et al. (1994) for the dielectric response. The treatment for the acoustic response is strictly analogous.

The Hamiltonian shown above in (5.3) is formally analogous to that of spin-1/2 particles in a magnetic field. The dynamics of such a system can be considered within the formalism of the Bloch equations familiar from NMR, with the energy splitting E playing the role of an effective dc magnetic field B_z. If Γ is the effective spin's gyromagnetic ratio, we can consider $H_{\rm ac}$ above and write down that $-\hbar\Gamma B_z = E$. Similarly, the measuring field $\boldsymbol{E}_{\rm ac}$, that oscillates with frequency ω, acts like the ac magnetic field $\boldsymbol{B}_{\rm ac}$ in an NMR experiment:

$$-\frac{1}{2}\hbar\Gamma \boldsymbol{B}_{\rm ac} = \left(\frac{\Delta_0}{E}, 0, \frac{\Delta}{E}\right) pE_{\rm ac}\cos\theta \cos\omega t. \tag{5.5}$$

In the absence of relaxation processes, the equation of motion for a spin S with gyromagnetic ratio Γ in a magnetic field $\boldsymbol{B} = B_0 \hat{z} + \boldsymbol{B}_1(t)$ is

$$\frac{d\boldsymbol{S}}{dt} = \Gamma \boldsymbol{S} \times \boldsymbol{B}. \tag{5.6}$$

In a frame of reference rotating with angular frequency ω around the z axis, that is, the direction of the static component of $\boldsymbol{B}$, the equation of motion becomes

$$\frac{d\boldsymbol{S}}{dt} = \boldsymbol{S} \times [(\omega_0 - \omega)\hat{z} + \omega_1 \hat{x}'], \tag{5.7}$$

where $\omega_0 = \Gamma B_0$, $\omega_1 = \Gamma B_1$, and $\hat{x}'$ is the new x axis.

Inserting both the τ_1 relaxation process and a τ_2 dephasing process, we can write the linearized Bloch equations. We write the spin operator $\boldsymbol{S}$ as a vector of Pauli matrices $\boldsymbol{S} = \boldsymbol{\sigma}/2$:

$$\frac{dS^x}{dt} - \Gamma(S^y B_z - S^z B_y) + \frac{1}{\tau_2}S^x = 0, \tag{5.8}$$

$$\frac{dS^y}{dt} - \Gamma(S^z B_x - S^x B_z) + \frac{1}{\tau_2}S^y = 0, \tag{5.9}$$

$$\frac{dS^z}{dt} - \Gamma(S^x B_y - S^y B_x) + \frac{1}{\tau_1}(S^z - \langle S^z \rangle) = 0. \tag{5.10}$$

Here, $\langle S^z \rangle = \frac{1}{2}\tanh(\beta \Gamma \hbar B_z(t)/2)$ is the thermal equilibrium value of S^z, where $\beta = 1/k_B T$ as usual.

For the case where B_{ac} described above is small, one can linearize the Bloch equations in a Taylor series about their unperturbed solution $\boldsymbol{S}^0(t)$. We define a few useful parameters:

$$\begin{aligned} \omega_0 &= \Gamma B_{z,\text{dc}} = -E/\hbar, \\ \delta &= -\frac{2\Delta}{\hbar E}pE_{\text{ac}}\cos\theta, \\ S_0^z(\infty) &= -\frac{1}{2}\tanh(\beta E/2), \\ \alpha &= \frac{\Delta_0}{\Delta}, \\ \lambda &= \hbar\beta[1 - 4(S_0^z(\infty))^2]/4. \end{aligned} \tag{5.11}$$

Introducing raising and lowering operators $S_\pm = S_1^x \pm \mathcal{I} S_1^y$ leads to two equations:

$$\frac{dS_+(t)}{dt} + \mathcal{I}\left(\omega_0 - \frac{\mathcal{I}}{\tau_2}\right)S_+(t) - \mathcal{I}\alpha\delta S_0^z(t)\cos(\omega t) = 0 \tag{5.12}$$

and its complex conjugate for S_-.

5.1 Interactions and Equilibrium Properties

Solutions to these equations are (Carruzzo et al. (1994))

$$S_0^z(t) = S_0^z(\infty) + [S_0^z(0) - S_0^z(\infty)]e^{-t/\tau_1}, \tag{5.13}$$

$$S_1^z(t) = \frac{\lambda\delta}{1+\tau_1^2\omega^2}[\cos(\omega t) + \tau_1 \sin(\omega t)], \tag{5.14}$$

$$S_+(t) = \frac{\alpha\delta[(\omega_0 - \mathcal{I}/\tau_2)\cos(\omega t) - \mathcal{I}\omega\sin(\omega t)]S_0^z(\infty)}{(\omega - \mathcal{I}/\tau_2)^2 - \omega^2}$$
$$+ [S_0^z(0) - S_0^z(\infty)]e^{-t/\tau_1}$$
$$\times \frac{\alpha\delta[(\omega_0 + \mathcal{I}/\tau_1 - \mathcal{I}/\tau_2)\cos(\omega t) - \mathcal{I}\omega\sin(\omega t)]}{(\omega_0 + \mathcal{I}/\tau_1 - \mathcal{I}/\tau_2)^2 - \omega^2}. \tag{5.15}$$

The solution for $S_-(t)$ is given by the complex conjugate of (5.15). The solutions above do not include the homogeneous solutions to the free spin precession (that happens at frequency ω_0) since we are only interested in dynamics at the driving frequency ω.

The first equation shows that an individual TS whose spin is out of thermal equilibrium will relax back to equilibrium with an exponential time dependence with characteristic time τ_1. The raising and lowering operators govern transitions between the given TS' energy levels due to the external ac field.

To extract measurable quantities from these equations, it is necessary to use the spin operators to find the susceptibility of the system. The electric dipole operator of a particular TS is, in the basis of (5.3),

$$\boldsymbol{\pi} = \left(\frac{\Delta}{E}\sigma_z + \frac{\Delta_0}{E}\sigma_x\right)\boldsymbol{p}. \tag{5.16}$$

Using our definition of $\boldsymbol{S}$ one finds that the average component of the dipole moment along the ac field, $p_\parallel$, is

$$p_\parallel = -\langle\boldsymbol{\pi}\rangle\cos\theta$$
$$= -p\cos\theta\left(\frac{2\Delta S_z^1(t)}{E} + \frac{\Delta_0(S_+ + S_-)}{E}\right), \tag{5.17}$$

where we have used the Bloch equation solutions to substitute for the averaged Pauli matrices.

Now finding the susceptibility $\tilde{\epsilon} = \epsilon' + \mathcal{I}\epsilon''$ is simply a matter of calculating $dp_\parallel/dE_{\text{ac}}$. Here ϵ' is the response in phase with the ac field, and ϵ'' is the out-of-phase response. It is convenient to break down ϵ' and ϵ'' as follows:

$$\epsilon' = \epsilon'_z + \epsilon'_{1\pm} + \epsilon'_{2\pm},$$
$$\epsilon'' = \epsilon''_z + \epsilon''_{1\pm} + \epsilon''_{2\pm} \tag{5.18}$$

and identify the terms on each right-hand side with the contributions of S_1^z and $S_\pm$. The results are

$$\epsilon'_z = \left(\frac{\Delta p\cos\theta}{E}\right)^2 \frac{\beta(1-4(S_0^z(\infty))^2)}{1+\tau_1^2\omega^2}, \tag{5.19}$$

$$\epsilon'_{1\pm} = -\frac{2}{\hbar}\left(\frac{\Delta_0 p \cos\theta}{E}\right)^2$$
$$\times \left(\frac{\omega_0+\omega}{1+(\omega_0+\omega)^2\tau_2^2} + \frac{\omega_0-\omega}{1+(\omega_0-\omega)^2\tau_2^2}\right)\tau_2^2 S_0^z(\infty), \quad (5.20)$$

$$\epsilon'_{2\pm} = -\frac{2}{\hbar}\left(\frac{\Delta_0 p \cos\theta}{E}\right)^2$$
$$\times \left(\frac{\omega_0+\omega}{1+(\omega_0+\omega)^2\tau_{12}^2} + \frac{\omega_0-\omega}{1+(\omega_0-\omega)^2\tau_{12}^2}\right)\tau_{12}^2$$
$$\times [S_0^z(0) - S_0^z(\infty)]e^{-t/\tau_1}, \quad (5.21)$$

$$\epsilon''_z = \left(\frac{\Delta p \cos\theta}{E}\right)^2 \frac{\beta(1-4(S_0^z(\infty))^2)\tau_1\omega}{1+\tau_1^2\omega^2}, \quad (5.22)$$

$$\epsilon''_{1\pm} = \frac{2}{\hbar}\left(\frac{\Delta_0 p \cos\theta}{E}\right)^2$$
$$\times \left(\frac{1}{1+(\omega_0+\omega)^2\tau_2^2} - \frac{1}{1+(\omega_0-\omega)^2\tau_2^2}\right)\tau_2 S_0^z(\infty), \quad (5.23)$$

$$\epsilon''_{2\pm} = \frac{2}{\hbar}\left(\frac{\Delta_0 p \cos\theta}{E}\right)^2$$
$$\times \left(\frac{1}{1+(\omega_0+\omega)^2\tau_{12}^2} + \frac{1}{1+(\omega_0-\omega)^2\tau_{12}^2}\right)\tau_{12}$$
$$\times [S_0^z(0) - S_0^z(\infty)]e^{-t/\tau_1}. \quad (5.24)$$

In the above, $\tau_{12}^{-1} = \tau_1^{-1} - \tau_2^{-1}$. The above equations include time-dependent terms that reflect the relaxation of the system to thermal equilibrium, as well as the standard equilibrium solution.

One should note the use of the Bloch equation to describe the dephasing essentially assumes the exponential character of this process. Sometimes this is incorrect for TS's (see e.g. Black and Halperin (1977) and Baier et al. (1988)). However, the results for the susceptibilities remain valid whereas $\omega\tau_{1T} \gg 1$, where τ_{1T} is the value of τ_1 for TS's with $\Delta_0 \approx E \approx k_B T$ as in (5.59). Our future considerations obey this condition, and we can therefore use the results of the Bloch approximation.

To obtain the susceptibility of the entire TS system, the above contributions must be averaged over possible dipole orientations θ and the distribution of TS parameters given in (2.21). Thus, the total TS contribution to the dielectric susceptibility is given by the integral:

$$\langle\tilde{\epsilon}\rangle = \frac{P_0}{2}\int_{-\Delta^{\max}}^{+\Delta^{\max}} d\Delta \int_{\Delta_0^{\min}}^{\Delta_0^{\max}} \frac{d\Delta_0}{\Delta_0} \int_{-1}^{+1} d(\cos\theta)\tilde{\epsilon}(\Delta,\Delta_0,\cos\theta). \quad (5.25)$$

In the linear response limit ($\mathbf{p}\cdot\mathbf{E}_{\mathrm{ac}}/k_B T \ll 1$ for electric drive, and $\gamma\epsilon_0/k_B T \ll 1$ for acoustic drive) both Kramers–Kronig and direct integration of the equations of motion lead to the same predictions for the form of the response. In the relaxational regime, where $\omega\tau_{1,\min} \ll 1$, one obtains a total reactive response (relaxational plus resonant) given by

$$\frac{1}{C}\frac{\delta v}{v}\bigg|_{\text{tot}} = -\frac{1}{C_d}\frac{\delta \epsilon'}{\epsilon'}\bigg|_{\text{tot}} = -\frac{1}{2}\ln\left(\frac{T}{T_0}\right), \qquad (5.26)$$

where

$$C = \frac{P_0\gamma^2}{\varrho v^2}, \qquad (5.27)$$

$$C_d = \frac{2}{3}\frac{P_0 p^2}{\epsilon_0 \epsilon'}, \qquad (5.28)$$

and T_0 is a reference temperature (not to be confused with the crossover temperature defined in Chap. 3). For use in later sections, we define two parameters, the "temperature slopes" of the equilibrium acoustic and dielectric behaviors:

$$S_{T,v} \equiv \frac{\partial(\Delta v/v)}{\partial \log_{10} T}, \qquad (5.29)$$

$$S_{T,\epsilon'} \equiv \frac{\partial(\Delta \epsilon'/\epsilon')}{\partial \log_{10} T}. \qquad (5.30)$$

In acoustic measurements, the dissipative response is characterized by Q^{-1}, the internal friction of the material. Similarly, in dielectric measurements one usually speaks in terms of the loss tangent, $\tan\delta \equiv \epsilon''/\epsilon'$. The dissipative response in the relaxational regime is

$$\frac{1}{C}Q^{-1} = \frac{2}{C_d}\tan\delta = \frac{\pi}{2}. \qquad (5.31)$$

For the resonant regime, when $\omega\tau_{1,\min} \gg 1$, we obtain

$$\frac{1}{C}\frac{\delta v}{v}\bigg|_{\text{tot}} = -\frac{1}{C_d}\frac{\delta \epsilon'}{\epsilon'}\bigg|_{\text{tot}} = \ln\left(\frac{T}{T_0}\right). \qquad (5.32)$$

The dissipative response in this regime is predicted to be (see Hunklinger and Arnold (1976))

$$\frac{1}{C}Q^{-1} = \frac{2}{C_d}\tan\delta = \frac{\pi^3}{24\omega\varrho\hbar^4}\left(\frac{\gamma_\ell^2}{v_\ell^5} + 2\frac{\gamma_t^2}{v_t^5}\right)k_B^3 T^3. \qquad (5.33)$$

Here the subscripts ℓ and t refer to longitudinal and transverse acoustic waves, respectively.

The crossover between the two regimes may be characterized by a temperature T_m:

$$T_m = \sqrt[3]{\frac{\omega}{Ak_B^3}}, \qquad (5.34)$$

where $A = (1/(2\pi\hbar^4\varrho))\left[\gamma_\ell^2/v_\ell^5） + 2(\gamma_t^2/v_t^5)\right]$, as in (2.18). This scaling, $T_m \propto \omega^{1/3}$, has been confirmed in several materials; for example, see Frossati et al. (1977).

To summarize the above predictions: With the standard noninteracting tunneling model, the reactive part of acoustic (dielectric) response is expected to vary logarithmically with T for temperatures well above and well below the crossover, T_m. The temperature slope $S_{T,v}$ or $S_{T,\epsilon'}$ ratio for the resonant ($T \ll T_m$) regime compared to the relaxational ($T \gg T_m$) regime is predicted to be -2:1. We choose to define S_T using the base-10 logarithm. Finally, because of the assumption that all TS's relax to thermal populations through one-phonon processes, the dissipation for $T \ll T_m$ is predicted to vary like T^3; see Table 4.1.

5.1.2 Interactions Between Tunneling Systems: Spectral Diffusion

We now turn our attention to the first experimentally observed consequence of interactions between tunneling states in amorphous solids, spectral diffusion.

The two-level nature of the tunneling excitations makes their mathematical description similar to that of spin-1/2 particles. Consequently many experimental methods that are useful for studying magnetic systems remain relevant for the analysis of TS's in amorphous solids. In particular it is possible to study echo phenomena in such materials, in analogy to similar nuclear magnetic resonance experiments. This idea has been realized in Hunklinger et al. (1973), Graebner and Golding (1979), Harrison (1979), Hu and Walker (1977), Hunklinger et al. (1973), and Hunklinger and Arnold (1976), for example, in various modifications of the dipole echo experimental technique. It is traditional to describe the relaxation of the echo signal in terms of the two relaxation times, i.e. the longitudinal time τ_1 and the transverse time τ_2. The longitudinal time is defined by the transition time of a two-level system or spin between its two states. The transverse time characterizes the phase memory of the wave function of an ensemble of TS's. If the TS's are isolated from each other and coupled to the phonon bath only, then both relaxation times are defined by the transition time (2.18). However, at sufficiently low temperatures ($T < 1$ K) the experimental data in glasses contradict this picture. In particular, measurements of the relaxation of the hole burned in the absorption [see Arnold and Hunklinger (1975) for example] show that the energy splittings of TS's slowly change with time, and the dephasing time extrapolated from echo measurements is shown to be remarkably shorter than the transition time (2.18). This anomalous behavior has been explained in terms of interactions between TS's. These interactions shift the energies of TS's in a chaotic manner, that gives rise to spectral diffusion, i.e. the random change of the energy of a specific TS due to the relaxation of its surroundings. As a first approximation we will focus our attention on a particular TS and treat the surrounding TS's as essentially independent.

5.1 Interactions and Equilibrium Properties

We will demonstrate how the coupling of TS's with phonons, (2.13), introduces a long-range interaction between TS's, and we will analyze the dephasing rate caused by this spectral diffusion.

In the foregoing discussion, we have considered the "resonant" interaction of a particular defect with the elastic medium. The term resonant means that the phonons participating in the interaction have energies close to the energy difference between the ground and excited levels of the tunneling defect. The processes of the resonant interaction with phonons involve the absorption or emission of phonons and the corresponding transition of the TS between its levels.

Another way in which the defect interacts with the medium is by perturbing the equilibrium positions of surrounding molecules with a strain field. Other defects feel this perturbation. Their energies become sensitive to the quantum state of the tunneling defect under consideration. This gives rise to an elastic interaction between TS's that is quite similar to the electric dipole–dipole interaction originating from the virtual exchange of photons.

Let us demonstrate the appearance of this interaction for two two-level defects in an amorphous medium separated by the distance R and coupled with phonons according to (2.13). We begin with a very general formulation, and proceed to tractable calculations by various simplifications. Our method of introducing the interaction of TS's is rather similar to that used by Grannan et al. (1990) in studying the interactions between rotating defects in the model-orientational glass KBr-KCN.

The Hamiltonian of the system including the phonon subsystem and the pair of defects can be rewritten as

$$\widehat{H} = H_{\text{ph}} + H_{\text{def}} + V_{\text{int}};$$
$$H_{\text{ph}} = \sum_{q,\mu}\left(\frac{|P_{q,\mu}|^2}{2M} + M\omega_{q,\mu}^2\frac{|U_{q,\mu}|^2}{2}\right), \quad (5.35)$$
$$H_{\text{def}} = -\Delta_1 S_1^z - \Delta_2 S_2^z,$$
$$V_{\text{int}} = -\sum_{\alpha,\beta}\left(\gamma_{1\alpha\beta}\frac{\partial U_{1\alpha}}{\partial x_{1\beta}}S_1^z + \gamma_{2\alpha\beta}\frac{\partial U_{2\alpha}}{\partial x_{2\beta}}S_2^z\right). \quad (5.36)$$

Here P and U represent momentum and displacement operators, respectively for phonon modes numerated by wave vector q and branch μ, and M is the mass of the elementary "cell" of the medium. For simplicity, we consider the long wavelength limit. Further, for clarity we have replaced the strain tensor u_{ij} by its definition. Note that $\gamma_{\alpha\beta} = \gamma_{\beta\alpha}$. Then the amorphous medium can be considered isotropic, and the phonon spectrum breaks up into one longitudinal and two transverse branches ($\mu = \ell, \text{t1}, \text{t2}$, respectively). In the long wavelength limit the spectrum of these modes is nothing other than the standard sound spectrum

$$\omega_{q,\ell} = v_\ell q,$$
$$\omega_{q,\text{t1},2} = v_t q, \quad (5.37)$$

where v_ℓ, v_t are the longitudinal and transverse velocities of sound, respectively. Displacements U_1 and U_2 are taken from the positions of defects x_1, x_2, and indices α, β enumerate the Cartesian coordinates. The internal degrees of freedom for each defect are represented by pseudo-spin-1/2 operators S_1^z, S_2^z. We neglect the tunneling amplitudes of the defects, assuming that the tunneling matrix elements are smaller than the energy of the relevant phonons. This assumption is analogous to the adiabatic approximation. The applicability of the adiabatic approximation will be discussed later in Sect. 5.2.5.

Consider the system described by the Hamiltonian in (5.36). The defects disturb the equilibrium phonon coordinates. The new equilibrium positions depend on the states of the defects S^z. One needs to proceed to the new equilibrium positions $U_{0q,\mu}$ where the defects and the phonon subsystem will be effectively separated. All we are doing here is finding the new normal modes of the system after accounting for the effect that the TS's have on the phonons.

To find the new equilibrium phonon coordinates one should make use of the definition of the local atom displacements in the positions of defects

$$U_{1\alpha}(\mathbf{x}) = \frac{1}{\sqrt{N}} \sum_{q,\mu} U_{q,\mu} e_{q,\mu,\beta} e^{(i\mathbf{q}\cdot\mathbf{x})}, \tag{5.38}$$

where N is the total number of elementary cells in the sample, and $e_{q,\mu}$ is the unit vector representing the direction of vibrations for mode μ. For the longitudinal mode

$$e_{q,\ell,\alpha} = q_\alpha/q, \tag{5.39}$$

whereas for the transverse modes $t1$, $t2$

$$e_{q,t1} \cdot \mathbf{q} = e_{q,t2} \cdot \mathbf{q} = e_{q,t1} \cdot e_{q,t2} = 0,$$
$$\sum_{\mu=t1,t2} e_{q,\mu,\alpha} e_{q,\mu,\beta} = \delta_{\alpha\beta} - \frac{q_\alpha q_\beta}{q^2}. \tag{5.40}$$

The equilibrium value of displacement for each mode $U_{0q,\mu}$ can be found by the minimization of the total energy (5.36) with respect to the displacements $U_{q,\mu}$, making use of the definition (5.38). After straightforward calculations one finds

$$U_{0q,\mu} = \frac{1}{\sqrt{N}M\omega_{q,\mu}^2} \sum_{\alpha\beta} iq_\beta (\gamma_{1\alpha\beta} e_{q,\mu,\alpha} S_1^z + \gamma_{2\alpha\beta} e_{q,\mu,\alpha} S_2^z). \tag{5.41}$$

Then we define the vibrational energy in terms of the displacements $\tilde{U}$ from equilibrium positions

$$\tilde{U}_{q,\mu} = U_{q,\mu} - U_{0q,\mu}. \tag{5.42}$$

The total energy of the system in terms of the redefined phonon coordinates reads

$$H_{\text{tot}} = \sum_{q,\mu}\left(\frac{|P_{q,\mu}|^2}{2M} + M\omega_{q,\mu}^2\frac{|\tilde{U}_{q,\mu}|^2}{2}\right.$$
$$\left. - M\omega_{q,\mu}^2\frac{|U_{0q,\mu}|^2}{2}\right) - \Delta_1 S_1^z - \Delta_2 S_2^z. \tag{5.43}$$

The first two terms represent the phonon bath, whereas the last term in parentheses contains the corrections to the energy caused by the presence of the defects. It is this term in which we are interested since it contains the part of the system energy that is dependent on the states of the defects. That term can be written as

$$H_{\text{def}} = -E_1 - E_2 - U_{12}S_1^z S_2^z,$$
$$E_{1,2} = \frac{1}{4NM}\sum_{q,\mu}\sum_{\alpha\beta}\sum_{\gamma\delta}\frac{1}{\omega_{q,\mu}^2}e_{\alpha,\mu}e_{\gamma,\mu}q_\beta q_\delta \gamma_{1,2\alpha\beta}\gamma_{1,2\gamma\delta},$$
$$U_{12} = \frac{2}{NM}\sum_{q,\mu}\sum_{\alpha\beta}\sum_{\gamma,\delta}\frac{1}{\omega_{q,\mu}^2}e_{\alpha,\mu}e_{\gamma,\mu}q_\beta q_\delta \gamma_{1\alpha\beta}\gamma_{2\gamma\delta}\cos(\boldsymbol{q}\cdot\boldsymbol{x}_{12}),$$
$$\boldsymbol{x}_{12} = \boldsymbol{x}_1 - \boldsymbol{x}_2. \tag{5.44}$$

The first two contributions, E_1 and E_2, describe the independent influence of defects on the total energy of the system. The last term contains the "interference" effect of both defect contributions. In essence, this term defines the interaction between the two TS's. Taking into account the relations (5.39) and (5.40), we can rewrite this term as

$$U_{12} = \frac{2}{N}\sum_{\alpha\beta}\sum_{\gamma\delta}\frac{\gamma_{1\alpha\beta}\gamma_{2\gamma\delta}}{M}\left(\frac{1}{v_t^2} - \frac{1}{v_\ell^2}\right)\sum_{q,\mu}\frac{q_\alpha q_\beta q_\gamma q_\delta}{q^4}\cos(\boldsymbol{q}\cdot\boldsymbol{R})$$
$$-\frac{2}{N}\sum_{\alpha\beta\delta}\frac{\gamma_{1\alpha\beta}\gamma_{2\alpha\delta}}{Mv_t^2}\sum_{q\mu}\frac{q_\alpha q_\beta}{q^2}\cos(\boldsymbol{q}\cdot\boldsymbol{R}),$$
$$\boldsymbol{R} = \boldsymbol{x}_1 - \boldsymbol{x}_2. \tag{5.45}$$

Assuming that the distance R exceeds the interatomic distance we can proceed from the summation to the integration over $\boldsymbol{q}$. After tedious calculations we arrive at the following final expression for the interaction between defects:

$$U_{12} = -\frac{1}{2\pi R^3}\sum_{\alpha\beta\delta}\frac{\gamma_{1\alpha\beta}\gamma_{2\alpha\delta}}{\varrho v_t^2}(\delta_{\beta\delta} - 3n_\beta n_\delta)$$
$$+\frac{1}{2\pi R^3}\sum_{\alpha\beta\gamma\delta}\frac{\gamma_{1\alpha\beta}\gamma_{2\gamma\delta}}{\varrho}\left(\frac{1}{v_t^2} - \frac{1}{v_\ell^2}\right)(-[\delta_{\alpha\beta}\delta_{\gamma\delta} + \delta_{\alpha\gamma}\delta_{\beta\delta} + \delta_{\alpha\delta}\delta_{\beta\gamma}]$$
$$+3[\delta_{\alpha\beta}n_\gamma n_\delta + \delta_{\alpha\gamma}n_\beta n_\delta + \delta_{\alpha\delta}n_\gamma n_\beta$$
$$+\delta_{\beta\gamma}n_\alpha n_\delta + \delta_{\beta\delta}n_\alpha n_\gamma + \delta_{\gamma\delta}n_\alpha n_\beta] - 15n_\alpha n_\beta n_\gamma n_\delta),$$
$$\boldsymbol{n} = \frac{\boldsymbol{R}}{R}, \quad \varrho = M/a^3. \tag{5.46}$$

Here ϱ is the density of the sample and a^3 is the volume of an effective elementary cell. When R is comparable to the interatomic distance the interaction becomes softer than $1/R^3$.

Consideration of the whole ensemble of TS's interacting with the strain field leads to the same interaction for all pairs of TS's. Thus we can write down the Hamiltonian of the interaction of the TS's as

$$\widehat{H}_{\text{int}} = -\frac{1}{2} \sum_{1,2} U_{12} S_1^z S_2^z, \quad (5.47)$$

where the interaction can be defined as

$$U_{12} = \frac{u_{12}}{R_{12}^3}. \quad (5.48)$$

Here R_{12} is the distance separating TS's 1 and 2 and the interaction constant u_{12} is defined using (5.46). The constant depends on the relative orientation of strain tensors $\widehat{\gamma}_1$ and is independent of the spatial separation of the TS's. It is important that the average value of the interaction constant is equal to zero:

$$\langle u_{12} \rangle = 0. \quad (5.49)$$

This is shown for the dipole–dipole interaction, where the averaging over the angles gives

$$\langle \delta_{\alpha\beta} - 3 n_\alpha n_\beta \rangle = \delta_{\alpha\beta} - 3 \cdot \frac{1}{3} \delta_{\alpha\beta} = 0, \quad (5.50)$$

and the same result can be obtained easily for the elastic interaction (5.46). For the average value of the modulus of the interaction constant we get from (5.46)

$$\langle |u_{12}| \rangle \equiv U_0 \approx \frac{\gamma^2}{\varrho v^2}, \quad (5.51)$$

where γ is the characteristic constant of the interaction of TS's with the strain field and v is the sound velocity. For simplicity below, we will treat the elastic coupling of TS's as dipolar in nature and neglect any correlations of the interaction constants for different pairs of TS's.

In this treatment the system of interacting TS's can be represented by the random Ising model of interacting spin-1/2 particles. The interaction varies with the inter-TS distance as $1/R^3$; i.e. it is long range. Since the average value of the interaction constant is equal to zero, these constants are of a random sign.

The relative strength of the interaction between TS's with respect to their own energies can be expressed by the dimensionless product

$$\chi = P_0 U_0 \approx P_0 \frac{\gamma^2}{\varrho v^2}. \quad (5.52)$$

Recall that P_0 is the density of TS's defined in (2.20). This parameter χ is universally small in all known glasses. It is of the order of 10^{-3}. It has a clear physical interpretation: In accordance with (5.26) and (5.31), it represents the characteristic value of the internal friction and the slope of the logarithmic temperature dependence of the speed of sound. Further, it is approximately equal to the ratio of the mean free path and the wavelength for the thermal phonons responsible for the thermal conductivity (see Freeman and Anderson (1986), Yu and Leggett (1988)). It has been shown by Freeman and Anderson (1986) using measurements of the heat capacity and the speed of sound in a wide variety of glasses that the parameter χ is almost material independent, despite the strong variation in parameters P_0, γ, ϱ and v. The universality of χ motivates an interaction-based scenario for the formation of TS's that we will discuss later in Sect. 5.3.

Now that we have seen the general description of elastic TS interactions and simplified it using the dipole approximation, let us consider the dephasing in a TS subsystem caused by these interactions (following Black and Halperin (1977)). The interaction between TS's, (5.47), gives rise to an additional term in the level asymmetry of a particular TS, labelled i,

$$\tilde{\Delta}_i = \Delta_i + \sum_j U_{ij} S_j^z . \tag{5.53}$$

The second term on the right-hand side of (5.53) depends on the states of the surrounding TS's. Since these TS's undergo chaotic transitions between levels because of the emission or absorption of phonons, the energy E_i of the TS of interest changes with time. This change is associated with the spectral diffusion of TS's. A cartoon of this sort of influence is shown in Fig. 5.4a.

This spectral diffusion also gives rise to fluctuations of the TS phase. The phase of the TS wave function $\Psi_i \propto \exp(\mathrm{i}\Phi_i(t))$ can be represented as

$$\Phi_i(t) \approx \int_0^t \mathrm{d}t' E_i(t') . \tag{5.54}$$

This integral has a random contribution

$$\delta \Phi_i(t) \approx \delta E_i(t)\, t , \tag{5.55}$$

because of the fluctuations of the energy E_i.

When the characteristic value of the random phase $\delta\Phi$ approaches unity, phase coherence is broken. Thus the dephasing time τ_2 caused by spectral diffusion can be estimated from the condition

$$\delta E(\tau_2)\tau_2 \approx 1 . \tag{5.56}$$

To estimate τ_2 we need to find $\delta E(\tau_2)$ first. We have to assume that the dephasing time from spectral diffusion is shorter than the relaxation time τ_1 given in (2.18); otherwise τ_1 also defines the dephasing time.

The fluctuation of energy can be estimated making use of (5.53):

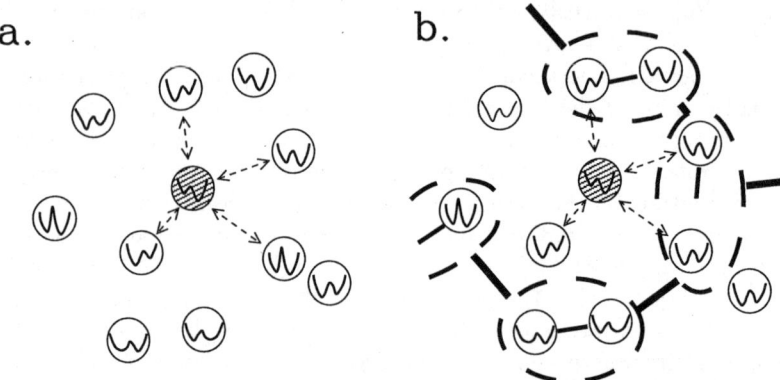

Fig. 5.4. The situation in (a) is traditional spectral diffusion: a particular TS's energy is stochastically fluctuated because of interactions with its neighbors, which are treated as independent TS's. In (b), the TS's environment is now dominated by the influence of TS participating in the cluster of resonant TS pairs, leading to dephasing behavior not predicted by the standard tunneling model (see Sect. 5.1.4).

$$\delta E_i(t) = \sum_j{}' U_{ij} S_j^z. \tag{5.57}$$

Here the prime means that the summation is made over TS's that undergo transitions during the time interval $(0, t)$. Those TS's that contribute the most to spectral diffusion are "thermal TS's" having $\Delta \sim \Delta_0 \sim k_B T$. TS's with energy splittings larger than the temperature are frozen in their ground states, and TS's with energy splittings much less than the temperature have a small phase volume. One chooses $\Delta_0 \sim k_B T$ because the TS's with very small Δ_0 have very long transition times ($\propto 1/\Delta_0^2$), so that the thermal TS's are more significant. The density of thermal TS's, n_T, can be found from the distribution (2.21)

$$n_T \approx P_0 \int_0^{k_B T} d\Delta \int_{k_B T/2}^{k_B T} \frac{d\Delta_0}{\Delta_0} \approx P_0 k_B T. \tag{5.58}$$

The transition time of thermal TS's is defined by (2.18) with $k_B T$ substituted for the energy and the tunneling amplitude of the TS's,

$$\frac{1}{\tau_{1T}} \approx \frac{1}{2\pi \varrho \hbar^4} \left(\frac{\gamma^2}{v^5}\right)(k_B T)^3. \tag{5.59}$$

For different TS's the random corrections to their energies are determined by the spatial distribution of TS's that have made transitions. The spatial density of TS's that have made transitions in a time $t < \tau_{1T}$ can be estimated as

$$n(t) \approx P_0 k_B T \frac{t}{\tau_{1T}}. \tag{5.60}$$

Note that this is the step in which we assume that the other TS's are the usual independent entities of the standard tunneling model.

As we have demonstrated above, the interaction of TS's scales with the distance between them as $1/R^3$, and their interaction constant is of a random sign. In this case the spatial distribution of $n(t)$, thermal TS's contributing to the energy shift (5.57), is uniform. For such a situation the distribution of the random energy shift has been calculated in Black and Halperin (1977). It has the Lorentzian form

$$G(\Delta) = \frac{1}{\pi} \frac{\Lambda}{\Lambda^2 + \Delta^2}, \qquad (5.61)$$

and the width Λ of the Lorentzian distribution is set by the characteristic interaction of TS at the average distance between them

$$\Lambda(t) = \frac{2\pi^2}{3} U_0 N(t) \approx k_B T P_0 U_0 \frac{t}{\tau_{1T}}. \qquad (5.62)$$

This width represents the characteristic value of the fluctuation of a TS's energy during the time t.

Now we are able to estimate the dephasing time τ_2 from the condition

$$\frac{\Lambda(\tau_2)\tau_2}{\hbar} \approx 1. \qquad (5.63)$$

Solving this equation one finds

$$\frac{1}{\tau_2} \approx \sqrt{\frac{k_B T P_0 U_0}{\hbar} \frac{1}{\tau_{1T}}}. \qquad (5.64)$$

One should check the condition for the validity of (5.64), $\tau_2 < \tau_{1T}$. Note that the time τ_2 increases with decreasing temperature as T^{-2}, whereas the relaxation time of thermal TS's τ_{1T} varies as T^{-3}. Thus, at sufficiently low temperatures, $T < 0.1$ K, (5.64) is valid. The crossover temperature appears to be of order 1 K.

5.1.3 Theoretical Approaches to the Relaxation of Tunneling Systems

Actually almost all experimental data demonstrate clear deviations from the model of noninteracting TS's at sufficiently low temperatures, as will be discussed below in Sect. 5.1.7 [see e.g Bernard et al. (1979), Kleiman et al. (1987), Baier et al. (1988), Enss et al. (1990), Classen et al. (1994), Esquinazi et al. (1992), Esquinazi et al. (1994), Rogge et al. (1996b)]. Both longitudinal and transverse relaxation rates show linear dependences on the temperature. Neither the linear temperature behavior for τ_2^{-1} nor that for τ_1^{-1} can be explained by the standard tunneling model's phonon-assisted mechanism described in Sect. 5.1.2.

For this reason, we are going to consider modifications to the above calculation. In particular we are going to look closely at the consequences of TS interactions and no longer treat the "environment" of a particular TS as an ensemble of noninteracting tunneling states.

Let us first try to obtain an intuitive understanding of the temperature dependences of the relaxation rates. If a certain type of excitation is responsible for the relaxation of the TS's, then it makes sense that the relaxation rate should vary like the number of such excitations. One can interpret the cubic temperature dependence of τ_{1T}^{-1} [see (5.59)] in terms of the temperature dependence of the heat capacity of the phonon subsystem $C_{\text{ph}} \approx T^3$. Similarly, the approximately linear temperature dependence for the heat capacity of the TS ensemble is known from Zeller and Pohl (1971). Therefore, if the transitions of the TS's occur due to an internal relaxation process in the TS's, one would expect a less steep T dependence for both τ_1^{-1} and τ_2^{-1} compared with the behavior in Eqs. (2.18) and (5.63); such a process does not involve the emission or absorption of phonons.

A useful analogy can be established with the relaxation of the magnetization of the nuclear spins in magnetic resonance experiments. It is well known that there exist two different mechanisms of relaxation. These are: spin lattice relaxation, very similar to the relaxation of TS's caused by their interactions with phonons; and the spin–spin relaxation, caused by the $1/R^3$ interaction between spins. Moreover, the spin-spin relaxation usually occurs much faster than the spin-lattice relaxation. It is certainly plausible that the interactions between TS's might be responsible for their relaxation at very low temperatures. We have obtained this interaction as the consequence of the virtual exchange of phonons in Sect. 5.1.2. We will consider its manifestation in TS dynamic properties. It will be shown that at very low temperatures the interactions between TS's lead to linearly temperature-dependent longitudinal and transverse relaxation rates. We then make a comparison between this theory and the experimental data.

We are going to consider the influence of TS interactions on the dynamics of the TS subsystem. This problem has been considered in detail in Burin and Kagan (1994a), and Burin and Kagan (1995) and we will follow these references in Sect. 5.1.4 below.

5.1.4 Many-Body Effects and Collective Excitations

To illustrate the effect of TS–TS interactions on the dynamics of the TS subsystem, consider a pair of the isolated TS's. The Hamiltonian of the pair reads

$$\widehat{H}_p = H_{01} + H_{02} - U_{12} S_1^z S_2^z,$$
$$H_{01} = -\Delta_{01} S_1^x - \Delta_1 S_1^z, \quad H_{02} = -\Delta_{02} S_2^x - \Delta_2 S_2^z, \qquad (5.65)$$

where U_{12} is the interaction caused by the exchange of virtual phonons, as introduced in (5.44). Below [(5.82), for example] we will see that this interaction is significant at length scales far exceeding the interatomic distance. Thus we can use the dipolar form given in (5.48).

Let us switch to the representation where the single TS Hamiltonians H_{01} and H_{02} are diagonal. This change corresponds to a rotation of the quantization axes in the spin space. The relation between the spin components in the old and new $(\tilde{S}^a)$ coordinate systems is given by

$$S_n^z = \frac{\Delta_n}{E_n}\tilde{S}_n^z - \frac{\Delta_{0n}}{E_n}\tilde{S}_n^x,$$

$$S_n^x = \frac{\Delta_{0n}}{E_n}\tilde{S}_n^z + \frac{\Delta_n}{E_n}\tilde{S}_n^x,$$

$$E_n = \sqrt{\Delta_n^2 + \Delta_{0n}^2}, n = 1, 2. \quad (5.66)$$

Then the total Hamiltonian reads

$$\widehat{H}_\mathrm{p} = -E_1\tilde{S}_1^z - E_2\tilde{S}_2^z - U_{12}\left[\frac{\Delta_1\Delta_2}{E_1 E_2}\tilde{S}_1^z\tilde{S}_2^z - \frac{\Delta_{01}\Delta_2}{E_1 E_2}\tilde{S}_1^x\tilde{S}_2^z\right.$$

$$\left.- \frac{\Delta_1\Delta_{02}}{E_1 E_2}\tilde{S}_1^z\tilde{S}_2^x + \frac{\Delta_{01}\Delta_{02}}{E_1 E_2}\tilde{S}_1^x\tilde{S}_2^x\right]. \quad (5.67)$$

In spite of the fact that each pair has four energy levels, only two of the levels can effectively contribute to the dynamics of the system. These levels correspond to the states of the pair when the first TS is in the ground state whereas the second one is in the excited state and vice versa. These are "flip–flop" (or up–down) pairs, considered by Burin et al. (1989) and Burin and Kagan (1995). The energy splitting between the two flip–flop levels of a pair, $E_\mathrm{p} = |E - E'|$, can be much less then T even if the energy splitting for each TS constituting a resonant pair is of the order of the temperature, $E \sim E' \sim k_\mathrm{B}T$. The two other levels of such a pair correspond to the case when both TS's are either in the ground or in the excited state. These levels are well separated from those giving rise to the flip–flop excitations and are weakly coupled to them. We shall see later that the pairs with $E_{1,2} \sim k_\mathrm{B}T$ and $|U_{12}| \ll k_\mathrm{B}T$ are most significant. The first term in brackets in (5.67) does not change at all in the up–down transition. Consequently, these transitions can be described by the Hamiltonian of a two-level system (2.1) with an asymmetry energy Δ_p and transition amplitude Δ_0p:

$$\Delta_\mathrm{p} \approx E_1 - E_2,$$

$$\Delta_\mathrm{0p} = \frac{1}{2}U_{12}\frac{\Delta_{01}\Delta_{02}}{E_1 E_2}. \quad (5.68)$$

The coherent coupling within the pair determines the transition amplitude Δ_0p. It decreases with distance as $1/R_{12}^3$. If this transition amplitude is less than the width of the level $\hbar/\tau_1$ defined in (2.18) of each TS in the pair, then the coherent coupling of the pair is actually destroyed. This is because the time needed for coherent oscillations of the pair becomes longer then the lifetime of the states of the TS constituents. This suggests a cutoff radius R_c for the formation of coherently coupled pairs. A direct estimate of $\tau_1(E)$ for $E \leq k_\mathrm{B}T$ gives [see (2.18) and (5.59)]

$$\frac{1}{\tau_1} \approx \left(\frac{\Delta_0}{k_B T}\right)^2 \frac{1}{\tau_{1T}}, \quad \frac{1}{\tau_{1T}} = \frac{\gamma^2}{\varrho v^5 \hbar^4}(k_B T)^3. \tag{5.69}$$

For the most significant thermal TS's, those with $\Delta_0, \Delta \sim k_B T$, the restriction for the radius of the coherent interaction reads

$$R_c \sim \lambda_T = \frac{\hbar v}{k_B T} \sim a \frac{\Theta_D}{T}. \tag{5.70}$$

Here Θ_D is the Debye temperature, a is the size of an elementary "cell" in the amorphous host, v is the sound velocity and λ_T is the wavelength of a phonon with the thermal energy.

We have shown that an interacting pair of TS's may, if the constituents are close enough together, be treated as a single dynamical entity. We have also obtained a finite radius within which coherent interactions between TS's are allowed. For the thermal TS's this radius is defined by the wavelength of thermal phonons [see (5.70)]. This means that our restriction is analogous to the electric dipole–dipole interaction (Landau and Lifschitz (1980a)) that is effectively static at distances less than the wavelength and is defined by the absorption or emission of real photons at larger distances.

Note that the most effective coupling between TS's appears in the resonant pairs where the asymmetry energy Δ_p and the pair transition amplitude Δ_{0p} are of the same order. These pairs contribute the most to the system's dynamics since their two up-down states overlap strongly.

Now that we know these pair excitations can exist, it is important to determine how many of them there can be, and what their effective TS parameters are. Let us estimate the probability w for an arbitrary TS to find a resonant up-down partner in the coherent region, i.e. to form a resonant pair with another TS. Let the given TS be described by asymmetry Δ and tunneling amplitude Δ_0. If the second TS is at a distance R ($R < R_c$) then its energy E_2 must satisfy the resonant condition

$$|E_2 - E| \leq \Delta_{0p} = \frac{U_0}{2R^3} \frac{\Delta_0 \Delta_{02}}{EE_2}, \tag{5.71}$$

where we have used our definition (5.51) for U_0.

The probability w is accumulated over large distances because of the long-range character of the TS interaction (see Fig. 5.5.). Therefore E_2 can be replaced by E in the right-hand site of (5.71). Then we can find w by counting TS's subject to the restriction (5.71):

$$w = \frac{P_0}{2} \int d\mathbf{R} \int_0^E \frac{d\Delta_{02}}{\Delta_{02}} \int_E^{E+\Delta_{0p}} \frac{E_2 dE_2}{\sqrt{E_2^2 - \Delta_{02}^2}} \Theta(R_c - R), \tag{5.72}$$

where $\Theta(x)$ is a step function that is equal to 1 for positive x and 0 otherwise. Taking advantage of the small value of Δ_{0p}, we find

$$w \approx \frac{\pi^2}{2} \frac{\Delta_0}{E} (P_0 U_0) \ln\left(\frac{\Theta_D}{T}\right). \tag{5.73}$$

Here the TS energy E is replaced by $k_B T$ in the argument of logarithm which is valid with logarithmic accuracy.

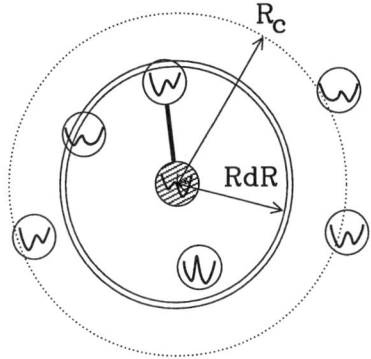

Fig. 5.5. This cartoon represents the process by which a TS finds a resonant partner. Because of finite temperature and the weakness of the TS interactions, a particular TS may only find a resonant partner within a distance R_c.

As we have already mentioned, the interaction of TS's is weak in all known glasses

$$P_0 U_0 \approx 10^{-3} \ll 1, \tag{5.74}$$

so that the probability w for a single TS to find a resonant partner is much less than unity even at very low temperatures, $T > \Theta_D \times \exp(-100)$. Therefore there are no delocalized excitations of the spin-wave type, in contrast to the assumptions of Yu (1985), Maleev (1989), and Continentino (1980). The statements about the presence of delocalized excitations were based on the long-range character of the $1/R^3$ interaction. However, as we have shown above, the interaction has a finite radius at nonzero T, and therefore it cannot lead to delocalized elementary excitations.

The result in (5.73) is essential for the future discussion of many-body excitations. It implies that since $w \ll 1$ an individual TS does not participate in more than one resonant pair.

Thus there exists a subsystem of resonant pairs with relatively low density, see the cartoon in Fig. 5.4b. This subsystem is responsible for the self-dynamics of TS's. To describe the resonant pairs, let us calculate their distribution as a function of the asymmetry energies Δ_p and tunneling amplitudes Δ_{0p}. Since we are considering excitations related to up-down transitions, we must introduce statistical Boltzmann factors to account for the fact that one TS in each pair must be in its excited state. Then the distribution of pairs is defined as

$$P_2(\Delta_p, \Delta_{0p}) = P_0^2 \int \frac{d\Delta_{01}}{\Delta_{01}} \int_{\Delta_{01}} \frac{E_1 dE_1}{\sqrt{E_1^2 - \Delta_{01}^2}} \frac{1}{1 + \exp(-E_1/(k_B T))}$$
$$\times \int \frac{d\Delta_{02}}{\Delta_{02}} \int_{\Delta_{02}} \frac{E_2 dE_2}{\sqrt{E_2^2 - \Delta_{02}^2}} \frac{\exp(-E_2/(k_B T))}{1 + \exp(-E_2/(k_B T))}$$
$$\times \int d\mathbf{R} \Theta(R_c - R) \delta(\Delta_p - E_1 + E_2)$$
$$\times \delta\left(\Delta_{0p} - \frac{U_0}{2R^3} \frac{\Delta_{01}\Delta_{02}}{E_1 E_2}\right). \tag{5.75}$$

After algebraic but tedious calculations [see Burin and Kagan (1995) for details], we arrive at the following distribution of up–down excitations as a function of their energies and tunneling parameters:

$$P_2(\Delta_p, \Delta_{0p}) = \frac{\pi^3}{12}(P_0 k_B T)(P_0 U_0) \frac{1}{\Delta_{0p}^2}$$
$$\times \Theta(k_B T - \Delta_{0p}) \Theta\left(\Delta_{0p} - U_0 \left(\frac{k_B T}{\hbar v}\right)^3\right). \tag{5.76}$$

It is significant that the most important contribution to the density of pair excitations comes from thermal TS, with energies and tunneling amplitudes of order of the temperature, that are separated by the distance R satisfying the condition

$$\Delta_{0p} \approx \frac{U_0}{R^3}. \tag{5.77}$$

The most essential feature of the distribution of pairs is the sharp increase of their density for small values of tunneling amplitude,

$$P_2 \propto \frac{1}{\Delta_0^2}. \tag{5.78}$$

Comparison of the distribution of pairs (5.76) with the distribution of the initial TS's (2.20) shows that for tunneling amplitudes small enough, $\Delta_{0p} < k_B T(P_0 U_0)$, the density of the pair excitations becomes larger than that of isolated TS's. Thus at sufficiently low temperatures the dominant low-energy excitations are up–down pairs.

We have demonstrated above [see (5.74) and following] that the resonant interaction between individual TS is weak and cannot lead to the delocalization of their excitations. However, this is not the case for up–down pairs, since the pair density is much larger at low energies. As we shall see, it is possible to obtain delocalized excitations due to interactions between different *pairs* of TS's.

We have seen that each pair excitation acts like an effective TS. Consider resonant pairs satisfying

$$\Delta_p < \Delta_{0p} \tag{5.79}$$

that have a tunneling amplitude Δ_{0p} that is the same order of magnitude as the tunneling amplitude of their individual constituent TS's,

$$\Delta_0 - \Delta_0/2 < \Delta_{0p} < \Delta_0 + \Delta_0/2. \tag{5.80}$$

For arbitrary values of tunneling amplitude $\Delta_0 < k_B T$, the density n_r of pairs satisfying (5.79) and (5.80) is roughly independent of Δ_{0p}:

$$\begin{aligned} n_r &\approx P_2(\Delta_p, \Delta_{0p})\Delta_{0p}\Delta_{0p} \\ &\approx \frac{\pi^3}{12}(P_0 k_B T)(P_0 U_0). \end{aligned} \tag{5.81}$$

We can define an analogous condition for an individual TS. We are interested in those particular resonant TS's that satisfy $\Delta < \Delta_0$. Their spatial density, given by $P_0\Delta_0$, decreases with decreasing tunneling amplitude, in contrast to the pair excitations discussed above.

Now we are interested in examining the interactions between resonant pairs of TS's; that is, we will consider each such pair as an effective TS itself, as described above in (5.75), and look at interactions between such pair excitations. Consider the subset of resonant pairs where each pair has a tunneling amplitude of the order Δ_{0p}. The average distance R_* between such pairs can be estimated by using their spatial density, (5.81),

$$\frac{1}{R_*^3} \approx (P_0 k_B T)(P_0 U_0). \tag{5.82}$$

The coupling between the resonant pairs in each energy interval $(\Delta_0 - \Delta_0/2, \Delta_0 + \Delta_0/2)$ can be estimated as

$$\Delta_* \approx \frac{U_0}{R^3} \approx k_B T (P_0 U_0)^2. \tag{5.83}$$

We have used the resonant character of the pairs (that means Δ_{0p}/E_p is of order unity), so the coherent coupling of pairs in (5.68) is set by their interaction U_0/R^3 if the distance between pairs is less than R_c. Under these conditions, the resonant pairs with $\Delta_{0p} \lesssim \Delta_*$ form an infinite coherent coupled cluster that enables the appearance of delocalized excitations. For $\Delta_{0p} \gg \Delta_*$ an infinite cluster does not form and excitations turn out to be localized, since their interaction is larger than the characteristic scale of their energy. Recall that the characteristic distance between pairs R_* must be less than the radius of interaction R_c defined in (5.70). This requirement leads to an upper bound for the temperature below that one expects to see delocalized excitations, $T < T_0$,

$$T_0 \approx \frac{1}{k_B}\sqrt{v^3\hbar^3/U_0}(P_0 U_0). \tag{5.84}$$

This restriction means that the interaction radius must be sufficiently large for the excitations to delocalize. Future references to "the resonant cluster" refer to this delocalized excitation comprising resonant pairs of TS's that are coupled to each other by interactions.

Hence we see that at sufficiently low temperatures, the TS's, rather than behaving independently, break up into pairs of TS's, each pair acting as an effective TS. The pairs interact with each other and are able to form delocalized excitations.

We now explore the consequences of such excitations. First, let us reexamine the spectral diffusion and dephasing problems analyzed in Sect. 5.1.2 in (5.55) and in the following text. Instead of the TS-phonon coupling setting the relevant time scale, it is necessary to consider the delocalized excitations. The characteristic time for the diffusive transport of the excitations is defined by the inverse energy:

$$\tau_* \approx \frac{\hbar}{\Delta_*}. \tag{5.85}$$

In this time the pairs with energy of order Δ_* change their state. The number density of such pairs is n_r [see (5.81)]. Their changes of state influence the energies of the other excitations. Statistically, this interaction effect causes the energies of the collective excitations to shift on the scale $U_0 n_r \sim \Delta_*$. This means that the resonant coupling between pairs will be broken for those pairs with $\Delta_{0p} \leq \Delta_p \ll \Delta_*$ in a time of order τ_*. The probability of a real transition is

$$(\Delta_{0p}\tau_*)^2 < 1. \tag{5.86}$$

Consequently, the transitions of TS's belonging to the resonant cluster shift the phase of the wave function for any arbitrary TS. Since for each TS this phase shift is unique and depends on that TS's specific surroundings, the phase coherency of the TS's will be destroyed. The wave function of an isolated TS with energy splitting E varies harmonically in time,

$$\Psi(t) \propto \exp(\Phi(t)), \quad \Phi(t) = iEt. \tag{5.87}$$

The interaction of this TS with its surroundings gives rise to a fluctuating part of the phase

$$\delta\Phi(t) \approx \frac{\delta E(t) t}{\hbar}, \delta E \approx \sum_i U_i. \tag{5.88}$$

Here δE is the shift of the TS energy caused by the transitions of the TS's i belonging to the resonant cluster. The TS forgets its initial phase when the phase shift $\delta\Phi$ becomes comparable to unity. We can easily see that this happens at $t \sim \hbar/\Delta_*$. Actually at this time n_r pairs of TS's undergo transitions, and the resulting shift of TS energy $U_0 n_r$ is nothing more than the characteristic energy scale Δ_*. Thus the rate of the dephasing in the ensemble of TS's, $1/\tau_2$, is set by the energy scale of the infinite cluster

$$\frac{1}{\tau_2} \approx \frac{\Delta_*}{\hbar} \sim \frac{k_B T (P_0 U_0)^2}{\hbar}. \tag{5.89}$$

Recall that the dephasing time has been measured in various echo experiments that are rather similar to the spin echo measurements for nuclear spins.

Since the dimensionless parameter $P_0 U_0$ is close to 10^{-3}, the final expression for the dephasing rate takes the form

$$\tau_2^{-1} = bT, \quad b \approx 10^6 \text{K}^{-1}\text{s}^{-1}. \tag{5.90}$$

The factor b defining the linear temperature dependence in (5.90) is in good agreement with that obtained via measurements of spontaneous echoes (Bernard et al. (1979)), where $b \approx 10^6$ K^{-1}s^{-1} has been observed, and the echo measurements in Suprasil-I (Baier et al. (1988)) where $b \approx 2 \times 10^6$ K^{-1}s^{-1} has been found.

Finally, let us apply our results to analyze recent experimental data obtained for the rotational echo in (KBr)$_{1-x}$(KCN)$_x$ ($x = 0.05; x = 0.08$) (Enss et al. (1995)). A linear temperature dependence for the dephasing rate was reported in that paper. To fit their data one should accept for the dimensionless parameter $P_0 U_0$ the value 10^{-2} for the concentration $x = 0.05$, and 0.7×10^{-2} for the concentration $x = 0.08$. To account for such a choice one should note that the heat capacity for the concentration x under consideration is larger than that in the glassy state that corresponds to concentrations $x \geq 0.3$. The heat capacity is known to decrease whereas x increases if the concentration x is small enough (see de Yoreo et al. (1986)). Therefore large dephasing rates can be accounted for by the larger value of the density of the low-energy excitations P_0 for small concentrations x.

We have described a new mechanism that causes dephasing in glasses. This mechanism is due to the long-range $1/R^3$ interaction between TS's. Such interactions result in a subsystem of strongly coupled excitations formed from resonant TS pairs. The relaxation of this subsystem leads to the linear temperature dependence of the dephasing rate seen at ultralow temperatures.

One should note that the phenomenon of delocalization has been analyzed in terms of the concepts in the theory of Anderson localization. Delocalization implies the possibility that a local perturbation may "go to infinity" in a diffusive way. This delocalization is essentially many-body in nature and appears at finite temperature only. It is interesting that this result is not restricted to systems with the long-range $1/R^3$ interaction, but it ensures delocalization for the short-range interaction case also. Our calculations for the density of up-down pairs (5.76) can be generalized easily for the case of arbitrary interaction $1/R^\alpha$ with $\alpha > 3$. The result is

$$P_\alpha(\Delta_0) \sim \frac{1}{\Delta_0^{(\alpha+3)/\alpha}}. \tag{5.91}$$

Consequently, the density of resonant pairs scales with their energy as

$$n_\alpha(E) \sim E^{(\alpha-3)/\alpha}, \tag{5.92}$$

and the coherent coupling of pairs behaves like

$$U_0(n_\alpha(E))^{\alpha/3} \propto \Delta_0^{(\alpha-3)/\alpha}. \tag{5.93}$$

For $\alpha < 6$ the coherent coupling decreases with E slower than linearly. At sufficiently small E the coherent coupling always becomes comparable to E, which causes the appearance of an infinite resonant cluster of pairs and the delocalization of the excitations. Thus the quadrupole interaction $1/R^5$ also leads to delocalized excitations. Generally, for a system of the dimensionality d the threshold is defined as $\alpha < 2d$. More generally, if the interaction in the d-dimensional system decreases with distance slower than $1/R^{2d}$, one always finds delocalization of the low-energy excitations for arbitrary disordering. Note that at zero temperature the delocalization always takes place for $\alpha < d$. Including more complex excitations (triples, etc) does not change the above results. This conclusion has been formulated by Burin et al. (1989).

5.1.5 Interaction-Stimulated Relaxation of Tunneling Systems

Consider the effect of delocalized excitations on the relaxation processes that govern the TS's. We will see that such collective modes can stimulate TS transitions more rapidly and with a different temperature dependence than the TS-phonon processes of the standard tunneling model.

In the delocalization scenario described in Sect. 5.1.4, n_r pairs per unit volume undergo transitions in each time interval τ_*. The resulting changes in local strain (and possibly electric) fields shift the energy levels of each TS. It follows that as the local environment changes, some intra- and interpair resonant couplings will be broken, whereas at the same time new resonant couplings will be formed. Relaxation processes in the TS system depend on such rearrangement.

Consider the evolution of the TS system caused by up-down transitions within the resonant cluster of pairs, beginning from some initial state. Let the density n of TS's that have undergone transitions be small in comparison with the total density of TS's n_0. The change in the asymmetry Δ_i of TS i caused by these transitions can be written as

$$\delta_i = \Delta_i' - \Delta_i = \sum_j{}' U_{ij}. \tag{5.94}$$

Here Δ_i is the initial asymmetry of TS 'i', Δ_i' is its final value and the sum is taken over all TS's undergoing transitions since the beginning of the time interval under consideration. When the interaction U_{ij} scales with distance as $1/R_{ij}^3$, the value of Δ_i is distributed in a Lorentzian law [see (5.61)] with the characteristic width

$$\Lambda(n) = \frac{2\pi^2}{3} n U_0. \tag{5.95}$$

This is exactly as we saw in our earlier calculation of spectral diffusion, (5.62), but in this case n has a time dependence due to the dynamics of the delocalized excitations, rather than the TS-phonon relaxation.

Thus the Lorentzian distribution for the level shift appears in the TS system because of the transitions of some of the TS's into new states. The width of the Lorentzian is proportional to the number of TS's that make such transitions. We have found earlier that about n_r pairs change their state in the time interval τ_*. Equation (5.95) implies that the characteristic scale $\Lambda(n_r)$ for the change in Δ_i exceeds $\Delta_* \sim U_0 n_r$. This change leads to the breakdown of the infinite resonant cluster in a time of order τ_*, since the majority of resonant pairs will have their levels shifted by more than the resonant energy Δ_*. However, the smoothness of the TS distribution versus energy leads to the appearance of the new resonant pairs and consequently a new infinite cluster of pairs. As some pairs "wander away" from resonant interactions with the rest of the cluster, new pairs "wander into" resonant interaction. This self-consistent picture arises as the number of TS's N (= twice the number of pairs) that have undergone the transition described by the simple equation

$$\frac{dN}{dt} = 2n_r \frac{\Delta_*}{\hbar} \tag{5.96}$$

increases. It follows from the calculations of the density of pairs [(5.75) and (5.76)] that the pairs with energy splittings less than the temperature are formed from thermal TS's with $\Delta \approx \Delta_0 \approx k_B T$. The density of the thermal TS's can be estimated as $P_0 k_B T$. As the self-consistent time evolution process proceeds all of these TS's will change their state after briefly sojourning within the infinite resonant cluster.

To establish the validity of this scenario let us assume first that the pairs that have left the resonant cluster do not return. This is not the case for individual thermal TS's that return to form resonant pairs with other partners. Then the relaxation time τ_0 of thermal TS's can be estimated from the requirement that during this time almost all thermal TS's change their state. Setting the solution of (5.96), $N(\tau_0) = 2n_r \tau_0 \Delta_* / \hbar$, equal to the density of thermal TS's, $P_0 k_B T$, we obtain

$$\frac{1}{\tau_0} = 2\Delta_* \frac{n_r}{\hbar P_0 k_B T} \approx 10 \frac{k_B T}{\hbar} (P_0 U_0)^3. \tag{5.97}$$

At this time the characteristic distribution width (5.95) reaches a value

$$\Lambda \approx \frac{2\pi^2}{3} \Delta_* \frac{\tau_0}{\tau_*} \sim k_B T (P_0 U_0). \tag{5.98}$$

Since this value corresponds to the interaction between thermal TS's at their average separation distance, it determines the maximum scale of Λ.

Let us verify the hypothesis that the TS does not return to the infinite cluster within the same resonant pair during the time τ_0. For this purpose we are going to estimate the probability for an individual pair to return back to a resonant state. Assume this pair was in resonance at time $t = 0$. The distribution of its energy, $F(\Delta)$, at time t is Lorentzian with the width defined by (5.98), where the final time τ_0 is replaced with t:

$$F(\Delta) = \frac{1}{\pi} \frac{\Lambda}{\Lambda^2 + \Delta^2},$$
$$\Lambda = \frac{2\pi^2}{3} \Delta_* \frac{t}{\tau_*} \approx k_B T (P_0 U_0). \tag{5.99}$$

The probability that the level asymmetry of a pair is within the resonant interval $(-\Delta_*, \Delta_*)$ within the elementary time window $(t, t + \tau_*)$ can be found making use of the distribution (5.99),

$$\delta w_b(t) \approx \frac{2}{\pi} \frac{\Delta_*}{\Lambda} = \frac{3}{\pi^3} \frac{\tau_*}{t}. \tag{5.100}$$

Taking the sum of all probabilities, (5.100), from $t = \tau_*$ up to $t = \tau_0$, we find the probability to return to the infinite cluster is

$$w_b = \frac{3}{\pi^3} \ln\left(\frac{\tau_0}{\tau_*}\right). \tag{5.101}$$

The argument of the logarithm is of order of $P_0 U_0$. If we assume this parameter has the characteristic glassy value 10^{-3} [see (5.74)], we find $w_b < 1$. It can be shown (within the above self-consistent consideration, taking into account the nonresonant pairs) that even for extremely small values of the parameter $P_0 U_0$, $\Lambda(t)$ increases with time at least as rapidly as $t/\ln(w_b t)$. This means that (5.100) holds.

In the above we have neglected the interaction of two level systems with phonons. That assumption is well justified if that interaction does not manifest itself during the characteristic transition time τ_0. In other words, the relaxation time τ_{1T} of the thermal TS's due to the interaction with phonons must be larger than τ_0. Since τ_0 increases with decreasing temperature as $1/T$ [see (5.97)] whereas the relaxation time caused by the interaction goes as T^{-3}, at sufficiently low temperatures, the relaxation mechanism induced by the interaction of TS's becomes more significant than that caused by their interaction with phonons. Comparing these relaxation times, we find that the crossover in relaxation mechanisms occurs at temperature T', defined as

$$T' \approx \frac{3}{k_B} \sqrt{\frac{v^3 \hbar^3}{U_0}} (P_0 U_0)^{3/2}. \tag{5.102}$$

The requirement $T < T'$ implies $T < T_0 = T'/(P_0 U_0)^{1/2}$ [see (5.84)] so that the interaction radius R_c is larger than the distance between resonant pairs in the infinite cluster. Thus, the crossover into interaction-dominated relaxation is consistent with the restriction on the radius of the interaction.

While the relaxation of thermal TS's is actually the fastest, there exists a whole ensemble of nonresonant TS with $\Delta_0/E < 1$, $E < k_B T$. They also relax, but more slowly than the thermal TS's. Let us estimate their relaxation time.

First, consider a resonant TS ($\Delta_0 \approx E$) with an energy splitting less than the temperature. Its relaxation depends on its ability to form a resonant pair with some other TS. This is easier if the second TS is resonant ($\Delta_{02} \approx \Delta_2$). Then a resonant pair will be formed if the difference in energies of both TS's is smaller than their interaction energy. The probability of finding a second TS satisfying that condition is determined by the interactions of the TS's, as described in (5.73). Each resonant TS has the same probability to form a resonant pair, and consequently each resonant TS has the same relaxation time τ_0.

A similar argument can be made for the nonresonant TS's ($\Delta_0 < E$). The effective tunneling amplitude for a pair will be less than that for the resonant TS's by a factor Δ_0/E. Consequently, starting with a given TS that forms a pair with tunneling amplitude Δ_* with some neighboring resonant TS, we can estimate the tunneling amplitude of the pair as $\Delta_* \Delta_0/E$. The probability of an up-down transition in such a pair during the time of the resonance contains an additional small factor $(\Delta_0/E)^2$, in accordance with (5.86). Thus the relaxation time for the nonresonant TS's can be estimated as

$$\tau(E, \Delta_0) \approx \tau_0 \left(\frac{E}{\Delta_0}\right)^2, \Delta_0 < E < k_B T. \tag{5.103}$$

Therefore, at sufficiently low temperatures, $T \ll T'$, the relaxation of TS is dominated by their interactions. However, even at higher temperatures, the relaxation time defined by (5.103) for TS's with $E < k_B T'$ is shorter than that caused by the interaction with phonons (2.18). It is interesting that despite the change in the relaxation mechanism at T', the estimation (5.103) actually remains valid even at temperatures higher than T'.

At temperatures $T > T'$ the relaxation of thermal TS's is dominated by their interactions with phonons. It can be characterized by the relaxation time $\tau_{1T} \propto T^{-3}$. The relaxation in the subsystem of thermal TS gives rise to a fluctuation of the energy splittings of the other TS. This fluctuation increases with time as [cf. (5.99)]

$$\delta E \approx k_B T P_0 U_0 \frac{t}{\tau_{1T}}, \quad t < \tau_1. \tag{5.104}$$

This spectral diffusion causes the excitations with low energies to undergo transitions. Consider an excitation with initial level shift $\Delta = 0$ and tunneling amplitude $\Delta_0 < \Delta_{1*} \equiv \sqrt{k_B T P_0 U_0 \hbar/\tau_{1T}}$. The probability of this TS making a transition can be estimated using the quantum-mechanical formula (see Landau and Lifschitz (1980b))

$$w_{\text{tr}} \approx \Delta_0^2/(k_B T P_0 U_0 \hbar/\tau_{1T}) \sim \frac{(\Delta_0 \tau_2)^2}{\hbar^2}, \tag{5.105}$$

where the "intermediate temperature" dephasing time τ_2 has been defined in (5.63). For a second transition to occur, the fluctuations must bring the level shift of the excitation back to small scales $\Delta \leq \hbar/\tau_2$. This will happen at a time $t \approx \tau_{1T}$, when all thermal TS governing the fluctuations of the excitation's energy will change their state several times. Thus we can estimate the average time between two successful transitions of the excitation as

$$\tau(\Delta_0) \approx \tau_{1T} w_{tr} \approx \frac{\hbar k_B T(P_0 U_0)}{\Delta_0^2}. \tag{5.106}$$

Thus we have found the relaxation time for excitations with very low-energy splittings. That time can be much shorter than the time defined by the excitations' interactions with phonons, see (2.18). Note, however, that our estimation is only applicable for TS's with very small tunneling amplitude $\Delta_0 < \hbar/\tau_2$.

Those TS's with intermediate tunneling amplitudes $\Delta_{1*} < \Delta_0 < k_B T$ can undergo transitions in pairs, provided that each pair's tunneling amplitude is smaller than the maximum energy scale for spectral diffusion $k_B T(P_0 U_0)$. That energy scale is defined by the interaction energy of two thermal TS's at the average distance between them. The level shift of such a pair must be less than the characteristic width of the level shift distribution caused by the transitions of thermal TS's $(w \sim k_B T(P_0 U_0))$. If this is the case, the transitions of neighboring thermal TS's can make Δ_p of the pair low enough to allow the transition. Thus, a given TS will form such a pair with the nearest TS having a level asymmetry Δ' close enough to Δ $(\Delta - \Delta' < k_B T U_0 P_0)$. The probability of finding such a neighbor within a sphere of radius R can be estimated as

$$w(R) \approx P_0 k_B T(P_0 U_0) R^3. \tag{5.107}$$

The pair will be formed within the distance $R_* \approx (P_0 k_B T(P_0 U_0))^{-1/3}$ where $w(R_*)$ approaches unity. The interaction of TS's forming such a pair can be estimated as $U_0/R_*^3 \approx \Delta_*$. If the TS's are resonant, then this interaction determines the tunneling amplitude of the pair, Δ_0. The relaxation time of this pair can be estimated with (5.106). It coincides with the time τ_0 found for the relaxation of the resonant thermal TS, (5.97). For nonresonant TS's, we can repeat the same argument as for low temperatures $T < T'$, arriving at the same result, (5.103).

If the relaxation rate from (5.106) is faster than the rate defined by the interaction with phonons, then the dominant relaxation mechanism is via the relaxation of neighboring TS's. Comparison of the relaxation time τ_0 with the phonon time, (2.18), shows that for TS's with energy splittings less than $k_B T'$ the relaxation is defined by (5.103).

Below, we consider the effect of the relaxation mechanism caused by the interaction between TS's on the equilibrium response of the system. For the acoustic and dielectric experiments the imaginary part of the dielectric constant ϵ'' as well as the internal friction Q^{-1} are mostly sensitive to the relaxation mechanism for the low-frequency external perturbation. According to Hunklinger and Raychaudhuri (1986), for example, at sufficiently low temperatures $T < T_m$ [see (5.34)], both ϵ'' and Q^{-1} are inversely proportional to the relaxation time of thermal TS's, τ_{1T},

$$Q^{-1} \approx P_0 \frac{\gamma^2}{\varrho v^2} \frac{1}{\omega \tau_{1T}},$$

$$\epsilon'' = P_0 p^2 \frac{1}{\omega \tau_{1T}}. \tag{5.108}$$

Here the measuring frequency ω is that of the external strain or electric field, and p^2 is the average TS electric dipole moment squared. In accordance with our results, at sufficiently low temperatures, $T < T'$, given by (5.102), the relaxation time τ_{1T} will be dominated by the interactions of TS's, and should be replaced by τ_0. Consequently, both the internal friction and the imaginary part of the dielectric constant (dielectric losses) must show a linear temperature dependence. Such behavior has been observed by Esquinazi et al. (1992) and Esquinazi et al. (1994) for the internal friction of SiO_2 at temperatures less than 30 mk [see Fig. 4.13B] and superconducting polycrystalline Nb-n, and in Rogge et al. (1996b) for the dielectric losses in a wide variety of Si-based dielectric glasses at temperatures less than 100 mK, and will be compared to the above theory in Sect. 5.1.7.

5.1.6 Equilibrium Acoustic and Dielectric Measurement Techniques

Before we compare the above theory with experimental results, we review the experimental techniques used to obtain the data in question. The techniques commonly used to measure the low-temperature dielectric and acoustic response of amorphous materials depend strongly on the frequency range of interest. Most of the work that we will discuss here has been carried out in the "acoustic frequency" range, 100 Hz $\leq \omega/2\pi \leq$ 100 kHz.

The most popular technique for high-resolution dielectric measurements in this frequency range is a bridge circuit, described in detail by Kibble and Rayner (1984). A typical schematic is shown in Fig. 5.6. The bridge consists of a ratio transformer acting as an inductive voltage divider, a reference capacitor, and decade resistors in series with the reference capacitor. Typically the moveable tap of the ratio transformer is grounded, and a lock-in amplifier is used to make a frequency and phase sensitive measurement of the bridge's off-balance signal. In variable-ratio bridges, the reactive phase of the off-balance signal is nulled by changing the tapping fraction of the ratio transformer, whereas in fixed-ratio bridges (such as the General Radio 1616) the tapping ratio is constant and balance is achieved by changing the reference capacitor. The resistive phase of the off-balance signal is nulled by adjusting the series resistance R_{ser} to balance the real part of the sample impedance. This series resistance also compensates for residual losses in the rest of the bridge circuit, making it difficult to obtain accurate absolute dielectric loss measurements.

One is typically interested in changes of the dielectric response rather than its absolute value. In this case the standard procedure is to null both

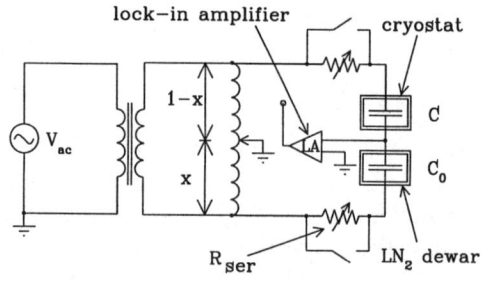

Fig. 5.6. A typical capacitance bridge for high resolution dielectric measurements.

capacitive and resistive components of the off-balance signal at the desired frequency and excitation voltage. The sensitivity of the bridge is then calibrated for each component by unbalancing the signal by a known amount and measuring the output of the lock-in. For relatively small deviations from balance, the off-balance voltage in the capacitive phase is linear in the change in sample capacitance. Similarly, the resistive off-balance signal is linearly proportional to the sample's change in dissipation away from the balance condition.

Changes in series resistance are related to the dielectric loss tangent by the following:

$$\Delta(\tan\delta) = \omega C(\delta R_{\text{ser}}). \tag{5.109}$$

Recall that $\tan\delta = \epsilon''/\epsilon'$.

Bridge techniques are impressively sensitive, allowing measurements of changes in $\tan\delta$ as small as 2×10^{-7} (Foote and Anderson (1987)) and of $\delta C/C$ as small as 3×10^{-8} (Rogge (1996)). It is possible to increase signal to noise in such bridge circuits by lowering capacitance to ground at the node leading to the input of the amplifier. The three ways to do this are (a) by incorporating the reference capacitor directly into the sample mounting scheme (Classen et al. (1994)); (b) by placing a cryogenic FET follower (for example, an InterFET IF1320 J-FET) inside the cryostat (Rogge (1996)) on the lock-in input lead; and (c) by using a current amplifier rather than a voltage amplifier. The upper limit on the bridge technique's frequency is set by both capacitive couplings and ratio transformer nonlinearities. The General Radio 1616 may be used up to 100 kHz before nonlinearity in the transformer becomes prohibitively important.

Another way to increase signal to noise is to boost the driving voltage. However, TS materials are inherently nonlinear dielectrics at low temperatures (especially for $T \ll T_m$). Thus one must be careful to ensure linear response. Typical measuring fields are on the order of $E_{\text{ac}} = 500$ V/m or lower in the linear regime.

One must also take care to heat sink samples well to avoid distortion of data by self-heating effects. Assuming that $\tan\delta$ is due entirely to dissipation within the sample, the heat produced per cycle of measuring excitation is given by $(\tan \delta) \times CV_{\text{ac}}^2/2$.

Note that bridge measurements give changes in sample capacitance. In the absence of changes in sample geometry, one may find the fractional change of the sample's dielectric susceptibility by working with the quantity $\Delta C/C = \Delta \epsilon'/\epsilon'$. This is an intensive quantity that allows comparisons between samples of differing geometries.

Details on the popular "vibrating reed" and "vibrating wire" methods of acoustic measurements are described at length in Sect. 4.3. Here we describe a modification to the reed technique, using magnetic rather than electrostrictive drive.

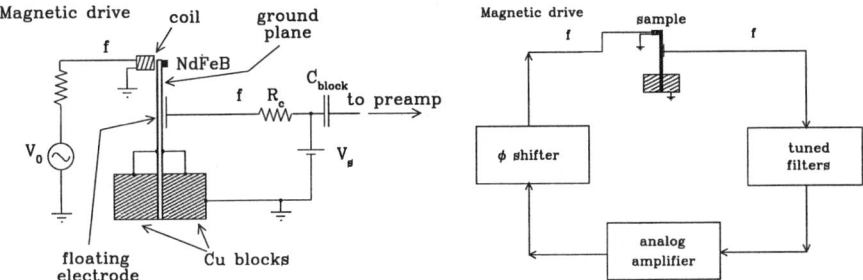

Fig. 5.7. Sample mounting and drive schemes for the vibrating reed technique. f denotes the driving frequency or the frequency of the detected signal.

For magnetic drive (see Fig. 5.7) a small (< 0.5 mm^3) piece of rare earth magnet is affixed with a tiny amount of epoxy to the end of the reed, oriented so that its magnetization is along the intended direction of reed motion. Since the part of the sample nearest the clamp experiences almost all of the strain during the oscillations, the magnet piece's only effect on the acoustic behavior of the reed is to lower the resonance frequencies by providing additional inertia. Rather than a driving electrode, a small coil is placed near the end of the reed, with the coil axis parallel to the magnetization. Because the magnetization is permanent rather than induced, the driving force is directly proportional to the current in the coil. One should also note that variation in the clamping method does not significantly alter the experimental results (see König et al. (1995)).

Figure 5.7 also shows a block diagram of the feedback method for the magnetic drive technique. The idea is to take the signal from the reed's motion, amplify it to a large enough amplitude that it may be fed into a frequency counter, and use that same signal to produce a driving force phase shifted from the signal by $\pi/2$ radians. In this way, the oscillator is made to self-resonate. To avoid frequency shifts due to nonlinearities in the oscillator, the drive amplitude is fed back to maintain constant oscillator amplitude. Magnetic drive allows one sample electrode to be electrically isolated from ground, permitting the study of dc electric field effects on acoustic response.

As in the capacitive measurements, one may obtain an intensive quantity independent of sample geometry by considering fractional changes; for resonance frequencies, $\Delta f/f = \Delta v/v$.

5.1.7 Equilibrium Acoustic and Dielectric Loss Data

As described above in (5.31) and (5.33), in the standard tunneling model at acoustic frequencies, the TS dissipative response is expected to be constant for $T \gg T_m$ and to vary as T^3 for $T \ll T_m$. The physical interpretation of the T^3 behavior is that single-phonon relaxations of the type described by (2.18) are the only source of TS relaxation out of phase with the driving ac field used to measure response. Thus, as the density of thermal phonons decreases as T^3, so should the dissipative response.

In practice, this T^3 loss behavior has never been seen at audio frequencies in either acoustic or dielectric behavior, for temperatures as low as $T/T_m \leq 0.01$. The loss seems to be better described by a linear variation in T in most cases. For example, Esquinazi et al. (1992) use the vibrating reed technique in equilibrium to measure internal friction in Suprasil I (SiO_2 with 1200 ppm OH^- impurities) and observe linear-T dependence between 3 mK and 30 mK. Similar work by Kleiman et al. (1987) in single crystal silicon shows the same T dependence over the same temperature range. Natelson et al. (1997) see linear T behavior in the acoustic dissipation in Suprasil 2 (also SiO_2 with 1200 ppm OH^-), shown earlier in Fig. 5.2, Suprasil 312 ($\sim$ 200 ppm OH^-), coverglass, and BK7, as well as in the dielectric loss, described below.

This systematic deviation from the predictions of the standard TS model is not limited to acoustic data. Dielectric losses also appear to vary linearly with temperature for $T \ll T_m$. Data for potassium-doped SiO_2 at two frequencies are shown in Figure 5.8 (Rogge et al. (1996b)). Again, the loss varies linearly with T. Further, when replotted on linear T axes, recent data of both Classen et al. (1994) and Enss et al. (1990) on Suprasil W and BK7, respectively, show the same qualitative behavior.

A speculation offered by Esquinazi et al. (1992) to explain the observed loss behavior was that perhaps the assumption of a three-dimensional phonon density of states was poor. However, the same linear temperature dependence of losses is observed in dielectric samples that may be well described by bulk three-dimensional phonons (Rogge et al. (1996b)), such as dielectric films on bulk crystalline substrates and samples immersed in liquid ^{3}He.

An alternate explanation for the linear-T behavior of the acoustic and dielectric dissipation is the existence of a new relaxation mechanism at very low temperatures. The relaxation process described above in Sect. 5.1.5 fits the observed facts well, as we now show.

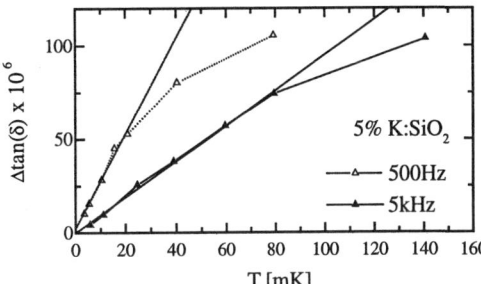

Fig. 5.8. Dielectric loss is well approximated by behavior linear in T at low temperatures, from Rogge (1996).

It should be emphasized that the experimental data we consider below are obtained at very low temperatures $T < 100$ mK. In this temperature regime, nonlinear effects are very significant and can strongly influence the experimental results [see e.g. Esquinazi et al. (1992), also Sects. 4.2.4 and 4.4.1]. Although the linear temperature dependence of the internal friction and dielectric losses are not sensitive to the external field amplitude, the characteristic value of the observed effect can be sensitive to nonlinearity. Thus we cannot completely exclude the possible influence of nonlinear effects.

Let us make a quantitative comparison between the theory outlined in Sect. 5.1.5 and the experimental data, starting with SiO_2. This experiment (Esquinazi et al. (1992)) is relevant since deviations in the internal friction from T^3 behavior have been observed at $T' \approx 30$ mK. Let us first compare our estimate for the crossover temperature T' with the experimental data. Recall that at that temperature the phonon relaxation time for the dominant (thermal) TS's, τ_{1T}, [see (2.18)] is equal to the interaction-induced relaxation time τ_0 [see (5.97)]. The relaxation time τ_{1T} shows an inverse cubic temperature dependence

$$\frac{1}{\tau_{1T}} \approx A(k_B T)^3. \tag{5.110}$$

The experimental data allow Esquinazi et al. (1992) (see Sect. 4.4.1 and Fig. 4.11) to fit the factor A as

$$A k_B^3 \approx 8 \times 10^7 \mathrm{K}^{-3} \mathrm{s}^{-1}. \tag{5.111}$$

Then the crossover temperature T' can be expressed as

$$T' \approx 3\sqrt{\frac{2}{\hbar A k_B^2}} (P_0 U_0)^{3/2}. \tag{5.112}$$

If one takes the dimensionless factor $\chi \equiv P_0 U_0 \approx 3 \times 10^{-3}$, one obtains a crossover temperature $T' \sim 30$ mK. This value of χ is in the characteristic range for that product in various glasses, though it is substantially larger than that obtained by measuring $P_0 \gamma_\ell^2/(\varrho v^2)$ using the plateau in the internal friction at $T > T_m$. To reconcile this discrepancy, recall that the interactions between TS's (described at length in Sect. 5.1.2) are strongly influenced by (virtual) transverse phonons, whereas the experiments measure the absorption of longitudinal sound. Consequently, the measured ratio $\gamma_\ell^2/(\varrho v^2)$ is substantially less than U_0 (Burin and Kagan (1995)).

Let us compare the observed internal friction with the theory (5.108). To make careful numerical comparison it is convenient to use the independent measurements of the product $P_0\gamma_\ell^2/(\varrho v^2)$ from the same experiment. One finds this by measuring the internal friction in the temperature-independent plateau observed by the experimenters at $T > 80$ mK. The ratio of the internal friction at the crossover temperature and in the plateau region was of order of 0.27; this should correspond to the factor $(\omega\tau)^{-1}$. If we check this by using the frequency of the measurement (400 Hz), using of (5.97) to define τ_0, and again assuming $P_0 U_0 \approx 3 \times 10^{-3}$, we find a satisfactory agreement between the above treatment and the data.

Consider the dielectric loss data. The experimental technique used by Rogge et al. (1996b) allows measurements at many different frequencies which is difficult for the vibrating reed method employed in the acoustic measurements. Therefore both the temperature dependence of the dielectric losses and their frequency dependence can be tested using the existing measurements. The linear temperature dependence for the dielectric losses has been observed at temperatures lower than 100 mK and over three orders of magnitude in frequency. Unfortunately, in all measurements the linear-T dependence has directly merged into the plateau region; a clear transition from linear-T to T^3 behavior has never been observed.

The data on K:SiO$_2$ of the experiment of Rogge et al. (1996b) are in satisfactory agreement with the theoretical predictions. Actually the parameter $P_0 p^2$ in (5.108) can be extrapolated from the high-temperature plateau region in ϵ''. Then the assumption $P_0 U_0 \approx 2 \times 10^{-3}$ results in quantitative agreement with the experimental value of the dielectric losses.

Let us turn to a qualitative discussion of the frequency dependence of the losses. In accordance with (5.108), the dielectric losses in the linear-T regime should be inversely proportional to the measuring frequency. The measurements have shown a decrease of the dielectric losses with increasing frequency, but this decrease is slower than $1/\omega$. A possible reason for this discrepancy is the presence of nonlinear effects at large frequencies. Actually, a decrease in the external field amplitude leads to a sharpening of the frequency dependence, but that dependence remains different from the predicted inverse proportionality. The source of this dependence on measuring field intensity is as follows: for low-intensity measuring fields (as are assumed in the above calculations) and low frequencies, the time required for TS's to couple to the measuring field is long compared to the TS-phonon relaxation time given in (5.137). For large measuring fields and high frequencies, this may no longer be the case, and a bottleneck can arise for energy from the measuring field to leave the sample via the phonons.

Another possible explanation is specific to the relaxation mechanism caused by the interactions between TS's. Recall that this relaxation process, presented in Sect. 5.1.5, involves TS's entering and leaving a continually shifting delocalized excitation (the resonant cluster). Those calculations assume that a TS's interaction with the cluster does not depend on previous such interactions. The presence of phase memory after interaction-driven transitions and the possibly nonexponential character of the TS relaxation can lead to a softer frequency dependence. This is a subject for future experimental and theoretical investigation.

5.1.8 Equilibrium Dielectric Saturation at Very Low Temperatures

Now that we have considered the experimental data on the low-temperature acoustic and dielectric dissipation of these materials in light of possible interaction effects, we turn our attention to the reactive component of the TS dielectric response.

According to the standard tunneling model, whereas the asymmetry Δ is allowed to be as small as zero in the case of degenerate TS's, a cutoff in the tunneling parameter Δ_0 is introduced at very low energy scales. The justification for this cutoff is to prevent an unphysical divergence in the distribution P_0/Δ_0 when $\Delta_0 \to 0$. Typical values of this cutoff, $\Delta_{0,\min}/k_B$, used in calculations of TS response, are of order $10^{-5} - 10^{-8}$ K [for acoustic data, see Fig. 4.12 and (4.30), for example], though interpretations of thermal measurements (Lasjaunias et al. (1978);Strehlow and Dreyer (1994)) find better fits to data with values of the order of 1 mK.

Within this framework, one expects to find logarithmic temperature dependence of both the sound speed and capacitance down to temperatures approaching $\Delta_{0,\min}/k_B$. In practice, however, for several samples the capacitance is found to saturate at the lowest temperatures, deviating substantially from $\log(T)$ behavior (Nishiyama et al. (1992); Rogge et al. (1997c)). Nishiyama et al. observed a saturation of the dielectric response of SiO_2 with 1200 ppm OH^- impurities, whereas Rogge et al. saw similar behavior in potassium-doped SiO_2, BK7, and SiO_x. These deviations from logarithmic temperature dependence were independent of measuring field. We define the saturation temperature T_{sat} by extrapolating the saturated dielectric response (as T approaches 0) onto the logarithmic temperature variation obeyed between T_m and the saturation regime.

It is important to distinguish intrinsic dielectric saturation due to TS behavior from other effects, such as poor heat sinking of samples, that would lead to qualitatively similar behavior. Both groups immersed their samples in liquid ^{3}He to ensure good thermal contact between their samples and the nuclear refrigerant. Similarly, both groups saw the saturation temperature to be independent of both frequency and measuring field for fields less than 10^3

V/m. Since heating effects should depend strongly on measuring field amplitude and frequency, these observations rule out heating due to the measuring field as a possible source of the saturation. It is more difficult to rule out possible saturation due to rf noise (see Rogge et al. (1997c)), though a purely analog bridge, optoisolated from digital multimeters was used for the measurements.

The lack of measuring field dependence also suggests that the low-temperature saturation is not an effect of TS nonlinearities.

Figure 5.9 shows the saturation temperature as a function of temperature slope $S_{T\epsilon'}$ [defined in (5.30) for $T < T_m$]. Recall that $S_{T\epsilon'} \sim P_0 p^2$. It is suggestive that the dielectric saturation temperature appears to be roughly proportional to TS density for several materials.

One speculative explanation for this behavior is that interactions between TS's lead to a cutoff in the density of states below energy scales on the order of $k_B T_{\text{sat}}$. The existence of long-time logarithmic relaxations in these systems [hole-burning experiments, dc dielectric polarization seen by Höhler et al. (1991), and the nonequilibrium dielectric and acoustic measurements described in Sect. 5.2.2] seems to preclude the possibility that the saturation is due to a low-energy cutoff in the tunneling matrix element Δ_0. The possible existence of such a gap in the density of states of TS's at low energies due to interactions is not a new idea; see Klein et al. (1978), Yu and Leggett (1988) and Lasjaunias et al. (1978), for example. A particular type of gap in the TS density of states is described in detail in the next section. It is difficult to understand the abruptness of this saturation with this explanation,

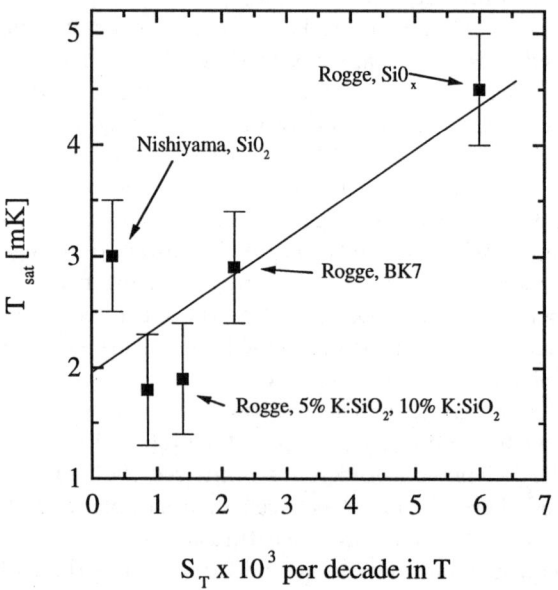

Fig. 5.9. Dielectric saturation temperature vs. low-temperature slope $S_{T,\epsilon'}$ (5.30). Data are from Nishiyama et al. (1992) and Rogge et al. (1997c).

however, since such gaps are smooth, continuous changes in the distribution of excitation energies.

Another possible explanation for the observed saturation is some sort of transition, a change in the TS dynamics that govern the dielectric response. Such a change, from the cooperative motion of pairs of TS's to motion of triplets, is discussed below, in Sect. 5.3.7.

5.2 Nonequilibrium Effects: Long-Time Relaxations and the Dipole Gap

We have discussed at some length the effects of weak dipolar TS interactions on the equilibrium dielectric and acoustic properties of amorphous insulators at low temperatures. Beginning with the virtual exchange of phonons we have approximated the elastic interaction between TS's as dipolar. The system of interacting TS's has an excitation spectrum that influences both dephasing rates (see Sect. 5.1.4) and longitudinal relaxation rates (See Sect. 5.1.5). It is a natural next step to consider what effects interactions between TS's may have on nonequilibrium behaviors.

A variety of nonequilibrium experiments have been performed on these materials at low temperatures. These include the acoustic hole burning and phonon/photon echo experiments mentioned in Sect. 5.0.2, as well as optical hole burning and dynamic heat release and heat-capacity measurements, discussed in Chaps. 6 and 2, respectively. Here, we choose to focus on a series of experiments performed recently at Stanford (Salvino et al. (1994),Rogge et al. (1996b),Rogge et al. (1997b)), and their theoretical interpretation in terms of the extension of the TS model to include weak elastic (or electric) dipolar TS-TS interactions.

We explain the experimental techniques and describe the resulting data, and then develop the theory of the nonequilibrium behavior of the interacting TS's. After comparing the theory's predictions with the experimental observations, we briefly describe some additional data, showing interesting hysteretic behavior, that are currently unexplained.

5.2.1 Nonequilibrium Experimental Techniques

The idea behind the Stanford experiments is a simple one: at a fixed temperature T and frequency $f = \omega/2\pi$, measure dielectric response while acting on the system with some external perturbation. In the first set of experiments (Salvino et al. (1994),Rogge et al. (1996b)), the perturbation was a large dc electric field applied across the sample. Figure 5.10 shows two possible methods of applying such a field bias to a capacitive sample. The output of a function generator capable of producing either dc or slowly (0.0001 Hz, for example) varying voltages was used as the input of a high voltage op-amp circuit. The large voltages with respect to ground thus produced were applied to one sample electrode through a charging resistor R_{ch}. The dc charging time of the capacitor was set by both sample capacitance C and capacitance to ground in the sample lead C_{gnd},

$$\tau_s = R_{ch}(C + C_{gnd}). \tag{5.113}$$

Typical values for the components are $C = 100$ pF, $C_{gnd} \simeq 1000$ pF, and $R_{ch} = 1$ MΩ, leading to a charging time of $\tau_s \simeq 40$ ms. Much shorter charging times were possible by using a relay to short R_{ch} with a small resistor (for example, 1 kΩ), leading to τ_s values in the microsecond regime. It was also possible to place an RC time constant between the function generator and the high voltage op-amp, if charging times on the order of seconds are required. To protect the lock-in amplifier during the application of the dc bias, a blocking capacitor was placed in series with the preamp input. For the fast charging experiments, a second relay was used to 'blank' the lock-in's input during the initial charging transient.

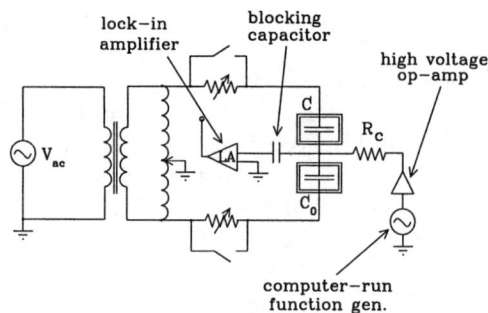

Fig. 5.10. Capacitance bridge set up for dc bias application.

In the second set of experiments (Rogge et al. (1997b)), a constant or slowly varying strain field ϵ_0 was applied to the sample. Figure 5.11 shows the method: the sample (commercial mylar film and BK7 glass were used) was mounted by its edges with epoxy (Stycast 1266) to pieces of silicon wafer, that were in turn affixed to piezoelectric plates (PZT8, 0.01" thick). Care was taken that ground planes shielded the sample from stray electric fields from the PZT that was biased by the same high voltage op-amp used for the electric field experiments. Again, a three-terminal bridge technique was used to measure sample capacitance and dissipation. In both the strain and electric field experiments, the samples were immersed in liquid ^{3}He with special heat exchangers for their leads (Rogge et al. (1997a)) to ensure good thermal sinking.

The overall elastic deformation of the sample under applied strain manifested itself as a temperature and frequency-independent capacitive response to bias strain, a $\delta C/C$ that scaled linearly with V_{PZT}. This purely geometric effect allowed the determination of how much strain was actually applied

5.2 Nonequilibrium Effects: Long-Time Relaxations and the Dipole Gap

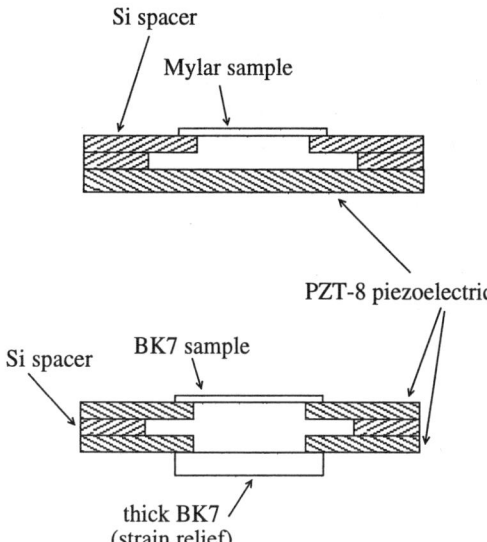

Fig. 5.11. Sample mounting scheme for strain application. After Rogge et al. (1997b).

to the sample, independent of any effects of epoxy joints. This geometric background, once characterized, was subtracted away, leaving behind true TS effects.

Finally, recent experiments have been performed that study the nonequilibrium acoustic response of glasses following their perturbation by a large dc electric field. Preliminary results of these investigations were reported by Osheroff et al. (1996).

The general experimental technique has been described above in Sect. 5.1.6. A glass reed was mounted by clamping one end between copper blocks. A small piece of ferromagnet was attached to the free end of the reed, and magnetic drive with capacitive sensing of reed position was used in a feedback circuit to make the reed self-oscillate at one of its natural frequencies. To enable the capacitive sense scheme, one side of the sample was coated with a ground electrode. Because of the magnetic drive technique, the other side of the sample could be coated with an electrode that was used to apply a bias electric field to the glass. The resonance frequency of the reed was observed during and following the application of the bias field. As described above, in the absence of electromechanical backgrounds, the fractional change in resonance frequency of the reed is essentially equal to the fractional change in the medium's sound speed. Similarly, by measuring the drive current necessary to maintain constant oscillator amplitude, the investigators measured the nonequilibrium response of the internal friction.

In both the dielectric and acoustic experiments, the nonequilibrium effects to be probed are $10^{-2} - 10^{-3}$, roughly the factor $\chi = P_0 U_0$, times smaller than the equilibrium TS temperature response. Thus signal to noise is of critical importance in all such experimental investigations. In particular, in the acoustic measurements the intrinsic width ($\propto Q^{-1}$) of the sample's resonance frequency is roughly two orders of magnitude larger than the size of the nonequilibrium contributions. Fortunately, averaging and the self-resonant feedback methods outlined previously make the sensitivity requirement achievable.

Further, in long-term experiments in particular, the experimental apparatus needs to have excellent long-term stability. Slow drifts in sample temperature or electronics performance can completely obscure tiny relaxations on the 10^4 second time scale. Again, in the acoustic measurements, where it is not possible to heat sink the sample by direct immersion in ^{3}He, this stability criterion is difficult to safeguard.

These three classes of experiments described above complement one another. In the first series, dielectric response after perturbation with an electric field, one is measuring a property due to those TS's with large electric dipoles in response to a perturbation that affects just those TS's. In the strain experiments, one again measures a property due to the TS's with large electric dipoles, but in this case the perturbation presumably affects a larger set of TS's, those with strong elastic couplings. Finally, by measuring sound speed in an amorphous insulator after perturbation with an electric field, one measures a property due to all elastically active TS's after perturbing only that subset that also have large couplings to electric fields. By comparing the nonequilibrium behaviors in these three sets of experiments with each other and with a theory developed to explain the results of the first set (see the following section), one can learn more about the distribution and dynamics of TS's in such materials.

To interpret the results of any of these experiments, it is important to consider the energy scale of such perturbations. As mentioned above, typical values of TS parameters are $\gamma \simeq 1$ eV, and $p \simeq 1$ Debye. Typical applied dc electric fields are on the order of $E_{\rm dc} = 1$ MV/m (100 V across a 100 μm thick sample), leading to an effective perturbation 'temperature' of

$$\frac{pE_{\rm dc}}{k_{\rm B}} \simeq 240 \text{ mK}. \tag{5.114}$$

Similarly, the largest strains that could be applied to the samples without detectable heating were $\epsilon_0 = 1.2 \times 10^{-5}$, and for a 1 eV TS-phonon coupling, this corresponds to an effective perturbation 'temperature' of 140 mK.

For each type of perturbation in the dielectric measurements, two experimental procedures were used, "jumps" and "sweeps". In a jump, the dielectric response was monitored as the perturbation was applied suddenly and maintained. For sweeps, the perturbation was ramped slowly over a broad range of values as a triangle wave.

In the acoustic measurements, electromechanical crosstalk between the biasing electrode and the sense electrode made the "sweep" experiments difficult to interpret. Large non-TS background frequency shifts were present.

5.2 Nonequilibrium Effects: Long-Time Relaxations and the Dipole Gap

These backgrounds were time and temperature independent, and so may be ignored if one is only interested in the system's nonequilibrium behavior.

5.2.2 Experimental Results

Figure 5.12 shows the ϵ' response of SiO$_x$ (a nonstoichiometric glass produced by reactive sputtering, with $x \approx 2.1$) following the application of a 10 MV/m dc electric field. At the moment of field application, ϵ' jumped up by a small amount $\delta\epsilon'(t = 0)$ and began to relax back downward. As shown in the inset, that relaxation was logarithmic in time. This behavior was observed in numerous samples, and may be characterized by both the size of the jump $\delta\epsilon'(0)$ and the slope of its relaxation, $S_{t\epsilon'} \equiv \partial(\delta\epsilon'/\epsilon')/\partial \log t$, with t measuring the time since dc bias application. Subsequent jumps to the same bias field were smaller and had relaxations with smaller slopes. This history dependence could be partially "annealed" away only by elevating sample temperature well above T_m for extended periods, and made obtaining precise $S_{t\epsilon'}$ values difficult.

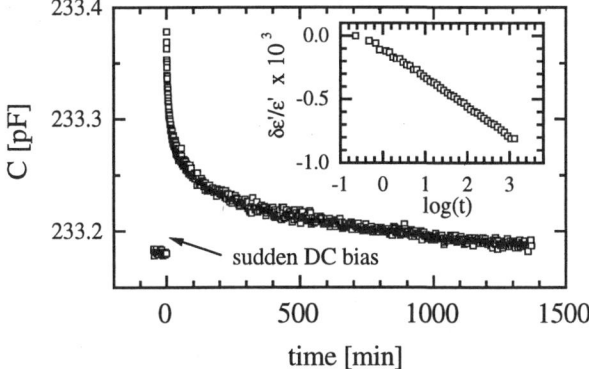

Fig. 5.12. Dielectric response at 1 kHz and 15 mK of a 1 μm thick SiO$_x$ sample to a 10 MV/m dc electric bias, after Salvino et al. (1994).

The dependence of this relaxation on temperature and measuring frequency is shown in Fig. 5.13 for two samples, the SiO$_x$ and a polymer photoresist. The data, taken at temperatures $T > T_m \geq 100$ mK, suggest that $S_{t\epsilon}$ is linear in $-\log(f)$ and $-\log(T)$. Error bars on the slope values are approximately 15 percent and are dominated by the history dependence discussed above.

Note that the nonequilibrium response does not directly scale with the equilibrium dielectric response. Rather than passing through a minimum at the crossover between relaxational and resonant regimes, $S_{t,\epsilon}$ monotonically

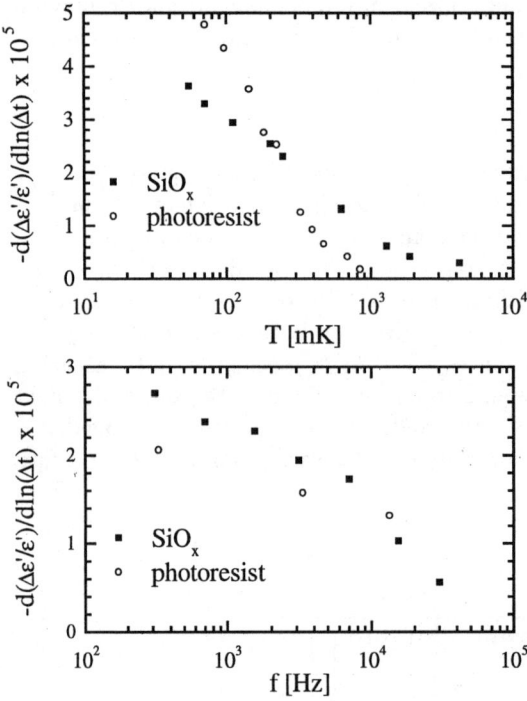

Fig. 5.13. Temperature dependence and frequency dependence of dielectric relaxation. Data in the upper graph were taken at a measuring frequency of 1 kHz. The SiO_x data in the lower graph is at 83 mK whereas the photoresist data is at 221 mK. After Salvino (1993).

decreased with rising temperature, even for temperatures well above T_m. This behavior is easily understood in the context of the model discussed below in Sect. 5.2.5.

Rogge et al. also studied the bias field dependence of the relaxation. To do so, at fixed temperature and frequency they performed a number of relaxations at different bias field amplitudes for 1000 s and fit the data to

$$\frac{\delta \epsilon'}{\epsilon'} = \delta \epsilon'(0) - S_{t,\epsilon'} \log_{10}(t), \tag{5.115}$$

where $t = 0$ is the time of the bias application. They found a constant value for $(\delta \epsilon'(0))/S_{t,\epsilon'}$ at all fields for a given sample. The size of the relaxation was symmetric in the applied field and could be written as

$$\frac{\delta \epsilon'}{\epsilon'} = f(|E_{dc}|) \left[\delta \epsilon'(0) - S_{t\epsilon'} \log(t - t_0) \right]. \tag{5.116}$$

They observed a quasilinear rise of the relaxation versus field for large fields and a "V"-shaped center section that was more pronounced at lower temperatures. Figure 5.14a shows the fit parameters versus field and Figure 5.14b the result of the fit versus field. The data were taken at one set of bias voltages, and then additional points were interleaved, explaining the corrugated appearance of the curves by history effects of neighboring relaxations. These effects are insignificant at longer times.

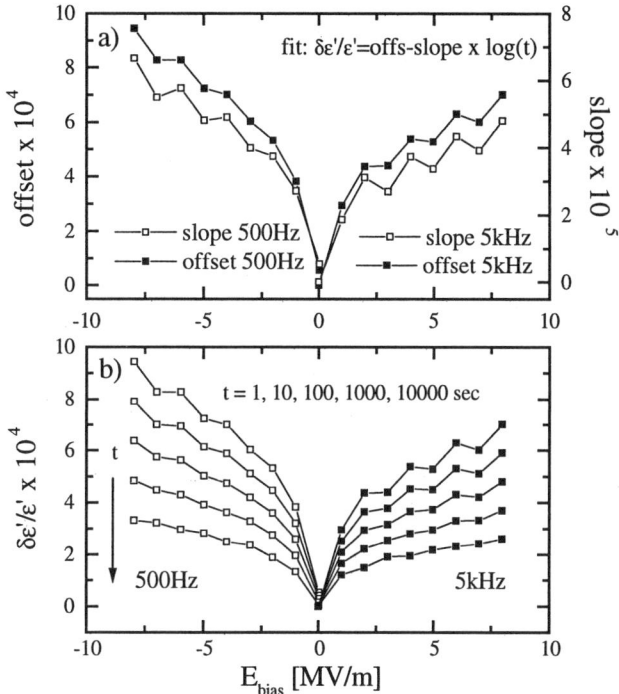

Fig. 5.14. Dielectric relaxation sizes and slopes for various E_{dc} for an SiO_x sample at 20 mK. The upper graph is a plot of parameters obtained by fitting logarithmic relaxations to $f(|E_{dc}|)(\delta\epsilon'(0) - S_{t\epsilon'}\log(t))$. The lower graph shows the resulting fits evaluated at specific values of $t - t_0$. After Rogge (1996).

By continuously varying the applied E_{dc} as a triangle wave with a period on the order of minutes, they traced out a curve of $\delta\epsilon'/\epsilon'$ vs. bias field very similar to that shown in Fig. 5.14. For example, consider the mylar data shown in Fig. 5.15. At higher temperatures, the curve looks more like a flat "V" and no longer shows the sharp center feature. Conceptually, this makes sense; in the limit that the continuous sweep experiment was performed quickly compared to the logarithmic relaxation seen in the jump experiments, one expects to see a map of the short-time $\delta\epsilon'/\epsilon'$ behavior as a function of bias field.

Figure 5.16 shows a bias sweep after a bias of 5 MV/m was applied for 1 hour for a photoresist sample. The sweep shows a new local minimum centered around 5 MV/m, and this minimum annealed itself back into the "background" in subsequent sweeps whereas the zero-bias "hole" regained its

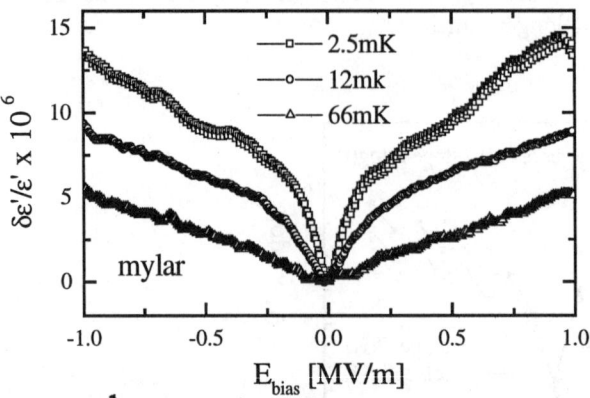

Fig. 5.15. Dielectric response to slowly sweeping electric bias field, after Rogge (1996).

original depth. Even if a bias is applied for 24 hours, the zero bias hole did not totally anneal out; rather, it decreased to about 2/3 of the equilibrium zero bias hole depth.

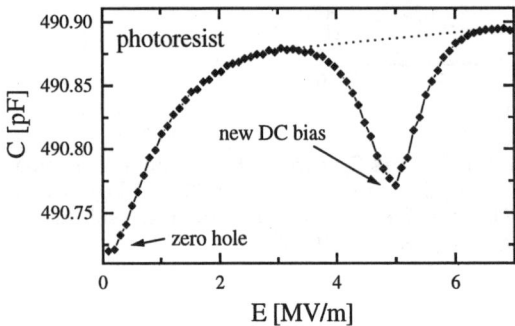

Fig. 5.16. "Hole" in dielectric response from sitting at fixed bias field, after Salvino (1993).

To probe the nature of this "V" feature around zero bias, Salvino et al. (1994) performed an experiment in which a sample was cooled in a nonzero bias field from 300 K down to below 1 K. In this case they observed a "hole" centered around the cooldown field instead of zero bias field. The "hole" would slowly anneal at low temperatures when the bias was set to a different value; in this way this cooldown field hole behaved exactly like the zero bias hole in the non-field-cooled sample. It seems reasonable to conclude that the only distinction between a new minimum created by perturbing the system and the "zero" hole is the fact that one was created at much higher temperatures, and cannot be completely annealed at low temperatures.

The schematic in Fig. 5.17 (Tigner et al. (1993)) links the creation of a new local minimum in $\epsilon'(\omega, T, E_{\rm dc})$ with the observed dielectric relaxation. The front plane shows dielectric response versus bias field (a sweep experiment) and the side plane shows dielectric response versus time (a relaxation experiment). When a bias field is applied at time $t = t_0$ the dielectric response suddenly increases and subsequently decays back into equilibrium (log relaxation). If we could do sweeps at the same time we would observe the creation of a new bias hole centered around the applied bias field. The drawing does not take into account that the zero bias hole fills in during this time.

An explanation for this sort of behavior may be found through an argument familiar from spin glasses. The capacitive measurements described above are most sensitive to those TS's whose local effective orienting fields (electric or elastic) are nearly zero; the ac measuring field dominates the dynamics of such TS's. Suppose interactions between TS's are important, and consider a ground state of the system in which a particular TS is in zero local electric field. If one of its interacting neighbors makes a transition, this may shift the energy splitting of the particular TS, causing a transition. Thus, the supposed ground state is unstable to perturbations. One would then expect the actual ground state of the system to have a diminished density of states near zero local electric field. The same would be true for elastic fields.

When an external perturbing field is applied, a new subset of TS's is moved into zero local field; thus the TS response jumps up until the hole described above has a chance to reestablish itself. Thermal broadening would smear the hole in question out to a width in energy of approximately $k_{\rm B}T$. The exact functional form of the bias field dependence and the numerical size of the response should depend on the nature of the TS interactions. A particular treatment employing dipolar elastic TS interactions, already discussed in Sects. 5.1.2 through 5.1.5 for the case of equilibrium measurements, is detailed below, and its predictions are found to agree well with experiment. The depressed response at zero local field will be referred to as the "dipole gap", in analogy to the Coulomb gap in semiconductors, whose origin is the Coulomb interaction between charge carriers (see Baranovskii et al. (1980)).

The frequency dependence of relaxations following the application of a dc electric field shows interesting temperature dependence beyond the behavior shown in Fig. 5.13. Figure 5.18 shows two data sets from Rogge et al. (1996b) on the SiO_x and mylar sample, one at 140 mK and the other at 20 mK and 5 mK using both 500 Hz and 5 kHz bridge excitations. The ratio of the slopes at the two different frequencies is very close to unity at the lower temperature (the two parallel lines). At the higher temperature the ratio of the slopes is about 1.4 for the SiO_x and 1.3 for mylar (the two intersecting lines). This behavior is common to all the samples examined by Rogge et al. (1996b), but the crossover temperature to the frequency-independent regime varies from sample to sample. In general, frequency-dependent relaxations after dc electric field application were seen when $T \gg T_{\rm m}$, and that frequency

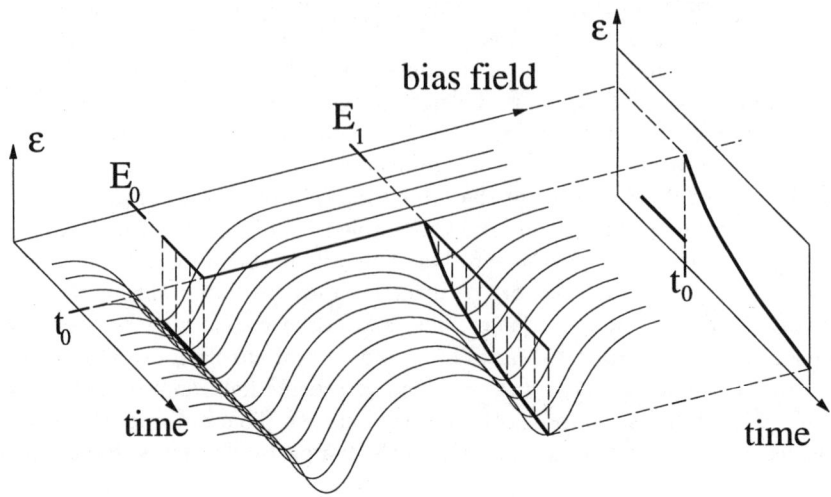

Fig. 5.17. The relationship between jump and sweep experimental results, from Tigner (1994).

dependence vanished for $T \ll T_{\mathrm{m}}$. Sample history did not markedly affect the slope ratios, and no ac excitation amplitude dependence was seen in any relaxation for all E_{ac} small enough to avoid heating the samples.

Rogge et al. observed no difference in relaxation slopes $S_{t\epsilon'}$ between jumps where the bias was applied in 3 μs and those in which the E_{dc} rise time was 2 s. Additionally, no deviation from logarithmic behavior was apparent up to 10^4 s after the perturbation, even in the case when (observation time)/(rise time) = 10^{10}.

Logarithmic relaxations were also reported in measurements of $\tan \delta \propto \epsilon''$. The fractional change in sample loss was up to 100 times larger than the nonequilibrium capacitive response.

We also consider dielectric response of a system following the rapid application of a dc strain field (Rogge et al. (1997b)). Figure 5.19 shows a logarithmic relaxation for the mylar sample up to 10^4 s. The capacitance instantaneously rose above the value that is due to the geometrical effect from straining the sample (as indicated by the dotted line) and the subsequent relaxation was logarithmic to within the noise. This behavior is analogous to that observed after a perturbation with an electric bias field. The ratio of the jump height to the log-slope of the relaxation $\delta\epsilon'(0)/S_{t\epsilon'}$ is the same for a perturbation with a strain field and an electric field. This suggests that the original dielectric relaxation is independent of the number of TS's perturbed, or that nearly all TS's in this material possess an appreciable electric dipole. The electric field sweeps also show interesting temperature dependence of the frequency dependence. Rogge et al. swept the dc electric field continuously

5.2 Nonequilibrium Effects: Long-Time Relaxations and the Dipole Gap 273

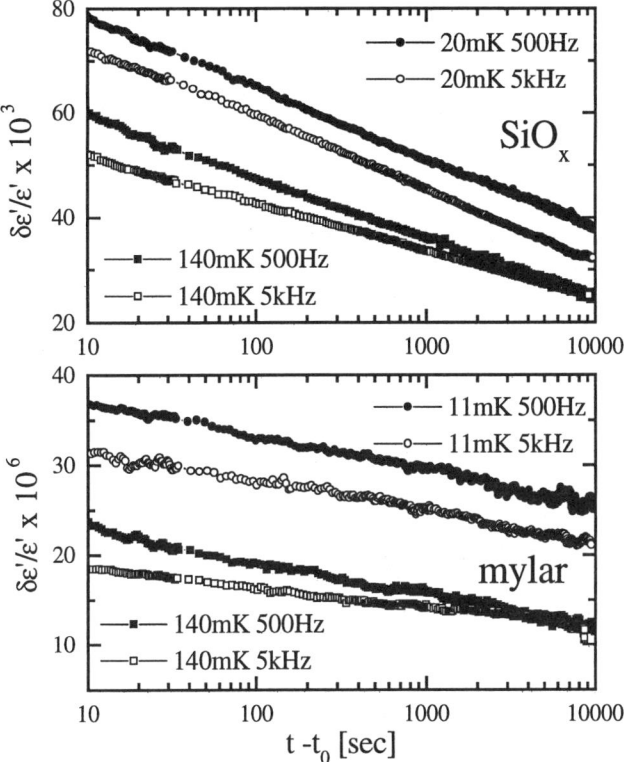

Fig. 5.18. Vanishing frequency dependence of relaxations at low temperatures. For each sample the lower pair of curves show relaxations after the application of a 3 MV/m bias field for $T \gg T_\mathrm{m}$, whereas the upper pair of curves show relaxations when $T \ll T_\mathrm{m}$. Well below T_m, the relaxation slope $S_{t\epsilon'}$ is independent of measuring frequency. From Rogge et al. (1996b).

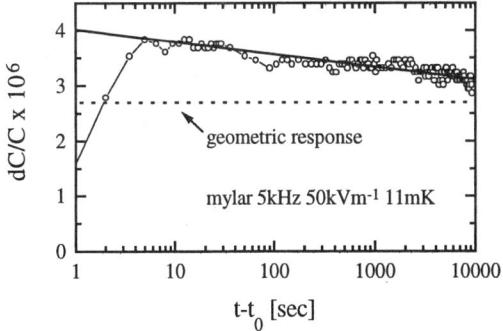

Fig. 5.19. Dielectric relaxation following application of strain field of $\Delta\ell/\ell = 2.7 \times 10^{-6}$, plotted as fractional change in capacitance.

using a triangular wave $(10\ \mathrm{kVm^{-1}s^{-1}})$, while measuring the capacitance. The change in $\epsilon'(\omega, \mathbf{E}_{\mathrm{dc}})$ with dc field is shown in Fig. 5.20 for the mylar sample. As described above, at low temperatures one can see a sharp hole in $\epsilon'(\mathbf{E})$ centered around zero bias field. The hole was ≈ 1 MV/m wide, similar to those reported by Salvino et al. (1994). This hole broadened and became shallower at higher temperatures, and is almost undetectable in the sweeps at 140 mK. At 140 mK the change in $\epsilon'(\omega)$ due to the bias field is 1.4 times larger at 500 Hz than at 5 kHz, whereas at low temperatures the traces were almost the same. This shows that although the sharp hole in $\epsilon'(\omega)$ is not visible at 140 mK, the same physics govern the broad minimum that still remains. Again, the change in $\epsilon'(\omega)$ due to the bias field was frequency dependent at high temperatures and not at low temperatures. This was true for all the samples examined, provided the low-temperature data were taken well below T_{m}.

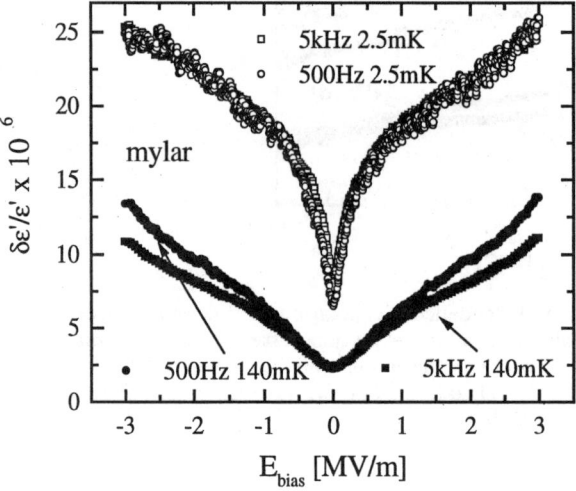

Fig. 5.20. Decreased frequency dependence of E_{dc} sweep response at low temperatures, after Rogge et al. (1996b).

Figure 5.21 shows strain field sweeps after subtracting the geometrical background. Again, as in the case of perturbation by an electric field, when the system is dominated by relaxational tunneling $(T > T_{\mathrm{m}})$, the response due to bias is frequency dependent, in contrast to the resonant tunneling regime $(T \ll T_{\mathrm{m}})$ where the frequency dependence of the bias effect vanishes. Figure 5.21 demonstrates this with the two overlying sweeps at 11 mK and the overall smaller but frequency-dependent sweeps at 140 mK.

The observed shape of the strain minimum in $\epsilon'(\omega, \Delta\ell/\ell)$ is different from that seen in the electrical field sweeps. In the strain sweeps there is a quadratic

5.2 Nonequilibrium Effects: Long-Time Relaxations and the Dipole Gap 275

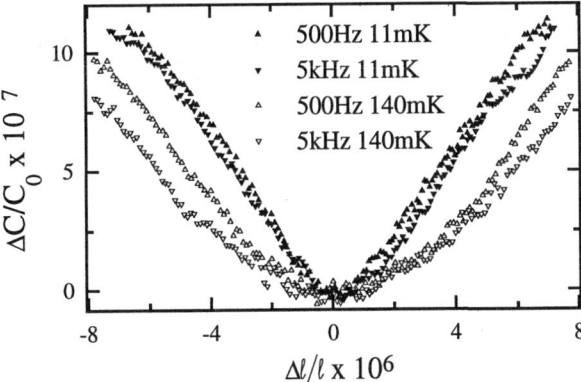

Fig. 5.21. Decreased frequency dependence of strain sweep response at low temperatures, plotted as fractional change in capacitance, of a mylar sample. After Rogge et al. (1997b).

shaped curve, whereas for the electric field sweeps $\epsilon'(\mathbf{E}, \omega)$ looked like a sharp cusp close to zero bias, with a broad "V"-shaped background. Rogge et al. performed sweeps at 12 mK up to 1.2×10^{-5} in relative strain, their maximum achievable ratio of perturbation energy compared to $k_B T$, and still saw no turnover into the broad background feature. This is consistent with the dc electric field perturbation data presented above if one considers the appropriate scaling for the x-axis, which is the energy shift due to the applied field.

For the mylar at 140 mK, the measured capacitive change due to strain is 13 times smaller at maximum strain field (8×10^{-6}) than at maximum electric field (3×10^6 Vm^{-1}). The data taken in the strain and in the electric field sweep experiments can be fit with the model discussed in Sect. 5.2.5 using one set of parameters: the electric dipole moment (1 Debye), the TS-phonon coupling (0.5 eV, consistent with the observed T_m), and the dimensionless interaction strength ($P_0 U_0$) multiplied by the measured equilibrium TS response (5×10^{-7}). The capacitive response is represented by this model very well, except for the 140 mK strain sweep, which is a factor of four larger than the prediction.

The logarithmic relaxations of $\epsilon'(\omega)$ as well as the temperature and frequency dependence of the strain field sweeps, especially the frequency independence at low temperature, are analogous to the behavior seen in the case of a perturbation with an electric field.

Jumps in TS response followed by decays that were logarithmic in time after the application of a large dc electric field bias were also seen in acoustic measurements. Recall that equilibrium reactive acoustic TS response is expected to differ from dielectric response only by multiplicative prefactors, one of which is -1. If nonequilibrium TS effects analogous to those seen above in dielectric behavior are present, they should manifest themselves as jumps downward in sound speed, followed by relaxations toward higher sound speeds.

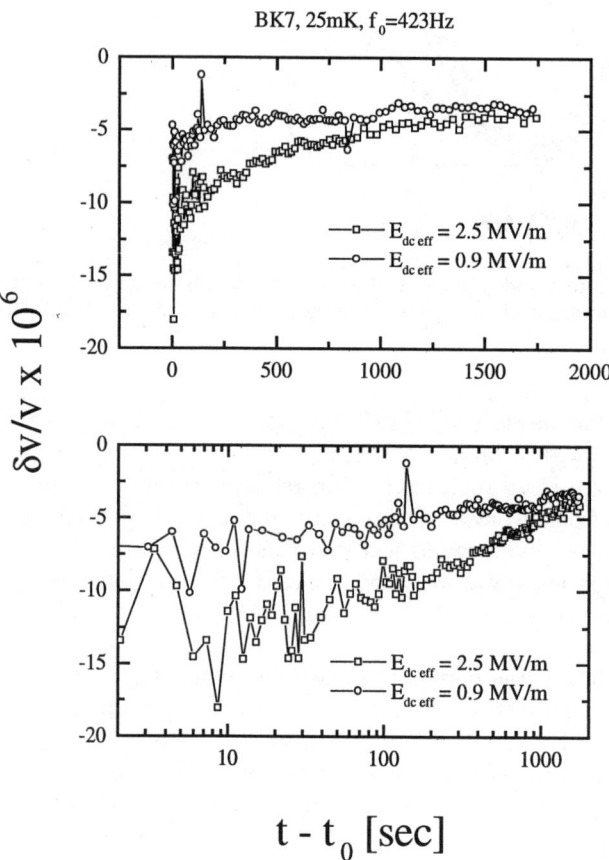

Fig. 5.22. Acoustic relaxations following application of large electric field in BK7 at 25 mK. Field values are effective dc fields. From Natelson et al., unpublished.

Figure 5.22 shows the results of two such jump experiments. The sample in question was a piece of BK7 glass 150 μm thick and approximately 2.5 cm in length. As in the dielectric measurements, the TS nonequilibrium response is logarithmic in time, and the response is larger at larger bias fields. Only the time-dependent portion of the bias-induced response is shown, and the quoted electric field values are effective fields due to electrode configuration.

The internal friction also exhibits a jump and logarithmic decay, as shown in Figure 5.23.

5.2 Nonequilibrium Effects: Long-Time Relaxations and the Dipole Gap   277

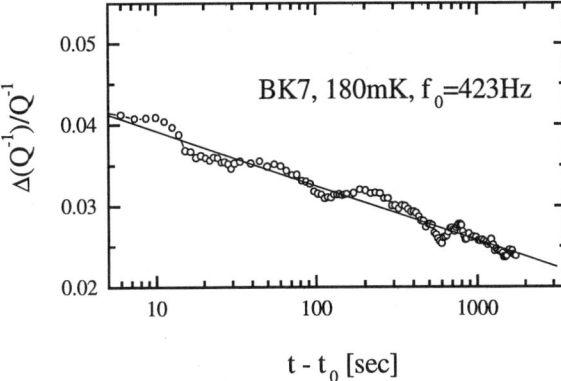

Fig. 5.23. 1/Q relaxation following the application of large electric field, from Natelson et al., unpublished.

Similar jumps and logarithmic decays in the dissipative component of susceptibility have been seen in spin glasses following the application of a large dc magnetic field (Fenimore and Weissman (1994), for example). Given all the above evidence, and the presence of such similar effects in systems known to have long-range interactions, we now move on to consider these data in the context of a theory of TS coupled to one another by dipolar interactions.

5.2.3 Nonequilibrium Behavior: General Remarks

When describing the nonequilibrium response of any quantum-mechanical system to a time-dependent perturbation, it is important to consider both the time scale over that the perturbation was applied and the time scale for the evolution of the unperturbed system. This is further complicated by finite-temperature effects; in addition to quantum mechanically coherent time evolution of the system in question, the system may be coupled to a thermal bath that can also drive relaxation processes.

First, consider a standard time-dependent perturbation theory problem: a given TS starting in a particular energy eigenstate has its energy splitting perturbed by an amount ϕ_0; the perturbation is applied with a rise time τ_s. The perturbation is considered to be *adiabatic* if the TS can adjust its wave function rapidly enough to follow the instantaneous energy eigenstate of the system; that is, under an adiabatic perturbation, a TS that starts in the ground state of the original Hamiltonian will end up in the ground state of the post-perturbation Hamiltonian. This is the familiar adiabatic approximation from undergraduate quantum mechanics. Similarly, the perturbation is considered to be *nonadiabatic* if the perturbation results in an appreciable probability of a transition of the TS to the other energy eigenstate; a TS that starts in the ground state that has been nonadiabatically perturbed has a non-negligible chance of ending up in the excited state of the post-perturbation Hamiltonian. This is equivalent to the "sudden" approximation.

The parameter in the TS Hamiltonian that effectively sets the pace at which the wave function can adjust is the tunneling matrix element Δ_0. For the parameters given above, there is a critical value of Δ_0 given by the Landau–Zener criterion (see Carruzzo et al. (1994), Burin (1995a), Shimshoni and Gefen (1991))

$$\Delta_{0c} \approx \sqrt{\frac{2\hbar\phi_0}{\tau_s \pi}}. \tag{5.117}$$

The subset of TS's with $\Delta_0 > \Delta_{0c}$ are perturbed adiabatically by ϕ_0.

The presence of a thermal bath leads to further complications by allowing TS's to change their state more rapidly than allowed by the pure quantum dynamics inherent in Δ_0. If thermal relaxation is rapid enough compared to the application time of the perturbation, then thermally driven dynamics determine the final state of the TS's, rather than the pure state perturbation theory described above. Since the relaxation rate τ_1 for a particular TS to the phonon bath is strongly temperature dependent [see (2.18)], the crossover to phonon-dominated dynamics also varies with T [see Burin (1995a) for more details]. A useful parameter to define is the crossover time τ_c, which is the value of τ_1 for TS's with energy splittings of $k_B T$ and tunneling matrix elements Δ_{0c}. Consider measuring the response of a system some time t after the application of the perturbation. At long observation times ($t \gg \tau_c$), the measured response is dominated by TS's that were perturbed nonadiabatically, whereas for short times ($t \ll \tau_c$) one sees primarily the response of adiabatic TS's.

5.2.4 Nonequilibrium Behavior Without Interactions Between Tunneling Systems

The dielectric response of an ensemble of noninteracting TS's to a perturbing dc field has been described in detail by both Carruzzo et al. (1994) and Burin (1995a). Here we outline the calculation and present the relevant results.

Beginning with the Hamiltonian for an individual TS, one can write, instantaneously following the application of a dc electric field, a new level asymmetry and energy splitting

$$\begin{aligned}\tilde{\Delta} &= \Delta + 2\,\boldsymbol{p} \cdot \boldsymbol{E}_{\text{dc}}, \\ \tilde{E} &= \sqrt{\tilde{\Delta}^2 + \Delta_0^2}.\end{aligned} \tag{5.118}$$

Since the energy splitting of each TS is altered by the dc field, the population difference between ground and excited states (that had been at its thermal equilibrium value prior to the perturbation) must adjust itself to

5.2 Nonequilibrium Effects: Long-Time Relaxations and the Dipole Gap

reestablish thermal populations. The population difference for a TS with new energy splitting $\tilde{E}$ then evolves according to the Bloch equations [see (5.13) and the following] like

$$\frac{\partial \Delta n}{\partial t} = -\frac{\Delta n - \tanh(\tilde{E}/(2k_B T))}{\tau_1}. \tag{5.119}$$

The issue of adiabatic or nonadiabatic perturbation affects the dielectric response by determining what initial conditions of the TS ensemble are used when solving (5.119). Knowing the time evolution of the TS population distribution, it is straightforward to compute the resulting dielectric response.

For noninteracting TS's that are perturbed *nonadiabatically* (= NINA case), one finds (Burin (1995a)) that the dielectric response $\delta\epsilon'_{\text{NINA}}(t)$ is

$$\delta\epsilon'_{\text{NINA}} \approx -\frac{\pi^{5/2}}{3\sqrt{2}} P_0 p^2 (\tau_{\min}/t)^{1/2} \left(1 - (k_B T/\phi_0) \right.$$
$$\left. \times \int_{-1}^{1} d(\cos\theta) \tanh(\phi_0(\cos\theta)/(2k_B T))/\cos\theta \right), \tag{5.120}$$

where the average over dipole direction has been left unevaluated. Here, $\tau_{\min}$ is defined to be τ_1 when $\Delta = 0$ and $\Delta_0 = k_B T$.

When the perturbation is adiabatic (= NIA case) and $\phi_0 > k_B T$, the predicted dielectric response is (Burin (1995a))

$$\delta\epsilon'_{\text{NIA}} = \frac{4\pi P_0 p^2}{3} \ln(\tau_{\max}/t). \tag{5.121}$$

$\tau_{\max}$ is τ_1 for TS's with energy splitting $= k_B T$ and the minimum tunneling matrix element, $\Delta_0 = \Delta_{0,\min}$. Note that the result (5.121) is frequency independent.

The predicted nonequilibrium dissipation (ϵ'') from such noninteracting TS's is not logarithmic in time (see Carruzzo et al. (1994)).

5.2.5 Weak Interactions: The Dipole Gap

To analyze the interaction-induced nonequilibrium response of the TS's, we begin by considering the effect of TS interactions on the TS density of states. The interaction effects lead directly to a change of the TS density of states caused by the application of an external dc field.

Consider an ensemble of weakly interacting TS's distributed according to the phenomenological law (Anderson et al. (1972), Phillips (1972)) [see (2.20)]

$$P(\Delta, \Delta_0) = \frac{P_0}{\Delta_0} \Theta(\Delta_0 - \Delta_{0,\min}),$$
$$\frac{\Delta_{0,\min}}{k_B} \simeq 10^{-7} K. \tag{5.122}$$

Both level asymmetries and tunneling amplitudes are effectively restricted to energies that are less than some crossover energy W because at larger energies, other excitations come into play.

The distribution of TS's (5.122) in terms of their energies and tunneling amplitudes can be written as

$$P(E, \Delta_0) = \frac{P_0}{\Delta_0} \frac{E}{\sqrt{E^2 - \Delta_0^2}} \Theta(E - \Delta_0) \Theta(\Delta_0 - \Delta_{0,\min}). \tag{5.123}$$

The Hamiltonian of the model reads [see (5.47) and (5.65)]:

$$\begin{aligned}
\widehat{H} &= \widehat{H_0} + \widehat{U}, \\
\widehat{H_0} &= -\sum_i (\Delta_i S_i^z + \Delta_{0i} S_i^x), \\
\widehat{U} &= -\frac{1}{2} \sum_{ij} U_{ij} S_i^z S_j^z.
\end{aligned} \tag{5.124}$$

The summation is taken over all TS's i. The operators S_i^z, S_i^x are the standard spin-1/2 operators, which are, respectively diagonal and off-diagonal on the quantum states of each TS' different wells. This is precisely the generalization of our earlier equation (5.44). According to Hunklinger and Raychaudhuri (1986) (see also Burin (1991a)) Δ is much more sensitive to the perturbations (atomic vibrations or fluctuations of the electric field) than the tunneling amplitude. That is why the only terms retained in the interaction Hamiltonian (5.124) are those with the products $S_i^z S_j^z$.

As discussed in Sect. 5.1.2, we treat the interaction between TS's as scaling with distance as $1/R^3$ [see Hunklinger and Raychaudhuri (1986), Yu and Leggett (1988), and (5.46)]. For dielectric glasses, our system of interest, this interaction is usually produced by the strain or electric dipole interaction (see Hunklinger and Raychaudhuri (1986), Tornow et al. (1994)). Recall that in both cases the angular average value of the interaction constant is zero. For simplicity we assume [see (5.51)]

$$U_{ij} = \pm U_0/R^3, \tag{5.125}$$

with equal probabilities for positive or negative sign of the interaction energy (see Sect. 5.1.2). As in that section, we assume that the dimensionless constant $\chi = P_0 U_0$ is always much smaller than unity [see (5.74)] for all known glasses. The small value of this product means that the interaction may be treated as a weak perturbation of the noninteracting TS Hamiltonian ($\widehat{H_0}$).

One should note that the excitation energies E_i should be defined taking into account the interaction between TS:

$$\begin{aligned}
E_i &= \sqrt{\Delta_i'^2 + \Delta_{0i}^2}, \\
\Delta_i' &= \Delta_i + \sum_{ij} U_{ij} S_j^z.
\end{aligned} \tag{5.126}$$

5.2 Nonequilibrium Effects: Long-Time Relaxations and the Dipole Gap

Let us discuss the influence of the interaction on the behavior of the density of states (5.123). We shall use the same stability arguments proposed by Baranovskii et al. (1980) for Coulomb gap theory.

The low-energy state of the system should be stable under perturbations with large energies. For single TS excitations of the type described by (5.124), this means that if the excitation energy E_i of TS i is larger than the thermal energy $k_B T$, then the probability that this TS is in its excited state is negligibly small. A similar condition should be satisfied for two-particle excitations. However, as we have seen in Sect. 5.1.4, the energy of a two-particle excitation (for two TS's i and j) differs from the sum of the energies of two isolated defects. The two-TS excitation energy can be approximately written as (Baranovskii et al. (1980), Burin (1991b))

$$E_{ij} \approx E_i + E_j - U_{ij}. \tag{5.127}$$

The existence of a "coupling" energy U_{ij} leads to a decrease in the density of low-energy excitations and even to the formation of a pseudogap (see Baranovskii et al. (1980), Efros (1985)).

Consider the TS density of states at low temperatures, with the energy of the ground state of the TS ensemble set at zero. For zero temperature, all two particle [see (5.127)] (and more complex) excitations must have positive energies. To enforce this restriction, the density P of such excitations in the pair approximation picks up the additional factor (see Baranovskii et al. (1980), Efros (1985))

$$P_i - P \prod_j \Theta(E_i + E_j - U_{ij}). \tag{5.128}$$

For finite temperature T, the product in (5.128) approximately contains only pairs i,j where the interaction exceeds the characteristic thermal energy,

$$U_{ij} > k_B T. \tag{5.129}$$

If the interaction is sufficiently weak [see (5.74)], then it is valid to keep only the correction to the density of excitations that is first order in $P_0 U_0$, originating from the product of step functions

$$\prod_j \Theta(E_i + E_j - U_{ij}) \Theta(U_{ij} - k_B T). \tag{5.130}$$

This factor ensures the decrease of the density of states in accordance with the theory of the dipole gap (Baranovskii et al. (1980)). After averaging (5.130) over the positions of TS j, one finds a decrease of the density of states of single-particle excitations

$$\begin{aligned} P'(E, \Delta_0) &\approx \frac{P_0}{\Delta_0} \frac{E}{\sqrt{E^2 - \Delta_0^2}} \Theta(E - \Delta_0) \Theta(\Delta_0 - \Delta_{0,\min}) \\ &\times (1 - \frac{P_0}{2} \int d\mathbf{R} \int_{\Delta_{0,\min}}^{W} \frac{d\Delta_0'}{\Delta_0'} \end{aligned}$$

$$\times \int_{\Delta_0'}^{W} dE' \Theta(-E - E' + U_0/R^3)\Theta(U_0/R^3 - k_B T)). \quad (5.131)$$

Equation (5.131) only accounts for the half of the neighbors with positive coupling energies U_0. This leads to a factor $\frac{1}{2}$ in the corrections to the density of states given by (5.131) [if $U_{ij} < 0$, then the argument of Θ-function in (5.130) is always positive]. A factor $E'/\sqrt{E'^2 - \Delta_0'^2}$ is omitted in (5.131) because the relevant tunneling amplitudes Δ_0' are much smaller than the relevant energy $E' \geq k_B T$.

After straightforward calculations, we obtain, with logarithmic accuracy,

$$P'(E, \Delta_0) \approx \frac{P_0}{\Delta_0} \frac{E}{\sqrt{E^2 - \Delta_0^2}} \Theta(E - \Delta_0)$$
$$\times \left(1 - \frac{2\pi}{3} P_0 U_0 \log\left(\frac{W}{E + k_B T}\right) \log\left(\frac{W}{\Delta_{0,\min}}\right)\right). \quad (5.132)$$

The additional contribution of order $\log(W/(k_B T))/\log(W/\Delta_{0,\min}) \approx 10^{-1}$ is neglected here. In other words, the formation of the pair excitations described in Sect. 5.1.4 leads to a depletion of the density of states of single-TS excitations, shown in (5.132). We will see that large energetic perturbations can disrupt the TS density of states in a way that leads to the observed nonequilibrium behaviors.

Consider the effect of the application of an external dc electric field $\boldsymbol{E}_{\rm dc}$. After this event occurs each TS will have its energy splitting randomly shifted by an amount within the scale ϕ that can be estimated as

$$\phi \simeq 2\boldsymbol{p} \cdot \boldsymbol{E}_{\rm dc}. \quad (5.133)$$

Here, $\boldsymbol{p}$ is the dipole moment of a typical TS. We assume for simplicity that the dipole moments of different TS's are oriented at random, and that all such dipoles have the same magnitude p. The TS energies are altered on the characteristic scale $\phi_0 = 2pE_{\rm dc}$.

The dc field, by shifting the energies of the TS's, changes those TS's that most strongly affect the stability condition given in (5.128). In the nonadiabatic regime (= INA case), TS's whose preperturbation energy splittings were $\leq \phi_0$ will now have to readjust their configurations. Therefore, the density of states (5.132) changes. Its variation can be described, with logarithmic accuracy, by the introduction of a new effective temperature

$$T_\phi \approx T + \frac{\phi_0}{k_B} \quad (5.134)$$

into (5.132). The redefined density of states P'' reads:

$$P'' \approx P' - \frac{2\pi}{3} \frac{P_0}{\Delta_0} \frac{E}{\sqrt{E^2 - \Delta_0^2}} P_0 U_0 \log\left(\frac{E + k_B T + \phi_0}{E + k_B T}\right)$$
$$\times \log\left(\frac{W}{\Delta_{0,\min}}\right),$$
$$E < \phi_0. \quad (5.135)$$

5.2 Nonequilibrium Effects: Long-Time Relaxations and the Dipole Gap

Consider the behavior in time of the density of states after DC field application at $t = 0$. According to (5.131) the hole in the density of states at low energies comes from certain specific pairs of TS's. For those pairs with pair excitation energies of order or less than ϕ_0 the conditions for stability (5.128) can be destroyed at the moment $t = 0$. For times $t > 0$ the pairs with relaxation times less than t will return to their equilibrium states. Therefore, the hole in the density of states will gradually be restored.

To estimate the density of states at the time t, one should exclude in (5.131) those pairs with energies of order or less than ϕ_0 and with relaxation times τ_p longer than t:

$$P(E, \Delta_0; t) \approx P'(E, \Delta_0; +\infty)$$
$$+ \frac{P_0}{\Delta_0} \frac{E}{\sqrt{E^2 - \Delta_0^2}} \frac{P_0}{2} \int d\mathbf{R} \int_{\Delta_{0,\min}}^{W} \frac{d\Delta_0'}{\Delta_0'} \int_{\Delta_0'}^{W} dE'$$
$$\times \Theta(-E - E' + U_0/R^3)\Theta(\phi_0 + E + E' - U_0/R^3)$$
$$\times \Theta(U_0/R^3 - k_B T)\Theta(\tau_p - t), \qquad (5.136)$$

where $P'(E, \Delta_0; +\infty)$ is the equilibrium density of states from (5.132), corresponding to an infinitely long time after the perturbation.

The relaxation time τ_p for a pair of TS's with energy splittings E_a, E_b and tunneling amplitudes Δ_{0a}, Δ_{0b} can be crudely estimated as a maximum relaxation time of these two TS's,

$$\tau_p \simeq \max(\tau_a, \tau_b). \qquad (5.137)$$

Substituting the definition of the pair relaxation time (5.137) into the equation for the time-dependent density of states (5.136) we obtain the logarithmic behavior of the density of states at intermediate times:

$$P(E, \Delta_0; t) \approx P(E, \Delta_0; +\infty) + \frac{\pi}{3} \frac{P_0}{\Delta_0} \frac{E}{\sqrt{E^2 - \Delta_0^2}} (P_0 U_0)$$
$$\times \log\left(\frac{\phi_0 + k_B T + E}{E + k_B T}\right) \log\left(\frac{t}{\tau_{\max}}\right);$$
$$E < \phi_0, \quad \tau_1 < t < \tau_{\max},$$
$$\tau_{\max} \simeq \frac{1}{A\Delta_{0,\min}^2 k_B T},$$
$$\tau_1 = \frac{1}{A\Delta_0^2 E \coth(E/(2k_B T))}, \qquad (5.138)$$

where we have repeated the definition of the parameter A, (5.4). If t becomes larger than $\tau_{\max}$, then the system returns to its equilibrium state.

The logarithmic relaxation law for the density of states, and consequently for the dielectric susceptibility (see below), is the main result to be compared with the experiments described above. It is evident that this behavior is rooted in the distribution of relaxation times of pairs, (5.137). To obtain this time evolution law we need only the trivial fact that the relaxation time of a pair is inversely proportional to the square of the minimum tunneling amplitude (Δ_0^2). Therefore our crude approximation for the relaxation time of pairs in (5.137) is sufficiently accurate.

To describe the time evolution of the TS density of states at very low temperatures $T < T' \leq 100$ mK [see (5.102)], one should substitute the relaxation time defined by (5.97) into (5.136). It can be shown that the time-dependent density of states takes the form in (5.138) with redefined minimum and maximum times:

$$\tau_{\min} \approx \frac{E^2}{\Delta_0^2 \zeta T}, \quad \tau_{\max} \simeq \frac{k_B^2 T}{\zeta \Delta_{0,\min}^2}, \tag{5.139}$$

where

$$\zeta \approx 10 \frac{k_B}{\hbar} (P_0 U_0)^3. \tag{5.140}$$

The new $\tau_{\min}$ replaces τ_1 in (5.138).

Consider the influence of the interaction between TS's on the dielectric response. Since the interaction is weak, the effects resulting from it must be small because of the small parameter $P_0 U_0$ of (5.74). That is why the corrections to the (already fairly small) noninteracting adiabatic nonequilibrium response (NIA above, jump up and subsequent logarithmic relaxation) cannot be important.

In fact, the predicted NIA response is never observed experimentally, even with a wide variety of perturbation switching times. It has been pointed out (see Burin (1996)) that when TS relaxations are driven by interactions [see (5.97)] the rethermalization predicted in (5.119) leads to a dielectric response that varies as $1/t$ rather than $\log t$. This rapid t dependence of the NIA response due to non-phonon-driven TS relaxation would lead to negligible NIA contribution to the dielectric measurements on experimental time scales.

For temperatures large enough that $T > T'$ the relaxation of thermal TS's is caused by their interaction with phonons. This is the case for the experiment of Salvino et al. (1994). The generalization of subsequent results for $T < T'$ is straightforward.

We restrict our investigation to the interacting nonadiabatic (INA) regime. This means that we consider relatively large times $t > \tau_c$.

Following the paper by Baranovskii et al. (1980), we restrict the investigation to the part of dielectric response caused by the nonequilibrium behavior of the density of states; we neglect for the moment the rethermalization of TS populations discussed above. For experimental times $t \gg \tau_{\min}$ the equilibrium value for the population difference of the relevant TS's ($\langle \tilde{S}_i^z \rangle = (1/2) \tanh(E/(2k_B T))$) can be assumed. In this approximation the contribution of the TS subsystem to the dielectric susceptibility reads [see (5.131) and (5.138)]:

$$\delta \epsilon(t) = \frac{4\pi p^2}{3} \int_{\Delta_{0,\min}}^{W} d\Delta_0 \int_{\Delta_0}^{W} dE (P(E, \Delta_0, t) - P(E, \Delta_0, +\infty))$$

5.2 Nonequilibrium Effects: Long-Time Relaxations and the Dipole Gap 285

$$\times \left[\frac{2\Delta_0^2}{E^3} \tanh\left(\frac{E}{2k_BT}\right) \right.$$
$$\left. + \frac{(E^2 - \Delta_0^2)}{k_BTE^2\cosh^2(E/2k_BT)(1 - i\omega\tau_1)} \right]. \tag{5.141}$$

In the zeroth order approximation the density of states of TS's can be considered in its noninteracting form, (5.123). Then we obtain the standard result for the TS contribution to the dielectric constant ($\delta\epsilon_0 = \delta\epsilon_0' + i\epsilon_0''$)

$$\delta\epsilon_0' = \frac{4\pi P_0 p^2}{3}\log(W^2 k_B T A/\omega), \quad \epsilon_0'' = \frac{2\pi^2 P_0 p^2}{3},$$
$$T > T_m. \tag{5.142}$$

The deviation of the system from its equilibrium state, caused by the external DC field application, leads to an instantaneous jump in the density of states at low energies, given by (5.135). To find the corresponding increase in the dielectric susceptibility, one should substitute (5.135) into the definition (5.141). After some calculations, we obtain, with logarithmic accuracy,

$$\delta\epsilon_{\text{jump}}' = \frac{8\pi^2}{9}(P_0 p^2)(P_0 U_0)\log(\phi_0/(k_B T))$$
$$\times \log(A(k_B T)^2\phi_0/\omega)\log(\phi_0/\Delta_{0,\min}),$$
$$\delta\epsilon_{\text{jump}}'' = \frac{4\pi^3}{9}(P_0 p^2)(P_0 U_0)\log(\phi_0/(k_B T))\log(\phi_0/\Delta_{0,\min}),$$
$$T > T_m. \tag{5.143}$$

Consider the relaxation of the dielectric constant after its instantaneous increase. Equation (5.138) for the time-dependent density of states can be used to calculate it. Substituting (5.138) into the definition (5.141), we find the logarithmic relaxation of the dielectric susceptibility

$$\delta\epsilon'(t) = \frac{4\pi^2}{9}(P_0 p^2)(P_0 U_0)\log(\phi/(k_B T))$$
$$\times \log(A(k_B T)^2\phi/\omega)\log(\tau_{\max}/t),$$
$$\delta\epsilon''(t) = \frac{2\pi^3}{9}(P_0 p^2)(P_0 U_0)\log(\phi/(k_B T))\log(\tau_{\max}/t),$$
$$\tau_{1T} < t < \tau_{\max}, \quad T > T_m, \tag{5.144}$$

where the maximum time $\tau_{\max}$ hase been defined in (5.138), whereas the minimum relaxation time $\tau_{\min}$ has been replaced with logarithmic accuracy by that of thermal TS's, $\tau_{1T} = 1/(A(k_B T)^3)$ [see (5.59)].

To analyze the experimental data we will need the logarithmic relaxation rates of the real and imaginary parts of the dielectric constant

$$S_{t,\epsilon'} = -\frac{\partial(\delta\epsilon'(t))}{\epsilon\partial\log(t)}$$
$$= \frac{4\pi^2}{9}\frac{(P_0 p^2)}{\epsilon}(P_0 U_0)\log(\phi/(k_B T))\log(A(k_B T)^2\phi/\omega),$$

$$S_{t,\epsilon''} = -\frac{\partial(\delta\epsilon''(t))}{\partial \log(t)}$$
$$= \frac{2\pi^3}{9}(P_0 p^2)(P_0 U_0)\log(\phi/(k_B T)). \tag{5.145}$$

The nonequilibrium dielectric response [(5.143), (5.144)] is in qualitative agreement with the experimental data described in Sect. 5.2.2. The dielectric susceptibility jumps up just after the dc field switches on and then decreases logarithmically to its equilibrium value. This theory explains such behavior for both the real and imaginary parts of the dielectric susceptibility. The theory for noninteracting adiabatic response does not predict a logarithmic relaxation for ϵ'' (see Carruzzo et al. (1994)).

The behavior of the INA logarithmic relaxation rate $S_{t,\epsilon'}$ (5.145) with ac field frequency and temperature is also in agreement with the experiment. Both the frequency and temperature dependences of $S_{t\epsilon'}$ are nearly logarithmic for $T > T_m$. The relaxation rate increases with decreasing frequency, as in (5.145). The rate also increases with decreasing temperature, at least for $T \gg T_m$ (see Fig. 5.13). One should note that the logarithmic relaxation rate for the noninteracting adiabatic case, (5.121) is always insensitive to frequency and temperature in the limit of large dc field. Thus the observed frequency dependence rules out the NIA response as an explanation for the data.

For the limit of very low temperatures, $T \ll T_m$, the increase in the characteristic relaxation time with decreasing temperature leads to the suppression of the relaxation contribution. This suppression is independent of the relaxation mechanism at low enough temperatures. The jump in the nonequilibrium dielectric susceptibility and its relaxation rate are defined only by the resonant contributions. Keeping only this contribution in (5.141), we obtain new relations for $\delta\epsilon_{\text{jump}}$ and $S_{t,\epsilon'}$:

$$\delta\epsilon'_{\text{jump}} = \frac{8\pi^2}{9}(P_0 p^2)(P_0 U_0)(\log(\phi/(k_B T)))^2 \log(\phi_0/\Delta_{0,\min}),$$
$$S_{T,\epsilon'} = \frac{4\pi^2}{9}\frac{(P_0 p^2)}{\epsilon}(P_0 U_0)(\log(\phi/(k_B T)))^2,$$
$$T \ll T_m. \tag{5.146}$$

Thus at very low temperatures we expect the logarithmic relaxation rate to be independent of frequency. This prediction has been confirmed in Rogge et al. (1996b), as shown above in Figure 5.18.

One may also evaluate the dielectric relaxation due to the dipolar gap for small perturbation fields, $\phi < k_B T$ (Burin (1995b)). For temperatures below T_m the relaxation is dominated by the resonant contribution

$$\delta\epsilon'_{\text{res}}(t) = -\frac{4\pi^2}{9\epsilon_0}P_0 p_0^2 P_0 U_0 \log^2\left[1 + \eta_1\frac{\phi_0}{k_B T}\right]\log\frac{\tau_{\max}}{t}, \tag{5.147}$$

for $\phi_0/k_B < T < T_m$, and the relaxational contribution is given by

5.2 Nonequilibrium Effects: Long-Time Relaxations and the Dipole Gap

$$\delta\epsilon'_{\rm rel}(t) = -\frac{2\pi^2}{9\epsilon_0} P_0 p_0^2 P_0 U_0 \log \frac{A(k_{\rm B}T)^3}{\omega} \log\left[1 + \eta_2 \left(\frac{\phi_0}{k_{\rm B}T}\right)^2\right]$$
$$\times \log \frac{\tau_{\max}}{t} \tag{5.148}$$

for $\phi_0/k_{\rm B} < T, T > T_{\rm m}$.

The two parameters η_1 and η_2 are of order unity. At temperatures around the temperature of the minimum, the relaxation consists of both contributions; the resonant and the relaxational ones. In this temperature range one uses the superposition of the responses in (5.147) and (5.148).

The dipole gap model also provides a framework in which to analyze the nonequilibrium sound speed data shown in Fig. 5.22. By comparing the observed nonequilibrium acoustic behavior with analogous nonequilibrium dielectric relaxations on the same materials, one can learn about the TS distribution of strain couplings. Assuming that f is the fraction of active TS's that have large electric dipoles, recall that the equilibrium temperature slopes scale like

$$S_{T,\epsilon'} = \frac{\partial \epsilon'}{\epsilon' \partial \log_{10}(T)} \approx f P_0 \langle p^2 \rangle,$$
$$S_{T,v} = \frac{\partial v}{v \partial \log_{10}(T)} \approx P_0 \langle \gamma^2 \rangle. \tag{5.149}$$

Also, for fixed temperature, frequency, and size of perturbation, the slopes of the logarithmic relaxations vary like

$$S_{t,\epsilon'} = \frac{\partial \epsilon'}{\epsilon' \partial \log_{10}(t)} \approx f P_0 \langle p^2 (P_0 U_{12}) \rangle,$$
$$S_{t,v} = \frac{\partial v}{v \partial \log_{10}(t)} \approx f P_0 \langle \gamma^2 P_0 U_{12} \rangle, \tag{5.150}$$

where the brackets $\langle \rangle$ denote an average over the TS distribution. If the TS interactions are dominated by elastic couplings, then we find that $\langle \gamma^2 U_{12} \rangle \sim \langle \gamma^3 \rangle \langle \gamma \rangle$. If one scales the nonequilibrium dielectric response by its equilibrium counterpart, and one does the same for the acoustic expressions, one arrives at

$$\frac{S_{t,\epsilon'}}{S_{T,\epsilon'}} \approx \langle \gamma \rangle,$$
$$\frac{S_{t,v}}{S_{T,v}} \approx f \frac{\langle \gamma^3 \rangle \langle \gamma \rangle}{\langle \gamma^2 \rangle}. \tag{5.151}$$

Making a ratio of ratios, one finds

$$\frac{S_{t,v}/S_{T,v}}{S_{t,\epsilon'}/S_{T,\epsilon'}} = \frac{f \langle \gamma^3 \rangle}{\langle \gamma^2 \rangle \langle \gamma \rangle}. \tag{5.152}$$

Thus, by comparing the sizes of acoustic and dielectric nonequilibrium effects (at fixed temperature, frequency, and perturbation size), one can extract information about both the fraction of TS's with large electric dipoles and about the distribution of TS strain couplings in a particular sample.

5.2.6 Discussion of the Experiments

There is quantitative agreement between the weakly interacting model theoretical predictions for the nonequilibrium dielectric response in adiabatic and nonadiabatic regimes and the experimental data of Salvino et al. (1994) and Rogge et al. (1996b) (some of which is reproduced in Sect. 5.2.2). Here we concentrate on the most investigated system, SiO_x.

It is important to use the correct quantities for the parameters P_0, p, U_0 in the calculations. We cannot use the parameters that are known for SiO_2 from other experiments to discuss the dielectric properties of SiO_x because the dielectric properties of these two materials are not similar at the experimental temperatures, despite the related chemical compositions. The slope of the temperature dependence of the equilibrium dielectric susceptibility $S_{T,\epsilon'}$, from (5.32), in SiO_x is 10 times larger than this parameter in SiO_2. The logarithmic relaxation rate $S_{t,\epsilon'}$ in SiO_x exceeds the same parameter in SiO_2 (see Salvino et al. (1994)) by a factor of 10^2.

Therefore, to analyze the nonequilibrium data in SiO_x it seems reasonable to use the results for the equilibrium dielectric response obtained in the same set of experiments. These are the low-temperature slope of the equilibrium temperature-dependent dielectric susceptibility $S_{T,\epsilon'} \sim 2.5 \times 10^{-3}$ and the imaginary part of the dielectric susceptibility $\epsilon'' \approx 10^{-2}$. Using (5.142) one can estimate the background dielectric susceptibility ϵ and the important dimensionless parameter $P_0 p^2$ as

$$\epsilon' \approx 5,$$
$$P_0 p^2 \approx 1.5 \times 10^{-3}. \tag{5.153}$$

Let us compare our predictions for the logarithmic relaxation rate with the data from Salvino et al. (1994) and Rogge et al. (1996b). According to (5.145) the relaxation rates for both real and imaginary parts of dielectric susceptibility $(S_{t,\epsilon'}, S_{t,\epsilon''})$ should be of the same order. Experimentally, both values are of order of 10^{-5} [see Rogge et al. (1996b) and Sect. 5.2.2]. The ratios of the relaxation rates and the equilibrium parameters: $S_{t,\epsilon'}/S_{T,\epsilon'}$, $S_{t,\epsilon''}/\epsilon''$ are defined by the the dimensionless product $P_0 U_0$ (with logarithmic factors, that are always of order of unity). This parameter describes the manifestation of the interaction between TS's. Using the experimental data from the papers one can estimate it as

$$\chi = P_0 U_0 \simeq 10^{-3}. \tag{5.154}$$

This value is in agreement with the magnitude of this parameter in other glass-like materials (see Yu and Leggett (1988), Hunklinger and Raychaudhuri (1986)).

5.2 Nonequilibrium Effects: Long-Time Relaxations and the Dipole Gap 289

While our treatment involving virtual phonons (see Sect. 5.1.2) assumed an elastic TS–TS interaction, the U_0/R^3 interaction can be caused by either electric or strain dipoles. If we assume that the dominant interaction in this material is the electric dipole–dipole one, then the coupling constant U_0 is related to the TS dipole moment by $U_0 \approx 4p^2/\epsilon$. In this case the estimation of $P_0 U_0 \approx P_0 4p^2/\epsilon \approx 10^{-3}$ is in agreement with the previous one, (5.154). Therefore one may conclude that the electric dipole–dipole interaction between TS's is important here and probably dominates in SiO_x (the elastic couplings between TS's are smaller than the electric ones).

Let us show that the proposed theory can describe the complex frequency and temperature dependences of the logarithmic relaxation rate $S_{t,\epsilon'}$ at large enough temperatures $T > T_m$. Taking $\phi_0/k_B \approx 450$ mK, $P_0 U_0 P_0 \mathbf{p}^2/\epsilon' \approx 1.25 \times 10^{-6}$, and $A \approx 0.6 \times 10^8$ for $T = 200$ mK, we obtain satisfactory agreement with the experimental data. The theoretical curve and the experimental points taken for the same temperature ($T \approx 200$ mK) are shown in Fig. 5.24. One should note that our estimation (5.145) for the relaxation rate $S_{t,\epsilon'}$ is valid far from the crossover temperature T_m. For temperatures close to T_m more careful consideration is necessary. The relaxation rate, and hence the hole in the dielectric response, has to be almost independent of frequency at larger frequencies or lower temperatures. This tendency clearly takes place for the experimental data at lower temperatures in a variety of materials (see Figs. 5.20 and 5.21).

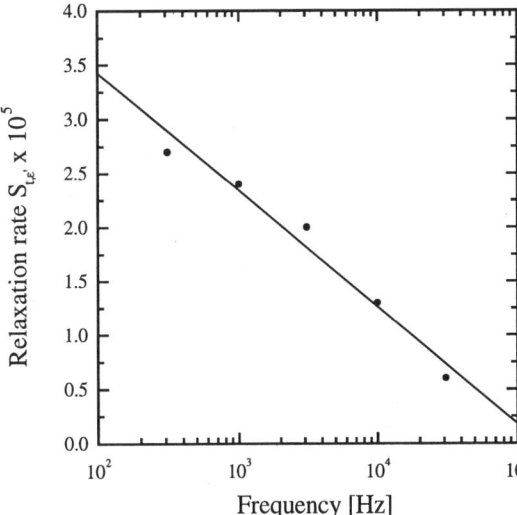

Fig. 5.24. Theory (solid line) and experimental points for frequency dependence of dielectric relaxation rate at 200 mK. After Burin (1995a).

To describe the temperature dependence of the SiO_x logarithmic relaxation rate $S_{t,\epsilon'}$, one should take $\phi/k_B \approx 0.17$ K to agree with the experimentally used electric fields, with the other parameters almost the same as for the frequency dependence of $S_{t,\epsilon'}$. The behavior $S_{t,\epsilon'}$ versus $\log(T)$ is shown in Fig. 5.25 simultaneously with the experimental points (Rogge er al. (1994), Salvino et al. (1994), Rogge (1994)) for frequency equal to 1000 Hz. Again, the theory describes the data well.

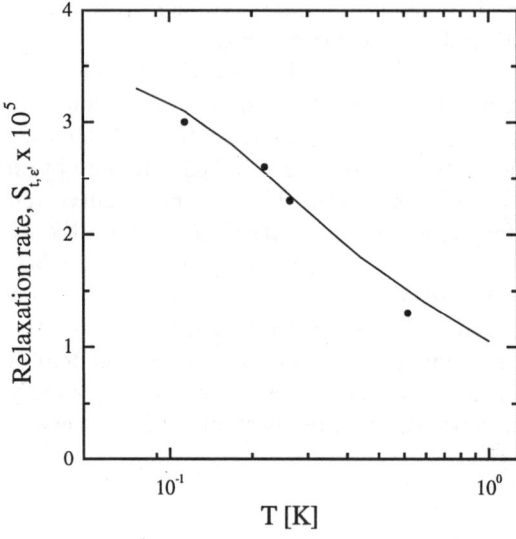

Fig. 5.25. Theory (solid line) and experimental points for temperature dependence of dielectric relaxation rate at 1 kHz. After Burin (1995a).

Consider the data for SiO_2 from Salvino et al. (1994). In SiO_2 the slope of the equilibrium dielectric susceptibility temperature dependence is $S_{T,\epsilon'} \approx 2 \times 10^{-4}$. The logarithmic relaxation rate is $S_{t,\epsilon'} \approx 10^{-7}$. The large difference between these two values strongly suggests that the observed nonequilibrium behavior is due to the interacting nonadiabatic response (in analogy with SiO_x). We estimate $P_0 U_0 \simeq 10^{-3}$. This estimation is in agreement with that in Burin and Kagan (1994a). The inequality $4p^2 < U_0$ seems to be obeyed. This means that strain interaction between TS's probably dominates in SiO_2.

The functional forms of the nonequilibrium dielectric response [given in (5.147) and (5.148)] adequately describe the "sweep" experiments discussed above for both electric and strain perturbations. For example, Fig. 5.26 shows theoretical fits to the data from Fig. 5.20 at $t = 10$ s. Because the functions depend only logarithmically on several parameters, it is difficult to perform careful nonlinear fits. However, both this data and the low-temperature strain data from Fig. 5.21 are consistent with a single set of fit parameters. The high-temperature strain data can be fit with the functional form, but this requires a value of γ almost a factor of two larger than that needed to fit the low-temperature data.

5.2 Nonequilibrium Effects: Long-Time Relaxations and the Dipole Gap

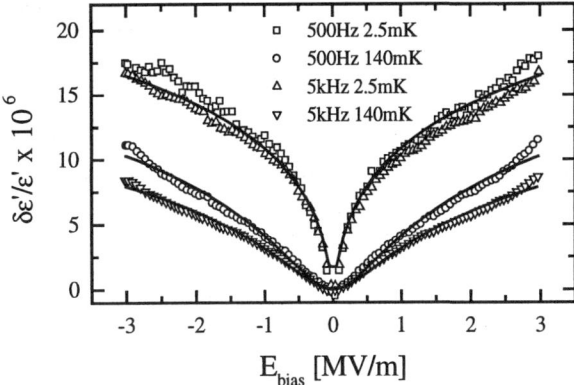

Fig. 5.26. Dipole gap model fit to sweep data on mylar sample. Data taken from Rogge et al. (1996a).

Now consider the nonequilibrium acoustic measurements discussed in the previous section. Carrying out the analysis described in the preceding section for two different materials, BK7 and Suprasil 2 (see Osheroff et al. (1996) and Natelson et al. (1997)), we find the following. Recall that all logarithms in the definitions of the S_T and S_t parameters are base-10, and the numbers quoted below are for $T < T_m$. In the limit that the distribution of the magnitude of the TS-phonon coupling γ is sharp (i.e. a δ-function at some particular value), then the ratio calculated in (5.152) should be a measure of f, the fraction of acoustically active TS's with significant electric dipole moments. Both phonon- and photon-echo experiments (Golding and Graebner (1976), Golding et al. (1979) for example) and other "crossexperiments" (Laermans et al. (1977)) have looked at this information.

For Suprasil 2 with E_{dc} =800 kV/m, $S_{T,\epsilon'} = 1.9 \times 10^{-4}, S_{t,\epsilon'} = 3.4 \times 10^{-7}, S_{T,v} = 3.6 \times 10^{-4}, S_{t,v} = 5 \times 10^{-7}$. Thus, the "magic number" defined in (5.152) is approximately 0.8. According to Golding et al. (1979), the ratio of the number of OH$^-$-based TS's that have an electric dipole moment of about 3.7D, to the number of intrinsic TS's in 1200ppm OH$^-$ SiO$_2$, is around 0.5. Therefore, the measured ratio of 0.8 is not terribly surprising, ignoring the issue of the distribution of strain couplings.

For BK7 with E_{dc} =800 kV/m, $S_{T,\epsilon'} = 2.4 \times 10^{-3}, S_{t,\epsilon'} = 8 \times 10^{-6}, S_{T,v} = 2.8 \times 10^{-4}, S_{t,v} = 1.8 \times 10^{-6}$. Thus, the "magic number" defined in (5.152) is approximately 2. Now, experiments by Laermans et al. (1977) indicate that f should be around 0.6. Thus, the measured ratio of 2 is rather surprising, possibly implying a very nonsharp distribution of strain couplings within the material.

Experiments by Natelson et al. using torsional oscillator techniques are underway to gather further data about this matter. It is hoped that such methods will be less prone to the electromechanical background effects discussed above in Sect. 5.2.1, and will thus allow other related experiments to be performed.

5.2.7 Anomalous Hysteretic Behavior and Ultralow Temperatures

One experimental fact that has not yet been discussed is the unusual hysteresis in rapid strain and electric field sweeps at low temperatures seen in dielectric measurements by Rogge et al. (1997b) in both BK7 and mylar. Figure 5.27 shows dielectric data on the mylar sample at 5 mK at 500 Hz and 5 kHz measurement frequencies at fixed sweep rate of the applied strain field. The low-frequency data show severe hysteresis whereas the higher-frequency (shallow "U"-shaped) curve is nonhysteretic. In addition to this frequency dependence, the authors observed strong temperature and sweep rate dependence of the hysteresis. The hysteresis was most pronounced at low temperatures, low measurement frequency, and high sweep rate.

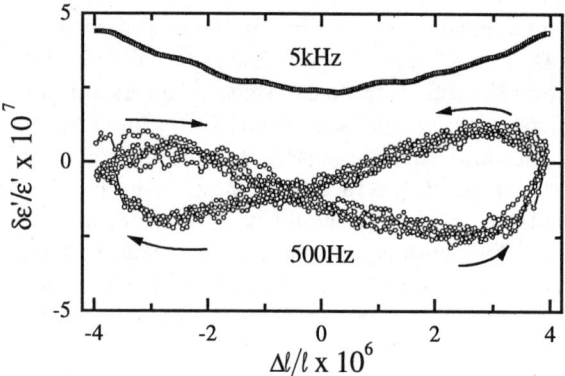

Fig. 5.27. Dielectric hysteresis in mylar when subjected to rapidly varying applied strain. From Rogge (1996).

Assuming that the hysteresis is due to TS dynamics rather than some artifact of the applied strain measurement scheme, one would expect to see similar behavior in dielectric response measured during rapid sweeps of a dc electric field. The authors performed this experiment, and the result is shown in Fig. 5.28. The figure clearly demonstrates the strong temperature dependence of the effect. Again, the effect is more dramatic at lower temperatures, lower frequencies, and higher bias field sweep rates.

The curves in the strain and electric bias figures look qualitatively rather different, but by altering sweep rates, the two cases can be made to resemble one another. Arguments in Rogge et al. (1997b) describe this and demonstrate that the observed behavior is not due to heating of the samples.

5.2 Nonequilibrium Effects: Long-Time Relaxations and the Dipole Gap   293

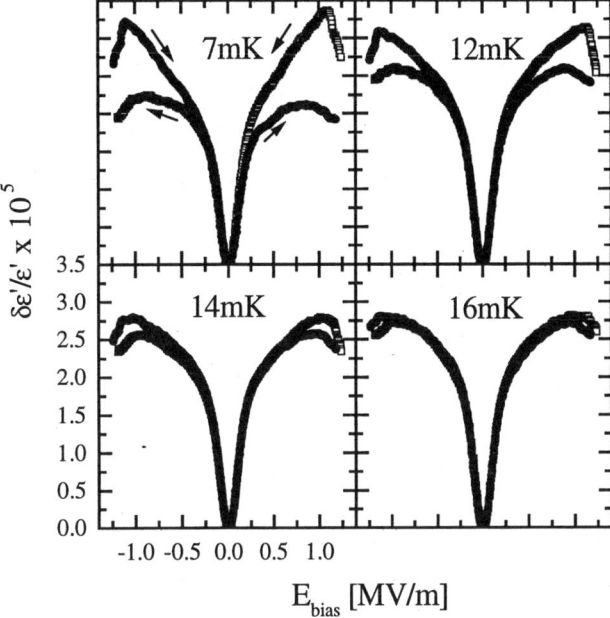

Fig. 5.28. Appearance of dielectric hysteresis in BK7 at 500 Hz under a rapidly varying dc electric field. After Rogge et al. (1997b).

In addition to the very strong temperature and frequency dependence noted above, other features of this hysteresis are worth noting. First, the directional sense of the curves in both Figs. 5.27 and 5.28 is unexpected. Within the context of dielectric jumps and relaxations developed above in Fig. 5.17, one would expect hysteresis to occur at sufficiently slow sweep rates due to the development of a hole in the dielectric response at the applied bias field. For this type of hysteresis, the dielectric response near the edge of the sweep should be lower upon returning to a value of bias field that was just visited; during the first visit, the response at a particular field begins to decrease as a hole is formed, and this hole does not have sufficient time to "refill" before the sweep has turned around and revisited that value of bias field. This behavior is indeed observed in very slow sweeps (Rogge (1996)). In the above figures, however, one observes dramatically enhanced, rather than decreased, response upon returning to a recently visited value of applied bias.

Secondly, the hysteresis is frequency dependent in a temperature regime where slower sweeps and equilibrium dielectric studies show no frequency dependence. It is clear from the formalism developed above to explain the long-time relaxations that this frequency dependence cannot come from the formation of the dipole gap, since that process simply modifies the effective TS density of states.

To compare the mechanical and electrical cases, the relevant parameter for the sweep rate is the amount of energy by which asymmetries of individual TS's are shifted per unit time. As shown in Fig. 5.28, the hysteresis in the electric field sweep on the BK7 sample started around 10 mK with an energy rate of $\partial \mathbf{E} \partial t \cdot (p/k_B) = 5$ mKs^{-1} (assuming one Debye for p). In the mechanical case there is already strong hysteresis for the same BK7 sample at 30 mK with a rate of $\gamma/(k_B \ell)(\partial \ell/\partial t) = 0.1$ mKs^{-1} (assuming $\gamma = 1$ eV). In BK7 the ratio of the number of TS's with just an elastic dipole to those with both large elastic and electric couplings is two to one (Laermans et al. (1977)). It is not clear why there is such a dramatic difference in the scale of the sweep rate needed to produce the hysteresis.

There is another fundamental difference between the strain and electric field hystereses. The two types of perturbations produce qualitatively different responses in "unipolar" sweep experiments (in which the bias field is swept from zero to some large value and back, rather than being symmetric about zero bias field). In the electrical case hysteresis was only seen at the maximum field turning point of the bias in a unipolar sweep. Conversely, a unipolar strain sweep causes hysteresis at both the zero field turn point and at the maximum field turn point. Again, this is surprising, especially since the model that appears to be very successful at describing jumps and slow sweeps would predict behaviors independent of the type of perturbing field.

One proposed explanation of the hysteresis (see Burin (1996)) suggests that the answer may involve the (noninteracting) adiabatic response of the TS. Again, a particular TS is considered to be adiabatically perturbed if its energy splitting is altered by the perturbation without its state being forced to make a transition. A TS in its ground state prior to an adiabatic perturbation will be found in its new ground state after the perturbation.

Such adiabatically perturbed TS's could lead to hysteresis due to their dielectric response as they rethermalize their populations after perturbation by the bias field (first discussed by Carruzzo et al. (1994)). Consider a set of nearly degenerate TS's in thermal equilibrium. As the bias field is swept, these systems' energy splittings are increased as they no longer experience zero local electric field. Assuming that this perturbation takes place adiabatically, the TS's are no longer in thermal equilibrium. If the TS's now rethermalize, there is a larger population difference between the ground and excited states than prior to the perturbation. Now suppose the bias field is swept back such that these TS's are adiabatically returned to zero local field. Because of their now athermal populations, their dielectric response is enhanced compared to its original preperturbation value. This would explain both the presence and the direction of the strong hysteresis observed at the extrema of the sweep.

The interacting rethermalization response after adiabatic perturbation is much larger than the relaxation due to the rearrangement in the density of states, but also much shorter lived, varying as $1/t$ according to Burin (1995a). This time dependence comes about because for temperatures less that T', the dominant relaxation mechanism is now the interaction-driven one described in Sect. 5.1.5. A $1/t$ dependence of the adiabatic response might explain why this contribution was not observed in the relaxation experiments, since the measuring electronics could not respond until a few seconds after the field application. At that point the adiabatic component might be very small compared to the logarithmic relaxation. One expects both mechanisms to be present after a perturbation of the system. In sweep experiments it may be possible to observe the adiabatic contribution if the sweep is fast enough.

The substantial differences between the two types of perturbations (differing temperatures for the onset of hysteresis, and differing behavior in the unipolar sweeps) raise the question of whether the adiabatic response model described above is correct. Another speculative explanation of the hysteresis involves spectral diffusion. The spectral diffusion of energy between the TS's driven by the external fields would cause athermally populated TS's, enhancing the dielectric response. In the case of strain perturbation all TS's are shifted, in contrast to the electric case where only the fraction with large electric couplings is perturbed. Since spectral diffusion is due to interactions, the differing spacing of perturbed systems in the two cases might explain the differences between electric and strain responses. It is far from obvious, however, whether either this approach or the NIA response can account for the observed frequency dependence.

This section's analysis demonstrates that the low-energy spectral properties of the TS subsystem are dominated by the temperature-dependent dipole gap. Consequently, one might wonder: at relatively high temperatures (50 K), the properties of the system differ markedly from those of the standard tunneling model; perhaps the standard TS distributions arise through substantial freezing of some system of initial defects because of interaction effects like the dipole gap. Self-consistent evolution of the density of states of excitations within the dipole gap theory suggests that the universality of the TS ensemble can result from this freezing. We now treat this problem using renormalization group in Sect. 5.3.

5.3 On the Universality of the Low-Temperature Properties

As we have seen, given the usual distribution functions of TS parameters from the standard tunneling model, and accounting for weak dipolar elastic interactions between TS's, it is possible to explain a wide variety of experimental results. The origin of these universal distribution functions in such a wide variety of materials remains an open question.

We consider the general problem of the nature of the universal properties found in amorphous solids at low temperatures. We will try to understand this universality within the model of interacting defects. Beginning with interacting defects with an arbitrary distribution of parameters, we shall see that, following a renormalization group analysis, the excitation spectrum of the renormalized system will be very close to that assumed in the standard tunneling model.

5.3.1 Basic Facts

The standard tunneling model of glasses assumes the presence of some specific subset of atoms or groups of atoms, each having an internal degree of freedom. This degree of freedom can be represented as particle moving in two-well configuration. The two wells are energetically offset on some scale Δ and are separated by a barrier, characterized by the tunneling matrix element Δ_0. These hypothetical excitations are called two-level tunneling systems.

The central computational point of theory is the particular choice of distributions of these parameters. Uniform distributions of the level shift and of the logarithm of the tunneling amplitude were assumed in the previous analysis. Calculations using these distributions show features similar to those in much of the existing experimental data on glasses at low temperatures. The assumption of the uniform distribution of level splittings seems quite natural. The second assumption is based on the hypothesis of a quasi-uniform distribution of semiclassical tunneling action, S, over a wide range of its variation:

$$\Delta_0 \propto \exp(-S), \quad P(S) = \text{const.} \tag{5.155}$$

The microscopic nature of TS's generally remains unknown.

During the last two decades the consequences of TS's have been seen in a wide variety of materials, including orientationally disordered crystals, superionic conductors, radiatively damaged crystals, polymers (Hunklinger and Raychaudhuri (1986), and references therein), cryocrystals of inert gases (Muromtzev et al. (1994)) and biological systems (Goldanskii et al. (1989)). Recently, data on sound attenuation and sound velocity measurements in a wide variety of polycrystalline metal and amorphous dielectric samples led to a statement made in Esquinazi et al. (1992) and Esquinazi et al. (1994) that TS's might be observed in crystals containing any kind of disorder; see Chap. 4 and Sect. 4.5.

These observations highlight the problem of the microscopic nature of TS's. Moreover, this problem might be closely related to the problem of the universal $1/f$ noise in solids at low temperatures, that is also currently unresolved.

There exist many theoretical models to describe the nature of the two-level systems. The tunneling model itself as well as the more detailed soft-potential model [see Chap. 9 or Parshin (1994a) for a review] are based on the hypothesis that the local-independent two-well configurations are frozen in the glassy host near the melting temperature. It is assumed that the barriers between wells have the properties needed to provide the universal distribution (5.122). These models, under some reasonable assumptions, ensure that the properties of the low-energy excitations are close to those in amorphous solids.

5.3 On the Universality of the Low-Temperature Properties

However, some peculiarities of the experimental data lead one to think about a more complex and general origin of two-level excitations.

One intriguing point is the quantitative universality of low-temperature glassy properties. One of the main aspects of this universality, found by Freeman and Anderson (1986), is the approximate conservation of the ratio of the phonon wavelength λ to the mean free path l for phonon scattering by TS's at low temperatures:

$$Q^{-1} \approx \frac{\lambda}{l} \approx \text{constant} \approx 10^{-3}. \tag{5.156}$$

It is remarkable that this ratio is almost unchanged in different glassy materials despite variations in λ and l over orders of magnitude.

The temperature independence of this value leads to T^2 thermal conductivity behavior:

$$\kappa \sim C_{\text{ph}} v l \sim \frac{l}{\lambda_T} \frac{\hbar v}{k_\text{B} T} C_{\text{ph}} v^2 \propto \frac{l}{\lambda_T} T^2, \tag{5.157}$$

where v is the sound velocity and C_{ph} is the phonon part of the specific heat. We label this ratio Q^{-1} because of its similarity to the internal friction in such materials that also has a fairly universal value.

The above universal ratio can be expressed in terms of the significant physical parameters of the system,

$$Q^{-1} \approx P_0 \frac{\gamma^2}{\varrho v^2}. \tag{5.158}$$

Recall that P_0 stands for the TS density of states, γ is the (assumed dipolar) coupling constant between TS's and phonons, and ϱ is the bulk density of the sample. Freeman and Anderson (1986) have shown that whereas all these parameters can vary markedly from sample to sample, the value of Q^{-1} changes much more weakly and remains close to 10^{-3}. This finding and earlier work by Klein et al. (1978) stimulated Yu and Leggett (1988) to propose their interaction-based scenario, discussed below, to explain this and other results.

5.3.2 Significance of $1/R^3$ Interactions

The factor $\gamma^2/(\varrho v^2)$ has a clear physical interpretation [see (5.46) and the following]. In addition to setting the scale for TS effects on sound propagation, it determines the scaling constant U_0 of the $1/R^3$ elastic interaction between TS's. As we have seen in Sect. 5.1.2, the same TS-phonon interaction that makes TS's strongly scatter thermal phonons leads to interactions between TS's through the exchange of relatively high-energy virtual phonons. This TS interaction is analogous to the electric dipole-dipole interaction caused by the exchange of virtual photons. It has the form

$$U(R) = \frac{U_0}{R^3}. \tag{5.159}$$

Combining the empirical statement of (5.156) with (5.158) leads to the conclusion

$$P_0 \propto \frac{1}{U_0}. \tag{5.160}$$

Both Klein et al. (1978) and Yu and Leggett (1988) argued that the universality of two-level excitations might be a result of strong interactions between some defects. This conclusion was essentially supported by the results of the investigation of low-temperature excitations in disordered systems of interacting centers, each center having an internal degree of freedom.

Consider a system of such centers, interacting via a $1/R^3$ law, with a random spatially varying potential of scale W playing the role of disorder. The stability arguments, proposed by in Baranovskii et al. (1980) and Efros (1985), result in the following low-temperature density of states for the excitations of such a system:

$$P_0' \propto \frac{3}{2\pi \ln(\frac{W}{k_B T}) U_0}. \tag{5.161}$$

Note that this density of states only weakly (logarithmically) depends on the original centers through the parameter W that represents the scale of the random external potential. We shall see this result again later in this section.

The very small value of product $P_0 U_0 \approx 10^{-3}$ is, however, difficult to reconcile with the last equation. Also, the above picture does not explain the origin of the logarithmically uniform distribution of tunneling amplitudes Δ_0, the existence of which is supported by many measurements of logarithmic dependences of various parameters on time, frequency or temperature.

This second issue can probably be addressed by considering many-center excitations (clusters) in the system of defects. Klein et al. (1978) showed that in a system of rotating molecular impurities dissolved in a crystal interacting via a $1/R^3$ law, pair excitations provide the same behavior of the heat capacity and thermal conductivity as that observed in glasses. The analysis, made in Burin (1991a) and Burin (1991b), of pair excitations in strongly disordered media demonstrates that for the interaction law $1/R^3$, pair excitations can be described by the same distribution as (5.122). However, the distribution of those pair excitations was not universal in that model. Coppersmith (1991) assumed that many-center clusters can be responsible for the logarithmically uniform distribution of the elementary excitations because of the evident behavior of the tunneling amplitude, $\sim \exp(-\alpha k)$, where k is the number of the defects involved.

The main purpose of this section is to study the low-temperature properties of a general model of interacting defects. These properties are determined by the elementary excitations with energies less than $k_B T$. Thus the problem reduces to the analysis of these excitations. This will be done within the framework of the renormalization group developed in Burin and Kagan (1996b) and Burin and Kagan (1996c). Earlier renormalization work studies of similar problems were made by Fisch (1980), based on scaling in energy space, and by Mogilyanskii and Raikh (1989), based on scaling in real space. These studies qualitatively reproduce the pseudogap in the density of states in the form found by Baranovskii et al. (1980).

The main idea of the method is that decreasing temperature leads to a decrease in the entropy of the system. The number of states available to the total system in thermal equilibrium is lowered. As the volume of phase space accessible by the system shrinks, the ways the system can make transitions between different available states are forced to become increasingly indirect. The different paths from state to state in phase space can be represented in terms of the elementary excitations of the system. These excitations can be described in terms of the total number of defects involved, i.e single-defect excitations, pair-excitations, ...k-defect excitations. When the temperature goes to zero, the characteristic number of defects involved in a typical excitation goes to infinity in systems with long-range interactions. These many-center excitations lead to the presence of very long relaxation times.

We will find that, for a general disordered system of interacting defects, the collective excitations involving many defects that dominate at the lowest temperatures possess the universal distributions seen in the tunneling model. These distributions of excitation parameters will depend only logarithmically on the original defect distributions. We will also discuss experimental tests of this model.

5.3.3 The Renormalization Group Model

First of all, we assume the existence of a set of primary defect centers, each possessing an internal degree of freedom. In orientational glasses these centers are obviously associated with impurity molecules, having rotational degrees of freedom in the crystal field of the sample. In traditional glasses the appearance of such centers can be linked to the local breakdown of spatial degeneracy. We want to look at the general properties of an interacting system of such defects, without assuming the standard tunneling model's distribution functions for the parameters of those defects.

Suppose the spatial density of the primary centers is n. In general such centers interact with one another via a potential given by $U(R) = u_{ij}/R^3$ by the exchange of virtual phonons (in dielectric glasses, described in Sect. 5.1.2) or due to virtual electron-hole pair exchange (in metals). In addition, those centers are influenced by external random disordering that provides each primary center i with a random level shift ϕ_i. The distribution of the random level shift is characterized by the energy scale $\phi_i \sim W$, and is not necessarily flat. The internal tunneling motion of center i can be described by the parameter Δ_{0i}. The distribution of this parameter is arbitrary (not necessarily flat) and has some characteristic scale $\Delta_{0i} \sim \Delta_{0*}$.

We make a few additional assumptions regarding the above parameters. First of all, we assume the internal tunneling motion of each primary center is very slow, that means

$$\Delta_{0*} \ll U_0 n, W. \tag{5.162}$$

Here the parameter $U_0 n$ represents the scale of the interaction energy at the average distance between TS's. This assumption seems quite reasonable from WKB considerations, keeping in mind that we are discussing the tunneling of "heavy" atoms.

We also assume that the interaction is weak in comparison with the energy scale of the external random field W. This assumption is necessary to pursue the theoretical description that follows. However, since the resulting behavior of the density of the low-energy excitations will be almost insensitive to this assumption, we hope that the spatial disordering of centers will be sufficient for the validity of our analysis.

Initially, we neglect the relatively weak tunneling dynamics of each TS. Then the model of the system reduces to the disordered Ising model with $1/R^3$ interactions. Arguments based on the static susceptibility will be presented to prove that a significant fraction of low-energy excitations can be represented as many-center clusters at very low temperatures.

The special threshold character of $1/R^3$ interactions (a volume integral of the potential in three dimensions is only logarithmically divergent) between long range (for example, the Coulomb interaction) and short range (for example, the Lennard–Jones potential), allows us to develop a renormalization group description for the low-energy excitations. Solving the renormalization group equations, we will find the density of states of the many-center clusters. The form of the final result will be universal, i.e. the resulting density of states is only logarithmically sensitive to the parameters describing the primary centers.

The weak tunneling dynamics terms will be taken into account through perturbation theory. The distribution of the low-energy excitations' level shifts and effective tunneling parameters will end up very close to the universal two-level system distributions.

We treat the lowest eigenstates of each defect as the two spin states of a spin-1/2 particle ($S^z = \pm\frac{1}{2}$, see Sect. 5.1.1), and we allow the particles to interact according to the $1/R^3$ law and have random level shifts ϕ_i. This results in the Hamiltonian of the Ising model in a random field,

$$\widehat{H} = -\sum_i \phi_i S_i^z - \frac{1}{2}\sum_{ij} U_{ij} S_i^z S_j^z, \quad U_{ij} = \frac{u_{ij}}{R_{ij}^3}. \tag{5.163}$$

This Hamiltonian has been considered by many authors (e.g. Nieuwenhuizen (1993), Kokshenev et al. (1996), and references therein) within various modifications of the mean field approach developed to study the spin-glass transition. We will see below that in the case of strong disordering the separate pairs of defects should be treated exactly and mean field theory cannot be applied. Moreover, even in the case of the absence of "diagonal" disorder ($\phi_i = 0$), the distribution of the molecular field $\sum_j U_{ij} S_j^z$ strongly differs from a Gaussian form. In general, mean field theory works when the interaction decreases with distance more slowly than $1/R^{3/2}$, as in the classical model of the spin glass, in which the interaction's average value is independent of distance (Sherington and Kirkpatrick (1975)).

Recall that the spatial density of primary defects i is n. We assume that the average value of the interaction constants u_{ij} is zero, and these constants are uncorrelated for different pairs of centers. The first assumption is valid for all known realizations of the R^{-3} interaction [see (5.49) and (5.50)]. The second assumption looks reasonable for metallic glasses, where the interaction constants contain the rapidly oscillating factor $\cos(k_F R_{ij})$. The assumption of uncorrelated interaction constants can also be justified for the elastic interaction in dielectric glasses using methods developed by Baranovskii et al. (1980) to study systems of interacting dipoles; we will not present the details here. The characteristic value of the interaction energy can be represented as the interaction at the average TS spacing, $U \approx U_0 n$, where U_0 is the characteristic magnitude of the interaction constant u_{ij}.

We characterize the random potential ϕ_i by some distribution function $g(\phi_i)$, and assume only that this function is even in ϕ and has no singularities near zero, where we assume $g(0) = 1/W$. The parameter W is the characteristic scale of the random potential fluctuations. Note that the standard tunneling model would assume g to be flat from $\phi = 0$ out to the cutoff W, and that our treatment is much more general.

Of course, for perturbation theory to be applicable, the interaction must be weak compared to the noninteraction parts of the Hamiltonian:

$$U_0 n \ll W. \tag{5.164}$$

5.3.4 A Key Identity

Before starting the renormalization group analysis, we give a simple illustration of future conclusions, that is based on the analysis of the static susceptibility of the system of defects at zero temperature.

Let us introduce the distribution function of spins (primary defects) at zero temperature. For $T = 0$ the system should be in its ground state. This means that the energies of all other states should be larger. Suppose that in the ground states all spins have the values

$$S_i^z = s_i. \tag{5.165}$$

Each excited state A_n can be described by the set of spins $S_1, S_2, ..S_n$ that have been flipped compared to their ground state configuration. The energy difference between this state and the ground state (or the energy of the excitation A) can be written as

$$E_{A;1..n} = 2\sum_{k=1}^{n} \phi_k S_k + U(A), \qquad (5.166)$$

where the interaction term $U(A)$ contains the interaction between spins (defects) in the set A with other spins. For example, for a two-spin excitation $S_1^z, S_2^z \to -S_1^z, -S_2^z$,

$$U(A) = 2 \sum_{i \neq 1,2} (U_{1i} S_1^z S_i^z + U_{2i} S_2^z S_i^z). \qquad (5.167)$$

Then we can describe the system in its ground state by the distribution function [cf. (5.128)]

$$P(..; S_i^z; ..) = \prod_A \Theta(E_A);$$
$$\Theta(x) = 0, x < 0;$$
$$\Theta(x) = 1, x \geq 0. \qquad (5.168)$$

The Θ function ensures that all excitations A have positive energies compared to our ground state.

The model under consideration possesses the average symmetry $S_i^z \to -S_i^z$ because of the symmetrical distribution of random energies $g(\phi) = g(-\phi)$ and the binary character of the interaction between defects. To calculate the linear response of the system even at zero temperature, we add a small external "magnetic" field to break the symmetry

$$-h \sum_i S_i^z \qquad (5.169)$$

that leads to a finite response

$$\langle S^z \rangle \sim \chi h. \qquad (5.170)$$

This response can be represented in terms of the density of the low-energy excitations for very small fields h. This allows us to derive the following significant identity.

Consider the differential operator

$$\sum_i S_i^z \frac{\partial}{\partial \phi_i}. \qquad (5.171)$$

Let us find its average value in the ground state:

$$\left\langle \mathrm{Tr}_{S_i} \frac{1}{2} \sum_i S_i^z \frac{\partial}{\partial \phi_i} P(..; S_i^z; ..) \right\rangle. \qquad (5.172)$$

Averaging is done over the disorder, and the trace is performed over possible spin configurations. Each derivative $S_i^z \partial/\partial \phi_i$ of $\Theta(E_A)$ gives $\delta(E_A)/2$ if the spin i participates in excitation A and gives zero otherwise. Then taking derivatives of all Θ-functions in the product (5.168), we obtain

5.3 On the Universality of the Low-Temperature Properties

$$\left\langle \langle \sum_A n_A \delta(E_A) \rangle \right\rangle, \tag{5.173}$$

where n_A is the number of spins participating in excitation A. Averaging is done over the ground state of the model and over disorder.

Since the average δ-function represents the density of excitations of the system having zero energy, one can rewrite the above expression as

$$\frac{V}{2} \sum_k k P_k(0). \tag{5.174}$$

Here V is the system volume, and $P_k(0)$ is the average density of states for k-spin excitations with zero energy.

Let us make another transformation of the expression in (5.172). Evaluating the trace over spins, and averaging each term i over all disorder excluding the random shift ϕ_i, we obtain

$$\sum_i \left\langle \frac{\mathrm{d}S_i^z(\phi_i)}{\mathrm{d}\phi_i} \right\rangle_i = N \left\langle \frac{\mathrm{d}S_i^z(\phi_i)}{\mathrm{d}\phi_i} \right\rangle_i. \tag{5.175}$$

Here $\langle .. \rangle_i$ denotes averaging over various configurations of ϕ_i,

$$\int \mathrm{d}\phi_i g(\phi_i) \left\langle \frac{\mathrm{d}S_i^z(\phi_i)}{\mathrm{d}\phi_i} \right\rangle. \tag{5.176}$$

Since, in the relevant domain $-U_0 n < \phi_i < U_0 n$ where $\langle S_i^z \rangle$ changes significantly, the distribution g is nearly constant, one can replace it with $g(0)$. Then, with the same accuracy, the integral can be evaluated to give

$$g(0)(\langle S_z(+\infty) - S_z(-\infty) \rangle) \approx g(0). \tag{5.177}$$

The last relation, together with the result (5.174), allows one to find a significant identity that is almost independent of the interaction between defects:

$$\sum_k k P_k = P_0(0) \equiv 2n g(0). \tag{5.178}$$

Now we will draw an important conclusion from this identity. As was shown by Baranovskii et al. (1980), the density of states at zero energy in the system with $1/R^3$ interactions decreases with decreasing temperature,

$$P(0) \propto \frac{1}{\ln(W/(k_\mathrm{B} T))}. \tag{5.179}$$

This tendency is general for excitations involving an arbitrary number of centers. Thus (5.178) can be satisfied only because of the dominant contribution of clusters containing a large number k of primary centers.

It is evident that the relaxation time for the complex excitations grows rapidly with increasing k. Consequently, long time relaxations become significant at low temperatures for various properties of the system. If we assume an exponential increase of the relaxation time with increasing k, then it follows from (5.178) that for a smoothly changing distribution P_k, the distribution of the relaxation times must be close to the logarithmically uniform one (Burin (1997)).

This foreshadows the results of the renormalization group calculation that make it possible to describe the universal low-temperature behavior of glasses.

5.3.5 General Model

To derive the renormalization group approach let us start by considering pairs of interacting defects in (5.163). They can be described by the Hamiltonian

$$\widehat{H}_{\text{pair}} = -\phi_i S_i - \phi_j S_j - U_{ij} S_i S_j. \tag{5.180}$$

The following analysis is very similar to that of Sect. 5.2.5, but the pairs discussed there and in Sect. 5.1.4 are coupled via an interaction that is much smaller then $k_B T$, and those pairs fluctuate and lose coherence dynamically due to thermal processes. Here, the pairs of initial states are coupled at energy scales strongly exceeding $k_B T$, and these pairs therefore exist permanently, in analogy to a diatomic molecule.

The following significant inequality is assumed to be valid:

$$|g(0) U_{ij}| \ll 1. \tag{5.181}$$

We are no longer assuming the standard tunneling model's distribution of parameters and are neglecting the tunneling matrix elements for now.

Consider the influence of the interaction on the density of low-energy excitations of Hamiltonian (5.180) for its eigenstate with energy $\Delta \ll |U_{ij}|$. Without interactions the spin density of states reads [explicitly writing out the contributions in (5.168)]:

$$P = \Theta(2\phi_i S_i)\Theta(2\phi_j S_j). \tag{5.182}$$

Then one can consider the single-spin excitations with energy

$$E_i = 2\phi_i S_i; \quad E_j = 2\phi_j S_j, \tag{5.183}$$

and the excitation of the pair

$$E_p = 2\phi_i S_i + 2\phi_j S_j. \tag{5.184}$$

The density of single-spin excitations with $\Delta > 0$ can be defined as

$$\begin{aligned}
P_1(\Delta) &= n \text{Tr}_{S_i} \langle \delta(2\phi_i S_i - \Delta) \rangle \\
&= n \text{Tr}_{S_i} \int d\phi_i g(\phi_i) \delta(2\phi_i S_i - \Delta) \Theta(2\phi_i S_i) \\
&\approx 2n g(0) \equiv P_1(0). \tag{5.185}
\end{aligned}$$

Again, n is the spatial density of primary defects, and the factor of two comes from the contribution of excitations with both $\phi_i = \pm\Delta$. Note that in this

5.3 On the Universality of the Low-Temperature Properties

section we are neglecting the tunneling internal to each defect (i.e. $\Delta_0 = 0$), so that a TS's level shift Δ is exactly its energy splitting.

Still neglecting interactions, the density of pair excitations is

$$\begin{aligned}
P_2(\Delta) &= \frac{n}{2}\text{Tr}_{S_i,S_j}\langle \delta(2\phi_i S_i + 2\phi_j S_j - \Delta)\rangle \\
&= \frac{n}{2}\text{Tr}_{S_i,S_j}\int d\phi_i g(\phi_i) \int d\phi_j g(\phi_j) \delta(2\phi_i S_i + 2\phi_j S_j - \Delta) \\
&\quad \times \Theta(2\phi_i S_i)\Theta(2\phi_j S_j) \\
&\approx \frac{n}{2}\text{Tr}_{S_i} 2g(0) \int d\phi_i g(\phi_i) \Theta(2\phi_i S_i)\Theta(\Delta - 2\phi_i S_i) \\
&= \frac{P_0}{2} g(0)\Delta.
\end{aligned} \quad (5.186)$$

The prefactor $1/2$ comes from the fact that each pair can be related to both the site i as well as the site j, but should be taken into account only once. The density of single-spin excitations (5.185) is negligibly small compared to the contribution from pairs in the limit of very low energy $\Delta \to 0$ in which we are interested.

Let us account for the interaction between spins and analyze its influence upon the density of the excitations. The energy of excitations of the interacting Hamiltonian can be represented as

$$E_i = 2\phi_i S_i + 2U_{ij}S_i S_j; \quad E_j = 2\phi_j S_j + 2U_{ij}S_i S_j, \quad (5.187)$$

for single spin excitations and

$$E_p = 2\phi_i S_i + 2\phi_j S_j \quad (5.188)$$

for pair excitations.

The energetic restrictions on the allowed spin configurations that may be considered are

$$\Theta(2\phi_i S_i + 2\phi_j S_j)\Theta(2\phi_i S_i + 2U_{ij}S_i S_j)\Theta(2\phi_j S_j + 2U_{ij}S_i S_j), \quad (5.189)$$

where the first Θ-function forces positive energies for pair excitations, and the remaining two Θ-functions force positive energies for single-spin excitations.

For the interacting spins, the density of single-spin excitations can be defined as ($\Delta > 0$)

$$\begin{aligned}
P_1'(\Delta) &= n\text{Tr}_{S_i,S_j}\int d\phi_i g(\phi_i)\int d\phi_j g(\phi_j) \\
&\quad \int dU f(U)\delta(\Delta - 2\phi_i S_i - 2U_{ij}S_i S_j) \\
&\quad \times \Theta(2\phi_j S_j + 2U_{ij}S_i S_j)\Theta(2\phi_i S_i + 2\phi_j S_j).
\end{aligned} \quad (5.190)$$

Here, $f(U)$ is the distribution of TS interactions. The evaluation of this integral gives

$$P_0\left(1 - \frac{U_0}{2}P_0\frac{1}{nR^3}\right), \quad U_0 = \langle |U|\rangle. \quad (5.191)$$

Thus, interactions lead to the dipole gap, described for the first time in Baranovskii et al. (1980).

Consider the density of pair excitations, now taking the interaction into account. Making use of the definition (5.186) we obtain

$$\begin{aligned} P_2(\Delta) &= \frac{n}{2}\text{Tr}_{S_i,S_j} \int d\phi_i g(\phi_i) \int d\phi_j g(\phi_j) \\ &\quad \times \int dU f(U)\Theta(2\phi_j S_j + 2U_{ij}S_iS_j)\Theta(2\phi_i S_i + 2U_{ij}S_iS_j) \\ &\quad \times \delta(2\phi_i S_i + 2\phi_j S_j - \Delta) \\ &\approx \frac{U_0}{2}\frac{P_0^2}{2nR^3}. \end{aligned} \quad (5.192)$$

This result leads to a finite density of pair excitations even as the excitation energy goes to zero. The contribution (5.186) can be neglected for $\Delta \ll U_0/R^3$.

Note that the identity in (5.178), that in the approximation taking pairs into account can be reduced to

$$P_1 + 2P_2 = P_0, \quad (5.193)$$

is satisfied exactly by our solution at small energies.

A pair can be understood as a bound state of two centers with a relatively large bond energy. A simple example is two spins without an external field, and with the interaction energy exceeding $k_B T$, that can be described by the Hamiltonian

$$\widehat{H} = -U_{12}S_1^z S_2^z. \quad (5.194)$$

At low temperatures, $k_B T \ll U_{12}$, only two states $S_1^z = S_2^z = \pm 1/2$ will be relevant and the dynamics of the pair is restricted to simultaneous flips of both spins.

The total correction (caused by the interaction) to the single excitation density of states can be written as the sum of the pair corrections (5.191) over j. This represents the first nonvanishing order of perturbation theory in the weak interaction U_{ij}:

$$\delta P_1 = -\left\langle \sum_j \frac{U_0}{2}\frac{P_0^2}{2nR_{ij}^3} \right\rangle. \quad (5.195)$$

We average over the spatial distribution of centers that allows us to replace the summation over j with a volume integration. After straightforward calculations similar to those used to derive (5.132), we find

$$\delta P_1 = -2\chi\xi, \quad \chi = \pi P_0 U_0 < 1, \quad \xi = \ln(L/r_{\min}). \quad (5.196)$$

Here L is the maximum radius of the interaction, corresponding to the system size, and $r_{\min}$ is some minimum size that can be defined as $(U_0/W)^{1/3}$ with logarithmic accuracy. The divergence of this correction at large distances L

5.3 On the Universality of the Low-Temperature Properties

means that the interaction remains significant despite its relative weakness. The density of pairs can be calculated in similar fashion. The total density of pairs increases as $\chi\xi$ with increasing L. It can be shown that the density of clusters containing k centers goes as $(\chi\xi)^k$. Thus $\chi\xi$ is the relevant parameter of the perturbation theory, and at $\chi\xi \approx 1$ the clusters containing many centers become extremely significant.

Let us turn to the derivation of the renormalization group theory. We will follow a procedure similar to that proposed by Levitov (1990) to study the localization in systems with long-range transition amplitudes. Let us restrict the interaction radius by some large size $L \gg r_{\min}$, but $L \ll r_{\min} e^{1/\chi}$. In this case the interaction can be treated perturbatively because of the small value of the characteristic parameter $\chi \ln(L/r_{\min}) < 1$. The density of states for single and two-particle excitations can be written as

$$P_1 = P_0(1 - 2\chi\xi), \quad P_2 = P_0 \chi\xi. \tag{5.197}$$

To derive the renormalization group equation, consider the evolution of the density of the low-energy excitations caused by increasing the interaction radius from R_1 to R_2 ($R_1 \ll R_2$, $R_1 e^{1/\chi} \gg R_2$). Since the density and the structure of the excitations changes significantly when the cutoff radius is increased by a factor of $e^{1/\chi}$, we can consider the excitations at this step as essentially pointlike objects, with a characteristic size of order $R_1 \ll R_2$. Then the same pair approximation can be applied to describe the evolution of the densities P_m corresponding to clusters containing m primary centers:

$$\begin{aligned} P_1(R_2) &= P_1(R_1) - 2\pi(\Delta\xi)P_1(R_1)\sum_{k=1}^{\infty}\langle| U_{1k} |\rangle P_k(R_1), \\ P_m(R_2) &= P_m(R_1) - 2\pi(\Delta\xi)P_m(R_1)\sum_{k=1}^{\infty}\langle| U_{mk} |\rangle P_k(R_1) \\ &\quad + \pi(\Delta\xi)\sum_{k=1}^{m-1}\langle| U_{m-k,k} |\rangle P_k(R_1) P_{m-k}(R_1); \\ \Delta\xi &= \xi_2 - \xi_1 = \ln(R_2/R_1). \end{aligned} \tag{5.198}$$

Here U_{mk} is the coupling constant between clusters m and k.

Going to the differential form of (5.198) we obtain

$$\frac{\partial P_m}{\partial \xi} = -2\pi P_m \sum_{k=1}^{\infty}\langle| U_{km} |\rangle P_k + \pi \sum_{k=1}^{m-1}\langle| U_{k,m-k} |\rangle P_k P_{m-k}. \tag{5.199}$$

The conservation law

$$\frac{\partial}{\partial \xi}\sum_{m=1}^{\infty} m P_m = 0 \tag{5.200}$$

ensures the validity of the identity (5.178). Initial conditions for P_m should obviously be taken as

$$P_m(\xi = 0) = P_0 \delta_{m1}. \tag{5.201}$$

The above relations are obtained for the ground state of the system, corresponding to zero temperature. To extend the description to finite temperatures T we can simply neglect the interaction if it is less than $k_B T$, since the binding energy should exceed the temperature in the coupled cluster. This allows us to introduce the maximum cutoff radius

$$R_T = (U_0/(k_B T))^{1/3}. \tag{5.202}$$

The solution of the renormalization group equation for $\xi_T = \ln(R_T/r_{\min}) = \frac{1}{3}\ln(W/(k_B T))$ will be responsible for the thermal properties of the system at finite temperatures.

To find the density of the excitations with some finite energy $\Delta > k_B T$, one can analogously introduce the energy-dependent cutoff radius

$$R_\Delta = (U_0/\Delta)^{1/3}. \tag{5.203}$$

If one neglects all clusters with $m > 1$, the renormalization group equations reduce to

$$\frac{\partial P_1}{\partial \xi} = -2\pi \xi P_1^2. \tag{5.204}$$

The solution of this equation for $\xi_T = \ln(R_T/r_{\min})$ results in the same behavior of the density of states as was obtained by Baranovskii et al. (1980):

$$P_1 = \frac{3}{2\pi} \frac{1}{U_0 \ln(W/(k_B T))}. \tag{5.205}$$

However, many-center clusters strongly affect this behavior (see below).

The interaction constants U_{km} depend on the number of centers in clusters k and m. Since the interaction between two pointlike excitations can be represented as

$$U_{km} = \sum_{i \in k, j \in m} \frac{U_{ij}}{R_{ij}^3} = \frac{1}{R_{km}^3} \sum_{i \in k, j \in m} U_{ij}, \tag{5.206}$$

and the different values of U_{ij} are supposed to be uncorrelated, we obtain

$$\langle |U_{km}| \rangle \approx U_0 \sqrt{km}. \tag{5.207}$$

Finally, the renormalization group equations take the form

$$\frac{\partial P_m}{\partial \xi} = -2\pi U_0 P_m \sum_{k=1}^{\infty} \sqrt{km} P_k + \pi U_0 \sum_{k=1}^{m-1} \sqrt{k(m-k)} P_k P_{m-k}. \tag{5.208}$$

Consider the asymptotic solution of (5.208) with coefficients (5.207) at $\chi \xi \gg 1$. It is easy to show that both P_1 and the total density $P(\chi \xi) = \sum_k P_k(\chi \xi)$ decrease with increasing "time" $\chi \xi$. However, the functions $P_k(0) = 0$ at $k > 1$. They increase at small $\chi \xi$ as $(\chi \xi)^{k-1}$. Therefore, the function $P_k(\chi \xi)$ at some $\xi_*(k)$ goes through a maximum and then decreases with subsequently increasing ξ. The inverse function $k_*(\xi)$ describes the position of the "wavefront", rapidly spreading to large k with increasing ξ. In other words, as interaction strength ξ is increased, the number of defects k_* in the dominant excitation increases.

The asymptotic solution of (5.208) with coupling constants (5.207) found in Burin and Kagan (1996b) and Burin and Kagan (1996c) has the form

$$P_k(t) \approx 4 \times 10^{-2} \frac{e^{-k/k_*}}{k^2 U_0 \xi} \ln(k(\chi\xi)^\alpha), \quad \alpha \approx 0.4. \tag{5.209}$$

It should be noted that the nonlinearity of (5.208) makes it possible to find the numerical coefficient in (5.209).

Taking into account the definitions of the cutoff radius and renormalization parameter ξ, respectively, we can rewrite the distribution of the excitations (5.209) in the approximate form

$$\begin{aligned} P_k &= 4 \times 10^{-2} \frac{e^{(-k/k_*)}}{k^2 U_0 \ln(W/(k_B T + \Delta))} \\ &\quad \times \ln\left(k(\chi \ln(W/(k_B T + \Delta)))^\alpha\right). \end{aligned} \tag{5.210}$$

The conservation law of the first moment, (5.178), can be used to estimate the wavefront "position", $k_*(\xi)$. Substituting the solution (5.209) into (5.178) and taking $k_*(\xi)$ as the upper limit in the integral form of (5.178), we obtain (with logarithmic accuracy)

$$k_*(\xi) \approx \exp\left(\sqrt{50(\chi\xi)}\right). \tag{5.211}$$

The numerical solution of (5.208) with coefficients (5.207) clearly demonstrates both the exponentially increasing wavefront "position" and the dependence of $P_k(\xi)$ on k close to that predicted in (5.209). Thus we see that many-center excitations do indeed become increasingly important in this self-consistent picture when the interaction lengthscale ξ is increased, as would happen if T is lowered [see Eq. (5.202)].

5.3.6 Tunneling Motion

Let us now include the intra-center motion in a primary set of defects. It can be introduced as a set of "transverse" fields Δ_{0i}:

$$\widehat{H}_{\mathrm{tr}} = -\sum_i \Delta_{0i} S_i^x. \tag{5.212}$$

Now the energy splitting of a particular primary defect is found by adding its effective fields in quadrature, $E_i = \sqrt{\Delta_i^2 + \Delta_{0i}^2}$, as usual.

Suppose first that the interaction radius is small so that the renormalization parameter $\chi\xi$ is less than unity. Then consideration can be restricted to single and pair excitations. Consider the dynamic properties of pairs of centers. The coherent tunneling amplitude connecting the excited state of a pair with its ground state is

$$\Delta_{0ij} \approx \frac{1}{2} \frac{\Delta_{0i}\Delta_{0j}}{\Delta_i \Delta_j} \mid U_{ij}(R) \mid, \quad \Delta_{0i,j} < \Delta_{i,j}. \tag{5.213}$$

This is rather similar to (5.68). As in that case, we can thus describe each pair excitation as an effective two-level entity, having an asymmetry energy Δ_{ij} and an internal tunneling degree of freedom Δ_{0ij}. The joint distribution function of the parameters Δ and Δ_0 for the pair excitations can be written as

$$\begin{aligned} P_2(\Delta, \Delta_0) &= \frac{P_0^2}{2} \int d\Delta_{01} P'(\Delta_{01}) \int d\Delta_{02} P'(\Delta_{02}) \\ &\times \int d\Delta_1 \int d\Delta_2 \int d\mathbf{R}_{12} \langle \delta(\Delta - \Delta_{12}) \\ &\times \delta(\Delta_0 - \Delta_{012}) \rangle_u, \end{aligned} \tag{5.214}$$

where $P'(\Delta_0)$ is the distribution of the tunneling amplitudes of the primary centers normalized to unity; the pair excitation energy Δ_{ij} is defined by (5.188). The integration in (5.214) gives

$$P_2(\Delta, \Delta_0) \approx P_0 \chi / (3\Delta_0). \tag{5.215}$$

The tunneling amplitude of pairs Δ_0 in this expression is defined over the interval

$$\Delta_{0*}^2 / W < \Delta_0 < \Delta_{0*}, \tag{5.216}$$

where Δ_{0*} is the characteristic scale for the distribution $P'(\Delta_0)$. Thus pair excitations already possess a uniform distribution of $\ln(\Delta_0)$, even for an arbitrary primary center distribution $P'(\Delta_0)$. This result, combined with the analysis of single defect excitations made in Ivanov (1985), explains the glass-like behavior found in solutions of rotating molecules in cryocrystals (see Bagatskii et al. (1992), Muromtzev et al. (1994)). A similar, though modified, analysis has been employed recently by Würger (see Chap. 3 of this volume and references therein, as well as Würger (1997a)) to consider pair excitations in model glasses like KBr-KCN.

We will suppose that

$$\Delta_{0*} < W e^{-\eta/\chi}, \quad \eta \simeq 1. \tag{5.217}$$

In this case at $t_m \equiv \xi \ln(R_{\max}/R_{\min}) \approx 1$ the tunneling amplitudes for pairs appear to be less than the effective interaction $U_0/R_{\max}^3$.

We want to generalize Eq. (5.213) to find the effective tunneling parameter Δ_{0m} of a cluster of m defects. If this cluster results from the coupling of k and $n-k$ clusters, then Δ_{0m} is defined by (5.213), with the replacement of Δ_{0i}/Δ_i and Δ_{0j}/Δ_j by Δ_{0k}/Δ_k and $\Delta_{0m-k}/\Delta_{m-k}$, respectively.

The analysis of (5.214) shows that the region $\Delta_k \sim \Delta_{m-k} \sim \mid U_{k,m-k} \mid /R^3$ gives the main contribution into the integral. This allows one to write

5.3 On the Universality of the Low-Temperature Properties

$L_m \approx L_m + L_{m-k} + 3\xi_R$, where $L_k = \ln(\Delta_{0k}/W)$. Continuing with an analogous procedure for L_k and L_{m-k}, we can approximately represent L_m through the primary defect tunneling amplitudes:

$$L_m \approx mL_* + 3m/\chi, \quad L_* = \ln(\Delta_{0*}/W). \tag{5.218}$$

According to the inequality (5.217) the second term in (5.218) can be neglected in comparison with the first one. Then using (5.209) and proceeding from the summation over m ($m > 1$) to the integration we find the general distribution function at $t_m \geq 1$ and finite temperature T:

$$\begin{aligned} P(\Delta, \Delta_0) &= \int dm P_m(\xi_{max}) \delta(\Delta_0 - We^{mL_*}) \\ &= 4 \times 10^{-2} \frac{3}{\pi U_0 \ln(W/(\Delta + k_B T))} \frac{\ln(W/\Delta_{0*})}{\ln^2(W/\Delta_0)} \frac{1}{\Delta_0}. \end{aligned} \tag{5.219}$$

Here R_{max} was taken in accordance with (5.202) with the substitution $\Delta \to E = \sqrt{\Delta_0^2 + \Delta^2}$. The condition $t_m > 1$ requires the inequality (5.217) to be satisfied not only for Δ_0 but for Δ and $k_B T$ as well.

5.3.7 Discussion of the Results

The distribution obtained above does not noticeably deviate from a uniform distribution of $\ln(\Delta_0)$. Since χ/P_0 does not depend on the distribution of primary centers, we come to the important conclusion that the resulting distribution function is not dependent on the primary centers' density or nature. It is interesting that the numerical factor in (5.219) also determines the quantitative scale of the distribution.

If one introduces the notation $P(\Delta, \Delta_0) = P_0/\Delta_0$, then the dimensionless parameter $\chi = P_0 U_0$ has a rather universal value for markedly different glasses. It is remarkable that (5.219) predicts that χ depends only logarithmically on the parameters of the primary defects, and its numerical value is close to the experimental value $\approx 10^{-3}$. An analysis of the absolute magnitude of the internal friction provides good support for the above conclusion.

Consider the internal friction Q^{-1} in the domain of the plateau, $T > T_m$, where it is actually independent of the temperature and frequency of sound. Let us restrict our consideration to dielectric glasses. Then we find

$$Q^{-1} \approx P_0 \Delta_0 U_{\text{eff}}, \tag{5.220}$$

where $U_{\text{eff}} = \gamma_{\text{eff}}^2/(\varrho v^2)$ and Δ_0 are the effective coupling constant and tunneling amplitude of the excitations that are responsible for the attenuation of sound with small enough frequency ω. The relevant TS's can be found from the condition

$$\omega \tau \approx 1. \tag{5.221}$$

The relaxation time τ in dielectric glasses at these temperatures is caused by the TS-phonon interaction [see (2.18)], and for thermal TS's such that $E = k_B T$,

$$\tau_1^{-1} = A\Delta_0^2 k_B T. \tag{5.222}$$

For vitreous silica the constant A can be estimated as $10^8 \text{s}^{-1}\text{K}^{-3}$ (Carruzzo et al. (1994), see also Chap. 4). For fixed ω and T one can use (5.221) to find the relevant tunneling parameter,

$$\Delta_0 \approx \sqrt{\frac{\omega}{Ak_B T}} \tag{5.223}$$

and $n_{\text{eff}} \approx \ln(W/\Delta_{0*})/\ln(W/\Delta_0)$. Substituting these values into the distribution (5.219), we obtain

$$Q^{-1} \approx 4 \times 10^{-2} \frac{1}{\ln(W/(k_B T)) \ln(W/\Delta_0)}. \tag{5.224}$$

Taking $\omega \approx 10^3 \text{ s}^{-1}$ and $W \approx 10^2$ K, we find

$$Q^{-1} \approx 10^{-3}, \tag{5.225}$$

which is close to the experimental value.

Thus the low-energy spectral properties of many-center excitations caused by $1/R^3$ interactions between defect centers demonstrate broad universality: Low-energy excitations that act like effective TS with a quasi-uniform distribution of $\ln \Delta_0$ and Δ; the absence of any strong influence of the density and distribution of primary centers on the final effective TS distribution function; and quantitative relevance to the experimental data. These results support the hypothesis discussed by Klein et al. (1978) and Yu and Leggett (1988) about the principal role of $1/R^3$ interactions in determining the nature of the universal properties of amorphous solids.

Let us summarize some recent data that are relevant to this model. Both experiments and numerical simulations support the significance of many-center clusters. There exists experimental evidence that the number of primary centers participating in each effective TS exceeds unity (for a discussion of this, see Hunklinger and Raychaudhuri (1986), Sects. 4.3 and 4.4). Recent numerical molecular dynamic simulations made by Heuer and Silbey (1996) also show that the number of atoms effectively participating in each TS can be essentially larger than unity. Moreover, Heuer and Silbey (1996) have found that the distribution of the number of atoms participating in tunneling excitations appears to be very close to that described by (5.209). This can be seen in Fig. 5.29, where the dots represent the data of the molecular dynamics taken from Heuer and Silbey (1996) and the solid line corresponds to (5.209).

Indirect arguments in favor of the significance of $1/R^3$ long-range interactions were proposed already by Yu and Leggett. In particular, they try to understand the superlinear temperature dependence of the heat capacity,

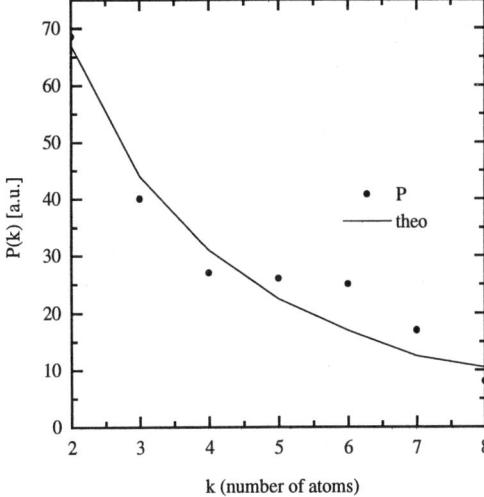

Fig. 5.29. Simulation results and analytic expression for distribution of number of atoms participating in each TS, after Heuer and Silbey (1997).

$C \propto T^{1+\alpha}$, as the result of the dipole gap in the density of states. The direct experimental evidence for the dipole gap has been described above in the discussion of the measurements of the Stanford group (described extensively in Sect. 5.2; see also Salvino et al. (1994), Rogge et al. (1996b), Rogge et al. (1997b), Natelson et al. (1997)).

However, recent data obtained by White and Pohl (1995) seems to be in contradiction with theories based on such long-range interactions. In these experiments, the internal friction of a very thin sample containing from three to 50 monolayers of vitreous silica attached to bulk crystalline Si has been measured. The authors conclude that the internal friction shows the same behavior as in bulk samples, which implies that $1/R^3$ interactions cannot be significant.

From the point of view of the theory outlined above, the interaction is significant if its energy scale exceeds $k_B T$. Only the three monolayer experimental point from the Cornell data involves sample dimensions that are comparable to the thermally limited interaction length scale given by (5.202), however, and for that point the internal friction already differs remarkably from the bulk sample. Because of this thermal limitation on the length scale over which interactions are important, it is not clear that this data actually conflicts with the interacting theory.

Another significant issue to be studied when analyzing these experiments is the strong acoustic mismatch between the amorphous film and the crystalline substrate; see Sect. 4.4.5. Note that the results of the renormalization group analysis are not sensitive to the dimensionality of the system. In a d-dimensional system, the exchange of virtual phonons produces a $1/R^d$ interaction between tunneling defects; this is a straightforward consequence of the

d-dimensional generalization of (5.45). Then all relations of the renormalization group analysis remain nearly unchanged. Thus, if the phonon spectrum of the surface layer is similar to that in an isolated two-dimensional film, no remarkable changes in the properties of the tunneling states are expected.

There may be quasi-isolation of the surface layer from the bulk crystalline substrate due to the markedly differing acoustic properties and weak coupling between the materials at their boundary. Indirect evidence of anomalies in the phonon spectrum of the thin amorphous surface layer at energies as low as 10^{-4} eV has been observed recently by Gaganidze et al. (1997) in sound velocity measurements in samples similar to those studied by the Cornell group.

It is important to consider experiments to verify the predictions of the renormalization group theory. The predictions of this theory for the time-dependent heat capacity or heat release can probably be checked directly. The renormalization group theory predicts the following time-dependent specific heat:

$$C(t) \propto \frac{1}{\ln(W/(k_B T))} - \frac{1}{\ln(W/(k_B T)) + \ln(t/t_{\min})/2}. \tag{5.226}$$

If the earliest experimental data are available at time t_0 after an initial heat pulse, then remarkable deviations from logarithmic time dependence of $C(t)$ might be observed at

$$t \approx \frac{t_0^2}{\tau_T} \frac{W^2}{(k_B T)^2}. \tag{5.227}$$

This crossover time can be made lower by decreasing the temperature and the initial time of the measurements. For current typical values $T \sim 1$ K, $\tau_T \sim 10^{-8}$ s and $t_0 \sim 1$ s, unrealizably long times are necessary. However, decreasing the initial time of observation to 10^{-3} s with the same other parameters gives one hour for t. Such measurements would be an important test for the interacting theory.

Another possibility is to look for evidence of the discreteness of the TS distribution when a small number of atoms n participate in one effective TS. In particular, the saturation of the equilibrium dielectric response at very low temperatures seen by Rogge et al. (1997c) and discussed above in Sect. 5.1.8 may be evidence for such discreteness. The observed saturation effect may be a signature of the change of those excitations that are the relevant effective TS's, for instance a change from pairs to triplets. In fact, since the density of triplets is several times less than the density of pairs [see (5.209)] the observed saturation could be caused by the replacement of pairs by triplets at $T \sim 10$ mK. Taking the characteristic tunneling parameter of single primary defects $\Delta_{0*} \sim 1$ K and the maximum bond energy of pairs about $W \sim 100$ K, we can estimate the minimum energy scale of the pair excitations as [see (5.216)] $\Delta_{0*}^2/W \approx 0.01$ mK, which is quite reasonable.

One should note that almost all measurements of relaxations including, for instance, time-dependent heat capacity and hole-burning experiments, show some anomalous behavior at early times, 1 ms $< t <$ 10 ms, that could be a manifestation of such discreteness.

5.4 Conclusion and Remarks

In this chapter we have attempted to provide an overview of recent experimental evidence for and theoretical treatments of interactions between TS's in amorphous solids. The remarkably universal applicability of the standard TS model and systematic deviations from the predictions of that model at low temperatures have motivated both the experimental and theoretical efforts described here.

We have discussed the standard noninteracting TS model and its predictions for equilibrium dielectric and acoustic properties. Further, we have analyzed one long-known consequence of interactions, spectral diffusion. From there, we presented several experimental techniques and results (equilibrium acoustic and dielectric loss as a function of temperature, low-temperature dielectric saturation, nonequilibrium dielectric relaxations after the application of electric and strain fields, nonequilibrium acoustic behavior after the application of dc electric fields) that are explained by a model employing dipolar inter-TS interactions. Profound interaction effects manifest themselves at very low temperatures $T < 0.1$ K because the phonons are substantially frozen out, and the kinetics of the TS system is dominated by TS-TS interactions rather than TS-phonon processes.

Finally, after describing the interacting model, we presented a renormalization group treatment of the system, showing that the very presence of interactions can lead to the observed universal distributions employed by Anderson et al. (1972) and Phillips (1972). Although the starting point of this model is a set of some unknown preliminary defects, the properties appearing at low temperatures are rather insensitive to those initial assumptions. Moreover, the scale-invariant solution found within the renormalization group analysis satisfactorily describes the low-temperature glassy universality. The elementary excitations at low temperature are clusters of coupled excitations of primary defects. Clusters containing many centers are responsible for the long time relaxation behavior seen in various experiments.

By relaxing one underlying assumption of the original TS model, it has been possible to explain a wealth of experimental data, and to gain insight into the problem of the universality seen in amorphous materials at low temperatures.

6. Investigation of Tunneling Dynamics by Optical Hole-Burning Spectroscopy

Hans Maier, Boris Kharlamov, and Dietrich Haarer

6.1 Introduction

Spectral hole burning originally evolved from the optical spectroscopy of intramolecular electronic and vibronic energy levels of organic molecules – mainly large dye molecules – with the solid surrounding these molecules merely serving as a matrix to handle them. The first observations of persistent hole burning were done by Kharlamov et al. (1974) in organic glasses and by Gorokhovskii et al. (1974) in organic crystals.

While the method became more sophisticated and the spectral resolution was increased, the necessity arose to understand the details of the interactions of the dye molecules with their surrounding matrix in order to describe the measured spectra. After that it was only the next logical step to turn the purpose of the experiment upside down and employ the molecules as probes for the properties of their surrounding matrix – spectral hole burning became a tool of solid state physics. This step was of course not as straightforward as described here. Efforts had to be made to understand the basics of what was observed. But finally spectral diffusion was discovered and confirmed in a number of investigations. Since then the technique of spectral hole burning has developed into a highly efficient standard method especially suited for the investigation of the low-temperature tunneling dynamics in organic amorphous solids.

Since this is the only chapter in this book which deals with optical spectroscopy, we first outline the basics of the optical spectroscopy of solids to provide the reader with the background necessary for the understanding of the techniques and experimental observations presented in this chapter. This outline also serves as a summary of many of the experimental results of previous decades. After an introduction to the experimental implementations of hole-burning spectroscopy, which also describes the special possibilities as well as the limitations of this technique, a section on the physics of photochemical and nonphotochemical hole burning is included.

The experimental results presented in the remainder of this chapter represent some of the very recent achievements in the field of spectral diffusion investigations of amorphous solids. Emphasis has been laid on the subject of long-term equilibrium dynamics – a field which cannot be readily investigated by conventional techniques of experimental solid state physics. But also nonequilibrium experiments are included – some of them illustrate the connection between spectral diffusion measurements and the heat-release experiments of Chap. 2, others represent a link to dielectric investigations.

6.2 Optical Spectra of Impurities in Solids

Optical spectroscopy of electronic transitions of impurities in solids is a powerful tool for the investigation of both the physical structure of impurity centers and, using the impurity as a spectroscopic probe, the physical properties of the matrix itself. Due to the very high sensitivity and spectral resolution of optical methods, the spectroscopy of impurity centers in solids serves as one of the most powerful methods for the investigation of insulating and semiconducting crystals. The last decades' development of new high-resolution optical methods – such as photon echo (PE), fluorescence line narrowing (FLN), persistent hole burning (HB), and single-molecule spectroscopy (SMS) – tremendously increased the ability of optical spectroscopy of solids, especially amorphous solids.

The aim of this chapter is to introduce and give a survey of the recent achievements in the field of the optical investigation of tunneling processes in amorphous solids. The original results discussed in the following sections were obtained in the investigation of organic glasses. Therefore, the following introductory review of optical impurity spectra in solids is restricted to organic systems. [A detailed list of references both for classical and new results of structure and dynamics of solids obtained via optical methods can be found in a recent review by Skinner and Moerner (1997) and other reviews cited below].

6.2.1 Crystals

Optical Bands of Impurities. In organic glasses optical impurities are mostly organic dye molecules, i. e. molecules which absorb light in the visible range and therefore create a specific color to the human eye. For this reason such molecules are also often referred to as chromophores. Here we consider only the lowest electronic transitions of such organic chromophore molecules in solids. In general, at least three energy levels of the impurity should be taken into account: The singlet ground state S_0 and the first excited singlet state S_1 as well as the lowest triplet state T_1. The corresponding energy diagram is shown in Fig. 6.1.

In absorption spectroscopy, the presence of a triplet state can usually be neglected, but this is not always true. The lifetime of the T_1-state of organic chromophores usually ranges between 10^{-4} and 10^{-1} seconds. This value is so high compared to a typical S_1 lifetime of nanoseconds because an optical transition $T_1 \to S_0$ involves a spin conversion. Nonradiating transitions

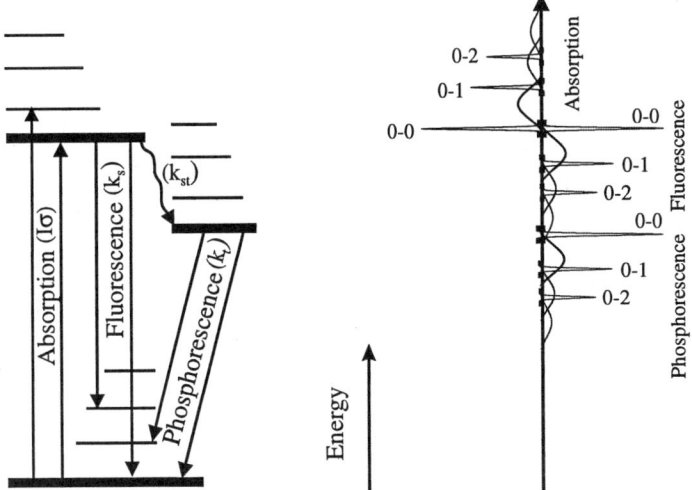

Fig. 6.1. Energy diagram of the three lowest electronic levels with vibrational sublevels of an organic molecule: The singlet ground (S_0) and excited (S_1) states and the triplet (T_1) state. The optical transitions and the absorption, fluorescence and phosphorescence spectra including phonon bands are also shown schematically. The numbers in the right-hand part denote the vibrational sublevels involved in the respective optical transitions, i.e. 0-0 denotes a transition from a vibrational ground state into a vibrational ground state.

from the excited singlet state S_1 to the triplet state T_1, so-called intersystem crossing processes, are however possible. Their yield can be as high as some tens of percent. This means, a saturation of the triplet state can be reached at rather moderate excitation conditions. This is considered in Sect. 6.3.

The spectra of individual chromophore molecules consist of optical bands, including a so-called 0-0 transition band which means a purely electronic transition (which will be of major interest in the further discussion) and vibronic bands, corresponding to transitions accompanied by the excitation of molecular vibronic modes. At sufficiently low temperatures (for organic crystals typically below 70–40 K) every optical band consists of two components: A sharp zero-phonon line (ZPL) and a broad phonon band (PB). [General aspects of impurity spectra in crystals have been discussed by Maradudin (1968) and Rebane (1968), detailed reviews of theoretical and experimental results were given by Osad'ko (1979), Osad'ko (1983), Skinner (1986), and Osad'ko (1991)]. The phonon band reflects an inelastic interaction of the impurity with the matrix upon electronic transitions: The creation or annihilation of phonons by the electronic transition. The relative magnitude of the zero-phonon line as compared to the total optical band intensity is characterized by the Debye–Waller factor (DWF)

$$\alpha = \frac{I_{\text{ZPL}}}{I_{\text{ZPL}} + I_{\text{PB}}}.$$

The DWF is temperature dependent and reaches its maximum value (determined by the electron–phonon interaction between impurity and matrix) at temperatures far below the Debye temperature Θ_D (typical values of Θ_D for organic crystals are 50–100 K). The maximum value of the DWF varies for different organic systems from 0 (many dye molecules) to 0.8 or even 0.9 (some polycyclic compounds in n-paraffin matrices). At temperatures of about Θ_D and higher, zero-phonon lines practically disappear and only broad phonon bands remain in the optical spectra. Therefore, not only high resolution but even moderate resolution spectroscopy of impurity centers in organic solids requires liquid-helium or at least liquid-nitrogen temperatures.

Homogeneous and Inhomogeneous Line Broadening. The zero–phonon lines of impurities are broadened via interactions with the host. These can be separated into static interactions causing an inhomogeneous broadening and dynamic processes which are responsible for homogeneous broadening.

Inhomogeneous broadening is a statistical distribution of the transition frequencies of impurity molecules due to imperfections of the crystalline structure, random variations of impurity implantation into a host crystal, mutual interaction of impurities, etc. It does not influence the spectral band shape of individual molecules, but increases the optical band width of the whole sample. The inhomogeneous broadening in inorganic crystals was thoroughly investigated both theoretically and experimentally (see Hughes (1968), Stoneham (1969), Orth et al. (1993), Jaaniso et al. (1994) and references therein). In organic crystals this is a more complicated problem than in inorganic ones due to the larger size of both host and guest molecules and the high sensitivity of the electronic spectra of organic molecules to the surrounding structure. There are special pairs of guest/host molecule combinations providing small inhomogeneous broadening (of the order of a few wave numbers), for example naphthalene in durene or pentacene in paraterphenyl. In addition, there is a rather universal class of host crystals – so-called Shpol'skii matrices – normal paraffins, where many complex organic molecules are incorporated into discrete sets of "well-defined" sites. Spectroscopically, this results in so-called Shpol'skii multiplets: Each component of a multiplet corresponds to a definite site. The inhomogeneous broadening within one of these sites is typically in the range of $1-10\,\text{cm}^{-1}$, depending on the host/guest pair. A spectral resolution of that magnitude is sufficient for many spectroscopic purposes, but not for high-resolution studies of the subtle details of impurity–host interactions, such as, for example, spectral diffusion. Many methods such as saturation spectroscopy, photon echo, hole burning, fluorescence line narrowing and single-molecule spectroscopy have been successfully employed to overcome the inhomogeneous broadening both in crystalline and amorphous solids. Some of these methods related to the topic of this chapter are described in more detail below.

The physical limit of the achievable spectral resolution is determined by the homogeneous width Γ of the zero-phonon line for a given guest/host combination. It is limited by two factors: The lifetime T_1 of the relevant energy level (which is largely temperature independent) and a temperature-dependent pure dephasing time T_2^* determined by elastic interactions with other excitations,

$$\Gamma = \frac{1}{\pi T_2} = \frac{1}{2\pi T_1} + \frac{1}{\pi T_2^*}.$$

The homogeneous width of vibronic lines at low temperatures is usually determined by the lifetimes of the involved vibrational modes which are very short. Therefore this width is in the range of $1\text{–}10\,\text{cm}^{-1}$ even at low temperatures. The lifetime of singlet electronic states of impurity molecules is usually inbetween $10^{-8}\,\text{s}$ and $10^{-10}\,\text{s}$, which corresponds to linewidths of 10^{-1} to $10^{-3}\,\text{cm}^{-1}$. That is why 0-0 transition lines are the main objects of high-resolution optical spectroscopy of impurities in solid.

At nonzero temperatures, however, there is still the second contribution to the linewidth, which is determined by the pure dephasing time T_2^*. The impurity lines in crystals are broadened by interactions with phonons: Acoustic and optic phonons as well as local modes. A detailed analysis of electron–phonon interactions and temperature-dependent line broadening can be found in reviews of Osad'ko (1979), Osad'ko (1983), Skinner (1986) and Osad'ko (1991). Here we give only a brief overview. In the case of weak electron–phonon interaction, the temperature dependent part of the homogeneous linewidth can be presented as (see for example Osad'ko (1991))

$$\frac{1}{T_2^*} \propto \int_0^\infty d\omega \left(\frac{\gamma(\omega)}{\sinh(\frac{\hbar\omega}{2k_B T})} \right)^2,$$

where ω is a frequency of the corresponding vibrational mode and $\gamma(\omega)$ is the phonon spectral function.

From this equation the limiting cases for the temperature dependence of homogeneous line broadening can be derived:

- Interactions of impurities with acoustic phonons: The Debye model yields for acoustic phonons:
  $\gamma(\omega) \propto \omega^3$ at $\omega \leq \omega_D$, and $\gamma(\omega) = 0$ if $\omega > \omega_D$. This results in $1/T_2^* \propto T^7$ if $T \ll \Theta_D$, and $1/T_2^* \propto T^2$ if $T \gg \Theta_D$.
- Interactions of impurities with optical phonons or local modes with an energy $\hbar\omega$:
  $1/T_2^* \propto \exp(-\hbar\omega/k_B T)$ for $\hbar\omega \ll k_B T$ and $1/T_2^* \propto T^2$ for $\hbar\omega \gg k_B T$.

All these cases were observed experimentally. In many inorganic crystals, the 0-0 line broadening can be described very well by interactions with acoustic phonons: McCumber et al. (1963), Yen et al. (1964), Thorne et al. (1985),

Powell et al. (1985), Babbit et al. (1989). The main source of optical dephasing in organic crystals is the interaction of impurities with pseudolocal modes, which results in an exponential line broadening in the low-temperature range as observed by Völker et al. (1977), Völker et al. (1978), Gorokhovskii and Rebane (1977), Korotaev and Kalitievski (1980), Dicker et al. (1981), Olson et al. (1982), Duppen et al. (1981), Molenkamp and Wiersma (1984) and Kikas et al. (1993).

In general, the behavior of homogeneous lines of impurities in crystals was successfully explained on the basis of existing theories of crystals. Some general properties of temperature-dependent optical line broadening in crystals should be noted, which are especially interesting in comparison with the analogous characteristics of glassy systems:

1. The temperature dependence of the linewidth is strongly nonlinear. Even at high temperatures a power law with an exponent of $\beta \geq 2$ can be measured.
2. At low temperatures the interaction with phonons is already very weak and usually $\Gamma(T) \approx 1/2\pi T_1 = \text{const}$ at $T \leq 5-6$ K. This is illustrated in Fig. 6.2 (taken from Kikas et al. (1993)) where comparing measurements of the homogeneous line broadening of chlorin in a special matrix were performed: Benzophenone is an organic molecule which exists as a crystalline solid and also in a glassy modification. Note the difference of the temperature-dependent line broadening in crystal and glass. This will be discussed in more detail in the following.

6.2.2 Amorphous Solids

Origin of the Inhomogeneous Line Broadening. The main distinction of amorphous solids is the absence of long-range order. This is the origin of many very different properties as compared to crystals. In optical spectroscopy, one of the most general features is a large inhomogeneous broadening of impurity spectra in glasses which is usually orders of magnitude larger than in crystals. In organic glasses, a typical value for the width of an inhomogeneously broadened absorption band is 100-200 cm^{-1}. The shape of such a band (inhomogeneous distribution function) in a glass is approximately Gaussian (Stoneham (1969), Kador (1991)). The large inhomogeneous broadening in glasses is the reason why their optical spectra are fundamentally different from those in crystals. It reduces the spectral resolution to such an extent that typical features of optical bands such as zero-phonon line/phonon band structures and even vibrational structures are "smeared out". The situation is depicted in Fig. 6.3 where different local guest-host configurations are shown: The impurity molecules are displayed as ellipses in a crystalline and an amorphous solvent schematically sketched as a regular and a distorted lattice. Due to their different environments, their spectra are located at different frequencies, as is symbolically depicted in the right-hand part of the figure. The spectrum of the crystal consists of a discrete number of lines, corresponding to different impurity sites, whereas the spectrum of the glass has a continuous distribution. If the inhomogeneous broadening is comparable with the vibrational frequencies of impurity molecules, all spectral features are smeared out. Only broad structureless bands can be observed in such electronic transition spectra. This is the case for organic molecules embedded in most organic glasses and polymers. For a long time, this circumstance was the general obstacle in optical spectroscopy of glasses and it was only overcome by the development of the high resolution techniques mentioned above.

Peculiarities of Homogeneous Broadening and its Temperature Dependence. Due to the large inhomogeneous broadening, the measurement of homogeneous linewidths of chromophores in glasses is impossible by conventional absorption or fluorescence spectroscopy. At the same time the general spectral features of individual impurity molecules in organic glasses are very similar to their features in crystals: Fine vibrational structure, zero-phonon line with phonon bands, etc. This became clear after the development of the

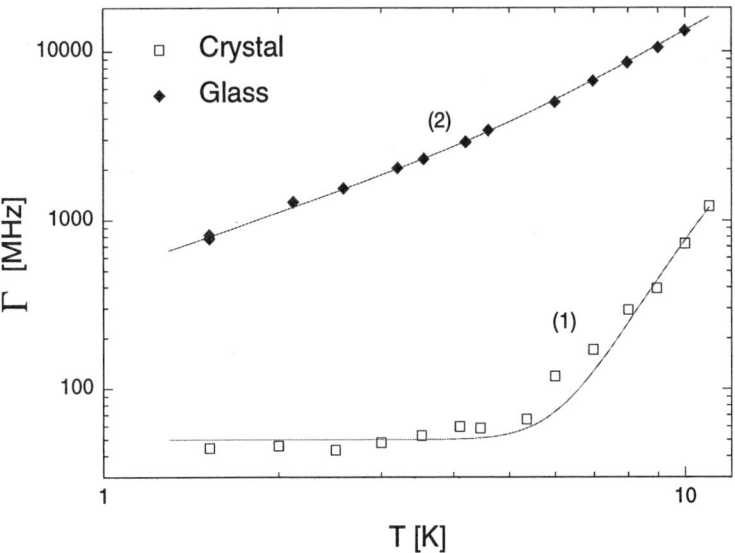

Fig. 6.2. Temperature dependence of the homogeneous absorption linewidth of chlorin in benzophenone. In the crystalline phase, the linewidth is essentially temperature independent below 5 K, whereas in the glassy modification the linewidth is larger by more than an order of magnitude and decreases monotonously with decreasing temperature. The solid lines are fit curves corresponding to: (1) $\Gamma = \Gamma_0 + A\exp(\hbar\omega/kT)/(\exp(\hbar\omega/kT)-1)^2$; (2) $\Gamma = \Gamma_0 + BT^{1.28} + C\exp(\hbar\omega_1/kT)/(\exp(\hbar\omega_1/kT)-1)^2$; where $\omega \simeq 36$ cm^{-1} and $\omega_1 \simeq 18$ cm^{-1} are the frequencies of the pseudolocal modes fitted for the crystalline and glassy phase, respectively. The reproduction of the experimental data and fit results has kindly been granted by Kikas et al. (1993).

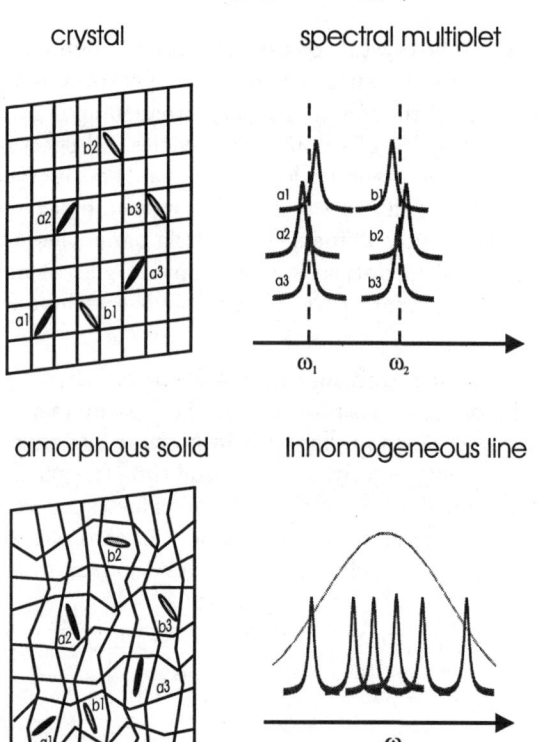

Fig. 6.3. A comparative sketch of crystal and glass with embedded impurity molecules. The impurities in the crystal build up two well-defined sites with different orientations which correspond to a doublet in the optical spectrum. The disorder of the glass causes a continuous Gaussian distribution of impurity transition frequencies in a broad spectral range, which is of the order of the distance between the multiplet components in the crystal.

first of a group of high resolution spectroscopy methods in glasses: Fluorescence line narrowing (FLN) (Personov et al. (1972), Personov (1983)). The first attempt to measure a homogeneous linewidth in a glass was made in 1974 by Kharlamov et al. (1974) introducing another new method – persistent hole burning (PHB). The measured 0-0 transition linewidth of perylene in ethanol at 4.2 K ($\approx 0.3\,\mathrm{cm}^{-1}$) appeared to be at least an order of magnitude larger than homogeneous linewidths in organic crystals. Later measurements confirmed that such a situation is typical, not only for organic but also for inorganic glasses (Selzer et al. (1976), Hegarty and Yen (1979)). For a long time, the widths measured via hole burning were treated as homogeneous in analogy to crystals. Much later it was ascertained that the situation in glasses is more complicated than in crystals and the linewidth of impurities in glasses measured via hole burning cannot be treated as truly homogeneous (see below). In the following we will call the linewidths measured by hole-burning, three-pulse photon echo, etc. "quasi-homogeneous". Another strange peculiarity was already found in very early measurements: The temperature dependence of the quasi-homogeneous linewidths as measured using hole burning by Thijssen et al. (1982) and using the photon echo technique by Walsh et al. (1986) is approximately linear. In the most extensively investigated organic glasses and polymers, an algebraic temperature dependence was found with an exponent of $\alpha \simeq 1.3$ (see Völker (1989), and references therein). The thermal line broadening in inorganic glasses shows a less uniform behavior: The exponent in the power law is distributed in the range $\alpha = 1 - 2$ for different systems (Macfarlane and Shelby (1987)), but these values cannot be explained either in the framework of "classical" homogeneous line broadening in crystals. As mentioned above, the temperature dependence of homogeneous linewidths in crystals cannot be "weaker" than T^2. So the properties of zero-phonon lines in glasses turned out to be very different from those in crystals (see the results of comparative measurements in Fig. 6.2).

These two features of impurity spectra in glasses – broad quasi-homogeneous lines and their quasi-linear temperature dependence – were interpreted in the same way as other "anomalous" low-temperature properties of glasses (specific heat, thermal conductivity, sound attenuation, etc.): As a result of an interaction of the impurities with two-level systems (TLS's) (Hayes et al. (1981), Small (1983)). Two-level tunneling systems are phenomenologically introduced local excitations with a very broad distribution of energies and relaxation rates [Anderson et al. (1972), Phillips (1972), Phillips (1987), see also Chap. 2].

Assuming a standard model distribution of TLS parameters as the one proposed by Anderson et al. (1972) and Phillips (1972), their interactions with impurities result in a linear temperature dependence of the zero-phonon line temperature broadening as was shown by Small (1983). A whole series of subsequent theoretical papers [see for example the list of references in the review of Osad'ko (1991)] – which differed in their calculation techniques but all employed the two-level tunneling model (with a few exceptions based on fractal models, for example) – was devoted to the explanation of the anomalous homogeneous broadening of zero-phonon lines in glasses. Definite problems were caused by the fact that the observed exponent of $\alpha \simeq 1.3$ deviated from the theoretical prediction $\alpha = 1$. This deviation was for example explained by simple modifications of the TLS's distribution function: The distribution of TLS's asymmetries within the standard model is $P(\Delta) = \text{const}$; if we assume $P(\Delta) = \Delta^{0.3}$, this results in the observed temperature dependence. It was shown by Reineker and Kassner (1986) that an averaging over interactions of an impurity with many TLS's also yields a linewidth $\Gamma(T) \propto T^{1.3}$. Jackson and Silbey (1983) explained the observed dependence by a combination of two different interaction mechanisms: (a) impurity–TLS, resulting in a linear temperature dependence and (b) impurity–local mode with an exponential temperature dependence. It is very surprising that only one group of authors, Hunklinger and Schmidt (1984), payed attention to the fact that due to the temperature dependence of the tunneling relaxation rate (2.18) of the tunneling systems (TS's)[1] even the standard tunneling model temperature dependence of optical linewidths should be slightly nonlinear (see below). This result was recently calculated in a more general form for two-pulse and three-pulse photon echo experiments by Geva and Skinner (1997a). The true origin of the nonlinearity of the zero-phonon line broadening in glasses is unclear even now, but we know that the influence of both TS's and local modes (Jackson and Silbey (1983)) should be taken into account for interpreting the line broadening.

Another surprising feature of zero-phonon lines in glasses was discovered in 1984 in the organic system quinizarin/alcohol glass by Breinl et al. (1984a), Friedrich and Haarer (1986): The time dependence of spectral linewidths, so-called spectral diffusion. This effect was explained in terms of interactions of impurities with slowly relaxing TS's and a theory of time-dependent spectral line broadening in glasses previously proposed by Reinecke (1979) was used for the description of the experimental results. This theory applied the earlier model of Klauder and Anderson (1962), developed for spin-1/2 systems, and its modification for phonon echoes in glasses (Black and Halperin (1977)) to optical transitions in glasses. Later, optical spectral diffusion was also discovered in inorganic systems in three-pulse photon echo experiments (Broer and Golding (1986), Broer (1986)).

At the beginning, the presence of spectral diffusion in optical spectra hardly influenced the interpretation of linewidths measured via hole burning. Spectral diffusion was conceived as a minor addition to the homogeneous linewidth. But this interpretation changed dramatically after the discovery of a huge discrepancy between the "homogeneous" linewidth of the same impurity in the same glassy host, as measured by two different techniques: Two-pulse photon echo and persistent hole burning (Molenkamp and Wiersma (1985), Walsh et al. (1986), Berg et al. (1987a)). The photon echo linewidths turned out to be 4-8 times narrower than the results of hole burning. The difference between these two methods is that the first one measures the coherence time of the electronic excited state of an impurity, so to speak the homogeneous linewidth per definition, whereas the second one is much slower: Persistent hole burning itself and the measurement of the spectrum take some tens of seconds. Nowadays, it is known that TS's relaxation times are distributed on a huge time scale, starting below the picosecond region and extending at least to 10^6 s. This resulted in a new interpretation of homogeneous linewidths in glasses as compared to crystals: The linewidth measured in hole-burning experiments at low temperatures is extremely broadened by spectral diffusion. This is the reason why the hole-burning linewidth is usually

---

[1] The abbreviation TS's will be used when the emphasis is given to the tunneling rate of the TLS's.

called "quasi-homogeneous". It follows from incoherent photon echo measurements for example by Narasimhan et al. (1991) and by Gruzdev and Vainer (1993) that even on the nanosecond time scale, the spectral diffusion contribution to the measured linewidth is readily observable. In terms of its quantum-mechanical definition, the homogeneous linewidth of an individual chromophore is determined by the width of its excited state energy level. Spectroscopically, the homogeneous linewidth is related to the coherence time of the molecule's excited state. According to this definition the homogeneous linewidths in glasses, where spectral line profiles are time dependent, can only be measured by coherent techniques, like two pulse photon echo or free induction decay. However, even such methods average in glasses over broad distributions of homogeneous widths of individual molecules. A detailed analysis of experimental results from different techniques in situations with broadly distributed and time-dependent linewidths has been performed by Vainer et al. (1996), Geva at al. (1996), Geva and Skinner (1997a), and Geva and Skinner (1997b).

6.3 Basic Methods of Hole-Burning Spectroscopy

6.3.1 Introduction

Spectral holes are created within inhomogeneously broadened optical bands by illuminating the band using a light source with a band-width smaller than the inhomogeneous width. This irradiation process selects those absorbers whose homogeneous lines overlap with the frequency of the incident light.

The simplest example of a spectral hole is the saturation of resonant two-level transitions in spin resonance experiments. Unlike spin resonance, not many systems in solid state optical spectroscopy can be treated as two-level transitions with long excited-state lifetimes. There are some examples of a successful application of this saturation technique to inorganic systems (Szabo (1975), Erickson (1977), Macfarlane and Shelby (1979), Schmidt et al. (1993), Schmidt et al. (1994)). In organic systems, however, there is another approach: populating the triplet state (Shelby and Macfarlane (1979), Jankowiak et al. (1993), Lindrum and Nickel (1990), Van der Zaag et al. (1990), Jahn et al. (1991), Wannemacher et al. (1992), Wannemacher et al. (1993), Kharlamov et al. (1994)). As already mentioned in the previous section, many large organic molecules have long-lived triplet states and high intersystem crossing yields (i. e. transitions between excited singlet and triplet states, see Fig. 6.1). Therefore, the triplet state of such molecules can easily be saturated and a transient spectral hole can be created. But in general the application of saturation methods is more difficult in optical solid state experiments than in spin resonance.

In the last two decades, however, a variety of systems were found which can be permanently bleached by light absorption at low temperatures: Upon excitation there is a nonzero probability for such systems to end up in a modified ground state with a different absorption frequency. This corresponds to a depletion of the number of states in the inhomogeneous frequency distribution of the absorbers in the vicinity of the excitation frequency: A persistent "hole" is created in this distribution function and, therefore, a spectral hole appears in the absorption spectrum. Besides inorganic optical absorbers like rare earth elements which undergo such modifications (see, for example, Macfarlane and Shelby (1981), Manson (1982) and Schmidt et al. (1994)), there are many organic systems which can perform photoreactions commonly utilized to create persistent spectral holes (Kharlamov et al. (1974), Gorokhovskii et al. (1974), Personov (1983), Small (1983), Friedrich and Haarer (1984), Personov and Kharlamov (1986), Moerner (1988), Völker (1989)). Low-temperature photoreactions in organic amorphous solids leading to persistent hole burning can be divided into two classes: Photochemical (PHB) and nonphotochemical hole burning. From the methodical point of view, however, it is usually not important which particular photoreaction is employed for burning a persistent spectral hole. The details of photoreaction mechanisms will be discussed in Sect. 6.4.

6.3.2 Experimental Techniques

A hole-burning spectrum always reflects the linewidths involved in its creation. When a hole is burnt at moderate light power, the result is basically the convolution of the spectral width of the employed light source and the intrinsic linewidth of the absorbers (for more details see reviews by Friedrich and Haarer (1984) and Personov and Kharlamov (1986)). To create a spectral hole representing the intrinsic (quasi-homogeneous) linewidth of an optical transition, a light source with a band width much narrower than this width is necessary. In most experimental situations, this can be achieved by using narrow-band lasers.

The second task after having created a hole is to measure its spectrum. This is performed by continuously scanning the frequency of a light source over the spectral range of the hole. This process again involves a convolution of the intrinsic absorber linewidth with the bandwidth of the employed light source. Therefore the most powerful tool for modern hole burning spectroscopy is a single-mode tunable laser such as a dye laser. Dye lasers provide monochromatic light with a bandwidth in the MHz range; their wavelength can be continuously changed by tuning optical elements inside the resonator. The stability of the emitted wavelength is generally assured by an active frequency control loop. Such systems can be employed for both of the above-mentioned steps: For step number one, hole burning, this provides absolute freedom for the frequency location within the inhomogeneous band, where the hole is to be created. For step number two, read-out of the hole spectrum, the necessity of continuous tuning is obvious. Since this second step also involves illumination of the sample, the light intensity has to be decreased by several

6.3 Basic Methods of Hole-Burning Spectroscopy

orders of magnitude between step one and step two to avoid distortion of the spectrum by additional hole burning.

In recent years, the development of tunable narrow-band diode lasers has led to a possible alternative to dye lasers for some experimental applications. Both kinds of light sources allow external access to wavelength tuning, which in turn allows computer-controlled conduction of the whole experiment. At elevated temperatures above the boiling point of liquid helium, the intrinsic linewidth of the sample is in many cases large enough for the use of high-resolution monochromators for hole detection.

As well as other features in optical spectroscopy the depth of a spectral hole is usually measured in terms of absorption by the dimensionless quantity "optical density", which is the negative common logarithm of the normalized transmitted intensity

$$\alpha(\lambda) = -\log_{10}\left(\frac{I(\lambda)}{I_0(\lambda)}\right). \tag{6.1}$$

Here I_0 represents the incident light intensity whereas I stands for the intensity transmitted by the sample; the quotient I/I_0 represents the transmission of the sample. Employing this definition, the depth of a spectral hole is given by the small change in absorption induced by the photoreaction at the burning wavelength λ as compared with the unperturbed value

$$\alpha(\lambda) - \alpha(\lambda_{\text{ref}}) = -\log_{10}\left(\frac{I(\lambda)}{I(\lambda_{\text{ref}})}\right),$$

where λ_{ref} is a reference wavelength outside of the spectral hole profile (see Fig. 6.5).

In cases where it is necessary to avoid the so-called saturation broadening, spectral holes have to be rather shallow. Usually their depth should be chosen in the range of a few percent of the optical density of the sample (Köhler et al. (1985), Kador et al. (1986)). Therefore, the required signal-to-noise ratio has to be rather high. In the case of a tunable single-mode laser, however, the incident light power is subject to rather large temporal fluctuations – technical noise caused by the frequency stabilization feedback circuit of such a laser system. This implies another basic technical aspect of hole-burning spectroscopy: Power normalization. The transmitted signal has to be normalized to the power of the incident light to eliminate the low-frequency noise or the incident light power itself has to be kept constant. Accordingly, there are two basic methods to overcome this technical problem:

One way is to put a power stabilization setup between light source and sample. This is a so-called "noise eater", which monitors the light power emitted by the laser and adapts its transmission in a feedback controlled fashion to counteract the laser's power fluctuations. For this purpose the output power of the noise eater of course has to be lower than its input power. But this loss of light power normally does not impose a serious restriction to hole-burning experiments.

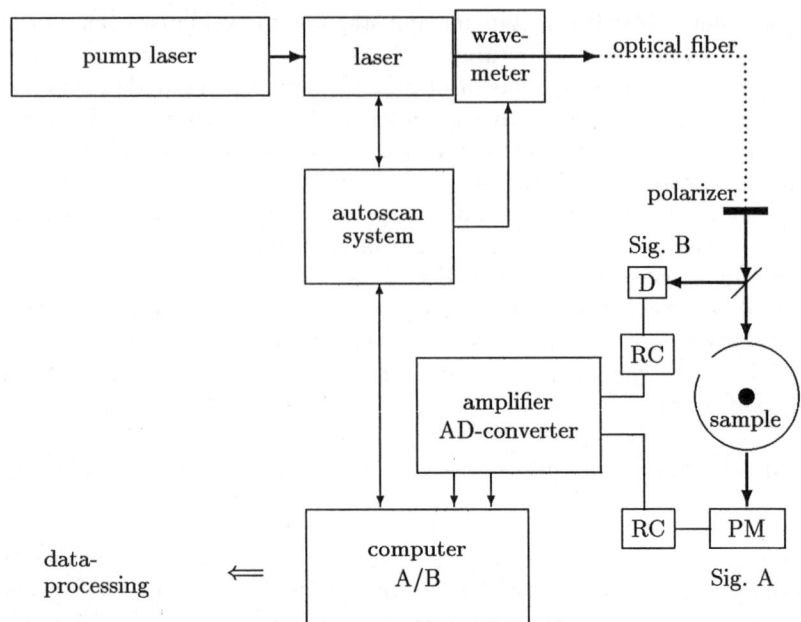

Fig. 6.4. Setup for transmission measurement of spectral holes; the computer-based data acquisition and frequency control are also displayed. A photomultiplier (PM) measures the transmitted light power (Sig. A). For monitoring the power of the incident light, a beam splitter is introduced into the optical path to direct a part of the laser beam onto a photodiode (D, Sig. B). "RC" represents integration circuits for bandwidth restriction. The polarizer is necessary to stabilize the splitting ratio.

Another approach is to include the temporal light power fluctuations into the above-mentioned power normalization process: Parallel to the measurement of the light power transmitted by the sample (signal A) a reference signal proportional to the incident power is monitored (signal B), while the laser is scanning the frequency range of the hole spectrum. Such a setup is displayed in Fig. 6.4 including the data acquisition system. After restricting the signal bandwidths to the necessary range by integrating both channels with identical overall time constants, the resulting signal-to-noise ratio can be close to the shot noise or quantum limit.

The advantage of the two-channel scheme as shown in Fig. 6.4 is the possibility of data processing after the actual measurement, e.g. digital filtering and signal normalization by a computer, which is more flexible and robust than the analog noise eater method. The disadvantages are the higher requirements of computer resources and the necessity to have absolutely identical analog responses in both registration channels. On the other hand the analog noise-eater method is not constrained by the actual signal registration circuit, but it is not as robust as the digital signal normalization of the two-channel scheme. Clearly, the most important advantage of a noise eater is that it provides a constant incident light power on the sample. This can be very important in the case of nonlinear photoreactions. In general, both methods yield identical results for the most part: The suppression of low-frequency noise. The higher frequency components of the noise spectrum can be efficiently suppressed by increasing the response time constant of the registration circuit(s) or, in the case of digital data processing, by performing multiple scans and data averaging.

In the limiting case of extremely weak absorption bands, it can be advantageous to utilize fluorescent light from the sample rather than measuring its transmission (Parker (1968)). This is necessary when the overall absorption of the band is so low that the achievable hole depth is close to the signal-to-noise ratio of a custom transmission spectroscopy setup. The so-called "fluorescence excitation" method typically scans the frequency range of the spectral hole using a narrow-band tunable laser and measures the red-shifted fluorescence intensity of the sample. This intensity is proportional to the respective number of excited absorbers at the corresponding frequency position. Nowadays, this is the only technique allowing the registration of single-molecule absorption spectra, which is the ultimate limit of detecting small numbers of absorbers (Moerner and Basché (1993), Orrit et al. (1993), Kador (1995), Plakhotnik et al. (1997)). In conventional hole-burning spectroscopy, however, where the overall sample absorption can usually be prepared as high as necessary, this technique has no special advantages but, on the contrary, two weak points. The first one is very general: Obviously, this technique can only be applied to luminescing compounds. The second point is specific for hole burning. In many cases it is necessary to reduce the incident light intensity as much as possible to avoid a distortion of the spectrum due to hole burning during read-out. But if the main source of noise is shot noise, the signal-to-noise ratio scales like $S/N \simeq \sqrt{S}$, where S is the signal registered by the detector. The signal contrast of shallow holes in moderately intense absorption bands is similar in transmission detection and fluorescence excitation. The respective signals are described by

$$\begin{aligned} S_{\text{tr}} &\simeq [I_0 \exp(-\ln(10)\alpha,)] \\ S_{\text{fl}} &\simeq [I_0 \left(1 - \exp(-\ln(10)\alpha)\right)]\theta\Phi\,. \end{aligned} \qquad (6.2)$$

The terms in square brackets in both equations are of the same order of magnitude, because the optical density α of the sample is usually of the order 0.3–0.5 in hole-burning experiments. But there are two additional factors in the right-hand side of (6.2), each of them being smaller than 1. The fluorescence quantum efficiency Φ determines the number of emitted photons per absorption process. It is close to 100 % only in a few special cases (for example for porphyrins which are widely used in hole-burning experiments $\Phi \leq 0.1$ in most cases). The registration aperture θ in conventional fluorescence detection schemes is usually < 0.1. Therefore, the detector signal in the case of fluorescence excitation will be at least an order of magnitude smaller than in a transmission experiment. This is why the fluorescence detection technique

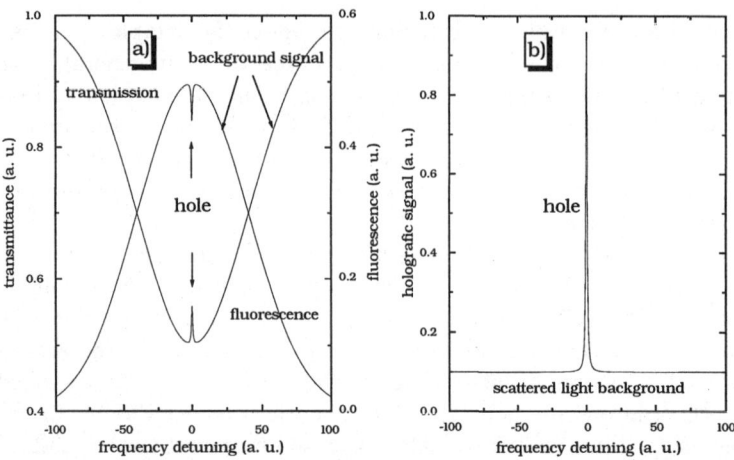

Fig. 6.5. Schematic representation of the registered signals for three different methods of optical absorption measurements (for a sample with an optical density at the absorption maximum of about 0.3): (a) The signals in transmission and fluorescence detection are mirror images of each other. Both have a background component. (b) The holographic signal contains only a diffracted component which is proportional to the change of absorption caused by hole burning. The background signal is determined only by stray light, the level of which depends mostly on the optical quality of the sample.

is not commonly used in hole-burning spectroscopy. In single-molecule spectroscopy with very sharp focusing of the incident light on a very thin sample, special parabolic mirrors are employed to cover up to 30–40 % of the full solid angle. With this method, fluorescence excitation has developed into the standard method in this field.

Both of the techniques described up to now have one disadvantage in common: There is a large background signal. In the case of transmission measurements this background is given by the initial transmission of the sample prior to hole burning. In the case of fluorescence excitation, the background signal is determined by the fluorescence quantum yield of those absorbers remaining within the spectrum after hole burning. This is schematically illustrated in Fig. 6.5a, where an absorption band with a spectral hole is shown as a transmission signal and as a fluorescence excitation signal [the conversion between transmission and absorption is defined in (6.1)].

There is, however, a technique which allows the registration of spectral holes without any background signal: Holographic hole burning (Meixner et al. (1989), Meixner et al. (1990), Renner et al. (1995)). This technique requires two coherent laser beams for hole burning. Their interference produces an absorption grating in the sample which only exists in a frequency range of the intrinsic linewidth of the absorbers. In a subsequent frequency scan with a single laser beam only the diffracted intensity is detected. The signal reaches its maximum at the center frequency of the spectral hole, where the contrast of the grating is most pronounced. This method has no transmission background. In spite of the nonlinearities involved in generating the signal, the shape of a spectral hole is undisturbed by this method (Meixner et al. (1989)). A holographic signal is schematically depicted in Fig. 6.5b. The only background appearing in this method is due to stray light; its intensity is determined by the optical quality of the sample and the optical elements of the setup. The main limitation of this method is the required very high optical quality of the samples. In addition, holographic hole burning is technically much more complicated than the transmission method and is therefore not commonly applied.

6.3.3 Technical Limitations

Temperature Limits. The high-temperature limit for hole-burning spectroscopy originates from two different sources. The first one has already been analyzed in Sect. 6.2: The temperature threshold for the existence of zero-phonon lines. For most organic molecules in organic matrices zero-phonon lines are observable at temperatures below 70–80 K. Porphyrins are commonly used as spectroscopic probes and tolerate higher temperatures than many other molecules. In their spectra zero-phonon lines can be observed at temperatures above 80 K.

The second limit is the thermal stability of the hole, i. e. the thermally activated back reaction of the photoproduct. Depending on the nature of the employed photoreaction, this limit can be as low as 20–30 K for nonphotochemical hole burning but also reach temperatures of 150 K (Furusawa and Horie (1991), Kümmerl et al. (1994), Ehrl et al. (1994)) in the photochemical case.

The low-temperature limit is only a technical one in the first place and therefore depends on the cryostat used. Conventional ^{4}He bath cryostats provide temperatures down to 1.5–1.6 K. A minimum temperature of 0.025 K was reached in a ^{3}He/^{4}He dilution refrigerator by Müller and Haarer (1991). In the sub-Kelvin range, the determination of the lowest working temperature also involves the incident light power which cannot be arbitrarily low. This aspect of the problem will be discussed below.

Power Limits. In hole-burning experiments there can be various reasons for an upper limit of the applicable light power when burning a hole. The first one to be discussed here is the limit of saturation broadening. Because of the nonzero homogeneous linewidth, absorption occurs also when the laser light and the absorption line are not exactly resonant, i. e. when the laser frequency is not centered at the peak of the absorption line. When the incident burning light power is beyond the saturation limit, the spectral hole will be artificially broadened because the fully resonant optical transitions are saturated while off-center absorption is not yet saturated. Therefore, nonresonant absorption is more efficient in the saturation regime and that causes the hole to braoden more the higher the burning power is.

For a two-level transition, the saturation limit is determined by the power density (notations of transition rates k_i here and in the following have been defined in Fig. 6.1):

$$I_s^{\max} = \frac{h\nu}{2\sigma}k_s, \qquad (6.3)$$

where σ is the absorption cross section and $h\nu$ is the photon energy. Depending on the oscillator strength and homogeneous linewidth, $I_s^{\max}$ can vary at liquid helium temperatures from about 0.1 to 100 W/cm² for most organic molecules. This two-level estimate mainly applies to short laser pulses where the population of triplet states can be neglected. Such power density levels, however, are not unusually high for pulsed lasers. This is the reason why in hole-burning experiments with pulsed lasers, holes are usually very broad as compared to CW excitation. In the case of CW excitation, the bottleneck of power saturation is the long-lived triplet state for the majority of complex organic molecules. The saturation limit for different compounds varies by orders of magnitude, depending on the intersystem crossing rates and triplet state lifetimes. A theoretical analysis of saturation via the bottleneck of triplet state population can be found in De Vries and Wiersma (1980). The saturation limit for this case can be expressed as

$$I_s = I_s^{\max} K \quad \text{with} \quad K = \frac{k_s + k_{st}}{k_s} \times \frac{2k_t}{2k_t + k_{st}}. \qquad (6.4)$$

Taking into account that for the majority of aromatic compounds $k_t \ll k_s$ and k_{st}, it follows $K \ll 1$. According to our current knowledge, the aromatic system with the lowest intersystem crossing rate into the triplet state is dibenzanthanthrene (DBATT), where K was estimated to be $K \simeq 0.9$ (Boiron et al. (1996)). This means the electronic excitation of this molecule practically corresponds to a two-level transition scheme. The saturation power for DBATT in n-hexadecane at 1.8 K was found to be about 100 mW/cm² (Boiron et al. (1996)). It should be noted that experimental values of I_s can be much higher than the estimate given by (6.4) due to a random orientation of the molecular transition dipole moments and other geometric factors [see the detailed analysis of this problem for the case of single-molecule excitation conditions by Plakhotnik et al. (1995)]. In CW experiments, such excitation levels can only be reached in single molecule spectroscopy. The opposite example are molecules like Zn-tetrabenzporphine (Zn-TBP), were intersystem crossing is the main deactivation channel of the S_1-state. The estimate for this molecule in a glassy matrix at helium temperatures gives $I_s^{\max} \simeq 100$ W/cm² for the two-level case but the high intersystem crossing rate causes a value of $I_s \simeq 10\,\mu$W/cm², which is in good agreement with experimental results (Khodykin et al. (1997)). For such molecules power broadening can readily be achieved even by CW excitation, which makes them very attractive for three-pulse echo and transient hole-burning experiments (Jahn et al. (1991), Zilker and Haarer (1997), Khodykin et al. (1997)).

Another power limitation can be the possibility of sample heating by light absorption during the hole-burning process. This depends on such factors as sample material, sample size and the efficiency of the chromophore to convert the light energy into heat. At conventional temperatures of 2–4 K this limit is in the range of 0.01–$10\,\mathrm{mW/cm^2}$, which is not as low as in the above example of triplet population. But the heating problem becomes more serious in the following two cases:

The first example are three-pulse echo and transient hole-burning experiments, which both employ triplet state population. The problem is that the total burning energy has to be high enough to create an observable change in the sample absorption and at the same time the pulse duration is much shorter than the thermal relaxation time of the sample. Therefore, heating of the sample by the burning laser pulse can easily occur. Taking a PMMA sample of 1 mm thickness as an example, at 3 K a simple estimate of the maximum burning pulse energy would be $1\,\mu\mathrm{J/cm^2}$. The experimental value is even an order of magnitude lower (Khodykin et al. (1997)). As a result of sample heating, strong distortions in the time dependence of spectral diffusion occur. Therefore the experimental conditions have to be checked very carefully in pulsed experiments to avoid heating artifacts. An analysis of sample heating consequences is given by Kharlamov et al. (1994), Khodykin et al. (1997) for the case of transient hole burning and by Zilker and Haarer (1997), Neu et al. (1997b) for three-pulse photon echoes.

The second critical case is the sub-Kelvin region, where a dramatic decrease of the specific heat reduces the power limit of sample heating by orders of magnitude. For example the widely employed porphyrins are a class of organic molecules which can release large amounts of their electronic excitation energy by radiationless processes. An estimate of the maximum burning laser power for phthalocyanine, which releases about 95 % of the absorbed energy nonradiatively, yields a value of $100\,\mathrm{nW/cm^2}$ at a temperature of 100 mK. This value is so low because almost all of the absorbed light energy is transferred into heat, which in turn has to be compensated by a sufficient cooling power of the cryo-setup of the respective experiment. For such compounds the burning light power in the extremely low temperature range below 1 K also has to be small compared with the cooling power at a given temperature (for low-temperature cooling methods, see Pobell (1992)).

Burning Fluence Limit. Hole burning can formally be treated as populating a long-living metastable state, the photoproduct state. Therefore the persistent saturation kinetics in hole burning is very similar to a saturation by triplet state population. The difference is that the lifetime of the photoproduct is usually much longer than the duration of an experiment. Hence, the limiting quantity is not the burning power but burning fluence, i.e. the total burning energy per unit area. If the hole becomes very deep, the ground state of molecules resonantly absorbing laser light is strongly depleted, which causes saturation broadening of the holes. This problem was investigated in detail by Rebane et al. (1982), Köhler et al. (1985) and Kador et al. (1986) [see also the detailed theoretical investigation of hole-burning kinetics of Jalmukhambetov and Osad'ko (1983) and the experimental data on the relation between hole width and burning fluence in the review by Völker (1989)]. In spectral diffusion experiments, however, the absolute width of holes is not investigated, but the hole broadening as a function of time is measured. This diffusional broadening is independent of the initial hole width, therefore a small amount of saturation broadening is acceptable. Absolute hole widths are usually determined by an extrapolation of a series of measurements with decreasing burning fluence.

Since the recording of the hole spectrum also involves irradiation, the light power has to be further reduced by 2–3 orders of magnitude, depending on the number of read-out processes and the quantum efficiency of the phototransformation, to prevent additional hole burning. These considerations finally yield a lower limit of light intensities in the pW range for recording hole spectra. For these reasons, normally photomultipliers have to be utilized for the detection of the transmitted light. If, however, a setup is employed where the detectors are located within the cryostat at liquid helium temperatures, the thermal noise of commercial photodiodes can be decreased to a noise equivalent power (NEP) value in the pW range. Using such a configuration, hole-burning experiments have been successfully conducted by Hannig et al. (1996). Such an unusual application, however, necessitates careful selection of the employed diodes.

Time Limits. As the above considerations have shown, the applicable light power is a crucial point in hole-burning spectroscopy. If an experiment is to be conducted at temperatures well below 1 K, the available cooling power, applicable light power, and therefore the achievable time resolution of an experiment, are not independent variables any more. Instead, light power and time resolution can be governed by the performance of the employed cryosetup. When an upper limit for the applicable light power is established, the achievable time resolution of the experiment can be estimated. Since the hole depth is a function of the accumulated burning energy, the minimum duration of the burning process is given by the quantum efficiency of the photoreaction. For experiments with porphyrin-doped amorphous solids in the vicinity of 1 K, these restrictions typically result in a minimum detection time of a hole spectrum of the order of 1–10 s after the beginning of the burning process. A higher time resolution can only be achieved with highly efficient transient hole-burning mechanisms (see references above). But as already noted, sample heating becomes an important problem in such experiments.

Currently, however, new experimental methods are emerging that allow an extension of the time resolution of the persistent spectral hole-burning method for the investigation of TS's dynamics to times shorter than those required for hole burning (Wunderlich et al. (1997)). At an arbitrary time after creating a spectral hole, a nonequilibrium situation can be induced by a quick variation of an external parameter such as an electric field. The fast relaxation process of the ensemble of TS's in response to this distortion can subsequently be detected. This rather new approach will be further described together with experimental results in Sect. 6.5.5.

Concerning the maximum observation time of persistent spectral holes, there is basically no upper limit, if the sample is stored at temperatures in the liquid helium range. For all practical purposes, the lifetime of a spectral hole in a porphyrin sample, for example, is essentially infinite, since thermally activated back reactions of the photoproduct are almost completely suppressed. Instead, the technical limitation is given by the achievable runtime of the employed cryo-setup. Here again, ^{3}He/^{4}He dilution refrigeration has to be mentioned in the first place, since it is a continuous cooling method. Employing this method, runtimes on the order of 100 days and total hole observation times of several weeks (Maier et al. (1996), Fritsch et al. (1996)) have been achieved. None of these times, however, necessarily constitutes an upper limit. As an alternative in the temperature range between 1.5 K and 4.2 K, a dipstick cryostat with integrated detectors can be used, providing a runtime of up to 50 days in a 100 l liquid helium container (Joyeux et al. (1993), Hannig et al. (1996)).

Spectral Resolution Limit. Finally, the spectral resolution of a hole-burning setup has to be addressed. As mentioned in the description of the experimental techniques, there are different options for the light source. Regarding the frequency resolution, however, nowadays, dye lasers are the superior choice in the range of visible light. While the actual bandwidth of light emitted from such lasers lies in the kHz range, electronically frequency stabilized commercial systems yield fast fluctuating time-averaged widths in the range of 1 MHz. This determines the accuracy of measuring the widths of spectral holes. Hole-burning spectroscopy, however, would usually not benefit much from higher resolutions, since the lifetime limited natural linewidth of most organic chromophores is $\geq$ 10 MHz. For porphyrins, for example, it is in the range of 50 MHz. The fact that the frequency resolution is of the order of 1 MHz yields a relative resolution of hole-burning spectroscopy of the order $1/10^9$, since the order of magnitude of optical frequencies is 10^{15} Hz. This means that relative changes in frequency of optical absorbers can be determined with an accuracy of one over one billion. Therefore hole-burning spectroscopy is termed a high-resolution method. It is this high sensitivity to extremely small frequency changes that makes spectral hole burning a suitable tool for investigating minute solid state processes such as TLS dynamics.

6.4 High-Barrier Versus Low-Barrier Tunneling

The development of high-resolution methods of laser spectroscopy of solids (fluorescence line narrowing, photochemical hole burning, two- and three-pulse photon echoes) opened up a new field of investigations: They enable the study of very infrequent tunneling processes accompanied by very small changes in optical spectra, hardly observable in inhomogeneously broadened spectra even in crystals, not to mention glasses. Some examples of such impurity–solvent interactions were already introduced above: Homogeneous line broadening and spectral diffusion in glasses. These processes are connected with phonon-assisted interactions of impurities with TLS's of the solid. The underlying tunneling systems (TS's) dynamics are spontaneous and usually involve low-energy TLS's. The phenomenon of spectral diffusion, caused by these processes, will be discussed in detail in the remainder of this chapter. But high-resolution optical experiments showed the existence of various photon-mediated tunneling processes of high-energy TS's: Low-temperature photoreactions. They were discovered via hole burning and became the basis of this technique. As mentioned in the previous section, they are commonly separated into two very inhomogeneous groups: Photochemical (PHB) and nonphotochemical (NPHB) reactions. In the following we briefly discuss the main characteristics of both, as investigated by means of optical hole-burning spectroscopy.

6.4.1 Photochemical Hole Burning

Solids at low temperatures are too rigid to provide much freedom for atomic and molecular movements necessary for "real" chemical reactions. The only mobile particles which can be responsible for phototransformations under such conditions are electrons and protons. Even for these particles, potential barriers are usually too high for spontaneous tunneling into new configurations. Such processes may be stimulated in some well-defined cases by photon absorption. In general, the process of such phototransformations can be formally represented by the four-level scheme shown in Fig. 6.6. The impurity center has two or more potential minima both in its ground and excited state. The difference between these two-level systems and TLS's of the matrix is that these two-level systems are created within the impurity itself or by impurity–matrix interaction. This is the reason why they are sometimes called "extrinsic" in contrast to the intrinsic solid-state TLS's of the matrix itself. In many cases the microscopic nature of "extrinsic" TLS's is known; they are well-defined systems with a very small dispersion of parameters.

6.4 High-Barrier Versus Low-Barrier Tunneling

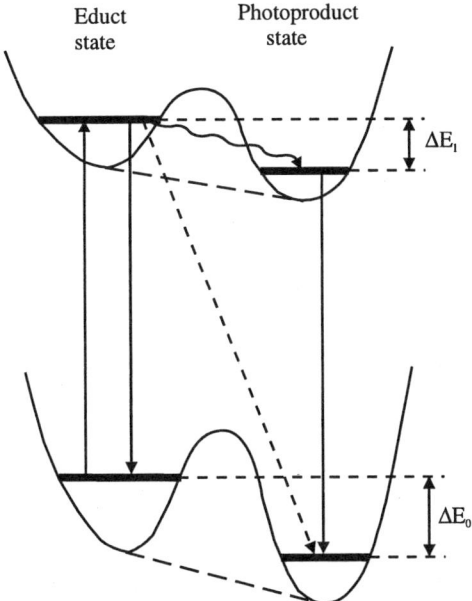

Fig. 6.6. General schematic representation of a phototransformation: The transition from the educt ground state to the product ground state by photon excitation, i.e. via an excited state.

The barrier shape and height in ground and excited states can differ substantially, in particular, the excited state can have only one minimum. The most probable transition channel from one minimum into another is tunneling while the impurity is in its excited state. Usually, the probability of tunneling is rather low, the molecule has to be excited many times. After tunneling, the transition frequency of the impurity is shifted, which can be detected in optical experiments. Examples of organic molecules which perform such photo-transformations are given in Fig. 6.7. The two different mechanisms of photochemical transformations indicated in the figure will be discussed in the following.

Proton Tautomerization. There is a whole family of organic molecules which exhibit proton tautomerization and which have been widely investigated by hole-burning spectroscopy: the porphyrins. The mechanism of photochemistry is the same for all members of this family. It is indicated in Fig. 6.7 for the molecule phthalocyanine, which also belongs to the porphyrins. The two possible orientations of the central protons are indistinguishable in a free molecule. But in a low symmetry matrix the tautomerization shown in the figure – formally equal to a rotation of the molecule – induces a small shift of the molecule's electronic states resulting in a modified absorption frequency. The existence of "dark" tautomerization of porphyrins at high temperatures was first noticed in NMR experiments by Storm and Teklu

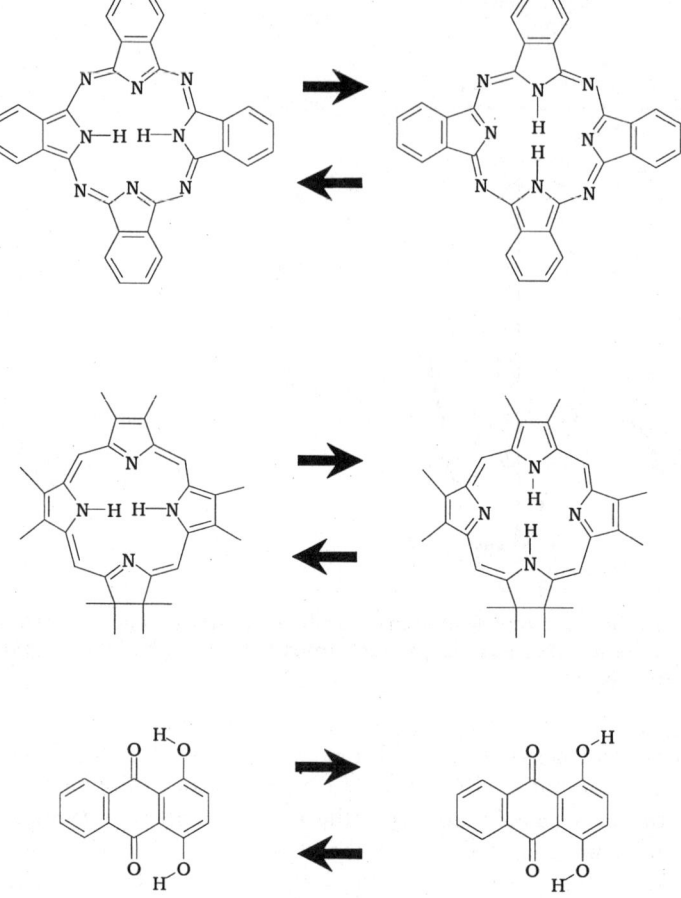

Fig. 6.7. Examples of organic molecules performing photochemical transitions, from top to bottom: phthalocyanine, chlorin, and quinizarin. The first two exhibit a so-called proton tautomerization whereas in the case of the third example the reaction is a rearrangement of a hydrogen bond. Note that in the case of phthalocyanine the surrounding nonsymmetric amorphous host breaks the symmetry so that the reaction does not merely correspond to a rotation of the molecule by 90° but results in a different electronic structure with a shifted absorption frequency.

(1972) in liquids and by Limbach et al. (1984) for the solid state. At low temperatures, however, the barrier is too high for spontaneous tautomerization. Light-induced low-temperature tautomerization of porphyrins was first discovered in n-paraffin matrices and displayed a redistribution of intensities of Shpol'skii multiplets under light irradiation (Solovjov et al. (1973), Völker and van der Waals (1976)). The proton tautomerization was also used for the first persistent hole-burning experiment in crystalline matrices (Gorokhovskii et al. (1974)). The hole-burning technique has become the main method for

6.4 High-Barrier Versus Low-Barrier Tunneling

the investigation of low-temperature proton tautomerization in porphyrins. Due to high barriers the backward tautomerization in the ground state at helium temperatures is negligibly slow. Therefore the burned holes persist virtually infinitely long if the temperature is sufficiently low. It should be noted that in many polymers and glasses both proton tautomerization (i.e. PHB) and NPHB can coexist simultaneously with comparable quantum efficiencies but very different other characteristics. In this section "pure" tautomerization will be considered, systems with mixed photoreactions will be discussed below.

The investigations of direct tautomerization reactions concentrated presumably on the measurement of the reaction efficiency and the redistribution of the photoproduct bands. It was found that in most porphyrins the low temperature tautomerization has a quantum efficiency of the order of $10^{-3} - 10^{-4}$ (see for example Kador et al. (1986)) phototransformations per excitation. In crystalline matrices with small inhomogeneous broadening, the two possible orientations of the central protons result in pairs of sites within the corresponding components of Shpol'skii multiplets that can be photo-converted into each other by appropriately selected light irradiation (Solovjov et al. (1973), Völker and van der Waals (1976), Völker and Macfarlane (1980)). In crystalline matrices with large inhomogeneous broadening and in glassy matrices, the photoproduct molecules are distributed within the inhomogeneous band showing no sharp spectral features (see, for example, Kharlamov et al. (1984)). There is however one interesting exception: chlorin. The structure of this molecule is also shown in Fig. 6.7. In contrast to other porphyrins, the photoproduct of chlorin is in fact chemically different from the educt. It is stable only at temperatures below 40–50 K. Its absorption band is shifted by about 1600 cm^{-1} towards the high energy side. Another interesting peculiarity of this compound is that in this case the back reaction has a quantum efficiency at least two orders of magnitude higher than the forward reaction as was found by Völker and Macfarlane (1980) in a crystalline matrix and Kharlamov et al. (1984) in a glassy one. This compound is one of the most popular objects for hole-burning experiments.

To investigate the back reaction thermal cycling can be employed. The scheme of such an experiment is usually the following: A hole is burned at a temperature T_b, then the sample is heated to a temperature T_c and cooled back down to T_b again. The measured quantity is the decrease of the hole area as a function of temperature (see for example Köhler et al. (1988), Zollfrank and Friedrich (1990)). If the back reaction from the product state is governed by thermal activation processes, the rate of reaction is given by

$$R = R_0 \exp(-V/k_B T), \quad (6.5)$$

where R_0 is the "frequency of attempts" to cross the barrier with a height of V. Assuming that during the annealing time τ at the cycle temperature T_c, roughly all molecules with $R \geq 1/\tau$ will come back into the initial state, we come to a limiting barrier height of

$$V_m = k_B T_c \ln(R_0 \tau). \tag{6.6}$$

The hole area remaining after the thermal cycle and the distribution of barrier heights are related by the proportionalities

$$A(T_c) \propto \int_{V_m}^{V\infty} P_v dV \quad \text{and} \quad P_v \propto -\frac{dA(T_c)}{dT_c} \frac{1}{\ln(R_0 \tau)}. \tag{6.7}$$

These relations were employed in thermally activated hole-filling experiments for evaluating P_v for some porphyrins in various matrices by Zollfrank and Friedrich (1990), Schellenberg et al. (1994) and Kümmerl et al. (1994) (here we consider only systems without NPHB). An approximately Gaussian, rather narrow distribution of barriers was found with the position of the maximum between 100 K and 150 K and a width of 10–30 K. Both parameters were found to be very weakly dependent on the matrix but fairly sensitive to the molecular structure. The reaction kinetics was also measured in real time at higher temperatures ($T \simeq 145 - 160$ K). The decay curves can be described by a stretched exponential

$$A(t) = A_0 \exp(-tR)^\beta, \tag{6.8}$$

which is typical for systems with a broad dispersion of relaxation rates and numerically resembles the more accurate integral equation

$$A(t) = \int_0^\infty P_v \exp[-tR(V)] dV. \tag{6.9}$$

Comparative measurements of the tautomerization of chlorin in the crystalline and the amorphous modifications of benzophenone were carried out by Schellenberg et al. (1994). In both cases only PHB, i.e. tautomerization, was present, but the difference in the barrier distribution is very pronounced. In both cases the distribution could be fitted with a Gaussian function with a maximum at about 30–45 K in the crystalline matrix and 50 K in the amorphous modification. But the width of the barrier distribution in the glass is about 22 K whereas the distribution in the crystal is very sharp with a width of no more than a few Kelvins.

Hydrogen Bond Rearrangement Reaction. For some molecules a light-induced creation or destruction of hydrogen bonds accompanied by intramolecular rearrangements is possible. Here the hydrogen bond can be intramolecular or a bond between the impurity and the surrounding matrix. A typical representative of the latter broadly used in hole burning spectroscopy is quinizarin. The chemical structure and the photoreaction between this molecule and the solvent is shown in Fig. 6.7. The peculiarity in this case is that matrix is directly involved in the photoreaction. The spontaneous refilling of spectral holes of quinizarin in alcohol glass samples (ethanol/methanol mixture) was investigated on a long-time scale by Breinl et al. (1984b). In the time range 1–10^3 min, a logarithmic hole refilling was observed. This would mean that the distribution of relaxation rates for the back reaction resembles that of the standard tunneling model. Indeed, the standard model distribution function of relaxation rates is given by

$$P(R) = \frac{P_0}{2R\sqrt{1-R}} \qquad (6.10)$$

with $R = r/r_{\max}$, where $r = \tau_1^{-1}$ is the relaxation rate of TS's and $r_{\max} = \tau_1^{-1}(\Delta_0 = E)$ is the maximum relaxation rate for TS's with a given energy E.

The change of the hole area due to tunneling processes of the photoproduct back to the initial state can be expressed as

$$A(t) = A_0 \left[1 - \int_0^\infty dE\, n(E) \int_0^1 dR \frac{\Delta}{E} P(R) \right. \\ \left. \times (1 - \exp(r_{\max} Rt)) \right]. \qquad (6.11)$$

The weight factor Δ/E reflects that symmetric TS's cannot participate in dipolar interactions. $n(E)$ is the probability that a "photoproduct TS" has the energy E. The inner integral can easily be solved (see Sect. 6.5.2, compare also to the time dependencies obtained in Sect. 2.2.3)

$$\int dR.... = \frac{1}{2} P_0 \ln\left[r_{\max}(E,T)t\right] + ... \qquad (6.12)$$

In hole-burning experiments, however, the time resolution is always limited by the time required for hole burning itself and the registration of the hole spectrum. Therefore the hole area at $t = 0$ s cannot be measured. What is measured in reality is the hole area at the experimentally achievable minimum time, $A(t_0)$. Therefore the measurable time dependence in a normalized representation is given by

$$A(t_0) - A(t) = A(t_0) \frac{P_0}{2} \langle n(E) \rangle_E \ln \frac{t}{t_0}. \qquad (6.13)$$

Two consequences follow from this equation:

- A logarithmic time dependence of hole filling, characteristic of the standard tunneling model.
- The impossibility to measure the parameter $r_{\max}$ due to the limited time resolution of the experiment.

Employing this analysis, we can argue that changing the mass of the tunneling particles can hardly account for the strong dependence of the hole-refilling rate on matrix deuteration. The only parameter which is massively affected by a change of the mass, $r_{\max}$, cancels due to the limited time resolution as long as $t_0 \gg 1/r_{\max}$ which is the case even at the lowest accessible temperatures in hole-burning experiments. This means that a modification of the tunneling mass cannot affect the observable rate of hole refilling, so that this change might have another reason. As (6.13) indicates, a modified

TS's density of states can be responsible for the modified hole-refilling rate instead.

Investigations of the back reaction via thermal cycling were performed on a set of systems with PHB and NPHB, including quinizarin in different matrices by Köhler and Friedrich (1987). The distribution of barrier heights was measured in these experiments using the above formalism for the interpretation of the data (6.5)–(6.7). The result was quite different from systems with proton tautomerization. For all systems exhibiting NPHB as well as PHB by impurity-matrix proton tunneling, the hole refilling as a function of the cycle temperature follows the law

$$A(T)/A_0 = 1 - (T^{1/2}\sqrt{k_B \ln(R_0 \tau)/V_{\max}}), \qquad (6.14)$$

where $V_{\max}$ is the maximum reaction barrier height. The above dependence corresponds to a barrier distribution

$$P_v \propto 1/\sqrt{V}, \qquad (6.15)$$

which is typical for NPHB in glasses (see below). This barrier distribution can be obtained from the standard tunneling model distribution in λ if we take into account that for proton tunneling, the tunneling mass is the same for all relevant TS's and the tunneling distance is expected to have only minor variations (Köhler and Friedrich (1987)). It is interesting to note that the experiments did not show any lower limit for the barrier height. This means that the hole recovery starts already at arbitrarily small temperature increases. The experiments were extended to temperatures as low as 500 mK by Zollfrank et al. (1989) with the same result.

The results of spontaneous hole-filling measurements (tunneling processes) and thermal cycles (activated barrier crossing) are in agreement, showing a "standard tunneling model-like" distribution of reaction barriers and relaxation rates. Actually, this is a rather surprising result (compared with intramolecular proton tautomerization). The TLS's representing hydrogen bonding of the impurity molecule with the matrix are not "intrinsic" TLS's. They are created by a rather special interaction of the impurity with the matrix, i.e. they are "extrinsic". Therefore they are not *a priori* expected to have the same distribution as "intrinsic" TLS's. This illustrates an important complication of photoreactions in low-temperature glasses: In many systems it is rather difficult to clearly separate the involved mechanisms. Therefore measurements on well-defined model systems are important for clarifying the elementary processes. Before introducing such a model system, we will summarize some experimental facts on nonphotochemical hole burning.

6.4.2 Nonphotochemical Hole Burning

Nonphotochemical hole burning (NPHB) is a term for describing puzzling low-temperature photoreactions, probably not a single one, and their mechanisms are not yet clear. Since its first observation in the system perylene in alcohol glass by Personov et al. (1972) and Kharlamov et al. (1974) many efforts

6.4 High-Barrier Versus Low-Barrier Tunneling

had been undertaken for understanding the origin of those processes (Kharlamov et al. (1975), Hayes and Small (1978), Klafter and Silbey (1981), Kharlamov et al. (1984), Al'shits et al. (1986), Shu and Small (1990), Lin et al. (1992), Shu and Small (1992), Takahashi et al. (1994)), but the problem has not been solved up to now. Nevertheless, NPHB is broadly used for high-resolution hole-burning spectroscopy of organic impurity molecules in glasses. Briefly summarized, the existing data concerning the area of its prevalence are the following.

A simple rule is that NPHB exists in polar glasses and polymers and that it is not observable in nonpolar glasses and in crystals. It is, however, more precise to say that in these types of matrices the output of the reaction is much lower than in polar glasses: Hole burning was observed even in Shpol'skii matrices [for example by Attenberger et al. (1991) and Boiron et al. (1996)], which are crystals, and direct observations of impurity zero-phonon line jumps in such systems under light irradiation have recently been reported by Ambrose et al. (1991) and Moerner et al. (1994). "Lower output" means that in "real" NPHB materials (according to the above rule), the majority of impurity molecules are photosensitive – holes of nearly 100% depth can be burned into the absorption spectrum – whereas in non-polar glasses and crystals, only very shallow holes can be generated via the NPHB effect; just a small fraction of the impurity molecules is sensitive to it.

The reaction is insensitive to the chemical nature of the impurity molecule but only sensitive to the type of matrix. This suggests light-induced rearrangements within the solid matrix as a mechanism of the reaction. The quantum output of the photoreaction is in the range of 10^{-3}–10^{-5} [see for example Moerner et al. (1984) and Takahashi et al. (1994)], which is sufficient for hole-burning spectroscopy with burning powers in the micro- and milliwatt range and exposition times of minutes. The photoreaction is of a single-photon type (Kharlamov et al. (1975)), i.e. its nonsaturated rate is a linear function of the excitation light power. The absorption of the photoproduct is distributed presumably in the same spectral region as the inhomogeneous band of the educt. The photoreaction is completely reversible: After heating the sample to a temperature of about 80–100 K, the absorption band of the educt recovers completely [see for example Kharlamov et al. (1984), and Al'shits et al. (1986)].

The very first idea concerning the nature of this reaction was the assumption that it is some type of impurity-matrix rearrangement (Kharlamov et al. (1974), Kharlamov et al. (1975)), which was formalized by Hayes and Small (1978) into a scheme of transitions of TLS's coupled to the impurity. This formal scheme works rather well for explaining many properties of NPHB. But questions still remain, the two most urgent being the following: What are the corresponding TLS's from the microscopic point of view? Why does NPHB not occur in all glasses?

Calculations of the hole-burning efficiency were already performed in early theoretical work by Klafter and Silbey (1981) using the assumption that NPHB is based on jumps of "internal" TLS's of the glass. This assumption, however, is very hard to reconcile with the observation that there is no correlation between the presence or absence of the NPHB effect and "genuine" properties of a glass like the anomalies of the specific heat and thermal conductivity or spectral diffusion. At the same time it was found that the existence of NPHB is often correlated with the presence of small amounts of other solvents in the matrix. For example it was found that NPHB can be observed in PMMA films if they are prepared via drying of a PMMA solution from a solvent. The presence of small amounts of such a solvent, which can hardly be avoided completely, results in the existence of the NPHB effect. If the film is dried very thoroughly, NPHB disappears. A similar observation was made in the case of ethanol. Alcohol glasses in general, ethanol in particular, are systems with a very pronounced NPHB effect. But ethanol normally contains about 2% of water. It is very hard to remove the water completely. A variation of the water content in ethanol in the percent range does not change the efficiency of NPHB. But in specially prepared extra dry ethanol NPHB disappears (Al'shits et al. (1987)).

The sum of all existing data supports the assumption that NPHB can be connected with hydrogen bond rearrangements within the solid matrix surrounding the impurity. There are definite correlations between the existence of NPHB and the presence of hydrogen bonds. NPHB was recently found in a classical hydrogen-bonding crystalline system: crystalline water doped with uranyl (Al'shits et al. (1996)). In spite of the crystalline structure of water, persistent hole burning in the uranyl absorption band was found to exhibit typical NPHB properties: a quantum efficiency on the order 10^{-4}, a broad distribution of barriers, and a relatively low stability (the holes disappear at sample annealing temperatures of 40–50 K). These findings support the idea of an involvement of hydrogen bonds in NPHB.

An investigation of the kinetics of spontaneous hole refilling, similar to the discussion of quinizarin in an ethanol/methanol glass in the previous section, was performed for the molecule tetracene in a matrix of glassy ethanol/methanol, a system exhibiting only NPHB, on a time scale from minutes up to a week by Köhler et al. (1987). A logarithmic time dependence of hole recovery was observed, similar to the case of quinizarin. It was also found that, in accordance with (6.13), there is no deuteration effect in hole recovery. This is especially remarkable, since in hole broadening the deuteration effect appeared to be very pronounced.

The distribution of barrier heights for the NPHB back reaction was extensively investigated using the above technique of thermal cycling (Köhler and Friedrich (1987), Köhler et al. (1987), Köhler and Friedrich (1988b), Köhler et al. (1989)). For systems with pure NPHB, a dependence of the hole area on the cycling temperature was found according to (6.14). As was shown above, this corresponds to a distribution of barrier heights $P(V) \propto 1/\sqrt{V}$. Special measurements devoted to the search of a minimum barrier height failed although the experimental temperature was as low as 500 mK (Köhler and Friedrich (1988b)). If we assume a constant tunneling mass and distance, as in the case of PHB, we can achieve agreement between this distribution and the postulate of the standard tunneling model $P(\lambda) = $ const, as was already discussed above. The question is whether it is possible to apply this assumption to the case of NPHB. Taking into account the numerical model calculations by Dab et al. (1995) which give an idea of the microscopic structure of TLS's, it is definitely impossible to assume a constant tunneling mass for "intrinsic" TLS's. As described in Chap. 8, we have to assume the mass of intrinsic TLS's to vary within a very broad range. On the other hand, if we assume some "extrinsic" TLS's to be responsible for NPHB, their nature is not clear. If the ideas about NPHB going along with hydrogen bond rearrangements are correct, the tunneling particles can be identified as protons and the above interpretations can be more or less reliable. If we deduce the properties of pure NPHB systems, we come to the conclusion that the distribution of barriers and other properties of the corresponding "extrinsic" TLS's are very similar to those of "intrinsic" TLS's. But at the same time NPHB does not exist in all glasses, which makes it impossible to identify "extrinsic" TLS's with the "intrinsic" ones which cause the well-known low-temperature anomalies of glasses.

The evidence presented here for tunneling of protons along hydrogen bonds within the solid matrix as a possible mechanism for NPHB illustrates the relationship between hole-burning spectroscopy and tunneling processes in low-temperature amorphous solids in general.

6.4.3 Hole Burning in a Model System: Benzoic Acid

The above discussion of NPHB in organic impurity/host systems has shown that the observed phenomena can be extremely complicated and puzzling. On the other hand tunneling along hydrogen bonds constitutes a well-defined model for tunneling in solids. In the following we will discuss a crystalline system which displays seemingly very complicated photoreactions with impurities along with well-defined proton transfer in hydrogen bonds: benzoic acid (BA).

Crystals of BA are built of dimers having two tautomeric forms. The structure of the BA dimer is shown in Fig. 6.8.

For an isolated dimer, both tautomers are energetically equivalent, but in a crystal, the interaction with the other molecules removes this degeneracy. In the frame of a model employing a double-well potential for the description of these two tautomeric forms and the transitions between them, this means that the corresponding double-well potential becomes nonsymmetric. In a crystal, this tunneling state has a splitting of about $35\,\text{cm}^{-1}$ (corresponding to 4.3 meV). At high temperatures the tautomerization process is thermally

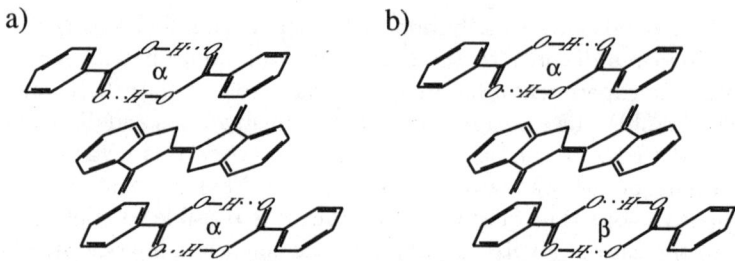

Fig. 6.8. A fragment of a benzoic acid (BA) crystal with an impurity molecule (thioindigo) substituting one of the BA dimers. (a) Both neighboring BA dimers are in the same state, designated as α ; (b) the BA dimers near the impurity molecule are in different states, correspondingly designated as α and β.

activated, but at temperatures below $\simeq 50$ K tunneling is the dominant proton exchange process. BA crystals proved to be a very good model system for the investigation of tunneling, dephasing and even phototransformations in glasses. In this section we will give only a short overview of the results which are interesting in a comparison with glasses. Detailed descriptions of the results on proton tunneling in BA dimers obtained via spectroscopic methods have been given by Clemens et al. (1984), Trommsdorff et al. (1984), Holtom et al. (1986), Trommsdorff (1986), Hegarty and Yen (1979), Skinner and Trommsdorff (1988b), and Oppenländer et al. (1989). Investigations of photoreactions in impurity/BA systems have been performed by Olson et al. (1982), Astilean et al. (1994), Barbara et al. (1995), and finally Neumann et al. (1996) gave a short review of the topic with a complete list of original references.

Very instructive spectroscopic results can be gathered by introducing suitable impurities into the BA crystal lattice. Some large organic molecules can substitute exactly one BA dimer within the crystal lattice. This is the case for thioindigo or pentacene. Hole burning using these molecules as impurities in benzoic acid is a very instructive example of the processes which take place in a well-defined system with small inhomogeneous broadening and at the same time exhibit a very complicated scheme of phototransformations.

Transfer of Localized Protons. Patterson et al. (1981) first observed hole burning on pentacene in BA. Many rather sharp photoproduct absorption lines were discovered as a result of intensive hole burning (Olson et al. (1982), Trommsdorff et al. (1984)). A strong deuteration effect was observed already in the first publications by Olson et al. (1982) and Walsh and Fayer (1985). Later Casalegno et al. (1992) investigated the influence of deuteration on the quantum efficiency of hole burning and on the decay rate of the photoproduct in detail. A careful investigation of the mechanism of the photoreaction was published by Astilean et al. (1994) and Barbara et al. (1995). Here we present only an overview of the reaction scheme.

The optical spectrum of pentacene in BA has only one 0-0 absorption line, because there is only one way of embedding the pentacene molecule into the crystal lattice. Laser irradiation creates more than 20 additional lines corresponding to altered metastable states of the lattice surrounding the pentacene molecules. The following mechanism has been proposed for the photoreaction by Astilean et al. (1994) and Barbara et al. (1995): The photoproduct corresponds to the molecular states of the lattice where the hydrogen atom of a BA dimer, which was abstracted in a first step of the reaction by a pentacene molecule, returns to the matrix, but not necessarily to its original place. The lifetime of these new metastable states varies from sub-second to tens of hours and can be increased up to 5 orders of magnitude by deuterating the host. Irradiation of these newly created defect sites causes a redistribution between the original stable state and other defect states, whereas the spontaneous decay of defect states partially populates other defect sites with lower ground state energy. In addition new defects can be created by irradiation, which cannot be accessed directly from the stable initial state.

Despite the fact that there is only one single mechanism – proton transfer from one hydrogen bond to another – the whole scheme of possible photoreactions looks very complicated. The complete energy diagram illustrating the scheme of transformations as proposed by Astilean et al. (1994) and Barbara et al. (1995) is presented in Fig. 6.9. The above example is the first carefully investigated case of a NPHB model with a clear spectroscopic localization of all photoproduct possibilities. This system displays many similarities to NPHB in glasses: Pentacene is a photostable compound and in many other matrices it shows no signs of photosensitivity. The photo-transformation is reversible and the spectral positions of the photoproduct states are distributed in the vicinity of the initial stable form with a dispersion of the order of a typical inhomogeneously broadened absorption band in glasses, which is $\approx 100\,\text{cm}^{-1}$. The main difference from the situation in glasses is the discrete number of photoproduct states, which is clearly due to the ordered structure of the crystalline matrix. Nevertheless this system can serve as a very instructive model for NPHB processes in general.

Tunneling of Delocalized Protons in Hydrogen Bonds. Thioindigo as an impurity exhibits very special hole-burning properties: The quantum efficiency is rather low, the burned holes decay spontaneously within minutes, and the burning mechanism is presumably a reversible photo-induced hydrogen abstraction which is supported by the fact that the effect is absent in deuterated BA (Clemens et al. (1984), Oppenländer et al. (1989)). If thioindigo is embedded in the crystal lattice of BA, the distortion created by this impurity molecule leads to the symmetrization of the double-well potential describing two neighboring dimers. Their tautomeric energy difference is lowered to less than $1\,\text{cm}^{-1}$. Four possible combinations of the states of neighboring dimers create four impurity states (two possible ways of inserting the impurity into the lattice double this number of states). This results in the appearance of a multiplet of lines in absorption and fluorescence spectra of thioindigo as observed by Clemens et al. (1984). As was shown in this reference, the optical excitation of the impurity strongly perturbs the double-well structure of neighboring dimers, removing the degeneracy of their TLS's. Therefore, the excited states of thioindigo in BA corresponding to the different states of neighboring BA dimers are spectroscopically well resolved, their splitting is of the order of $30\,\text{cm}^{-1}$, which is very close to the tautomeric energy spitting in the pure crystal. If thioindigo is in its optically excited S_1 state, the tautomerization of the BA dimers occurs on the time scale of the S_1 lifetime. This causes a very strong dependence of the fluorescence multiplet intensity distribution on the excitation scheme, i.e. which component of the multiplet is selectively excited. The kinetics of tautomerization was investi-

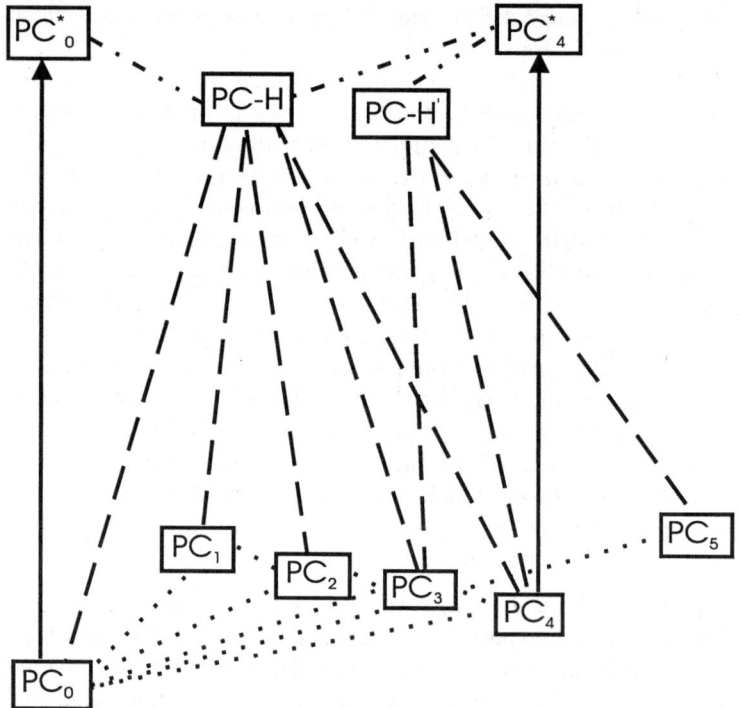

Fig. 6.9. Energy diagram indicating the processes involved in the creation and evolution of different sites observed for Pentacene (PC) in BA. PC_0 corresponds to PC in the stable state. After optical excitation ($PC_0 \Rightarrow PC_0^*$), hydrogen/proton abstraction from the matrix leads to the intermediate PC-H (dash-dotted line). PC-H dissociates but the H does not necessarily return to the original position so that defect sites PC_i are created (dashed lines). These defect sites decay on a longer time scale (dotted lines) and finally repopulate PC_0. Optical excitation of PC in defect sites leads to analogous processes. From Astilean et al. (1994) and Barbara et al. (1995).

gated in detail using selective excitation of different absorption components of two thioindigo sites (Clemens et al. (1984)). The tunneling rate in the excited state was found to be $4.5 \times 10^8\,\text{s}^{-1}$ at an excited state lifetime of the lowest component of the multiplet of $\tau = 12.04\,\text{ns}$. Tunneling in BA dimers when the thioindigo is in its ground state was investigated in by Clemens et al. (1984) using the hole-burning technique. Employing the measured hole widths for all components of the multiplet as well as the data obtained from the kinetic measurements on the excited state, the tunneling rate in the ground state was deduced. A rate in the range of $\simeq (2.4\text{–}5.4) \times 10^8\,\text{s}^{-1}$ for different transitions at $T = 1.6\,\text{K}$ was obtained. The investigation of tunneling within BA dimers with the thioindigo in its ground state was continued by Hochstrasser and Trommsdorff (1987) and Oppenländer et al. (1989) employing high-resolution fluorescence line narrowing and hole-burning spectroscopy. As a result, the whole structure of energy levels of BA-thioindigo-BA sandwiches in the thioindigo ground state was reconstructed. In particular, the value of the tunneling matrix element was calculated to be $J = 8.4 \pm 0.1\,\text{GHz}$. The asymmetry of the tautomeric double-well potential is of the same order of magnitude. This means that the tautomeric states of a BA dimer neighboring a thioindigo molecule are indeed strongly delocalized.

6.4.4 Conclusion

The experimental facts about photochemical and nonphotochemical hole burning summarized in this section show that the behavior of organic glasses doped with a variety of dye molecules can be very complex. The example of benzoic acid has also shown that there are organic model systems which are especially instructive for studying solid-state tunneling processes by optical methods. Doping can introduce new degrees of freedom in the form of the dopant's intrinsic molecular properties or even initiate complicated interactions between dopant and solid as the example of hydrogen bond rearrangements has shown.

Of course the possible influence of the impurities has to be taken into account when the application of such dye-doped systems for the investigation of the tunneling dynamics of "intrinsic" TLS's in amorphous solids is considered. With the help of careful investigations of the combined dye-matrix system, perturbations of the intrinsic solid-state dynamics as well as mutual interactions between dye molecules can be ruled out. The disadvantage that optical spectroscopy requires the introduction of impurities to investigate glasses, is compensated by the high precision and sensitivity of the method as well as by the huge time range accessible. A remarkable benefit of optical spectroscopy is the possibility to investigate the model situation of TS dynamics in thermal equilibrium on time scales from picoseconds to months without the necessity to perturb any thermodynamic parameters of the system. At the same time, it is also possible to introduce well-defined deviations from equilibrium to observe equilibration processes similar to the time dependence of the specific heat or heat release. Several examples of the potential of optical hole-burning spectroscopy for investigating solid state TS dynamics are presented in the remainder of this chapter.

6.5 Spectral Diffusion: Low-Barrier Tunneling

6.5.1 Spectral Diffusion

Experiments with external perturbations such as electric and magnetic fields (Personov and Kharlamov (1986), Maier (1986), Personov (1992), Kohler et al. (1995)) or hydrostatic pressure (Richter et al. (1984), Jankowiak et al. (1993), Ellervee et al. (1993), Pschierer et al. (1994), Den Hartog et al. (1996)) have proven the high sensitivity of hole profiles and positions to rather small changes of the external conditions. Since hole-burning spectroscopy is a high-resolution optical method, the dynamics of tunneling states in glasses can be investigated by inserting suitable optical absorbers as probes into the solid. The observation of an inhomogeneous broadening due to the solvent shift itself already illustrates the fact that the absorption frequency of such an absorber probe is very sensitive to the specific structure of its surrounding matrix (see Sect. 6.2.2). If a given configuration is altered, the probe will immediately react by reconfiguring its electronic structure – causing a shift of its absorption frequency. In terms of TS dynamics, this means that every time a TS flips, i.e. changes its state, optical absorbers interacting with it exhibit a jump in frequency space.

In single-molecule spectroscopy, this process has been observed directly by monitoring the frequency position of an individual absorption line as a function of time not only in glasses but even in crystals (Ambrose et al. (1991), Basché et al. (1992), Moerner et al. (1994)). There have been observations of single molecules strongly coupled to a small number of neighboring TLS's; the absorption lines of such molecules consequently accessed a corresponding small number of discrete frequency positions. An example of such frequency jumps of a single chromophore due to interactions with fluctuating TLS's is presented in Fig. 6.10, which was made available by Moerner et al. (1994). The 0-0 absorption frequency of the impurity molecule has 5 discrete values, which can be interpreted as the result of the interaction with at least 3 neighboring TLS's. A qualitative scheme of such a TLS–impurity interaction is sketched in Fig. 6.11 (see also Fig. 5.4). A particular chromophore with a transition frequency $\hbar\omega$ is coupled to an ensemble of TLS's via electric or strain fields. The energy splitting and, therefore, the transition frequency of the chromophore depends on the configuration of the TLS ensemble.

This is the mechanism of spectral diffusion: Ideally, the hole-burning process is a method to take a snapshot of the current distribution of absorption frequencies at a given instant of time. After a hole has been burned into the inhomogeneous absorption band of an amorphous solid doped with probe

6.5 Spectral Diffusion: Low-Barrier Tunneling 353

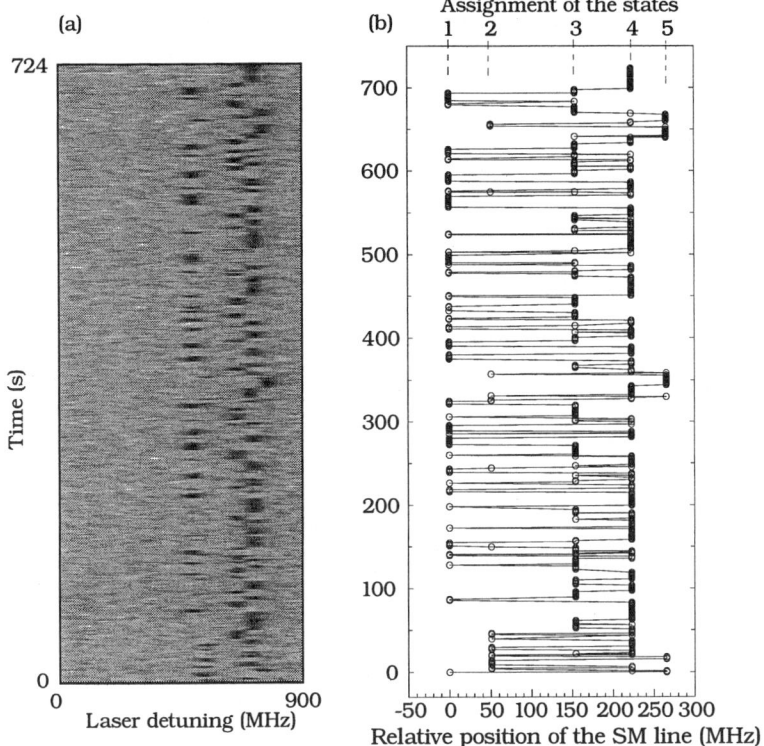

Fig. 6.10. Frequency positions for a single terrylene molecule in a Shpol'skii matrix of hexadecane as a function of time. The vertical axis denotes the time in seconds, the horizontal axis the frequency position in MHz. (a) Fluorescence intensity plotted as gray scale (from 566 scans of 256 points each, 5 ms per point, laser intensity 0.36 W/cm^2). (b) The assignment of five frequency positions with 0 MHz being the initial frequency. The lines are guides to the eye to show the time sequence. From Moerner et al. (1994).

molecules, TLS flips cause the remaining absorbers to perform frequency jumps. Since the molecules constituting a hole spectrum are distributed all over the sample volume, the time evolution of a hole spectrum reflects the dynamics of the whole TLS ensemble within the sample. As the surrounding matrix is different for every individual probe molecule, coupling to a different number of TLS's with various strengths occurs. The ensemble behavior of the complete hole spectrum is therefore statistical. Since the hole-burning process reduces the number of absorbers within the hole spectrum, there are on the average more jumps of absorption lines from the edges towards the center of the hole spectrum than the other way. The result of this ensemble dynamics is therefore a diffusional broadening process of the hole spectrum – the so-called "spectral diffusion".

The shape of a spectral hole is in general Lorentzian (Friedrich and Haarer (1984)). In the limit of low-burning energy and intensity its width consists of two parts given by

$$\Gamma_{\text{hole}} = 2\Gamma_H(T) + \Gamma(t,T). \tag{6.16}$$

Here $\Gamma_H(T)$ represents the homogeneous linewidth of the optical transition. The factor of 2 appears because this width is involved in both burning and recording of the hole spectrum (see Sect. 6.3.2). $\Gamma(t,T)$ describes the hole broadening by spectral diffusion. The experimental verification of this time-dependent term was originally performed on organic systems via hole-burning experiments in the time range from minutes to days by Breinl et al. (1984a) and Friedrich and Haarer (1986) and on the other hand on inorganic glasses via three-pulse photon echo experiments on a time scale from micro- to milliseconds by Broer and Golding (1986), Broer (1986). But a clear understanding of the decisive role of spectral diffusion for line broadening in organic systems already on short time scales was achieved via comparing hole-burning linewidths measured in the range of seconds with photon-echo data acquired in the nanosecond range (Molenkamp and Wiersma (1985), Walsh et al. (1986), Berg et al. (1987a), Walsh et al. (1987), Berg et al. (1987b)). This comparison yielded a large difference for the values of the respective

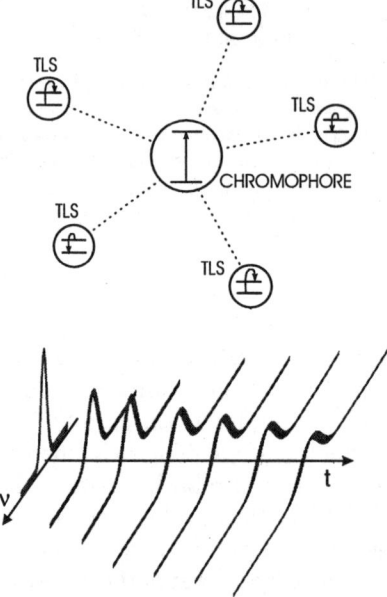

Fig. 6.11. A probe molecule interacting with TLS's. Every TLS flip slightly modifies the absorption frequency of the molecule. The statistical result of diffusional hole broadening for a whole ensemble of chromophores constituting a spectral hole is shown in the lower part.

linewidths obtained with these two different methods (up to factor of 4–8). This was accounted for by introducing the time-dependent term in 6.16. It is now experimentally proven that spectral diffusion plays a significant role even on time scales as short as nanoseconds (Bai and Fayer (1988a), Bai and Fayer (1988b), Narasimhan et al. (1991), Gruzdev and Vainer (1993)) and that the dynamics extends at least up to weeks (Maier and Haarer (1995), Maier et al. (1996)).

In the framework of the standard tunneling model the thermodynamical nonequilibrium nature of the glassy state is not considered. In the following, a glass is considered to be in equilibrium when the TS's are in equilibrium with the phonons, which are assumed to equilibrate infinitely fast. Due to the extremely broad distribution of TS relaxation rates (its long-time limit was not yet reached experimentally), even in the framework of this model, a glass at low temperatures is never completely in equilibrium. Nevertheless, it is possible to distinguish between situations where either equilibrium or nonequilibrium dynamics dominates. One of the advantages of the optical detection of TS dynamics via spectral diffusion is the possibility to monitor not only nonequilibrium relaxations, but also unperturbed equilibrium TS dynamics on a broad time scale (see below). Historically, the investigation of equilibrium phenomena was begun earlier and only in the last years extensive studies of nonequilibrium spectral diffusion have been started. In the following sections we present a generalized theoretical description and experimental data on equilibrium and nonequilibrium spectral diffusion.

6.5.2 Theoretical Description of Spectral Diffusion

Diffusional line broadening was first analyzed for spin systems in terms of spin–spin interaction by Klauder and Anderson (1962). In particular they found that in the limit of a low spin density and a dipolar spin–spin interaction, such a diffusion leads to a Lorentzian line shape. Later a similar formalism (TS–TS dipolar interaction) was used by Black and Halperin (1977) for the description of time-dependent anomalies in ultrasonic absorption and phonon echoes. Reinecke (1979) extended this model to optical transitions interacting with TS's. We will basically follow his model description.

Two-level systems flips cause shifts of the transition frequencies of the chromophores. An initially monochromatic ensemble of impurities will be broadened with time. This is the so-called diffusion kernel. It is the theoretical representation for the time-dependent broadening of a spectral hole described above. Assuming a low spatial TS density and a dipole–dipole interaction of the TS's with the probe, Reinecke obtained a Lorentzian diffusion kernel

$$D(\omega - \omega_0) = \frac{1}{\pi} \frac{\Gamma(t,T)}{(\omega - \omega_0)^2 + (\Gamma(t,T))^2} \,. \tag{6.17}$$

A dipolar interaction term can always be regarded as a good first approximation if the spatial extensions of the interaction partners are small as compared to their distance. A dipole–dipole chromophore–TS interaction and a low spatial number density of TS's can, in general, not be assumed *a priori*. The dependence of the diffusion kernel on the interaction mechanism is not completely clear. The dependence of the line profile on the TS density was already investigated in the first publications devoted to spectral diffusion in spin echoes (Klauder and Anderson (1962)) and phonon echoes (Black and Halperin (1977)). It was shown that in the limit of high TS concentrations the line profile becomes Gaussian. The diffusional kernel for optical spectra at arbitrary TS concentrations was calculated by Kador (1991) (a criterion for the limiting cases of "high" and "low" concentrations is also given in this reference). Up to now no distinct deviation from a Lorentzian line shape was observed.

The time- and temperature-dependent width of the Lorentzian diffusion kernel is given by

$$\Gamma(t,T) = \frac{2\pi^2}{3\hbar} \langle C_{ij} \rangle \left\langle \frac{\Delta}{E} n(t,T) \right\rangle_{\Delta_0, E}, \tag{6.18}$$

Δ and E are the energetic asymmetry and the total energy splitting of a TS, respectively. For a single TS $n(t,T)$ would describe the probability to have left its initial state after a time t at a given temperature T. In our representation, for a whole TS ensemble, it is a number density. The brackets denote averaging over a distribution of TS parameters Δ_0 and E. The proportionality constant $\langle C_{ij} \rangle$ represents an averaged coupling strength between probe and TS.

Let us calculate the quantity $n(t,T)$. The initial ($t=0$) and final ($t=\infty$) states of TS's in equilibrium are represented by the probability vector

$$n_{0,\infty} = \begin{pmatrix} p \\ 1-p \end{pmatrix}, \text{ where } p = 1/\left(1 + \exp\left(\frac{E}{k_B T}\right)\right) \tag{6.19}$$

is the probability that a TS is in its *upper* state and $1-p$ in its *lower* state. All possible events in the intermediate time are represented by the matrix

$$n_0 \otimes n_\infty = \begin{pmatrix} p^2 & p(1-p) \\ p(1-p) & (1-p)^2 \end{pmatrix}, \tag{6.20}$$

where only the off-diagonal elements are related to changes of the TS state. The sum of these elements multiplied with a time-dependent relaxation term for the TS transition probability yields

$$n(t,T) = \frac{1}{2}\text{sech}^2\left(\frac{E}{2k_B T}\right)[1 - \exp(-r(E,\Delta_0,T)t)]; \tag{6.21}$$

$r(E, \Delta_0, T)$ is the TS relaxation rate. Now we can explicitly rewrite the averaging of (6.18) in an integral form and insert $n(t,T)$:

6.5 Spectral Diffusion: Low-Barrier Tunneling

$$\Gamma(t,T) = \frac{2\pi^2}{3\hbar}\langle C_{ij}\rangle k_B T \int_0^\infty dx\, \text{sech}^2(x) \int_0^{\Delta_{0,\max}} d\Delta_0 \frac{\Delta}{E} P(E,\Delta_0)$$
$$\times [1 - \exp(-r(x,\Delta_0,T)t)]\,. \tag{6.22}$$

Here we have substituted $x = E/2k_B T$. At this stage a few comments on the choice of the integration limits are necessary: First of all the averaging should of course be performed over the distribution of TS parameters, i. e. the limits should be the extremes $E_{\min}$, $E_{\max}$, $\Delta_{0,\min}$, and $\Delta_{0,\max}$ respectively. These, however, are unknown *a priori*. Therefore we have to choose reasonable values. A first reasonable lower limit for the TS energy is $E_{\min} \to 0$. Recalling that $E = \sqrt{\Delta^2 + \Delta_0^2}$, this implies $\Delta_0 \to 0$. Δ_0 is the tunneling matrix element: If it vanishes, then no tunneling, i. e. no relaxation occurs. In other words, TS's with $E \to 0$ would only contribute at infinitely long times. The external integrand $\text{sech}^2(x)$ decreases very fast at $x > 1$, therefore the upper integration limit may be taken as ∞ – large values of x do not contribute significantly to the result of the integration. In a similar manner the lower limit of the internal integral can be put to zero due to the tendency $r(x,\Delta_0,T) \to 0$ at $\Delta_0 \to 0$ – the integrand vanishes in this limit. Only the upper integration limit for Δ_0 has to be chosen rigorously as: $\Delta_{0,\max} = \min(\hbar\omega_0, x2k_B T)$. Taking into account that in the framework of the standard tunneling model the distribution function is given by

$$P(E,\Delta_0) = P_0 \frac{E}{\Delta\Delta_0}, \tag{6.23}$$

the internal integral can easily be solved for the low-temperature case, where only one-phonon-assisted TS jumps need to be taken into account. In this case the relaxation rate is given by (2.18), i. e.

$$r = A\Delta_0^2 E \coth\left(\frac{E}{2k_B T}\right). \tag{6.24}$$

If we insert this expression, the integration over Δ_0 can be performed analytically without any approximations, yielding

$$\int_0^{\Delta_{0\max}} d\Delta_0 \ldots = \frac{1}{2}\left[\ln(r_{\max}(x,T)t) + Ei(1,r_{\max}(x,T)t) + \gamma\right], \tag{6.25}$$

where $r_{\max}$ is defined as $r_{\max}(x,T) = A(2k_B T)^3 x^3 \coth(x)$; $Ei(1,y) = \int_y^\infty e^{-y}/y dy$ is the integral exponent, and $\gamma \simeq 0.577\ldots$ is Euler's constant. For $r_{\max}t \gg 1$, this result can be approximated very well by

$$\int_0^{\Delta_{0\max}} d\Delta_0 \ldots \simeq \ln(r_{\max}(x,T)t). \tag{6.26}$$

This result is a general prediction of the standard tunneling model for the time dependence of spectral diffusion. An estimation of the parameter A in (6.24) for organic polymers (Maier and Haarer (1995), Fritsch et al. (1996)) yields $Ak_B^3 \simeq 10^{10} s^{-1} K^{-3}$ (for comparison $Ak_B^3 \simeq 10^8 s^{-1} K^{-3}$ for SiO$_2$ obtained from acoustic experiments, see Sect. 4.4.1). Therefore the above condition is fulfilled with very high precision in hole-burning experiments having a time resolution ≥ 1 s, even at the lowest temperatures reached.

Qualitatively, a time dependence of TS's dynamics as given by (6.26) agrees rather well with most experimental results. This statement holds for almost all measurement techniques applied in this field of research. Especially in the case of persistent spectral hole burning, however, where experiments can be conducted at very long times and over several orders of magnitude in time, pronounced deviations from this logarithmic prediction have recently been observed. These experimental results as well as their theoretical implications for our current understanding of TS dynamics will be discussed in the next section.

A common way of experimentally obtaining information about densities of states is the measurement of temperature dependencies. In the case of an inhomogeneous line broadening by spectral diffusion, however, the interpretation of such data is absolutely not straightforward. The temperature dependence of a diffusionally broadened linewidth is not simple: Due to the time dependence of a diffusional linewidth, the result may be very sensitive to the experimental procedure. Let us illustrate this with the following examples:

1. If the whole linewidth is measured (we assume for simplicity that the phonon-induced line broadening is negligible), the observed temperature dependence should be slightly superlinear instead of the linearity implied by $P(\Delta) = $ const. This is due to the temperature dependence of $r_{\max}$. This superlinearity was analyzed by Hunklinger and Schmidt (1984) and Kharlamov et al. (1994). It was shown that a temperature dependence of $r_{\max}$ leads to the phenomenological relation $\Gamma(T) \propto T^{1+\alpha}$, with $\alpha \simeq 0.1 - 0.3$. Recently the more general analysis, which is in agreement with the above resuts, was done by Geva and Skinner (1997a).
2. If only diffusional hole broadening is measured during a well-defined time interval $\Delta t = t_1 - t_0$, i.e. $\Delta \Gamma(T, \Delta t) = \Gamma(T, t_1) - \Gamma(T, t_0)$, it is easy to show that in this case a logarithmic time evolution leads to $\Delta \Gamma(T, t_1) \propto T \ln(t_1/t_0)$, where the temperature dependence of the relaxation rates disappears completely.

In general, the temperature dependence of spectral linewidths in glasses is even more complicated by additional phonon-induced line broadening which becomes a dominating factor at temperatures above 3–4 K.

6.5.3 Equilibrium Glass Dynamics

In the case of hole-burning spectroscopy a large number of absorbers is involved in constituting the spectral shape of a hole. The fact that these absorbers are neighbors in frequency does not, however, imply that they are also located in the same spatial position in the sample. There are, on the contrary, many spatial sites distributed all over the sample volume that initially cause more or less degenerate energetics for the incorporated absorption probes. In addition it has to be taken into account that the photochemically transformed molecules are not involved in the spectral diffusion process. Instead, absorption spectroscopy observes the majority of remaining absorbers. The time-resolved observation of the diffusional broadening of a single hole spectrum is therefore capable of yielding an averaged measure of the TS dynamics in the whole sample volume.

Such an experiment yields the same amount of information as a "macroscopic" measurement of, for instance, the time dependence of the heat capacity, see Sect. 2.2.3 and Sect. 2.3.1. A detailed comparison of the heat capacity and the optical linewidth as a function of temperature has in fact been performed for two inorganic systems by Schmidt et al. (1993), (1994). The result for one of them (Schmidt et al. (1994)) is shown in Fig. 6.12. These results can be taken as direct evidence that both methods are actually sensitive to the same degrees of freedom. There is one major difference remaining, however: A measurement like that of the heat capacity monitors the response of a macroscopic thermodynamic parameter to a disturbance of the thermodynamic equilibrium of the TS ensemble in the sample – in this example –

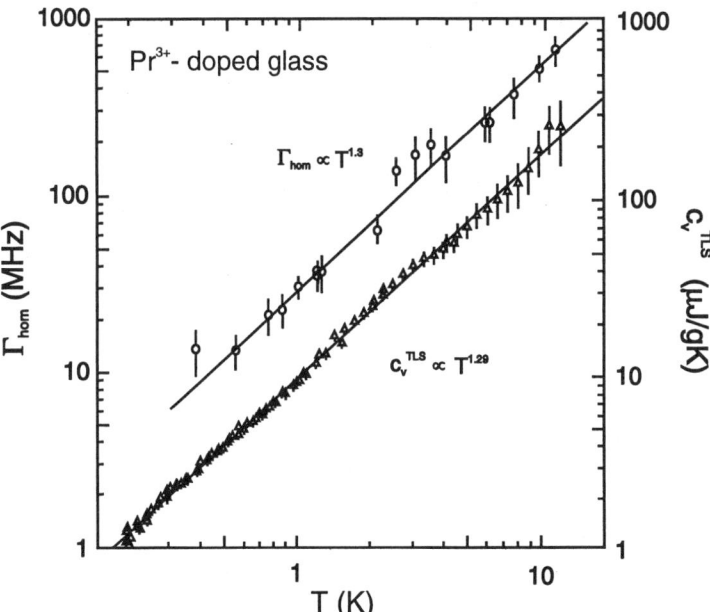

Fig. 6.12. Comparison of the quasi-homogeneous hole-burning linewidth (denoted as Γ_{hom}) and the specific heat for a Pr^{3+}-doped silicate glass. Both quantities display the same temperature dependence. From Schmidt et al. (1994).

by measuring the time evolution of the temperature after imposing a heat pulse on the system. The hole-burning approach is different in this context. At every instant of time the absorption frequency of a probe is given by the configuration of the TS's interacting with it. What we do when we burn a hole is indirectly marking this TS configuration by memorizing the distribution of absorption frequencies in the spectrum at the time of hole-burning. Changes of the marked TS configuration, given by the term $n(t,T)$ in (6.18), are reflected by frequency shifts of the probes, which, on the average, yield a broadening of the initial hole. The occurrence of such changes, however, is not restricted to thermal relaxations in nonequilibrium situations. Since thermal equilibrium is only a global statistical condition, transitions of individual TS's occur also in an equilibrium situation by random absorption and emission of thermal phonons without affecting the global statistics. In this way, the time evolution of an initially marked TS configuration can be monitored by observing the width of a spectral hole without the experimental necessity of preparing a nonequilibrium situation and inducing a thermal relaxation process.

6.5.4 Long-Time Equilibrium Dynamics: Nonclassical Distribution of Tunneling States

Motivation. In this section we describe recent experiments on the long-time behavior of equilibrium spectral diffusion ranging from seconds to weeks. As will be shown later on in this chapter, such experiments are extremely time consuming. Before starting measurements on this topic, a long waiting time is required in order to avoid nonequilibrium contributions from thermal relaxation. This is because cooling a sample from high temperatures to the liquid helium range initiates a long-lasting relaxation process (see Chap. 2). For these reasons the experiments described in this section require low-temperature run times on the order of three months. In spite of this extreme effort, such experiments are particularly interesting from the viewpoint of the theoretical description of TS dynamics. The standard tunneling model as described in Chap. 2 as well as in other chapters of this book suffers from two major problems:

The first one is the fact that there is no microscopic consideration within this model which would be able to establish a distribution function for the parameters employed to model the TS ensemble describing an amorphous solid. Instead the original publications *a priori* assumed a flat distribution of the parameters they chose, namely, the TS asymmetry Δ and the dimensionless tunneling parameter λ. The latter is connected with the tunneling matrix element Δ_0 via $\Delta_0 = \hbar\omega \exp(-\lambda)$, which results in a distribution $P(\Delta_0) \propto 1/\Delta_0$ which was previously used in this chapter. There is, however, no physical justification for these specific distributions. If we look at the mathematical consequences of this choice described in Sect. 6.5.2, we can conclude that especially the assumption $P(\Delta_0) \propto \Delta_0^{-1}$ represents a serious

6.5 Spectral Diffusion: Low-Barrier Tunneling

bias for the prediction of the time evolution of TS dynamics. It is essentially this choice that leads to the logarithmic time dependence given by (6.26). Any deviation of the exponent from 1 will turn the result into an algebraic law. Mathematically, it is even irrelevant how large this deviation should be, only an exponent of exactly 1 results in a logarithmic dependence. Experimentally, the situation is unfortunately not as clear. If an experiment is to distinguish a logarithmic behavior from a power law, the necessary effort does not only depend on the magnitude of the deviation of this exponent from 1 but also on the quality of the data. Without prior knowledge of the deviation even a very low noise data set will necessarily have to extend over as many orders of magnitude in time as possible. This is one of the main reasons for a large experimental time scale.

The second problem is the fashion in which the standard model treats the interaction of a TS with its surroundings. In the original work a weak perturbative coupling of TS's to phonons is the only interaction included. But a look at experimental data and theoretical considerations in the literature shows that two more mechanisms may actually be present:

- First, there is the question of how strong TS's interact with phonons. A model has been put forward by Kassner and Silbey (1989), which explicitly includes TS-phonon interaction in the system's Hamiltonian instead of treating this interaction perturbatively as in the original model. Such an explicit inclusion is necessary if the interaction is too strong to be treated in a perturbative manner. According to these considerations, deviations from the time dependence predicted by the standard model should occur at very long observation times only if they are due to a strong coupling of TS's and phonons.

- Second, there is experimental evidence that TS's do indeed interact with each other. This necessarily had to be assumed by Black and Halperin (1977) for the interpretation of early phonon echo experiments by Arnold and Hunklinger (1975). As we will show below, the presence of such an interaction leads to a time evolution of spectral diffusion which is different from the prediction of the standard model.

From an experimental point of view, spectral diffusion is a suitable tool for addressing the above questions: In contrast to heat release or acoustic experiments, spectral diffusion is an "integrating method" of observing TS dynamics. As has been mentioned in the previous section, hole burning indirectly marks a given TS configuration at a certain time. Any change of this configuration occurring by TS dynamics will result in a broadening of the spectral hole. If the dynamics are slightly different from our expectation in a given short-time interval, then the diffusional hole broadening during this time will also be slightly different. But if we look at the following time interval, this small deviation will not be lost. Instead, if we monitor a hole spectrum as long as possible, then such small deviations will accumulate more and more – the longer the observation time, the clearer small differences in the rate of diffusional hole broadening will be visible. In addition, the measurement of spectral diffusion is an experimental method with no inherent limits for the observation time of TS dynamics – very long-lasting measurements of TS dynamics are possible. Given that the technical requirements can be fulfilled, spectral diffusion experiments are therefore the method of choice to shed light on the specific questions raised here.

Experiments. The long-time experiments described here have been performed at 0.5 K, 1 K and 2 K. The low-temperature parts of these measurements were carried out in a continuously operating ^{3}He/^{4}He dilution refrigerator. ^{3}He is pumped, recondensed, and brought back to the low-temperature stage via heat exchangers in a closed circle. Therefore, this method allows us to keep a sample continuously at low temperatures for very long times. For a detailed description of this method, we refer the reader to Pobell (1992). The experiments at 2 K were performed in a special dipstick cryostat, designed for insertion in a normal He storage vessel. With a 100 l He vessel, Hannig et al. (1996) achieved a runtime of more than 50 days.

The cryostat employed for the experiments in the sub-Kelvin region was especially designed for transmission spectroscopy (Müller and Haarer (1991), Maier et al. (1997)). Since a hole-burning experiment necessarily requires the sample to absorb energy from a laser beam (Sect. 6.3), the sample was placed directly into the mixing chamber to optimize its cooling by direct contact to the cooling fluid. In our design this was achieved by using an optically transparent mixing chamber made of glass. A cross-section of the mixing chamber and the optical window system is shown in Fig. 6.13.

The experiments described here were performed to monitor long-time TS dynamics under thermal equilibrium conditions . The subject of spectral dif-

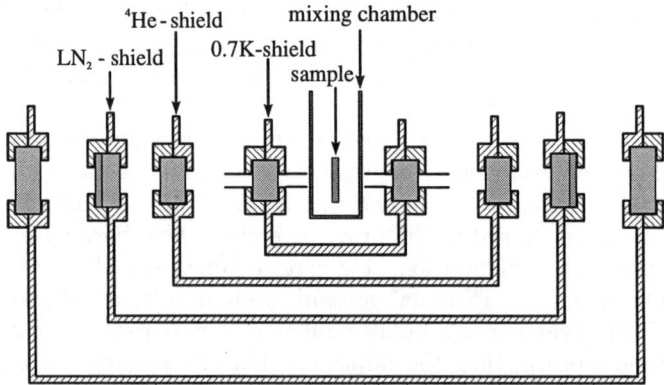

Fig. 6.13. Cross-section of the optical part of our ^{3}He/^{4}He dilution refrigerator. To minimize thermal radiation on the sample, there is an optical window at every temperature stage. The windows in the LN$_2$ shield additionally have an IR-reflection coating. The sample is located inside the mixing chamber in direct contact with the cooling fluid.

6.5 Spectral Diffusion: Low-Barrier Tunneling

fusion in nonequilibrium situations will be addressed in Sect. 6.5.5. To avoid any contributions to the hole broadening from thermal relaxations of the TLS ensemble, the sample has to be kept at low temperatures for a sufficiently long time before starting the experiments. This time must be long as compared to the desired observation time. Since our purpose was to observe spectral diffusion for a time of 10^6 seconds, the following time sequence was used for the dilution refrigerator experiments:

- T below 0.5 K for six weeks;
- T constant at 0.5 K for two weeks;
- burning of the first series of holes at 0.5 K;
- measurement of spectral diffusion for 10 days at 0.5 K;
- T raised to 1.0 K;
- T constant at 1.0 K for two weeks;
- burning of the second series of holes at 1.0 K;
- measurement of spectral diffusion for 10 days at 1.0 K.

This schedule required a total runtime of the cryosystem of 90 days. While the upper time limit of 10^6 s was determined by the time schedule, the lower limit depended on the time required for hole burning. The experiments were performed on a sample of polymethylmethacrylate (PMMA) doped with the organic dye molecule phthalocyanine (see Fig. 6.7). The optical density of the sample was 0.4 at the maximum of the lowest absorption band with a sample thickness of 1 mm. At 0.5 K the power used for hole burning was about 40 nW with a total burning energy of 300–350 nJ. At 1.0 K both values were about twice as large. Hole-burning and scanning of the spectra were performed using a single-mode cw dye laser. For read-out the light power was decreased by a factor of about 2000. The employed laser system possesses an integrated wavelength measurement device with an accuracy of 30 MHz. This feature allowed us to monitor several holes in parallel at every temperature.

Examples of hole spectra from the experiment are shown in Fig. 6.14. The left-hand part shows the spectrum of a hole immediately after burning, the right-hand part shows the same hole at the end of the experiment – after a waiting time of 10^6 s. They are given in units of optical density; i.e. the signal amplitude at the edges represents the initial absorption of the sample and the absorption decreases towards the center of the hole. Increasing waiting time, the depth of the hole decreases. This is due to the fact that on the one hand the width of the hole increases as a function of time, but on the other hand the integrated area of the hole is conserved, since it is a measure of the number of photochemically transformed absorbers.

The shapes of *both* hole spectra are perfectly Lorentzian. For this outcome two conditions have to fulfilled: The hole spectrum without diffusional broadening as well as the diffusion kernel have to be Lorentzian (6.16–6.18). While the former condition addresses a rather general issue of optical spectroscopy, the latter already yields some qualitative insight into the properties of our sample material regarding the tunneling model itself: The hole broadening of our experiment can be described by a Lorentzian diffusion kernel as given by the theoretical description of Reinecke (1979). This implies a low spatial density of tunneling centers, as is generally assumed within the standard tunneling model.

The measurements at 2 K were performed using a PMMA sample doped with a different chromophore: H_2-tetra-phenyl-porphin (TPP). The influence of thermal relaxation of TLS's on spectral diffusion at higher temperatures is less pronounced (see the next section); therefore it is not necessary to keep the sample at the working temperature for as long as in the sub-Kelvin experiments before starting the measurements. Still, this waiting time was about one month. The results of the experiments are displayed in Fig. 6.15 (Maier et al. (1996), Hannig et al. (1996)).

Here and in the following hole widths are given in terms of the half width at half maximum (HWHM). The two lower curves represent the results obtained in the dilution refrigerator at 0.5 K and 1.0 K. The upper curve was measured at 2 K in the dipstick cryostat. In Fig. 6.15a the hole broadening data $\Delta\Gamma(T,t) = \Gamma(T,t) - \Gamma(T,t_0)$, hole width at time t minus the initial hole width immediately after burning, are displayed on a semilogarithmic scale. Fig.ure reflong-time-datab shows the absolute hole widths in a double logarithmic plot. The data points in the figure are averaged values of five spectral holes each.

As described above, the measurements were performed to reveal a possible deviation from a logarithmic time dependence. The results display important features:

The hole broadening is a nearly logarithmic function of time only in the short-time range of the experiment (cf. linear increase in the left-hand part of Fig. 6.15a, semilog plot). For longer times, the functional dependency

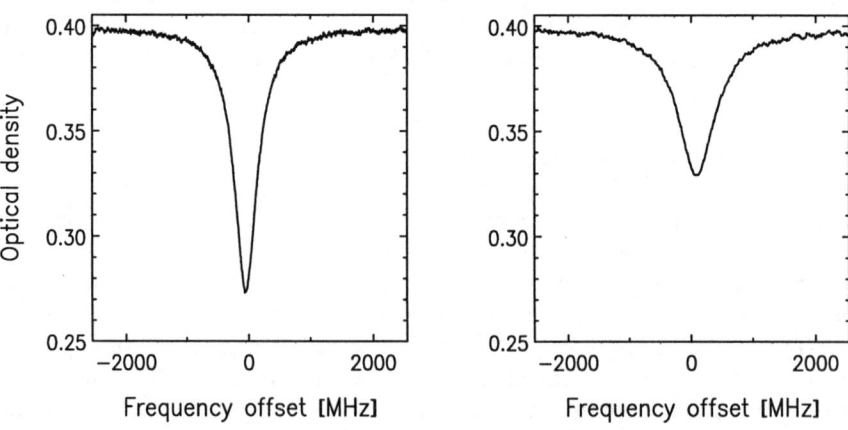

Fig. 6.14. Hole spectra from the experiment described in this section ($T = 1.0$ K). *Left*, spectrum of a hole immediately after burning; *right*, spectrum after a waiting time of 10^6 s.

is clearly nonlogarithmic. The asymptotic behavior for $t \to \infty$ is a square root law. This is not consistent with the standard model, which predicts a logarithmic increase of the hole width on all time scales.

The experiments described here covered 5 orders of magnitude in time. To summarize, we can conclude that the results clearly disagree with the predictions of the standard model: The overall time evolution of the diffusional hole broadening is definitely nonlogarithmic. On the other hand, however, the short-time part of all data sets is nearly logarithmic. This may indicate that on short time scales (i.e. below 10^3 s at 1 K) the standard model could provide a useful description for the observed hole broadening. It must be modified only for very long observation times.

Model Considerations. The results of the experiments described above clearly indicate that the long-time behavior of TS dynamics substantially deviates from the standard model's predictions. In principle, this can be caused by many reasons, for example, by a nonvalidity of the tunneling model for the description of long-time spectral diffusion. But it is quite reasonable to try to describe this phenomenon in terms of the tunneling model at first. Only if such attempts fail, a more "exotic" explanation may be necessary. In the framework of the tunneling model, two reasons may be responsible

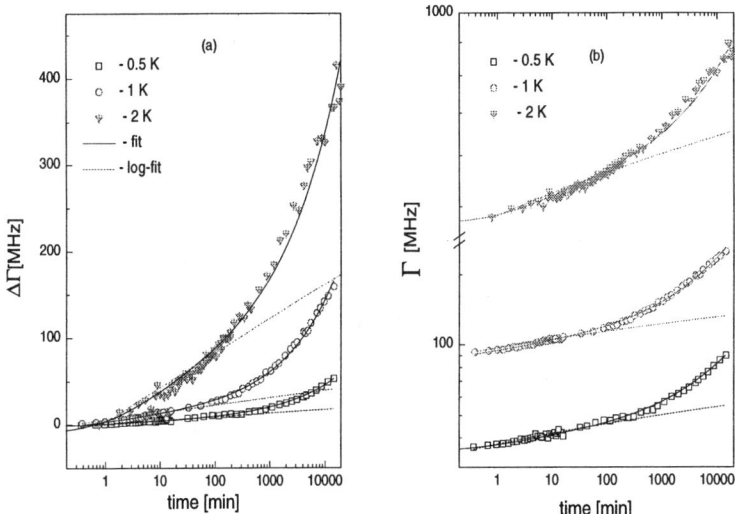

Fig. 6.15. Observed long-time hole broadening at three different temperatures. The solid lines represent the results of calculations based on our model distribution function 6.27, the dashed lines are logarithmic fits of the short-time part of the data (see the next section). The error is typically 1 MHz in the entire time range displayed: (a) semilog plot of the hole broadening, (b) double-log plot of the full hole width. The sample material is PMMA doped with H_2-tetra-phenyl-porphin in the case of the 2.0 K data and doped with phthalocyanine for the two other data sets. From Hannig et al. (1996).

for the experimentally observed deviation of diffusional hole broadening from the logarithmic prediction:

- A deviation of the TS parameter distribution from the standard model assumptions. Some possible reasons for such a deviation have been given in the introductory part of this section.
- A nonequilibrium TS population. This will be treated in Sect. 6.5.5. Qualitatively, this would lead to the same result as displayed by the above experimental data, the hole width increases faster than predicted by the standard model.

There are quite clear criteria to distinguish whether or not the TS ensemble is in thermal equilibrium; they will be discussed in detail in the next section. Here we mention only that all necessary conditions were fulfilled in the above experiments. We can be sure that an excess hole broadening caused by thermal relaxations, which cannot be completely avoided, was negligibly small. This means that the observed long-term nonlogarithmic line broadening is essentially an equilibrium phenomenon.

Let us discuss the modification of the TS distribution function necessary to account for the observed algebraic asymptotic behavior of spectral diffusion.

In the time range accessible to usual hole-burning experiments, the temporal development of spectral diffusion is completely dominated by the TS distribution parameter Δ_0, whereas the TS energy distribution is responsible for the temperature dependence. This clear separation can be demonstrated by inserting numbers into the expression for the TS relaxation rate (6.24): If we use the estimate of the coupling constant for TS–phonon interactions $A \simeq 10^{10}$ s^{-1} K^{-3} at $T \simeq 1$ K and assume an experimental time resolution of $t \geq t_{\min} = 1$ s and an average TS energy of the order 1 K, we obtain $\Delta_0 \lesssim 10^{-5}$ K for TS's with relaxation rates $r \leq 1/t_{\min}$. According to the relation $E = \sqrt{\Delta^2 + \Delta_0^2}$, this means $E \simeq \Delta$ for most of the active TS's in such an experiment and therefore $P(E, \Delta_0) \simeq P(\Delta)P(\Delta_0)$.

As this discussion shows, an examination of $P(\Delta_0)$ is sufficient to obtain an estimate for the time dependence of spectral diffusion. Let us now generalize $P(\Delta_0)$ in the following way: $P(\Delta_0) = 1/\Delta_0^\alpha$. To perform the analysis in a simplified picture we recall that the TS relaxation rates scale with Δ_0^2 (6.24) and that a "rate" corresponds to the inverse of time, i.e. $\Delta_0 \sim t^{-1/2}$. In this picture the integral over the generalized form of $P(\Delta_0)$ yields

$$\int P(\Delta_0) d\Delta_0 \sim \int \frac{1}{\Delta_0^\alpha} d\Delta_0 \sim \Delta_0^{1-\alpha} \sim t^{\frac{1}{2}(\alpha-1)}.$$

This shows that the assumption $\alpha = 1$, which corresponds to the standard model distribution function, is a singular case – it yields a logarithmic time dependency. Any other exponent results in an algebraic time evolution. Now it is easy to choose an exponent which provides the observed asymptotic $\Delta\Gamma \propto \sqrt{t}$, namely, $\alpha = 2$. Interestingly, it is exactly this exponent which has

6.5 Spectral Diffusion: Low-Barrier Tunneling   367

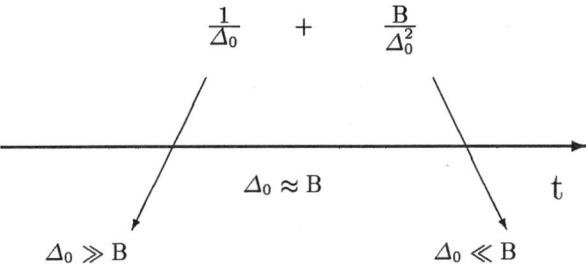

Fig. 6.16. Schematic illustration of the crossover behavior provided by the distribution function (6.27). At short times, corresponding to values of the tunneling parameter $\Delta_0 \gg B$, the dominating first term results in a logarithmic time dependency. On the other hand, at long times, the second term produces an algebraic behavior.

been found by models dealing with TS–TS interactions (Klein (1990), Burin and Kagan (1994a), Burin and Kagan (1995)). While Klein (1990) analyzed defect crystals in a completely microscopic fashion, Burin and Kagan (1993) present an extension of the standard model of amorphous solids which includes additional very low temperature excitations arising from the formation of interacting pairs of TS's. This is discussed in detail in Chap. 5.

Although the existence of such models is an encouraging fact, they cannot be employed to account for the observed time evolution of spectral diffusion in a straightforward fashion: Both of these models yield $P(\Delta_0) = 1/\Delta_0^2$, which results in an algebraic behavior $\Delta\Gamma \sim \sqrt{t}$ on all time scales. Our data, however, clearly display a crossover from a nearly logarithmic to an algebraic behavior. Therefore we have constructed an ad hoc distribution function, which includes both types of Δ_0 dependencies, i. e. both types of time evolutions:

$$P(\Delta_0) = P(1/\Delta_0 + B/\Delta_0^2). \qquad (6.27)$$

The first term in this distribution function leads to the "classical" logarithmic part of diffusional line broadening, the second one provides the algebraic square root dependency. The constant B has to be introduced for dimensionality reasons, it corresponds to the tunneling energy Δ_0 at which both terms become equal. The nice feature of this specific choice is the fact that it yields a "natural" separation of the two terms in the time domain: For large values of Δ_0, i. e. $\Delta_0 \gg B$, the first term dominates whereas in the other extreme the second term becomes larger. Since large values of Δ_0 correspond to short times and vice versa, this distribution therefore yields the observed crossover behavior with the location of the crossover being determined by the value of B. This is illustrated in Fig. 6.16.

Now, if we insert the distribution function (6.27) into (6.22), a sum of two time-dependent terms results:

$$\Gamma(t,T) = Ak_\mathrm{B}T \int_0^\infty dx \, \mathrm{sec}\, h^2(x) \{G_\mathrm{st}(t,x) + G_\mathrm{ad}(t,x)\},$$

where

$$G_\mathrm{st}(t,x) = \frac{1}{2}[\ln(r_\mathrm{max}(x,T)t) + Ei(1, r_\mathrm{max}(x,T)t) + \gamma]$$

is the standard model part [see the designation of the variables in the previous section and (6.25)], and

$$G_\mathrm{ad}(t,x) = \int_0^{\Delta_{0\,\mathrm{max}}} \frac{B d\Delta_0}{\Delta_0^2}[1 - \exp(-r(x,T,\Delta_0)t]$$

$$= \frac{B}{\Delta_{0\,\mathrm{max}}}\left[\sqrt{\pi}\sqrt{r_\mathrm{max}(x,T)t}\, \mathrm{erf}\left(\sqrt{r_\mathrm{max}(x,T)t}\right)\right.$$

$$\left. + \exp(-r_\mathrm{max}(x,T)t) - 1\right]$$

is the additional term responsible for the square root behavior of spectral diffusion at long times. Here erf(y) designates the error function, all other variables are the same as in the previous section.

These expressions can be further simplified taking into account the given experimental conditions. In the limit $r_\mathrm{max}t \gg 1$, which always applies in typical hole-burning experiments, and with a time resolution of t_0, we arrive at:

$$\Gamma(t,T) \simeq Ak_\mathrm{B}T \int_0^\infty dx \, \mathrm{sech}^2(x) \left\{ \frac{1}{2} \ln \frac{t}{t_0} \right.$$

$$\left. \times \frac{B\sqrt{\pi}}{\Delta_{0\,\mathrm{max}}} \left[\sqrt{r_\mathrm{max}(x,T)t} - \sqrt{r_\mathrm{max}(x,T)t_0} \right] \right\}. \quad (6.28)$$

Now we can numerically fit this result independently to the two experimental data sets. This is performed by using P and B from (6.27) as free parameters. The resulting good agreement over the whole investigated time range – five orders of magnitude – is demonstrated in Fig. 6.15. As expected from our analysis, the Δ_0-distribution does indeed model the two limiting cases of the time evolution, logarithm and square root, as well as the transition between them and therefore provides a good fit to the data. Consistently, the resulting fit parameter $B = k_\mathrm{B}10^{-7}$ K is the same for 0.5 K and 1 K. For one-phonon relaxation, this corresponds to crossover times of 10^3 s and 10^4 s for 1.0 K and 0.5 K, respectively. The results for the second parameter P deviate by 10 % for the two data sets. This could be interpreted as a small increase of the total density of states, but keeping in mind the ad hoc fashion of our modeling, this result should not be overinterpreted. The fitted value of the parameter B at 2 K is smaller: $B = k_\mathrm{B}0.3 \times 10^{-7}$ K. The value of P obtained from the 2 K data cannot be compared with that of the low-temperature data because they correspond to different chromophore molecules.

6.5 Spectral Diffusion: Low-Barrier Tunneling

As was already mentioned before, the distribution function $P(\Delta_0) \propto 1/\Delta_0^2$ has recently been predicted by models of interacting TS's (Klein (1990), Burin and Kagan (1994a), Burin and Kagan (1995)). Burin's and Kagan's model is especially of interest, because it includes dynamical interactions of TS's and offers a scenario of effective interactions of TS's with nonresonant energies (see Chap. 5). But this model, in its original form, cannot describe our data since the expected direct TS–TS coupling should be very weak. This means that the coherence of a TS ensemble should be destroyed due to interactions with phonons at rather low temperatures. The estimate of the temperature region where TS coherence is supposed to exist is in the milli–Kelvin region. Hence, the model can qualitatively account for the decrease of the parameter B in our distribution function (2 K data) at temperatures above some threshold by the loss of TS coherence due to phonons. On the other hand, the quantitative discrepancy between the values of experimentally measured parameters and their theoretical estimates is orders of magnitude.

These discrepancies are analyzed by Neu et al. (1997a). The authors introduce a modified model of interacting TS's based on earlier ideas of Kassner and Silbey (1989) about strong interactions of TS's with phonons, which had been employed to account for less pronounced deviations from standard model behavior (Maier and Haarer (1997)). This inclusion of a strong TS–phonon coupling improved the agreement between experiment and theory substantially. But still a quantitative difference between the measured temperature dependence and its theoretical prediction remains.

Another explanation of the nonlogarithmic spectral diffusion we observed was recently proposed by Heuer and Neu (1997). Based on our data and on the results of sound attenuation measurements in various polymers, the authors proposed the existence of two subensembles of TS's. The first one more or less corresponds to the standard model. The second one is associated with degrees of freedom specific for particular materials. To model this, the authors chose a Gaussian distribution as a function of the tunneling parameter λ. This secondary subensemble of TS's is assumed to be responsible for the nonlogarithmic hole broadening on long time scales. By numerically fitting the parameters of this Gaussian distribution, the authors obtained good agreement with our data.

We wish to emphasize that up to now there is no unambiguous interpretation for the clear deviation of the time dependence of spectral diffusion from the predictions of the standard model. The validity or nonvalidity of the above models can only be verified by more experimental data on even longer time scales or preferably at higher temperatures. In addition, detailed long-term data of more materials are necessary for clarifying the situation: There are indications that the behavior we observed in PMMA is not universal for glasses. Recent measurements of the line broadening in an alcohol glass by Fritsch et al. (1996) with time scales and temperatures comparable to our experiments coincide with the predictions of the standard model. At the same time measurements of long-term spectral diffusion in proteins by the same group, Fritsch et al. (1997), show a well-pronounced $\sqrt{t}$ behavior similar to our data.

Both of the above models promise to bring new features into the theoretical treatment of low-temperature dynamics in glasses, which has essentially been unchanged for the past 20 years. For example, the double distribution function of Heuer and Neu (1997) bears some similarities to the model of hierarchical barrier distributions in proteins, which has in the last years been employed for explaining many properties of proteins (see Ansari et al. (1985), Frauenfelder and Wolunes (1994)). The model of interacting TS's can also be interpreted as a first approximation of weakly communicating local basins in a multibasin barrier hierarchy.

6.5.5 Nonequilibrium Glass Dynamics

After a brief introduction to the nature of nonequilibrium states of TS's in this section, we first wish to outline the general treatment of spectral diffusion phenomena under nonequilibrium conditions and then apply this formalism to several experimental examples.

The necessary equations and formulas are more complicated than in the previous sections: More than one temperature value can be involved at the same time or even individual TS energies can be changed during the experiment. For the sake of simplicity we will therefore restrict the theoretical treatment to the application of the standard tunneling model. As we shall see in the following, this will be sufficient for the description of some cases. It will in general be sufficient for the understanding of the physics presented in the following examples. In the cases where nonstandard behavior is observed, this will of course be discussed.

Introduction. If an amorphous sample is cooled rapidly from high temperatures to the liquid helium range, then some of the tunneling centers within this sample will be caught in their high energy level. The reason for this is the strong temperature dependence of the TS relaxation rates, see Chap. 2. At sufficiently high temperatures, the dominating relaxation mechanism is thermal activation, which can be a very fast mechanism (Köhler and Friedrich (1989), Khodykin et al. (1996)). Its rate, however, is strongly temperature dependent: $r \sim \exp(-V/k_B T)$, where V denotes the barrier height. Therefore, upon cooling below some temperature it becomes inefficient as compared to tunneling. If the cooling rate is large in comparison with tunneling relaxation rates of the TS's, the populations of their energy levels will not follow the temperature change. Simplifying the picture, we can say that the TS populations freeze at the temperature, where activation relaxations become comparable to or slower than tunneling. A nonequilibrium state is created. This is also the starting point of the heat-release experiments described in Chap. 2 of this book: The energy stored in a nonequilibrium TS population is released

6.5 Spectral Diffusion: Low-Barrier Tunneling

as a function of time and thereby the sample is heated. The amount of heating decreases with time as the population approaches its thermal equilibrium distribution.

In a spectral hole-burning experiment the situation is quite similar. It is illustrated in Fig. 6.17 (Maier and Haarer (1995)). In the figure, the time origin is set to the moment when the final temperature of 300 mK is reached after cooling from liquid nitrogen temperature (77 K). At this time a hole is burnt and its broadening with time is observed. This is the data set labelled as **A**. It is in full correspondence to the heat-release data in Chap. 2 except for one point: Where the differential technique of heat-release experiments observes a time dependence as, for example, $1/t$, the integral technique of spectral diffusion correspondingly observes a $\log(t)$ dependence. If a new hole is burnt after some time delay t_d, then a weaker broadening with time is observed reflecting the fact that meanwhile the ensemble has already come closer to its equilibrium population; the relaxation becomes weaker. This is illustrated by the next two data sets in Fig. 6.17. Finally, after a sufficiently long waiting time t_d, we can observe a time evolution which is independent of t_d, if the observation time is restricted to values much smaller than t_d

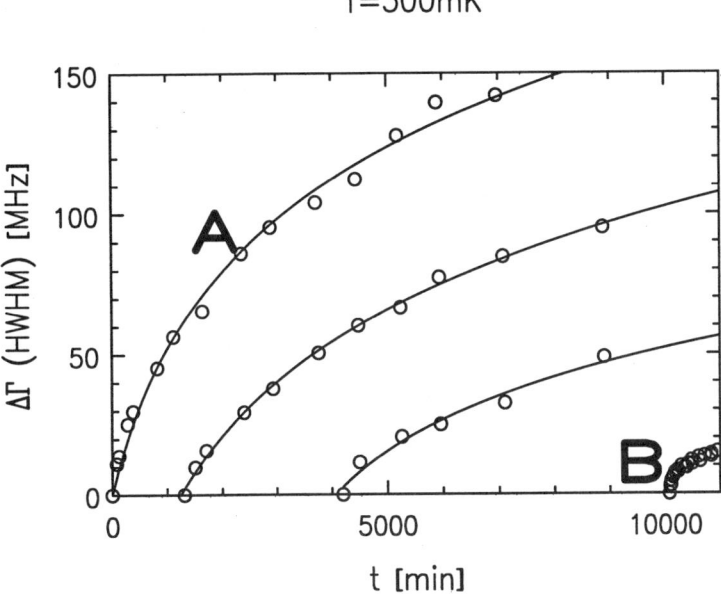

Fig. 6.17. Spectral hole broadening as a function of the time after cooling of the sample (phthalocyanine in PMMA). The time origin represents the moment when a temperature of 300 mK was reached after cooling from 77 K. Then the temperature was kept constant. The four data sets show the broadening of spectral holes burnt at approximately 0, 1000, 4000, and 10.000 min after reaching 300 mK. The solid lines are logarithmic fits. From Maier and Haarer (1995).

(Maier and Haarer (1995)). This is illustrated by the dataset labelled as **B** in Fig. 6.17. In this time interval, the thermal relaxation has become too weak to dominate the observed hole broadening. This is the limiting case of equilibrium dynamics described in the previous section. These different time regimes will subsequently be analyzed in the following theoretical treatment.

The relaxation behavior illustrated above explains why long-term equilibrium experiments as described in Sect. 6.5.4 are so extremely time consuming: The sample must be given a relaxation time t_d which is longer than the duration of the subsequent experiment in order to capture only equilibrium dynamics. In this section the theory of spectral diffusion as introduced in Sect. 6.5.2 is first formally extended to the general treatment of nonequilibrium and then applied to several nonequilibrium situations.

Theory of Nonequilibrium Phenomena in Spectral Diffusion. The treatment of a glass as an equilibrium system introduced in Sect. 6.5.2 has strict boundaries even in the framework of the standard tunneling model. In a rigorous mathematical sense, an ensemble of TS's will never reach its thermal equilibrium state after cooling. This is also true for the description of many experimental situations. Indeed, no upper limit for TS relaxation times was found experimentally up to now. Therefore, all phenomena at low temperatures in glasses are more or less influenced by a nonequilibrium population of the TS ensemble. Nevertheless, the long-time nonequilibrium aspects of TS dynamics were ignored both theoretically and experimentally for a long time (with the exception of heat release, where the physical situation necessarily involves a nonequilibrium state, see Chap. 2). The first numerical modeling of some consequences of a nonequilibrium TS population for spectral diffusion was performed by Köhler and Friedrich (1988a), dynamical aspects were briefly analyzed by Bai et al. (1989), Littau et al. (1990), Narasimhan et al. (1990) and in a more general way by Kharlamov et al. (1994), Khodykin et al. (1996), Fritsch et al. (1996).

We will follow the analysis of Fritsch et al. (1996) and Khodykin et al. (1996). The equilibrium situation is easily generalized for the nonequilibrium case. Consider an ensemble of TS's in thermal equilibrium at an initial temperature $T = T_i$ with a population vector of

$$n_0 = \begin{pmatrix} p_0 \\ 1 - p_0 \end{pmatrix}. \tag{6.29}$$

Then at $t = 0$ the temperature is changed to $T = T_f$ infinitely fast. After an infinite waiting time, the TS population will have adapted to the temperature change and will again be in thermal equilibrium with the temperature now being T_f:

$$n_\infty = \begin{pmatrix} p_\infty \\ 1 - p_\infty \end{pmatrix}. \tag{6.30}$$

The matrix of possible events during the intermediate time is similar to (6.20)

$$n_0 \otimes n_\infty = \begin{pmatrix} p_0 p_\infty & p_0(1-p_\infty) \\ p_\infty(1-p_0) & (1-p_0)(1-p_\infty) \end{pmatrix}. \tag{6.31}$$

As in the equilibrium case, only the off-diagonal elements are related to TS flips. Substituting the two equilibrium population values given by $p_0 = 1/(1+\exp(E/k_B T_i))$ and $p_\infty = 1/(1+\exp(E/k_B T_f))$ into (6.31), we obtain the time-dependent probability for a TS jump:

$$\begin{aligned} n(t, T_i, T_f) &= \frac{1}{2}\left[1 - \tanh\left(\frac{E}{2k_B T_i}\right)\tanh\left(\frac{E}{2k_B T_f}\right)\right] \\ &\quad \times [1 - \exp(-r(E, \Delta_0, T_f)t)]. \end{aligned} \tag{6.32}$$

For a better comparison with the equilibrium case it is useful to split the last expression into two parts:

$$n(t, T_i, T_f) = [n_e(T_f) + n_{\text{ne}}(T_i, T_f)][1 - \exp(-r(E, \Delta_0, T_f)t)],$$

with:

$$n_e(T_f) = \left[1 - \tanh^2\left(\frac{E}{2k_B T_f}\right)\right] = \frac{1}{2}\text{sech}^2\left(\frac{E}{2k_B T_f}\right), \tag{6.33}$$

$$\begin{aligned} n_{\text{ne}}(T_i, T_f) &= \tanh\left(\frac{E}{2k_B T_f}\right) \\ &\quad \times \left[\tanh\left(\frac{E}{2k_B T_f}\right) - \tanh\left(\frac{E}{2k_B T_i}\right)\right]. \end{aligned} \tag{6.34}$$

The physical meaning of this separation is quite obvious: The probability for a TS jump would be equal to $n_e(T_f)$ if the TS ensemble were in equilibrium at $T = T_f$. The term $n_{\text{ne}}(T_i, T_f)$ is essentially the nonequilibrium part of the TS transition probability. This quantity depends on two temperatures: T_f is the temperature of the phonon bath during the whole process of equilibration discussed here; T_i is the temperature that describes the equilibrium populations of the TS energy levels prior to switching the phonon bath temperature at $t = 0$. We wish to emphasize that $n_{\text{ne}}(T_i, T_f)$ can be either positive or negative, depending on whether T_i is higher or lower than T_f.

In the following we apply the above formalism to the description of some experimental nonequilibrium scenarios.

Thermal Relaxation. The first scenario corresponds to the most typical situation found in laboratory experiments: The sample is cooled down at the beginning of the experiment. This is also the situation described in the introduction of this section yielding diffusional hole broadening as shown in Fig. 6.17.

Due to the extremely broad distribution of TS relaxation rates, extending over many weeks at low temperatures, the TS ensemble in the sample never really reaches thermal equilibrium populations. Let us analyze the consequences of this fact for spectral diffusion.

Suppose that the sample is cooled down very fast. For convenience we shift the time origin and let this happen now at $t = -t_d$ instead of $t = 0$ s as above. As described in the introduction, the TS relaxation rate can decrease strongly with decreasing temperature, therefore, during the cooling process we have: At high temperatures, the main relaxation mechanisms are thermally activated processes with an exponential dependence of the relaxation rate on temperature (Köhler and Friedrich (1989), Khodykin et al. (1996)). Below a specific temperature, the transition rate determined by thermal activation becomes smaller than the cooling rate for TS with a given set of parameters. From then on, relaxations can only occur via tunneling. We assume the populations of such TS to remain constant for the rest of the cooling process. When crossing this threshold, the TS populations will be therefore frozen at their current values in our analysis. Although this specific temperature varies within the whole ensemble of TS – it is determined by the barrier height – we further assume that it is equal for all TS's in the sample. Because of these simplifications and the finite duration of a real cooling process, the "initial" temperature from above T_i has to be replaced by an *effective* temperature T_eff.

The physical interpretation of T_eff is the following: When reaching the temperature T_f, the initial TS populations at $t = -t_d$ are described by a Boltzmann factor, $p_0/(1 - p_0) = \exp(-E/k_B T_\text{eff})$. A relaxation process towards the new thermal equilibrium described by T_f sets in and the TS populations become time dependent. Since the process of hole burning indirectly "marks" a given configuration of a TS ensemble by marking the absorption frequencies of probe molecules interacting with the TS (cf. Sect. 6.5.1), the important quantity is the population at the time of hole burning $t = 0$ s:

$$p(T_\text{eff}, T_f, t = 0) = p(T_f) + [p(T_\text{eff}) - p(T_f)] \exp[-r(T_f) t_d].$$

Substituting this expression into (6.31) and taking advantage of (6.33) and (6.34), we can split the spectral diffusion linewidth into its equilibrium and nonequilibrium parts according to $\Gamma = \Gamma_e + \Gamma_\text{ne}$. For the equilibrium part, this of course yields the same result as deduced in Sect. 6.5.2 for the thermal equilibrium situation

$$\Delta \Gamma_e(t, T_f) = A k_B T_f \ln\left(\frac{t}{t_0}\right) \int_0^\infty dx \left[1 - \tanh^2(x)\right], \qquad (6.35)$$

with $x = E/2k_B T_f$; the actual temperature here being given by $T = T_f$. As before, t_0 is the time required to determine the initial hole width after burning.

For the nonequilibrium contribution to spectral diffusion, we now obtain the additional term

$$\Gamma_\text{ne}(t, T_\text{eff}, T_f) = A k_B T_\text{eff} \ln\left(\frac{t + t_d}{t_d}\right) \qquad (6.36)$$
$$\times \int_0^\infty dx \tanh\left(x \frac{T_\text{eff}}{T_f}\right) \left[\tanh\left(x \frac{T_\text{eff}}{T_f}\right) - \tanh(x)\right].$$

6.5 Spectral Diffusion: Low-Barrier Tunneling

Here we have substituted $x = E/2k_B T_{\text{eff}}$. In the above two equations we can distinguish three time regimes:

- $t \ll t_d$. The time interval between sample cooling and hole burning is much longer than the experimental observation time t. In this limit $\Gamma_{\text{ne}} \ll \Delta\Gamma_e$. Thin means that we can neglect the nonequilibrium part of spectral diffusion; the observed diffusional line broadening will practically be dominated by thermal equilibrium TS dynamics with the temperature given by T_f. We obtain a broadening behavior which is independent of t_d. Within the standard model, the hole broadening is proportional to the phonon bath temperature and increases logarithmically with time. In practice this limiting case is what we measure when we investigate "equilibrium dynamics" at long waiting times after cooling as described in Sect. 6.5.4: The long waiting time after cooling strongly reduces nonequilibrium contribution to the hole broadening.
- $t \approx t_d$. In this intermediate case the equilibrium and nonequilibrium term are comparable. The hole broadening in this time interval has a complicated temperature dependence and is nonlogarithmic in time. Within the standard model it is essentially described by a superposition of (6.35) and (6.36).
- $t \gg t_d$. In this limit Γ_{ne} becomes the dominating term. It corresponds to the experimental situation that holes are burnt at short waiting times after cooling of the sample.

Experimental data for protoporphyrin IX in an alcohol glass including these three cases are displayed in Fig. 6.18 (Fritsch et al. (1996)). The data are presented in a semilog plot. The decrease of the nonequilibrium contribution with increasing t_d is clearly visible. The data were fitted using (6.35) and (6.36). First the parameter A (the product of the chromophore-TS coupling constant and the TS density of states) was fitted using only the equilibrium (short-term) part of the data. Then T_{eff} was fitted to the long-term data. It should be noted that 14 data sets with t_d varying from 19 h to 190 h, measured at temperatures of 100 mK and 800 mK were fitted with the same set of parameters A and T_{eff} (see Fritsch et al. (1996)). This shows that our simple model describes the aging effect in glasses at low temperatures quite satisfactorily. The fitted value of the "effective temperature" in the above data is $T_{\text{eff}} \simeq 12$ K. Comparing this value with the experimental temperature of $T = 100$ mK, we come to the conclusion that nonequilibrium spectral diffusion is governed by T_{eff}.

Indeed, for $T_{\text{eff}} \gg T_f$, it can be shown that (6.36) is practically independent of T_f, since in this limit the integral yields

$$\lim_{T_{\text{eff}}/T \to \infty} \int_0^\infty dx \, \tanh\left(x \frac{T_{\text{eff}}}{T_f}\right) \left[\tanh\left(x \frac{T_{\text{eff}}}{T_f}\right) - \tanh(x)\right] = \ln(2).$$

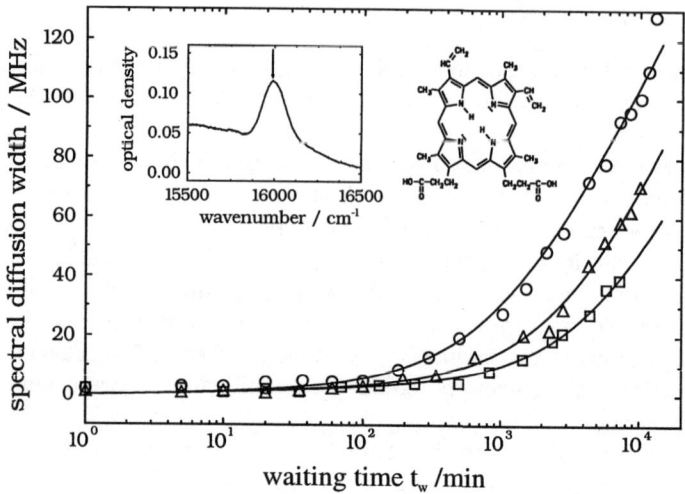

Fig. 6.18. Aging experiment at T = 100 mK. Spectral diffusion and the respective fit curves for three holes burned at different delay times t_d after the final temperature of 100 mK was reached. The delay times t_d were 19 h for the upper, 66 h for the middle and 114 h for the lower trace. The inserts show the inhomogeneous absorption spectrum and the structure of protoporphyrin IX. The arrow indicates where the hole burning was performed (15988 cm^{-1}). From Fritsch et al. (1996).

This means that the rate of spectral diffusion becomes *independent* of the actual phonon bath temperature T_f! We expect a logarithmic curve in time with a slope exclusively given by T_{eff}. Experimentally, this limiting case can be achieved by fast cooling the sample and starting the measurements immediately after T_f has been reached. We performed such experiments on thermal relaxation in a temperature range $T_f = 100$ mK to 700 mK (Maier and Haarer (1995)): The system was cooled down from 77 K to 4 K within one hour and then cooled down further by ^{3}He/^{4}He dilution refrigeration to T_f in slightly varying times of several hours – the duration of this step depends on T_f, but a temperature below 1 K is reached more or less immediately. The results of these experiments are shown in Fig. 6.19. In this plot the time origin for each run is identical, namely, the time at which the cooling process from 77 K was started. For better visibility the data sets corresponding to 500, 300, and 100 mK have been shifted by 0.5, 1.0, and 1.5 orders of magnitude along the logarithmic time axis, respectively. In spite of the large variation of the final temperature T_f over almost one order of magnitude, the four data sets in Fig. 6.19 display the same hole broadening behavior. Since the cooling procedure was similar for all of these experimental runs, we can expect the effective TS "charging temperatures" T_{eff} to vary only within a narrow range. This explains why all four curves in Fig. 6.19 are parallel: The

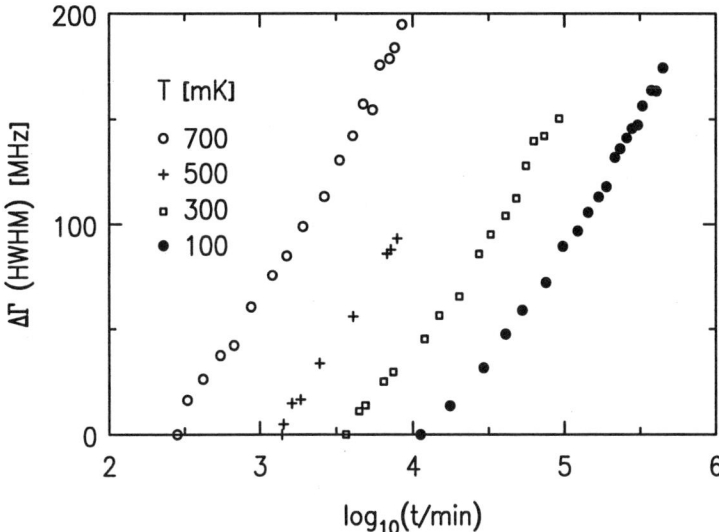

Fig. 6.19. Broadening of holes burnt immediately after reaching the displayed temperatures T in a semilogarithmic plot. The data are shifted for better visibility (see text). The sample material is phthalocyanine in PMMA. From Maier and Haarer (1995).

large amount of spectral diffusion observed immediately after fast cooling of the sample reflects the relaxation of the TS to thermal equilibrium which is insensitive to the actual phonon bath temperature in the limit $T_{\text{eff}} \gg T_f$ as discussed above. The observable slight deviations from linearity in the semi-logarithmic plots of Fig. 6.19 can be attributed to the influence of equilibrium dynamics as well as to deviations from the parameter distribution function $P(\Delta_0)$ of the standard tunneling model. The above results, however, suggest that thermal relaxation experiments do not seem to be the appropriate instrument for investigating TS–TS coupling phenomena as discussed in Sect. 6.5.4 because of the fairly high "charging temperatures" involved.

Concluding this section we can say that spectral diffusion induced by thermal relaxation phenomena of TS's can be described in the framework of an extended standard tunneling model introducing two more parameters:

1. The time interval t_d between sample cooling and hole burning. On the one hand, this is an independently measurable parameter. On the other hand, however, the sample-cooling procedure is usually not very fast; therefore t_d is not a very accurate quantity on a linear time scale. But Γ_{ne} depends on t_d logarithmically and therefore the accuracy of t_d is sufficient.
2. The effective temperature T_{eff} which is formally related to the initial TS populations. Due to the complexity of the processes during sample cooling, T_{eff} cannot be calculated. Actually, the introduction of T_{eff} is

a strongly simplified way to describe the complicated distribution of TS populations after sample cooling. But, as our results have shown, this parameter describes the observed aging effect very well within a broad time and temperature range and can easily be determined from the experimental data. Detailed measurements by varying the starting temperature may yield valuable information about high-temperature relaxations of TS and about boundaries of the applicability of the TS model itself.

Thermal Cycling. Thermal cycling is an external perturbation of the TS ensemble, meaning a change of the sample temperature for a given time interval. This opens up the possibility to extend optical experiments into a high-temperature region which cannot be reached by "passive" spectral diffusion experiments. Above temperatures of 6–10 K, the spectral-diffusion-induced broadening of a spectral hole becomes negligibly small as compared to the initial hole width. The time-dependent broadening cannot be resolved in this region because the width of a spectral hole at these temperatures is completely dominated by the large phonon-induced homogeneous linewidth of the chromophores, see Sect. 6.2. Therefore, the following method is applied in optical spectral diffusion experiments (Fritsch et al. (1996)).

Suppose the sample is in thermal equilibrium at $T = T_f$ before hole burning. A hole is burned at $t = 0$. At $t = t_1$, we start a thermal cycle by increasing the temperature to $T = T_c$ until $t = t_2$, then the temperature is lowered to its initial value T_f. Let us consider all three time regions in this scenario, calculating separately the equilibrium and nonequilibrium parts of diffusional linewidth Γ as before.

1. The time before the thermal cycle: $0 < t < t_1$.
In this time interval there is regular spectral diffusion in thermal equilibrium. The equilibrium part of Γ is described by (6.35); the nonequilibrium part is zero.

2. The thermal cycle: $t_1 < t < t_2$.
The equilibrium part is almost the same as before but with a modified relaxation rate corresponding to the higher temperature T_c:

$$\Gamma_e(t, T_c, T_f) = Ak_B T_f \int_0^\infty dx \left[1 - \tanh^2(x)\right] \qquad (6.37)$$
$$\times \int_0^{\Delta_{0,\max}} \frac{d\Delta_0}{\Delta_0} \Big[1 - \exp[-r(x, \Delta_0, T_f) t_1$$
$$- r(x, \Delta_0, T_c)(t - t_1)]\Big].$$

The above equation reduces to (6.35) if we keep the temperature constant and set $T_c = T_f$. The only consequence of the temperature increase (a decrease is also possible) during the thermal cycle is an acceleration of spectral diffusion. At low temperatures, where the TS relaxation rates depend very weakly on

6.5 Spectral Diffusion: Low-Barrier Tunneling

temperature, this effect is negligible. If the cycle temperature is in a range where relaxations by thermally activated barrier crossing become relevant, this acceleration may, however, be an important factor.
The nonequilibrium part now reads:

$$\Gamma_{\rm ne}(t, T_c, T_f) = Ak_{\rm B}T_f \int_0^\infty dx \tanh(x) \left[\tanh(x) - \tanh(x\frac{T_f}{T_c})\right]$$
$$\times \int_0^{\Delta_{0,\max}} \frac{d\Delta_0}{\Delta_0} \left[1 - \exp\left(-r(x, \Delta_0, T_c)(t - t_1)\right)\right]. \quad (6.38)$$

This provides either an increase ($T_c > T_f$ as in our example) or a decrease ($T_c < T_f$) of the diffusional hole broadening. As mentioned above, this increase is hardly measurable if $T_c > 6$–$10\,\mathrm{K}$ because of the large increase of the homogeneous linewidth. Therefore, the most interesting part from the experimental point of view is the last step of the scenario.

3. The time interval after the thermal cycle: $t_2 < t$.
At $t = t_2$, the temperature is reduced to its original value T_f. The equilibrium part of the hole broadening returns to its original state with the relaxation rate from now on again determined by T_f. The increased relaxation rates during the cycle interval, however, cannot simply be dropped in some situations, as already mentioned.

If we evaluate the Δ_0-integral for the nonequilibrium part, we can now write this contribution to the hole broadening in the following way:

$$\Gamma_{\rm ne}(t, T_c, T_f) = Ak_{\rm B}T_f \int_0^\infty dx \tanh(x) \left[\tanh(x) - \tanh(x\frac{T_f}{T_c})\right]$$
$$\times \ln\left[1 + \frac{r_{\max}(x, \Delta_0, T_c) \times (t_2 - t_1)}{r_{\max}(x, \Delta_0, T_f) \times (t - t_2)}\right]. \quad (6.39)$$

From this result some immediate conclusions can be drawn:

- Thermal cycling gives rise to an excess hole broadening as in the case of the previous example of thermal relaxation. The magnitude of this excess broadening is again determined by the two involved temperatures T_c and T_f. This can allow conclusions about the distribution of the TS energy splittings E.
- The excess hole broadening induced by a temperature cycle is reversible as can be seen from the logarithmic term in (6.39): $\Gamma_{\rm ne} \to 0$ for $r(x, \Delta_0, T_f) \times (t - t_2) \gg r(x, \Delta_0, T_c) \times (t_2 - t_1)$. Immediately after the thermal cycle there is a hole-narrowing regime, whose duration is determined by the "memory time" of the TS's ensemble. This memory time can roughly be estimated as

$$t_m \approx (t_2 - t_1)\frac{r(x, \Delta_0, T_c)}{r(x, \Delta_0, T_f)}. \quad (6.40)$$

In the low-temperature region, where the relaxation rates are only weakly temperature dependent, we can set $t_m \approx (t_2 - t_1)$. On the other hand, if the temperature dependence of the relaxation rates is strong, which is the case when the cycle temperature is sufficiently high for two-phonon and especially thermally activated processes to be dominating, the memory time can be orders of magnitude longer than the cycle duration $t_2 - t_1$. This is probably the reason why the hole-narrowing process after a temperature cycle had not been observed experimentally for a long time in conventional thermal cycling experiments with cycle durations of 10–100 min. The memory time depends on the cycle duration. Therefore, in short time scale measurements with pulsed sample heating and a sufficiently high time resolution, the effect should be observable. The first observation of hole narrowing after thermal cycling was performed and interpreted in terms of nonequilibrium spectral diffusion by Kharlamov et al. (1994). Figure 6.20 shows the result of transient hole measurements at different hole-burning energies. The hole burned with low laser energy (lowest trace) shows regular spectral diffusion broadening. High burning energies cause sample heating which can be treated as a thermal cycle. Recooling the sample after the laser pulse results in hole narrowing.

The above experiments had some methodical imperfections for quantitative measurements of the influence of thermal cycles on spectral diffusion: Hole burning and sample heating were not separated in time; the cycle temperature could not be measured, etc. Later thermal cycling measurements

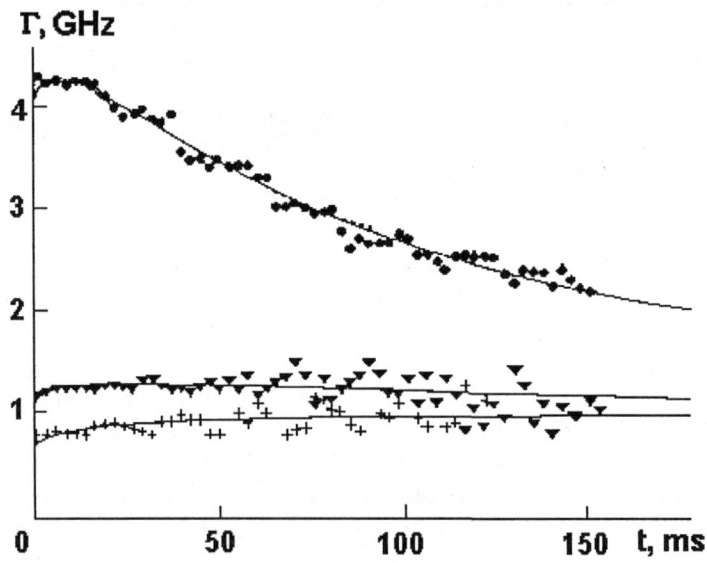

Fig. 6.20. Temporal variation of the transient hole width at three burning pulse energies $E = 0.1$, 1 and 10 μJ/cm^2 for the bottom, middle and upper curves, respectively. The solid lines are results of model calculations taking the influence of the heating pulses on hole broadening into account. The sample is Zn-tetrabenzporphin in PMMA. The cryostat temperature is $T = 1.9$ K. From Kharlamov et al. (1994).

6.5 Spectral Diffusion: Low-Barrier Tunneling

under well-defined conditions at low temperatures provided new evidence of diffusional hole narrowing. The corresponding data (Fritsch et al. (1996)) are presented in Fig. 6.21.

The experiment was performed on the same sample as the previously described aging experiment. The important point is that all fit parameters taken from the aging data fit except for the parameter describing two-phonon TS relaxations (so-called Raman processes, see Sect. 2.2.5). The inclusion of the additional relaxation mechanism for describing the data was necessary because the cycling temperature was rather high, and two-phonon processes and even thermally activated relaxation are expected to be important in this temperature regime, as observed in heat-release experiments, see Sect. 2.5.2. The data in Fig. 6.21 clearly show the very long "memory time": The hole does not come to its equilibrium width even a week after the cycle, which lasted only one hour (the calculated equilibrium hole broadening is also displayed). Actually, thermal cycling measurements are a promising technique for investigating the temperature dependence of TS relaxation rates because this temperature dependence is manifested unambiguously in the post-cycle

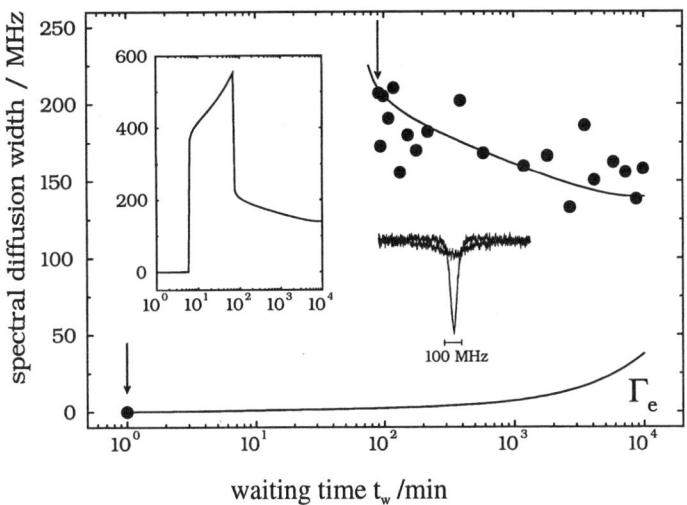

Fig. 6.21. Temperature cycling experiment (100 mK – 4 K – 100 mK). Shown is the spectral diffusion linewidth together with the respective fit curve of a hole burned at $T = 100$ mK directly before performing the cycle. Before burning, the sample was allowed to relax for 10 days at 100 mK. In addition, the figure shows a model calculation for the spectral diffusion width when no cycle is performed (Γ_e). The time origin is given by the hole-burning time. The cycle duration was about 75 min. The right-hand insert shows the hole immediately before and after the cycle (arrows). The left-hand insert shows a model calculation of the spectral diffusion width for the whole cycle experiment. From Fritsch et al. (1996).

memory time. The data presented here are only a first indication of these opportunities.

Electric Field Effects. As illustrated above, the creation of nonequilibrium populations provides an additional degree of freedom in experimental investigations of low-temperature dynamics in amorphous solids. By observing relaxation phenomena, essential information about the dynamics of the system under investigation can be obtained, as in heat-release experiments (Chap. 2). In the following, we will however introduce a special way of creating such relaxations with high time resolution: External fields can be employed to create nonequilibrium populations without modifying other parameters characterizing the system such as the temperature and, as a consequence of that also the relaxation rates, a method also introduced in Chap. 5.

It was shown both experimentally and theoretically by Maier et al. (1995) that applying an external electric field to a sample causes perturbations in a TS ensemble, which can be observed in spectral diffusion experiments. Let us describe the situation analytically employing a simplification: Suppose all TS's in the sample possess the same electric dipole moment μ and are in thermal equilibrium. When we apply an electric field F to the sample, the energy of every specific TS will be shifted from its original value E to $E_f = E + \vec{\mu} \cdot \vec{F}$, see also Sect. 5.1 (to avoid confusion between the tunneling energy E and electric field, we use in this chapter the symbol F for the electric field). This means that at the instant when we turn on the field, all TS's are immediately out of equilibrium because we have changed their energies at constant temperature T. The TS ensemble will start to relax to the new equilibrium populations being defined by the now modified energies E_f and the temperature T. The number of additional TS transitions can be described in analogy to (6.34) with the modification that we now have changed the TS energies instead of the temperature

$$n_{\mathrm{ne}}(T, E, \vec{F}) = \tanh\left(\frac{E}{2k_{\mathrm{B}}T}\right)\left[\tanh\left(\frac{E}{2k_{\mathrm{B}}T}\right) - \tanh\left(\frac{E + \vec{\mu}\cdot\vec{F}}{2k_{\mathrm{B}}T}\right)\right]. \qquad (6.41)$$

Since $\vec{\mu}$ and $\vec{F}$ are vector quantities, there is, however, one more complication: Due to the random orientations of TS electric dipoles with respect to the external field, an angular averaging procedure has to be performed and we obtain

$$\widetilde{n_{\mathrm{ne}}}(T, E, F) = \tanh\left(\frac{E}{2k_{\mathrm{B}}T}\right)\cdot\left[\tanh\left(\frac{E}{2k_{\mathrm{B}}T}\right) - \frac{k_{\mathrm{B}}T}{\mu F}\ln\frac{\cosh\left(\frac{E+\mu F}{2k_{\mathrm{B}}T}\right)}{\cosh\left(\frac{E-\mu F}{2k_{\mathrm{B}}T}\right)}\right]. \qquad (6.42)$$

6.5 Spectral Diffusion: Low-Barrier Tunneling

Because any orientation between an individual TS dipole moment and the external field is possible, the averaging has two major consequences:

- A decrease of the effect: After averaging, the field-induced energy shift enters $\widetilde{n_{\mathrm{ne}}}$ only logarithmically. This is the reason why this phenomenon can practically only be observed at sub-Kelvin temperatures.
- In contrast to the general expression for spectral diffusion induced by a temperature change (6.34) the averaged quantity $\widetilde{n_{\mathrm{ne}}}$ is always positive. This means that the application of an external electric field always causes an increase of the diffusional hole broadening.

Inserting this term into the expression for the nonequilibrium part of spectral diffusion results in an equation similar to (6.38). It should be noted that while the external field is applied, spectral diffusion cannot be observed. The reason is a Stark effect hole broadening caused by the direct interaction of the electric field with the chromophores which, however, is completely reversible (Meixner et al. (1986)). Kador and Haarer (1987) showed that the applied electric field strength must be kept sufficiently low so that charge carrier injection into the sample is avoided.

Because of this reversibility, the result of the electric field action on the TS dynamics can be observed after the field is switched off. The TS ensemble, taken out of equilibrium by the field-induced energy shifts, starts to relax back to its initial equilibrium state as the individual TS energy splittings E are restored. This leads to a decrease of $\varGamma_{\mathrm{ne}}$. The result is similar to the previous case of thermal cycling. But, since, in the experiments discussed here, a nonequilibrium population is created without changing the temperature, we can assume the relaxation rates to be constant and completely evaluate the Δ_0-integral,

$$\varGamma_{\mathrm{ne}}(t) = AT \frac{1}{2} \ln \left[\frac{t-t_1}{t-t_2} \right] \int_0^\infty \mathrm{d}x\, \widetilde{n_{\mathrm{ne}}}(T, E, F) \,. \tag{6.43}$$

We expect a logarithmic decrease of the field-induced excess broadening with time.

Figure 6.22 shows the results of such experiments at a temperature of 100 mK for four different values of the field strength F (Wunderlich et al. (1997)). The experiments were performed on a sample of PMMA doped with the dye tetraphenylporphin. The sample had a thickness of 50 μm and was placed between two semitransparent glass plates coated with indium tin oxide (ITO) which served as electrodes. The holes were burnt with an energy of 10 μJ within a time of about 10 s. As already mentioned, the spectral hole cannot be observed while the electric field is switched on. In the sub-Kelvin temperature range, the hole spectrum is completely invisible due to the Stark effect caused by the direct interaction of the probe molecules with the external electric field.

After switching the field off, an additional hole broadening can be observed, see Fig. 6.22. It is induced by the interaction of the TS with the

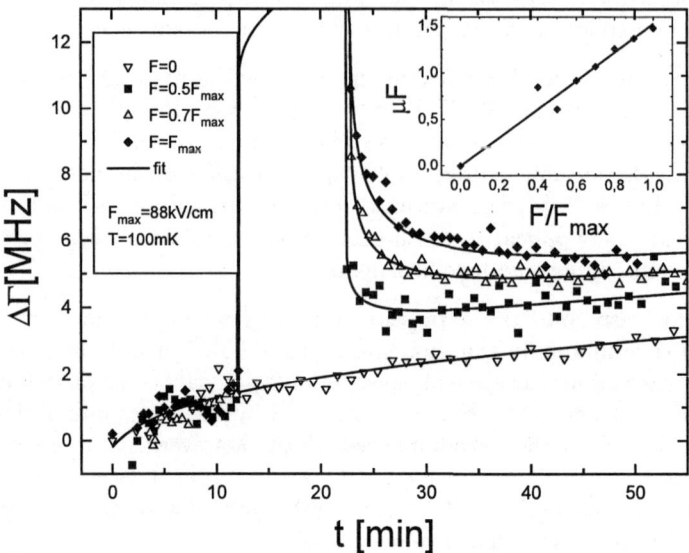

Fig. 6.22. Time-resolved broadening of spectral holes before and after the application of an external electric field for several values of the field strength F at a temperature of 100 mK. The field was applied for a period of 10 minutes in each of the measurements. During this time no data can be obtained, as is discussed in the text. All solid lines are theoretical fit curves. The insert shows the dependence of the fitted value of μF on the field strength. From Wunderlich et al. (1997).

external field and scales with the value of F. The effect, however, is very small: 10 MHz at the highest field strength of 88 kV/cm in our case, that is only about 10% of the initial half width of the holes at 100 mK. This means that the field-induced perturbation of the thermal equilibrium populations of the TS is small even at such extremely low temperatures; see also Sect. 5.2.2.

As expected, the excess broadening decreases in an approximately logarithmic fashion with time after the field is switched off again. The TS ensemble relaxes back to its initial equilibrium state and only the thermal equilibrium hole broadening prevails at long times. This specific time evolution as well as the observed dependence on temperature T and electric field strength F are in good agreement with the theoretical description developed in this section.

The temperature dependence of the field effect is shown in Fig. 6.23. These data are normalized with respect to the temperature. In accordance with our model, the maximum relative effect is observed at the lowest temperature, its magnitude decreasing very fast with increasing temperature. The field effect practically disappears at $T = 700$ mK. This shows the high temperature limit for the observation of this effect.

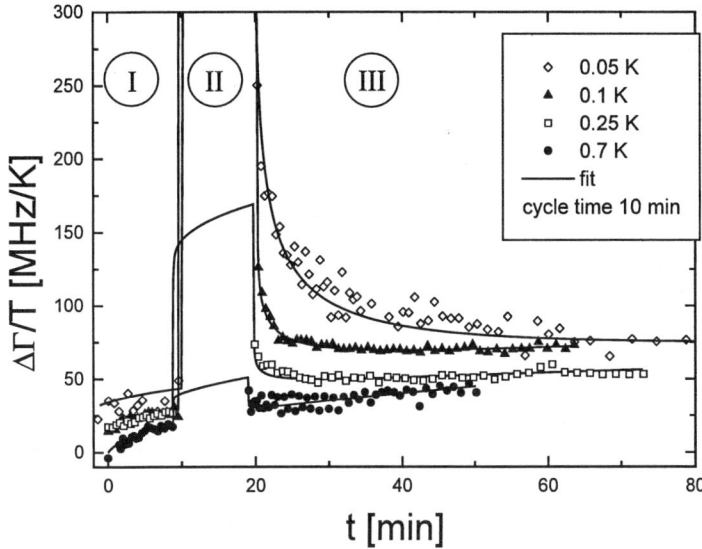

Fig. 6.23. Temperature dependence of the field-induced nonequilibrium spectral diffusion process. All measurements were carried out at a maximal electrical field change of $F = 88$ kV/cm during the field cycle. The hole broadening in the curves is normalized on temperature. The solid lines are fit curves (see text). The data sets are shifted with respect to each other along the vertical axis for clarity. From Wunderlich et al. (1997).

It must be mentioned, however, that the standard tunneling model is not strictly sufficient here. As was shown in Sect. 6.5.4, the TS distribution function in PMMA differs from that assumed in the standard model. Deviations from the logarithmic law in equilibrium hole broadening are noticeable already at 10^2 s. Therefore, the observed deviations from the logarithmic time evolution according to the standard model necessarily have to be included into the description of the electric field effects: In a consistent theoretical treatment, the distribution function $P(\Delta_0)$ must be the same for evaluating the equilibrium and the nonequilibrium contribution to the hole broadening. For fitting the data, the distribution function (6.27) is employed, which yields good agreement.

Measurements of field-induced spectral diffusion yield the following information:

– The average value of the static electric dipole moments of the TS's can be measured. This was done for the two systems PMMA and PS by Wunderlich et al. (1997). The obtained values are $\mu = 0.2$ D and 0.1 D, respectively. The electric dipole moment of the TS's can also be measured in dielectric experiments, but dielectric experiments yield a value of μ at nonzero frequency, that is, however, in the same range. The optical method yields the

average static dipole moment; it can be measured for TS's with relaxation rates as small as 10^{-2}–10^{-3} s^{-1}. Moreover, Wunderlich et al. (1997) observed that there is a correlation between the TS dipole moments and their relaxation rates. This means that measurements of the TS electric dipole moment on various time scales can yield valuable information about their microscopic structure.
- The field-induced nonequilibrium spectral diffusion is governed by the same TS distribution as the equilibrium dynamics. But the time resolution in measurements of the field-induced effect is not limited by the usually slow hole-burning process. Instead at an arbitrary time after hole burning, an instantaneous nonequilibrium situation can be created by switching an external field. Therefore, the time resolution is formally only limited by the rate at which the field can be switched on and off. This allows an increase of the time resolution in hole-burning experiments by orders of magnitude. This particular advantage of electric field experiments was already demonstrated by Wunderlich et al. (1997): This way, a time resolution of milliseconds was realized in persistent hole-burning measurements.

6.6 Conclusion

This chapter evolves around the technique of optical hole burning and clearly shows that the introduction of this rather new technique has boosted the optical spectroscopy of solids. The method is able to enhance the optical resolution far beyond the limit given by the inhomogeneous linewidth. This leads, at low temperatures, to an increase of resolution of up to six orders of magnitude.

The most attractive aspect of the hole-burning technique is the fact that the method works equally well for amorphous solids like polymer glasses. It also works for crystals but here the gain in resolution is less.

Combining the aspects of low temperatures and the concomitant enhancement of resolution, this chapter focuses on the spectroscopy of polymer glasses that are doped with probe molecules. The concentration of these probe molecules that mostly absorb in the visible spectral range is rather low, namely, 10^{-3} to 10^{-5} molar. It is known that these concentrations do not change the bulk properties of the glasses, since polymers have a free volume between 10% and 15% and thus, can easily accommodate the guest molecules. There are also many experimental evidences for the fact that the probe molecules are suitable for obtaining information on the unperturbed host matrix. (This has, for instance, been proven for the compressibility parameters of the solids.)

One important aspect of this chapter is that hole-burning experiments allow a thorough reexamination of the model of the low-temperatures properties and anomalies of glasses that is now a quarter of a century old. This model is based on the assumption of tunneling states as carriers of the dynamical properties of glasses and has proven to be extremely useful as a first step for an understanding of the low-temperature properties of glasses. With the much more precise optical experiments at hand, the old two-level system model, that was put forward by Anderson and Phillips, needs to be modified, but can still be looked upon as a very good first approach. In view of new experiments, a modified model involving TS–TS coupling and TS–phonon coupling has to be considered, as is done in previous chapters.

The optical methods provide a magnificently broad time scale from subpicoseconds (for echoes) to months and, thus, allow a very thorough analysis of the glass dynamics at low temperatures. However, there is still a large need to correlate the low-temperature properties of glasses with their properties in the vicinity of the glass transition temperature, where mode coupling theories have lately been applied quite successfully.

Naturally, the chapter is also influenced by the various experiments of the authors even though a broad overview of the relevant literature is given. We believe that the study of spectral diffusion is extremely helpful for the further development of a molecular model of the tunneling systems that are still unknown to a large extent. Particularly interesting are the electric field experiments. They offer the possibility to measure the static electric dipole moments of the TS's. These experiments show that the dipole moments are correlated with other TS properties such as the TS relaxation rates.

We are not unhappy and not surprised that there are still many open questions. This has to be expected, since the physics of glasses can be considered as one of the remaining challenges of solid state physics. Finally, the chapter opens analogies between glasses and other related fields such as the field of the dynamic properties of proteins. Looking from this view angle, the study of glasses gains a breadth that goes well beyond that of a special subsection of solid state physics.

Acknowledgments

The authors would like to thank R. Wunderlich for his help during the preparation of this chapter, and R. I. Personov, J. L. Skinner, L. Kador, S. Zilker, A. Bard and J. Beier for carefully reading and improving the manuscript. They would also like to thank all the authors mentioned in the text who kindly granted the permission to reproduce their figures for this chapter.

This work was financially supported by the Volkswagen Foundation (Grant No. AZ I/70 526), the Deutsche Forschungsgemeinschaft (Sonderforschungsbereich 213) and the Russian Foundation of Basic Research (Grant No. 96-02-17566).

7. Tunneling of H and D in Metals and Semiconductors

Gaetano Cannelli, Rosario Cantelli, Francesco Cordero, and Francesco Trequattrini

7.1 Introduction

Due to their low mass, interstitial H and its isotopes may exhibit marked quantum effects in solids. These appear both as coherent delocalization of the H wave function over two or more interstitial sites, and as a mobility that is not described by the classical Arrhenius law. The earliest and most extensive investigations have been carried out in *bcc* metals, notably in Nb and Ta and recently in the *hcp* rare earths Y and Sc. Quantum effects in *fcc* metals are found to be less marked, and a coherent tunneling system (TS) of H has never been found, possibly because in these lattices, the distance between the interstitial sites occupied by H is greater than in *bcc* metals, hence reducing the overlap between the H wave functions in adjacent sites.

The main difference between the treatment of tunneling of interstitial H in crystalline solids and tunneling in disordered solids is the absence of broad distributions of the tunneling parameters in the former case. The main source of inhomogeneous broadening in a diluted solid solution of H trapped by impurities is the long-range elastic interaction among the impurities and the H atoms. In the limit of very high dilution, the distribution functions for parameters of the TS's that are supposed to be linearly coupled to local strain can be approximated by Lorentzians, as discussed by Stoneham (1969).[1] In the most studied cases of TS's of H trapped by heavier interstitial atoms in Nb and Ta, it turns out that the tunneling energies are of the order of a few Kelvin or less, with no or little dispersion of the values, whereas the perturbations to the site energies reach tens of Kelvin, depending on the defect concentration. These cases can be treated with the same formalism adopted for the TLS's in glasses, the only difference being in the distribution functions of the asymmetry and tunneling energies Δ and Δ_0 that are Lorentzians or Gaussians instead of uniform distributions. This situation is midway between TLS's in disordered solids and the multilevel TS's of the off-centre impurities

---

[1] When the width of the distribution increases, the use of a Lorentzian becomes questionable, due to the high wings that require a cutoff; indeed, the Lorentzian curve is obtained without imposing a lower limit to the reciprocal distance between defects, and the highest deviations of the distributed parameter from the mean value come from interactions of pairs of defects arbitrarily close to each other.

in alkali halides (Narayanamurti and Pohl (1970)), where the influence of random internal strains is generally neglected both on the tunneling energies and off-centre site energies.

Coherent delocalized states of H in *bcc* metals are clearly observed only for H trapped at some impurity, but there are indications that untrapped H may also delocalize over more sites or at least perform extremely rapid jumps within a set of relaxed interstitial sites, with a rate much faster than that for hopping to another localized configuration. Unfortunately, the solubility of H in the gaslike α phase becomes vanishingly small at low temperature (see Sect. 7.4), where these delocalized states should be easily recognizable as coherent states. The situation is different for the *hcp* rare earths, where no precipitated hydride phase is found with H concentrations up to 30 at%; in these metals the H tunnel states can be studied also in samples in which the influence of trapping impurities is negligible.

The picture that emerges from the body of experimental data is that H is delocalized within a restricted number of sites among which it forms coherent tunneling states below about 10 K and performs fast hopping at higher temperature. The number of sites over which H is delocalized depends on the lattice type and impurity-H symmetry in the case of trapped H. The hopping between different sets of delocalized configurations occurs at a much slower rate and is responsible for the long-range diffusion and reorientational dynamics around impurities. Also, this slower hopping dynamics does not follow the classical rate theory, valid for heavier atoms.

Figure 7.1 shows most of the hopping and tunneling rates of H and D in *bcc* and *hcp* metals, obtained from experiments of anelastic relaxation at acoustic frequencies and below (AR), ultrasonic attenuation (UA), and quasi-elastic neutron scattering (QNS).

Reviews of the properties of the metal-H systems, also including quantum dynamics of H, can also be found in Alefeld and Völkl (1978), Wipf (1997) and Fukai (1993).

7.2 Solid Solutions of Hydrogen

Interstitial hydrogen is an impurity always present in metals; at room temperature, its solubility varies from a few at ppm in Mo, W, Fe, Cu, Al up to several at % in Pd, Ti, Zr, Cr, Nb,Ta, V (Schober and Wenzl (1978)) and even as much as 20 at % H in the rare earths Y and Sc (Vajda (1995)). The lattice structure alone does not determine such different capabilities of absorbing hydrogen, because the *bcc*, *fcc* and *hcp* lattices are represented in all the three groups of metals mentioned above.

7.2 Solid Solutions of Hydrogen   391

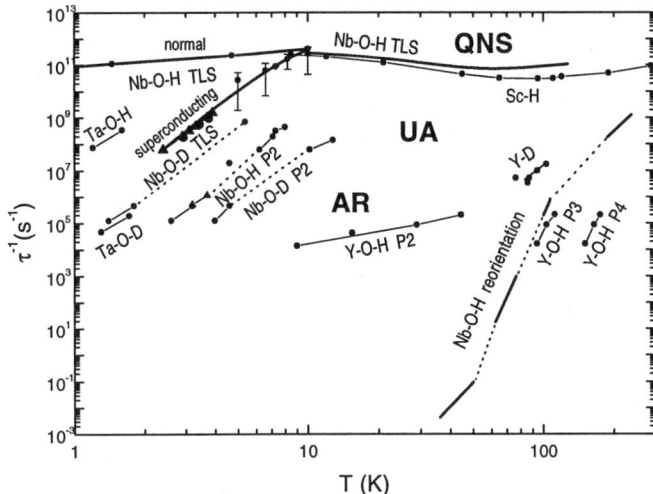

Fig. 7.1. Relaxation rates of H tunnel systems and for H reorientation around impurities in some *bcc* and *hcp* metals (from Cannelli et al. (1993)). Note that the reorientation rate of the OH pair in Nb is a straight line from 1 to 10^9 s^{-1} in a plot versus reciprocal temperature. Also indicated are the experimental techniques adopted in the different frequency ranges: AR = anelastic relaxation (resonating sample), UA = ultrasonic attenuation, QNS = quasi-elastic neutron scattering.

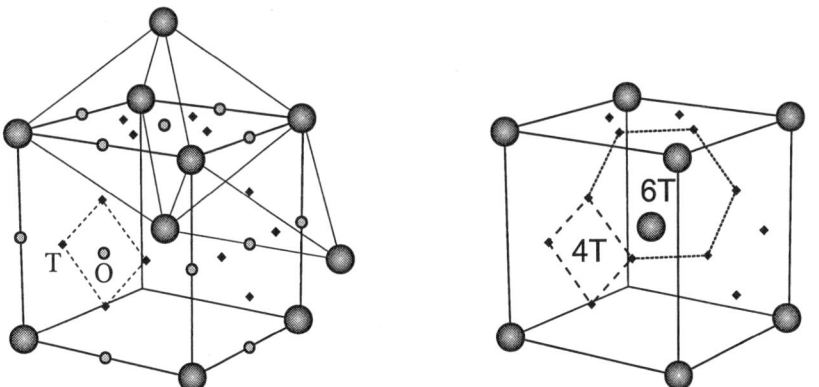

Fig. 7.2. Unit cell of *bcc* lattice, with tetrahedral (T) and octahedral (O) sites. Also shown are two possible types of delocalization of H over four or six T sites.

7.2.1 The bcc Metals V, Nb and Ta

The transition metals of the group 5, V, Nb and Ta have a *bcc* unit cell, as shown in Fig. 7.2. Interstitial atoms may occupy octahedral (O) sites or tetrahedral (T) sites. Both the O and T sites are equivalent among themselves, but have three different crystallographic orientations, depending on the direction of the nearest neighbor lattice atoms. The heavy impurities like O, N and C occupy the O sites, whereas H occupies the T sites.

At sufficiently low H concentration and high temperature, H occupies randomly the T sites in a gaslike phase, called α phase. Lowering temperature, the (mainly elastic) interactions among the H atoms cause its precipitation in ordered hydride phases, analogously to the condensation of a gas phase into liquid and solid phases. The phase diagram of a metal-H system is generally represented using only two variables: temperature T, and concentration c expressed as atomic ratio of the number of hydrogen atoms n_H and the number of metal atoms n_M. Below 300 °C, the third variable, the external pressure, is not necessary, as the metal surface constitutes a barrier for the in- and out-diffusion of hydrogen (except for Pd). In Fig. 7.3, the phase diagram of the Nb-H system is shown, that is representative for most metals (both *bcc* and *fcc*). The boundary lines "T versus c" separate the single phases (or two coexisting phases), which are commonly named α, α', β, δ, ϵ, γ, ζ.

The dynamics of H in all the ordered hydride phases simply consists of

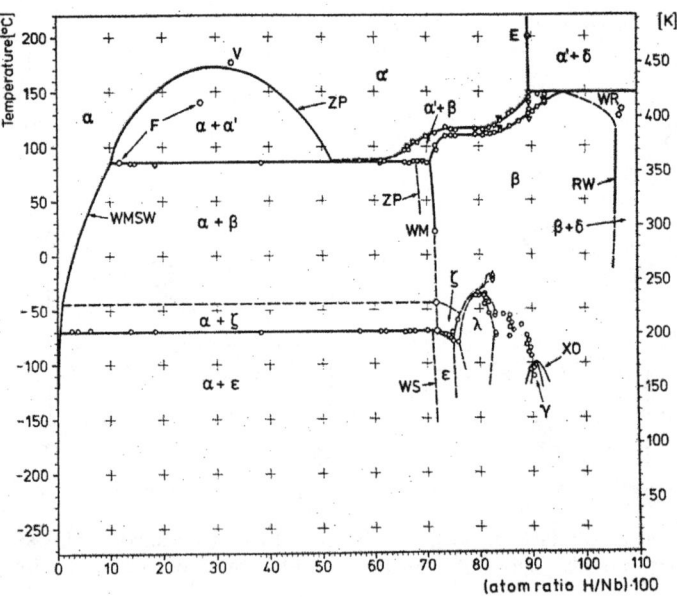

Fig. 7.3. Phase diagram of NbH$_x$, from Schober and Wenzl (1978).

localized nearly harmonic vibrations, without any delocalization over tunnel sites; therefore we are only interested in the solvus line between the α and the precipitated phases, that is given by $c_\alpha = c_0 \exp\left(-H^\alpha/k_\mathrm{B}T\right)$, where H^α is the enthalpy of solution; c_α becomes negligibly small at low temperature. The H exceeding the solubility limit precipitates in the hydride phases, which have different lattice constants and therefore constitute misfitting particles and generate plastic strains (dislocations or even cracks).

It should also be mentioned that the positions of the boundary lines of the M-H phase diagrams may depend on the isotope mass, so that the M-D phase diagrams are not identical with the M-H ones (Schober and Wenzl (1978)); few experiments have been carried out with tritium.

7.2.2 The Rare Earths Sc, Y and Lu

The rare earths Sc, Y, La, Lu and others have a *hcp* unit cell, as shown in Fig. 7.4. The properties of the rare earth-H systems have been recently reviewed by Vajda (1995). The phase diagram of the Y-H system, that is representative of other rare earths, consists essentially of three phases: α, β and γ. In the solid solution α-phase, the H atoms populate mainly the tetrahedral (T) sites, with little octahedral (O) occupancy.

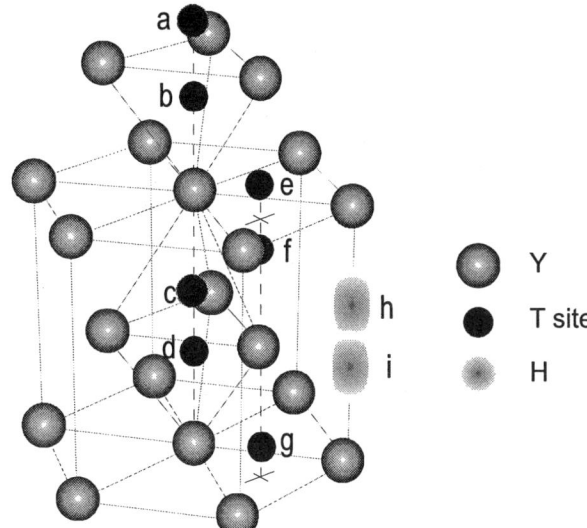

Fig. 7.4. Unit cell of *hcp* lattice, with tetrahedral (T) sites. Also shown is a TLS of H delocalized over the two nearest neighbor T sites h and i.

The mobility of hydrogen in the rare earths is lower than in *fcc* and *bcc* metals and, as temperature is lowered, the α^* metastable phase appears in which hydrogen tends to order in pairs occupying next-nearest neighbor T sites along the *c*-axis with a bridging metal atom between two H atoms. This ordering prevents the hydride precipitation so that this phase presents the striking feature of retaining H in solid solution down to the zero absolute temperature at concentration as high as 20 at%. The dihydride or β-phase crystallizes in the fcc fluorite type structure with an ideal composition YH_2. The trihydride or γ-phase can retain up to three H atoms per atom Y, with both T and O occupancy; its structure is hcp but with a larger unit cell than the metal. The formation of this phase gives rise to a metal–insulator transition accompanied by drastic changes in the optical properties.

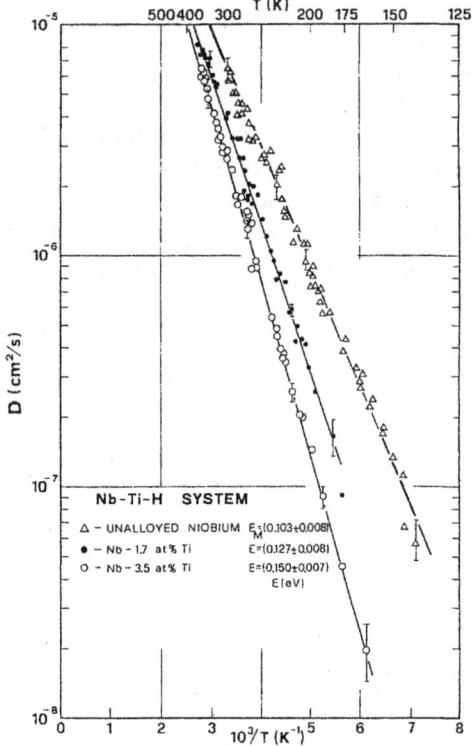

Fig. 7.5. Influence of the substitutional Ti on the diffusion coefficient of H in Nb (from Cannelli and Cantelli (1977)).

7.2.3 Trapping of Hydrogen by Impurities

The M-H phase diagrams refer to pure metals (generally having a 99.99% atomic purity), but interstitial and substitutional impurities can drastically change the phase diagram of a M-H system by trapping one or more H atoms in stable impurity-H complexes. As a consequence, the solvus line $c_\alpha(T)$ is shifted to higher values of concentration. Experimental evidence of H trapping by interstitial and substitutional impurities is given by the enhancement of solubility in α-phase in M-H systems, the decrease of long-range diffusion coefficient with the impurity content, the anelastic or magnetic relaxation processes associated with the reorientational hopping of H around the impurity.

Figure 7.5 shows an example from Cannelli and Cantelli (1977) in which the diffusion coefficient of H in Nb, determined by the Gorsky technique, is affected by the presence of substitutional Ti impurities. The experimental data can be well described (continuous lines) by a model formulated by Koiwa (1974) for the apparent diffusivity of interstitial impurities in a bcc lattice in which randomly dispersed substitutional atoms are assumed to act as trapping centers only for the nearest neighboring sites. In the case of tetrahedral occupancy for H and supposing that the saddle-point energies are not modified, the apparent diffusion coefficient D' is given by the expression

$$D' = \frac{D}{1 - 4c + 4c(\nu_0/\nu_1)\,\exp(E_\mathrm{B}/k_\mathrm{B}T)}, \tag{7.1}$$

where D is the normal diffusion coefficient, c the atomic fraction of the substitutional Ti, ν_0 and ν_1 are the frequency factors of H in the normal and trap sites, E_B is the binding energy. The fit to the experimental data allows the determination of the binding energy of H to Ti, $E_\mathrm{B} = (0.061 \pm 0.009)$ eV.

In a simple approach, the trapping is explained by the lowering of the ground state energy (binding energy) of the interstitial sites surrounding the trapping center; the values of the binding energies, determined by different methods, are generally included between 50 and 200 meV (Matsumoto et al. (1975)). In most cases, one impurity can trap one H atom. Vargas and Kronmüller (1980) and Vargas et al. (1985) have proposed a phenomenological model that describes satisfactorily the trapping mechanism of hydrogen by substitutional impurities in several host metals Nb, V, Pd, Ni, Fe. In the model it is assumed that both electronic and elastic interactions contribute to the formation of stable S-H complexes, where S is the substitutional atom. The elastic energy contribution is proportional to the volume change of the host metal produced by the S impurity. The electronic component depends on the electric charges transferred from the host atom to the substitutional impurity; this latter quantity can be evaluated using concepts developed by Miedema (1973).

7.3 Experimental Techniques Revealing the Tunneling of Hydrogen

7.3.1 Specific Heat

The contribution of a TLS to the specific heat is easy to calculate, since it only depends on the energy levels of the TLS. The fraction n_α of TLS occupying the α-th level with energy E_α is $n_\alpha = Z^{-1}\exp(\beta E_\alpha)$, with $\beta = 1/(k_\mathrm{B}T)$,

$Z = \sum_\alpha \exp(\beta E_\alpha)$, and the average energy of the TS at a temperature T is simply $\langle E \rangle = \sum_\alpha n_\alpha E_\alpha$. The specific heat is the derivative of the internal energy with respect to temperature. For a TLS with $E_2 - E_1 = E$ is[2]

$$C_{\text{TLS}} = \frac{\partial \langle E_{\text{TLS}} \rangle}{\partial T} = k_B \left(\frac{E}{2k_B T}\right)^2 \text{sech}^2 \left(\frac{E}{2k_B T}\right). \tag{7.2}$$

It is peaked at $T = 0.42 E/k_B$, and vanishes as $[T \exp(E/k_B T)]^{-2}$ at lower temperatures, due to the fact that the upper level becomes less and less populated. Even if a broad distribution of asymmetries Δ exists, the lower limit for the energy separation between the levels of the TLS's is Δ_0, so that $C_{\text{TLS}}(T)$ on a double logarithmic scale has a cutoff at $k_B T \simeq 0.4\Delta_0$. This is different from the case of TLS's in disordered systems, where also Δ_0 has a nearly uniform distribution down to very small values, causing the famous linear term in the low-temperature specific heat (see Sect. 2.2.3). Indeed, Wipf and Neumaier (1984) used the low-temperature cutoff of the excess specific heat due to O-H and O-D pairs in Nb to estimate Δ_0 for these TLS's.

The $C(T)$ curves do not provide any information on the dynamics of the TS, since they are a thermodynamic property; however, at very low temperatures, the relaxation time of the TS can become so long that it can be estimated from the slow temperature response of the sample on cooling (see Chap. 2).

7.3.2 Acoustic Measurements

The influence of TS's to the acoustic properties of solids has been described in Chap. 4 (see also Chap. 5). Here we will reconsider the contribution of a TS to the elastic response from different points of view; both for gaining further physical insight and because a different approach is needed, in order to put in evidence the dependence of the acoustic properties on the symmetry and geometry of the TS in a single crystal.

We consider first the interaction between a TS and stress or strain from a purely thermodynamic point of view, introducing later the dynamic part. The contribution to strain of a TS changes with its occupation state, and therefore it depends on temperature and time. Such a contribution is called *anelastic*, to distinguish it from the instantaneous elastic one; it modifies the elastic response of the material causing elastic energy dissipation and softening of the moduli.

Since we will be dealing also with multiwell TS's, the notation must be somewhat more complex than that used in the rest of the book, dealing with TLS's. The states of the TS's in the diagonal representation will be labelled

---

[2] Equation (7.2) is the specific heat at constant volume, whereas the quantity actually measured is the specific heat at constant pressure. The quantities are practically coincident at low temperatures, because the volume change with temperature is negligible. See also Sect. 2.2.3.

with μ, ν, while the indexes α, β will be used for the localized representation, i.e. α, β refer to the sites among which the H atom tunnels; in the absence of tunneling, the two representations coincide. The indexes $i, j, \ldots$ will refer to the components of the elasticity tensors. The treatment of the influence of a TS on the acoustic properties that follows is based on the localized representation, because it is intuitive and allows the construction of simple models of tunneling systems that explicitly include their geometry.

Elastic Dipole, Double Force Tensor and Defect Elastic Energy. A defect can be characterized from the elastic point of view in two equivalent ways, by the elastic dipole and the double force tensor. The elastic dipole λ_{ij} is defined through the strain ϵ_{ij} created by a homogeneous distribution of c defects per mole (when the chemical formula of the host material consists of only one atom, the mole concentration coincides with the atomic ratio),

$$\epsilon_{ij} = c\lambda_{ij} \to \lambda_{ij} = \frac{\partial \epsilon_{ij}}{\partial c}, \tag{7.3}$$

and is adimensional. On the other hand, the double force tensor is defined through the stress due to a homogeneous distribution of c defects per mole; since σ usually represents an external stress applied to the crystal, a minus sign is necessary to convert it to the internal stress

$$\sigma_{ij} = -cp_{ij} \to p_{ij} = -\frac{\partial \sigma_{ij}}{\partial c}; \tag{7.4}$$

with this definition, p has the dimensions of an energy per volume. The two quantities are simply connected to each other by the Hooke's law (summation over repeated indexes is understood)

$$p_{ij} = C_{ijhk}\lambda_{hk}, \qquad \lambda_{ij} = S_{ijhk}p_{hk}, \tag{7.5}$$

where C and S are the elastic stiffness and compliance constants. A detailed treatment of the elastic properties of defects is provided in Leibfried and Breuer (1978), Nowick and Berry (1972).

A defect can assume different configurations, which we label with the index α, e.g. H in a bcc lattice can occupy three types of T sites with the major axis of the elastic dipole oriented along $\alpha = x, y, z$.

The elastic energy of a defect in state α can be expressed in terms of λ^α or p^α as

$$E_\alpha(\sigma) = E_\alpha(0) - v_0 \lambda_{ij}^\alpha \sigma_{ij}, \qquad E_\alpha(\epsilon) = E_\alpha(0) - v_0 p_{ij}^\alpha \epsilon_{ij}, \tag{7.6}$$

where v_0 is the unit cell volume. These relationships can be obtained by taking the second derivatives of the appropriate thermodynamic potentials. The differential of the internal energy u per unit volume (all the extensive variables are expressed per unit volume) is

$$du = Tds + \sum_{ij} \sigma_{ij} d\epsilon_{ij} + \frac{1}{v_0} \sum_\alpha E_\alpha dc_\alpha, \tag{7.7}$$

where $\sigma_{ij}d\epsilon_{ij}$ is the work done on the crystal, the last term is the contribution from the defects, and $\sum_\alpha c_\alpha = c$. The Helmholtz free energy per unit volume is $f = u - Ts$ and its differential is

$$df = \sum_{ij} \sigma_{ij}d\epsilon_{ij} - sdT + \frac{1}{v_0}\sum_\alpha E_\alpha dc_\alpha; \qquad (7.8)$$

Eq. (7.6) is obtained by combining (7.4) and (7.8):

$$p_{ij}^\alpha = -\frac{\partial \sigma_{ij}}{\partial c_\alpha} = -\frac{\partial^2 f}{\partial c_\alpha \partial \epsilon_{ij}} = -\frac{1}{v_0}\frac{\partial E_\alpha}{\partial \epsilon_{ij}}. \qquad (7.9)$$

Similarly, using the Gibbs free energy $g = u - \sum_{ij}\epsilon_{ij}\sigma_{ij} - Ts$, whose differential is

$$dg = -\sum_{ij}\epsilon_{ij}d\sigma_{ij} - sdT + \frac{1}{v_0}\sum_\alpha E_\alpha dc_\alpha, \qquad (7.10)$$

and (7.3), one obtains the other form of (7.6):

$$\lambda_{ij}^\alpha = \frac{\partial \epsilon_{ij}}{\partial c_\alpha} = -\frac{\partial^2 g}{\partial c_\alpha \partial \sigma_{ij}} = -\frac{1}{v_0}\frac{\partial E_\alpha}{\partial \sigma_{ij}}. \qquad (7.11)$$

From (7.9) it appears that the dipole force tensor is simply related to the deformation potential, that is generally used in the literature on tunneling systems [see also (4.10)]:

$$\gamma_{ij}^\alpha \simeq \frac{1}{2}\frac{\partial \Delta_\alpha}{\partial \epsilon_{ij}} = -v_0 p_{ij}^\alpha. \qquad (7.12)$$

Analogous relations can be written for describing the elastic behavior of the eigenstates of a TS, instead of the particle localized in site α; in that case the quantities p_{ij}^μ and λ_{ij}^μ depend also on strain, as shown later.

Contributions of a TS to the Static Elastic Constants. In order to calculate the contribution of a TS to the static elastic compliance S_{ijhk}, we apply an external stress σ_{hk} and determine the resulting strain ϵ_{ij}. The stress σ_{hk} perturbs the energy levels E_μ of the TS according to (7.6), and this in turn causes a change in the equilibrium populations $\langle c_\alpha \rangle$, according to the Boltzmann distribution ($\langle ... \rangle$ indicates the thermal average); it is also convenient to introduce the occupation fractions n_α,

$$\langle n_\alpha \rangle = \frac{\langle c_\alpha \rangle}{c} = \frac{e^{-\beta E_\alpha}}{Z}, \quad \text{where } Z = \sum_\alpha e^{-\beta E_\alpha}. \qquad (7.13)$$

A change in the TS populations is reflected in the strain, since, according to (7.11), a concentration c_α of TS in state α contributes to the strain ϵ_{ij} with

$$\epsilon_{ij}^{an} = c_\alpha \lambda_{ij}^\alpha; \qquad (7.14)$$

this strain is indicated as *anelastic*, to distinguish it from the elastic response $\epsilon_{ij}^{el} = S_{ijhk}\sigma_{hk}$. The contribution of the TS to the compliance is

7.3 Experimental Techniques Revealing the Tunneling of Hydrogen

$$\delta S_{ijhk} = \frac{\mathrm{d}\langle\epsilon_{ij}^{\mathrm{an}}\rangle}{\mathrm{d}\sigma_{hk}} = c\frac{\mathrm{d}}{\mathrm{d}\sigma_{hk}}\left(\sum_\alpha n_\alpha \lambda_{ij}^\alpha\right) =$$

$$= c\sum_{\alpha\beta}\frac{\partial n_\alpha}{\partial E_\beta}\underbrace{\frac{\partial E_\beta}{\partial \sigma_{hk}}}_{-v_0\lambda_{hk}^\beta}\lambda_{ij}^\alpha + c\sum_\alpha n_\alpha \frac{\partial \lambda_{ij}^\alpha}{\partial \sigma_{hk}} = \delta S_{ijhk}^{\mathrm{para}} + \delta S_{ijhk}^{\mathrm{dia}},$$

and consists of paraelastic and diaelastic parts:

$$\delta S_{ijhk}^{\mathrm{para}} = -v_0 c \sum_{\alpha\beta}\frac{\partial n_\alpha}{\partial E_\beta}\lambda_{ij}^\alpha \lambda_{hk}^\beta, \tag{7.15}$$

$$\delta S_{ijhk}^{\mathrm{dia}} = c\sum_\alpha n_\alpha \frac{\partial \lambda_{ij}^\alpha}{\partial \sigma_{hk}}. \tag{7.16}$$

In a completely analogous way it is possible to calculate the contribution of a TS to the static elastic stiffness C_{ijhk}:

$$\delta C_{ijhk}^{\mathrm{para}} = \frac{c}{v_0}\sum_{\alpha\beta}\frac{\partial n_\alpha}{\partial E_\beta}p_{ij}^\alpha p_{hk}^\beta, \tag{7.17}$$

$$\delta C_{ijhk}^{\mathrm{dia}} = c\sum_\alpha n_\alpha \frac{\partial p_{ij}^\alpha}{\partial \epsilon_{hk}}. \tag{7.18}$$

Another way of expressing (7.17), (7.18) is, using (7.9),

$$\delta C_{ijhk}^{\mathrm{para}} = \frac{c}{v_0}\sum_{\alpha\beta}\frac{\partial n_\alpha}{\partial E_\beta}\frac{\partial E^\alpha}{\partial \epsilon_{ij}}\frac{\partial E^\beta}{\partial \epsilon_{hk}}, \tag{7.19}$$

$$\delta C_{ijhk}^{\mathrm{dia}} = c\sum_\alpha n_\alpha \frac{\partial^2 E^\alpha}{\partial \epsilon_{ij}\partial \epsilon_{hk}}, \tag{7.20}$$

and similarly for (7.15) and (7.16). This form of δC_{ijhk} can also be obtained by directly applying the definition $\delta C_{ijhk} = \partial^2 f/\partial\epsilon_{ij}\partial\epsilon_{hk}$ with the free energy of the defects expressed as $f = -\frac{c}{v_0}k_\mathrm{B}T\ln(Z)$ [Granato et al. (1985)], and puts in evidence that the paraelastic response is due to the slope of the defect energy vs. strain, whereas the diaelastic one is proportional to the curvature of the defect energy versus strain.

Paraelastic Polarization. The term paraelastic polarizability describes the elastic counterpart of the paraelectric reorientation of dipoles in an electric field. In fact, a typical situation in which one observes paraelastic polarization is that of an interstitial atom that can occupy sites that are energetically equivalent, but with different crystallographic orientations, like T or octahedral sites in the bcc lattice, that may have $\alpha = x, y, z$. In that case the elastic dipoles have the expressions

$$\lambda^x = \begin{pmatrix} \lambda_1 & 0 & 0 \\ 0 & \lambda_2 & 0 \\ 0 & 0 & \lambda_2 \end{pmatrix}, \quad \lambda^y = \begin{pmatrix} \lambda_2 & 0 & 0 \\ 0 & \lambda_1 & 0 \\ 0 & 0 & \lambda_2 \end{pmatrix},$$

$$\lambda^z = \begin{pmatrix} \lambda_2 & 0 & 0 \\ 0 & \lambda_2 & 0 \\ 0 & 0 & \lambda_1 \end{pmatrix}, \tag{7.21}$$

representing dilatations along the three crystal directions; if $\lambda_1 > \lambda_2$, the application of a uniaxial stress $\sigma_{ij} = \sigma_0 \delta_{i1} \delta_{j1}$ lowers the energy of the atoms in sites with $\alpha = x$ by $-v_0 (\lambda_1 - \lambda_2) \sigma_0$ with respect to those with $\alpha = y, z$, producing an increase of the population n_x at the expenses of n_y and n_z, or equivalently, it reorients (polarizes) some elastic dipoles in order that their major axis is parallel to the uniaxial stress.

As shown by Cordero (1993), the paraelastic term can be written in a more transparent form by observing that

$$\frac{\partial n_\alpha}{\partial E_\beta} = \beta n_\alpha (n_\beta - \delta_{\alpha\beta}) \quad \text{and} \quad \sum_\alpha n_\alpha = 1,$$

and the sum can be decomposed as

$$\begin{aligned}
\delta S^{para}_{ijhk} &= -cv_0\beta \sum_{\alpha\beta} \left[\lambda^\beta_{ij} \lambda^\alpha_{hk} n_\alpha (n_\beta - \delta_{\alpha\beta}) \right] \\
&= -cv_0\beta \sum_\alpha \left[\lambda^\alpha_{ij} \lambda^\alpha_{hk} n_\alpha (n_\alpha - 1) + \lambda^\alpha_{hk} n_\alpha \sum_{\beta \neq \alpha} \lambda^\beta_{ij} n_\beta \right] \\
&= -cv_0\beta \sum_\alpha \left[-\lambda^\alpha_{ij} \lambda^\alpha_{hk} \sum_{\beta \neq \alpha} (n_\alpha n_\beta) + \lambda^\alpha_{hk} \sum_{\beta \neq \alpha} \lambda^\beta_{ij} (n_\alpha n_\beta) \right] \\
&= -cv_0\beta \sum_\alpha \sum_{\beta < \alpha} \left(-\lambda^\alpha_{ij} \lambda^\alpha_{hk} - \lambda^\beta_{ij} \lambda^\beta_{hk} + \lambda^\alpha_{hk} \lambda^\beta_{ij} + \lambda^\beta_{hk} \lambda^\alpha_{ij} \right) n_\alpha n_\beta \\
&= \sum_\alpha \sum_{\beta < \alpha} cv_0 \beta n_\alpha n_\beta \left(\lambda^\alpha_{ij} - \lambda^\beta_{ij} \right) \left(\lambda^\alpha_{hk} - \lambda^\beta_{hk} \right).
\end{aligned} \tag{7.22}$$

In a completely analogous way,

$$\delta C^{para}_{ijhk} = \sum_\alpha \sum_{\beta < \alpha} \frac{c\beta}{v_0} n_\alpha n_\beta \left(p^\beta_{ij} - p^\alpha_{ij} \right) \left(p^\beta_{hk} - p^\alpha_{hk} \right). \tag{7.23}$$

In the above equations the paraelastic term has been decomposed as the sum of the contributions from all the pairs of defect levels. This decomposition will be useful for analyzing the contributions to the relaxation intensity from states with different symmetries in a multilevel tunnel system.

The energy levels E_μ of a TS differ in general from the site energies E_α, and when describing the effects of transitions between eigenstates of a TS, the indexes α, β running over the sites should be substituted with μ, ν running over the eigenstates; then, p^μ becomes the strain derivative of E_μ, according to (7.9), and similarly for λ^μ. The energy levels of a TLS of an H atom tunneling between sites a and b with site energies $E_{a,b} = \pm\frac{1}{2}\Delta = \pm\frac{1}{2}\gamma\epsilon$ are given by

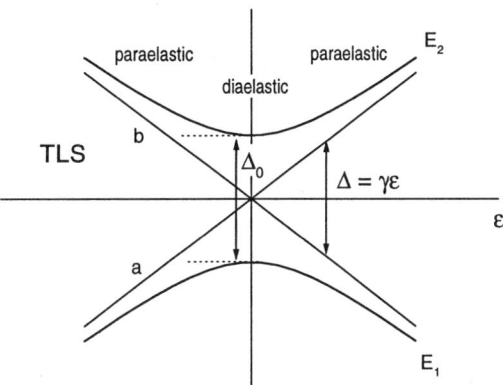

Fig. 7.6. Energy levels of a TLS within two sites a and b with site energy $\pm\frac{1}{2}\Delta$ as a function of strain. The elastic response of nearly symmetric TLS's ($\epsilon \simeq 0$) is of the diaelastic type, while with increasing the asymmetry it becomes more and more of the paraelastic type.

$$E_{1,2}(\epsilon) = \mp \frac{1}{2}\sqrt{\Delta_0^2 + \Delta^2} = \mp \frac{1}{2}\sqrt{\Delta_0^2 + \left(\sum_{ij}\gamma_{ij}\epsilon_{ij}\right)^2}, \tag{7.24}$$

(neglecting the influence of strain on Δ_0, as explained in the introduction) and are shown in Fig. 7.6. At large strains the TLS energy levels coincide with the two site energies, but at $\epsilon = 0$, the slope of $E_\mu(\epsilon)$ is zero and therefore also p^μ and the paraelastic polarization vanish. Sometimes, an effective deformation potential for paraelastic relaxation is introduced [see also (4.9)]:

$$D = 2\frac{\partial E}{\partial \epsilon} = 2\frac{\Delta}{E}\gamma. \tag{7.25}$$

There is still another effect that reduces the paraelastic relaxation amplitude: from (7.22) and (7.23), it appears that the contribution of each pair of states is proportional to the product of the respective populations divided by temperature; therefore $\delta C(T) \sim 1/T$ only for $k_B T \gg E$, and it goes to zero at $k_B T \ll E$, because the upper level is unpopulated, passing through a maximum at $k_B T \sim E$.

Diaelastic Polarization. The diaelastic terms, (7.18, (7.16), are due to the change of the elastic dipole or double force tensor of the defect under the action of the external stress or strain. In the absence of tunneling, the magnitude of the diaelastic effect is generally much smaller than the paraelastic one. In fact, if the polarization of a defect in configuration α is a function of strain, $p^\alpha(\epsilon)$, then the associated elastic energy is, according to (7.6),

$$E^\alpha(e) = -v_0\left(p^\alpha(0) + \frac{\partial p^\alpha}{\partial \epsilon}\epsilon\right)\epsilon,$$

from which it appears that the paraelastic contribution is linear with strain, whereas the diaelastic one is quadratic. On the other hand, we have just seen that the curves of the energy levels of a TS as a function of strain may become flat at $\epsilon = 0$, suppressing the paraelastic response (see Fig. 7.6). As noted by Granato et al. (1985), this implies that the response of a nearly symmetric TS is mainly diaelastic. Essentially, the elastic constants are the second derivatives or curvatures of a thermodynamic potential with respect to strain; each occupied level of the TLS contributes with its own curvature and, since the populations are temperature dependent, this diaelastic contribution is also temperature dependent. Starting from $T = 0$ K, the fundamental energy level is populated and lowers the elastic moduli with its negative curvature; increasing T, part of the TLS's go to the upper levels with positive curvature, and therefore the negative diaelastic contribution is reduced and tends to zero at $T \gg E$. The step in the diaelastic contribution occurs at $T \simeq E$. The diaelastic contribution to C_{ijhk} from a TLS can be written using (7.20), (7.13) and (7.24) as

$$\delta C_{ijhk}^{\text{dia}} = -\frac{c}{2} \gamma_{ij} \gamma_{hk} \left(\frac{\Delta_0}{E}\right)^2 \frac{1}{E} \tanh(\beta E/2). \tag{7.26}$$

The relative variation of the elastic constants or sound velocities provides a good estimate of the tunneling energy because the temperature dependence, $\tanh(E/2k_\mathrm{B}T)$, is directly related to $E \simeq \Delta_0$. In fact, even though a distribution $P(\Delta)$ of asymmetries exists, the nearly symmetric TLS's contribute much more than the others, because they are weighted with the $(\Delta_0/E)^2$ factor and because the distribution function of E arising from $P(\Delta)$ becomes infinite at $E = \Delta_0$ (see Fig. 7.21).

Relaxation at $\omega \neq 0$. Up to now, we have considered the effect of TS's on the static elastic constants from a purely thermodynamic point of view. Let us consider now the frequency dependence of the dynamic susceptibility. The diaelastic response has a resonant character, i.e. the frequency dependence is maximum when $\hbar\omega \approx E$; however, the ultrasound measurements that have been made up to now on TS's of H in metals have frequencies well below 1 GHz, and therefore the static condition $\hbar\omega \ll E$ is valid.

Instead, the paraelastic contribution has a relaxational character that produces frequency dependence of the elastic susceptibilities in the whole experimentally accessible range of frequencies; this dependence can be calculated as follows. Consider for simplicity a set of TLS's with $E = E_2 - E_1$ and with instantaneous occupation fractions n_1 and n_2, that obey

$$\begin{cases} n_1 + n_2 = 1 \\ \dot{n}_1 = -\nu_{21} n_1 + \nu_{12} n_2 \end{cases}, \tag{7.27}$$

where ν_{12} and ν_{21} are the transition rates due to the interaction of the TLS with the phonon bath or with the conduction electrons. Consider also a particular type of strain symmetry, so that we can treat all the elasticity tensors as scalars. If we apply a periodic stress

7.3 Experimental Techniques Revealing the Tunneling of Hydrogen

$$\sigma(t) = \sigma_0 e^{-i\omega t}, \tag{7.28}$$

we perturb all the energy levels according to (7.6), and therefore also the TLS populations and the corresponding anelastic strain, according to (7.11). It is our aim to find an expression for the anelastic strain as a function of the applied stress, Eq. (7.28), from which we can deduce the dynamic compliance. Let us divide the instantaneous occupation fraction $n_1(t)$ as follows:

$$n_1(t) = \underbrace{\overline{n}_1^0 + \overbrace{\Delta \overline{n}_1(t) + \delta n_1(t)}^{\Delta n_1(t)}}_{\overline{n}_1(t)}, \tag{7.29}$$

where $\overline{n}_1^0$ is the equilibrium value in the absence of stress, $\Delta \overline{n}_1(t)$ is the perturbation to the equilibrium value due to the applied stress,

$$\Delta \overline{n}_1(t) = \frac{d\overline{n}_1}{dE}\frac{\partial E}{\partial \sigma}\sigma_0 e^{-i\omega t} = \Delta \overline{n}_1 e^{-i\omega t}, \tag{7.30}$$

$\overline{n}_1(t)$ is the instantaneous equilibrium value, $\delta n_1(t)$ is the deviation from the instantaneous equilibrium due to the delayed response of the defects, and $\Delta n_1(t)$ is the deviation from the static equilibrium in the absence of applied stress,

$$\Delta n_1(t) = \Delta n_1 e^{-i\omega t}; \tag{7.31}$$

note that the amplitude of $\Delta n_1(t)$ and similarly that of $\delta n_1(t)$ are complex because they include a phase lag between the excitation and the response, due to the finite time needed for reaching the instantaneous equilibrium according to (7.27). The anelastic strain that we need for computing the dynamic compliance is that due to Δn_1, and, according to (7.3), is

$$\Delta \epsilon = \lambda_1 \Delta n_1 + \lambda_2 \Delta n_2 = \frac{c}{v_0}\frac{\partial E}{\partial \sigma}\Delta n_1. \tag{7.32}$$

The next step is to write the rate equation, (7.27), in a convenient form and to express it in terms of $\Delta n_1(t)$. In order to do that, we use the condition for the equilibrium values of the populations, i.e.

$$\frac{d\overline{n}_1(t)}{dt} = 0 \rightarrow \overline{n}_1(t)\nu_{21} = \overline{n}_2(t)\nu_{12}, \tag{7.33}$$

also known as detailed balance condition. With this relation, it is easy to transform (7.27) into

$$\frac{dn_1(t)}{dt} = -\frac{\delta n_1(t)}{\tau}, \tag{7.34}$$

where

$$\tau^{-1} = \nu_{21} + \nu_{12} = \tau_1^{-1}; \tag{7.35}$$

(7.34) indicates that the rate of change of the population is proportional to the deviation from the instantaneous equilibrium (note that in this chapter we will always deal with longitudinal relaxation rate τ_1^{-1}, i.e. the relaxation rate of σ_z in the pseudospin notation of the TLS Hamiltonian, and therefore we will omit the subscript 1). We can now express (7.34) in terms of Δn_1 by noting that

$$\frac{dn_1(t)}{dt} = \frac{d\Delta n_1(t)}{dt} = -i\omega \Delta n_1(t) \tag{7.36}$$

and

$$\delta n_1(t) = \Delta n_1(t) - \Delta \bar{n}_1(t)$$

and using (7.31), we obtain

$$\Delta n_1 = \frac{\Delta \bar{n}_1}{1 - i\omega\tau} = \Delta \bar{n}_1 \left[\frac{1}{1 + (\omega\tau)^2} + i\frac{\omega\tau}{1 + (\omega\tau)^2} \right]. \tag{7.37}$$

Equation (7.37) contains the dynamics of the relaxation process. When $\omega\tau \ll 1$, we are in the quasi-static limit in which the actual response Δn_1 follows the instantaneous equilibrium $\Delta \bar{n}_1$, when $\omega\tau \gg 1$, the system is too slow to respond, and the transition between the two situations occurs at $\omega\tau = 1$, where the out-of-phase part of the response has a peak. In order to obtain the dynamic part of the elastic susceptibility, we write Δn_1 in terms of the resulting anelastic strain through (7.32), and $\Delta \bar{n}_1$ in terms of the stress excitation through (7.30)

$$\Delta S(\omega) = \frac{\Delta \epsilon}{\sigma} = \frac{c}{v_0} \frac{d\bar{n}_1}{dE} \left(\frac{\partial E}{\partial \sigma} \right)^2 \frac{1}{1 - i\omega\tau}. \tag{7.38}$$

An equivalent expression can be derived for the dynamic modulus with the roles of ϵ and σ exchanged; by comparing with (7.15)–(7.18), it appears that the dynamic response function is obtained multiplying the static one by $(1 - i\omega\tau)^{-1}$. In the case of many levels, the response functions coincide with (7.15)–(7.23) at $\omega = 0$, but they cannot be simply muliplied by $(1 - i\omega\tau)^{-1}$ for $\omega \neq 0$, because several relaxation modes with different values of τ are involved. However, if the relevant relaxation occurs only between the levels α and β with relaxation time τ, remembering (7.23) and (7.9), we can write the relative change of the elastic stiffness as:

$$\delta C_{ijhk}(\omega) = \frac{c}{v_0} \beta n_\alpha n_\beta \left(\frac{\partial E^\beta}{\partial \epsilon_{ij}} - \frac{\partial E^\alpha}{\partial \epsilon_{ij}} \right) \left(\frac{\partial E^\beta}{\partial \epsilon_{hk}} - \frac{\partial E^\alpha}{\partial \epsilon_{hk}} \right) \frac{1}{1 - i\omega\tau}. \tag{7.39}$$

The imaginary part of the relative change of the elastic constant, (7.39), is peaked at $\omega\tau = 1$ and gives rise to the acoustic absorption (e.g. Nowick and Berry (1972)),

$$\text{Im}(\delta C/C) = \text{Im}(\delta S/S) = Q^{-1} = \frac{2v}{\omega}\alpha, \tag{7.40}$$

where Q^{-1} is the elastic energy loss coefficient or internal friction, α is the sound attenuation expressed in Np/cm and v is the sound velocity; the real part of the r.h.s of (7.39) gives a negative (positive) step in the relative change of the modulus (compliance) with an amplitude $2Q_{\max}^{-1}$ in correspondence with the absorption peak. Since τ is generally a decreasing function of temperature, the condition $\omega\tau = 1$ is satisfied at higher temperatures when measured at higher frequencies and the relaxation curves are correspondingly shifted to higher temperatures. The frequency-independent part of the relative change of $\delta C/C$ or $\delta S/S$ is generally called *relaxation strength*, $R(T)$, so that

$$\begin{bmatrix} Q^{-1} \\ \delta C_{ijhk}/C_{ijhk} \end{bmatrix} = \begin{bmatrix} \mathrm{Im} \\ \mathrm{Re} \end{bmatrix} \left(\frac{\delta C_{ijhk}}{C_{ijhk}} \right)$$

$$= R_{ijhk}(T) \frac{1}{1+(\omega\tau)^2} \begin{bmatrix} \omega\tau \\ -1 \end{bmatrix}. \qquad (7.41)$$

For a TLS with the energy levels given by (7.24), and $\frac{\partial E_{1,2}}{\partial \epsilon_{ij}} = \mp \frac{1}{2}\frac{\Delta}{E}\gamma_{ij}$, (7.39) becomes

$$\delta C_{ijhk}(\omega) = \frac{c}{v_0}\left(\frac{\Delta}{E}\right)^2 \gamma_{ij}\gamma_{hk}\beta n_1 n_2 \frac{1}{1-i\omega\tau}.$$

Substituting $n_1 n_2 = \frac{1}{4}\mathrm{sech}^2(\beta E/2)$, with $\beta = 1/k_\mathrm{B}T$, the absorption and dispersion are

$$\begin{bmatrix} Q^{-1} \\ \delta C_{ijhk}/C_{ijhk} \end{bmatrix} = \begin{bmatrix} \mathrm{Im} \\ \mathrm{Re} \end{bmatrix}\left(\frac{\delta C_{ijhk}}{C_{ijhk}}\right)$$

$$= \frac{c}{v_0}\left(\frac{\Delta}{E}\right)^2 \gamma_{ij}\gamma_{hk}\frac{\beta}{4}\mathrm{sech}^2(\beta E/2)\frac{1}{1+(\omega\tau)^2}\begin{bmatrix}\omega\tau \\ -1\end{bmatrix}, \qquad (7.42)$$

see also Sect. 4.2.2. Equation (7.42) is equivalent to the classical relaxation due to the reorientation of defects, except for the factor $(\Delta/E)^2$ that comes from the fact that the energy levels of a TLS become flat at $\Delta = 0$ (see Fig. 7.6). The factor $\frac{1}{4}\mathrm{sech}^2(\beta E/2)$ comes from the product of the populations, $n_1 n_2$ in (7.23), and reduces to zero the relaxation strength $R(T)$ at $k_\mathrm{B}T \ll E$, when the upper level is not populated.

Relaxation Rate. The relaxation rate is given by (7.35), i.e. by the sum of the hopping (transition) rates between the two sites (energy levels). These rates, when not given by the simple Arrhenius law, can be calculated in terms of transitions involving absorption or emission of one or more phonons, or scattering with the conduction electrons, or other mechanisms, see Chaps. 3 and 4. The actual rate is the sum of all the contributions:

$$\tau^{-1} = \tau_{1p}^{-1} + \tau_{2p}^{-1} + \tau_e^{-1} + \ldots$$

The one-phonon rate can be evaluated with the time-dependent perturbation theory, where the perturbation is the change in site energy $\gamma_{ij}u_{ij}$ (7.12) due to the local strain u_{ij} expressed in terms of phonon operators (Sussman (1967), Jäckle (1972)), see also the discussion on (7.59) or with other methods, see also (3.110), and is

$$\tau_{1\mathrm{p}}^{-1} = \sum_\lambda \frac{1}{2\pi\hbar^4 \varrho v_\lambda^5} \left(\frac{\Delta_0}{E}\gamma_\lambda\right)^2 E^3 \coth\left(\beta E/2\right), \qquad (7.43)$$

where the sum is over the acoustic phonon modes with sound velocity v_λ, ϱ is the mass density. The rate is linear in T at $k_\mathrm{B} T > E$, and tends to a finite value at $T=0$; compare with (2.18). The values of γ can be deduced from the relaxation intensity in (7.42), which therefore determines the contribution of $\tau_{1\mathrm{p}}^{-1}$ to the total rate.

At higher temperatures the multiphonon processes and/or incoherent tunneling (see Chaps. 3 and 4) also contribute, and finally give an Arrhenius-like T dependence. There is a vast literature on the T dependence of multiphonon processes (see Sect. 7.4.1). Essentially, the rate is predicted to follow a T^n law with $n > 2$, or $\exp\left(-E/k_\mathrm{B}T\right)$ with some corrections.

In metals, it is found that the scattering of the conduction electrons from the tunneling particle is far more effective than the phonons in inducing transitions of the TS's; this is also discussed in Chaps. 3 and 4, and later we will provide many experimental evidence that it is the case for tunneling of H in metals. Regarding the expression of τ_e^{-1}, besides the simpler approach based on first-order time-dependent perturbation theory, there are other theories that, starting from Kondo (1976), take into account the possibility that the interaction of the electrons with the TS's is not fully adiabatic, i.e. that there is a finite density of electron excitations that do not follow adiabatically the tunneling particle, see the discussion in Sect. 4.2.3.

7.3.3 Neutron Spectroscopy

The neutron scattering experiments provide many types of information on the dynamics of interstitial H. The quantity that is measured is the partial differential cross-section, $\partial^2\sigma/\partial\Omega\partial\hbar\omega$, i.e. the number of neutrons that are scattered by the sample in the solid angle $\partial\Omega$, changing their momentum from $\hbar\vec{k}_i$ to $\hbar\vec{k}_f = \hbar\vec{k}_i + \hbar\vec{q}$ and their energy from E_i to $E_f = E_i + \hbar\omega$, normalized by the incident flux. Assuming that the neutrons are plane waves, and a little fraction of them is scattered as spherical waves by the nuclei, the differential scattering cross-section can be written as (a complete treatment of the subject can be found for example in Marshall and Lovesy (1971))

$$\frac{\partial^2\sigma}{\partial\Omega\partial\omega} = N\frac{k_f}{k_i}\int_{-\infty}^{\infty}\frac{dt}{2\pi}e^{-i\omega t}\sum_{mn}\left\langle \overline{b_m^* b_n} e^{-i\vec{q}\cdot\vec{R}_m(0)} e^{i\vec{q}\cdot\vec{R}_n(t)}\right\rangle, \qquad (7.44)$$

where N is the number of the nuclei, b_m is the scattering length, i.e. the scattering strength and phase, for a neutron interacting with the m-th nucleus, that depends on the nucleus type and reciprocal spin orientation; the average is both thermal and over the isotope and spin configurations of the nuclei, that are simply at random. For static and identical nuclei (all with the same b), (7.44) reduces to the elastic Bragg diffraction (coherent scattering). In general, $\overline{b_m^* b_n} = \overline{b^*}\,\overline{b} = \left|\overline{b}\right|^2$, when $m \neq n$, since the nuclei are uncorrelated, whereas $\overline{b_m^* b_m} = \overline{|b|^2}$. Therefore it is possible to write $\overline{b_m^* b_n} = \left|\overline{b}\right|^2 + \delta_{mn}\left(\overline{|b|^2} - \left|\overline{b}\right|^2\right)$ and separate the sum over the nuclei in a sum over all pairs of nuclei that is proportional to $\left|\overline{b}\right|^2 = \sigma_c/(4\pi)$ and is called coherent, and a sum with $m = n$, that is proportional to $\left(\overline{|b|^2} - \left|\overline{b}\right|^2\right) = \sigma_i/(4\pi)$ and is called incoherent. The nucleus of H has spin $\frac{1}{2}$ and the two scattering lengths corresponding to the collisions with spin parallel or antiparallel to that of the neutron have values such that σ_i is about 50 times larger than σ_c. Therefore, the scattering from H is

$$\frac{\partial^2 \sigma}{\partial \Omega \partial \omega} = N \frac{\sigma_i}{4\pi} \frac{k_f}{k_i} \int_{-\infty}^{\infty} \frac{dt}{2\pi} e^{-i\omega t} \sum_m \left\langle e^{-i\vec{q}\cdot\vec{R}_m(0)} e^{i\vec{q}\cdot\vec{R}_m(t)} \right\rangle \quad (7.45)$$

$$= N \frac{\sigma_i}{4\pi} \frac{k_f}{k_i} S_i(\vec{q},\omega), \quad (7.46)$$

where $S_i(q,\omega)$, also called incoherent scattering law, is the spatial and temporal Fourier transform of the correlation function $G_s(r,t)$ of a same nucleus at the two times 0 and t, and therefore gives information on the single-particle dynamics. On the contrary, the cross-section of deuterium is mainly coherent, so that experiments on deuterated samples are suitable for determining the structure of the hydrides or collective dynamical properties. Note that the terms "coherent" and "incoherent" when referred to the scattering of neutrons have a different meaning from that referred to tunneling; in the first case, they distinguish between the interference of neutrons scattered in the same manner from all the atoms as a whole and the scattering from the uncorrelated single atoms.

The two extreme cases of interest here are: (i) the H atom jumps among different sites without correlation between successive jumps with mean hopping rate ν (incoherent hopping or tunneling); (ii) the H atom forms a coherent tunnel system with levels separated in energy by E.

Inelastic Scattering from Coherent TS's. The incoherent inelastic scattering law for a TLS is essentially constituted by two peaks at $\hbar\omega = \pm E$, corresponding to neutrons that stimulate a transition of the TLS to the other level and absorb or loose the energy E, according to the energy conservation law. The peaks are broadened by the finite lifetime of the levels, and, as also shown by Grabert et al. (1986), a good approximation is to assume the response function of a harmonic oscillator resonating at E and with damping $\Gamma = \hbar\nu$; the near equivalence between the dynamic response of a coherently tunneling particle and a damped harmonic oscillator is the reason why much of the theoretical literature concerned with TLS's uses the term "damping" instead of relaxation rate. The expression adopted in the analyses of the data is

$$S_i(q,\omega) \propto \int_{-\infty}^{\infty} d\Delta \frac{P(\Delta)}{\pi} \frac{\nu(E/\hbar)^2}{\left[\omega^2 - (E/\hbar)^2\right]^2 + (\nu\omega)^2}$$

$$\times \left(\frac{\Delta_0}{E}\right)^2 \frac{2}{1+e^{\hbar\omega/k_B T}}. \tag{7.47}$$

The harmonic oscillator function reduces to two delta functions at $\hbar\omega = \pm E$, for $\nu = 0$; the last factor ensures that the detailed balance condition,[3] $S(q,-\omega) = S(q,\omega)e^{-\hbar\omega/k_B T}$, is satisfied. As already noted for the diaelastic contribution to the moduli, (7.26), the factor $(\Delta_0/E)^2$ weighs the symmetric TLS's more than the asymmetric ones, and the distribution $P(\Delta)$ centered at $\Delta = 0$ causes a singularity in $P(E)$ at $E = \Delta_0$ [see (7.80) below], so that the low-energy transfer sides of the inelastic peaks provide a good estimate of the tunneling energy (Magerl et al. (1986b)).

Expressions for the scattering from a TLS both in the coherent and incoherent regimes have been calculated by Grabert et al. (1986), including the nonadiabatic interaction of the electrons with the TLS.

Quasi-elastic Scattering from Incoherent Hopping (QNS). The above result (7.47) holds in the coherent tunneling regime, that means that the H wave function remains undisturbed for at least a few tunneling periods, and corresponds to the picture in which the H atom oscillates between the two sites with frequency $\Delta_0/\hbar$. With increasing temperature, this (coherent) motion is so frequently disturbed by the phonon bath and electron scattering that it becomes stochastic hopping between the two sites, with a mean hopping time comparable to or lower than the tunneling period. The scattering law (7.44) is the Fourier transform of the self-correlation function $G_s(\vec{r},t)$, that is also the probability of finding the hopping atom at $\vec{r}$ at time t if it was at $\vec{r} = 0$ at $t = 0$; this probability obeys a rate equation describing the jumps with frequency ν to and from the neighboring sites at $\vec{r} + \vec{l}$. The Fourier transform of the solution of the rate equation is a Lorentzian or a sum of Lorentzians (see e.g. Bée (1988)),

$$S_i(\vec{q},\omega) = \frac{1}{\pi} \frac{\nu f(\vec{q})}{\omega^2 + \left[\nu f(\vec{q})\right]^2}, \tag{7.48}$$

whose width is given by the hopping rate ν multiplied by the structure factor

$$f(\vec{q}) = \sum_{\vec{l}} \left[1 - e^{-i\vec{q}\cdot\vec{l}}\right].$$

---

[3] Essentially, if there are two states that differ in energy by $\hbar\omega$, the higher-energy state is less populated than the lower-energy one according to the Boltzmann factor; the probability that the neutron acquires the energy $\hbar\omega$ from the upper level, $S(q,-\omega)$, and the probability for the opposite transition $S(q,\omega)$, are in the same ratio as the populations of the initial states.

The peak given by (7.48) is called quasi-elastic because it is centered at $\omega = 0$, and it must be distinguished from the intense elastic peak (the neutrons that are not scattered by the sample). The expression adopted by Steinbinder et al. (1991) for interpreting the data of H hopping within a two-site TS is

$$S_i(q,\omega) \propto \int_{-\infty}^{\infty} d\Delta \frac{P(\Delta)}{\pi} \frac{2\nu}{\omega^2 + 4\nu^2} \frac{1}{\cosh^2(E/2k_BT)} \frac{2}{1+e^{\hbar\omega/k_BT}}, \quad (7.49)$$

where $2\nu = \nu_{21} + \nu_{12}$ corresponds to the relaxation rate of the preceding paragraphs, an integration over the distribution $P(\Delta)$ of asymmetries is performed, it is taken into account that hopping occurs between sites that differ in energy by E, and the last factor ensures detailed balance.

Inelastic Scattering from the Local H Vibrations. At higher values of $\hbar\omega$, the $S_i(q,\omega)$ has peaks due to the transitions of H between the localized vibration modes within the same site. These modes are only as a first approximation those of a harmonic quantum oscillator with energy spacing of the order of 100–150 meV, and their actual spacing, broadening and structure give information on the potential felt by the H atom, the dynamical coupling with the lattice and tunnel splittings of the excited states.

Figure 7.7 represents schematically the main contributions to the low-temperature inelastic neutron spectrum from a sample containing interstitial H that forms a TLS, as described above.

7.3.4 Nuclear Magnetic Resonance

Nuclear magnetic resonance (NMR) is a flexible technique that can be used to extract several types of microscopic information, from the local environment of certain atoms to the hopping rate of diffusing atoms. The principles of the technique can be found in Abragam (1961) and its application to the study of the metal-H systems is discussed by Cotts (1978) and Messer et al. (1986). Here, only a brief description is provided, in order to discuss the small number of papers devoted to the local fast motion of H in metals.

What is generally measured is the spin-lattice relaxation rate T_1^{-1} or Γ_1 of the proton, i.e. the relaxation rate of the component of the proton spin parallel to the applied magnetic field (sometimes called ''longitudinal" for this reason). The relaxation of the spin whose resonance is measured is due to the interaction of that spin with the local magnetic field fluctuations from the spins of the conduction electrons (Korringa relaxation, T_{1e}^{-1} proportional to temperature) of the nuclei of the lattice or of the other nuclei of the same type (spin-spin relaxation). The case of interest is the spin-lattice relaxation rate of the protons, when the fluctuating field is caused by the hopping of the H atoms in the field created by the other spins. If the magnetic moments of the host nuclei are not much larger then those of the proton (as for ^{45}Sc) and the H concentration is sufficiently high, then the H–H interactions are also important.

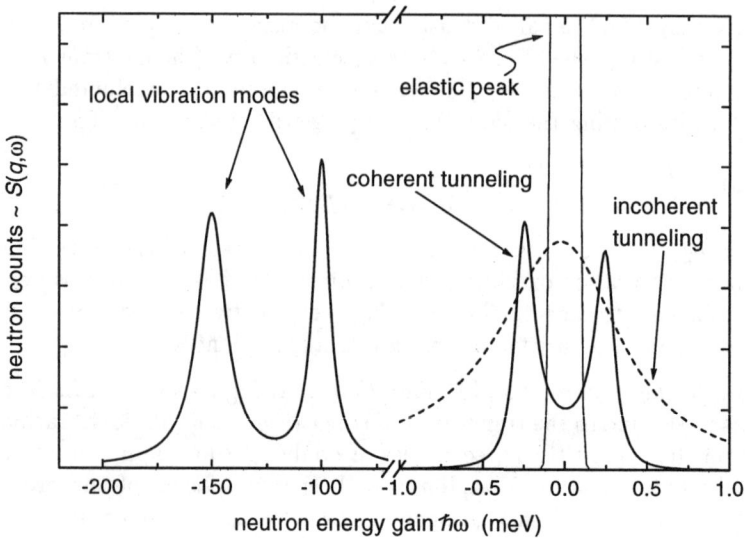

Fig. 7.7. Schematic diagram of the main components of the low temperature neutron spectrum from a sample containing interstitial H that forms TS's. The energy scale is roughly valid for the case of NbO$_x$H$_y$, whereas the ordinates scale is arbitrary. Increasing temperature, the double inelastic peak due to coherent tunneling transforms into the quasi-elastic peak due to incoherent hopping between the same sites.

The relaxation rate T_1^{-1} can be expressed as a sum of power spectra of randomly varying dipolar fields, $J^{(q)}(\omega)$ roughly corresponding to the $S(q,\omega)$ introduced above for the neutron scattering cross-section that are the Fourier transforms of the correlation functions $g(t)$ of the spins that participate in the relaxation, analogous to the $G(r,t)$ (see Chap. VIII.C of Abragam (1961)). The diffusive motion is generally modelled with exponential decays of the dipolar correlation functions, $g(t) \propto \exp(-t/\tau_c)$, where the *correlation time* τ_c corresponds to the hopping time of the H atom, and the resulting spectral functions are of the type

$$J^{(q)}(\omega) = g^{(q)}(0) \frac{2\tau_c}{1 + (\omega\tau_c)^2},$$

that can be compared with the quasi-elastic peak width (7.48) with $\nu = 1/\tau_c$. In the simplest case, the spin-lattice relaxation rate is then

$$1/T_1 \propto \frac{2\tau_c}{1 + (\omega\tau_c)^2}. \tag{7.50}$$

The measuring frequency is that of the resonance between the two energy levels of the proton with spin parallel or antiparallel to the applied magnetic field H, $\omega = \gamma H$, where γ is the gyromagnetic ratio of the proton, and can be changed by varying H. Note that the dependence of T_1^{-1} on the correlation time is identical with that of the elastic energy loss due to anelastic relaxation, (7.41), that is maximum at $\omega\tau = 1$. Therefore, as in the acoustic case, the contribution to $T_1^{-1}(T)$ from the hopping of H with a mean hopping time $\tau_c(T)$ is a peak at the temperature where $\omega\tau_c = 1$; this temperature increases with the measuring frequency. The intensity is proportional to the strength of the interaction between the spin of the proton and the spins of its environment.

7.4 Long-Range Diffusion and Incoherent Hopping of Hydrogen in bcc Metals

7.4.1 Theories of Quantum Diffusion

The literature on the quantum diffusion of light particles in solids started to grow at least from the late sixties and is now vast, especially on the theoretical side; it is also partially overlapped with that on the TLS in amorphous solids. We will not present here a review, but only mention the works that have been adopted for interpreting the data on H tunneling presented below. The theory of quantum diffusion has been treated, for example, by Kehr (1978), Kagan (1992), Fukai (1993), Grabert and Schober (1997).

The Arrhenius law for the atomic diffusivity can be derived within the classical rate theory (e.g. Vineyard (1957)) that calculates the flux of mobile particles through the saddle-point configuration in a phase space containing all the atomic coordinates relevant to the particle diffusion (including the neighboring lattice atoms). The well-known result for the hopping rate ν is

$$\nu = \nu_0 \exp\left(-E_a/k_B T\right), \tag{7.51}$$

where the exponential essentially comes from the probability that the particle and environment are in the saddle-point configuration with an energy higher by E_a than that with the particle in the initial site. The activation energy E_a is considered to be independent of the mass of the particle. The so-called attempt frequency ν_0 is of the order of the vibration frequency of the particle within the site and should be inversely proportional to the square root of the particle mass.

In most metals the diffusion coefficient of H depends on temperature according to the Arrhenius law, but the values of E_a and ν_0 do not depend on the isotope mass as expected. Moreover, in Nb, Ta and V the activation energy is comparable or even lower than the vibrational energy of H (Alefeld and Völkl (1978)), indicating that the classical picture is inadequate. Various corrections to (7.51) have been proposed, like taking into account the ground state vibration energy of the H atom in the site and at the saddle point. The next step is to consider the motion of the light particle from a quantum-mechanical point of view. Initially, only the interaction with the phonon bath was considered relevant for describing the dynamics of a mobile particle in a solid. A simple treatment is to not consider explicitly the relaxation of the lattice around the mobile atom (self-trapping), and to suppose that it feels a static double-well potential plus a time-dependent perturbation due to the thermal phonons. The perturbation is the cause of the transitions of the mobile atom between different states, whose rate is calculated in first-order perturbation theory and contains the square of the matrix element of the perturbation between the initial and final states of the particle (Δ_0 in the TLS literature). This approach has been discussed, for example, by Sussman (1967) and Jäckle (1972), and is that which yields the rate for one-phonon transitions, (7.43), and nearly power-law dependences like T^7 for two-phonon transitions (see also Doussineau and Robin (1980)), depending on the details of the potential and matrix elements. A formalism of this type has also been recently adopted by Svare (Leisure et al. (1993a), Svare et al. (1991)) to describe the relaxation rates of H tunnel systems in *hcp* rare earth metals.

Other theories, like those of Flynn and Stoneham (1970) and Kagan and Klinger (1976), take explicitly into account the self-trapping or polaron effect due to the relaxation or "polarization" of the lattice atoms around the diffusing one. In the high-temperature limit that is generally identified with $T > \theta_D$, almost all theories predict an Arrhenius-like temperature dependence of the hopping rate, where, however, the activation energy has a different meaning with respect to the classical rate theory, and the pre-exponential factor contains the square of some matrix element between ground or excited states of the particle in the two sites. At lower temperatures, a transition to a power law, typical of transitions involving few phonons, is found. Therefore, all these theories have in common a change of the slope of the logarithm of the hopping rate versus the reciprocal of temperature around θ_D. We will cite the expression derived by Flynn and Stoneham (1970) and put in a computable form by Stoneham (1972) because it has often been used to interpolate the temperature dependence of the long-range diffusion rate and reorientational hopping rate around impurities of H in Nb and Ta. Recently, it has also been used for the reorientation rate of H around B in Si. The expression proposed by Stoneham (1972) is

$$\tau^{-1} = \frac{|J|^2}{\hbar k_B \theta_D} e^{-f(\theta_D/(2T))} \int_{-\infty}^{+\infty} dt \left[e^{g(\theta_D/(2T),t)} - 1 \right], \quad (7.52)$$

$$g(a,t) = \frac{20 E_a}{k_B \theta_D} \int_0^{+\infty} dx \, x^3 \cosh(ax) \cos(xt),$$

$$f(a) = \frac{20 E_a}{k_B \theta_D} \int_0^1 dx \, x^3 \coth(ax),$$

where θ_D is the Debye temperature, E_a is an activation energy taking into account an antisymmetric local lattice mode that overcomes the self-trapping in the initial site, and J is a tunneling matrix element between the wave functions of the atom localized in the initial or in the final site. In the fits to the experimental H hopping rates, J has values much larger than the tunneling matrix elements Δ_0 found for various H tunnel systems; this large value should be due to the intervention of a symmetric local mode that lowers the barrier and promotes tunneling. As pointed out also by Flynn and Stoneham (1970), the influence of these symmetric lattice modes is expected to be important in the *fcc* lattices, where the trajectory of a jump between two octahedral sites passes through two close lattice atoms. In the case of hopping between T sites in a *bcc* lattice, however, there are no atoms that obstruct the jump.

More complete calculations on the same line, using empirical potentials similar to that adopted by Sugimoto and Fukai (1980) for the H–Nb interaction, have been made by Schober and Stoneham (1988) for the diffusion rate of H and D in Nb, and by Klamt and Teichler (1986) for H and D in Nb and Ta.

The interaction between a tunneling particle and the conduction electrons has been studied intensely, after Golding et al. (1978) showed that it is the dominant interaction in metallic glasses, like in PdSiCu; however, see the discussion in Sects. 4.4.2 and 4.4.3. Below, we will show several experimental evidence that this is true also for fast tunneling of H in Nb and Ta; this type of interaction, however, has not been considered when discussing the slower diffusional or reorientational motion of H.

7.4.2 The Gorsky Effect: Long-Range Diffusion

The paraelastic relaxation described in Sect. 7.3.2 is due to the relaxation between different orientations α and β of point defects whose symmetry is lower than the crystal symmetry, and is therefore proportional to the square of the so-called *shape factor* of the distortion due to the defect, $(\lambda^\alpha - \lambda^\beta)^2$. There is another type of relaxation, known as Gorsky relaxation (Gorsky (1935)), that arises when all point defects representing a center of dilatation, whatever their symmetry, interact with a dilatational stress like that produced by bending a crystal. When a rectangular plate is subjected to bending, the deformation generates a dilatation stress gradient across its thickness. Consequently, the defects migrate from the less energetically favorable compressed region to the dilated region, giving rise to a concentration gradient that causes the additional time-dependent anelastic strain of the crystal. For a diluted concentration c, the relative change of strain or relaxation strength Δ_G, is

$$\Delta_G = c v_0 M_u (\text{tr}\lambda)^2 / (9kT) = \Theta/T , \tag{7.53}$$

where v_0 is the atomic volume, M_u is the unrelaxed modulus, $\text{tr}\lambda = \lambda_1 + \lambda_2 + \lambda_3$ is the *size factor* of the defect. The process is characterized within 1% by a single relaxation time given by, for the case of a rectangular plate of thickness h,

$$\tau_G = \left(\frac{h}{\pi}\right)^2 D(T) , \tag{7.54}$$

where D is the long-range diffusion coefficient of the defect. This equation allows the diffusion coefficient to be determined by the measurement of the relaxation time without any model-dependent factor for the conversion of the relaxation time into the diffusion coefficient D.

In principle, any type of point defect could give rise to the Gorsky relaxation, but in practice, this process involves long-range diffusion over the macroscopic distance h, so that the effect can be observed only in the case of very fast diffusing interstitial H, otherwise, the relaxation time becomes dramatically long. The effect predicted by Gorsky was first experimentally observed by Schaumann et al. (1970) with the quasi-static technique (elastic after-effect) and by Cantelli et al. (1969) with dynamic measurements. In the static technique, the sample is bent and then the exponential response of the anelastic strain, characterized by the single relaxation time τ_G, is recorded isothermally as a function of time. With particular geometries like wires wound into coiled springs (Schaumann et al. (1970)), some difficulties may arise from the impossibility of annealing the introduced dislocations. The dynamic technique consists in measuring the elastic energy loss or internal friction Q^{-1} as a function of temperature, at constant angular vibration frequency ω, that is described by a Debye curve

$$Q^{-1} = \Delta_G \frac{\omega \tau_G}{1 + (\omega \tau_G)^2} . \tag{7.55}$$

By combining (7.53)–(7.55), the diffusion coefficient can be obtained by the following relation containing all measurable quantities:

$$\ln D(T) = \ln(\omega h^2/\pi^2) - \cosh^{-1}(\Theta/TQ^{-1}) .$$

The dynamical method allows the determination of the diffusion coefficient through one experiment, over a wide temperature range (from 150 to 700 K), where the relaxation curve extends.

A large amount of experimental data on the diffusion coefficients D_H of H and D_D of D have been collected by the Gorsky technique in fcc and bcc metals (Schober and Wenzl (1978)), in hcp metals (Vajda (1995)), in substitutional alloys (Cannelli and Cantelli (1977)), and in metallic glasses (Berry and Pritchet (1981)). For low H concentrations, these data have an excellent consistency with the results from other surface-independent methods (QNS, resistivity relaxation, Mössbauer effect) and indicate a diffusivity of hydrogen higher in the bcc structure than in the fcc one (Schober and Wenzl (1978)). This behavior can be explained with the smaller distance between interstitial sites (tetragonal or octahedral) of the bcc lattice.

Contrary to what was observed for the localized mobility of H, confined around trapping centers or in pairs of sites in hcp metals, the long-range diffusion coefficient shows a less marked deviation from the classical behavior in most of the investigated systems. This may also be a consequence of the relatively high temperatures at which the experimental data refer, where marked quantum effects are not expected. The most relevant findings, also

7.4 Long-Range Diffusion and Incoherent Hopping of Hydrogen in bcc Metals

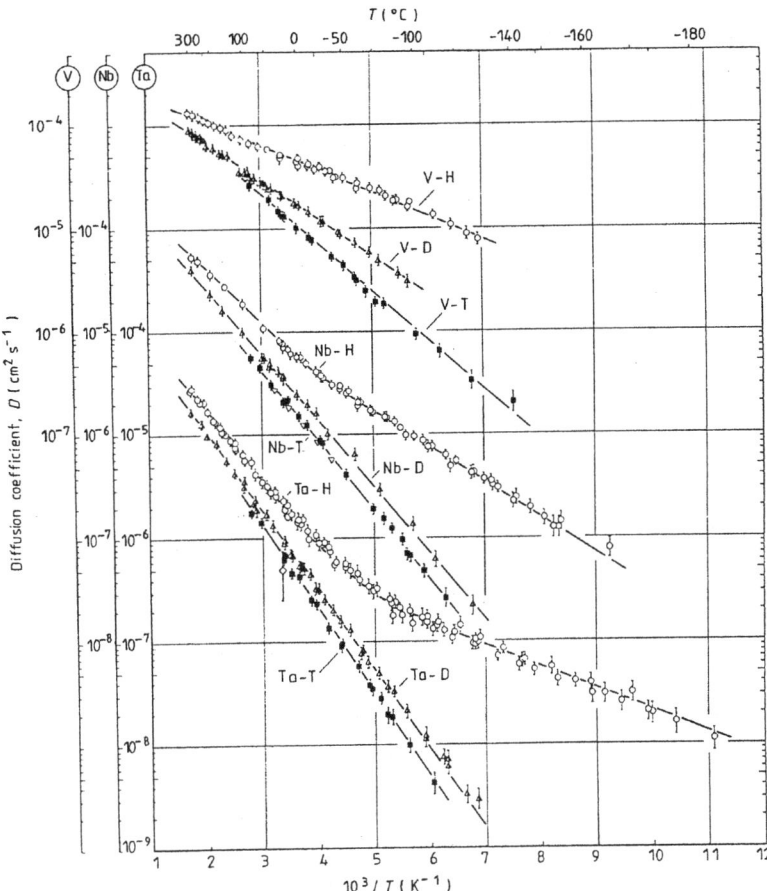

Fig. 7.8. Diffusion coefficient of H, D and T in V, Nb and Ta from Gorsky effect experiments, from Qi et al. (1983).

visible in Fig. 7.8 (Qi et al. (1983)) for the case of Ta, are: the isotope effects in bcc Nb, Ta, V (the ratio D_H/D_D can reach values as high as 20); a significant deviation (below 220 K) of the H mobility from the classical behavior reported by the Alefeld group in very pure Nb (Schaumann et al. (1970)) and Ta (Kokkinidis (1977)), and observed to a less extent by Cantelli et al. (1977) in UHV annealed Ta.

The change of the slope of $\ln D$ vs T^{-1} would be due to the crossover to the transitions assisted by few phonons. The existence of this intrinsic change in the temperature dependence of the diffusion coefficient has been questioned in terms of the influence of the hydride precipitation (Westlake (1972)), and also because it was not reproduced in Nb samples containing about 300 at ppm O (Matusiewicz et al. (1974)). Further confirmation of the deviation has been given by the Alefeld group by electrical resistivity relaxation (Wipf and Alefeld (1974)) and by QNS measurements (Richter et al. (1977)). Alefeld and co-workers emphasized that the deviation from the high-temperature Arrhenius-like dependence can be observed only in very pure materials where trapping effects are negligible; in their Nb samples, the total amount of interstitial trapping impurities was below 20 at ppm.

The data showing the change of the slope of $\ln D$ vs T have been interpreted in terms of models of polaronlike hopping of H and we cite, among others, the works by Schober and Stoneham (1988) and Lagos and Cerón (1988).

In the α^*-phase of R–H(D) (R =Sc,Y, Lu), the experimental data of the Gorsky effect indicate that the long-range diffusivity of H follows the Arrhenius law (Völkl et al. (1987)). In these systems, the hopping rate deduced from the Gorsky relaxation coincides with that of the relaxation process observed around 300 K (Vajda et al. (1991)), and attributed to stress-induced ordering of the H pairs in T sites bridging the R atom. Thus, local ordering and long range diffusion involve the same jump of hydrogen.

7.4.3 Hopping of Hydrogen near Interstitial Impurities

A clear evidence that the mobility of interstitial hydrogen could not be described by the classical rate theory dates from the sixties, with the first anelastic relaxation experiments conducted in Ta and Nb doped with hydrogen and deuterium (Cannelli and Verdini (1966a) and (1966b)). The hopping rates deduced from the peak temperatures of the relaxation processes at different frequencies (around 100 K in the kHz range, the samples being resonating discs and bars) did not display any appreciable deviation from the classical Arrhenius law, due to the small temperature range explored, but it appeared immediately surprising that the ratio of the pre-exponential factors ν_0 in (7.51) of the two isotopes is $\nu_0^H/\nu_0^D \sim 80$ instead of $(m_D/m_H)^{1/2} = \sqrt{2}$ as predicted by the classical theory.

At that time, it was believed (Nowick and Berry (1972)) that those relaxations were the analogous of the well-known O, N, C Snoek peaks observable above room temperature in bcc metals, i.e. hopping of H in the α-phase among the T sites, with the associated reorientation of the elastic dipole. The large isotope effect and the fast relaxation rate of hydrogen also stimulated the first experiments on the Gorsky relaxation (see Sect. 7.4.2) that however yielded different values of the H hopping rates. The discrepancy was explained by Alefeld (1970) who hypothesized that the low-temperature relaxation peaks are due to hopping of H around heavy impurities like O and N, as confirmed by numerous anelastic experiments in $NbO_x(H/D)_y$ (Schiller and Schneiders (1975), Baker and Birnbaum (1973), Zapp and Birnbaum (1980)).[4] Those

---

[4] The exact geometry of these jumps has not been ascertained, but the jumps should occur between tunnel systems like that represented in Fig. 7.10.

7.4 Long-Range Diffusion and Incoherent Hopping of Hydrogen in bcc Metals

experiments were carried out at frequencies included from 10^{-3} Hz (anelastic after effect) up to 200 MHz, allowing us to determine the hopping rate τ^{-1} of H around O and N in a wide temperature range (35–250 K), and revealing a significant deviation from the classical Arrhenius law.

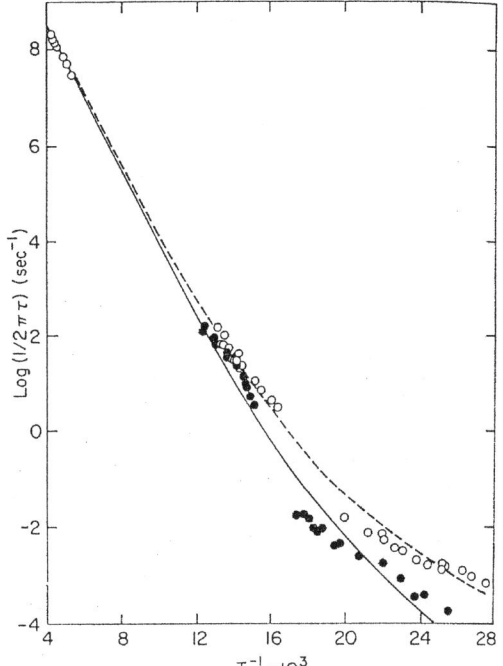

Fig. 7.9. Arrhenius plot of the hopping frequencies of H around O (open symbols) and N (closed symbols) in Nb. The lines are fits with the Flynn–Stoneham theory (from Zapp and Birnbaum (1980)).

Figure 7.9 shows the $\tau^{-1}(T)$ curves fitted by Chen and Birnbaum (1976) with the model of Flynn and Stoneham (1970) for polaronlike hopping, (7.52), with $E \simeq 0.2$ eV and $J = 0.24$ eV (0.11 eV) for the N(O)-H pair. Note that J is three orders of magnitude larger than Δ_0 of the TLS's among which H jumps, that makes the application of (7.52) doubtful, even when taking into account a lowering of the barrier due to phonon fluctuations. In such cases, a more complex approach is required, as in Schober and Stoneham (1988) or Klamt and Teichler (1986).

Further confirmation of the large isotope effect of hydrogen was then given by Hanada (1981) who reported also the relaxation process caused by tritium trapped by interstitial O(N) in Nb. The relaxation due to the H reorientation around interstitial impurities has been found also for V (Cannelli et al. (1983), Cannelli and Mazzolai (1973)), and recently also in the hcp Y (Cannelli et al. (1991b)); since the available data span a restricted region of temperature, no crossover to a faster relaxation has been found.

7.4.4 Hopping of Hydrogen near Substitutional Impurities

In bcc transition metals, substitutional impurities may trap H within the 24 nearest neighboring T sites of the cubic cell containing the substitutional atom [consider Fig. 7.2 with the substitutional (S) atom in the center of the cell]. In spite of the simpler geometry (all 24 T sites around the S atom are equivalent), the anelastic relaxation spectrum of samples with S-H pairs is much more complex than that with I-H pairs. This is also due to the possibility of reaching much higher concentrations of S-H pairs than I-H pairs: the solubility of many metal atoms in Nb, Ta and V is complete, whereas in order to obtain concentrations of O in solid solution higher than ~ 1 at%, the sample must be quenched from high temperatures, and it is likely that part of O is precipitated.

The H dynamics within S-H complexes has been mostly studied in Nb and V with substitutional Ti or Zr (see e.g. Cannelli et al. (1994b), and references therein). A relaxation peak similar to that due to the hopping of H near O is found in all these diluted alloys (with less than 5 at% substituted atoms), and analogously attributed to the slow thermally activated reorientation of H around the S atom. The experimental data are limited to the acoustic frequencies, and it is not possible to reveal deviations from the Arrhenius law in the correspondingly restricted range of temperature. Instead, marked effects on the apparent activation energy are observed by increasing the concentration of the S atoms above 0.5 at%, that have been explained in terms of distributions of the interstitial site energies caused by the S atoms (Cannelli et al. (1985)). Additional peaks are also observed between that of the S-I reorientation and the low-temperature tunneling processes. The geometry of the S-I pair has been investigated by acoustic measurements on single crystals by Cannelli et al. (1994b) and Yoshinari et al. (1996).

7.5 Coherent Tunneling and Fast Local Motion of Hydrogen

7.5.1 Hydrogen Trapped by Interstitial O,N and C in Nb and Ta: A Two-Level System

The first evidence of coherent tunneling states of interstitial H in metals were found in Nb containing both H and heavier interstitial impurities, O, N and C, that trap H. Much work has been done on this system with specific heat, ultrasound and lower-frequency acoustic experiments, and neutron spectroscopy. At present, there is a large body of evidence that at liquid He temperatures, H forms a TLS in a pair of equivalent sites, presumably tetrahedral, near the trapping impurity. Since the vibrational energies of H trapped by O and N in Nb are very close to those of untrapped H, Magerl et al. (1983) conclude that H occupies sites not too close to the trapping impurity, like the third-nearest neighbor T sites. The configuration generally accepted for the O-H pair is shown in Fig. 7.10. The sites are equivalent

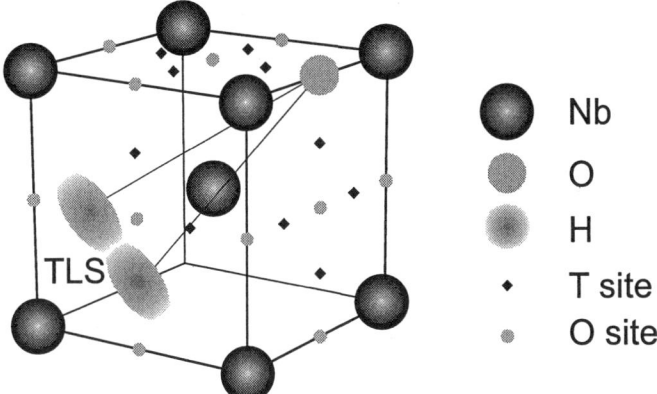

Fig. 7.10. Nb unit cell with the proposed configuration of a H-TLS near an O atom.

with respect to the trap, so that the TLS is in principle symmetric, but the elastic interactions among the complexes induce asymmetries in the TLS's that may exceed the tunneling energies already at concentrations as low as 1000 at ppm. The tunneling energy, on the contrary, is relatively insensitive to the random strains, as demonstrated by specific heat (Wipf and Neumaier (1984)) and neutron spectroscopy (Magerl et al. (1986b)) on samples with concentrations of impurity-H pairs ranging from 150 to 14000 at ppm. The strong influence of the long-range elastic interactions on the H site energies (TLS asymmetries) also implies that coherent tunneling is confined within only two trap sites. In fact, any additional site would be nonequivalent with respect to the O atom, and would therefore differ in energy with respect to the first pair by an energy that is about two orders of magnitude larger than the measured tunneling energy; the formation of a coherent state equally delocalized over such different sites would be impossible. On the other hand, the delocalization over a ring of four T sites or other symmetric configurations is possible and indeed observed when H is trapped by a substitutional atom, thanks to the higher symmetry.

Table 7.1 summarizes the values of the parameters characterizing the TLS associated with the O-H pair in Nb, found by various techniques. There is reasonable agreement among all the data, indicating that the tunneling model provides a good description of the H-TS's.

Specific Heat. Specific-heat measurements have provided the first evidence of atomic tunneling both in the amorphous solids and for interstitial H in metals. Sellers et al. (1974) first observed excess specific heat below few Kelvin

Table 7.1. Values of the parameters describing the TLS associated with H(D) trapped by O(N,C) in Nb and Ta, deduced from various techniques: INS = Inelastic neutron scattering, QNS = quasi-elastic neutron scattering, UA = ultrasound attenuation, SV = sound velocity, SH = specific heat. $\overline{\Delta}$ is the width of a Lorentzian distribution of the asymmetries, $P(\Delta) = \pi\overline{\Delta}/(\overline{\Delta}^2 + \Delta^2)$. $\gamma(K)$ is the coupling constant of TS's and phonons (electrons).

complex	$c_{H/D}$ (at ppm)	$c_{O/N}$ (at ppm)	Δ_0 (K)	$\overline{\Delta}$ (K)	γ (K)	K
O-H[a]	800	1000	2.2	20		
O-H[a]	2400	3000	2.2	60		
O-D[a]	500	800	0.24	5		
O-D[a]	2600	5000	0.24	30		
O-D[a]	1400	19000	0.24	100		
O-H[b]	150	150	1.97±0.06[m]	0.7		0.04
O-H[b]	2200	2200	2.7±0.17[m]	–[n]		0.04
O-H[b]	11000	11000	3.1±1.2[m]	–[n]		0.04
O-H[c]	200	200	2.6	–		0.05–0.06
O-H[d]	2000	2000	2.6	23		
C-H[d]	200	200	1.9	2.3		
O-H[e]	200	200	2.6	1.6		
N-H[e]	500	500	1.91	2.3		0.055
N-H[e]	4000	4000	2.00	25		0.055
N-H[e]	4000	4000	2.04	35		0.055
O-H[f]	2000	2000	2.4[o]	2.3		0.055[o]
O-H[g]	17000	700	1.5		150	
O-H[h]			1.5		230–750	
N-H[i]	3000	1500	1.4	3	270	0.07
N-D[i]	2400	1500	0.18	3	370	0.07
O-H[j]	700	100	1.7	0.42	590	
O-D in Ta[k]	2500	800	1.0±0.5	15±5	≥900	$(1-3)\times 10^{-4}$
H in Sc[l]	160000		3.9	<50		0.039

Technique and Reference
[a] SH, Wipf and Neumaier (1984).
[b] INS, Magerl et al. (1986b).
[c] INS, Wipf et al. (1987).
[d] INS, Neumaier et al. (1989).
[e] INS+QNS, Steinbinder et al. (1991).
[f] QNS, Steinbinder et al. (1988).
[g] SV, Bellessa (1983).
[h] UA+SV, Bellessa (1985) based on the data of Poker et al. (1984).
[i] UA, Morr et al. (1989).
[j] UA+SV, Drescher-Krasicka and Granato (1985).
[k] UA, Cannelli et al. (1986).
[l] QNS, Anderson et al. (1990).
[m] Width of a Lorentzian distribution of Δ_0, $P(\Delta_0) = \pi^{-1}\overline{\Delta_0}/\left(\overline{\Delta_0}^2 + \Delta_0^2\right)$.
[n] $\overline{\Delta}$ is large, but the fit is insensitive to its exact value.
[o] Taken from the low-temperature INS data and used in the QNS fit.

due to interstitial H in Nb; moreover, the contribution to the $C(T)$ curves shifted to lower temperature when D was added instead of H. This isotope effect demonstrated the presence of tunnel split states of H, whose energy scale was reduced by increasing the isotope mass. The data were initially interpreted by Birnbaum and Flynn (1976) in terms of polaronlike self-trapping of interstitial H in a multisite tunnel system, but later Morkel et al. (1978) made experiments on pure Nb samples and N doped samples and showed that the tunneling states were observed only for H trapped by the N impurity.

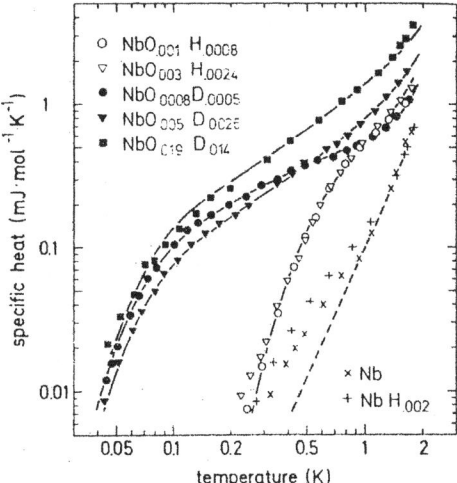

Fig. 7.11. Specific heat of $NbO_x(H/D)_y$ (from Wipf and Neumaier (1984)). The broken line is the calculated specific heat of pure Nb; the full lines are fits with the tunneling model.

Additional measurements on $NbO_x(H/D)_y$ were made by Wipf and Neumaier (1984), who were finally able to quantitatively explain the data in terms of TLS's. The $C(T)$ curves of Nb, $NbH_{0.002}$, NbO_xD_y and NbO_xD_y are shown in Fig. 7.11. The O-free sample, both before and after charging with H, has a specific heat little above the expected contribution due to phonons and electrons in the superconducting state. This fact is also in accordance with the negligibly small solubility of H at those temperatures, and demonstrates that H in the β-phase precipitates does not form tunneling states. Instead, the O-H and O-D pairs cause broad peaks in the $C(T)$ curves, with a low temperature cutoff depending on the isotope but not on the concentration of the complexes. The data can be fitted assuming that H forms a TLS, using (7.2). An additional integration is required over a distribution of the asymmetries Δ, that has been assumed to be Lorentzian with width $\overline{\Delta}$, but still a large contribution comes from TLS's with energy difference $E = \sqrt{\Delta_0^2 + \Delta^2}$ just above Δ_0; as discussed in more detail in Sect. 7.5.2, the large number of TLS's with $E \simeq \Delta_0$ produces a relatively sharp cutoff at low temperatures. The values of the tunneling energy deduced from the fits are $\Delta_0 = 2.2$ K for H and $\Delta_0 = 0.24$ K for D, with no apparent influence from the impurity concentration. On the contrary, the width of the distribution of asymmetries increases strongly with c_O and c_N, as shown in Table 7.1.

Acoustic Measurements. After the discovery of TS's due to the O-H pair in Nb by specific-heat measurements, Poker et al. (1984) found a peak in the ultrasound attenuation (10-200 MHz) in the same system around 2 K, that we label P1. That peak was recognized as due to the anelastic relaxation between the levels of the TS, according to the mechanism of (7.42). The sound velocity, below the temperature of the softening accompanying the relaxation peak, exhibited an additional decrease that was interpreted as the diaelastic or resonant effect due to the finite curvature of the energy levels vs strain [(7.20) and (7.26)].

The measurements of Poker et al. (1984) contained also some features that have been later confirmed but are not yet fully explained. The experiments were made on single crystals with different polarizations of the sound wave, and showed relaxation and dispersion of the $C' = (C_{11} - C_{12})/2$ elastic constant but not of C_{44}. This fact was at first explained in terms of a multilevel tunneling system, with H delocalized over a ring of 4 T and 4 triangular sites of a face of the unit cell (the triangular sites are halfway between the T ones); in that case, thanks to the high symmetry of the wave function, no relaxation is expected for the shear of the C_{44} type. On the contrary, there are no symmetry reasons for not observing relaxation of the C_{44} mode for a TLS; the only reason for a reduced influence of the ϵ_{44} shear on the asymmetry of the TLS is a relatively large distance between the H and the O atom. This will be discussed in more detail in Sect. 7.5.2.

The second observation that still remains largely unexplained is an additional peak around 6 K, which we label P2, whose intensity seemed to increase faster than linearly with the H concentration. First we will consider P1, that is identifiable as due to the TLS of the O-H pair also observed by specific heat and neutron spectroscopy experiments; Sect. 7.5.1 will be devoted to P2.

After the observation by Morkel et al. (1978) that H could hardly be delocalized over more than two equivalent sites near an O atom, the low-temperature experiments on the O-H pair have been interpreted in terms of the standard tunneling model. With acoustic experiments, it has been possible to determine all the parameters that define a TLS and its interaction with the environment, i.e. Δ_0, Δ, and the coupling constants to phonons γ and to electrons K. The tunneling energy can be estimated from the relative variation of the elastic constants or sound velocities, $\Delta v/v = \frac{1}{2}\Delta C/C$, as a function of temperature through (7.26). As explained above, the diaelastic response is little affected by the asymmetric TLS's, so that a good estimate of Δ_0 is possible. This has been first done by Bellessa (1983), who measured the sound velocity of $NbO_{0.007}H_{0.017}$ at 200 MHz down to 0.1 K, finding $\Delta_0 = 1.5$ K; from the amplitude of the step in $\Delta v/v$ the coupling

7.5 Coherent Tunneling and Fast Local Motion of Hydrogen

to strain was estimated to be $\gamma \geq 150$ K. Drescher-Krasicka and Granato (1985) repeated the ultrasonic absorption and sound-velocity measurements in Nb with a very small concentration of O-H pairs, in order to minimize the strain broadening, and deduced $\Delta_0 = 1.7$ K and $\gamma = 590$ K. Finally, Morr et al. (1989) found $\Delta_0 = 1.4 \pm 0.1$ K and $\gamma = 270$ K for the N-H pair in Nb and $\Delta_0 = 0.18 \pm 0.01$ K and $\gamma = 370$ K for the N-D pair. There is a remarkable agreement between all these values obtained from samples with various H and impurity concentrations and prepared under different conditions, and this confirms the reliability and consistency of the interpretation of the data. Some discrepancies appear in the estimates of the coupling to strain γ, but this is due to the fact that this parameter is extracted from the amplitude of the diaelastic effect (mainly asymmetric TLS's with local strain $\epsilon \to 0$); this in turn also depends on the TLS concentration, which is difficult to determine especially when part of H precipitates, and on the distribution of asymmetries $P(\Delta)$ that is assumed to fit the data.

Additional information on the TLS can be obtained from the relaxational paraelastic response, (7.42) (mainly symmetric TLS's with large local strain). The peak of the elastic energy loss coefficient, Q^{-1}, due to the O(N)-H(D) pair in Nb, has been measured by Drescher-Krasicka and Granato (1985), Cannelli et al. (1986) and Morr et al. (1989). The $Q^{-1}(T)$ curve depends also on the relaxation rate of the TLS, τ^{-1}, and Wang et al. (1984) provided the first strong evidence that τ^{-1} is dominated by the interaction with the conduction electrons. Niobium is superconductor below $T_c = 9.2$ K, and the $Q^{-1}(T)$ peak P1 is in the superconducting state up to the highest frequencies used. Wang et al. (1984) forced the sample in the normal state by applying a magnetic field and observed the disappearance of P1. This fact, later confirmed by Drescher-Krasicka and Granato (1985) and Morr et al. (1989), was an evidence that $\tau^{-1} \simeq \tau_e^{-1}$, as already found for the TLS's in metallic glasses (Black and Fulde (1979)). In fact, the superconducting transition does not affect the asymmetries of the TLS's, since these are due to elastic interactions, but has some influence on the tunneling matrix element Δ_0 [see (4.19) in Chap. 4 and later in Par. 7.5.1]. The largest effect, however, is on the relaxation rate due to the scattering of the conduction electrons from the TLS. Various expressions have been proposed for τ_e^{-1} in the normal state, from the Korringa-like interaction (7.57) (Sect. 7.5.1) to more elaborated expressions including nonadiabatic coupling to the electrons, like (7.64). For the moment, it is sufficient to use the approximated expression proposed by Black and Fulde (1979) for metallic glasses in the superconducting state

$$\tau_e^{-1} = \frac{2\pi}{\hbar} K \left(\frac{\Delta_0}{E}\right)^2 \frac{k_B T}{e^{\Delta_s/k_B T} + 1} \quad , \tag{7.56}$$

where $\Delta_s(T)$ is the superconducting energy gap. This expression is valid for $E < \Delta_s$, which is verified for not too asymmetric TLS's already few Kelvin below T_c, where Δ_s approaches 17 K. Equation (7.56) follows nearly an Arrhenius law with an activation energy equal to the superconducting energy gap, since relaxation occurs through the electrons that are thermally activated above the gap. In the normal state, peak P1 is not observed because $\tau_e^{-1} \sim 10^{11}$ s^{-1}, as determined from the width of the QNS peak, and varies very slowly with temperature; as a consequence, only a small high-temperature tail of the Debye peak is observed. When the sample becomes superconducting, τ_e^{-1} rapidly decreases according to (7.56), and passes through the region in which $\omega\tau \sim 1$, where the maximum occurs. Peak P1 has indeed an apparent activation energy close on the energy gap of Nb, as already appeared from the measurements of Poker et al. (1984), and this is in accordance with (7.56). In the ultrasound attenuation measurements, peak P1 is superimposed to the absorption due to the interaction between the sound wave and the electrons.

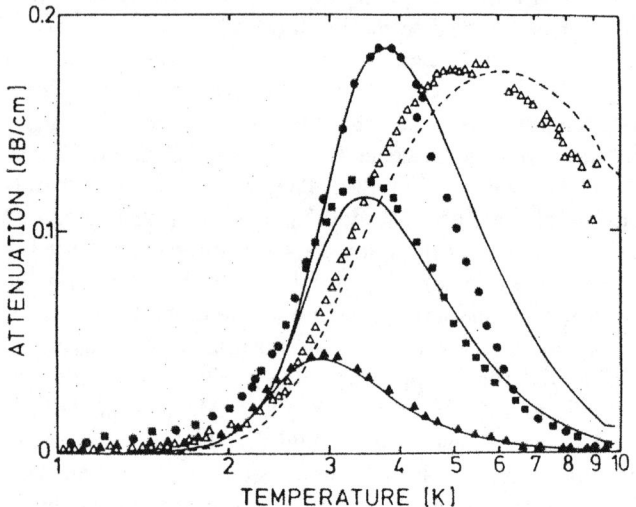

Fig. 7.12. Ultrasonic attenuation of NbN$_{0.0015}$ doped with H and measured at 30 MHz (closed triangles), 90 MHz (squares), 150 MHz (circles) and doped with D at 110 MHz (open triangles). The lines are fits with the tunneling model (from Morr et al. (1989)).

Morr et al. (1989) were also able to fit their attenuation curves in NbN$_{0.0015}$H$_{0.003}$ and NbN$_{0.0015}$D$_{0.0024}$ in terms of the TLS model, with (7.26), (7.42) and (7.56). The fits are shown in Fig. 7.12, and the values of the parameters are in fair agreement with the rest of the data shown in Table 7.1.

O-H Pairs in Ta and V. The research on the TS's formed by H in metals has been concentrated on niobium, but it is expected that, also in other metals, coherent TS's are formed. The only existing data on metals other than Nb are acoustic measurements on Ta and V.

7.5 Coherent Tunneling and Fast Local Motion of Hydrogen

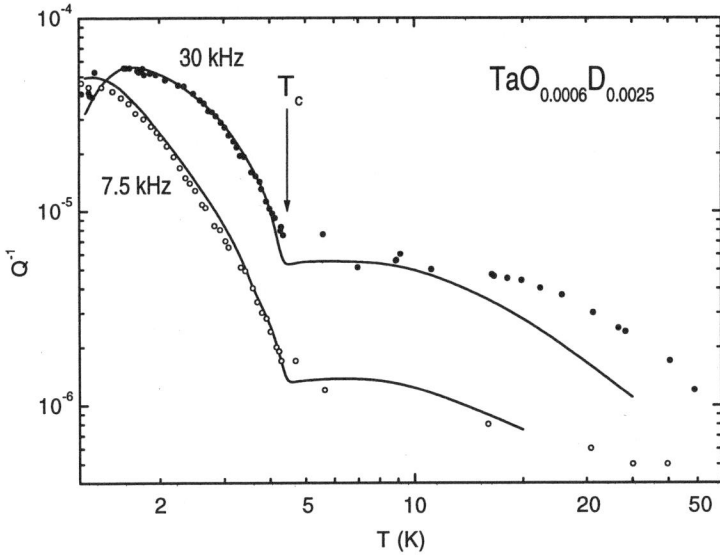

Fig. 7.13. Elastic energy loss (or internal friction) of $Ta_{0.0008}D_{0.0025}$ measured at 7.5 and 30 kHz. The arrow indicates the temperature of the superconducting transition. The lines are a fit with the tunneling model (from Cannelli et al. (1987)).

Cannelli et al. (1987) measured $Q^{-1}(T)$ of $Ta_{0.0008}D_{0.0025}$ at 7.5 and 30 kHz down to 1.2 K, and found a peak below the superconducting transition at $T_c = 4.5$ K. The data are shown in Fig. 7.13, after subtracting the small background of the D free sample. The $Q^{-1}(T)$ curves have two distinct behaviors above and below T_c, which itself is a demonstration that the relaxation rate of the TLS is determined by the interaction with the conduction electrons. In the normal state, there is a very broad curve that roughly scales as the reciprocal of the vibration frequency. This means that we are observing the tail of the Debye peak (7.42) at $\omega\tau \ll 1$, that gives $Q^{-1}(T) \simeq R(T)(\omega\tau)^{-1}$; if the relaxation time τ varies slowly with temperature, as is the case of a strong interaction with the electrons, then the T dependence is mainly given by the relaxation strength $R(T)$, and a change in ω only changes the amplitude. Below T_c, the relaxation rate rapidly drops to the region where $\omega\tau \sim 1$, producing the peak below 2 K. The continuous lines are a fit using the standard TLS model and assuming that the relaxation rate is dominated by the conduction electrons above and below T_c, i.e. $\tau^{-1} = \tau_e^{-1}$, as is already seen for niobium.

For τ_e^{-1}, the expression proposed by Golding et al. (1978) and Black and Fulde (1979) was adopted, that in the normal state,

$$\tau_e^{-1} = \frac{\pi}{2\hbar} K \left(\frac{\Delta_0}{E}\right)^2 E \coth(\beta E/2) \quad , \tag{7.57}$$

whereas in the superconducting state, it becomes (7.56); K is the coupling strength to the electrons adopted also in the rest of the book. This type of interaction between the TS's and the conduction electrons corresponds to a simple treatment in first-order perturbation theory; the interaction Hamiltonian consists of a term that perturbs the site energies when an electron is scattered, analogously to the term $\gamma\epsilon$ of (7.6) and (7.12) that describes the TLS–phonon interaction.

As explained by Black (1981), the local potential $U(r)$ felt by the electron (a plane wave with vector k) is slightly different when the tunneling atom is in the left or right site. If $\delta U(r)$ is the difference between the two situations, the matrix element $\langle k|\delta U(r)|k'\rangle$ for the scattering of an electron from $\mathbf{k}$ to $\mathbf{k}'$ with a transition of the atom to the other site can be taken as the interaction potential between the electron and the TLS, that has an expression of the type (3.5). In the localized representation of the TLS in pseudospin notation, the Hamiltonian becomes [see also (3.2)]

$$H_{\text{TLS}}^{\text{loc}} + H_{\text{TLS}-e}^{\text{loc}} = \Delta \frac{\sigma_z}{2} + \Delta_0 \frac{\sigma_x}{2} + K_\parallel \frac{\sigma_z}{2} \sum_{k,k'} c_k^+ c_{k'} \quad , \tag{7.58}$$

where the interaction $\langle k|\delta U(r)|k'\rangle$ is averaged to a constant coupling $K_\parallel$, (the subscript "longitudinal" refers to the coupling through the pseudospin σ_z). After diagonalizing $H_{\text{TLS}}^{\text{loc}}$ into $H_{\text{TLS}} = \sigma_z E/2$, the off-diagonal part of the interaction that produces transitions between the TLS eigenstates is

$$H_{\text{TLS}-e} = \frac{\Delta_0}{E} K_\parallel \frac{\sigma_x}{2} \sum_{k,k'} c_k^+ c_{k'} \quad . \tag{7.59}$$

The coupling strength in (7.59) is often indicated as

$$v_\perp = \frac{\Delta_0}{E} K_\parallel,$$

and contains the Δ_0/E factor due to the diagonalization of the TLS Hamiltonian. The strength is maximum for a symmetric TLS, since in that case the diagonalization of $H_{\text{TLS}}^{\text{loc}}$ requires a rotation in the pseudospin space that transforms the diagonal terms into off-diagonal ($\Delta_0/E = 1$).[5] Instead, for $\Delta_0 \ll \Delta$, $H_{\text{TLS}}^{\text{loc}}$ is almost diagonal and the effect of the scattering of the electrons remains a change of the TLS asymmetry also after diagonalizing.

---

[5] The Hamiltonian $H_{\text{TLS}}^{\text{loc}} = \frac{1}{2}\begin{pmatrix} -\Delta & -\Delta_0 \\ -\Delta_0 & \Delta \end{pmatrix}$ in the localized basis (Ψ_L, Ψ_R) (see Sect. 2.1) can be diagonalized by a rotation $T = \begin{pmatrix} \cos\phi & \sin\phi \\ -\sin\phi & \cos\phi \end{pmatrix}$ in the pseudospin space, that transforms the TLS basis (Ψ_L, Ψ_R) into $\Psi_- = \cos\phi\, \Psi_L + \sin\phi\, \Psi_R$ and $\Psi_+ = -\sin\phi\, \Psi_L + \cos\phi\, \Psi_R$, and $H_{\text{TLS}}^{\text{loc}}$ into

$$H_{\text{TLS}} = T H_{\text{TLS}}^{\text{loc}} T^{-1} = \frac{1}{2}\begin{pmatrix} \Delta\cos 2\phi + \Delta_0 \sin 2\phi & \Delta_0 \cos 2\phi - \Delta \sin 2\phi \\ \Delta_0 \cos 2\phi - \Delta \sin 2\phi & -\Delta\cos 2\phi - \Delta_0 \sin 2\phi \end{pmatrix}.$$

7.5 Coherent Tunneling and Fast Local Motion of Hydrogen

Black and Fulde (1979) proposed an expression for the relaxation rate (7.35) due to the interaction (7.59) in the first-order perturbation theory (Fermi's golden rule), that is valid both in the normal and superconducting state:

$$\tau_e^{-1} = \frac{\pi}{4\hbar}\left|\rho(\epsilon_F)\frac{\Delta_0}{E}K_\parallel\right|^2 I(E,T,\Delta_s(T)), \qquad (7.60)$$

$$I(E,T,\Delta_s(T)) = \int d\epsilon \int d\epsilon' g(\epsilon) g(\epsilon') f(\epsilon) f(1-\epsilon') \left(1 - \frac{\Delta_s^2}{\epsilon\epsilon'}\right)$$
$$\times [\delta(\epsilon - \epsilon' - E) + \delta(\epsilon - \epsilon' + E)], \qquad (7.61)$$

where the sum over k and k' becomes an integration over the electron energies with a density of states $\rho(E_F)$ that in the superconducting state acquires an additional dependence $g(\epsilon) = |\epsilon|/(\epsilon^2 - \Delta_s^2)$ for $\epsilon > \Delta_s$ and 0 below, $\Delta_s(T)$ being the superconducting energy gap ($\Delta_s = 0$ in the normal state). The factor $(1 - (\Delta_s^2/\epsilon\epsilon'))$ is the coherency factor that takes into account the wave function of the Cooper pairs, and $f(\epsilon) = [1 + \exp(\beta\epsilon)]^{-1}$ is the Fermi–Dirac distribution function. In the normal state, $I(E,T,0) = \coth(\beta E/2)$, so that (7.60) reduces to (7.57) with $K = \frac{1}{2}|\rho(\epsilon_F) v_\perp|^2 = \frac{1}{2}|\rho(\epsilon_F)(\Delta_0/E)K_\parallel|^2$. As explained above, the factor $(\Delta_0/E)^2$ is due to the assumption of an influence of the electron scattering only on the site energy, and for an asymmetric TLS, only a fraction of this perturbation produces transitions between the TLS eigenstates. That factor is present for the same reason in the expression of the relaxation rate due to one-phonon transitions and in more elaborated models of the TLS–electron interaction [see e.g. (3.175)]. The expression 7.60 produces a nearly linear rise of the rate with temperature, in contrast to theories on nonadiabatic coupling to the electrons (see Chap. 3) that predict a T^{2K-1} dependence, as first proposed by Kondo (1976). However, it must be noted that the more recent calculations of nonadiabatic interaction with the electrons contain correction factors for asymmetric TLS's that reduce the difference between the two cases for large asymmetries; for instance, (3.175) for nonadiabatic coupling reduces to (7.60) for $\Delta \gg \Delta_0$ and $K \ll 1$.

The best fit of the experimental curves shown in Fig. 7.13 has been done with $\Delta_0 = 1$ K, $K = 1.5 \times 10^{-4}$, and a Gaussian distribution of asymmetries was assumed with a width $\overline{\Delta} = 15$ K. With these values both the conditions $\Delta \gg \Delta_0$ and $K \ll 1$ are fully satisfied; in addition, the anelastic relaxation strength $R(T)$ is proportional to $(\Delta/E)^2$ (7.42), so that the nearly symmetric TLS's do not contribute.

Below T_c, $I(E,T,\Delta_s(T))$ in (7.60) cannot be expressed in analytical form, and can be approximated to (7.56) only for $E \gg \Delta_s$. For the fit

---

H_{TLS} is diagonal for $\tan 2\phi = \Delta_0/\Delta$, i.e. $\cos\phi = \sqrt{E + \Delta/2E}$, $\sin\phi = \sqrt{E - \Delta/2E}$ with $E = \sqrt{\Delta_0^2 + \Delta^2}$. If the interaction term is purely diagonal in the localized representation, like (7.59), the transformation $TH_{\text{TLS}-e}T^{-1}$ will introduce nondiagonal terms proportional to $\sin 2\phi = \Delta_0/E$. These terms are responsible for the transitions between the TLS eigenstates and are null for $\phi = 0$, or $\Delta \gg \Delta_0$, and maximum for $2\phi = \pi/2$, or $\Delta = 0$.

of Fig. 7.13, however, the exact expression (7.60) was integrated, since the $\Delta_s(0) = 8$ K, half than in Nb, and especially near T_c, the energy split of the TLS is larger instead of much smaller than Δ_s.

Regarding the coupling of the TLS with strain, assuming that it is mainly through the shear of E-symmetry like in Nb [$C' = \frac{1}{2}(C_{11} - C_{12})$ elastic constant], from the intensity of the peak and (7.42), it is deduced $\gamma \geq 900$ K, larger than in Nb. Since the coupling to strain γ is larger than in Nb and that with the electrons, K is smaller, it has also been checked that the one-phonon relaxation rate τ_{1p}^{-1}, given by (7.43) and proportional to γ^2, does not exceed τ_e^{-1}.

The TLS associated with the O-H pair in Ta was also revealed by Maschhoff and Granato (1985), who found an ultrasound absorption peak around 1.2 K with an apparent activation energy close to the superconducting energy gap of Ta, as expected for the relaxation of a TLS coupled with electrons.

In conclusion, in Ta, there is also evidence of the delocalization of H and D in TLS's near interstitial O. The tunneling energies and asymmetries of the TLS found by Cannelli et al. (1986) are comparable with those in Nb, and again the transition rates are governed by the interaction with the conduction electrons; nonetheless, the coupling parameter K is considerably smaller than in Nb and the coupling to strain is larger. The effects due to the nonadiabaticity of the interaction with the electrons, essentially the T^{2K-1} dependence of the relaxation rate in the normal state, were not observed; this is also expected, since the anelastic relaxation comes mainly from very asymmetric TLS's and K has a very small value.

Vanadium has been very little studied. The tail of an absorption peak around 1 K has been found by Cannelli et al. (1982) at acoustic frequencies in samples containing O-D pairs. That peak is most likely due to a TLS similar to that which exists in Nb and Ta, but the available data are not yet sufficient to analyze it.

A Second Tunnel System of Trapped H in Nb? Although the specific heat, neutron scattering and acoustic experiments are well interpreted assuming that H trapped by O delocalizes within a symmetric TLS, the anelastic relaxation spectrum of NbO_xH_y clearly exhibits two distinct peaks at the liquid He temperatures (see Fig. 7.14). In order to distinguish them, we label P1 the peak associated with the TLS and P2 the second one.

The first observation of P2 was made as early as in 1967 by Kramer and Bauer (1967) with the composite resonator technique at 80 and 240 kHz. At that time it was not known that H is always present in Nb, unless a thorough purification treatment is made, and the authors attributed the peak to the motion of dislocations, because the intensity increased after deformation. Kramer and Bauer (1967) also observed a shift of the peak from 3.2 to 2.2 K when the sample was forced in the normal state by a magnetic field, and

7.5 Coherent Tunneling and Fast Local Motion of Hydrogen

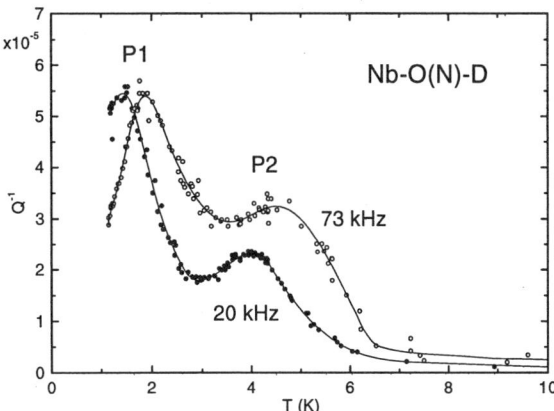

Fig. 7.14. Low-temperature anelastic relaxation spectrum of Nb containing 0.14 at% O and 0.14 at% D measured at two vibration frequencies (from Cannelli et al. (1986)).

determined an apparent activation energy of 18 and 22 K in the normal and superconducting state. The process was later observed by the group of Urbana by ultrasound attenuation (Poker et al. (1984), Huang et al. (1985) and Drescher-Krasicka and Granato (1985)) and at lower frequency (20-75 kHz) by Cannelli and Cantelli (1982) with H and by Cannelli et al. (1986) with D.

According to the Urbana group, peak P2 can be observed after quenching a highly hydrogenated sample, and the intensity decreases if the sample is aged at temperatures above 90 K, where H can migrate. This behavior was explained by assuming that P2 is due to a TS formed by H pairs or O-nH complexes with $n > 1$. Drescher-Krasicka and Granato (1985) also confirmed the shift to lower temperature when the sample passes in the normal state, and attributed it to the influence of the electrons on the tunneling energy. An attempt to fit P2 in terms of the TLS model was made by Cannelli et al. (1986), for a D-doped sample that contained D precipitates and therefore also dislocations. As already observed by Cannelli and Cantelli (1982), P2 is very narrow and does not require any distribution of parameters. The fits shown in Fig. 7.15 were obtained assuming a TLS with $\Delta_0 = 11$K, and assuming either a prevalent interaction with conduction electrons according to (7.60) and with a small coupling parameter $K \sim 10^{-4}$, or through two-phonon processes, as calculated by Orbach (1961). Both assumptions give comparable results, even though the effect of the suppression of superconductivity on the peak temperature indicates an involvement of the conduction electrons. Fits similar to those of Fig. 7.15 were also made for the same sample containing H instead of D, with the peak temperatures at the same vibration frequencies of Fig. 7.15 shifted to 2.7 and 3.2 K (Cannelli et al. (1986)). The result is that the tunneling energy of H is even slightly lower than that of D, $\Delta_0 \sim 9$ K, and the peak shift to lower temperature has to be accounted for by a smaller value of the interaction strength with the electrons (or phonons). The value of Δ_0 is essentially determined by the higher intensity of P2 at higher temperatures, that requires $E \geq 9$ K (7.42), and by the narrowness of the peak that excludes considerable asymmetries of the TLS, implying $\Delta_0 \sim E$. The fact that the two isotopes yield the same value of the tunneling energy and different coupling strengths to the electrons, means that the standard TLS model does not provide a satisfactory interpretation of P2.

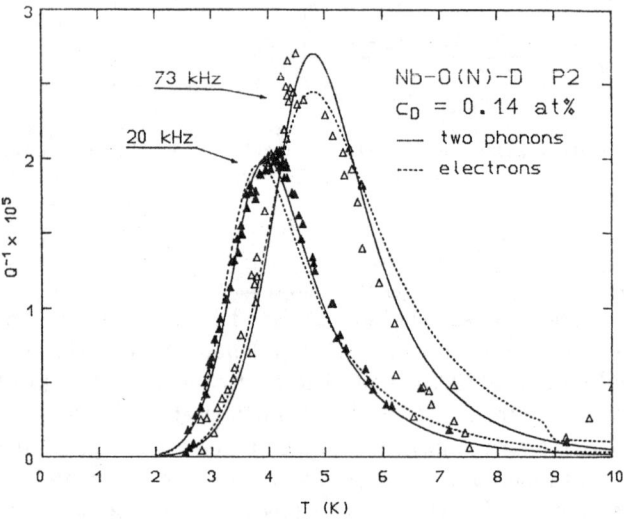

Fig. 7.15. Peak P2 in $NbO_{0.0013}$ doped with 0.14 at D after subtraction of the contribution of peak P1 (from Cannelli et al. (1986)). The lines are fits with the TLS model.

Pal-Val et al. (1993) repeated anelastic measurements at 90 kHz on a very pure (RRR = 10000) single crystal, again finding a peak like P2; according to these authors, H cannot be involved, since it was not introduced in their sample, and dislocations may have a role. We repeated additional measurements at different levels of deformation, H and D concentrations and cooling conditions, but the intensity of P2 did not change enough to provide a clear clue to the mechanism.

It can be concluded that, even though in all the samples investigated, dislocations were present, either due to H precipitation or deliberately introduced, the marked shift of P2 when D is introduced instead of H, demonstrates that it is essentially due to H tunneling.

Inelastic Neutron Scattering: Coherent Tunneling. The neutron scattering experiments by Wipf and co-workers (Wipf et al. (1981), Wipf et al. (1987),

7.5 Coherent Tunneling and Fast Local Motion of Hydrogen

Magerl et al. (1986b), Steinbinder et al. (1991)) have greatly contributed to demonstrate that H trapped at an interstitial impurity forms a TLS, and also allowed the study of the TLS dynamics at high temperatures, where the fast incoherent hopping cannot be investigated by other techniques.

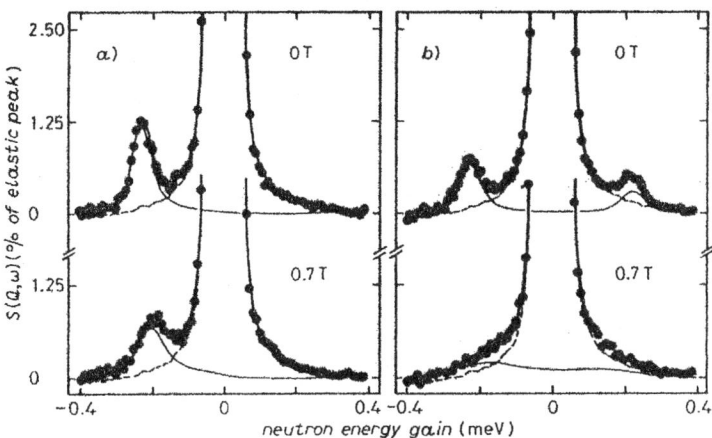

Fig. 7.16. Neutron spectra of Nb(OH)$_{0.0002}$ at 0.2 K (a) and 4.3 K (b) in the normal (0 T) and superconducting (0.7 T) states (from Wipf et al. (1987)). The solid lines are fits with the TLS model.

Wipf et al. (1981) measured for the first time the inelastic scattering from Nb containing O-H pairs below 5 K, and found a small elastic energy loss peak at $\hbar\omega \simeq 2$ meV, that was in accordance with a TLS with $\Delta_0 \simeq 2.2$ K, as previously determined by specific heat experiments (Morkel et al. (1978)). Later, the quality of the measurements was much improved, and it has been possible to obtain good estimates of the tunneling energy from the low-energy transfer edge of the inelastic peak, of the distributions of asymmetries from the ratio between the inelastic and elastic intensity, and of the relaxation rate of the TLS from the width of the inelastic peaks. In all cases, the inelastic spectrum of the coherent tunnel system has been described by (7.47). An example of the spectra in the coherent regime is shown in Fig. 7.16 for Nb(OH)$_{0.0002}$. The two inelastic peaks are centered at about the tunneling energy, and the peak corresponding to the neutron energy loss is more intense by a factor $\exp(E/k_\mathrm{B}T)$, which is the ratio of the populations of the lower level of the TLS to that of the upper level (detailed balance). The figure shows the spectra both in the normal and superconducting state, and it appears that in the normal state the tunneling energy is smaller (the peaks are shifted toward zero energy) and the TLS relaxation rate is higher (the width or damping factor of the peaks is increased). The rates deduced in this way have a rather large error, but extend the ultrasonic data at higher temperatures; the rates from neutron scattering appear about an order of magnitude smaller then those from ultrasound attenuation, but the agreement can be considered fair.

The effect of the conduction electrons on the tunneling matrix element had also been estimated by Yu and Granato (1985) in the framework of perturbation theory starting from the interaction Hamiltonian (7.59) of Golding et al. (1978) and Black and Fulde (1979); for H in Nb, a reduction of Δ_0 was predicted by 20% in the normal state with respect to the superconducting state. Wipf et al. (1987) found a decrease by 10%, and interpreted it in the framework of the nonadiabatic theory of Kondo (1986) and Grabert et al. (1986) (see also Chap. 3). According to these theories, the renormalized tunneling matrix element Δ_0^n in the normal state and Δ_0^s in the superconducting state at $T = 0$ are related to each other by (Wipf et al. (1987))

$$\frac{\Delta_0^n}{\Delta_0^s} = \sqrt{\Gamma(1-2K)\cos(\pi K)} \left(\frac{\Delta_0^n}{2\Delta_s(0)}\right)^K, \qquad (7.62)$$

where Γ is Euler's gamma function and K is the coupling parameter, also appearing in (7.57) and (7.56). From (7.62) and the values of Δ_0^n and Δ_0^s deduced from the inelastic spectra, Wipf et al. (1987) determined $K \simeq 0.055$, the same value needed to interpolate the rate τ_e^{-1} determined from the width of the inelastic peaks by means of (7.57) and (7.56). Since these equations come from the weak coupling limit ($K \ll 1$) of the nonadiabatic theory, their quantitative verification together with (7.62) is a confirmation of the validity of the nonadiabatic approach.

The tunneling energy of the N-H pair has been shown to be lower than that of the O-H pair (Steinbinder et al. (1991)), and the progressive broadening of the inelastic peaks by increasing the O-H concentration from 10^{-4} to 10^{-2} (Magerl et al. (1986b)) demonstrated an influence of the elastic interactions between defects also on Δ_0. Table 7.1 summarizes the values of the parameters of the O-H and N-H TLS's deduced from neutron scattering and other experiments.

Quasi-elastic Neutron Scattering: Incoherent Tunneling. With rising temperature, the neutron scattering spectra of the O-H pairs in Nb transform from the two inelastic peaks (like the response of a damped oscillator) into a central Lorentzian peak (like the response of a stochastically hopping particle). This is due to the increased disturbance of the tunneling motion from the scattering of electrons and phonons, and the formula used to interpret the data is (7.49). For the incoherent hopping rate, $\tau^{-1} = 2\nu$, Steinbinder et al. (1988) and (1991) adopted the expressions of the nonadiabatic TLS–electron interaction provided by Grabert et al. (1986):

$$\tau_e^{-1} = 2\nu(T, \Delta = 0) = \frac{\Gamma(K)}{\Gamma(1-K)} \frac{\Delta_0}{\hbar} \left(\frac{2\pi k_B T}{\Delta_0}\right)^{2K-1} \qquad (7.63)$$

for symmetric TLS, and

$$\tau_e^{-1} = 2\nu(T, \Delta)$$
$$\simeq 2\nu(T,0) \frac{(\pi K)^2 \Delta/(2k_B T)}{(\pi K)^2 + [\Delta/(2k_B T)]^2} \coth[\Delta/(2k_B T)] \quad (7.64)$$

for asymmetric TLS's in the limit $K \ll 1$ [where $\Gamma(K)/\Gamma(1-K)$ becomes K^{-1}]. According to (7.63), the rate is proportional to T^{2K-1}, as first predicted by Kondo (1976), which is opposite to the linear dependence predicted by first order perturbation theory, (7.57). However, for asymmetric TLS's with $\Delta \simeq E$ and for $K \to 0$, (7.64) becomes the expression of first-order perturbation theory, (7.57). Figure 7.17 shows the data of the H hopping rate ν within O-H pairs in Nb from Steinbinder et al. (1988); the closed circles refer to a sample with 2000 at ppm O-H pairs that was fitted with a particularly low value of the average asymmetry, $\overline{\Delta} = 2.2$ K, which is comparable with Δ_0.

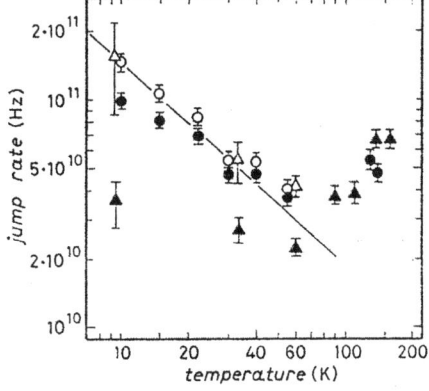

Fig. 7.17. Jump rate of H in two Nb(OH)$_x$ samples with $x = 0.002$ (circles) and $x = 0.011$ (triangles) from Steinbinder et al. (1988). The open symbols are the jump rate estimated for symmetric TLS's, $2\nu(T,0)$. The line is the theoretical prediction for $2\nu(T,0)$, with the model of nonadiabatic interaction between TLS and electrons. The straight line corresponds to a TLS-electron coupling constant $K = 0.055$.

Between 10 and 60 K, $\nu(T)$ decreases, instead of increasing linearly, whereas above 60 K multiphonon processes become predominant. The continuous line is obtained from (7.63), valid for symmetric TLS's, with $K = 0.055$ and $\Delta_0 = 2.4$ K, as determined by other methods. The triangles are from a sample with a higher concentration of O-H pairs and therefore whose TLS's are on the average much more asymmetric than in the former case; the rate is depressed for this reason, as results from (7.64). Svare (1989) questioned the above interpretation, showing that the quasi-elastic spectra can be also interpreted in terms of (7.57) plus phonon contributions when the TLS's have a broad distribution of asymmetries, $\overline{\Delta} \gg \Delta_0$, that is verified unless the concentration of O-H pairs is particularly low. Indeed, in such a situation both the first-order perturbation theory and nonadiabatic descriptions of the TLS–electron interaction give the same expression for the hopping rates, so that there would not be any need to split (7.57) into (7.64) and (7.63). This argument, however, is no more valid for the data of Steinbinder et al. (1988) at the lower impurity concentration, where the TLS's should not be too asymmetric; in that case the decrease of the rate between 10 and 60 K is not explainable in terms of interaction with phonons or with electrons within the first-order perturbation theory. It should also be mentioned that the T^{2K-1} dependence of the hopping rate has also been observed for the positive muons in Al by Hartmann et al. (1988) and in Cu by Luke et al. (1991).

A fit to the data of Fig. 7.17 including the phonon contribution is provided by Würger (1998b), see also Chap. 3. An additional verification of the tunneling model has been provided by the comparison between the hopping rates within the O-H and the N-H pairs, that have different values of Δ_0; Steinbinder et al. (1991) verified that the hopping rate, determined as explained above, is proportional to the square of the tunneling matrix element.

NMR. Most of the NMR experiments on hydrogenated Nb and Ta have been made at temperatures higher than 100 K, and give information on the long-range diffusion mechanism and trapping from the O atoms; they have been reviewed by Messer et al. (1986). These experiments are in substantial accordance and are complementary with the anelastic relaxation ones. More recently, the NMR measurements in $NbO_{0.016}H_{0.04}$ have been extended to 2 K by Pfiz et al. (1989). The spin-lattice relaxation rate Γ_1 at 72 MHz exhibits a clear peak at 200 K following (7.50), due to the reorientation of H around the O atoms. Below 100 K, a broad shoulder in Γ_1 was observed, that has been assigned to the fast incoherent hopping within the TLS, but no quantitative information could be obtained.

7.5.2 Hydrogen Trapped by Substitutional Ti and Zr in Nb: Two- and Four-Level Systems

The case of H trapped by a substitutional atom in Nb presents remarkable differences from that of the interstitial-H pair. In spite of the simpler geometry, the anelastic relaxation spectrum is much more complex both at liquid He and intermediate temperature and the specific heat is qualitatively different from that measured in Nb containing H trapped by interstitial O, N and C.

Four-level system. In this paragraph we find the energy levels and eigenstates of a four-level system (FLS) like that of Fig. 7.19 and their dependence on strain, in order to evaluate their contribution to the specific heat and to the elastic susceptibility. The influences of strain and electric field on perfectly symmetric multisite TS's have been treated several times with the help of symmetry group theory (see e.g. Würger (1997a)) to describe the paraelastic and paraelectric behavior of impurities in alkali halides that tunnel within off-center positions. In the present case however, this type of treatment is certainly inadequate: the case of tunneling within the O-H pair in Nb demonstrates that the site energy shifts due to the elastic interactions with the other complexes in Nb are larger than the tunneling energies already at concentrations of the order of a few hundreds at ppm, and therefore the tunnel system cannot be considered as symmetric even if it spans sites that are equivalent with respect to the first neighbor trap. The elastic response of a centrosymmetric multilevel system has been analyzed by Granato et al. (1985) in connection with the O-H pair, although it now appears that the geometry of the O-H pair does not allow the formation of a TS over more than two equivalent sites. Here, the method adopted by Cannelli et al. (1994a) will be followed, that assumes that the perturbation of strain on the FLS is mainly centrosymmetric, as expected from a homogeneous strain; the perturbation due to the sample vibration, instead, is let to be arbitrary, and can take into account the actual symmetry of the impurity-H pair (Fig. 7.19).

The parameters of the FLS are expressed in the localized representation, i.e. in terms of four H wave functions, each localized in one of the four sites, analogously to what is generally done for the TLS. This allows an easy and intuitive parametrization of the influence of strain on the FLS. The indices $\alpha, \beta = a, b, ...$ will indicate the sites of the TS, whereas $\mu, \nu = 1, 2, ...$ will label the states in the diagonal representation. In analogy with the TLS, we call $\frac{1}{2}\Delta_\alpha$ the energy of the α-th site and $\frac{1}{2}\Delta_{0,\alpha\beta}$ the tunneling matrix element between the localized wave functions in sites α and β. The Hamiltonian of the FLS in the localized representation is then

$$H_{\text{FLS}}^{\text{loc}} = \frac{1}{2} \begin{bmatrix} \Delta_a & \Delta_{0,ab} & 0 & \Delta_{0,ad} \\ \Delta_{0,ab} & \Delta_b & \Delta_{0,bc} & 0 \\ 0 & \Delta_{0,bc} & \Delta_c & \Delta_{0,cd} \\ \Delta_{0,ad} & 0 & \Delta_{0,cd} & \Delta_d \end{bmatrix}, \qquad (7.65)$$

where, for comparison, $H_{\text{TLS}}^{\text{loc}}$ for tunneling within the pair of sites a and b is also written. The basis is

$$|a\rangle = \begin{bmatrix} 1 \\ 0 \\ 0 \\ 0 \end{bmatrix}, \quad |b\rangle = \begin{bmatrix} 0 \\ 1 \\ 0 \\ 0 \end{bmatrix}, \quad |c\rangle = \begin{bmatrix} 0 \\ 0 \\ 1 \\ 0 \end{bmatrix}, \quad |d\rangle = \begin{bmatrix} 0 \\ 0 \\ 0 \\ 1 \end{bmatrix}. \qquad (7.66)$$

The tunneling matrix elements between nonadjacent sites have been put equal to zero, because the overlap of the wave functions over a distance $\sqrt{2}$ times longer is certainly much reduced. Anyway, they can be easily put into $H_{\text{FLS}}^{\text{loc}}$, and produce a relative shift between the pair of levels 1 and 4 and the pair 2 and 3, changing little the physical picture. The perfectly symmetric FLS has $\Delta_\alpha = 0$ (assuming as zero energy the mean site energy) and $\Delta_{0,\alpha\beta} = \Delta_0$; as mentioned before, this is not a suitable Hamiltonian for the FLS, exactly like a symmetric TLS does not describe H tunneling within the O-H pair. Still,

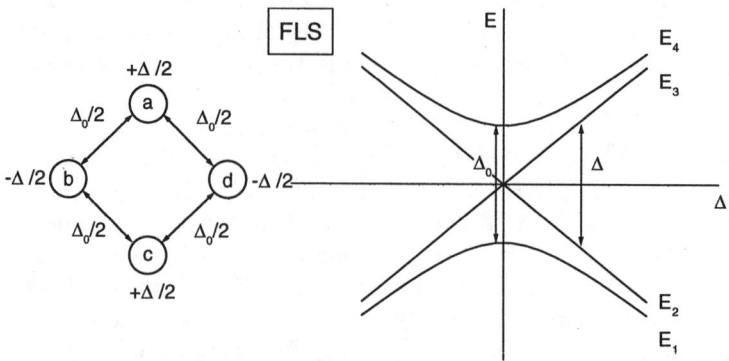

Fig. 7.18. Energy levels of a centrosymmetric FLS as a function of the asymmetry Δ between the pairs of opposite sites.

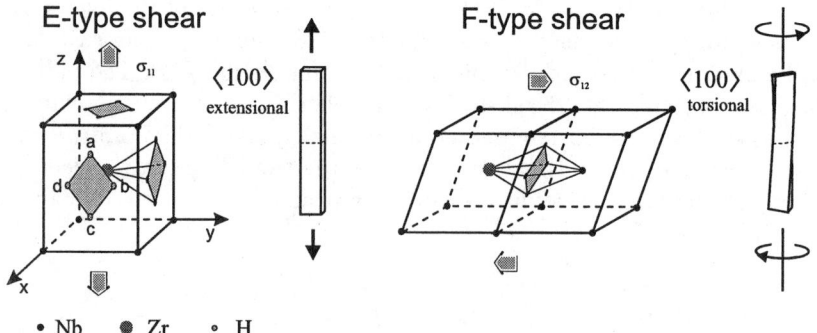

Fig. 7.19. Effect of the two types of shear on the bcc cell and on the FLS's. The deformation of the sample in the two cases is also shown.

a Hamiltonian with completely free values of the site and tunneling energies would not account for the marked dependence of the relaxation intensity on the symmetry of the applied strain, as shown later.

The first approximation in which the effect of the local strain is taken into account is to assume that the unperturbed (by the externally applied stress) FLS is centrosymmetric, since its asymetry is caused by the long range strain interactions with the other TS's, and strain is a centrosymmetric tensor. A perturbation caused by a homogeneous strain has the same symmetry of the strain ellipsoid and consists of a differentiation of the two pairs of opposite sites, leaving equivalent the two sites within a same pair. The resulting Hamiltonian is

7.5 Coherent Tunneling and Fast Local Motion of Hydrogen

$$H_{\text{FLS}}^{\text{loc}} = \frac{1}{2} \begin{bmatrix} \Delta & \Delta_0 & 0 & \Delta_0 \\ \Delta_0 & -\Delta & \Delta_0 & 0 \\ 0 & \Delta_0 & \Delta & \Delta_0 \\ \Delta_0 & 0 & \Delta_0 & -\Delta \end{bmatrix}. \tag{7.67}$$

The eigenvalues of (7.67) are

$$\begin{aligned} E_{1,4} &= \mp \tfrac{1}{2}\sqrt{4\Delta_0^2 + \Delta^2} = \mp \tfrac{1}{2} E, \\ E_{2,3} &= \mp \tfrac{1}{2}\Delta, \end{aligned} \tag{7.68}$$

and are plotted in Fig. 7.18 as a function of Δ. The above eigenvalues are the same as those adopted by Granato et al. (1985). The eigenstates expressed in the basis (7.66) are

$$|1,4\rangle = \frac{1}{2\sqrt{E(E\pm\Delta)}} \begin{bmatrix} 2\Delta_0 \\ -(\Delta \pm E) \\ 2\Delta_0 \\ -(\Delta \pm E) \end{bmatrix},$$

$$|2\rangle = \frac{1}{\sqrt{2}} \begin{bmatrix} 0 \\ 1 \\ 0 \\ -1 \end{bmatrix}, \quad |3\rangle = \frac{1}{\sqrt{2}} \begin{bmatrix} 1 \\ 0 \\ -1 \\ 0 \end{bmatrix}. \tag{7.69}$$

There are analogies between a centrosymmetric FLS and a TLS. The lowest and highest levels of the FLS, E_1 and E_4, have the same dependence on Δ as the levels of a TLS; the $4\Delta_0^2$ instead of Δ_0^2 in the expression of the energy split is due to the fact that the wave function of H in each site is overlapped with those in the two neighboring sites instead of only one. When $\Delta = 0$, the H wave function is equally delocalized over all four sites, and it is intuitive that the intensity of the paraelastic relaxation (7.22) vanishes at $\Delta = 0$, exactly as for a TLS. In fact, the local distortion is the same for the two completely symmetric states: $\lambda^1 = \lambda^4$. This can be seen by assuming that the elastic dipole λ^μ of the eigenstate μ is the average of the elastic dipoles of all the sites weighted with the probability of finding the atom in each site

$$\lambda^\mu = \sum_\alpha c_{\mu,\alpha}^2 \lambda^\alpha, \tag{7.70}$$

where the $c_{\mu,\alpha}$'s are given by (7.66) and (7.69), $|\mu\rangle = \sum_\alpha c_{\mu,\alpha} |\alpha\rangle$. From this point of view, it is clear that paraelastic relaxation, occurring between states μ and ν for which $\lambda^\mu \neq \lambda^\nu$, requires a change of the spatial distribution of the H probability density. When $\Delta = 0$, the $c_{\mu,\alpha}$'s of $|1\rangle$ and $|4\rangle$ differ only for the sign, so that a transition between these states causes no redistribution of the H wave function (only at the nodes of the wave function that are symmetrically arranged) and no change of the elastic dipole: $\lambda^1 = \lambda^4 = \tfrac{1}{4}\left(\lambda^a + \lambda^b + \lambda^c + \lambda^d\right)$. The same holds for a symmetric TLS. Instead, in the eigenstates $|2\rangle$ and $|3\rangle$, H is equally delocalized over either pair of opposite sites for any value of Δ, with no density in the other pair, so that the energy of these eigenstates is simply the energy of the occupied pair of sites, and transitions between these two levels are completely equivalent to a paraelastic reorientation.

Let us now consider the effect of the applied strain on the energy levels (7.68) to deduce the paraelastic relaxation strength from (7.23) and (7.9). In doing this, we will relax the condition that the perturbation to the FLS should be centrosymmetric. Indeed, that hypothesis is equivalent to neglecting the influence of the trapping impurity, and supposing that the four site energies E_α are affected by strain according to (7.6), with the λ^α's of the T sites given by (7.21). In that case, the elastic energy changes only on application of stresses that contain the diagonal uniaxial components, i.e. hydrostatic and E-type shear, whereas the F-type shears σ_{12}, σ_{23}, and σ_{13} do not change E_α. Since the E-type shears differentiate perpendicular pairs of opposite sites along x, y and z, the resulting Hamiltonian must be of the form (7.67). This fact can also be seen by the inspection of Fig. 7.19, representing a Zr-H pair in Nb under the application of a F-type shear. In the perfect lattice, all the sites would be affected by the F-type shear in the same way, because when the distance between a site and an atom changes, there is another atom symmetrically arranged, whose distance changes in the opposite way. However, if one of the two atoms is substitutional, then the net effect is that certain sites are brought closer to the substitutional atom (like site c in Fig. 7.19 that becomes closer to the Zr atom), whereas others (site a) more far apart from it. The presence of the trapping impurity cannot be completely discarded, as implicit in the fact itself that it acts as a trap, and that intense anelastic relaxation processes due to the H motion are observed near the impurity but not in the pure lattice.

The easiest way to parametrize the effect of the trapping impurity on the site energy E_α is to suppose that it contains a component $e(d_\alpha)$ that depends on the distance d_α between site α and the trap. Therefore, the application of strain ϵ_{ij} causes a perturbation

$$\delta E_\alpha = -v_0 p^\alpha_{ij}\epsilon_{ij} + \frac{\partial e}{\partial d_\alpha}\frac{\partial d_\alpha}{\partial \epsilon_{ij}}\epsilon_{ij} = \gamma^\alpha_{ij}\epsilon_{ij} \quad , \tag{7.71}$$

where the first part is that in the absence of the trap, (7.9), with p^α_{ij} related to λ^α_{ij} of a T site by (7.5), and the deformation potential γ^α_{ij} for site α has been introduced (7.12). As noted above, λ^α_{ij} and therefore also p^α_{ij} are diagonal and do not couple to F-type shears, but $\partial d_\alpha/\partial \epsilon_{ij}$ is different from zero also for ϵ^F, as is shown in Fig. 7.19. Therefore, the second term in (7.71) adds off-diagonal elements to the double force and elastic dipole tensors of the T sites near the trap. Introducing $P = v_0 p$, the double force tensor of a single site, and

$$\frac{\partial e}{\partial d_\alpha}\frac{\partial d_\alpha}{\partial \epsilon_{ij}} = \eta^\alpha_{ij} \quad , \tag{7.72}$$

we can separate the effective deformation potential of site α as

$$\gamma^\alpha = -P^\alpha + \eta^\alpha \ . \tag{7.73}$$

Another possible approach to parametrize the effect of the trapping impurity is to suppose that the strain of the impurity-H pair must have a component with one of the principal axes oriented like the axis of the pair; also in this case, off-diagonal elements corresponding to η^α arise if the impurity-H pair is not oriented exactly along one of the crystal axes. The deformation potentials γ^α are not known for the impurity-H pairs, and even the elements of P^α for H in a T site are not well known, as discussed later in Sect. 7.5.4. On the one hand, it is expected that the weight of the interaction term η^α between trapping impurity and H decreases with their reciprocal distance d; for this reason, the interaction term should be more important for the substitutional-H pair in Nb, where $d = \sqrt{5}/4a$, than for the interstitial-H pair, where $d = \sqrt{17}/4a$ or $3/4a$, depending on whether H occupies the third or second neighbor T sites to the octahedral impurity. On the other hand, the trap site for H must be close to a minimum of the function $e(d)$, that makes η^α small and could also make it possible that η^α is larger for a more distant site.

We can now evaluate the derivatives with respect to strain of the energy levels of a FLS, (7.68), with the eigenstates $|\mu\rangle$ given by (7.69). Since the strain ϵ_{ij} due to the sample vibration or sound wave is extremely small, the calculation can be done at the first order of perturbation theory. The perturbation can be written in the localized representation in terms of the deformation potentials γ^α_{ij}'s given by (7.71) as

$$\delta H^{\text{loc}}_{\text{FLS}} = \frac{1}{2} \begin{bmatrix} \gamma^a_{ij} & \gamma^{ab}_{0,ij} & 0 & \gamma^{ad}_{0,ij} \\ \gamma^{ab}_{0,ij} & \gamma^b_{ij} & \gamma^{bc}_{0,ij} & 0 \\ 0 & \gamma^{bc}_{0,ij} & \gamma^c_{ij} & \gamma^{cd}_{0,ij} \\ \gamma^{ad}_{0,ij} & 0 & \gamma^{cd}_{0,ij} & \gamma^d_{ij} \end{bmatrix} \epsilon_{ij} \ , \tag{7.74}$$

where the changes of the tunneling energies have also been introduced because they can be very easily treated. The perturbation to the energy levels can be evaluated as

$$\delta E_\mu = \langle \mu | \delta H^{\text{loc}}_{\text{FLS}} | \mu \rangle \ , \tag{7.75}$$

resulting in

$$\delta E_{1,4} = \left[\overline{\gamma_{ij}} \mp \left(\frac{\Delta}{2E} \delta_{ij} + 4 \frac{\Delta_0}{E} \overline{\gamma_{0,ij}} \right) \right] \epsilon_{ij} \ , \tag{7.76}$$

$$\delta E_{2,3} = [\overline{\gamma_{ij}} \mp \delta_{ij}] \epsilon_{ij} \ ,$$

where

$$\begin{aligned} \overline{\gamma_{ij}} &= \tfrac{1}{4} \left(\gamma^a_{ij} + \gamma^b_{ij} + \gamma^c_{ij} + \gamma^d_{ij} \right) \ , \\ \delta_{ij} &= \tfrac{1}{4} \left(\gamma^a_{ij} + \gamma^c_{ij} - \gamma^b_{ij} - \gamma^d_{ij} \right) \ , \\ \overline{\gamma_{0,ij}} &= \tfrac{1}{4} \left(\gamma^{ab}_{0,ij} + \gamma^{ad}_{0,ij} + \gamma^{bc}_{0,ij} + \gamma^{cd}_{0,ij} \right) \ . \end{aligned} \tag{7.77}$$

The anelastic relaxation intensity contains only the differences between the double force tensors of pairs of states [(7.23) and (7.22)], and therefore the mean change of all the site energies, proportional to $\overline{\gamma_{ij}}$, does not appear and we can write (neglecting $\overline{\gamma_{0,ij}}$ as usual)

$$\frac{\partial E_{1,4}}{\partial \epsilon_{ij}} = \mp \frac{\Delta}{2E} \delta_{ij},$$
$$\frac{\partial E_{2,3}}{\partial \epsilon_{ij}} = \mp \delta_{ij}.$$
(7.78)

Due to the symmetry of the eigenfunctions in (7.69), the relaxation strength does not contain the variations of the energies of the individual sites, but only the asymmetry δ_{ij} between pairs of opposite sites. The parameter δ corresponds to the deformation potential γ of a TLS.

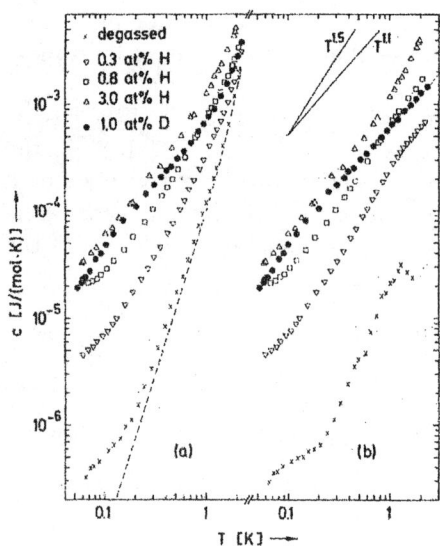

Fig. 7.20. Specific heat of $Nb_{0.95}Ti_{0.05}H_x$ from Neumaier et al. (1982). The broken line is the calculated specific heat of pure Nb.

Specific Heat . Neumaier et al. (1982) measured the specific heat of $Nb_{0.95}Ti_{0.05}H_x$ ($x = 0.003, 0.008, 0.03$) and $Nb_{0.95}Ti_{0.05}D_{0.01}$ and found the rather unexpected result that the excess specific heat due to H had finite values down to the lowest temperature, instead of the cutoff below $k_B T \simeq 0.4\Delta_0$, with a temperature dependence reminding that of the disordered solids containing TLS's (Fig. 7.20). When interpreted in terms of TLS's, this yielded the consequence that in $Nb_{0.95}Ti_{0.05}H_x$, there are TLS's with tunneling energies distributed according to a broad distribution function extending to very small values of Δ_0. Indeed, the first interpretation was that Nb containing a high concentration of Ti-H pairs behaves like a glass. Later, Cannelli et al.

(1989) and (1993) recognized that there is no need for a glasslike distribution of the values of Δ_0 to explain the observation of excess specific heat down to very low temperatures, since this is expected from a four-level system with fixed Δ_0 and a distribution of Δ.

The difference between the contribution to the specific heat at the lowest temperature of TLS's and FLS's can be understood by looking at the density of states of the lowest excitation energy. This is $E = \sqrt{\Delta_0^2 + \Delta^2}$ for a TLS. Assume that only the asymmetries Δ are distributed according to a distribution function $P_\Delta(\Delta)$, for example a Lorentzian

$$P_\Delta(\Delta) = \frac{1}{\pi} \frac{\overline{\Delta}}{\Delta^2 + \overline{\Delta}^2}; \tag{7.79}$$

since the contribution of each TLS to the specific heat depends on E [see (7.2)], we are interested in the distribution of E that can be obtained by writing $P_\Delta(\Delta)\,d\Delta = P_\Delta(\Delta)\,(d\Delta/dE)dE$, or

$$P_{\text{TLS}}(E) = 2\frac{P_\Delta(\Delta)}{|dE/d\Delta|} = \frac{2}{\pi} \frac{\overline{\Delta}}{E^2 - \Delta_0^2 + \overline{\Delta}^2} \frac{E}{\sqrt{E^2 - \Delta_0^2}}, \tag{7.80}$$

where the factor 2 is due to the fact that both $+\Delta$ and $-\Delta$ produce the same E. The curve $P_{\text{TLS}}(E)$ is shown in Fig. 7.21 for $\overline{\Delta} = 10\Delta_0$; it diverges when E approaches Δ_0 and is zero below Δ_0 because that is the minimum value for E. Note that $\int_{\Delta_0}^\infty dE\,P(E) = 1$, because $P_\Delta(\Delta)$ is normalized to 1. The large weight of the distribution for E just above Δ_0 is responsible for the relatively sharp cutoff of $C_{\text{TLS}}(T)$ below $T \sim 0.4\,\Delta_0/k_{\text{B}}$. Instead, the excitation energy E of a FLS does not have a lower limit because the first two levels approach each other when the asymmetry increases [see (7.68)]: $E = \frac{1}{2}\left(\sqrt{4\Delta_0^2 + \Delta^2} - \Delta\right)$. This is reflected in the density of states $P_{\text{FLS}}(E)$, and we show the case of a Lorentzian and of a Gaussian distribution of the asymmetries. In the case of a Lorentzian, (7.79), one has, substituting $\left(\Delta_0^2 - E^2\right)/E$ for Δ,

$$P_{\text{FLS}}^{\text{Lorentz}}(E) = \frac{P_\Delta(\Delta)}{|dE/d\Delta|} = \frac{2}{\pi} \frac{\overline{\Delta}\,\Delta_0^2}{\left(E^2 - \Delta_0^2\right)^2 + \left(E\overline{\Delta}\right)^2}. \tag{7.81}$$

The resulting distribution of E, also shown in Fig. 7.21 for the particular case $\overline{\Delta} = 10\,\Delta_0$, extends to $E = 0$, where it tends to the finite value $P_{\text{FLS}}^{\text{Lorentz}}(0) = \frac{2}{\pi}\overline{\Delta}/\Delta_0^2$. The weight to the smallest values of E depends much on the distribution of asymmetries; in fact, if this is a Gaussian,

$$P_\Delta(\Delta) = \frac{1}{\overline{\Delta}\sqrt{\pi}} e^{-(\Delta/\overline{\Delta})^2}, \tag{7.82}$$

then $P(E)$ becomes

$$P_{\text{FLS}}^{\text{Gauss}}(E) = \frac{2}{\sqrt{\pi}}\overline{\Delta}\left(\frac{\Delta_0}{E}\right)^2 \exp\left[-\left(\frac{E^2 - \Delta_0^2}{E\overline{\Delta}}\right)^2\right], \tag{7.83}$$

that now goes to zero at small E. In both cases however, the weight for the FLS's is predominantly at $E < \Delta_0$, so contributing to the excess specific heat down to very low temperatures. It can be concluded that the experiment of Neumaier et al. (1982) is an indication that H trapped by substitutional Ti in Nb forms FLS's.

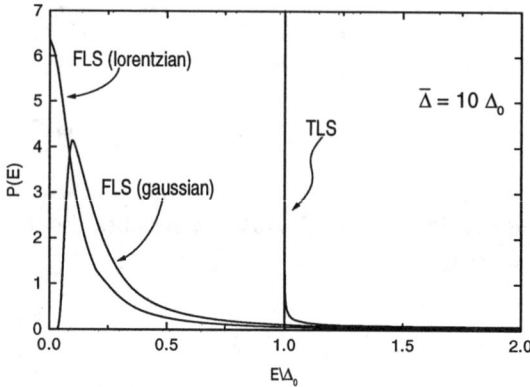

Fig. 7.21. Density of states of the separation E between the first two levels of: (i) a set TLS's with a Lorentzian distribution of the asymmetries Δ of width $\overline{\Delta} = 10\Delta_0$. (ii) A set of FLS's with the same distribution of Δ. (iii) A set of FLS's with a Gaussian distribution of Δ with standard deviation $\overline{\Delta} = 10\Delta_0$.

Anelastic Relaxation in $Nb_{1-x}Ti_x$ Polycrystals. The complex evolution of the AR spectrum can be partially justified by the fact that it is possible to reach much higher concentrations of substitutional-H pairs than interstitial-H pairs (actually there is complete solubility of Ti and Zr with Nb). Indeed, in order to obtain concentrations of O in solid solutions higher than ~1 at%, the sample must be quenched from high temperatures, and it is likely that part of it is precipitated.

Figure 7.22 shows the low-temperature anelastic relaxation spectrum of $Nb_{1-x}Ti_xH_y$ with $x \leq 0.05$ that is the highest value at which it is meaningful to interpret the data in terms of substitutional-H complexes. It appears, however, that already at $x = 0.025$, the spectra are qualitatively different from those at lower impurity concentrations; the H content has always been kept lower than that of the traps, and it was verified that H precipitation never occurred.[6] At the lowest impurity content, $x = 0.001$, there is a relatively narrow peak at 3-5 K, comparable to that due to the O-H pair, that shifts to lower temperatures by increasing the concentration of H. At $x = 0.005$, the peak is considerably broadened, and at $x = 0.025$, it spreads from 2 to 50 K.

A complete analysis of the rather complex phenomenology, see Fig. 7.22, has not been attempted yet, but a few conclusions can be drawn. The most evident consideration is that the rate of the TS of H around Ti are dominated by the interaction with the conduction electrons (Cannelli et al. (1989)),

---

[6] Both the Young's modulus and the elastic energy loss are very sensitive to the onset of H precipitation, and no anomaly was observed in the anelastic spectra.

7.5 Coherent Tunneling and Fast Local Motion of Hydrogen   443

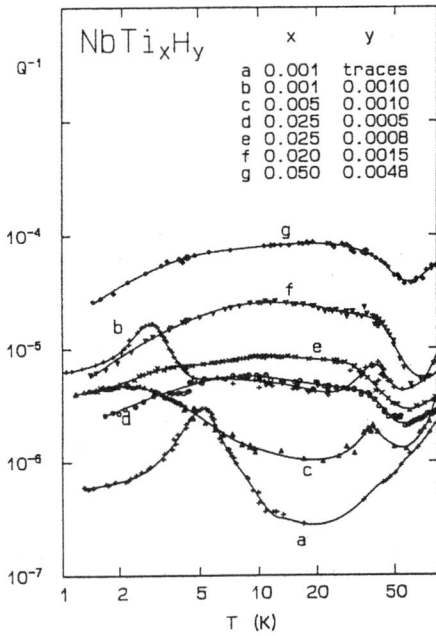

Fig. 7.22. Elastic energy loss of polycrystalline $Nb_{1-x}Ti_xH_y$ measured at 70 kHz (from Cannelli et al. (1993), Cannelli et al. (1981)).

as already established for the TLS of H trapped at interstitial impurities. This is apparent from the change of the slope of the $Q^{-1}(T)$ curves at $T_c = 9.25$ K (indicated by a dashed line in Fig. 7.23 and less evident with the logarithmic temperature scale in Fig. 7.22). The change of the $Q^{-1}(T)$ curves was so marked in the early measurements by Cannelli et al. (1981) on $Nb_{0.95}Ti_{0.05}H_y$, especially at the lower vibration frequency, that the data seemed to indicate two separate peaks, below and above T_c. Figure 7.23 shows an example of these curves at both frequencies in linear scale. The shapes of these $Q^{-1}(T)$ curves can be qualitatively understood as a rapid squeezing of the abscissas scale below T_c, due to the change of the temperature dependence of $\tau(T)$ from an expression of the type of (7.57) in the normal state to (7.56) in the superconducting one.

The fact that the anelastic relaxation spectrum in the presence of the Ti-H pairs appears so different from that with the O-H pairs, seems a consequence of the different geometry of the pair. Indeed, the transformation of the narrow peak between 2 and 5 K into a broad maximum by increasing the impurity level, has been tentatively explained in terms of FLS by Cannelli et al. (1993): the narrow peak would be due to transitions between the intermediate levels, 2 and 3 in Fig. 7.18, that cause paraelastic relaxation also for a symmetric FLS ($\Delta = 0$); the broad peak, that becomes predominant at higher levels of internal strains, would be due to transitions between the 1st and 4th levels,

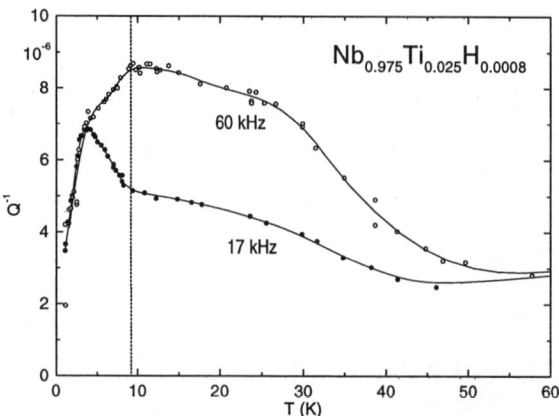

Fig. 7.23. $Q^{-1}(T)$ curves of hydrogenated $Nb_{0.975}Ti_{0.025}$ at the superconducting transition (indicated with a dashed line), from Cannelli et al. (1990).

whose slopes versus strain become appreciable when the asymmetry is higher than the tunneling energy.

Some doubts on this interpretation come from recent results by Cannelli et al. (1995) on the corresponding narrow peak in $Nb_{1-x}Zr_x(D/H)_y$ single crystals: as shown in the next paragraphs, the dependence of the intensity of that peak on the isotope mass can be very well explained in terms of relaxation between levels that cause no paraelastic relaxation at zero asymmetry, that is in contrast with the above interpretation. At present, there is no quantitative and comprehensive interpretation of the anelastic effects of the TS's associated with substitutional-H pairs.

Anelastic Relaxation in $Nb_{1-x}Zr_x$ Single Crystals. Recently, Cannelli et al. (1994a) measured the anelastic relaxation spectrum of $Nb_{1-x}Zr_xH_y$ single crystals with $x \sim 0.005$ and $y < x$, in order to prevent H precipitation and minimize the sources of internal strains. The crystals were bars with the longer dimension oriented along $\langle 100 \rangle$ and $\langle 111 \rangle$, and were excited on their extensional and torsional modes, so that the two types of shears, $\epsilon_4 = \epsilon_{12}$ (C_{44} elastic constant, F symmetry) and $\epsilon_{11} - \epsilon_{22}$ [$C' = (C_{11} - C_{12})/2$, E symmetry] could be excited on the same sample (see Fig. 7.19). Similar to the case of $Nb_{1-x}Ti_xH_y$, at the lowest H concentration, $y = 600$ at ppm, the spectrum contains a peak at $T \simeq 150$ K that can be attributed to H hopping with an apparent activation energy of 0.14 eV around the Zr atom and another peak at $T \simeq 2$ K, due to paraelastic relaxation within a tunnel system. By increasing the H concentration, new intermediate peaks appear, whereas the peak at 2 K saturates. Cannelli et al. (1994b) established that the slow reorientation around the Zr atom includes two different types of occupations with different symmetries, presumably FLS's and TLS's or other types of occupations, but an explanation of the whole spectrum has not yet

7.5 Coherent Tunneling and Fast Local Motion of Hydrogen

been provided. Here we will consider two features of the peak at 2 K, the dependence of its intensity on the symmetry of the applied stress and on the isotope mass.

Dependence on the Stress Symmetry. The intensity of the whole spectrum of $Nb_{1-x}Zr_xH_y$ is smaller under excitation of C_{44} with respect to that of C'; the ratio of the intensities is between 1.5 and 3 for the peaks above 10 K, but > 20 for the peak at 2 K (Fig. 7.24).

As discussed above, the smaller effect on the C_{44} mode is expected, due to the absence of off-diagonal elements of the elastic dipole describing H in a T site, and the finite intensity of the processes also under excitation of the C_{44} mode is a measure of the perturbation of the Zr atom on the symmetry of the trap sites. In fact, without the interaction term η in (7.73), the relaxation of C_{44} should be exactly null. We show now how that symmetry of a FLS can explain the much stronger reduction of the intensity of the peak at 2 K with respect to the other peaks.

The effect of a strain on the energy levels of the FLS can be evaluated with (7.71), (7.73) and (7.78). Let us first consider the effect of a deformation of E symmetry, $\epsilon^E = \epsilon_0/\sqrt{6}\left(2\epsilon^3 - \epsilon^1 - \epsilon^2\right)$, with

$$\epsilon^1 = \begin{pmatrix} 1 & 0 & 0 \\ 0 & 0 & 0 \\ 0 & 0 & 0 \end{pmatrix}, \quad \epsilon^2 = \begin{pmatrix} 0 & 0 & 0 \\ 0 & 1 & 0 \\ 0 & 0 & 0 \end{pmatrix}, \quad \epsilon^3 = \begin{pmatrix} 0 & 0 & 0 \\ 0 & 0 & 0 \\ 0 & 0 & 1 \end{pmatrix}, \quad (7.84)$$

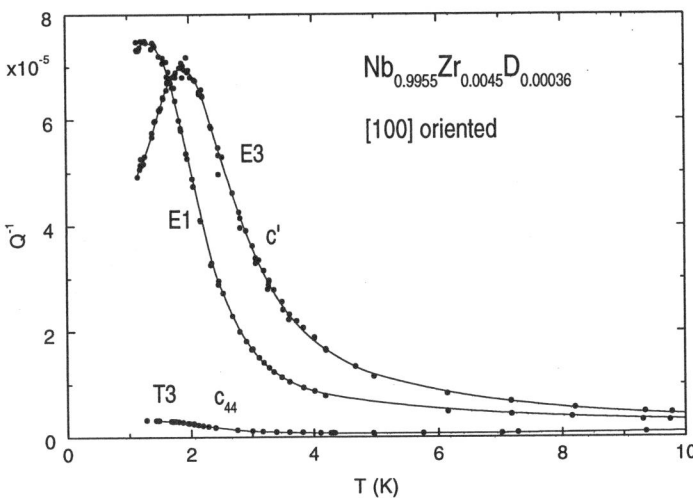

Fig. 7.24. Temperature dependence of the low-temperature anelastic process in a [100] oriented single crystal of $Nb_{0.9955}Zr_{0.0045}D_{0.00036}$ on the symmetry of the applied strain. E1 and E3 are the 1st and 3rd extensional modes (E-type shear), T3 is the 3rd torsional mode (F-type shear).

and amplitude ϵ_0, on the FLS in the cube face perpendicular to the y axis, as in Fig. 7.19a. For the T sites, in the perfect lattice, it is [using the dipole force tensor p that is directly related to the deformation potential γ, instead of the λ tensor, see (7.9) and (7.11)]

$$v_0 p^a = v_0 p^c = \begin{pmatrix} P_1 & 0 & 0 \\ 0 & P_1 & 0 \\ 0 & 0 & P_2 \end{pmatrix},$$

$$v_0 p^b = v_0 p^d = \begin{pmatrix} P_2 & 0 & 0 \\ 0 & P_1 & 0 \\ 0 & 0 & P_1 \end{pmatrix}, \qquad (7.85)$$

so that $-P_{ij}^a \epsilon_{ij}^E = \frac{2}{\sqrt{6}} \epsilon_0 (P_2 - P_1)$ and $-P_{ij}^b \epsilon_{ij}^E = -\frac{1}{\sqrt{6}} \epsilon_0 (P_2 - P_1)$. Regarding the change of the perturbation due to the trapping atom, by the inspection of Fig. 7.19a, we have $\eta^a = \eta^c = \eta$ and $\eta^b = \eta^d = -\frac{1}{2}\eta$ (the factor $1/2$ comes from the choice of the strain that causes deformations twice larger in the z direction than in the x one). We can therefore write

$$\gamma^{a,E} = \gamma^{c,E} = \delta E_a/\epsilon_0 = \frac{2}{\sqrt{6}}(P_2 - P_1) + \eta,$$
$$\gamma^{b,E} = \gamma^{d,E} = \delta E_b/\epsilon_0 = -\frac{1}{\sqrt{6}}(P_2 - P_1) - \frac{1}{2}\eta, \qquad (7.86)$$

where superscript E indicates the strain symmetry, and finally obtain the parameter describing the relaxation intensity of the FLS, (7.77), as

$$\delta^E = \frac{\sqrt{6}}{4}(P_2 - P_1) + \frac{3}{4}\eta. \qquad (7.87)$$

The result is that a strain with E symmetry differentiates the energy levels of a FLS according to (7.78) and (7.87), and therefore causes paraelastic relaxation.

Let us now consider the effect of a deformation of F symmetry, $\epsilon^F = \epsilon_0 \frac{1}{\sqrt{2}} \epsilon^5$, where

$$\epsilon^5 = \begin{pmatrix} 0 & 0 & 1 \\ 0 & 0 & 0 \\ 1 & 0 & 0 \end{pmatrix}. \qquad (7.88)$$

The effect of such a strain on the cell is shown in Fig. 7.19b. The terms $-P_{ij}^a \epsilon_{ij}^F$ are null and we remain with the $\eta^\alpha \epsilon_0$'s. The fact that $P_{ij}^a \epsilon_{ij}^F = 0$ certainly reduces the relaxation intensity under this type of shear; nonetheless, the terms $\eta^\alpha \epsilon_0$, that reflect the change of the symmetry of the T site near the trap, should not be negligible at all. In fact, the distortions of the impurity-H pairs have anisotropies that differ from that of the unperturbed T sites, as demonstrated by the relaxation peaks due to the slow reorientation of H around the impurities, that do not disappear under excitation of F-type strains. This type of anisotropy must come from η^α. Inspecting Fig. 7.19b, we have $\eta^a = -\eta^c = \eta$ and $\eta^b = \eta^d = 0$ that yield (7.77):

$$\delta^{\text{F}}_{\text{FLS}} = 0 \, . \tag{7.89}$$

Equation (7.89) indicates that strains of the F-type do not change the energy levels of a centrosymmetric FLS; this is due to the symmetry of the wave function that always cancels the opposite contributions η and $-\eta$. Instead, if H is delocalized over only two adjacent sites, this cancellation does not occur, and the relaxation strength is proportional to the square of

$$\gamma^{\text{F}}_{\text{TLS}} = \eta \, . \tag{7.90}$$

We conclude that, in order to explain the strong reduction of the intensity of the peak at 2 K under excitation of the C_{44} mode, we must suppose that it is due to relaxation of a FLS rather than of a TLS.

Dependence on the Isotope Mass. Cannelli et al. (1995) found a remarkable effect of the isotope mass on the peak at 2 K: when passing from H to D the peak intensity increases about 50 times. This is shown in Fig. 7.25, presenting the peak on the same sample doped with 360 at ppm D and ~ 600 at ppm D. On the other hand, the peak around 150 K due to the hopping reorientation of H around Zr behaves as expected: when caused by D, it is at slightly higher temperature but the intensity does not change. The enhancement of the intensity of the tunneling peak with D can be explained in a natural way, by supposing that it is due to transitions between the 1st and 4th levels of FLS's or those of TLS's whose asymmetries are distributed according to a Lorentzian $P(\Delta)$ of width $\overline{\Delta}$, and that the isotope mass only affects the tunneling energy Δ_0.

The continuous lines below 10 K if Fig. 7.25 are not the best fit, but serve only to demonstrate that it is possible to reproduce the isotope dependence of the 2 K peak simply by adopting the same parameters that Morr et al. (1989) used to interpolate the $Q^{-1}(T)$ curves due to the N-H and N-D pairs in Nb. The theoretical curves of Fig. 7.25 are calculated assuming anelastic relaxation of a TLS, but are also valid for the relaxation between levels 1 and 4 of a FLS with half the tunneling energy [see the discussion after (7.69)]. The coupling parameter γ refers to the excitation of $C' = \frac{1}{2}(C_{11} - C_{12})$ and enters both in the expression of the peak intensity and in that of the one-phonon relaxation rate, (7.43); in the present case, the contribution of τ_{1p}^{-1} had to be included, because at lower temperatures it becomes comparable to τ_e^{-1} (7.56). Table 7.2 compares the values of the TLS parameters that produce the curves of Fig. 7.25 with those used by Morr et al. (1989) for the N-H and N-D pairs in Nb.

The main reason for the dependence of the peak intensity on the isotope mass is that the relaxation intensity $R(\Delta, E, T)$ is proportional to the square of the deformation potential $D = 2\gamma\Delta/E$ [see (7.25), (7.41) and (7.42)], i.e. the slope of the energy levels versus T, and vanishes for those TS's with $\Delta \ll \Delta_0$ [see (7.25)]. If all the other parameters of the TS remain constant, a decrease of Δ_0 results in a narrowing of the range of the Δ values for which

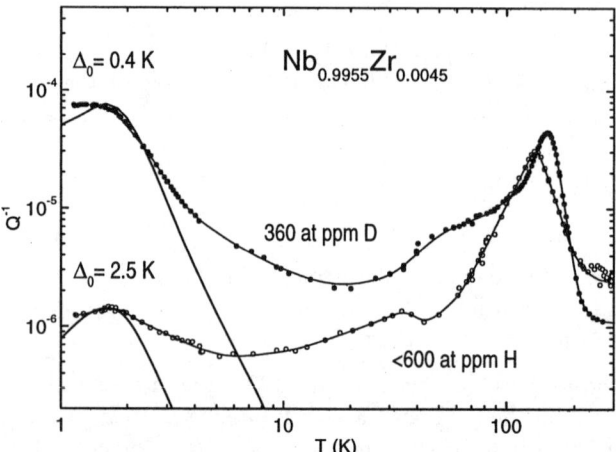

Fig. 7.25. Elastic energy loss of a Nb$_{0.9955}$Zr$_{0.0045}$ single crystal charged with about the same amount of H and D, vibrating at 40 kHz. For the theoretical curves below 10 K, see text. The continuous lines passing through all the experimental points are a guide for the eye.

Table 7.2. Comparison between the values of the TLS parameters that roughly reproduce the peak below 2 K due to H tunneling within the Zr-H(D) pair (Fig. 7.25) and those used by Morr et al. (1989) for the N-H(D) pair in Nb.

	Zr – H pair in Nb TLS/FLS	N – H pair in Nb TLS
Δ_0^H (K)	2.5 (TLS) 1.25 (FLS)	1.4
Δ_0^D (K)	0.4 (TLS) 0.2 (FLS)	0.18
$\overline{\Delta}$ (K)	10 (0.45 at%Zr)	3 (0.15 at%N)
γ^H (K) γ^D (K)	250	270 370
K	0.07	0.07

the relaxation is reduced (those for which $\Delta < \Delta_0$) and therefore corresponds to an increase of the TS's contributing to relaxation.

The reason why the peak due to the O-D pair measured at 100 MHz does not show an enhancement like that of the Zr-D pair measured at 40 kHz is connected with the Debye factor in (7.42), that is peaked at $\omega\tau \simeq 1$. One consequence is that the peak measured at higher frequency is shifted to higher temperature, where the depression of the relaxation strength at $\Delta < \Delta_0$ is less important, because at higher temperatures the weight of $R(\Delta, E, T)$ is shifted to higher values of Δ. A further depression of the dissipation due to H with respect to D in Fig. 7.25 is due to the fact that at the peak temperature the D-TS's with the maximum relaxation strength happen to have also $\omega\tau \simeq 1$, whereas practically all the H-TS's have $\omega\tau$ smaller by at least one order of magnitude (therefore for the H-TS's the condition $\omega\tau \simeq 1$ is satisfied at lower temperatures, where R is strongly depressed).

The entity of the effect depends also on the relaxation rate and frequency of measurement, since the Debye factor in (7.42) weighs more than those TS's with $\omega\tau \sim 1$. For this reason, the peak due to the O-D pair measured at 100 MHz does not show an enhancement like that of the Zr-D pair measured at 40 kHz.

Neutron Spectroscopy. Very recently, an indication that H delocalizes over more than two sites when trapped by a substitutional atom, has been provided also by neutron spectroscopy experiments on $Nb_{0.999}Ti_{0.001}H_{0.001}$. Hauer et al. (1996) found at 1.5 K two small energy loss peaks at 0.2 and 0.4 meV, indicating that the ground state level of H trapped by Ti is split in more than two levels. They were also able to see the splitting of the excited H vibrational state in a sample with a higher concentration of Ti-H pairs. The analysis of the peak showed that it consists of probably four lines, in accordance with the hypothesis of a delocalization of H over a FLS.

7.5.3 Tunneling of H in *hcp* Rare Earths

There is an important difference between the *bcc* transition metals and the *hcp* rare earths: the rare earths retain large amounts of H in solid solution down to 0 K as H pairs, with a fraction of the order of 5 − −10% of H that remains unpaired; this allows the observation of the tunneling dynamics of isolated H atoms.

Neutron Spectroscopy. Most of the neutron spectroscopy experiments on *hcp* rare earths have been made to obtain information on the local vibrational dynamics of H and H pairs, and to estimate the diffusion coefficient at high temperatures from the quasi-elastic peak. Nonetheless, QNS measurements of H in Y above room temperature showed that H performs a rapid local motion within pairs of nearest neighbor T sites (sites a, b or c, d or h, i in Fig. 7.4); Anderson et al. (1989) deduced an activation energy of 0.09 eV for the TT motion, compared with an average 0.55 eV energy barrier for the OT and TO jumps. Later, Anderson et al. (1990) extended the quasi-elastic measurements on ScH_x ($0.05 < x < 0.20$) down to 10 K, to study this fast local motion. The Sc metal was chosen instead of Y, because it has a smaller separation between the two nearest neighbor T sites (1.0 Å instead of 1.3 Å in Y), and therefore a higher tunneling energy is expected. Anderson et al. (1990) verified that the q independence of the width of the Lorentzian describing the incoherent hopping of H in Sc below 300 K indicates localized motion of H, therefore what is observed is the incoherent hopping between the two nn T sites. The temperature dependence of the quasi-elastic spectrum of ScH_x qualitatively reproduces the behavior of the spectrum of $Nb(OH)_x$, and accordingly, Anderson et al. (1990) interpreted the data in terms of nearly symmetric TLS's interacting nonadiabatically with the conduction electrons: the peak was fitted with (7.49) with ν given by (7.64) and (7.63). The tunneling energy is deduced to be $\Delta_0 = 3.7$ K and the coupling to electrons $K = 0.039$, that are comparable to the values found for the O-H pair in Nb. The incoherent hopping rate for the nearly symmetric TLS's shows a T^{2K-1} dependence between 10 and 80 K, and a rise at higher temperatures due to multiphonon contributions. Regarding the comments of Svare (1989) on the alternative interpretation of the linewidths in terms of asymmetric TLS with rates increasing with temperature, Anderson et al. (1990) observed that, in spite of the very high concentrations of H, their data indicate TLS asymmetries of the order of few tenths of meV, comparable with those of the dilute $Nb(OH)_x$ system.

Nuclear Magnetic Resonance. Several NMR investigations of H in *hcp* rare earths exist, mainly devoted to the long-range diffusion and to the type of occupancy of the H atom in the α and hydride phases. Lichty et al. (1989) found two peaks in the proton spin-lattice relaxation rate in ScH_x ($x = 0.11, 0.27$); the intense peak at 500 K (40 MHz) has been identified as due to the diffusive motion of H, whereas the smaller peak at 60 K as due to fast local motion of unpaired H atoms within the nearest neighbor T sites. Lichty et al. (1989) observed that the low temperature peak does not follow (7.50) with a single correlation time τ_c, and interpreted the data in terms of classical overbarrier hopping of the H atoms between the pair of T sites, with an extremely broad distribution of the barrier heights, as in disordered systems.

The unconventional shape of the peak at 60 K was reinterpreted by Svare et al. (1991) in terms of a model more closely related to the dynamics of H in a *hcp* lattice. Instead of a broad distribution of activation energies, Svare et al. (1991) adopted two distributions of the pair asymmetries $P^{(p)}(\Delta)$, corresponding to the paired H atoms (centered at $\Delta \sim 800$ K), and $P^{(u)}(\Delta)$ for the unpaired hydrogens that feel the perturbation from other H atoms in the neighboring chains (centered at $\Delta < 100$ K). The transition rates have been calculated as the superposition of one-phonon transitions (7.43), electron-assisted transitions (7.57), and various types of two-phonon transitions with T^5 and T^7 dependence. The experimental data can be interpolated with various sets of TLS parameters, rather in line with those adopted for the O-H pair in Nb; in particular, the assumption of $K = 0.06$ as TLS-electrons coupling was found to be adequate. Only the adopted tunneling energy, $\Delta_0 = 0.37$ K, is smaller than expected; the value estimated by Anderson et al. (1990) from quasi-elastic neutron scattering is 10 times larger. Although it is not strictly true that the maximum of the asymmetric peak of the NMR relaxation rate occurs at $\omega\tau = 1$, the relaxation rates of the TLS's contributing to such a relaxation must be at least two orders of magnitude lower than those probed by QNS. Even though different techniques probe TLS's with different ranges of asymmetries, the fact that the NMR and QNS experiments cannot be explained with the same set of parameters for the TLS, means that they are indeed probing different types of TLS's or there is some problem with the interpretation.

Acoustic Measurements. Cannelli et al. (1991b) measured the low temperature elastic energy loss of $YH_{0.1}$ and $YO_{0.0027}H_{0.016}$ between 2 and 30 kHz, and found that the sample with the lowest H content presents several relaxation peaks that are unobservable in $YH_{0.1}$. Some of these processes have been ascribed to O-H pairs, that are blocked by further addition of H (Cannelli et al. (1997a)). The peak labelled P3 in Fig. 7.26 corresponds to a single relaxation time with an activation energy of 0.15 meV and is attributable to reorientational motion of H around O. Peak P2 instead, cannot be interpolated by any meaningful classical expression, and is due to a TLS formed by H. Since its maximum shifts considerably in temperature with increasing frequency, the condition $\omega\tau = 1$ must roughly hold at the maximum of P2, and this means that the relaxation rate of this TLS is much slower than those probed by NMR and QNS (see also Fig. 7.1). This fact suggests that P2 is due to a different type of TLS that is associated with an O trap. From the rise of the intensity with temperature, determined at the three vibration frequencies, it was deduced that relaxation occurs between H states that differ in energy of $\Delta \geq 80$ K; this is also expected, since, whatever the occupation of interstitial O is, any pair of T sites in its neighborhood has to be asymmetric. No fit to the $Q^{-1}(T)$ curves was attempted, although it appears that multiphonon transitions prevail. In fact, both the one-phonon and electron-assisted transitions produce a too slowly varying $\tau(T)$ to reproduce peak P2; in that case the $Q^{-1}(T)$ curve, (7.42), would be dominated by the shape of the relaxation strength $R(T)$ instead of the Debye peak, and would not shift in temperature with varying frequency as P2 does.

The acoustic absorption of single crystal samples at higher frequencies has been measured by Leisure et al. (1993a) in $ScD_{0.18}$ and $ScH_{0.25}$ and by Leisure et al. (1993b) in $YD_{0.1}$ with the resonant ultrasound spectroscopy technique. The peaks they found seem to correspond to those of the NMR experiments, rather than to peak P2 in YO_xH_y, as expected from the high H content. The $Q^{-1}(T)$ curves were fitted in a manner analogous to the analysis of the NMR relaxation rates by Svare et al. (1991). Equation (7.42) was adopted, assuming (7.60) for τ_e^{-1} and a phenomenological power law expression for the multiphonon contributions at higher temperatures. Leisure et al. (1993a) also discussed the possible configuration of the TLS; in fact, the jump of an isolated H atom between a pair of nearest neighbor T sites along c does not change the elastic dipole because of its centrosymmetry. Still, the observed anelastic relaxation implies that the energy of one of the two sites is changed by strain with respect to the other site; this is possible if the two sites are made inequivalent for example by the presence of another H atom. Analyzing the intensity of the peak at the different vibration modes, Leisure

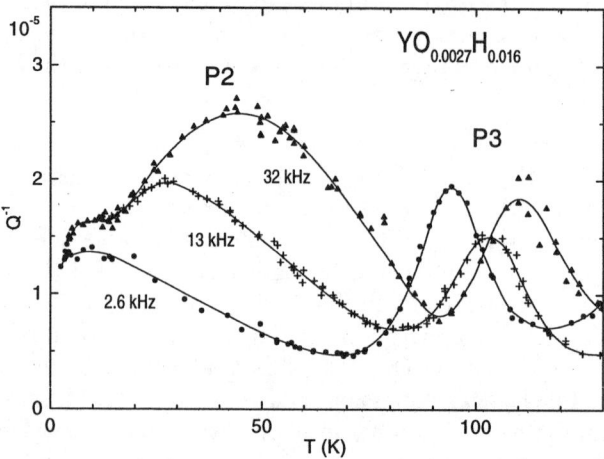

Fig. 7.26. Elastic energy dissipation of $YO_{0.0027}H_{0.016}$ at three different vibration frequencies. The continuous lines are only a guide to the eye (from Cannelli et al. (1991b)).

et al. (1993a) concluded that the anelastic relaxation is due to the modulation of the distance between two H atoms occupying T sites in adjacent c axes (see Fig. 7.4). This was modelled with a distribution of asymmetries centered at $\Delta = \pm 100$ K for Sc and $\Delta = \pm 60$ K for Y. Note that these values of the mean asymmetry of the TLS are in accordance with the estimate $|\Delta| \geq 80$ K by Cannelli et al. (1991b) for the peak in YO_xH_y, although they presumably refer to different types of pairs (H-H and O-H).

7.5.4 Motion and Delocalization of Untrapped Hydrogen in Nb, Ta and V

The Local Vibration Modes of H. In the previous sections, the issue of the actual potential felt by the H atom has not been addressed. The relaxation of the lattice around the H atom or the possible coupling of the H vibrations with local lattice vibrations have not been explicitly taken into account; similarly, the nature of the interaction between the trapping impurity and H is not considered. Nonetheless, these simple tunneling models provide a fairly good description of many experimental data, so that we can consider that parameters like Δ_0, Δ and γ are effective tunneling matrix elements, asymmetries and deformation potentials that somehow take into account the complications of the actual H-lattice interaction. In addition, detailed information on delocalized states has been found only for H trapped at impurities; this is probably a consequence of the small solubility of H, that allows the observation of coherent delocalized states of H at low temperatures only in the presence of traps that retain a sufficient concentration of H atoms in solid solution.

The shape of the potential felt by the H atom can be probed by neutron spectroscopy through the positions and shapes of the peaks corresponding to H transitions among the localized vibration modes. These vibrations are approximately those of a harmonic oscillator, two of them are degenerate, as required by the tetragonal symmetry of the T site, and the vibration frequencies of the three isotopes H, D and tritium are nearly in the expected ratios $1 : 2^{-1/2} : 3^{-1/2}$. For example, in Nb, the departure from these values due to potential anharmonicities does not exceed 7% even in the hydride phase (Rush et al. (1981)). There are, however, various features without an obvious explanation. The width of the lines corresponding to the H vibrations in V, Nb and Ta are much larger than expected. Magerl et al. (1986a) suggest that this can be due to H delocalization over more sites with a consequent tunnel splitting of the degeneracy between the vibrational levels in the different sites. The splitting of the excited states is expected to be larger than that of the ground state, due to the larger spread of the H wave function; therefore, if the ground state splitting is of few tenths of meV, as for Δ_0 of the O-H TLS, the splitting of the higher levels can account for the widths of the lines of several meV.

Cubic Distortion and Fast Local Motion in Nb . Another puzzling issue about interstitial H in *bcc* metals is that it occupies T sites, whose symmetry is tetragonal, but it causes a cubic distortion. That the actual symmetry is tetragonal results from the splitting of the local vibration modes of H, determined by neutron spectroscopy. That the long-range distortion is cubic is deduced not only from the absence of a detectable reorientation of the distortion associated with the H diffusion (Snoek effect), but also from diffuse X-ray and neutron scattering; this matter has been discussed in detail by Dosch et al. (1987) and (1992). Dosch et al. (1992) proposed that the distortion around H is cubic because H can jump at an extremely fast rate, of the order of 10^{14} s^{-1}, among a restricted number of T sites that includes all three possible orientations (e.g. three T sites or a ring of six T sites). The local motion of H would be so fast that the lattice cannot follow it and relaxes with an averaged cubic symmetry. This fast motion would remain localized in the relaxed set of sites and therefore would not contribute to the long-range diffusion; the latter would consist of jumps from a delocalized state to another. Note that the diffuse scattering experiments are done at or above room temperature, in order to have enough H in solid solution. Since the explanation proposed by Dosch et al. (1992) is of dynamical nature and requires jump rates exceeding 10^{13} s^{-1}, its relevance below 100 K is not clear, where the highest hopping rate actually measured is that of H within the O-H TLS and is more than two orders of magnitude smaller. Still, a slow lattice relaxation and fast local H motion as those proposed are not predicted by any of the present models of H diffusion, and therefore would require some revision also at lower temperatures. Later, Elsässer et al. (1994) calculated the forces exerted by an (immobile) H atom on the surrounding lattice atoms in Nb and Pd, and found that these forces are short ranged and with very small anisotropy for tetrahedral occupancy in Nb and octahedral in Pd. Although these calculations allow the explanation of the nearly isotropic force-dipole tensor in terms of static forces, they do not rule out the dynamic model of Dosch et al. (1992) that can explain other experimental observations, like the $\vec{q}$ dependence of the incoherent inelastic scattering of H in Nb.

Delocalization of Untrapped H in the 4T Configuration. In order to determine the energetically more favorable configuration of H in a *bcc* lattice, Sugimoto and Fukai (1980) solved the Schrödinger equation for the H atom interacting with the neighboring atoms through an empirical potential; the surrounding atoms were allowed to relax and the solution was minimized with respect to energy. The potential was chosen in order to reproduce the experimental double force tensor and vibrational energies of H in the α-phase.

Four types of occupation were explored: a single T site (T), a ring of four T sites in a cube face (4T), octahedral (O) and a ring of six T sites (6T); these configurations are shown in Fig. 7.2. The result of the calculation is that the self-trapping energy of H is higher in a T site than in an O site, as expected. However, increasing the distance of the two lattice atoms above and below the cube face, as under uniaxial tension, the 4T configuration becomes more favorable and the H wave function shifts progressively toward the O site. The relative stability of these configurations depends also on the isotope mass, and the result is that for the lightest muon, the tendency to occupy the O site is greater (Sugimoto and Fukai (1980)).

The above ideas have been verified by Yagi et al. (1986), who performed channeling experiments on $VH_{0.01}$ at room temperature applying a uniaxial compression along [100]. The yield profiles were observed to change under a stress of 7 kg/mm^2, in a manner compatible with a T or 4T configuration displaced toward the O site (that kind of experiment cannot determine whether each H atom is delocalized over the four sites). Additional evidence of a displaced T configuration come from X-ray diffuse scattering (Suzuki et al. (1983)) and neutron scattering (Kajitani et al. (1989)) measurements. Suzuki et al. (1983) also found in a series of experiments that the application of a uniaxial tension along [111] increases the H diffusivity by more than 50 times in V and in Nb, and explained this fact in terms of the formation of a network of displaced 4T rings. The existence of this superdiffusion, however, has been the object of controversy (see Koike et al. (1990), and references therein).

7.6 Nonclassical Motion of Hydrogen in Doped Semiconductors

The semiconductor-H systems are quite different from the metal-H systems. The solubilities and mobilities of interstitial H in Si, Ge and GaAs are very small, compared to those in the *bcc* and *hcp* metals considered above. In addition, H can be found in different ionization states, depending on the material doping, and form stable H_2 molecules that prevent its bulk diffusion. A recent review of the properties of semiconductor-H systems can be found in Pearton et al. (1992). The structure of the IV-type (Si, Ge) and III-V-type (GaAs, InP) semiconductors is the diamond structure, with each atom coordinated with other four atoms forming a regular tetrahedron. Hydrogen may form stable pairs with substitutional atoms, like B in Si or Si in GaAs, and these complexes are of interest here. A substitutional foreign atom S in a semiconductor drastically changes the electronic properties of its environment, and the stability and geometry of the S-H pair depend on the valence of S. Usually, acceptors trap a H atom in a stable *bond-center* (BC) configuration, i.e. in a site in the middle of one of the bonds between the S atom and the surrounding lattice atoms (see Fig. 7.27). There are theoretical predictions

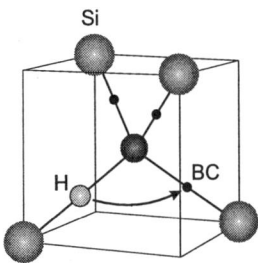

Fig. 7.27. Geometry of an acceptor-H complex in Si doped with B.

(Amore Bonapasta et al. (1992)) from density-functional calculations that the BC site may be unstable if the size of the S atom is sufficiently large, in which case an *off-axis* BC configuration is the stable one. In the case of *off-axis* BC occupation, the formation of a tunnel system over the three or six minima around the bond is expected.

There is little experimental work connected with the possible existence of coherent tunneling states of H in semiconductors. Measurements of infrared (IR) absorption on Ge containing Si-H pairs (Joòs et al. (1980)) and Si containing Be-(H,D,Li) pairs (Muro and Sievers (1986)) have been interpreted in terms of a coherent tunnel system of H over the four BC sites around the foreign atom. This type of measurements provide indirect and strongly model-dependent information on the possible tunnel system, since they rather probe the electronic states of the acceptor complex substitutional-H-hole; indeed, the data on Ge have been revised later (Kahn et al. (1987)), and no more evidence for tunneling was found. The experiments on the Be-H pair in Si provide rather high values of the tunneling matrix elements, $\Delta_0 = 28$ K for H and $\Delta_0 = 11.6$ K for H,[7] but no information on the transition rates.

On the other hand, the relaxational dynamics of H within B-H pairs in Si has been measured by two types of experiments: anelastic relaxation at acoustic frequencies (Cannelli et al. (1991a)) and decay of the stress-induced IR dichroism for relaxation times between 10^3 and 10^6 s (Cheng and Stavola (1994)). The latter technique consists in determining the populations of the H atoms among the four possible types of bonds from the infrared absorption lines of the H vibrations polarized in different directions. The sample is cooled under uniaxial stress down to the temperature of the measurement, in order to change the equilibrium populations of H in the four types of occupation, and then the relaxation of the H populations is followed after releasing the stress. This technique is equivalent to the elastic after-effect, where the dichroism is monitored instead of the anelastic strain. Similar to the case of the reorientation of O-H pair in bcc metals, by joining the anelasticity and infra-red data that span 12 decades of the relaxation rate, a weak deviation from the Arrhenius dependence is found (see Fig. 7.28), that Cheng and Stavola (1994) interpolated with the Flynn–Stoneham formula for the polaronlike hopping, (7.52). The resulting values of the activation energy and tunneling energy are $E = 0.27$ meV and $J = 56$ meV, and there is little departure from the Arrhenius law with ν_0 of the order of the frequency of the local vibration of H. This relaxation rate results from the hopping of H among the four BC sites around a B atom. The anelastic measurements have been extended down to 1.1 K, searching for a faster relaxation process associated with tunneling states, but no trace of additional relaxation has been found (Cannelli et al. (1997b)).

A fast relaxation process has been found by Cannelli et al. (1996) in GaAs containing Zn-D pairs. The elastic energy loss peak around 20 K (between 1 and 13 kHz) is far too wide to be interpreted in terms of classical overbarrier hopping, and the relaxation rate deduced from the peak temperature is tens of orders of magnitude faster than the extrapolated reorientation rates of H around substitutional atoms in Si (Fig. 7.28). It would be tempting to ascribe this fast relaxation to a tunnel system over *off-axis* BC sites; in this case, however, a slower relaxation process due to the reorientation among different bonds is expected, but not yet observed.

---

[7] Assuming that the Hamiltonian of the tunnel system has all the off-diagonal elements equal to $\Delta_0/2$, for compatibility with the notation used here.

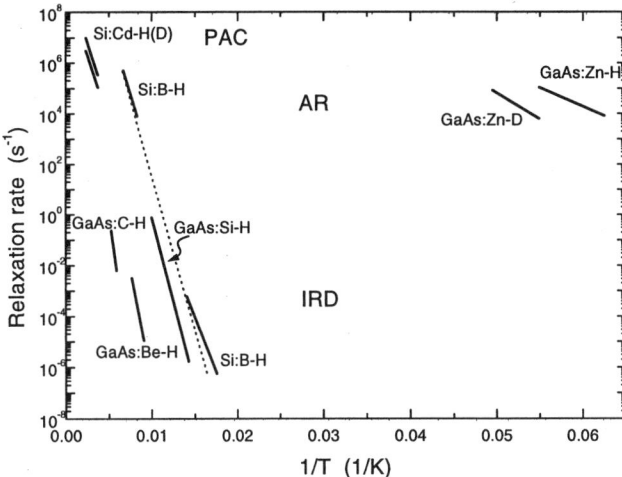

Fig. 7.28. Relaxation rates for the reorientation of H around substitutional atoms in Si. Low frequency data (IRD = infrared dichroism) from Cheng and Stavola (1994) and Kozuch et al. (1993), acoustic frequencies (AR = anelastic relaxation) of Si:B from Cannelli et al. (1991a), Si:Cd (PAC = perturbed angular correlation) from Gebhard et al. (1991), GaAs:Zn-D from Cannelli et al. (1996), GaAs:Zn-H from Cannelli et al. (1997b).

7.7 Conclusion

Hydrogen trapped by heavier interstitial or substitutional impurities in the *bcc* metals Nb and Ta constitutes an excellent model system for testing the present theories of quantum diffusion. Below about 10 K, where coherent tunneling states are observable, the comparison between theory and experiment allows a substantial confirmation, sometimes quantitative, of the presently adopted models of tunneling systems and of their interaction with phonons and conduction electrons. The TLS formed by H near O in Nb has been studied by specific heat, acoustic, inelastic and quasi-elastic neutron scattering measurements, which can all be explained with the same set of parameters of the TLS. For this tunnel system the test of the theory has been extended up to 100 K, for example checking by QNS experiments that the hopping rate is proportional to the square of the tunneling matrix element. Also delocalization over four sites is observed, when the symmetry of the impurity-H pair allows it.

Purely quantum effects are certainly essential in determining also slower types of motion, like the H long-range diffusion or its reorientation around the trapping impurities, through hopping between different tunnel systems. These processes, however, are measured above 30 K up to room temperature, where all theories, even if based on different hypotheses, yield similar Arrhenius-like temperature dependences of the hopping rate on temperature. A few models have been used to fit the temperature dependence of the rate of the H diffusion or reorientation around impurities, but a thorough comparison between the various available models has not been done. Moreover, all these theories contain parameters that are very difficult to estimate theoretically or measure with independent experiments, so that their ability to fit some experimental data is not sufficient to clearly discriminate the one which best describes the H mobility. It should also be mentioned that recent neutron scattering studies in Nb, Ta and V have revealed anomalously broad widths of the peaks of the local vibrations of H, tentatively interpreted as due to delocalized states of H also in the α phase, whereas the diffuse scattering of H in Nb can be explained by an extremely fast motion of H within a set of relaxed sites that cause an averaged cubic distortion. These phenomena should be included in theories of H diffusion.

In the *hcp* rare earths, there are clear evidences of fast tunneling motion of H between the close pairs of tetrahedral sites along the c axis from NMR, QNS and acoustic experiments. It seems that there are at least two types of TLS's that are formed by H atoms that are isolated or are paired with interstitial O atoms; it is not clear whether the H pairs formed at low temperatures exhibit a fast local dynamics too. The differences between results from the different experiments can be partially explained with the fact that QNS mainly probes symmetric TLS's whereas the acoustic absorption experiments probe the asymmetric ones; still, it has not yet been possible to explain all the experimental data with the same set of parameters for the TLS's and their coupling to strain.

In semiconductors, the first clear manifestation of quantum dynamics has been found in the relaxation rate of H within Zn-H complexes in GaAs, that is tens of orders of magnitude faster than the reorientation rates of H within any other dopant-H pair in Si and GaAs. The available data of the relaxation rate of H trapped by substitutional atoms in Si show little deviation from the behavior expected from overbarrier hopping, similar to the case of the reorientation of H around trapping impurities in *bcc* metals. Regarding the existence of coherent H tunnel systems, no clear evidence exists up to now, apart from indirect indications from infrared absorption spectra of Si containing Be-H pairs.

8. Microscopic View of the Low-Temperature Anomalies in Glasses

Andreas Heuer

8.1 Introduction

As shown in previous chapters, most of the low-temperature anomalies of structural glasses are well described by the tunneling model. Experimentally, strong evidence for the presence of tunneling systems (TS's) is provided, e.g. by the possibility of echo experiments (Golding et al. (1979)). Although the presence of TS's with a broad distribution of energies and relaxation times was originally only related to glasses, in the meantime, even in polycrystalline materials, TS's have been observed [Esquinazi et al. (1992); see Sect. 4.5].

The tunneling model by Phillips (1972) and Anderson et al. (1972) is based on the assumption that in amorphous materials a wide range of local environments exists. Then the distribution of TS's can be derived from simple statistical arguments and no specifications on the nature of TS's have to be made. This TS distribution allows quantitative fits of low-temperature anomalies like the specific heat below 1 K that approximately increases linearly with temperature. Most experimental deviations from predictions of the tunneling model can be removed by fine-tuning the tunneling model, e.g. by slightly modifying the distribution of TS's or by taking into account the small but finite interaction among TS's; see e.g. Rogge et al. (1996b) or Enss and Hunklinger (1997).

The generality of this approach is part of the explanation why the tunneling model is adequate for a variety of structural glasses like silicate glasses or polymers. Due to the phenomenological nature of the tunneling model, adjustable parameters have to be introduced. These are the density of TS's, denoted as P_0, and the deformation potential γ, that is a measure of the coupling strength between TS's and phonons. No statements are made about the microscopic origin of the TS's. Unfortunately, only few experiments contain information about the microscopic structure of TS's. An example are neutron scattering experiments, which to some degree are sensitive on the length scale of dynamic processes (Buchenau et al. (1988)).

In a review of Phillips (1987) on TS's in glasses, written nearly ten years ago, the author refers to this lack of information by concluding: *However, two basic questions remain: Why is the number of tunnelling states comparable in all glass-forming systems, and what (in the absence of impurities) is the microscopic structure of a tunneling state?* These two questions will not be answered in a single all-embracing theory.

Both questions touch the basis of the tunneling model. The first question is related to the universality of the low-temperature anomalies. Most, if not all, glasses show a linear temperature dependence of the specific heat in the sub-Kelvin regime. This universality is a direct consequence of the basic ideas of the tunneling model. However, surprisingly, the prefactor is also very similar for a variety of glasses. This *quantitative universality* is the real surprise when comparing different glasses. This chapter attempts to elucidate the underlying reason for this quantitative universality.

From a theoretical perspective, an appropriate way to tackle the second question is to start with a microscopic description of a model glass-forming system and then to identify the TS's. This would give unique information about TS's not accessible by experiments. Formally, this identification reduces to decomposing the full Hamiltonian $\mathcal{H}$ of the glass according to

$$\mathcal{H} = \mathcal{H}_{TS} + \mathcal{W} + \mathcal{H}_B, \tag{8.1}$$

where the first term describes the TS's, the third term the phonon bath and the second term the interaction between TS's and phonons. Such a decomposition would allow, on the one hand, to check whether the predictions of the tunneling model are correct for this specific glass and, on the other hand, to relate the adjustable parameters entering the tunneling model to the microscopic properties of this glass. This approach will guide the analysis presented in this chapter.

The outline of this chapter is as follows. In Sect. 8.2 the tunneling model as well as its generalization – the soft-potential model – are discussed from the perspective of the simulations described further below. A detailed discussion of both models is presented elsewhere in this book. Section 8.3 presents a thorough discussion of what computer simulations can tell about the nature of $\mathcal{H}_{TS}$. Among other things, the tunneling model is derived from first principles for a model glass. Section 8.4 is devoted to the analysis of the coupling between TS's and phonons, i.e. $\mathcal{W}$. First, the deformation potential γ is determined from computer simulations, second, its relation to the structure of the glass as well as to the nature of TS's is elucidated. In Sect. 8.5, some approaches are presented which yield information about the nature of tunneling systems (TS's) from analytical considerations rather than computer simulations. Section 8.6 connects the simulation results with experimental data on the low-temperature anomalies. It is verified that the parameters of the low-temperature anomalies show only a weak dependence on the microscopic properties. Furthermore, a possible reason for this quantitative universality is discussed. In Sect. 8.7, selected experiments are briefly summarized containing specific information about the microscopic nature of TS's. Section 8.8 contains a summary and an outlook.

8.2 Phenomenological Description of the Low-Temperature Anomalies

8.2.1 The Tunneling Model

One basis of the tunneling model is the postulate that via some collective rearrangement of atoms (or molecules) the glass can locally switch between two stable configurations corresponding to two local potential energy minima. One of the goals of this chapter is to clarify this aspect. Let the positions of the atoms in both configurations be described by the set of coordinates $\{r_{i,L}\}$ and $\{r_{i,R}\}$. The motion of the individual atom is expressed by the shift vector

$$d_i \equiv r_{i,R} - r_{i,L} \,. \tag{8.2}$$

The total distance, moved by the particles, can be conveniently defined via

$$d^2(\{r_{i,L}\}, \{r_{i,R}\}) = \sum_{i=1}^{N} d_i^2 \,, \tag{8.3}$$

where the sum is over all N particles of the glass and $d_i \equiv |d_i|$. The effective mass p is introduced via

$$p = d^2/d_{i,\max}^2 \,, \tag{8.4}$$

where "$i, \max$" denotes the index of the atom that moves the largest distance between both wells. Hence p is a measure for the number of particles that are involved in the transfer.

In general the Hamiltonian of the glass can be written as

$$\mathcal{H} = \frac{m}{2} \sum_{i=1}^{N} \left(\frac{\mathrm{d}}{\mathrm{d}t} r_i\right)^2 + V(\{r_i\}) \,, \tag{8.5}$$

where, for reasons of simplicity, all atoms possess the identical mass m. In order to describe the dynamics between both local potential energy minima, it is convenient to introduce the reaction coordinate via

$$r_i(x) = r_{i,L} + \frac{x}{d}(r_{i,R} - r_{i,L}) \,, \tag{8.6}$$

where $x = 0$ and $x = d$ belong to the two minimum-energy configurations. Then the Hamiltonian, parameterized along the reaction coordinate, reads:

$$\begin{aligned}\mathcal{H}(x) &\equiv \frac{m}{2} \sum_{i=1}^{N} \left(\frac{\mathrm{d}}{\mathrm{d}t} r_i(x)\right)^2 + V(\{r_i(x)\}) \\ &= \frac{m}{2} \left(\frac{\mathrm{d}}{\mathrm{d}t} x\right)^2 + V_{\mathrm{DWP}}(x) \,. \end{aligned} \tag{8.7}$$

The potential energy term $V_{\mathrm{DWP}}(x) \equiv V(\{r_i(x)\})$ describes a double-well potential (DWP) with minima at $x = 0$ and $x = d$. Implicitly, it has been assumed that all particles move along straight lines between both minima. However, in general, the dynamics may occur along curved trajectories in order to reach the saddle point between both minima and hence to cross a potential barrier with minimum energy. The simulations discussed below indicate that on average the length along this curved trajectory is 10% longer than the Euclidean distance d as defined above. Due to the smallness of the curvature any complications related to this aspect will be neglected.

According to (8.7), the collective dynamics between two potential energy minima can be represented by the dynamics of a single particle (with coordinate x) in a DWP. A typical DWP is shown in Fig. 8.1.

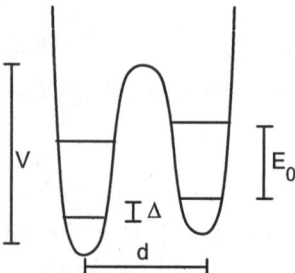

Fig. 8.1. Schematic representation of a DWP with barrier height V, asymmetry Δ, distance d, and excitation energy E_0.

Whereas for $k_B T \geq E_0$, the dynamics in a DWP can be explained in classical terms and displays an Arrhenius behavior, for $k_B T \ll E_0$, the quantum-mechanical nature of the dynamics is relevant [see e.g. Neu and Heuer (1997) for a discussion of this crossover]. If, furthermore, $\Delta \leq k_B T$, tunneling between the lowest two energy levels may occur (otherwise the particle is localized in the lower well) and the DWP effectively acts as a TS. Since typical experiments are performed in the Kelvin temperature regime, it is appropriate to consider DWP's with $|\Delta|/k_B \leq 1$ K as TS's. Apart from the asymmetry it is characterized by the tunneling matrix element Δ_0 that basically denotes the overlap of the eigenfunctions of both levels. In the WKB approximation it can be expressed as (Phillips 1987)

$$\Delta_0 = E_0 \exp(-d(2mV/\hbar^2)^{1/2}) \equiv E_0 \exp(-\lambda) . \tag{8.8}$$

The energy splitting E of the two eigenvalues of the TS is given by $E = \sqrt{\Delta^2 + \Delta_0^2}$. The dynamics of a TS coupled to a heat bath is elucidated in Chap. 3. It turns out that the transition rate Γ between both wells is mainly determined by the tunneling matrix element Δ_0, since $\Gamma \propto \tilde{\Delta}_0^2$. The tilde takes into account that one has to consider a renormalized tunneling matrix element.

Apart from the existence of DWP's a second important ingredient of the tunneling model is the assumption of a broad distribution of DWP parameters. This translates into a broad distribution of TS parameters (Δ, Δ_0), given by

8.2 Phenomenological Description of the Low-Temperature Anomalies

$$P(\Delta, \Delta_0) = P_0/\Delta_0 \tag{8.9}$$

and via $\Gamma \propto \tilde{\Delta}_0^2$ into a broad distribution of relaxation times.

Experiments like sound absorption probe the interaction of phonons with TS's [Tielbürger et al. (1992), see Chap. 4]. The strain field u of the sound waves modulates the parameters of the TS's, i.e. Δ and Δ_0. Generally, it is assumed that the modulation of the asymmetry is dominant. In the linear response regime the coupling constant can then be defined as

$$\gamma_\eta \equiv \frac{1}{2} \frac{\partial [\mathcal{H}(x=d) - \mathcal{H}(x=0)]}{\partial u}, \tag{8.10}$$

with $\eta = l$ (coupling to longitudinal phonons) or $\eta = t$ (coupling to transverse phonons). γ_η is denoted as *deformation potential*.

8.2.2 Determination of Tunneling Parameters from Experiments

Experimental observables like the thermal conductivity [see e.g. Freeman and Anderson (1986)] or the specific heat are a convenient source of information about the low-temperature anomalies. In order to obtain theoretical expressions for these observables, the standard procedure is first to calculate the observable for a single TS and then to average over the TS distribution. One basic physical process is the scattering of phonons on TS's. Knowledge of the phonon scattering rate enables the calculation of the thermal conductivity and other important physical observables at low temperatures. The resonant phonon scattering rate $\tau_\eta^{-1}(E)$ for a specific polarization η with phonon frequency ω and velocity of sound v_η from a single TS with tunneling matrix element Δ_0 and energy E can be easily determined from Fermi's golden rule and reads:

$$\tau_\eta^{-1}(E) = \frac{\pi \gamma_\eta^2 \omega}{\varrho v_\eta^2} \frac{\Delta_0^2}{E^2} \tanh(E/2k_\mathrm{B}T); \tag{8.11}$$

ϱ is the density of the glass. The total scattering rate is obtained by averaging over the distribution of the TS's. We first consider TS's with fixed energy E. Then $\tau_\eta^{-1}(E)$ is proportional to

$$P_\mathrm{eff}(E) \equiv \left\langle \frac{\Delta_0^2}{E^2} \delta(E - \sqrt{\Delta_0^2 + \Delta^2}) \right\rangle, \tag{8.12}$$

where the brackets denote the average over all TS's. This average can be easily performed for the distribution $P(\Delta, \Delta_0) = P_0/\Delta_0$ of the tunneling model. The result is surprisingly simple and reads: $P_\mathrm{eff}(E) = P_0$. Hence in the tunneling model the average scattering rate is independent from the energy of the TS's. Since the thermal conductivity at low temperatures is determined by resonant scattering processes, knowledge of τ_η^{-1} allows us to express the thermal conductivity $\kappa(T)$ as

$$\kappa(T) = \frac{\varrho k_B^3}{6\pi\hbar^2} \left(\sum_\eta \frac{v_\eta}{P_0 \gamma_\eta^2} \right) T^2 \ . \tag{8.13}$$

Note that $\kappa(T)$ depends on the product $P_0\gamma_\eta^2$. It is a general feature that observables referring to resonant scattering processes contain the parameters of the tunneling model in the form of this product. Experimentally, $P_0\gamma_\eta^2$ is often determined either from sound absorption experiments or the shift of the sound velocity (Berret and Meißner (1988)). In order to determine both parameters individually one may measure the acoustic properties, i.e. sound velocity and internal friction. From the maximum in the sound velocity, its slope and the value of the plateau in the internal friction one obtains P_0 and γ_η^2 individually.

Experimentally, it turns out that for most glasses the temperature dependence of the thermal conductivity is described by $T^{2-\beta}$ with $0.05 \leq \beta \leq 0.2$, indicating some deviations from the predictions of the tunneling model. In general, one may expect some energy dependence $P_{\text{eff}}(E) \propto E^{\delta_1}$ of the TS density. As will be shown in Sect. 8.4, the deformation potential γ_η also depends typically on energy, i.e. $\gamma_\eta^2(E) \propto E^{\delta_2}$. These small energy dependences translate into $\kappa(T) \propto T^{2-\delta_1-\delta_2}$, i.e. $\beta = \delta_1 + \delta_2$. Experiments that monitor the time dependence of optical linewidths are also very sensitive on the energy dependence of $P_{\text{eff}}(E)$; see e.g. Silbey et al. (1996) and Chap. 6.

The specific heat $C(T)$ depends on the total number of TS's $P_{\text{total}}(E)$. Tunneling systems with relaxation times τ_1 much larger than typical experimental times τ_{exp} do not contribute to the specific heat; see Chap. 2. Hence an appropriate definition is

$$P_{\text{total}}(E, \tau_{\text{exp}}) = \left\langle \delta(E - \sqrt{\Delta^2 + \Delta_0^2})\Theta(\tau_{\text{exp}} - \tau_1) \right\rangle \ . \tag{8.14}$$

In the tunneling model and in the limit of infinite τ_{exp}, the density (8.14) is independent of E, yielding $C(T) \propto T$. More generally, for $P_{\text{total}}(E, \tau_{\text{exp}}) \propto E^{\delta_3}$, one has $C(T) \propto T^{1+\delta_3}$. Experimental values for δ_3 range between 0.1 and 0.3 for most glasses. In Sect. 8.3 the δ_i will be determined from computer simulations.

8.2.3 Soft-Potential Model

The standard tunneling model predicts the linear temperature dependence of the specific heat $C(T)$ but fails to account for the bump of the scaled specific heat $C(T)/T^3$, see for example Fig. 9.16. The strong temperature dependence of $C(T)$ above a few Kelvin indicates the occurrence of a new class of low-temperature defects that are not included in the tunneling model. Many authors have identified these defects as local low-frequency vibrations. They must not be mixed up with delocalized low-frequency vibrations (phonons) that exist in glasses as well as in crystals.

The presence of these localized low-frequency modes is captured by the soft-potential model as originally suggested by Karpov et al. (1983). In Buchenau et al. (1991) and (1992), and Gil et al. (1993), as well as in Chap. 9, relevant applications are discussed. It proposes a unified description of the TS's and the localized vibrations, both denoted as *soft modes*. Since these localized vibrations can be related to the occurrence of specific single-well potentials, we abbreviate them as SWP's. Here, just a few basic facts are summarized that are relevant to the subsequent analysis.

For the description of the dynamics along the reaction path between two adjacent minima, the potential energy $V_{\text{DWP}}(x)$ has been introduced in (8.7) where the DWP is characterized by (Δ, d, V). One can represent the potential of the DWP by a polynomial of fourth order

$$V(x) = B[w_2(x/a)^2 - w_3(x/a)^3 + w_4(x/a)^4] \,. \tag{8.15}$$

For given energy scale B and length scale a of the glass the $w_{2,3,4}$ may be viewed as adjustable parameters, reproducing the values of (Δ, d, V). One minimum corresponds to $x = 0$, i.e. $w_2 > 0$. For reasons of stability, one always has $w_4 > 0$. In the soft potential literature many different parametrizations can be found. For reasons outlined below, the present parametrization is somewhat different from that used in Chap. 2 where $V(x) = \epsilon[\eta(x/a)^2 + \xi(x/a)^3 + (x/a)^4]$ [see (2.56)].

The general idea of the soft-potential model is to claim that the statistical nature of the glassy structure leads to an *independent* statistical distribution of the w_i described by distribution functions $p_i(w_i)$. Whereas for $w_4 w_2/w_3^2 < 9/32$, the potential $V(x)$ describes DWP's, for the other limit, this potential only contains a single minimum. Hence (8.15) allows a unified description of DWP's and SWP's. For general reasons [see (9.9)], one may require $p_2(w_2) \propto w_2$ for small w_2 and $p_3(w_3) \approx \text{const}$ for small w_3. For reasons of simplicity, the coefficient of the quartic term is chosen as a constant, i.e. $w_4 = w_4^0$.

For fixed $w_4 = w_4^0$, there exists a well-defined energy E_{SPM} such that SWP's can only have excitations energies $E \geq E_{\text{SPM}}$. The density of TS's roughly agrees with the distribution function predicted by the tunneling model. However, due to the additional presence of SWP's for $E \geq E_{\text{SWP}}$, the strong increase of the total density of states, as reflected by the bump of $C(T)/T^3$, is contained in the soft-potential model. Of course, in analogy to the tunneling model, the soft-potential model is purely phenomenological.

8.3 Double-Well Potentials in Computer Simulations

8.3.1 The Scope of Computer Simulations in the Present Context

Atomistic computer simulations are predestined to yield detailed information about the nature of the low-temperature anomalies in glasses. However, despite the dramatic progress in computer technology in the last decades, there

still exist basic problems impeding a direct evaluation of the low-temperature anomalies. Specifically, four problems are mentioned:
(i) Experimentally, the number of TS's that in Sect. 8.2 have been defined as DWP's with $|\Delta| \leq 1K$, is very small (approximately 1 TS per 10^6 particles). Thus, even for very extensive computer simulations, it will be difficult to observe a sufficiently large number of TS's.
(ii) The low-temperature anomalies correspond to slow processes that are furthermore dominated by quantum-mechanical effects. Already, for purely classic simulations only fast dynamic processes beyond the μs timescale are accessible. Furthermore, the computational expense gets larger when considering additional quantum-mechanical processes. Hence it is obvious that via simulations the low-temperature anomalies cannot be characterized via analysis of the time evolution of the glass at temperatures in the Kelvin regime.
(iii) The cooling rates in real experiments are orders of magnitudes smaller than those accessible by computer simulations. A priori it is not obvious how strongly the low-temperature anomalies depend on the cooling rate. In order to estimate the size of this effect, Vollmayr et al. (1996) have performed careful computer simulations with cooling rates varying by a factor of 300 (in the case of amorphous silica). An extrinsic quantity like the volume changed by approximately 3%. Also, from experiments (see Sect. 8.7.2 for more details), one may conclude that a major variation of the cooling rate has indeed some influence on the low-temperature anomalies that, however, does not qualitatively change their nature.
(iv) Very often computer simulations have to deal with finite-size effects when attempting to simulate bulk properties of real systems. This effect is dominant if one wants to simulate phase transitions where the length scale of fluctuations may become arbitrarily large. In this field very elaborate methods have been developed exploiting the additional finite-size information [see e.g. Binder and Stauffer (1987)]. As already discussed in Sect. 8.2, the TS's are expected to correspond to rather localized rearrangements so that the properties of TS's are not expected to depend critically on the size of the system. In most simulations of glasses the number of particles N is between 100 and 10000 where the chosen size depends on the application in mind.

From (i) it is evident that one has to concentrate on the analysis of DWP's in order to learn something about TS's. Important questions refer to the absolute number of TS's, the distribution of TS parameters, and some insight into their microscopic realization. Later on, how to predict properties of TS's, i.e. very symmetric DWP's, from knowledge of the distribution of DWP's in general, will be discussed. The problem of timescale, mentioned in (ii), suggests that it may be advantageous to develop schemes beyond straightforward low-temperature MD simulations in order to localize DWP's.

A concept that became popular in recent years and which is particularly relevant for the characterization of the low-temperature properties of glasses is that of the *energy landscape* (Goldstein (1969)). The basic idea is to con-

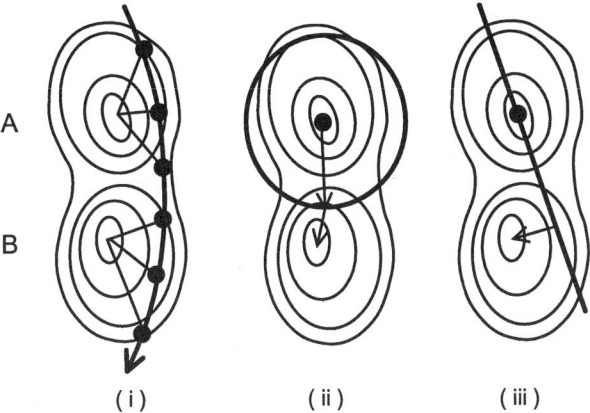

Fig. 8.2. Sketch of the different algorithms used in this text to locate nearby minima. They are described in detail in the text.

centrate on the potential energy regarded as a function in the 3N-dimensional configuration space of glasses, given by the set of coordinates $\{r_i\}$. Mainly, the energy landscape contains information about the local energy minima as well as their mutual saddles, characterizing their connectivity. In this picture, DWP's can be characterized as nearby pairs of minima of the energy landscape. Except for the rare saddle-point configurations, all configurations $\{r_i\}$ can be uniquely mapped on a local energy minimum. This minimum can be reached by a quenching procedure. Alternatively, one can say that the configuration space can be divided in a number of attraction basins related to the individual local energy minima. It is evident that this picture is particularly helpful for characterizing the low-temperature anomalies. However, as already outlined by Goldstein (1969), the properties of the energy landscape may be useful for the description of glasses even somewhat above the glass transition. Of course, in the deeply liquid regime the dynamics is no longer characterized by the nature of the minima of the energy landscape.

The complexity of the problem to locate DWP's on the energy landscape is related to the fact that the reaction path, i.e. the set of $\{d_i\}$ (see Sect. 8.2), may point in some *a priori* unknown direction of the high-dimensional configuration space. A typical DWP in a 2D energy landscape is shown in Fig. 8.2. In what follows three different approaches are described which allow us to locate DWP's.

1. The straightforward approach is to analyze molecular dynamics (MD) runs. In Stillinger and Weber (1983) this idea was systematically exploited for the first time. Afterwards Barnett et al. (1985) applied this method, revealing the presence of at least four DWP's in a metallic glass. During a MD run of a simple model glass the glass is regularly quenched to the next local energy minimum, corresponding to the present attraction basin; see Fig. 8.2i. Afterwards, the MD run is continued with the high-temperature configuration. In this way the present position in configuration space is projected on the corresponding local minimum. Keeping the information about the quenched minima, Stillinger and Weber (1983) observed some dichotomic sequences where the system switched between two energy minima. This has been interpreted as the dynamics within a DWP. In Sect. 8.3.4 some recent results along these lines are presented. Unfortunately, as discussed below, the class of DWP's found in this way is not representative of the DWP's relevant for the low-temperature anomalies.

2. Starting from a relaxed glassy structure one attempts local moves with the ultimate goal to find the reaction path leading to a second nearby local minimum of the potential energy. Starting from minimum A in Fig. 8.2, the goal is to localize the adjacent minimum B. In general, all particles will be somewhat involved in the dynamics between both minima. Minimum B could be simply located via a quenching process if one were able to reach some points of its attraction basin. Hence the main problem is to find a configuration belonging to the attraction basin of minimum B. This is a difficult task, since, from the perspective of minimum A, only a very limited choice of the $\{d_i\}$ leads to this attraction basin. In Heuer and Silbey (1993a) an algorithm has been presented which in most cases allows us to reach the attraction basin of a nearby minimum. The basic idea is to consider in the first step a set of configurations $\{r_{i,S}\}$ with fixed distance d_0 from minimum A, i.e. $d(\{r_{i,A}\}\{r_{i,S}\}) = d_0$. This sphere is defined in the 3N-dimensional configuration space and the value of d_0 is of the order of the expected distance between the two energy minima. Then, via a simulated annealing routine (Kirkpatrick et al. (1983)), an energy minimum $\{r^0_{i,S}\}$ in this subspace is determined. In case a nearby second minimum exists, the resulting configuration tends to be part of the attraction basin of this minimum. In Fig. 8.2ii, this step corresponds to the determination of the energy minimum on the circle around the minimum A.

For a practical realization of this idea one has to take into account that it is hard, if not impossible, to find a global energy minimum in a (approximately) 3N-dimensional space reliably. However, this problem can be circumvented. Already, from the MD simulations by Stillinger and Weber (1983), it became obvious that DWP's correspond to some localized dynamics in space. Although all particles perform some motion during the transfer between both minima, the main dynamics is restricted to a small number of adjacent particles. Hence, one may hope that by restricting the dynamics on the sphere to this limited number of particles, one may nevertheless reach the attraction basin of minimum B. In practice, for a given $T = 0$ glass configuration, the algorithm selects one initial particle and its m nearest neighbours with m chosen between 15 and 20 and chooses an energy minimum under the constraint $d(\{r_{i,A}\}\{r_{i,\tilde{S}}\}) = d_0$. $\tilde{S}$ denotes the sphere in the subspace of the selected

8.3 Double-Well Potentials in Computer Simulations

$(m+1)$ particles. Starting from this configuration $\{r^0_{i,\tilde{S}}\}$ a minimization routine, involving all particles, is performed which, in case of the presence of a nearby minimum, is supposed to reach this second minimum. Otherwise the minimization routine brings the system back to the original minimum. This procedure is repeated until all particles have been selected as the initial particle. In this way a systematic search for DWP's is achieved. Of course, starting from a sufficiently large glass, several DWP's can be detected, being connected to different parts of the glass. In Sect. 8.3.3 the results and the quality of this algorithm are discussed.

3. One might guess that the presence of a nearby second energy minimum is reflected by the *local* properties around the original minimum (here: minimum A). Qualitatively, the reaction coordinate is related to an escape direction with small restoring forces. Formally, this information is contained in the *force matrix* $\mathbf{K} \equiv (\partial^2 U / \partial r_i^\alpha \partial r_j^\beta)$ [U: potential energy, r_i^α: coordinate of i-th particle in direction α ($\alpha = x, y, z$)], evaluated in minimum A. All eigenvalues of $\mathbf{K}$ are positive. The eigenmodes with small eigenvalues are possible candidates. Since DWP's correspond to rather localized dynamics, one may concentrate on localized eigenvectors (phonon modes are unlikely to contain information about local relaxation processes). A practical procedure would be to calculate the potential energy along these eigenvectors. In case of a very good agreement between some eigenvector and the DWP reaction coordinate, one expects the occurrence of a second energy minimum if moving the configuration along this eigenvector. Starting from this second minimum, a final energy minimization would end up in minimum B. In case the agreement between some eigenmode and reaction coordinate is somewhat weaker, one might still expect that this eigenvector points to the attraction basin of the second minimum. Hence, even if the energy along the eigenvector only increases in a monotonic way, subsequent minimization, of the potential energy would lead to the second minimum if starting from some configuration in the new attraction basin. In Sect. 8.3.4 this approach will be tested.

Of course, the analysis of the force matrix itself yields information about SWP's (see Sect. 8.2.2). Work along this line has been mainly carried out by the group of Schober for a variety of model glasses (soft sphere glass, amorphous selenium, binary metallic glass). Important references are Laird and Schober (1991), Schober et al. (1993), Oligschleger and Schober (1995), Schober and Oligschleger (1996). Since this chapter mainly deals with the properties of tunneling systems and hence with DWP's, we only mention some major results:

– The SWP's give rise to a bump in $C(T)/T^3$. Hence it is indeed justified to identify the occurrence of this bump with the presence of soft modes as done in the soft-potential model. For frequencies corresponding to the maximum of $C(T)/T^3$, the modes strongly interact with each other.

- The soft modes are typically due to the correlated dynamics of more than 10 particles (with a large spread in number). They can be related to irregularities of the glass structure. Particles, that are strongly involved in a soft mode, tend to have a nearest-neighbor shell with a somewhat reduced distance together with a reduced density. No significant dependence of the distribution of the localized modes on aging has been observed.
- The number of localized low-frequency modes is much larger for soft spheres than for LJ glasses.

8.3.2 Summary of Earlier Simulations

In the first approaches designed to learn something about the low-temperature anomalies from computer simulations, it was common to study the effect of single-particle motions. Starting with some realistic amorphous structure of a computer glass several methods have been applied.

Amorphous Si and Ge were simulated by Smith (1978). The presence of possible adjacent local minima was probed by shifting one atom a unit bond length and then relaxing this atom by a steepest-descent procedure. In this way approximately four additional energy minima per atom were located. Although, due to very high energy, most states are globally unstable, states with low asymmetry were also found. Their number appears to be sufficiently high in order to account qualitatively for the experimental low-temperature specific heat. In a similar way Harris and Lewis (1982) searched for tunneling states in metallic glasses. They found out that single-particle motions between two local minima are only possible in the presence of voids in which a single particle can move, giving a first hint about the microscopic nature of DWP's.

A different numerical approach was chosen by Brawer (1981). He performed molecular-dynamics simulation of a BeF_2 glass for temperatures around $0.3T_g$ and monitored the ion dynamics. Some jump events by 1Å were detected during runs of 10 ps. Two important properties related to these mobile ions could be observed. First, these ions tend to be either three- or five-fold coordinated as compared to the regular four-fold coordination. Furthermore, quite often correlated back- and forthjumps were observed indicating the existence of DWP's. The author concludes that these ions in atypical environments may be responsible for the low-temperature anomalies of BeF_2, showing a connection between the occurrence of DWP's and structural anomalies.

Guided by the soft-potential model, Dyadyna et al. (1989) modelled a fragment of the SiO_2 structure as a quasi-molecule with nine atoms. They produced different realizations of this fragment so that the structural parameters correspond as closely as possible to the structure of SiO_2. Then they analyzed the potential of the central oxygen atom for these different realizations. After parametrizing this potential by a fourth-order potential in analogy to (8.15), strong fluctuations of the potential parameters were observed, reflecting one of the premises of the soft-potential model. Furthermore, the density of TS's resulting from this analysis is in agreement with typical experimental values.

All simulations have in common the fact that a possible cooperativity of relaxation processes has not been taken into account. An exception is the simulation of SiO_2 presented by Guttman and Rahman (1985). By chance, they observed two adjacent potential energy minima. The transition between both minima could be described as a coupled rotation of several SiO_4 tetrahedra. This specific interpretation of DWP's in SiO_2 is also derived from neutron scattering experiments; see Sect. 8.7.3.

An interesting approach has been developed by Brandt and Kronmüller (1987). Starting from one local minimum they devised a way to escape by adding an external potential chosen as an inverted harmonic potential centered around that minimum. It is parametrized such that the original minimum now corresponds to a maximum so that the system can easily escape. The important step of the algorithm is to estimate when the attraction basin of the original minimum is left. From that time on the modification is removed and the system can relax into an adjacent minimum on the basis of its intrinsic energy. In this way several DWP's have been found. However, it has not been checked whether these DWP's are relevant to the properties at low temperatures.

In the following sections simulations are presented along the lines discussed in Sect. 8.3.1. Here, cooperativity is automatically included and will turn out to be an essential ingredient for a realistic description of DWP's.

8.3.3 Systematic Search of Double-Well Potentials for a Model Glass

In what follows simulations by Heuer and Silbey (1993a) are discussed, yielding detailed information about DWP's.

The Model System. The model glass is similar to the system proposed by Weber and Stillinger (1985). It is a binary glass with the pair potential

$$f_{kl}(r) = A_{kl}[(\alpha_{kl}r)^{-12} - 1]\exp[(\alpha_{kl}r - a_c)^{-1}] \qquad (8.16)$$

for $0 < \alpha_{kl}r \leq a_c$ and zero else. $k, l \in \{1, 2\}$ describes which pair of particles is considered. The pair potential is constructed such that it resembles a pure LJ potential $(A/r^{12} - B/r^6)$ as close as possible and displays a single minimum at the equilibrium distance of two adjacent particles. It has the technical advantage of having a cutoff (at a_c/α_{kl}) for which nevertheless all derivatives are continuous. For reasons of simplicity, we nevertheless denote the model glass as a LJ glass. Anyhow, we expect that all results presented below are independent from the details of the interaction potential. 80% type 1 and 20% type 2 particles are taken. The cutoff distance a_c is chosen $a_c = 1.652\sigma$, where σ is the unit length. Values for the potential parameters are $A_{12} = 1.5 A_{11}, A_{22} = 0.5 A_{11}, \alpha_{11} = 1, \alpha_{12} = 1.05, \alpha_{22} = 1.13$. The parameters are chosen in a way that for $\sigma = 2.2$ Å and $A_{11} = 8200$ K, the model glass is a good representative for an amorphous nickel phosphorous alloy. All simulations are performed at constant density of $\varrho = 8350$ kg m^{-3}. Periodic boundary conditions are used in order to minimize boundary effects. The simulation box contains at most $N = 500$ particles.

The Search for DWP's. The algorithm has already been sketched in Fig. 8.2ii and discussed in Sect. 8.3.1. In what follows the set of all DWP's with parameters $0.1\sigma < d < 1.0\sigma$, $V < 600$ K, $|\Delta| < 800$ K will be denoted as $\mathcal{D}_0$. Only DWP's $\in \mathcal{D}_0$ have been recorded during the simulation run. This limitation is sensible since only for this subensemble of DWP's with *reasonable* potential parameters a systematic search may be attempted. In principle there are many more DWP's not captured by $\mathcal{D}_0$. A trivial example is the exchange of two nearly identical particles, corresponding to $d \approx 2\sigma$. This process is not relevant at very low temperatures. The radius d_0, defined in Sect. 8.3.1, was chosen as $d_0 = 0.4\sigma$. Starting from different initial configurations and systematically selecting all particles as the initial particle, approximately 300 DWP's have been found.

Reliability of the Algorithm. It is impossible to check whether the search algorithm indeed found most DWP's $\in \mathcal{D}_0$. However, indirect arguments for evaluating the quality of this algorithm can be found. In principle a single DWP can be found starting from different initial particles. A necessary condition is that the initially selected particles have a significant overlap with the central part of the DWP. The number of times a specific DWP $\in \mathcal{D}_0$ is found by the algorithm may be used to judge how difficult it is to locate this DWP. It turns out that this number varies between 1 and approximately 20. The fraction of DWP's which has only been found once or twice is rather small (less than 20%). This indicates that most DWP's are reliably found by the algorithm. Furthermore, whether the properties of the DWP's which were only found once or twice are different as compared to the average properties, has been checked. First, it turns out that the average value of the effective mass p for this subensemble is approximately 20% larger than the average value for the whole ensemble. This is in agreement with intuition since DWP's with larger effective masses can be expected to be more difficult to locate. Fortunately, the effect is not dramatic. Furthermore, it turns out that the average potential height of this subensemble is nearly identical to the average potential height of the whole ensemble. This observation is very promising since it indicates that the average energy scale of the DWP's which are hard to locate is similar to the energy scale of all DWP's. Hence one may conclude that even if a significant number of DWP's has been missed, the distribution of DWP's, found by the above algorithm, is representative for all DWP's.

Distribution of DWP's. From the simulations many features about the nature of DWP's can be extracted. First of all, it turns out, that on average, one

8.3 Double-Well Potentials in Computer Simulations   473

DWP $\in \mathcal{D}_0$ per 105 particles is found. Later on, the density of TS's will be estimated on the basis of this number. At first, the distribution of the DWP parameters Δ, d, and V is discussed. For the determination of V it is essential to determine the saddle point between both minima. This reflects the fact that the reaction coordinate in the multidimensional configuration space may correspond to a curved line. Actually, the potential height along the straight line between both minima is often twice as high as the true saddle point. This reduction is typically achieved by a very mild bending of the reaction path. The value of the distance along the curved reaction path across the true saddle is subsequently taken as the distance d. The projection of the (Δ, d, V)-distribution on the V, Δ-plane is displayed in Fig. 8.3. Note that only DWP's $\in \mathcal{D}_0$ are considered. First of all one can see a large scatter over the whole V, Δ-plane. This directly reflects the statistical nature of the formation of DWP's and hence strongly supports the conceptual basis of the tunneling model. Obviously, small values of V and Δ are preferred and both quantities are strongly correlated. DWP's with a small asymmetry tend to have a smaller potential height. It turns out that these DWP's on average also

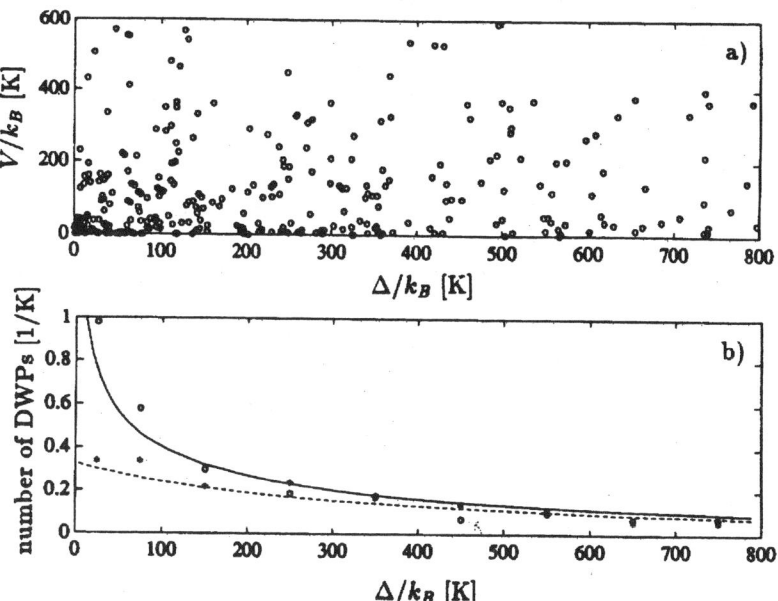

Fig. 8.3. (a) Distribution of 310 DWP's $\in \mathcal{D}_0$ with respect to the asymmetry Δ and the potential height V. (b) Same distribution for different regions of V [$V \leq 100$ K (o), $V \geq 100$ K (*)] as compared to the distribution derived from the DWP's $\in \mathcal{D}_0$ generated from the p_i (*straight line* and *dashed line*, respectively). The good agreement between circles and the straight line as well as between stars and the dashed line shows that the generated DWP's have a similar statistics as compared to the DWP's found numerically. Taken from Heuer and Silbey (1993a).

have very small values of d. This observation is in agreement with intuition since two nearby configurations naturally tend to have similar energies.

As expected, most DWP's are highly asymmetric and one has to apply some appropriate statistical procedure in order to predict properties of TS's from information about asymmetric DWP's. As was evident from Fig. 8.3, the individual parameters of the DWP's are strongly correlated. Hence, a straightforward extrapolation on the basis of the distribution in V, d, Δ is not possible.

The decisive step is to find a new representation of the DWP's so that all three parameters are statistically independent. The relevance of the statistical independence is clarified for a simple example; see Fig. 8.4. In this figure a statistical sample of $N = 100$ measurements has been taken, obeying the underlying (*a priori* unknown) distribution $p(x_1, ..., x_d)$, defined for $x_i \in [0, 1]$ (here $d = 2$). The goal is to estimate $\alpha \equiv \int_0^{\epsilon_1} ... \int_0^{\epsilon_d} dx_1...dx_d p(x_1, ..., x_d) \ll 1$. For reasons of simplicity it is assumed $p(x_1, ..., x_d) = \text{const}$, i.e. $\alpha = \epsilon_1 \cdot ... \cdot \epsilon_d$. The value of α is estimated by the number of measurements falling in the box $[0, \epsilon_1] \times ... \times [0, \epsilon_d]$. The estimation will be exact for $N \to \infty$. For finite N the number of measurements in this interval may show some variation which is approximately given by $\alpha N \pm \sqrt{\alpha N}$. Hence the relative error of the estimation is $1/\sqrt{\alpha N}$. The situation dramatically changes if the total probability function factorizes, i.e. $p(x_1, ..., x_d) = p_1(x_1)...p_d(x_d)$. Then one may first consider all projections on the d dimensions and hence individually estimate the values α_i, which correspond to the probability that $x_i \leq \epsilon_i$. In analogy to the above, one has $\alpha_i = \epsilon_i(1 \pm 1/\sqrt{\epsilon_i N})$. Due to the statistical independence, one may now estimate α via the product $\epsilon_1(1 \pm 1/\sqrt{\epsilon_1 N}) \cdot ... \cdot \epsilon_d(1 \pm 1/\sqrt{\epsilon_d N})$. For the simple case $\epsilon_1 = ... = \epsilon_d$, this estimation shows the relative error $d\alpha^{-1/2d}/\sqrt{N}$. One directly sees that for $\alpha \ll 1$, the quality of the estimation is much better than without using the statistical independence ($d\alpha^{-1/2d}/\sqrt{N} \ll 1/\sqrt{\alpha N}$).

Heuer and Silbey parametrized the DWP's by the fourth-order polynomial $V(x) = B[w_2(x/a)^2 - w_3(x/a)^3 + w_4(x/a)^4]$, where $a = 1.1\sigma$ is the average equilibrium distance between two particles. Furthermore, $B = mv^2 = 4 \times 10^4$ K for the present case, where the average sound velocity 2600 ms^{-1} of NiP

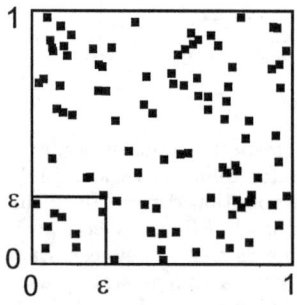

Fig. 8.4. Random distribution of points in 2D. In the text this sketch is used to explain why statistical independence may dramatically increase the information content of a given statistical sample.

has been used (see Sect. 8.4.1 for an exact definition of the average sound velocity v). This is the polynomial of minimum size which may reproduce the original parameters of the DWP Δ, d, and V. Choosing both minima of a DWP as $x = 0$, one obtains two triplets (w_2, w_3, w_4) per DWP and hence the distribution $p_w(w_2, w_3, w_4)$. In analogy to the soft-potential model, one might anticipate that the coefficients are statistically independent and are described by the probability distributions $p_i(w_i)$ (i=2,3,4). Now the analysis proceeds in two steps. First, the $p_i(w_i)$ are determined by the condition that $g_1 \equiv \langle [p_w(w_2, w_3, w_4) - p_2(w_2)p_3(w_3)p_4(w_4)]^2 \rangle$ is a minimum. The brackets denote the sum over the w_i axes which for practical reasons have been discretized. Here it is essential to restrict the summation to $\mathcal{D}_0$ (which corresponds to a deformed volume in (w_2, w_3, w_4)-space), since only for this parameter regime systematic information about DWP's is available. Second, one checks that the factorization assumption is consistent with the actual distribution of DWP's. This has been done in two ways: (i) The same number of DWP's has been randomly generated according to the p_i's. For this set of DWP's, the same statistical analysis has been performed, yielding a value g_2. Since, by construction, the generated set of triplets is based on independent distribution functions p_i, the ratio g_1/g_2 is a good measure for the statistical independence of the $\{w_i\}$ in the original set of DWP's. Indeed, it turned out that $g_1 \approx g_2$, hence confirming that possible residual statistical dependences are not relevant. (ii) The factorization assumption can be alternatively checked by comparing the original distribution of DWP's with the distribution of generated DWP's in the (V, Δ, d)-space. In Fig. 8.3b the results for the (V, Δ) plane are presented. Taking into account the finite size of the sample, the original distribution is reasonably well approximated by the generated distribution.

The parametrization, suggested by the soft-potential model, is $V(x) = \epsilon(\eta(x/a)^2 - \xi(x/a)^3 + (x/a)^4)$. In this case, every DWP can be expressed in terms of the parameter triplet (ϵ, η, ξ). However, it turns out that the probability function $\tilde{p}(\epsilon, \eta, \xi)$ cannot be factorized. Hence for the present purpose, the parametrization in terms of (w_2, w_3, w_4) is preferred.

In this new (w_2, w_3, w_4)-representation, DWP's are symmetric whenever $w_2 w_4/w_3^2 = 1/4$. In the (w_2, w_3, w_4)-space, this relation defines some curved 2D subspace. Due to the independence of the ω_i, information about this subspace can also be obtained from the distribution of asymmetric DWP's. Hence, symmetric DWP's do not arise as an *extrapolation* of asymmetric DWP's as in the case of the (V, D, Δ)-representation. Rather, this subspace is fully embedded in the set of all DWP's. Therefore reliable statistical information about TS's can be obtained without having simulated a large number of symmetric DWP's.

The numerically determined distribution functions p_i are displayed in Fig. 8.5. Note that $p_i(w_i)$ is significantly different from the probability function $q_i(w_i)$ denoting the probability that a DWP found from the simulation

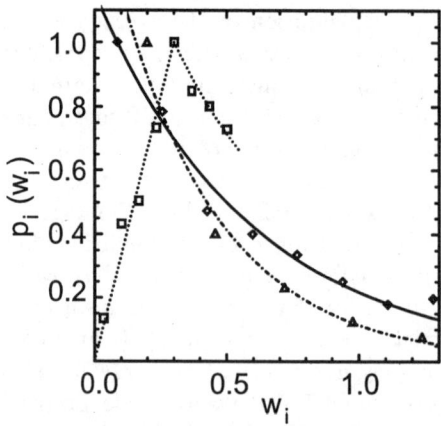

Fig. 8.5. The distribution $p_i(w_i)$ as determined from the analysis of the DWP's. The fitting functions are specified in the text (squares: $i = 2$, diamonds: $i = 3$, triangles: $i = 4$). Adapted from Heuer and Silbey (1996).

has some specific value w_i. This apparent contradiction has a very simple reason. The set $\mathcal{D}_0$ of DWP's only corresponds to a small and nontrivial volume in the w_2, w_3, w_4-space. Only if this volume were a simple cuboid, one would have $p_i = q_i$.

To a very good approximation the p_i can be approximated as

$$p_2(w_2) \propto (w_2/A_2)\Theta(A_2 - w_2) + \Theta(w_2 - A_2)\exp[-(w_2 - A_2)/2A_2],$$
$$p_3(w_3) \propto \exp(-w_3/A_3),$$
$$p_4(w_4) \propto \exp(-w_4/A_4), \tag{8.17}$$

with $A_2 = 0.3, A_3 = 0.6, A_4 = 0.4$. Initially, the p_i were not restricted to any specific form. The distribution functions p_i and hence the A_i slightly depend on the discretization chosen for the w_i axis. From the variance of the A_i, the statistical error of A_2 and A_4 can be estimated to be around 20%. The error of A_3 is somewhat larger since DWP's typically have large values of w_3 so that the determination of p_3 for small values of w_3 has a larger statistical error. Furthermore, only few DWP's with very large values of w_2 have been found since large w_2 are typically related to SWP's rather than DWP's. Therefore, the final decrease of p_2 with w_2 is only estimated qualitatively.

This result may serve as a microscopic verification of the soft-potential model in the regime of DWP's. In extension to the standard version of that model as described in Chap. 9, a distribution of the fourth-order parameter is also obtained.

Determination of Low-Temperature Parameters. For given distributions $p_i(w_i)$ the low-temperature parameters can be numerically determined in a straightforward way. The density of tunneling systems $P_{\text{eff}}(E)$ has been defined in (8.12) where the brackets denote the average over all DWP's. For each DWP, generated by the distributions $p_i(w_i)$, the tunneling matrix element Δ_0 as well as the excitation energy E has been calculated via the solution of the Schroedinger equation. For the present purpose, $E/k_B = 1$ K is chosen. Since

8.3 Double-Well Potentials in Computer Simulations

the number of DWPs's has been estimated on a quantitative basis, an absolute value of P_{eff} can be calculated. Using the material constants of NiP, it turns out $P_{\text{eff}}(E/k_B = 1 \text{ K}) = 1.6 \times 10^{46}$ J^{-1}m^{-3}.

For the calculation of the deformation potential, one has to take into account that $\gamma_\eta^2 \propto d^2$. This relation is motivated and derived in Sect. 8.4. Qualitatively, this relation indicates that energy minima which are further away are more susceptible to variations due to sound waves. Whereas the proportionality constant will be estimated in Sect. 8.4, the average value of d^2, denoted as $d_{\text{eff}}(E)^2$, can be already derived from the distribution of DWP's. Formally, this quantity can be written as

$$d_{\text{eff}}^2(E) \equiv \frac{\langle \delta(E - \sqrt{\Delta^2 + \Delta_0^2})(\Delta_0^2/E^2)d^2 \rangle}{P_{\text{eff}}(E)} \ . \tag{8.18}$$

In analogy to the definition of P_{eff}, it is essential to include the factor (Δ_0^2/E^2). Basically, it takes into account that the scattering of sound waves is more efficient for symmetric DWP's. Numerical evaluation of d_{eff}^2 around $E = 1$ K yields $d_{\text{eff}}/a \approx 0.35$. Together with the results for the proportionality factor in Sect. 8.4, one finally obtains $\gamma_t = 0.25$ eV, and $\gamma_l = 0.38$ eV.

For the estimation of $T_{C,1}$ and $T_{C,2}$ (temperatures related to the minimum and the maximum of the bump of $C(T)/T^3$) one has to calculate the specific heat on the basis of the distribution of soft modes. In the limit that for all soft modes only the lowest two energy levels are kept, the specific heat is given by the well-known expression

$$C(T) = \left\langle \frac{E^2}{4k_B T^2} \cosh^{-2}(E/2k_B T) \right\rangle \ . \tag{8.19}$$

In extension to the calculation of the tunneling parameters, the SWP regime also has to be taken into account. It turns out that the incorporation of higher energy levels does not influence the value of $T_{C,1}$, whereas $T_{C,2}$ is shifted downwards by approximately 10%. Numerically, one obtains for NiP $T_{C,1} = 2.8$ K and $T_{C,2} = 10.2$ K (Heuer and Silbey (1996)). Similar calculations with adjustable parameters have already been performed in Buchenau et al. (1991) in the framework of the soft-potential model and are presented in detail in Chap. 9. One also finds that the temperature dependence of the specific heat only slightly changes if $p_4(w_4) = \exp(-w_4/A_4)$ is replaced by $p_4(w_4) = \delta(w_4 - A_4)$. Therefore, for many practical purposes, it is possible to skip the integration over w_4 when averaging over all soft modes, as is done in the soft-potential model. If one wants to describe the outcome of heat release experiments, the total distribution of w_4 is relevant [see Parshin and Sahling (1993), and Sect. 2.3.1]. A comparison of the absolute values of the four low-temperature parameters with experimental data will be performed in Sect. 8.6.

Deviations from the Tunneling Model. The above values of the tunneling parameters were evaluated for $E/k_B = 1$ K. Within the tunneling model these values should not depend on E. However, numerically, some dependence on E can be found. In Sect. 8.2.2 the deviations from the tunneling model have been expressed in terms of $P_{\text{eff}}(E) \propto E^{\delta_1}$, $\gamma_\eta^2 \propto d^2 \propto E^{\delta_2}$, and $P_{\text{total}}(E, \tau_{\text{exp}}) \propto E^{\delta_3}$. These exponents can be easily calculated by repeating the above calculations for different values of E. One obtains for E around 1 K, $\delta_1 = 0.28, \delta_3(\tau_{\text{exp}} = \infty) = 0.09$, and $\delta_3(\tau_{\text{exp}} = 5\text{s}) = 0.29$. The energy dependence of the deformation potential is described by $\delta_2 = -0.14$. This value is negative because with decreasing d, typical tunneling matrix elements Δ_0 and hence typical energies E become larger. Taking into account $\delta_1, \delta_2 \neq 0$, one obtains $\kappa(T) \propto T^{1.86}$, i.e. $\beta = 0.14$, in excellent agreement with typical experimental values. Here, it was essential to take into account the energy dependence of γ_η^2 because otherwise $\beta = 0.28$ which would indicate too strong deviations from the T^2 behavior.

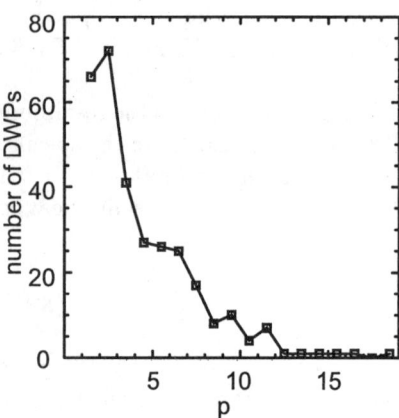

Fig. 8.6. The distribution of the effective mass p. Adapted from Heuer and Silbey (1996).

Microscopic Properties of DWP's. Apart from the distribution of potential parameters, the simulations contain information about microscopic characteristics of DWP's. In Fig. 8.6 the distribution of the effective mass p [see (8.4)] is displayed. Basically, p denotes how many particles significantly move when switching between both energy minima of a DWP. The distribution of p is rather broad with an average value of approximately four. This is exemplified for two DWP's (which for pedagogical reasons were calculated for a monatomic 2D LJ glass) shown in Fig. 8.7. Collective dynamics is important to explain the occurrence of the DWP shown on the right side of Fig. 8.7. However, the DWP's can still be regarded as rather localized. By comparing results from the simulations of glasses with $N = 150$ to $N = 500$, no significant dependence on the size of the box has been observed. The nature of

8.3 Double-Well Potentials in Computer Simulations   479

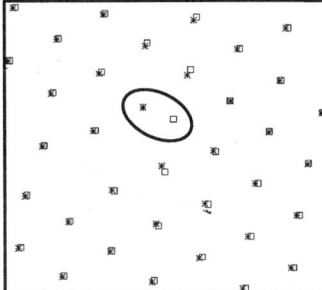 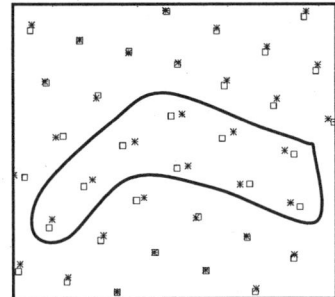

Fig. 8.7. Microscopic realization of two DWP's in a 2D glass. The stars and the squares correspond to the configurations at the two energy minima, respectively. The solid lines indicate which region of the glass is mainly involved in the reorganization process, when switching between both minima.

the present algorithm may tend to oversee DWP's with very large values of p ($p > 10$).

Interestingly, it turns out that the effective mass is strongly correlated with the potential parameters. The dependence of the average values of w_i on p is shown in Fig. 8.8. In a first approximation, typical values of w_i are inversely proportional to p. Qualitatively, this result is not surprising. In the limit of small p, the environment of a DWP is very stiff and hence typical energies involved in the transition between the two energy minima tend to be rather large. Of course, this results in a significant dependence of the potential height on the effective mass. DWP's with a small effective mass, i.e. localized DWP's, tend to have high barrier heights. This statement has been quantified by splitting the above sample of DWP's in two halves in dependence of their effective mass. The potential height V of the low-p subset of DWP's is on average nearly twice as high as V of the high-p subset. As already noted before, the distance d is also strongly correlated with the potential height V

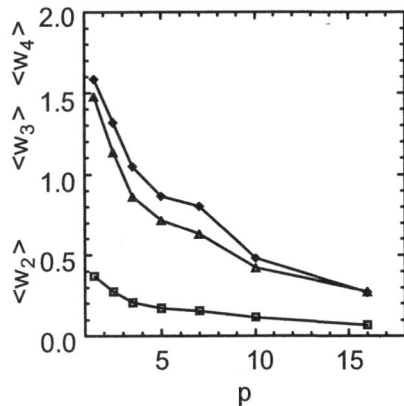

Fig. 8.8. The dependence of the average values w_i on the effective mass p. Adapted from Heuer and Silbey (1996).

(to a rough approximation $V \propto d^2$). Nearby minima are typically separated by a smaller saddle. This correlation is frequently used for a quantitative analysis of experimental data (Tielbürger et al. (1992)).

On first sight, it might seem surprising that, e.g. typical values of $\langle w_4 \rangle$ are of the order of one whereas A_4 is close to 0.5 [remember $p_4(w_4) \propto \exp(-w_4/A_4)$]. One reason for this apparent contradiction is that there exist a few DWP's with very large values of w_4 which are not captured by the exponential function $\exp(-w_4/A_4)$. The occurrence of such DWP's somewhat increases the average value of w_4. Furthermore, one has to keep in mind that in general the distribution p_4 is different from q_4, i.e. the number of DWP's with some specific value w_4 (see discussion above).

A DWP is defined by the vectors $\{d_i\}$ and the intermediate positions $\{r_{i,0}\}$ which are defined as

$$r_{i,0} = \frac{r_{i,L} + r_{i,R}}{2}. \tag{8.20}$$

The interesting question arises as to whether the $\{d_i\}$, characterizing the reaction coordinate within a DWP, are correlated among each other or correlated with the $\{r_{i,0}\}$, i.e. the local structure of the glass. For this purpose, the following terms have been analyzed:

$$\chi_1 \equiv \sum_{i<j}{}' \frac{d_{ij}^2}{12d^2},$$

$$\chi_2 \equiv \sum_{i<j}{}' \frac{(d_{ij} r_{ij,0})^2}{r_{ij,0}^2} / \sum_{ij} \frac{d_{ij}^2}{3}, \tag{8.21}$$

using the abbreviations $d_{ij} = d_i - d_j$, etc. The prime indicates that the sum only extends over nearest neighbors. The terms have been defined such that $\chi_i = 1$ in case that no correlations exist among the $\{d_i\}$ or between the $\{d_i\}$ and the $\{r_{i,0}\}$. For the numerically obtained DWP's, it turns out $\chi_1 = 0.7$ and $\chi_2 = 0.4$. $\chi_1 < 1$ implies that adjacent particles move along the same direction during the transfer between both minima. This can be envisaged as a chainlike motion in a tube formed by the other particles. The second relation implies a strong correlation between d_{ij} and $r_{ij,0}$. It means that two nearest neighbors favor a relative movement perpendicular to the vector which connects both particles. In this way, the change in distance tends to be small, hence decreasing the change in energy. Both effects can be observed in the 2D DWP's shown in Fig. 8.7. The wormlike structure is very obvious. Furthermore, it never occurs that two adjacent particles move antiparallel to each other. This gives rise to the second correlation. As will be discussed in Sect. 8.4, the value of the deformation potential depends in a sensitive way on these correlations. A further analysis of 2D DWP's can be found in Dab et al. (1995).

8.3.4 Application of Different Search Strategies

As discussed in Sect. 8.3.1, two further approaches may be chosen to identify DWP's.

Using Information about Low-Frequency Eigenmodes. Separate simulations were performed in order to check whether the reaction path is correlated with some low-frequency eigenmode resulting from diagonalization of the force matrix in one local energy minimum (Heuer (1997c)). For this purpose, a rather small glass with $N = 100$ particles is taken so that, on average, only a single DWP is present per glass configuration. Starting from one minimum of DWP's found by the algorithm discussed in Sect. 8.3.3, the results can be summarized as follows:

1. For a given DWP characterized by the $\{d_i\}$ and for the eigenvectors $v_{i,j}$ (shift of the i-th particle in the j-th mode) calculated for the initial energy minimum, the correlation coefficient

$$c_j = |\sum_{i=1}^{N} \hat{d}_i \hat{v}_{i,j}| \tag{8.22}$$

is calculated and finally the maximum $c_{\max} \equiv \max c_j$ taken. In case of perfect correlation, one would obtain $c_{\max} = 1$. The result indicates significant correlations. (i) On average it turns out $c_{\max} \approx 0.5$. In case no correlations are present (numerically this is achieved by random variation of the indices of the d_i), one obtains $c_{\max} \approx 0.1$. (ii) In 40% of all cases, the eigenmode with the largest correlation corresponds to the eigenmode with the lowest frequency. The correlation coefficient $c_{\max}$ is in agreement with earlier simulations (Schober et al. (1993), Oligschleger and Schober (1995)). The authors have concluded that these correlations are weaker than anticipated within the framework of the soft-potential model.

2. The effective mass of the selected eigenmode is, on average, a factor 2–3 larger than for the respective DWP. This shows that the cooperativity is much larger in the case of SWP's as compared to the case of DWP's. By increasing N, this difference gets even larger.

3. Calculating the potential energy along the eigenmode with the best correlation to the $\{d_i\}$, for more than 95% of all DWP's, a continuous increase of the potential energy in both directions of the minimum is observed. Hence the correlation of the eigenmode with the $\{d_i\}$ is not close enough in order to directly locate the second minimum. This result is consistent with the observation that the straight line between two minima of a DWP does not meet the saddle point and hence anharmonic contributions strongly influence the potential energy even on short length scales.

4. Finally, whether final relaxation from different points along the path constructed from the eigenmode allows localization of the second minimum is checked. Here, it turns out that the probability to end up in the second minimum strongly depends on the nature of the DWP. This is shown in Fig. 8.9. Whereas for small d/a, the second minimum could be reliably found, for larger d/a this probability dramatically decreases. Qualitatively, it is not surprising that the eigenmode with the highest correlation coefficient goes through the attraction basin of the second minimum in case the second minimum is very close.

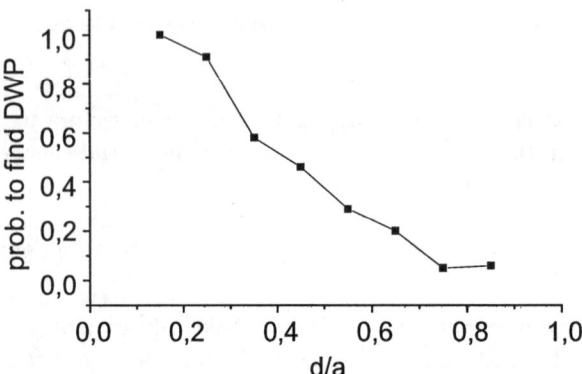

Fig. 8.9. The probability that knowledge of the most similar eigenmode allows us to find the second minimum of the DWP.

In summary, one may conclude that there is some correlation between SWP's and DWP's, mainly expressed by the fact that the related eigenmode tends to be the eigenmode with the lowest frequency. However, the correlation is not strong enough in order to allow a systematic search of DWP's from the knowledge of the force matrix. Stated differently, the *local* information from one well (expressed by the force matrix **K**) is only sufficient to locate very nearby minima ($d < 0.4$) and hence does not allow us to obtain information about the energy landscape on a more *global* scale.

Using Information from MD Runs. Finally, the information which can be obtained from MD simulations is discussed, as sketched in Fig. 8.2i and already discussed above. Due to the excessive need for energy minimizations, rather small systems have been simulated ($N = 40, 80$); see Buechner and Heuer (1997). As a result of the simulations, one obtains a list of minima which were successively visited during the MD run. From this list, information about the nature of transitions between adjacent local energy minima can be extracted.

For three different temperatures, the average value of d^2 is shown in Fig. 8.10. The units of the temperatures are determined by the energy of the potential minimum of the pair potential. It turns out that at lower temperatures, typical jump distances are significantly smaller. The underlying reason

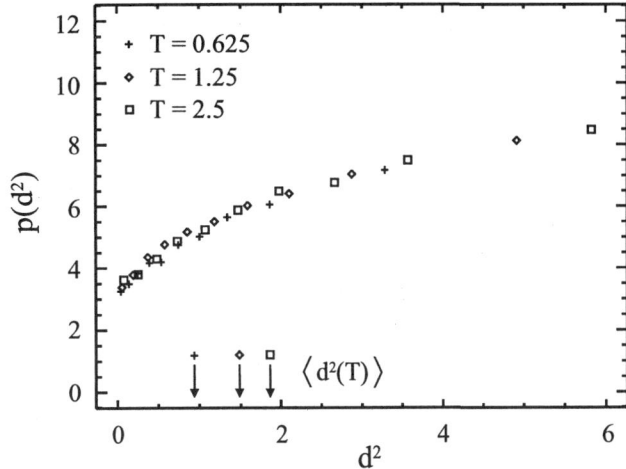

Fig. 8.10. The dependence of the effective mass on d^2, obtained from MD runs with $N = 40$ at different temperatures. The average values of d^2 are indicated for all temperatures. For $N = 80$, the $p(d^2)$ curve is shifted to higher values of p by approximately 20%. d is given in units of σ which approximately corresponds to the average distance between particles [see also the definition of σ in the discussion of the pair potential (8.16)].

is that DWP's with small distances tend to have lower potential heights (see discussion in Sect. 8.3.3), which are therefore preferred at low temperatures. Having in mind that for TS's, typical values of d are less than 0.5, this result clearly shows that only limited information can be obtained from MD simulations if one is really interested in the physics of the low-temperature anomalies.

It may be interesting to check the correlation of the effective mass with distance d. One observes a slight dependence of p on d^2 which, however, is much weaker than $p \propto d^2$. Hence the decrease of d^2/p with increasing p shows that, for motion with large effective masses, the individual particles themselves move less. Qualitatively, this effect can already be seen from the comparison of both DWP's in Fig. 8.7. It turns out that this correlation is, to a first approximation, independent of the simulation temperature, as shown in Fig. 8.10. Hence the nature of DWP's with respect to the $p(d^2)$ dependence does not change with temperature. Of course, for this kind of simulation, one expects large finite-size effects. However, upon going from $N = 40$ to $N = 80$, the typical effective mass $p(d^2)$ only increases by approximately 20%.

Finally, it is mentioned that due to the dependence of d^2 on temperature, one also obtains an increase of the effective mass with temperature. This again shows that one has to be careful in predicting properties of TS's from high-temperature MD simulations.

These results are in qualitative agreement with the work by Oligschleger and Schober (1995). In that work, a significant correlation of the effective mass with temperature and jump distance has been observed for a soft sphere glass and selenium too.

The attentive reader might be startled about the above-mentioned relation between p and d. As stated in Sect. 8.3.3, DWP's with larger potential heights, on the one hand, are related to DWP's with small effective masses p and, on the other hand, to DWP's with large values of d. However, the straightforward conclusion that d should be inversely correlated to p is *not* true, as shown in Fig. 8.10. This example shows that one has to be very careful when predicting statistical properties of strongly correlated quantities.

8.3.5 Tunneling Systems in the Presence of Impurities

One of the basic assumptions of the tunneling model is the broad distribution of DWP parameters, reflecting the somewhat random structure in glasses. Furthermore, coooperativity is an important ingredient for the dynamics at low temperatures. This situation dramatically changes if one considers the properties of very small particles dissolved in a glass. Qualitatively, one expects a diffusive motion of the impurities in the fixed structure of the glass. This implies that (i) the impurities display single-particle motion rather than cooperative dynamics and (ii) typical distances between local minima of the DWP will be of the order of typical nearest-neighbor distances in the glass and thus are no longer characterized by some broad distribution. The DWP's related to the dynamics of impurities are called *extrinsic* DWP's. The other DWP's are denoted as *intrinsic*.

By considering a single particle with variable size, diluted in a glass, it is possible to interpolate between both types of DWP's (Heuer and Neu (1997)). The main results of this work are presented below.

Model Glass with an Impurity. To the LJ glass, defined by (8.16), a single particle (formally called type 3) is added. Its size is characterized by the parameter α. Remember that the distance between Ni atoms corresponds to $\alpha_{1,1} = 1$ and the distance between Ni and P atoms to $\alpha_{1,2} = 1.05$. The fraction of adjacent P-P pairs is very small. Upon variation of α, one expects that for $1 \leq \alpha \leq 1.05$, the impurity behaves like the rest of the glass whereas otherwise deviations may occur (α large: small impurity; α small: large impurity). The energy parameter A in (8.16) is chosen as for the Ni atoms. The simulations have been performed in the range $0.7 \leq \alpha \leq 1.6$.

The search algorithm for DWP's is identical to the one employed in Sect. 8.3.3. Since one is interested in the region around the impurity, only a single cluster of particles is selected for the initial minimization routine, containing the impurity as well as its 15 nearest neighbors; see Sect. 8.3.1. This procedure is repeated for approximately 3000 different initial configurations. For all DWP's, the particle that moves most during the transition between both local energy minima is determined. If this particle is the impurity, this DWP is defined as extrinsic, otherwise intrinsic.

8.3 Double-Well Potentials in Computer Simulations

Simulation Results. First, the probability for the existence of a DWP close to the impurity is determined. Figure 8.11 shows the probability that an extrinsic DWP is detected. One can see that the number of extrinsic DWP's dramatically increases for large values of α. For the smallest particles analyzed in the simulation runs, nearly 80% of all impurities are connected with a DWP. Of course, this number has to be viewed as a lower limit since it is possible that the algorithm may miss some DWP's. For $\alpha = 1$, the impurity is identical to the other particles of the glass. Hence, the probability at $\alpha = 1$ has to be the same as the probability to find a DWP in the whole glass, expressed as a per-particle probability. In agreement with the results discussed in Sect. 8.3.3, one DWP exists per 100 particles. It turns out as a surprise that already for $\alpha = 1.2$, a significant number of extrinsic DWP's can be observed. Hence, only small differences between the size of the impurity and the rest of the particles strongly enhance the probability for the formation of a DWP. For $\alpha = 0.7$, no extrinsic DWP's have been found. A huge particle is locked in a cage formed by the surrounding small particles without a chance to escape.

In Fig. 8.11, the probability to find intrinsic DWP's close to the impurity is displayed too. This probability is again calculated on a per-particle basis. By definition, the formation of extrinsic and intrinsic DWP's for $\alpha = 1$ has the same probability. Interestingly, the number of intrinsic DWP's close to the impurity strongly depends on the size of the impurity. The number is largest for $\alpha \approx 1$, but strongly decreases for larger or smaller impurities. This result can be understood on a qualitative level. For α, very different to unity, the impurity does not participate in the dynamics of the glass because it does not move at all (small α) or because it only moves by itself (large α). Hence, from the viewpoint of the glass particles, there exists an "alien element" which forms a barrier for the surrounding particles. Thus, close to the impurity, the degrees of freedom and hence the chance to form a DWP are significantly reduced. This effect is irrelevant only for $\alpha \approx 1$, since the impurity may simply participate in the dynamics.

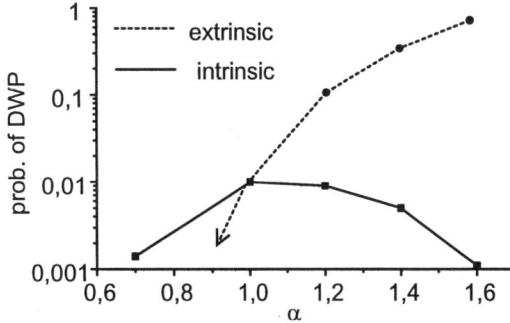

Fig. 8.11. Probability that extrinsic DWP's are formed by the impurity and intrinsic DWP's are formed close to the impurity. Note that α is inversely proportional to the radius of the impurity. The array indicates that for $\alpha = 0.7$, no extrinsic DWP's have been detected. Adapted from Heuer and Neu (1997).

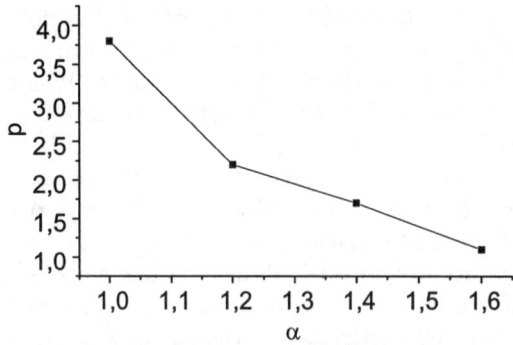

Fig. 8.12. The dependence of the average effective mass p of the extrinsic DWP's on α. Adapted from Heuer and Neu (1997).

In order to characterize the nature of the extrinsic DWP's somewhat closer, the average effective mass p of the DWP's in dependence on α is determined; see Fig. 8.12. For $\alpha = 1.6$, the extrinsic DWP's are related to motions of the impurity alone. In the limit of very small particles, this result could have been expected. Basically, the impurity jumps between different interstitial positions formed by a fixed environment. However, it is again surprising that already for $\alpha = 1.2$, the effective mass of extrinsic DWP's is significantly reduced as compared to the effective mass of intrinsic DWP's.

In Fig. 8.13, the distribution of barrier heights of intrinsic DWP's without a nearby impurity as well as the distribution for extrinsic DWP's (i.e. related to impurities) with $\alpha = 1.6$ is plotted. In both cases, a broad distribution of barrier heights is observed, reflecting the statistical nature of the formation process of a DWP. However, obviously the extrinsic DPW's in case of small impurities are significantly shifted to higher potential heights. Since the DWP's related to small impurities have small effective masses (as shown in Fig. 8.12), the resulting correlation between small effective masses and high barrier heights is in qualitative agreement with the general conclusions drawn in Sect. 8.3.3.

In summary, these simulations indicate that the presence of small impurities may give rise to additional (extrinsic) DWP's with properties different

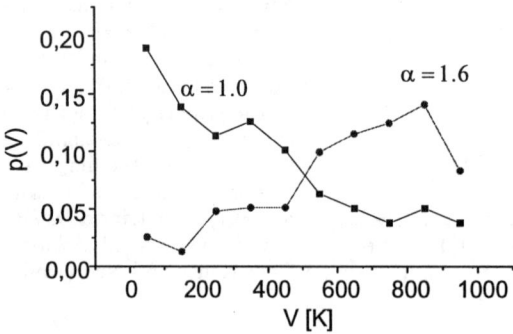

Fig. 8.13. Distribution of barrier heights of intrinsic ($\alpha = 1.0$) and extrinsic DWP's in the case $\alpha = 1.6$. Adapted from Heuer and Neu (1997).

from the DWP's already present in the original amorphous system (intrinsic DWP's). Thus, for the present case, a description of the low-temperature anomalies only in terms of the tunneling model would not be sufficient. Due to the high probability of the formation of extrinsic DWP's, already small concentrations of defects may lead to significant additional contributions (see also the discussion in Sect. 8.7.4).

8.3.6 Total Energy Landscape of a Glass-Forming System

In Fig. 8.2, the existence of DWP's has already been sketched as a property of the high-dimensional energy landscape of the glass. The DWP's, which are relevant at low temperatures, can be viewed as nearby pairs of local energy minima. In this sense, they are characterized by the fine structure of the energy landscape. At higher temperatures, additional parts of the energy landscape, unaccessible at low temperatures, become relevant. Basically, the energy landscapes contain information about all local energy minima as well as all saddles between contiguous minima. As already mentioned before, it has become a convenient approach to describe properties of glass-forming systems in terms of its multidimensional potential energy landscape in configuration space [see e.g. Goldstein (1969), Stillinger (1995), Ball et al. (1996), and references therein].

For several reasons, the knowledge of the total energy landscape is advantageous as compared to the knowledge of the DWP's obtained from the simulations described above: (i) The identification of DWP's from the knowledge of the total energy landscape is definitely unbiased. As discussed above, the systematic search algorithm somewhat tends to prefer DWP's with slightly smaller effective masses. (ii) How the DWP's are embedded in the total energy landscape is clarified. Qualitatively, DWP's can be viewed as small bumps in a very rugged energy landscape. The overall structure of the energy landscape is thought to contain information about the glass transition. In this way, the DWP's can be regarded as precursors to the glass transition. Of course, the energy landscape also contains information about the melting of crystals.

It is *a priori* not evident how to define a DWP. Considering the (one-dimensional) energy landscape in Fig. 8.14a, one would tend to call the pair of minima D and E a DWP. Two conditions are fulfilled for this pair. First, both minima are very close in configuration space so that they have a common saddle, and second, the energy at their common saddle is smaller than the energy of all other saddles which can be reached from the left or the right minimum. The second condition guarantees that at sufficiently low temperatures, the system can switch between both minima without escaping to a third minimum.

The determination of the total energy landscape for glass-forming systems with a typical size of 1000 particles is not possible since the number of minima exponentially grows the number of particles. For example, a small noble gas cluster with 13 particles already has the order of 100 local energy

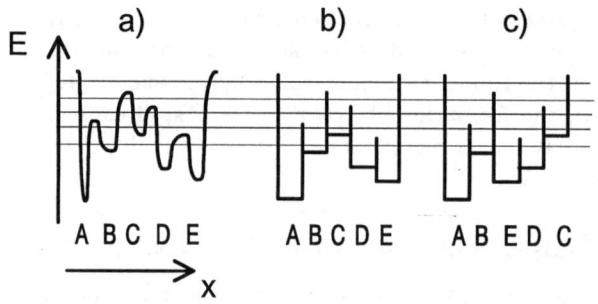

Fig. 8.14. (a) A simple 1D energy landscape; (b) schematic representation of (a); (c) the potential minima are rearranged such that the transfer matrix remains the same. Adapted from Heuer (1997a).

minima; see Berry (1993). The extension to larger clusters is not possible in a complete way and the energy landscape has to be analyzed in statistical terms (Ball et al. (1996)). The physical properties of small clusters are largely dominated by surface effects. Thus, if one is interested in the simulation of bulklike properties like the low-temperature anomalies for small systems, it is essential to use periodic boundary conditions. In Heuer (1996) (2D system) and Heuer (1997a) (3D system), the total energy landscape of a glass-forming system with periodic boundary conditions has been determined. The system is identical as described above except that only one type of particle is present. The system contains 32 particles. By choosing such a small system of identical particles, the number of minima is still accessible from simulations (see below). First results for the identical model are already given in Stillinger and Weber (1983). The simulated densities are $\varrho = 1$ and $\varrho = 1.075$ in units of the nearest-neighbor distance a and unit mass. For a polymer glass the density difference corresponds to an applied pressure of approximately 4 kbar (Grace and Anderson (1989)). The energy of the fcc crystal $E_{\text{cryst}}(\varrho)$ has its minimum for $\varrho = 1$ [$E_{\text{cryst}} \equiv E_{\text{cryst}}(\rho = 1) = -192$ in LJ units, Stillinger and Weber (1983)]. The analysis of different densities is motivated by the experimental observation that the density of tunneling systems in a glassy polymer significantly decreases upon the application of pressure, Grace and Anderson (1989); see Sect. 8.7.5.

In a first step ca. 10^5 conjugate gradient minimization procedures, starting from arbitrarily chosen initial configurations, were performed in order to get a (hopefully) complete list of energy minima $E(k)$, where k denotes the number of the minimum. 367 minima were found with different energy for $\varrho = 1$ and 75 for $\varrho = 1.075$. Their distribution is plotted in Fig. 8.15. Whereas the crystalline minimum turns out to be stable upon increasing density this does not hold for most amorphous minima. This is already reflected by the observation that the absolute number of minima decreases by more than a factor of four when going from $\varrho = 1$ to $\varrho = 1.075$. Performing the minimization with *variable* density, it was observed that the number of different minima dramatically increases so that a systematic search is no longer possible. This partly explains the observation why the number of energy minima

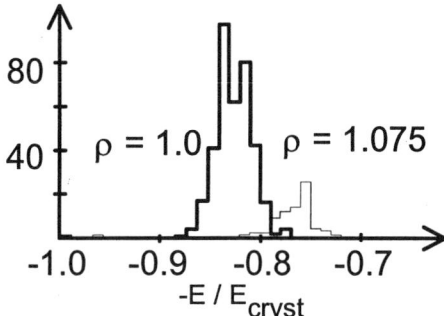

Fig. 8.15. Distribution of energy minima for two different densities ϱ. Adapted from Heuer (1997a).

of 13 particles in a cluster with no constraints on density is of the same order as that of 32 particles in a fixed volume.

In a second step, the distances of all pairs (k_1, k_2) of minima in configuration space has been determined. If the positions of the N particles are given by $\{r_{i_1,k_1}\}$ and $\{r_{i_2,k_2}\}$, one can define the Euclidean distance by

$$[d(k_1, k_2)]^2 = \sum_{i_1=1}^{N} (r_{i_1,k_1} - r_{i_2(i_1),k_2})^2 \ . \tag{8.23}$$

The notation $i2(i1)$ indicates that *a priori* it is not evident which particle of configuration k_2 corresponds to which particle of k_1 so that several mappings have to be checked. For the application of (8.23), two further aspects have to be considered. First, due to the periodic boundary conditions, the configuration $\{r_{i,k_j}\} + a_j$ with arbitrary vector a_j also belongs to energy minimum k_j. Hence, for an appropriate definition of $d(k_1, k_2)$, one additionally has to determine the value of $a_1 - a_2$ which minimizes $d(k_1, k_2)$. Second, for a 3D cube, one minimum corresponds to 48 different configurations which are related by symmetry operations like 90° rotations. This number results from 3! permutations of axes and 2^3 reflections. Hence a comparison of two energy minima in reality corresponds to a comparison of one configuration belonging to k_1 with 48 symmetry related configurations belonging to k_2. For the determination of $d(k_1, k_2)$, the minimum distance for all symmetry related configurations has been chosen. Since the MD simulations in Stillinger and Weber (1983) revealed that for ambient temperatures, only configurations with $E(k) < (5/6)E_{cryst}$ are relevant, the analysis was restricted to these 223 minima. For $\varrho = 1.075$, all minima were analyzed. From Fig. 8.11, which resulted from a MD study, it became clear that contiguous amorphous minima in configuration space tend to have a distance of $d \leq 2$ (in LJ units).

Properties of Adjacent Minima. In Sect. 8.3.3 properties of adjacent pairs of minima of the energy landscape were extracted. Here, we specifically refer to properties for which knowledge of the total energy landscape is essential. Although DWP's are only a subset of pairs of adjacent minima, knowledge of the general structure of the energy landscape also allows a closer understanding of the nature of DWP's.

For a given configuration k_1, one defines $d_{min}(k_1) = \min_{k_2} d(k_1, k_2)$. This value is a measure of how close the relevant configurations are in the high-dimensional configuration space. In Fig. 8.16, the distribution of d_{min}^2 is plotted for both densities.

In agreement with intuition, in both cases, the configuration with the largest value of d_{min} corresponds to the crystalline structure. Interestingly, for $\varrho = 1.075$ the whole d_{min} distribution is shifted to larger values, yielding a gap for $d_{min} < 0.6$. Hence, for denser systems, the different configurations are further away from each other in configuration space. Since the DWP's which dominate the low-temperature properties correspond to $d \approx 0.3 - 0.4$ (see Sect. 8.3.3), the present calculations, at least qualitatively, predict a significant decrease of the tunneling systems with increasing pressure (see Sect. 8.7.5 for comparison with experiment).

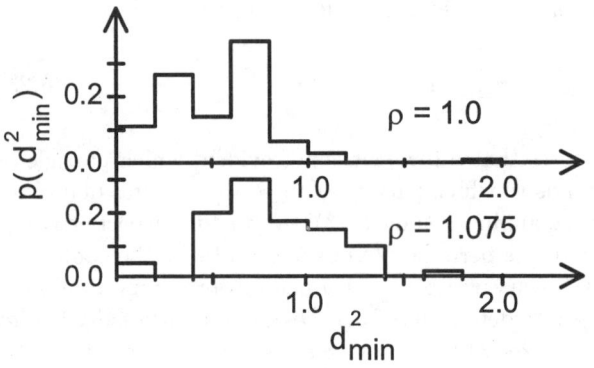

Fig. 8.16. The distribution of d_{min}^2 for $\varrho = 1$ (upper curve) and $\varrho = 1.075$ (lower curve). Note the depopulation of small values of d_{min}^2 for the higher density. Adapted from Heuer (1997a).

One-Dimensional Representation of the Energy Landscape. As a step towards identification of the DWP's, it is convenient to find a 1D projection of the total energy landscape which keeps the information about DWP's. Such a projection scheme has been presented in Heuer (1997a). For the characterization of DWP's, the energy $V(k_1, k_2)$ at the saddle of contiguous minima k_1 and k_2 has to be known. In principle, it is possible to determine the saddle between two minima in the high-dimensional configuration space. For the present purpose, it is sufficient to have a rough estimate of its value. Here, the approach of the soft-potential model is applied, which parametrizes the potential along the reaction path between adjacent energy minima by quartic polynomials of the type $B[w_2(x/a)^2 - w_3(x/a)^3 + w_4^0(x/a)^4]$ with constant w_4^0 determined above and independently distributed w_2 and w_3. Postulating that w_4^0 describes the quartic term of the transition between all pairs of adjacent minima, the values of w_2 and w_3 and thus of $V(k_1, k_2)$ can be directly estimated from the knowledge of $d(k_1, k_2)$ and $E(k_1) - E(k_2)$. For pairs of

8.3 Double-Well Potentials in Computer Simulations

minima with $d(k_1, k_2) > 2$, one formally sets $V(k_1, k_2) = \infty$. The subsequent results are insensitive to the precise value of w_4^0 and to the definition of adjacent minima ($d \leq 2$).

The information about the energy landscape is expressed by the energies $E(k)$, the distances $d(k_1, k_2)$ and the potential heights $V(k_1, k_2)$. In the spirit of the work of Stillinger (1990), one may define the *transfer matrix* $\tilde{V}(k_1, k_2)$ which contains the minimum saddle-point energy for all (indirect and direct) paths from minimum k_1 to minimum k_2. This matrix expresses the connectivity among all minima. As discussed by Stillinger (1990), $\tilde{V}(k_1, k_2)$ contains important information about the dynamics in glass-forming systems and is the basis for the definition of metabasins. For a 1D potential, the values of $\tilde{V}(k_1, k_2)$ are easily determined because minima k_1 and k_2 are connected by only a single path. For arbitrary multidimensional potential, it its possible to construct a 1D potential with *identical* transfer matrix. This enables visualization of important information in 1D and furthermore gives a strict recipe of how 1D representations of multidimensional potentials may be interpreted.

The algorithm can be outlined as follows. For given energy E_0, one defines groups of minima such that $\tilde{V}(k_1, k_2) \leq E_0$ for all members of one group and $\tilde{V}(k_1, k_2) > E_0$, otherwise ($k_1 \neq k_2$). For $E_0 \to \infty$, one has a single group which, during a decrease of E_0, continuously splits into smaller groups. For $E_0 \to -\infty$, no group is left. From checking all different E_0, indicated in Fig. 8.14 as horizontal lines, the different groups for the potential in Fig. 8.14 read as (A,B,C,D,E), (A,B), (C,D,E), (D,E). One can easily convince oneself that it is possible for arbitrary multidimensional potential to sort all minima $k_1, ..., k_N$ such that all the members of any group are contiguous. For the potential of Fig. 8.14, this is fulfilled, for example ABCDE and ABEDC but not for example ABECD, since, here, the members of the group (D,E) are not contiguous. Based on this sorting, the schematic potentials, shown in Figs. 8.14b and 8.14c, can be constructed with identical $\tilde{V}(k_1, k_2)$ as the original (possibly multidimensional) potential. In order to have a unique representation, it is further required that, starting from the left, the sequence of energy minima increases as monotonically as possible, see Fig. 8.14c. Note that bins which are adjacent in the 1D projection may have a large distance in the real potential landscape like minima B and E in Fig. 8.14c. A conceptually similar representation of a high-dimensional potential energy landscape can be found in Becker and Karplus (1997).

Another example for a 1D projection is shown in Fig. 8.17. Again, it is evident that the 1D projection keeps most of the relevant information of the original high-dimensional potential. First, the separation of minima A and E from the rest is conserved. Second, the effective saddle between minima B and C is small as compared to the high direct saddle. This reflects the fact that at sufficiently low temperatures, the system will move from B to C via D, hence avoiding the direct saddle.

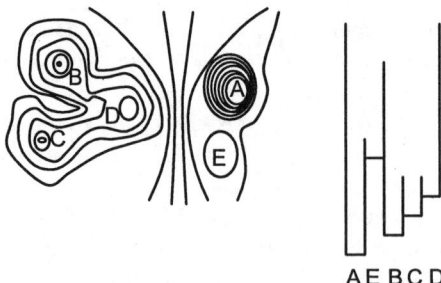

Fig. 8.17. A 2D energy landscape and its 1D projection.

In Fig. 8.18, this schematic potential is shown for the energy landscape of the LJ glass with $\varrho = 1$. The minimum on the very left side is the crystalline minimum, corresponding to a pure fcc structure. A very high energy has to be reached before the crystal can "melt". Here, melting means that the system leaves the first minimum. For the noncrystalline minima, basically two regions I and II can be distinguished; region I containing minima 3–6 [only minimum 3, having an energy of $E(3)/E_{\text{cryst}} \approx 0.895$, is relevant], and region II the other relevant amorphous minima. From the knowledge of the energy landscape, one may predict that the longest timescale of relaxation at low temperatures is related to the transition between both noncrystalline regions I and II and that the time to leave minimum 3 is longer than for other minima. This can already be seen from the MD simulations in Stillinger and Weber (1983).

Estimation of the Number of DWP's. In the remaining part of this section, the number of DWP's is estimated from the knowledge of the energy landscape. For the multidimensional potential, the DWP's have been defined as

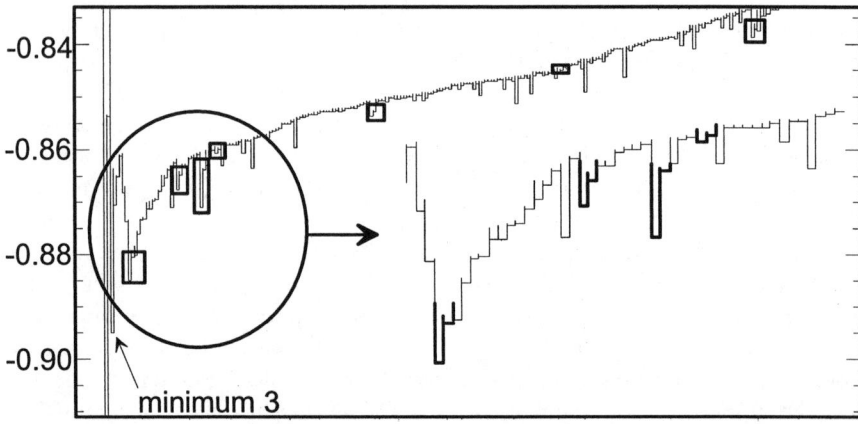

Fig. 8.18. The 1D projection of the energy landscape for $\varrho = 1$. The DWP's and minimum 3 are highlighted. All minima with energy $E < (5/6)E_{\text{cryst}}$ are included.

stable entities at low temperatures (see the strict definition in the discussion of Fig. 8.14). The projection scheme presented above has been devised such that exactly those pairs of minima which would be identified as DWP's in the original multidimensional potential also turn out to be DWP's in the projected 1D potential. Hence, they can be identified just from having a close view on Fig. 8.18. They are marked by squares. Most of them have a distance $d^2 \ll 1$, and hence correspond to very nearby configurations. Having found 7 DWP's for 223 minima, one has a probability of 14/223 per minimum that it belongs to a DWP. Hence, a configuration with $32 \times 223/14 \approx 500$ particles on average contains 1 DWP.

This number is by a factor of 5 smaller than the number of DWP's found from the simulations discussed in Sect. 8.3.3. Partly, this discrepancy may be due to the fact that the disorder in glasses with only identical particles is smaller than in glasses with different types of particles. Anyhow, these calculations show that, already for very small systems, DWP's naturally arise as a general feature of the energy landscape.

8.4 Coupling Between Tunneling Systems and Heat Bath

The coupling between TS's and phonons can be expressed in terms of the deformation potentials γ_η ($\eta = l, t$); see Sect. 8.2. The index expresses the experimental observation that the coupling to longitudinal phonons is different than the coupling to transverse phonons. Typically, one has $G_{lt} \equiv (\gamma_l/\gamma_t)^2 = 2.6 \pm 0.4$. For most glasses, the value of G_{lt} is smaller than 3. This is demonstrated in Fig. 8.19 where data of the work of Berret and Meißner (1988) are collected. The data indicate that, on average, the silica-based glasses have a somewhat smaller value of G_{lt}.

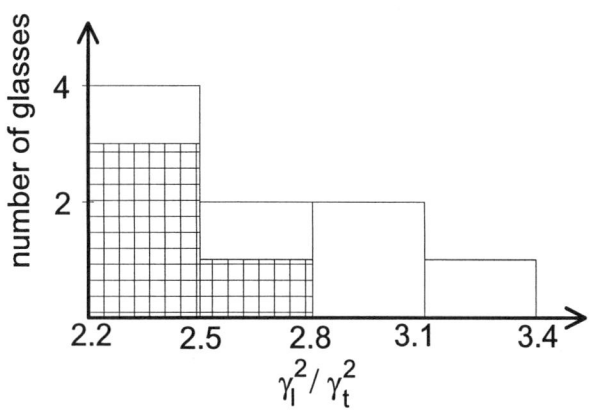

Fig. 8.19. The experimental value of G_{lt} for the set of glasses presented in the work of Berret and Meissner (1988). The shaded areas denote the contributions of the silica-based glasses. Adapted from Heuer (1997b).

A rough estimation of the absolute value of the deformation potential has been given within the soft-potential model (Galperin et al. (1985b), Buchenau et al. (1992)). In Heuer and Silbey (1993b) and Heuer (1997b) it has been demonstrated that the absolute value of γ_η as well as the ratio G_{lt} contain important information about the nature of TS's and the structure of the glass. This work is outlined below.

8.4.1 Microscopic Origin of the Deformation Potential and the Velocity of Sound

The microscopic nature of a TS is fully characterized by the coordinates of the left and right position $\{r_{i,R}\}$ and $\{r_{i,L}\}$ or, alternatively, by the shift vectors $d_i \equiv r_{i,R} - r_{i,L}$ and the average positions $r_{i,0} \equiv (r_{i,R} + r_{i,L})/2$. For reasons of simplicity, a system with only one type of particle interacting via some pair potential $f(r)$ is considered. The first goal is to calculate γ_η for a given TS. This can be achieved by expressing γ_η in terms of $\{d_i\}$, $\{r_{i,0}\}$, and the pair potential $f(r)$. Since the deformation potential turns out to be strongly related to the velocity of sound v_η, this quantity is also discussed. Throughout the analysis it is assumed that $d_i^2/a^2 \ll 1$ (a: equilibrium distance).

First, how the presence of sound waves changes the positions of the individual particles is discussed. A sound wave is considered with wave vector q and polarization vector p, with p chosen along the x-direction. For longitudinal sound waves, one has $q \propto p$, for the transverse case, the wave vector is oriented along the y-direction. Then the shift of the particle positions by a single sound wave can be expressed as

$$r_i - r_{i,0} = -i[\exp(iqv_{i,0})b - \exp(-iqv_{i,0})b^*]\frac{1}{\sqrt{N}}e_x, \qquad (8.24)$$

where $q \equiv |q|$ denotes the wave number and N the absolute number of particles. b and b^* are proportional to the creation and annihilation operator, respectively. In case of a longitudinal wave $v_{i,0}$ is identified with $x_{i,0}$ (the x-component of $r_{i,0}$), in case of a transverse wave with $y_{i,0}$. In the limit of long wavelenghts, the value of q is much smaller than typical nearest-neighbor distances. It follows that

$$r_{ij} - r_{ij,0} = v_{ij,0}qQ\frac{1}{\sqrt{N}}e_x, \qquad (8.25)$$

where $r_{ij} \equiv r_i - r_j$ and $[\exp(iqv_{i,0})b + \exp(-iqv_{i,0})b^*]$ has been identified with the position operator Q of the corresponding phonon. For small displacements, one obtains from (8.25)

$$r_{ij} \approx r_{ij,0} + \frac{v_{ij,0}x_{ij,0}}{r_{ij,0}}u, \qquad (8.26)$$

where the strain field u was introduced via

$$u \equiv qQ/\sqrt{N}. \qquad (8.27)$$

8.4 Coupling Between Tunneling Systems and Heat Bath

It is a measure of the intensity of the sound wave. Of course, the same relation also holds for deviations within the left or right well where the index 0 is substituted by L or R.

Determination of the velocity of sound. First, small fluctuations around the metastable configuration $\{r_{i,L}\}$ are analyzed, giving rise to an energy shift

$$\Delta E = \frac{1}{2} \sum_{i<j} f^{(2)}(r_{ij,L})(r_{ij} - r_{ij,L})^2 . \tag{8.28}$$

$f^{(m)}(r)$ denotes the m-th derivative of $f(r)$.

Combining (8.26) and (8.28), one obtains

$$\Delta E = \frac{1}{2N} q^2 \mathcal{Q}^2 \sum_{i<j} f^{(2)}(r_{ij,L}) \frac{(x_{ij,L} v_{ij,L})^2}{r_{ij,0}^2} . \tag{8.29}$$

In this way the deviations from the local energy minimum are expressed by the sound wave coordinate $\mathcal{Q}$. This expression can be compared with $\Delta E = m\omega^2 \mathcal{Q}^2/2$ which relates the energy change of a sound wave to its frequency. Since $\omega = v_\eta q$ (v_l: longitudinal, v_t: transverse velocity of sound) in the long-wavelength regime, this comparison finally yields

$$mv_\eta^2 = \frac{1}{N} \sum_{i<j} f^{(2)}(r_{ij,L}) \frac{(x_{ij,L} v_{ij,L})^2}{r_{ij,0}^2} . \tag{8.30}$$

Due to the lack of long-range order in amorphous systems, one does not expect any dependence of physical observables on the direction of the sound wave. Therefore, it is appropriate to average (8.30) over all directions of the wave vector and the polarization vector. Alternatively, one may average over all directions of the vector r_{ij}. This boils down to calculate $s_l = \langle x^4 \rangle$ and $s_t = \langle x^2 y^2 \rangle$, where the brackets denote the average over the unit sphere. A straightforward calculation yields $s_l = 1/5$ and $s_t = s_l/3$. Now (8.30) can be rewritten as

$$mv_\eta^2 = \frac{s_\eta}{N} \sum_{i<j} f^{(2)}(r_{ij,L}) r_{ij,0}^2 . \tag{8.31}$$

In a further step, we take into account the structure of the LJ glass. To a good approximation, each particle is on average surrounded by n-nearest neighbors with equilibrium distance a. For close packed structure like 3D LJ glasses, the coordination number n is 12. The contribution of terms corresponding to pairs which are not nearest neighbors is rather small because $f^{(2)}(r)$ strongly decreases for larger values of r. Therefore, one finally obtains

$$mv_\eta^2 \approx \frac{ns_\eta}{2} f^{(2)}(a) a^2 . \tag{8.32}$$

A direct evaluation of (8.30) on the basis of the generated $T = 0$ glass configurations yields on average 0.95 times the right side of (8.32) so that the above approximations are indeed reasonable. (8.32) can be rewritten in terms of the average phonon velocity v which is defined via $3v^{-3} = 2v_t^{-3} + v_l^{-3}$. Incorporating the numerical factor of 0.95, one finally obtains

$$mv^2 = \frac{2n}{51} f^{(2)}(a) a^2 . \tag{8.33}$$

For the case of a two-component LJ glass one simply has to average $f^{(2)}(a)$ over the three different pairs of particles with the correct statistics. m is the average mass. For the parameters of NiP, given in Sect. 8.3, one has $f^{(2)}(a)a^2 \approx 91000$ K and $m \approx 56 m_p$. Inserting these values in (8.33), one obtains $v = 2530$ ms^{-1} which compares well with the experimental value of 2600 ms^{-1}, from Bellessa (1980).

Determination of the Deformation Potential. The definition of γ_η in (8.10) can be rewritten as

$$\gamma_\eta = \frac{1}{2} \sum_{i<j} \frac{\partial [f(r)|_{r=r_{ij,R}} - f(r)|_{r=r_{ij,L}}]}{\partial u} . \tag{8.34}$$

With the help of [see (8.26)]

$$\frac{\partial r_{ij,0}}{\partial u} = \frac{v_{ij,0} x_{ij,0}}{r_{ij,0}} , \tag{8.35}$$

(8.34) can be approximated as

$$\gamma_\eta \approx \frac{1}{2} \sum_{i<j} \frac{v_{ij,0} x_{ij,0}}{r_{ij,0}} [f^{(1)}(r_{ij,R}) - f^{(1)}(r_{ij,L})] . \tag{8.36}$$

Since

$$[f^{(1)}(r_{ij,R}) - f^{(1)}(r_{ij,L})] \approx f^{(2)}(r_{ij,0})(\hat{r}_{ij,0} d_{ij}) , \tag{8.37}$$

γ_η is proportional to $d \times f^{(2)}$. In (8.36), terms proportional to $(d/a) \times f^{(1)}$ have been neglected because of $f^{(1)}(r_{ij,0}) \ll a f^{(2)}(r_{ij,0})$ for $r_{ij,0} \approx a$.

In analogy to the velocity of sound, an isotropic average should be performed in order to get rid of any direction dependence. It is essential to specify for which power of γ_η this average is performed. One is guided by the observation that experimentally accessible quantities like the thermal conductivity depend on γ_η^2 (see Sect. 8.2). Hence, one has to determine the isotropic average of the term $W_\eta \equiv \langle v_{ij,0} x_{ij,0} \rangle \langle v_{kl,0} x_{kl,0} \rangle$ by considering all possible orientations of the polarization vector and the wave vector. We introduce three orthogonal unit vectors h_i ($i = 1, 2, 3$) such that $p \parallel h_1$ and $q \parallel h_1$ (h_2) in the longitudinal (transverse) case. For reasons of simplicity, it is chosen $r_{ij} = (1, 0, 0)$ and $r_{kl} = (\cos\varphi, \sin\varphi, 0)$. This leads to

$$\begin{aligned} W_l &= \langle h_{1,x}^2 (h_{1,x} \cos\varphi + h_{1,y} \sin\varphi)^2 \rangle \\ &= A_2 + (A_1 - A_2) \cos^2\varphi \end{aligned} \tag{8.38}$$

and

8.4 Coupling Between Tunneling Systems and Heat Bath

$$W_t = \langle h_{1,x} h_{2,x}(h_{1,x}\cos\varphi + h_{1,y}\sin\varphi)(h_{2,x}\cos\varphi + h_{2,y}\sin\varphi)\rangle$$
$$= A_4 + (A_3 - A_4)\cos^2\varphi, \qquad (8.39)$$

with $A_1 = \langle h_{1,x}^4 \rangle$, $A_2 = \langle h_{1,x}^2 h_{1,y}^2 \rangle$, $A_3 = \langle h_{1,x}^2 h_{2,x}^2 \rangle$, $A_4 = \langle h_{1,x} h_{2,x} h_{1,y} h_{2,y}\rangle$ (the brackets again denote the isotropic average). The terms proportional to $\cos\varphi\sin\varphi$ disappear due to symmetry reasons. The A_i can be evaluated in a straightforward manner:

$$A_1 = 1/5, \qquad (8.40)$$

$$A_1 + 2A_2 = \langle h_{1,x}^2 \rangle = 1/3 \Rightarrow A_2 = 1/15, \qquad (8.41)$$

$$A_3 = A_2, \qquad (8.42)$$

and, using $\boldsymbol{h}_1 \cdot \boldsymbol{h}_2 = 0$,

$$A_3 + 2A_4 = 0 \Rightarrow A_4 = -1/30. \qquad (8.43)$$

Hence one obtains

$$W_l = (1/15)(1 + 2\cos^2\varphi) \qquad (8.44)$$

and

$$W_t = (1/30)(-1 + 3\cos^2\varphi). \qquad (8.45)$$

Summarizing these results, one can finally write for the averaged deformation potentials ($\hat{\boldsymbol{r}}_{ij,0} \equiv \boldsymbol{r}_{ij,0}/r_{ij,0}$),

$$\gamma_\eta^2 = \frac{s_\eta}{96} \sum_{i,j}\sum_{k,l} r_{ij} r_{kl}$$
$$[f^{(1)}(r_{ij,R}) - f^{(1)}(r_{ij,L})][f^{(1)}(r_{kl,R}) - f^{(1)}(r_{kl,L})] A_{ijkl,\eta}, \quad (8.46)$$

with

$$A_{ijkl,l} = 2\left[1 + 2(\hat{\boldsymbol{r}}_{ij,0} \cdot \hat{\boldsymbol{r}}_{kl,0})^2\right] \qquad (8.47)$$

and

$$A_{ijkl,t} = 3\left[-1 + 3(\hat{\boldsymbol{r}}_{ij,0} \cdot \hat{\boldsymbol{r}}_{kl,0})^2\right]. \qquad (8.48)$$

8.4.2 Numerical Evaluation of the Deformation Potential

In Heuer and Silbey (1993b) the deformation potential has been calculated on the basis of (8.46) for DWP's which were found from simulations of LJ glasses. Due to (8.37), one expects from (8.46) that $\gamma_\eta^2 \propto (d/a)^2$. Hence the coupling is stronger for DWP's whose minima are further away in configuration space. In the derivation of the relation it was assumed that all particles only move small distances so that only the first expansion term was relevant. Defining

$$\Gamma_\eta^2 \equiv \gamma_\eta^2/(d/a)^2, \qquad (8.49)$$

in Fig. 8.20, the values of Γ_l for approx. 200 DWP's are shown in dependence of d/a. As a dominant feature, one observes very large fluctuations among different DWP's. Hence the value of the deformation potential depends in a very sensitive way on the microscopic realization of the DWP. Of course, without having performed the isotropic average, the variations would even be larger. To a good approximation, the average value of Γ_l is indeed independent of d, hence confirming the relation $\gamma_\eta \propto d$. Averaging over all DWP's one obtains $\Gamma_l = 1.07\text{eV}$ and $\Gamma_t^2 = \Gamma_l^2/2.5$, i.e. $G_{lt} = 2.5$. As mentioned before this value of G_{lt} agrees very well with typical experimental data.

Together with (8.33), these results can be rewritten as

$$\Gamma_l \approx \frac{mv^2}{\sqrt{n}} \tag{8.50}$$

and

$$\Gamma_t \approx \frac{2mv^2}{3\sqrt{n}}, \tag{8.51}$$

where n is the coordination number.

8.4.3 Relation Between the Deformation Potential and the Structure of DWP's

The simulations do not reveal why $G_{lt} \approx 5/2$ and which properties of DWP's determined the value of γ_η.

For the subsequent analytical estimations we will use

$$\gamma_\eta^2 = \frac{s_\eta}{96} \sum_{i,j,k,l} r_{ij} r_{kl} f^{(2)}(r_{ij}) f^{(2)}(r_{kl}) A_{ijkl,\eta} (d_{ij} \hat{r}_{ij})(d_{kl} \hat{r}_{kl}) \tag{8.52}$$

that can be obtained from (8.36) and (8.37). For reasons of simplicity the index '0' has been omitted.

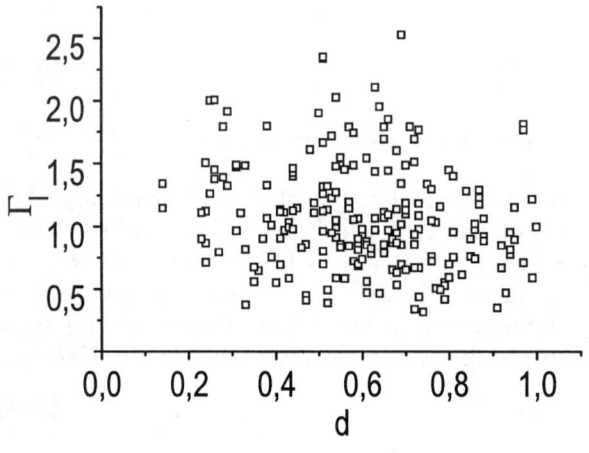

Fig. 8.20. The distribution Γ_l for simulated DWP's in dependence of d where $\Gamma_\eta^2 \equiv \gamma_\eta^2/(d/a)^2$.

8.4 Coupling Between Tunneling Systems and Heat Bath   499

The deformation potentials can be abbreviated as $\gamma_l^2 = 2(u^2 + 2v)$ and $\gamma_t^2 = -u^2 + 3v$ with $u \propto \sum_{ij} f^{(2)}(r_{ij}) r_{ij} (\hat{\mathbf{r}}_{ij} \mathbf{d}_{ij})$ and a similar expression for v. It is evident that G_{lt} has its minimum value when $u^2 = 0$, yielding $G_{lt} = 4/3$. For a given glass structure $\{\mathbf{r}_i\}$, it is always possible to find vectors $\{\mathbf{d}_i\}$ such that $u = 0$. On the contrary, for $u^2 \approx 3v$, one obtains $G_{lt} \to \infty$. Hence, for a single DWP, the value of G_{lt} may vary between $4/3$ and ∞.

Experimentally, one observes an average over many different DWP realizations. As discussed several times in this chapter, the statistical nature of the formation of DWP's is an essential ingredient for understanding the low-temperature anomalies in glasses. Hence one may attempt to find an appropriate statistical description of DWP's. The average over the possible realizations of a DWP will then fix the value of γ_η^2 and G_{lt}. It will be shown that, after this average, the value of G_{lt} is much more restricted than one might guess from the general relation $G_{lt} \in [4/3, \infty]$.

For a LJ glass, the second derivative of $f(r)$ strongly decreases for distances larger than the nn shell. Hence, it is a reasonable approximation to restrict the sum in (8.52) over nn's. As a technical simplification, we assume that for all nn's $r_{ij} = a$. Then (8.52) boils down to

$$\gamma_\eta^2 = \frac{s_\eta}{96} a^2 [f^{(2)}(a)]^2 \sum_{i,j,k,l} A_{ijkl,\eta} (\mathbf{d}_{ij} \hat{\mathbf{r}}_{ij})(\mathbf{d}_{kl} \hat{\mathbf{r}}_{kl}) \; . \tag{8.53}$$

On this basis, G_{lt} can be evaluated if the $\{\mathbf{r}_i\}$ as well as the $\{\mathbf{d}_i\}$ are given.

Averaged values of γ_η^2 and G_{lt} can be obtained after the specification of reasonable configurations $\{\mathbf{r}_i\}$ and reaction paths $\{\mathbf{d}_i\}$. This will be done on three different levels.

(1) *Minimum Correlations.* At first, it is assumed that all properties of DWP's are as uncorrelated as possible. One only takes into account that a particle is surrounded by n other particles on the nn shell. The directions of the $\{\mathbf{d}_i\}$ are assumed to be uncorrelated among each other as well as to the glass structure $\{\mathbf{r}_i\}$. This choice only contains very little information about DWP's. Neglecting correlations in the structure implies that the particle on the nn shell are randomly distributed. After performing the average over the directions of the $\hat{\mathbf{r}}_{ij}$, $A_{ijkl,\eta}$ is nonzero only for $(i,j) = (k,l)$ and $(i,j) = (l,k)$ ($A_{ijij,\eta} = A_{ijji,\eta} = 6$). Furthermore, one has to average over the directions of the $\{\mathbf{d}_i\}$, keeping their lengths d_i fixed. Since they are assumed to be uncorrelated among themselves and uncorrelated to the $\{\mathbf{r}_i\}$, the term $(\hat{\mathbf{r}}_{ij} \mathbf{d}_{ij})^2$ transforms into $(1/3)(d_i^2 + d_j^2)$. Taking into account that $d^2 = \sum d_i^2$ and that each particle has n particles on the nn shell, one obtains the transformation

$$\sum_{ij} (\hat{\mathbf{r}}_{ij} \mathbf{d}_{ij})^2 \Rightarrow (2/3) n d^2 \; . \tag{8.54}$$

Finally this results in

$$\begin{aligned} \gamma_\eta^2 &= \frac{n s_\eta}{12} [f^{(2)}(a)]^2 a^4 (d/a)^2 \\ &\equiv \Gamma_{\eta,\text{uncorr}}^2 (d/a)^2 \; . \end{aligned} \tag{8.55}$$

One immediately obtains $G_{lt} = 3$. Specification to the case of NiP yields $\Gamma_{l,\text{uncorr}} = 3.5$ eV. In comparison to the simulations (see Sect. 8.4.2) the value of $\Gamma_{l,\text{uncorr}}$ is by more than a factor of 3 too large. Since experimental observables are proportional to γ_η^2, this would result in predictions which are wrong by an order of magnitude. In contrast, the ratio G_{lt} is close to the value of 2.5 found in experiments and simulations.

(2) *Correlations Connected with the DWP's.* A major simplification in the previous analysis was the absence of any correlations. As discussed in Sect. 8.3.3, the numerical observation $\chi_{1,2} < 1$ indicates that (i) adjacent $\boldsymbol{d}_i$ are correlated and (ii) the direction of the tunneling dynamics expressed by the $\boldsymbol{d}_{ij}$, are correlated with the local structure of the glass, defined by the $\boldsymbol{r}_{ij}$. Taking into account these correlations, one obtains in generalization to (8.54),

$$\sum_{ij}(\hat{\boldsymbol{r}}_{ij}\boldsymbol{d}_{ij})^2 \Rightarrow (2/3)nd^2\chi_1\chi_2 \ . \tag{8.56}$$

Hence both correlations may lead to a strong reduction of the absolute value of γ_η^2. Note that the ratio G_{lt} remains unchanged so that these correlations do not explain why the value of $G_{lt} \approx 2.5$, appearing in simulations as well as experiments, is smaller than three.

(3) *Correlations Connected to the Glass Structure.* In a final step of the statistical description, one may take into account that the direction vectors $\boldsymbol{r}_{ij}$ of the particles of a nn shell are correlated itself. Even in glasses, there exists a significant amount of local order. Hence, one may expect that the nn shell somehow reflects the symmetry of the crystalline counterpart. Rewriting (8.36) together with the approximation in (8.37) and $\boldsymbol{d}_{ij} = \boldsymbol{d}_i - \boldsymbol{d}_j$ as

$$\gamma_\eta = 2\sum_{i,j}\frac{v_{ij}x_{ij}}{r_{ij}}\boldsymbol{d}_i\hat{\boldsymbol{r}}_{ij} \ , \tag{8.57}$$

it is evident that for strict inversion symmetry $\gamma_\eta = 0$. Hence residual local order further reduces the absolute value of γ_η^2. More important, the value of G_{lt} also turns out to depend on the degree of local order.

In order to estimate this dependence quantitatively, all correlations related to the DWP's are neglected for reasons of simplicity. Anyhow, as shown above, they alone do not influence the value of G_{lt}. First, G_{lt} is rewritten as [see (8.53)]

$$G_{lt} = \frac{\sum_{ijkl} B_{ijkl,l}}{\sum_{ijkl} B_{ijkl,t}} \tag{8.58}$$

with

$$B_{ijkl,l} = 2\left[1 + 2(\hat{\boldsymbol{r}}_{ij}\hat{\boldsymbol{r}}_{kl})^2\right](\hat{\boldsymbol{r}}_{ij}\boldsymbol{d}_i)(\hat{\boldsymbol{r}}_{kl}\boldsymbol{d}_k) \tag{8.59}$$

and

$$B_{ijkl,t} = \left[-1 + 3(\hat{\boldsymbol{r}}_{ij}\hat{\boldsymbol{r}}_{kl})^2\right](\hat{\boldsymbol{r}}_{ij}\boldsymbol{d}_i)(\hat{\boldsymbol{r}}_{kl}\boldsymbol{d}_k) \ . \tag{8.60}$$

8.4 Coupling Between Tunneling Systems and Heat Bath

In analogy to above, the average over the directions of the d_i is first taken. Since all d_i are uncorrelated, only terms with $i = k$ contribute. The orientational average of $(\hat{\mathbf{r}}_{ij}\mathbf{d}_i)(\hat{\mathbf{r}}_{il}\mathbf{d}_i)$ over d_i can be easily performed and yields $(d_i^2/3)\hat{\mathbf{r}}_{ij}\hat{\mathbf{r}}_{il}$.

For the statistical analysis one may assume that, on average, the structure of all nn shells display identical properties. Therefore, the analysis may be restricted to a single nn shell surrounding particle i. Choosing the origin as $r_i = 0$, one can finally write

$$G_{lt} = \frac{\sum_{jk} C_{jk,l}}{\sum_{jk} C_{jk,t}} \tag{8.61}$$

with

$$C_{jk,l} = 2\left[(\hat{\mathbf{r}}_j \hat{\mathbf{r}}_k) + 2(\hat{\mathbf{r}}_j \hat{\mathbf{r}}_k)^3\right] \tag{8.62}$$

and

$$C_{jk,t} = \left[-(\hat{\mathbf{r}}_j \hat{\mathbf{r}}_k) + 3(\hat{\mathbf{r}}_j \hat{\mathbf{r}}_k)^3\right] . \tag{8.63}$$

Now, one may specify the correlations among the $\{\hat{\mathbf{r}}_j\}$, i.e. the nn shell of particle i. First, the case of a regular structure (tetrahedral and inversion

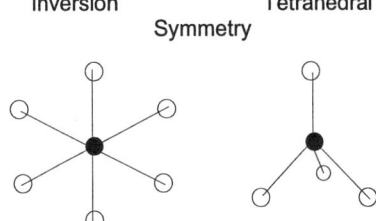

Fig. 8.21. Sketch of inversion symmetry (displayed in 2D for reasons of simplicity) and tetrahedral symmetry.

symmetry, respectively; see Fig. 8.21) is discussed. It is relevant for describing glasses like SiO_2 where, to first approximation, the individual SiO_4-tetrahedra are surrounded by 4 other tetrahedra, sharing identical O-atoms. In the extreme limit of a regular structure, one has as $\sum_j \hat{\mathbf{r}}_j = 0$ so that the first term of $C_{jk,\eta}$ does not contribute. Hence one directly obtains $G_{lt} = 4/3$. This result is noteworthy. It shows that the order effect indeed results in a strongly reduced value of G_{lt}. Furthermore, it turns out that for tetrahedral symmetry, the smallest possible value of G_{lt} can indeed be reached even if averaged over all realizations of the DWP's. As already shown in the discussion of (8.57), for the case of inversion symmetry, the value of G_{lt} is not defined since $\gamma_\eta = 0$.

In Heuer (1997b) the full transition between the fully ordered and fully disordered nn shell has been determined. Disorder has been defined such that the directions $\hat{r}_j$ of the n particles of the nn shell (LJ glass: $n = 12$, tetrahedral symmetry: $n = 4$) deviate from the regular directions $\hat{\mathbf{u}}_j$ by angles which are randomly drawn from a Gaussian distribution with variance ϵ^2. The dependence of G_{lt} on ϵ can be calculated analytically for tetrahedral as well as inversion symmetry. The result is shown in Fig. 8.22a. In agreement with the results mentioned before one obtains $G_{lt}(\epsilon \to \infty) = 3$. For inversion symmetry, one observes a transition from 7/4 to 3, for tetrahedral symmetry from 4/3 to 3.

The calculations leading to Fig. 8.22a were based on the assumption that all distances to the center particle are equal to a. In real systems, one rather expects a distribution of nn distances. In order to estimate the effect of such variations, the simple case that one particle of the nn shell is missing has been considered. This calculation has been performed in analogy to the above and the result is displayed in Fig. 8.22b (for the case of inversion symmetry $n = 12$ was used). Increasing the disorder by shifting away one particle leads to an increase in G_{tl}. This is consistent with the idea that disorder in general tends to shift G_{lt} to the value of 3. Interestingly, for small ϵ, the increase of G_{lt} upon removing one particle is larger than for tetrahedral symmetry. The underlying reason is that in this regime the individual values of γ_η for a full nn shell are so small (see above) that the disorder introduced by shifting away one particle dominates the individual γ_η and hence G_{lt}. In contrast, for tetrahedral symmetry, γ_η remains finite if the disorder is reduced ($\epsilon \to 0$). From the general formulas given in Heuer (1997b), it can be estimated that, e.g. γ_l^2 is approximately reduced by 50%. Therefore, the effect of missing particles on the nn shell in the ordered limit is weaker.

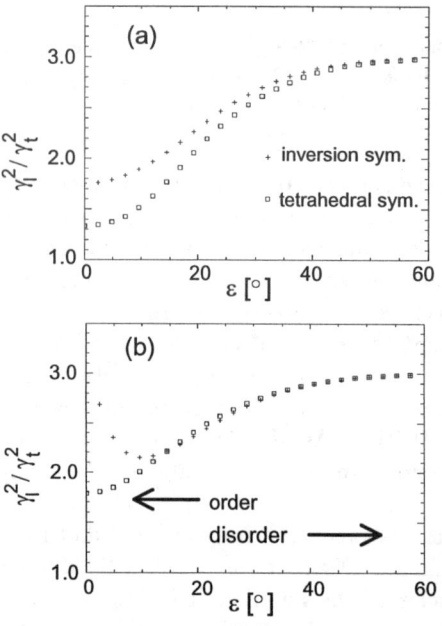

Fig. 8.22. (a) The value of G_{lt} in dependence on disorder, expressed by the average deviation ϵ (in degrees) of the individual bonds from the regular structure for inversion symmetry (crosses) and tetrahedral symmetry (squares). (b) Same as in (a) except that one particle of the nn shell (for inversion symmetry $n = 12$, corresponding to LJ glasses) is missing. This additional increase of disorder translates into an increase of G_{lt}.

In summary, the statistical approach chosen to describe the nature of DWP's revealed that the actual range of G_{lt} is much smaller than the range $[4/3, \infty]$ which theoretically can be achieved for individual DWP's. Furthermore, it turned out that the value of G_{lt}, may be interpreted as a measure for the local disorder of glasses close to TS's.

This analysis also allows one to speculate why the experimental G_{lt} is smaller for silica-based glasses. Two reasons may be relevant. First, due to the formation of networks, local disorder is weaker as compared, e.g. to polymer glasses. Second, even for comparable weak local disorder, G_{lt} is smaller for tetrahedral symmetry, see Fig. 8.22.

It is interesting to compare the above approach with older results by Buchenau and Schober (1989) which were obtained on the basis of elasticity theory. In that reference, it is shown that for a specific TS, related to a deformed octahedron, one obtains $G_{lt} = 4/3$. Via additional compressional contributions, it is possible to obtain arbitrary large values of G_{lt}. Although this approach is consistent with the present analysis, it is limited to a specific kind of DWP and hence the actual range of G_{lt} as found experimentally is not obtained.

8.5 Nature of Tunneling Systems Beyond Computer Simulations

For some further understanding of the low-temperature anomalies, it would be helpful to obtain information beyond experiments or computer simulations (which to some degree can also be viewed as "experiments" except that the information content is somewhat different). An important question deals with the relation between low-temperature parameters and properties of the glass in general. In the framework of the soft-potential model, one might be specifically interested to know the values of P_0, γ_η, and A_4. Starting from the Hamiltonian of, e.g. a LJ glass, it is not evident how these values can be predicted from first principles. In two ways this unsatisfactory situation may be circumvented.

First, one may analyze appropriate toyglasses which are supposed to mimick real structural glasses. Below, two examples are presented for which the low-temperature anomalies could be derived on analytical grounds.

Second, postulating some features of soft modes (which may be motivated by the simulation results), it is possible to estimate the low-temperature parameters analytically. For the example of Γ_η, this already has been demonstrated in Sect. 8.4. In this section some approaches in this direction are presented.

8.5.1 1D Model Glass

A simple model glass with an analytically accessible distribution of TS's has been presented by Reichert and Schilling (1985). Later on, the model was extended (Häner and Schilling (1989)) by considering the influence of external pressure.

In this model a chain of N particles is considered with the energy

$$V = \sum_n \sum_{l=1}^{2} V_l(u_{n+l} - u_n) - p \sum_n (u_{n+1} - u_n) . \tag{8.64}$$

u_n is the position of the n-th particle and p mimicks the external pressure. V_1 is chosen to describe a DWP with minima at a_1 and a_2:

$$V_1(x) = \frac{C_1}{2}[x - a_+ - a_-\sigma(x)]^2 , \tag{8.65}$$

where $a_\pm = (1/2)(a_2 \pm a_1)$, and $\sigma(x) = \text{sgn}(x - a_+) = \pm 1$. In contrast, $V_2(x)$ describes a purely harmonic interaction

$$V_2(x) = \frac{C_2}{2}(x - b)^2 , \tag{8.66}$$

with $C_2 \neq 0$

The important step is the determination of *all* local minima. The following results can be strictly derived.

1. For $N \to \infty$ the solutions of $(\partial V / \partial u_n) = 0$ can be written as

$$v_n(\sigma, p) = \alpha_1 + \alpha_2 \sum_{i=-\infty}^{\infty} \eta^{|i|} \sigma_{n+i} + \alpha_3 p , \tag{8.67}$$

where $\sigma_i = \pm 1$ and α_i and η only depend on the specific choice of the model parameters. According to this formula, all 2^N possible sets $\sigma = \{\sigma_i\}$ correspond to a local energy minimum, respectively. The situation is slightly more complicated since, additionally, the self-consistency condition

$$\sigma_n = \text{sgn}(v_n(\sigma, p) - c) \tag{8.68}$$

has to be fulfilled for all n, where c is another model-dependent constant. However, the parameters of the model can be chosen such that this condition is fulfilled for all 2^N choices of σ.

2. DWP's can be identified with single-particle transitions over a barrier. An exact condition that the potential around particle i contains a barrier is given by $\sigma_{i-1} \neq \sigma_i$. The new metastable configuration after the transition of particle i is characterized by a sequence $\{\sigma_i'\}$ for which σ_{i-1} and σ_i are exchanged. Thus, for a random sequence $\{\sigma_i\}$ (corresponding to an amorphous structure), on average, half of the particles are connected with a DWP. Due

to the simple structure of DWP's, their distribution can be determined analytically. The resulting low-temperature specific heat reads as $C(T) \propto T^d$, where the precise value of $0 < d < 0.63$ depends on the model parameters.

3. The number of metastable configurations has been determined in dependence of pressure in Häner and Schilling (1989), starting from a stress-free model. In agreement with the simulations of Sect. 8.3.6, this number decreases with increasing pressure. Interestingly, this decrease is not continuous (even for $N \to \infty$) but occurs in a stairlike manner. For sufficiently large pressure, one ends up with a single metastable configuration.

A related but more complex model is the 2D Frenkel–Kontorova model with piecewise harmonic substrate potential in Uhler and Schilling (1988). A microscopic derivation of the properties of such DWP's, involving only a single-particle movement, is possible. The density of DWP's depends on the quenching temperature related to the glass formation and on the type of quenched disorder, inherent in this model. The model displays a constant density of low-energy states in dependence of energy.

8.5.2 Spin Glass Like Model Glass

In structural glasses, the disorder is self-induced and is due to the complexity of the energy landscape. A different type of disorder as encountered in spin glass theory is the case of quenched disorder. It is expressed via some random coupling constants J_{ij} in the potential energy. In a recent work by Kirkpatrick and Thirumalai (1989), the fundamental similarity between both kinds of disorder has been stressed.

Motivated by these similarities, Kühn and Horstmann (1997) have analyzed a model which is devised to describe structural glasses but for which the disorder is introduced in analogy to spin glasses. The potential energy of their model reads as

$$U(v) = -\frac{1}{2}\sum_{i,j}^{N} J_{ij} v_i v_j + \frac{1}{\gamma}\sum_{i}^{N} G(v_i) \tag{8.69}$$

with some on-site potential

$$G(v) = \frac{1}{2}v^2 + \frac{a}{24}v^4 \ . \tag{8.70}$$

The coupling constants J_{ij} are independent Gaussian random numbers with mean J_0/N and variance $1/N$. The v_i can be viewed as the coordinate of the i-th particle where $v_i = 0$ corresponds to some reference position. The amorphous nature of the model comes in through the random linear contribution proportional to $v_i v_j$. For reasons of simplicity, the additional contributions, expressed in $G(v)$, are nonrandom. The consideration of $G(v)$ is important in order to stabilize the potential. Evidently, by the parameter γ, the number of modes with a negative second derivative can be tuned.

The formal treatment of this problem is borrowed from spin-glass theory. The basic idea is to consider an ensemble of systems with independent realizations of the J_{ij} (different *replica*). Using the replica symmetric approximation, one finally ends up with an effective replica-symmetric single-site potential

$$U_{\rm RS}(v) = -h_{\rm RS} v - \frac{1}{2} C v^2 + \frac{1}{\gamma} G(v) \qquad (8.71)$$

with the local field

$$h_{\rm RS} = [J_0 m + \sqrt{q} z] \,. \qquad (8.72)$$

z denotes a zero-mean univariance Gaussian variable and q, m, C denote coefficients which are determined self-consistently. It can be viewed as a mean-field potential where, however, the influence of the random parameter J_{ij} is still present in the random variable z. Whenever $\gamma C > 1$ the above potential contains two local minima (at least for sufficiently small z). By variation of the random variable z, the distribution of TS's, and hence of Δ and λ, can be easily obtained.

In contrast to the standard tunneling model, Δ and λ turn out to be highly correlated since both are functions of a single Gaussian variable. However, at low temperatures nevertheless, the omnipresent relation $C(T) \propto T$ is observed. Taking into account also the contributions of SWP's and the upper energy levels in DWP's at higher temperatures (the harmonic contributions), even a bump in the $C(T)/T^3$-plot occurs.

An important conclusion of this work is related to the expected degree of universality in the tunneling regime in comparison to the harmonic regime [bump of $C(T)/T^3$]. The linear temperature dependence at low temperatures is related to the global part of the potential mediated by the J_{ij} and hence may be viewed as a collective effect which thus tends to display universal properties. In contrast, the harmonic excitations giving rise to the bump in $C(T)/T^3$ are mainly related to the on-site potentials $G(v)$ so that a stronger material dependence can be expected. This aspect of the degree of universality will be discussed in Sect. 8.6 from a different point of view.

8.5.3 Simple Models of Soft Modes

Coupled Rotations. As will be discussed in Sects. 8.7.3 and 9.2.3, the neutron scattering on SiO_2 indicates the presence of coupled rotations of SiO_4 tetraedra. Buchenau (1985a) has proposed a simple model for soft modes which takes into account these rotations. It is a coupled rotation of two adjacent molecular units in a glass as shown in Fig. 8.23, see also Sect. 9.2.3.

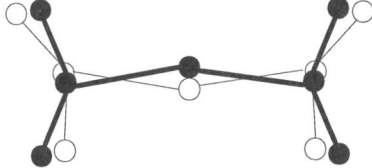

Fig. 8.23. Sketch of two adjacent molecules performing a coupled rotation. Adapted from Buchenau (1985a).

The potential energy function which is related to the coupled rotation is essentially split in two parts. The first part describes the intramolecular bond bending potential of the two central bonds. The second part contains the energies which arise from the interaction of this molecule with the elastic dipole and the elastic quadrupole of the surrounding glass. Averaging over all possible elastic dipole and quadrupole configurations with an appropriate Boltzmann weighting, a distribution of potentials for the coupled rotation is obtained. Knowledge of this distribution allows one to estimate the absolute number of TS's in good agreement with the experiment.

Wormlike Soft Modes. As already mentioned in Sect. 8.3.3, the analysis of DWP's and SWP's suggests that the soft modes have a low-dimensional structure. Here the extreme limit is treated such that they are strictly one-dimensional. The realization of such a soft mode is shown in Fig. 8.24. For reasons of simplicity, the soft mode is realized by a parallel motion of p particles along one line. The distance, covered by one particle, is denoted as d_p. If the soft mode corresponds to a DWP with a distance d between both minima, one has $d_p = d/\sqrt{p}$. Each atom along the chain is surrounded by

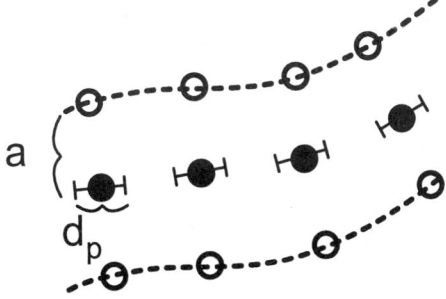

Fig. 8.24. Sketch of a wormlike DWP as discussed in the text.

approximately $(n-2)$ atoms (n: coordination number) which for this simplistic model do not participate in the motion. Let x denote the collective coordinate varying between 0 and d. Then the potential energy $V(x)$ reads to a good approximation as

$$V(x) \approx p(n-2)f(\sqrt{a^2 + (x-d/2)^2/p}), \qquad (8.73)$$

where f denotes the pair potential. The factor p in front of the expression takes into account that in this model, all p particles experience the identical interaction.

A straightforward calculation of the fourth derivative yields

$$\frac{1}{4!}\frac{\partial^4 V(x=d/2)}{\partial x^4} = \frac{3(n-2)f^{(2)}(a)}{4!(a^2 p)}, \qquad (8.74)$$

that can be identified with Bw_4/a^4 in (8.15). Comparison with (8.33) finally yields

$$w_4 \approx 2.5/p. \qquad (8.75)$$

First of all, this relation confirms the numerical observation that w_4 approximately scales with $1/p$. Furthermore, even the absolute value of $w_4(p)$ agrees reasonably well with the numerical data (compare Fig. 8.7). Hence this simple DWP model contains basic features of the numerical DWP's and allows a semiquantitative prediction of the value of w_4. A somewhat different order of magnitude estimation of w_4 can be found in Parshin (1994b), yielding $w_4 \approx 1$.

8.6 Universality of the Low-Temperature Parameters

8.6.1 Corresponding States

In this section the first question posed by Phillips about the comparison of different glasses (see Sect. 8.1) is elucidated. Experimentally, it turns out that the omnipresent product $P_0\gamma_\eta^2$ for most glasses, analyzed in the review by Berret and Meißner (1988), only varies within a factor of 10. For example, one obtains $P_0\gamma_t^2$ (SiO$_2$) / $P_0\gamma_t^2$ (Se) ≈ 9. Taking into account the very different microscopic nature of different glasses, it may come as a surprise that the variations are less than an order of magnitude. As pointed out by Leggett (1991), the agreement is even better if one considers the dimensionless quantity $P_0\gamma_t^2/(\varrho v_t^2)$. Here the variations among most glasses are smaller than a factor of 3. Intuitively, it is not surprising that the variation of a dimensionless quantity tends to be smaller.

This example shows that it is essential to compare appropriately scaled quantities in order to learn something about possible universal properties of the low-temperature parameters. It may be instructive to discuss this aspect for the simple example of the melting of xenon and argon. Experimentally, it turns out that the melting temperature T_m of xenon is 1.9 times larger than that of argon. However, despite this large difference, the microscopic nature of melting is basically identical. The difference of the melting temperatures is exclusively ascribed to different internal energy scales describing the pair interaction of the atoms. The *reduced* melting temperature, which is corrected for the different energy scales, is hence identical for both systems. For example, one would expect that the ratio $T_m/(mv^2)$ is identical for all noble gases, since mv^2 is a unique measure for the energy scale (atomic mass m, velocity of sound v as introduced in Sect. 8.4). Qualitatively, this means that a film of a molecular dynamics simulation of the microscopic dynamics of argon cannot be distinguished from that of xenon if one allows the adjusting of the speed of the film as well as the distance between projector and screen. This is denoted as the law of *corresponding states*. The remaining differences would directly hint of the existence of real microscopic differences. In general, noble gases can be classified by m, v, and the density ϱ, hence fixing the length, mass, and energyscale. Then any observable like T_m or the expansion coefficient should be identical if normalized by an appropriate combination of these three parameters.

The situation becomes more complicated if quantum-mechanical effects are effective as in the case of the melting of He (which may also be described as a noble gas). For He, the ratio $T_m/(mv^2)$ turns out to be smaller by more than a factor of two as compared to argon or xenon (see DeBoer (1948)). Qualitatively, the presence of quantum-mechanical effects changes the microscopic nature of melting since, due to zero-point fluctuations, the instability of the crystal already occurs at lower temperatures as expected from purely classical considerations. From a formal point of view, quantum-mechanical effects give rise to a new dimensionless quantity (remember $B \equiv mv^2$):

$$\mu \equiv (0.011B)^{8/3} v^{-10/3} \varrho^{-2/3} \hbar^{-2} \ . \tag{8.76}$$

Apart from the numerical prefactor which will be motivated below, μ is the only dimensionless combination of $(m, v, \varrho, \hbar)$. In the classical limit, one has $\mu \to \infty$. On the basis of the law of corresponding states, the relation $T_m \propto mv^2$ has to be generalized to $T_m = mv^2 f(\mu)$ with some *a priori* unknown function $f(\mu)$. Since, in the classical limit, both relations have to be identical, one may require $\lim_{\mu \to \infty} f(\mu) = \text{const}$. The function $f(\mu)$ contains the information on how quantum-mechanical processes influence the quantity under consideration. Expressing μ in terms of a, m, and B (length, mass, and energyscale) one obtains $\mu \propto m$. Hence μ is a measure of the mass of the particles.

8.6.2 Universal Relations for LJ Glasses

In Sect. 8.3.3, the four low-temperature parameters $(P_0, \gamma_\eta, T_{C,1}, T_{C,2})$ have been calculated for NiP which corresponds to one special choice of (m, v, ϱ).

Since all results related to the nature of DWP's have been expressed in a dimensionless form (leading e.g. to the definition of the w_i), all low-temperature quantities can, of course, be calculated for arbitrary (m, v, ϱ). Based on the above arguments these calculations can be dramatically simplified by taking into account the concept of corresponding states. Rather than determining the three-dimensional function $T_{C,1} = T_{C,1}(m, v, \varrho)$, one only has to determine a function $f_3(\mu)$ via ($k_B \equiv 1$):

$$T_{C,1} = B f_3(\mu) . \qquad (8.77)$$

Hence, the analysis of the set of LJ glasses, involving three different parameters, is reduced to the variation of a single parameter for the determination of $f_3(\mu)$. In analogy, one has

$$T_{C,2} = B f_4(\mu) , \qquad (8.78)$$

$$\gamma_t = B \tilde{f}_1(\mu) , \qquad (8.79)$$

$$P_0 \gamma_t^2 = \varrho v^2 \tilde{f}_2(\mu) . \qquad (8.80)$$

Later on, these expressions will be applied to glasses different than LJ glasses. This allows one to check whether, e.g. SiO_2 behaves different than LJ glasses at low temperatures. One essential difference between SiO_2 and LJ glasses is the different average coordination number ($n = 4$ in the case of SiO_2). However, as already discussed for the deformation potential, this quantity possesses a well-defined dependence on coordination number. Hence, in order to write the above equations as general as possible, it is advantageous to explicitly consider the expected dependence on n rather than absorbing this factor in the functions $\tilde{f}_i(\mu)$. From $\gamma_\eta \propto n^{-1/2}$ [(8.50)], it follows that

$$\gamma_t = B n^{-1/2} f_1(\mu) , \qquad (8.81)$$

$$P_0 \gamma_t^2 = \varrho v^2 n^{-1} f_2(\mu) . \qquad (8.82)$$

Performing the same calculations as for NiP for different values of μ (e.g. by scaling the mass of all particles), one obtains $f_i(\mu) = b_{i,\text{sim}} \mu^{-c_i}$. The c_i and $b_{i,\text{sim}}$ are given in Table 8.1 (Heuer and Silbey (1996)).

All four exponents c_i given above can in principle also be obtained in the framework of the soft-potential model with the standard assumptions (i) $p_2(w_2)$ is linear in w_2 for small w_2 and finally levels off, (ii) $p_3(w_3)$ is constant, and (iii) $p_4(w_4)$ is a delta-function for some value w_4^0. Apart from logarithmic corrections, c_1 and c_2 can be estimated by analyzing the Jacobi determinants for the transformation to the tunneling parameters; see Parshin (1994b). One obtains $c_1 = 1/6$ and $c_2 = 1/2$, which compares rather well with the results of the numerical analysis. Differences are due to the logarithmic corrections and the use of the more general distribution functions p_i. Anyhow, for most glasses (see Sect. 8.6.3), the value of μ varies at most by a factor of ten so that even for the estimation of $P_0 \gamma^2$, where the deviations between

8.6 Universality of the Low-Temperature Parameters

Table 8.1. The exponent parameter c_i and the prefactor $b_{i,\text{sim}}$ as obtained from the simulation, the exponent parameter $c_{i,\text{SPM}}$ as calculated within the soft-potential model, and the prefactors $b_{i,\text{exp}}$ and deviations σ_i as obtained from the comparison with experimental data in Sect. 8.6.3 for the deformation potential ($i=1$), the product $P_0\gamma^2$ ($i=2$), the minimum $T_{C,1}$ ($i=3$) and the maximum $T_{C,2}$ ($i=4$) of $C(T)/T^3$. The quality of the predictions of the low-temperature parameters will be discussed in Sect. 8.6.3.

i	$b_{i,\text{sim}}$	$b_{i,\text{exp}}$	c_i	$c_{i,\text{SPM}}$	σ_i
1	22	25	0.2	0.16	0.23
2	0.0035	0.003	0.4	0.5	0.43
3	0.008	0.011	0.67	0.67	0.16
4	0.027	0.033	0.5	0.5	0.11

simulations and soft-potential model are largest, both predictions would at most differ by 25% ($10^{0.5-0.4} = 10^{0.1} \approx 1.25$). This similarity indicates that the values of the exponents are rather insensitive to the exact structure of the distribution functions and hence do not depend on the details of the numerical simulations. The relation $c_1 > 0$ has a simple intuitive interpretation. In the limit of large masses, i.e. large values of μ, tunneling is only possible for small values of d. Since $\gamma \propto d$, the average value of γ decreases for larger masses, hence $c_1 > 0$.

The exponents $c_{3,4}$ are related to the temperature dependence of the specific heat and hence to the energy distribution of soft modes. As explicitly discussed in the framework of the soft-potential model in Chap. 9, the soft modes below some crossover energy E_1 are mainly due to DWP's, and above due to SWP's. This value can be estimated from setting $w_2 = w_3 = 0$, $w_4 = A_4$ and solving the corresponding Schroedinger equation for this quartic potential. One obtains $E_1 \propto BA_4^{1/3}\mu^{-2/3}$ (Parshin (1994b)), yielding $c_3 = 2/3$, in good agreement with the numerical result. In the regime $E > E_1$ the density of SWP's and, correspondingly, the specific heat, dramatically increases due to the linear increase of p_2. Since the distribution p_2 levels off around $w_2 = A_2$, the strong increase of the density of SWP's slows down around $E = E_2$, yielding the maximum in $C(T)/T^3$. The eigenvalue of the corresponding harmonic SWP is proportional to $A_2^{1/2}B\mu^{-1/2}$, yielding $c_4 = 1/2$, once more in agreement with the simulations. Note that the estimated value of $T_{C,2}$ is proportional to the Debye temperature of the glass.

The good agreement of the simulated c_i and those obtained in the framework of the soft-potential model reflects the similarity of both approaches, albeit the soft-potential model is purely phenomenological whereas the simulation results are based on a microscopic reasoning. The b_i are treated as adjustable parameters in the soft-potential model. Therefore, a comparison with the simulated b_i is only possible with experimental results.

As outlined in Parshin (1994b) near the temperature $T_{C,2}$ the interaction between the soft modes starts to dominate the dynamics. This effect may also lead to a maximum in $C(T)/T^3$. Hence it seems that the existence of a maximum temperature $T_{C,2}$ can also be explained without the final decay of p_2. Since both mechanisms yield the same μ-dependence of $f_4(\mu)$, they cannot be distinguished.

The values of $T_{C,1}$ and $T_{C,2}$ basically depend on A_2 and A_4, respectively. However, these values have to be treated with some caution. Originally, these parameters were derived from a systematic analysis of DWP's. As discussed in Sect. 8.3.4, the comparison of DWP's and SWP's revealed that the effective mass of SWP's are at least by a factor of four larger than the effective mass of DWP's. Due to the significant correlation between effective mass and energy parameters (see Fig. 8.8), one might expect that the values of A_2 and A_4 are too large to correctly describe the distribution of SWP's. However, since the temperatures $T_{C,1}, T_{C,2}$ only mildly depend on A_2 and A_4, respectively, no dramatic effect for the temperature dependence of the specific heat is expected. Interestingly, the value of A_3 does not enter into the low-temperature parameters. However, it is easy to check that the number of DWP's relative to the number of SWP's depends on A_3. This ratio basically corresponds to the height of the bump of $C(T)/T^3$. The larger the A_3, the larger the relative number of DWP's.

In summary, on the basis of the microscopic approach it is possible to calculate all four low-temperature parameters for *all* LJ glasses without adjustable parameters. Unfortunately, until now, no comparison with experimental data on LJ systems has been performed. These systems have a strong tendency to crystallize, hampering the determination of bulk low-temperature properties for nonmetallic LJ glasses.

8.6.3 Application for Different Types of Glasses

The results presented above are valid for the set of all LJ glasses. Now we are in a position to tackle the important question as to which degree the physics of the low-temperature properties quantitatively depends on the microscopic nature of the glass. The procedure is simple. First, one assumes that all low-temperature properties do not depend on the microscopic structure at all. Then all glasses must behave like LJ glasses and one can use (8.78)–(8.82) to predict the low-temperature parameters of glasses. In case this prediction does not agree with the experiment, the degree of deviation is a direct measure of how strongly the actual microscopic structure of the glass influences the physics at low temperatures; see also Heuer and Silbey (1994), Parshin (1994b), and Heuer and Silbey (1996).

For LJ glasses, the low-temperature parameters have been expressed in terms of the velocity of sound v, the density ϱ, and the mass of the elementary units m. Furthermore, the influence of the coordination number n has been included. Trivially, for a LJ glass, the elementary units can be identified with

8.6 Universality of the Low-Temperature Parameters

the individual particles. For a more complex glass the elementary units are typically the molecular units. For some glasses the molecular units can be reliably identified. For example, for SiO_2 the molecular units are the SiO_4-tetrahedra, since at low temperatures, its tetrahedral structure is very stable whereas the tetrahedra can be moved relative to each other. Unfortunately, for some glasses, e.g. silicate blends, it is hard to say which molecular unit should be taken in order to define a mass m.

In Heuer and Spiess (1994) a number of glasses have been analyzed for which the value of m could be guessed. For these glasses, to a good approximation, m could be related to the glass transition temperature T_g via

$$T_g = c_g m v^2 \tag{8.83}$$

with $c_g \approx 0.011$. This criterion can be considered as a generalization of the Lindemann melting criterion. The underlying relation $T_g \propto v^2$ has been experimentally verified for polymer glasses where the change of T_g and v was monitored upon variation of pressure by Pietralla et al. (1996).

This empirical relation motivates one to treat T_g rather than m as one of the three basic parameters characterizing a glass. Therefore, one can express the low-temperature parameters on (T_g, v, ϱ), without resorting to the somewhat arbitrary choice of m. Of course, one could have started from the very beginning with characterizing a LJ glass by (T_g, v, ϱ) instead of (m, v, ϱ) as done in Sect. 8.6.2. However, for LJ glasses, the particle mass is a more natural choice. It should be noted that the glass transition temperature is not a thermodynamic quantity but depends for example on the cooling rate or on the molecular weight in case of polymers. However, these dependences are relatively small and can be neglected for the subsequent analysis. The only remaining uncertainty is the choice of n. The choices are $n = 4$ for tetrahedral glasses and $n = 6$ else.

Now the theoretical predictions are compared with experimental data, using (8.78)–(8.82). The experimental data are listed in Tables 8.2. and 8.3.

The tunneling parameters have been exclusively taken from the data collection in the work of Berret and Meißner (1988). The value of the deformation potential of $LiCl\text{-}7H_2O$ has been omitted since this value is hard to access experimentally [see the discussion in Berret and Meißner (1988)]. Furthermore, the data for the $Se_x Ge_{1-x}$ semiconductors will be discussed separately below. The polymer polyethylene has not been analyzed since, due to its semicrystallinity, the glass transition temperature is no well-defined quantity.

Experimental uncertainties are mainly connected with the tunneling parameters. The quantity $P_0 \gamma^2 / \varrho v^2$ can be derived either from the plateau of the sound absorption, or from the temperature dependence of the sound velocity [see Phillips (1987) and Chap. 4]. It turns out that, e.g. for LASF-7, both values vary by a factor of 2 indicating some experimental or theoretical inconsistencies; see Berret et al. (1986). Since, in the data collection of Berret and Meißner (1988), all data have been evaluated in the same manner, at least the relative error should be smaller.

Table 8.2. The experimental tunneling parameters as used in this work. They are taken from Berret and Meissner (1988).

	ϱ [kg m^{-3}]	T_g [K]	v 10^3 [m s^{-1}]	γ_t [eV]	$P_0\gamma_t^2$ 10^7 [J m^{-3}]	n
SiO2	2200	1473	4.1	0.65	0.92	4
Se	4300	304	1.17	0.14	0.10	6
PMMA	1180	374	1.70	0.27	0.11	6
PS	1050	355	1.67	0.13	0.09	6
Epoxy	1200	350	1.66	0.22	0.08	6
BK7	2510	836	4.19	0.65	1.19	4
As2S3	3200	444	1.69	0.17	0.12	6
LaSF7	5790	957	3.95	0.92	1.05	6
SF4	4780	693	2.48	0.48	0.64	4
SF59	6260	635	2.13	0.49	0.63	4
V52	4800	593	2.51	0.52	1.13	6
BALNA	4280	520	2.59	0.45	1.09	6
LAT	5250	723	3.1	0.65	1.52	6
Zn-Glass	4240	570	2.58	0.38	0.81	6
PC	1200	418	1.86	0.18	0.14	6
LiCl-7H2O	1200	139	2.5		0.46	6

Table 8.3. The experimental temperature extrema $T_{C,1}$ and $T_{C,2}$ of the specific heat. The corresponding references are listed in Heuer and Silbey (1996).

	ϱ [kg m^{-3}]	T_g [K]	v 10^3 [m s^{-1}]	$T_{C,1}$ [K]	$T_{C,2}$ [K]
SiO2	2200	1473	4.1	2.1	10
Se	4300	304	1.17	0.7	3.1
PMMA	1180	374	1.70		3.5
PS	1050	355	1.67	0.9	3.4
Epoxy	1200	350	1.66	1.0	3.7
LiCl7H2O	1200	139	2.5	3.3	10.7
GeO2	3600	830	2.6	2.1	8.1
B2O3	1800	523	2.06	1.3	5.4
PB	930	186	1.69	1.4	5.1
(SiO2)$_{0.75}$(NaO)$_{0.25}$	2440	735	3.5	3.5	13.5
Glycerole	1300	185	1.7	2.2	8.0

The result of the comparison with experimental data is presented in Figs. 8.25 and 8.26. Using the c_i and $b_{i,\text{sim}}$ the comparison could have been attempted without adjustable parameters. For practical reasons the prefactors are treated as adjustable parameters $b_{i,\text{exp}}$ and it is only finally checked that $b_{i,\text{sim}}$ and $b_{i,\text{exp}}$ agree within reasonable limits. In order to quantify the scattering of the data the relative deviation σ_i are defined as the average value of $|x_{i,\text{sim}} - x_{i,\text{exp}}|/x_{i,\text{exp}}$, where x_i denotes the corresponding low-temperature

8.6 Universality of the Low-Temperature Parameters   515

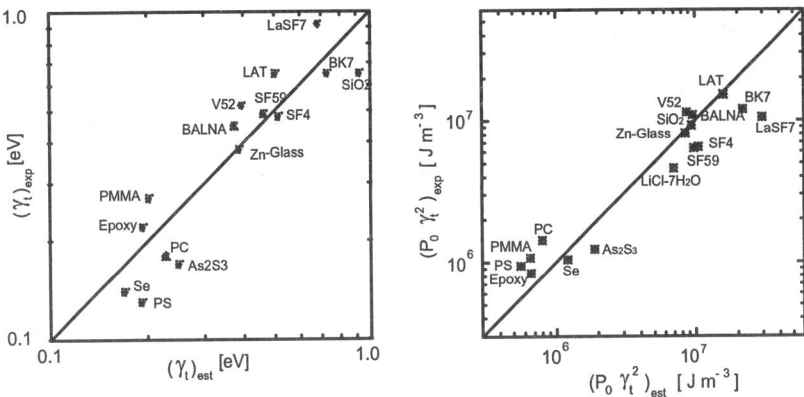

Fig. 8.25. A comparison of the estimated and the experimental tunneling parameters. A single adjustable parameter of the order of one has been used to scale the estimated data relative to the experimental data, respectively. Adapted from Heuer and Silbey (1996).

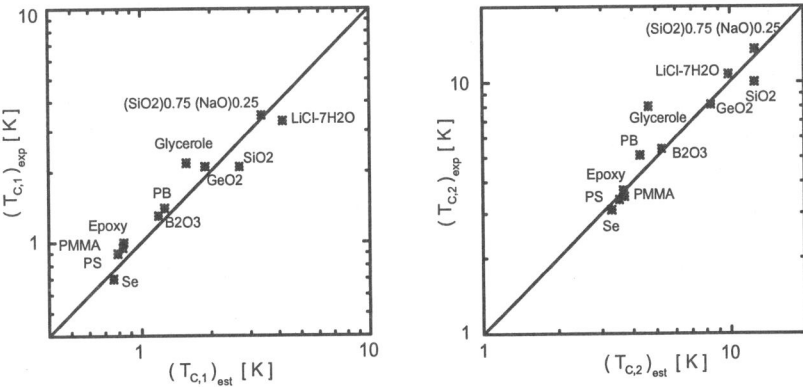

Fig. 8.26. Comparison of estimated and calculated values for the extrema of $C(T)/T^3$. Adapted from Heuer and Silbey (1996).

parameter. The results for the fitted proportionality constants $b_{i,\mathrm{exp}}$ and the average relative deviations σ_i are also listed in Table 8.1.

For all four quantities significant correlations can be observed. Hence one can already conclude that the individual microscopic structure does not dominate the actual low-temperature behavior. From the σ_i-values it is obvious that the correlations for the extrema of the specific heat are better than those for the tunneling parameters. Finally, it turns out that the $b_{i,\mathrm{exp}}$ agree very well with the simulated values $b_{i,\mathrm{sim}}$. Hence the low-temperature behavior as derived from the simulations of LJ glasses *quantitatively* agrees with the experimental data of most glasses.

One might argue that the above results are just manifestations of some trivial correlations. For example, the mere fact that the $T_{C,i}$ for SiO_2 are higher than those for, e.g. polystyrene (PS), can be simply related to the fact that silicate glasses are much stiffer than polymers as expressed by the higher glass transition temperature. However, obviously this picture breaks down in the case of $LiCl-7H_2O$ which among all glasses analyzed above has the smallest glass transition temperature but one of the highest values of $T_{C,2}$. Of course, the reason for this behavior is the strong dependence of $T_{C,i}$ on μ. In this sense the correlations in Figs. 8.25 and 8.26 contain more information than guessed intuitively. To stress this point, the correlation of $T_{C,1}/T_g$ and $T_{C,2}/T_g$ with μ is explicitly checked. The corresponding plot can be seen in Fig. 8.27. One can clearly see that the predicted behavior $T_{C,1}/T_g \propto \mu^{-2/3}$ and $T_{C,2}/T_g \propto \mu^{-1/2}$ is very well fullfilled. In principle the exponents c_i could have been derived directly from this plot with good accuracy.

As further discussed in Sect. 8.7.2, the properties of many glasses only mildly depend on the thermal history. Otherwise it would have been impossible to find correlations of the quality presented above. Interestingly, it turns out that metallic glasses significantly change their low-temperature properties upon annealing; see also Sects. 8.7.2 and 4.4.5.

The previous analysis indicates that to a large degree all low-temperature properties are insensitive to the microscopic structure. Hence it might be surprising that there exists another low-temperature parameter which cannot be analyzed in analogy to the quantities discussed until now. The temperature dependence of the absorption of sound displays a maximum around 50 K (for SiO_2), hence defining a temperature T_s. At these elevated temperatures the dynamics of the DWP's can already be described in classical terms [Tielbürger et al. (1992); however, see Rau et al. (1995a) and Chap. 3

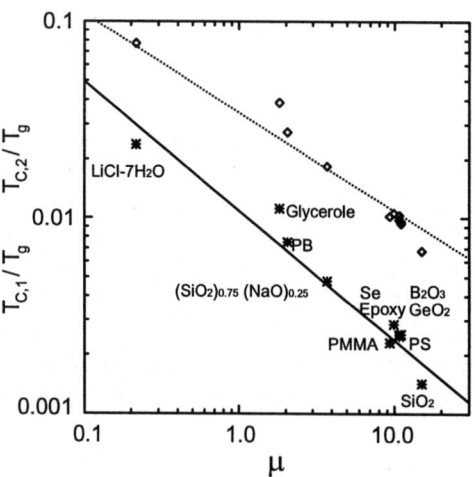

Fig. 8.27. A plot of $T_{C,1}/T_g$ and $T_{C,2}/T_g$ vs. μ where $T_{C,1}$ and $T_{C,2}$ correspond to the extrema of the specific heat and T_g to the glass transition temperature. $\mu = \mu(T_g, v, \varrho)$ has been defined in (8.76). The solid lines correspond to a scaling with $\mu^{-1/2}$ and $\mu^{-2/3}$, respectively.

for some recent new aspects of the sound absorption]. Since the relaxation of DWP's determines the sound absorption for these temperatures, one can already conclude from dimensional arguments that $T_s \propto B \propto T_g$ if measured in approximately the same frequency region (T_s only depends logarithmically on frequency). The values of T_s and T_s/T_g for four different glasses are shown in Table 8.4. Obviously, the ratio T_s/T_g is far from being constant.

Table 8.4. The ratio of T_s (temperature of the sound absorption maximum) and T_g (glass transition temperature) for four glasses.

	T_s [K]	T_s/T_g	ω [MHz]	Ref.
SiO2	45	0.03	20	Krause and Kurkjian (1968)
LASF7	110	0.11	30	Berret et al. (1986)
B2O3	83	0.16	20	Krause and Kurkjian (1968)
GeO2	170	0.20	20	Krause and Kurkjian (1968)

This means that the temperature of the peak of sound absorption strongly depends on the microscopic structure of the glass.

Recently, Parshin (1994b) has extensively applied the soft-potential model to the explanation of the low-temperature properties. Although his approach has some similarities with the analysis presented above, there exist some major differences. First, in contrast to the phenomenological approach of the soft-potential model, the above analysis starts from a microscopic description of the glass. In this way the low-temperature parameters can be obtained in a quantitative way. Second, for the estimation of P_0 and γ, Parshin (1994b) used the atomic mass as a value for m rather than the molecular mass which seems to be more appropriate. For the case of amorphous $As_x Se_{1-x}$ Parshin et al. (1993) were able to explain the dependence of the low-temperature specific heat on the composition in terms of the above-mentioned scaling relations.

8.6.4 Quantitative Universality: What Does it Express?

The surprising scaling property of all four low-temperature quantities already results from the knowledge of the exponents c_i relating the material constants to the experimental low-temperature observables. Since these values are insensitive to details of the simulations (see discussion above) the scaling results are of very general nature. In contrast, the prefactors $b_{i,\text{sim}}$ strongly depend on details of the computer simulation like the ability to systematically detect DWP's. The good agreement between $b_{i,\text{sim}}$ and $b_{i,\text{exp}}$ for all i demonstrates that the computer simulations indeed gave a reasonable picture of the low-temperature anomalies.

The above analysis has revealed that for most structural glasses the low-temperature properties, except for T_s, are insensitive to the exact microscopic structure. This implies that at low temperatures, silicate glasses or polymers do not behave very differently as LJ glasses. For a simple polymer like polystyrene (PS), a monomer may be identified as the elementary unit serving as a generalized LJ particle. The main difference between PS and a LJ glass is the presence of covalent forces among monomers which fix their distance along a polymer chain. However, from the systematic analysis of the microscopic structure of the DWP's, it became obvious (see Sect. 8.3) that two adjacent particles tend to keep their relative distance fixed during the transition between the two wells. Hence, additional forces fixing the distance of some particles only weakly influence the nature of DWP's. Therefore, the presence of some covalent forces are not particularly relevant for the nature of soft modes. Another argument supporting the notion of quantitative universality refers to the observation that a soft mode typically corresponds to the collective motion of a number of adjacent units, implying an average over microscopic details. This argument is similar to the explanation of universal behavior near phase transitions, where the correlation length far exceeds the microscopic length scale; see e.g. Ma (1987).

The latter argument implies that the degree of universality should be correlated with the effective mass p of the soft modes. This observation can be used to explain the different degrees of universality for the low-temperature parameters. As shown in Sect. 8.3, SWP's have much larger effective masses than DWP's. Since the extrema of $C(T)/T^3$ are related to the energy distribution of SWP's, it is not surprising that the degree of universality is very high for $T_{C,1}$ and $T_{C,2}$. In contrast, the values of $\gamma, P_0\gamma^2$, and T_s depend on the properties of DWP's. In Tielbürger et al. (1992), it has been shown for SiO_2 that the relaxation behavior near T_s is dominated by DWP's with barrier heights of the order of 500K. It is easy to check from the application of the WKB-formula that potentials with barrier heights of this order have negligible tunneling matrix elements. Stated differently, at temperatures near 1 K, only DWP's with significantly smaller barrier heights are relevant. According to the simulation results (see Fig. 8.8), this implies that DWP's relevant for the sound absorption near T_s are much more localized than those which are important at 1K. This argument may explain why the degree of universality for γ and $P_0\gamma^2$ is still rather high whereas there is no universal behavior at all for T_s.

An alternative explanation of the quantitative universality has first been proposed by Yu and Leggett (1988) and Leggett (1991) and has been further outlined by Burin and Kagan (1996a). The basic idea is that the TS's visible in low-temperature experiments are different from the primary defects but rather correspond to coupled clusters of primary defects. Among others this theory predicts a universal density of TS's; see Sect. 5 for more details. The strong-interaction approach still has to explain why (i) the other low-temperature parameters also display quantitative universal behavior, and (ii) why the degree of universality seems to be related to the effective mass of the TS's. Furthermore, the nature of the primary defects and the relation of this approach to numerical simulations of DWP's is not clear. In simulations the interaction cannot be removed so that it is impossible to observe the primary defects. Maybe the simulated DWP's already correspond to the coupled clusters. Anyhow, these aspects have to be further clarified in the future.

The number of glasses analyzed in the work of Berret and Meißner (1988) is sufficiently large in order to identify some generic behavior of structural nonmetallic glasses at low temperatures. This implies that for any major deviations one should be able to point out microscopic peculiarities. One example for glasses which do not follow the standard behavior is $Se_{60}Ge_{40}$ (Duquesne and Bellessa (1985)), which forms a very strong tetrahedral network. It turns out that the deformation potential is more than twice as small as expected from the correlation expressed in Fig. 8.25. This can be understood from the results of Sect. 8.4 where a correlation between the absolute value of the deformation potential and the disorder has been established. According to this analysis, glasses with less disorder are expected to display significantly reduced values for the deformation potential. For strongly bonded tetrahedral glasses, this behavior can be expected.

8.7 Experimental Hints about the Microscopic Nature of the Soft Modes

In this section some selected experimental results are summarized containing information about the microscopic structure of glasses. This set of experiments just represents a subjective selection and different experiments could be mentioned as well.

8.7.1 Relation to Strong and Fragile Glasses

It has been pointed out by Sokolov et al. (1993) that the height of the bump of $C(T)/T^3$ is related to the fragility of glasses. The fragility characterizes the temperature dependence of the α-relaxation around the glass transition; see e.g. Ediger et al. (1996). The α-relaxation in strong glasses like SiO_2 is Arrhenius-like whereas for fragile glasses like low-molecular glass formers or polymers, an apparent divergence of the α-relaxation time close to the glass transition is observed. Experimentally, it turns out that strong glasses show a larger bump. This directly shows that properties of the glass transition and the low-temperature anomalies have a strong connection.

Qualitatively, for strong glasses like network glasses, the harmonic vibrations around the metastable positions are important. Actually, the dominance of harmonic behavior in strong glasses has been recently verified in computer simulations of a-H_2O where it was shown that the so-called fast β-relaxation above the glass transition is exclusively due to dephasing of harmonic excitations. Hence the observation of a large bump in $C(T)/T^3$ for strong glasses directly indicates that the bump has to be related to some harmonic contributions (SWP's). Actually, this is just the statement of the soft-potential model.

Formally, the ratio of SWP's to DWP's, i.e. the relevance of harmonic contributions, is determined by the parameter A_3, as defined in (8.17). The above-mentioned observations indicate that A_3 should be smaller for strong glasses. Interestingly, the strong glasses analyzed in Fig. 8.26 (SiO_2 and GeO_2) do not show anomalous behavior with respect to their values of $T_{C,1}$ and $T_{C,2}$. Therefore, one may conclude that in contrast to A_3, the values of A_2 and A_4 do not seem to be significantly correlated with the degree of fragility.

8.7.2 Cooling Rate Dependence of TS's

In Lasjaunias et al. (1981) and (1986), and several related papers, Lasjaunias and co-workers studied to which degree the density of TS's is influenced by the conditions of preparation. They compared *sputtered* samples with *melt-spun* samples. One can consider sputtering as a quenching process with an enormous quench rate in comparison to melt-spinning. For a sputtered sample there is no memory of an initial liquid state in terms of structural short-range order.

They observed for a Zr-Cu alloy that the density of TS's of the sputtered sample is by a factor of 1.5 to 4 (in dependence on the composition) higher than the melt-spun sample. However, upon annealing of the sputtered sample, the density of TS's is reduced and brings the sample to a state closer to the melt-spun ones. Furthermore, it turns out for Zr-Ni that the sputtered sample displays a sublinear specific heat at low temperatures ($C(T) \propto T^{0.55}$). The specific heat decreases after annealing and approaches the standard linear dependence. In contrast, annealing does not have a significant effect on melt-spun samples. However, see Sect. 4.4.4 for somewhat different results on melt-spun amorphous metals.

These experimental results indicate that the presence of TS's is strongly related to the disorder of glasses which itself depends on the cooling rate. Due to fast quench rates of computer simulations the absolute number of TS's of computer glasses probably also tends to be somewhat too large. Experimentally, the comparison between two very different cooling mechanisms only leads to variations of the TS density which is of the order of two (except for metals). Thus one may be optimistic that the "cooling-rate-deficiency" of computer simulations does not hamper a semiquantitative analysis of the low-temperature anomalies.

8.7.3 The Microscopic Nature of Soft Modes in SiO_2

Neutron Scattering. From neutron scattering important information about the microscopic origin of dynamical processes can be obtained. For the case of SiO_2, Buchenau et al. (1984) measured the inelastic structure factor at 290 K. In order to model the underlying dynamics they first guessed the expected nature of the low-frequency harmonic vibrations. SiO_2 is built from SiO_4 tetrahedra sharing each of their corners with one other tetrahedron in a continuous three-dimensional network. The corner linkage between tetrahedra can be viewed as the softest spring in the system. Hence it is suggestive to model the dynamics by the rotation of a single or of coupled tetrahedra. Calculating the expected wave-vector dependence of the inelastic structure factor, it turned out that only the model of coupled rotations agrees with the experimental data. It is not possible to tell how far these modes extend from the central tetrahedron. Furthermore, Buchenau et al. (1988) have performed quasi-elastic neutron scattering experiments for SiO_2 in order to probe the microscopic orgin of the structural relaxation processes. Also, these processes can be consistently described in terms of coupled rotations of SiO_4 tetrahedra through distances of about 0.5 Å. Hence, for SiO_2, it is possible to show that the relaxation modes as well as the low-frequency harmonic vibrations can be related to the same microscopic origin. This proposal for the dynamics of SiO_2 agrees with the notion of Sect. 8.6 to identify the individual molecules rather than the atoms as elementary units in order to obtain the surprising quantitative universality among different glasses.

Analysis of Neutron-Irradiated Quartz. Ultrasonic experiments on neutron-irradiated quartz have been performed by Laermans and Keppens (1995) with different doses and different orientations of the quartz sample. The effect of the neutron irradiation is to generate so-called Dauphiné twins: groups of SiO_4 tetrahedra which are rotated 180° with respect to each other. Their rotational axes point in the crystallographic z-direction of the quartz crystal so that the dynamics is restricted to the xy-plane.

Recently, strong evidence was put forward that the TS's which are found for neutron-irradiated quartz are due to such twins. Their signature was the anisotropy of the longitudinal deformation potential. The coupling of longitudinal sound waves to this type of TS's is larger if the wave vector lies in the xy-plane as compared to the z-direction because of the better coupling to the dynamics. Experimentally, for large doses the value of γ_l^2 in both cases differs by a factor of three. Based on these results one may speculate that this coupled rotation of SiO_4-twins may constitute a general mechanism to explain tunneling also in amorphous SiO_2.

8.7.4 The Properties of Defects

Defects play an important role for several experiments in glasses. In several standard glasses they are always present as a minority component. For example the SiO_2-type glass Suprasil I approximately contains 1000 ppm OH-impurities. In comparison for Infrasil the OH-content is less than 5 ppm. In experiments like hole burning [see e.g. Maier et al. (1996) and Chap. 6] some optical active molecules are diluted in the glass. These defects are essential for these experiments since they serve as a probe for the amorphous material.

In several experiments on materials containing OH-impurities some observables display significant changes as compared to the pure glass. For example electric echo experiments by Golding et al. (1979) show a dramatic increase in echo intensity when going from Infrasil to Suprasil I. A microscopic model has been proposed by Phillips (1981b) which explains the extrinsic DWP's by a model involving rotation of the OH group about a SiO bond. A similar dramatic effect can be observed in sound absorption experiments on B_2O_3. Comparing the sound absorption for a "dry" (130 ppm) and "wet" (1.6% OH-content) probe in Rau et al. (1995a), again significant changes in the absorption peak can be observed. This should be contrasted with the observation that hardly any change occurs in the tunneling regime. The mere fact that the number of *extrinsic* DWP's related to the OH-molecules is of the same order as that of *intrinsic* DWP's shows that OH-molecules are very efficient in forming DWP's. Furthermore the insensitivity in the tunneling regime implies that these additional DWP's tend to have rather high potential barriers (basically, these DWP's contribute to the sound absorption).

As indicated by the simulations in Sect. 8.3.5, small defects in LJ glasses display the properties found for the OH-molecules in B_2O_3, namely a high probability to form extrinsic DWP's with high barrier heights.

8.7.5 Pressure Dependence

The pressure dependence of the low-temperature parameters of epoxy-resin, an amorphous polymer, has been analyzed by Grace and Anderson (1989). Interestingly, they observed a significant decrease of tunneling systems upon increasing pressure. For $p = 3.9$ kbar, which corresponds to an increase of the density by 8%, the value of P_0 has decreased by more than 20%. This change may be related to a change in the material constants density and velocity of sound via $P_0 \propto \varrho^{10/9} v^{7/3}$ (mass is unchanged upon pressure). On this basis Parshin (1994b) has estimated that the reduction should be only as large as 10%. Hence the stronger reduction hints towards structural changes of the glass, as extensively discussed in Sect. 8.6. The additional reduction of P_0 with pressure is in qualitative agreement with the simulations in Sect. 8.3.6. which also indicated major structural changes with pressure.

In this respect it is interesting to note that the sound absorption of SiO_2 displays a more complicated dependence on pressure; see Rau et al. (1995b). It has been explained in terms of a broadening of the DWP parameter distribution upon the application of pressure.

8.7.6 Length-Scale Dependence

In a careful study of the low-temperature internal friction of SiO_2-films, White and Pohl (1995) searched for the dependence of $P_0\gamma^2$ on the thickness of the films. The thickness was varied between 0.35 nm and 108.5 nm. Surprisingly for values as small as 0.75 nm they did not find any major dependence of the sound absorption on the film thickness in the tunneling regime (below 5 K). The authors concluded that any elastic interactions between the defects are not significant for the nature of the tunneling states on length scales as small as 0.75 nm, and thus the origin of these states is short range. Of course, the size of the tunneling entity is also bounded by this length scale. Interestingly, it turned out that at higher temperatures, where classical relaxation processes prevail, a significant dependence on the film thickness is observed. Furthermore, it was observed for this regime that wet thermal oxide films behave differently as compared to films deposited by electron beam evaporation.

Recently, it has been shown by Gaganidze et al. (1997) that in contrast to the internal friction plateau the sound velocity maximum significantly changes with thickness. An interpretation of this result is given in Gaganidze et al. (1997) in terms of the reduced dimensionality of the interaction between TS's in the amorphous films, see also Sect. 4.4.5.

These results do not contradict the notion that, despite the cooperative nature of the tunneling dynamics and the presence of interaction between different TS's, the TS's are rather localized in space. Furthermore, the variation of the absorption close to the relaxation peak might indicate that the states involved in the tunneling dynamics and those involved in the relaxation dynamics at somewhat higher temperatures (30 K) are not directly related.

8.8 Summary and Outlook

The main goal of this chapter was to show that the phenomenological tunneling model as well as the soft-potential model can be justified on a purely microscopic basis. This was achieved by decomposing the glass Hamiltonian according to

$$\mathcal{H} = \mathcal{H}_{\text{soft modes}} + \mathcal{W} + \mathcal{H}_{\text{B}}, \tag{8.84}$$

yielding information about the soft modes as well as their interaction with the phonon bath. Presently, it is not possible to perform this decomposition from a first principle analysis of a realistic glass Hamiltonian in analytical terms. However, on the basis of computer simulations, reliable information about $\mathcal{H}_{\text{soft modes}}$ and $\mathcal{W}$ can be obtained. Some major results are:

– *Microscopic Nature of Soft Modes.* Due to the atomistic information, microscopic properties of the soft modes can be deduced. Examples are their spatial extension and their degree of cooperativity.

- *Deviations from the Tunneling Model.* A detailed analysis of the distribution of DWP's allows one to specify deviations from the tunneling model like some energy dependence of the density of TS's and the deformation potential. These deviations agree with experimental observations. The statistical approach of the tunneling model breaks down for the case of impurities which may have a different distribution of DWP parameters.
- *Qualitative Universality.* The distribution of soft modes gives rise to low-temperature anomalies like the linear temperature dependence of the specific heat. This property is mainly based on the statistical independence of the expansion parameters w_i and basically reflects the notion that the formation of soft modes can be described in statistical terms. Furthermore, the simulations yield a microscopic justification of the assumptions inherent in the soft-potential model.
- *Quantitative Universality.* The simulation results motivated a specific way to compare experimental data of different structural glasses by appropriate scaling with material constants. A comparison has shown that these low-temperature parameters only show small deviations if compared among different glasses. This surprising observation implies that for many glasses the microscopic structure only has a weak influence on the nature of the soft modes. The underlying reason is still under debate. Some researchers relate this observation to the dominance of interaction effects between primary defects, triggering the formation of coupled clusters, appearing as the experimentally accessible TS's. Alternatively, the present simulations suggest that the quantitative universality can be regarded as a manifestation of the cooperative nature of the low-temperature dynamics. This approach also explains why the degree of quantitative universality is correlated with the size of the relevant soft modes.

From a more general perspective the low-temperature anomalies can be related to the fine structure of the energy landscape in the multidimensional configuration space of a glass-forming system. In this sense, the presence of DWP's can be viewed as the precursor to the glass transition, which is governed by properties of the energy landscape on a somewhat larger scale. Indeed, first indications of such a correlation between the low-temperature anomalies and the glass transition are found by relating the height of the bump of the specific heat to the fragility. It would be helpful to strengthen this connection and to characterize the microscopic nature of the glass transition in a way similar to the low-temperature anomalies. Approaching the glass transition from the low-temperature side may complement present theories which analyze the glass transition from the liquid side; see e.g. Götze and Sjörgen (1992).

Acknowledgments

I would like to thank Prof. R. J. Silbey for introducing me to this field and for a fruitful collaboration over many years. Special thanks are to Dr. P. Neu, S. Büchner, Prof. S. Hunklinger, Prof. R. Schilling and Prof. H.W. Spiess for many helpful discussions and the support of this work.

9. Beyond the Standard Tunneling Model: The Soft-Potential Model

Miguel A. Ramos and Ulrich Buchenau

9.1 Introduction

As already pointed out in previous chapters, the standard tunneling model introduced by Phillips (1972) and by Anderson, Halperin and Varma (1972), is able to explain many of the anomalous features of glasses found below 1 K to ~ 0.05 K, at least.

There are, however, two strong reasons to extend the standard tunneling model. The first is a theoretical argument on the nature of these tunneling states (see Chap. 8). In order to obtain a tunnel splitting of the order of 1 K, one needs a low energy barrier. This holds in particular if more than one atom participates in the tunneling motion. One estimates barrier heights of about 100 K for these tunneling states, lower than the thermal energy at the glass transition where the disordered structure freezes in. There is an even more stringent requirement for the asymmetry between the two wells, which has to be smaller than 1 K. It is difficult to imagine such a potential without postulating a large number of similar potentials with a whole distribution of asymmetries and barrier heights in the disordered structure. One would expect barrier heights down to zero, the potential distribution continuing into a broad set of single-well potentials with a small restoring force. In fact, numerical simulations of model glasses seem to confirm this picture (see Chap. 8).

The second strong reason is given by the clear deviations from the predictions of the standard tunneling model at temperatures above 1 K (or even below 1 K, if one extends the measurements to frequencies above 100 GHz).

This chapter discusses an extension of the standard tunneling model capable of explaining these marked deviations. This is the soft-potential model, proposed by Karpov, Klinger and Ignat'ev (1982) and extended by Il'in, Karpov and Parshin (1987). As we will see in more detail below, it postulates that the tunneling states in glasses are a vanishingly small part of a large number of local modes with a very small or even negative restoring force constant, at the boundary between local stability and local instability. In this introduction, we begin with a short survey of the development of the model.

Historically, thirty years ago, the low-temperature thermal properties of amorphous solids were expected to be very similar to those of crystals, well understood in terms of Debye's theory. It was thought that the structural disorder characteristic of glasses or amorphous materials would become unimportant at low temperatures, when the wavelength and the mean free path of acoustic phonons increases. This belief was destroyed by Zeller and Pohl (1971), who after some previous experimental indications by other groups, presented unambiguous evidence for thermal properties of amorphous solids which differ remarkably from those of their crystalline counterparts. The glass always shows a higher specific heat and a much lower thermal conductivity than the corresponding crystal, with a striking universality of the glass behavior. The specific heat at constant pressure $C_p(T)$ of glasses exhibits a quasi-linear temperature dependence below 1 K (see, for example, Fig. 2.6), and a broad maximum in the heat capacity plotted as C_p/T^3 at a few K, in sharp contrast to the T^3 dependence of C_p found in dielectric crystals at low temperature in agreement with Debye theory. The thermal conductivity $\kappa(T)$ varies approximately as T^2 below 1 K, followed at higher temperatures by a plateau, and a subsequent rise above 10 K.

While the most conspicuous deviations from the Debye sound-wave model occur below 1 K and are successfully accounted for by the standard tunneling model, the broad peak in C_p/T^3 and the plateau in the thermal conductivity above 1 K are still very marked deviations from the Debye picture. They are not accounted for by the standard tunneling model. It was generally felt that one needed an explanation of these two features, if possible, a combined explanation of both. Zeller and Pohl (1971) already proposed a mechanism for the scattering of sound waves in order to explain the plateau in the thermal conductivity in terms of Rayleigh scattering from the atomic disorder in the glass. However, Zaitlin and Anderson (1975) and Jäckle (1976) estimated the strength of that scattering, considering every atom of the glass as a scattering defect. They came to the conclusion that the mechanism was two orders of magnitude too weak to explain the plateau.

Since then, there has been a continuous controversy between two different groups of hypotheses. The soft-potential model belongs to the first group, which postulates a common origin of the anomalies below and above 1 K (but is by no means the only model in that group, see for instance the description of the ideas of Yu and Leggett, and Burin and Kagan in Chap. 5, as well as the concept of incoherent tunneling by Würger in Chap. 3). As we will see in Sect. 9.4, the soft-potential model explains the plateau in terms of resonant scattering of the sound waves from local (more precisely quasi-local or resonant) vibrational modes similar in nature and having a common origin with the tunneling states. The second group of explanations for the plateau postulates the scattering from different kinds of static disorder like clusters, fractals or disorder in the force constants. The broad maximum in C_p/T^3 is then explained in terms of vibrational localization of the strongly scattered sound waves. That controversy is still going on; at present, the soft-potential model is not a generally accepted model for the low-temperature anomalies of glasses. On the other hand, it does indeed explain consistently the glassy anomalies above and below 1 K with the same set of parameters, as will be shown in Sect. 9.4.

Within the soft-potential model itself, there is agreement on the assumptions concerning the region of crossover between tunneling states and soft vibrations. As one goes to higher vibrational frequencies, however, there are divergent views on the nature of the peak in C_p/T^3. One view assumes a central role of the interaction between different soft modes at these higher frequencies [Parshin (1994a,b), Gurevich, Parshin, Pelous and Schober (1993)], the other neglects that interaction and explains the peak in terms of a Gaussian distribution of linear potential terms around zero [Gil, Ramos, Bringer and Buchenau (1993), Ramos, Gil, Bringer and Buchenau (1993)]. These approaches will be discussed in more detail in Sect. 9.3.

One important information on the modes responsible for the broad peak in C_p/T^3 came from Raman and neutron scattering measurements, which indicate that these modes are no longer tunneling modes. If one measures the temperature dependence of the signals in that frequency region, one finds the increase with temperature expected for the scattering from harmonic vibrational modes (at least for not-too-high temperatures and not too low frequency). This was first seen in Raman scattering [Shuker and Gammon (1970), Winterling (1975)], and the broad peak in the scattering corresponding to the peak in C_p/T^3 was consequently called the "boson peak", in order to stress its harmonic vibrational nature. This somewhat infelicitous name has remained.

At first, there was some controversy as to whether the peak indeed reflected a true maximum in the vibrational density of states divided by frequency squared, thus really corresponding to a peak in C_p/T^3, or whether it corresponded to a maximum in the coupling factor of sound waves in the Raman technique (Martin and Brenig (1974)). Later, neutron scattering measurements (Buchenau et al. (1986)) were able to clarify this point. They showed that one deals with a true maximum in the vibrational density of states over frequency squared. It turned out to be possible to calculate the broad hump in C_p/T^3 quantitatively from the neutron data. In addition, the measured dynamic structure factor showed unambiguously that these modes are not sound waves, at least in the prototype glass vitreous silica, thus strongly supporting the soft-potential model postulate of local modes coexisting with the sound waves.

In the following Sect. 9.2, we will first give a short overview of the anomalous properties of glasses above 1 K which cannot be explained by the tunneling model alone, such as the rise of the specific heat beyond the linear low-temperature term and the plateau in the thermal conductivity. We further describe some of the neutron scattering experiments showing the boson peak. The soft-potential model will be described in detail in Sect. 9.3. In Sect. 9.4, the predictions of the model for the thermal and acoustic properties of glasses are worked out and compared to experimental results. Section 9.5 summarizes the conclusions.

9.2 Tunneling States and Soft Modes in Glasses

9.2.1 Specific Heat

The crossover between tunneling states and soft vibrations is seen most clearly in measurements of the specific heat. Figure 9.1 shows this for vitreous silica, the most thoroughly studied glass [Phillips (1981a), Phillips (1987), Hunklinger and Raychaudhuri (1986), Hunklinger and Arnold (1976)]. Its specific heat is shown in Fig. 9.1 as C_p/T^3 vs. the temperature T in a double logarithmic scale. The Debye contribution of the sound waves (in this figure a constant, which can be calculated from the known sound velocities) has been subtracted. Therefore only the contribution of the additional excitations appears in Fig. 9.1. The measurements were done by different authors [Zeller and Pohl (1971), Hunklinger and Raychaudhuri (1986), von Löhneysen and Platte (1979), von Löhneysen et al. (1985), Buchenau et al. (1986)] on different samples and do in fact differ appreciably. Nevertheless, they all have at the lowest temperatures a section where C_p increases essentially linear with temperature and, also, all of them show a rapid increase between 3 and 5 K. We will see in Sect. 9.3 that the soft-potential model does in fact predict an increase of the number of soft vibrations per frequency interval with the fourth power of the frequency (at least under certain circumstances), which would imply an increase of the corresponding specific heat with the fifth power of the temperature. The two sections – extrapolated low-temperature part and extrapolated higher-temperature part – intersect at a crossover temperature $T_{\min}$ in the minimum of C_p/T^3 [in suprasil W there are still additional excitations, see (von Löhneysen and Platte (1979), von Löhneysen et al. (1985), centered at 1 K which shift the minimum to higher temperatures; the extrapolated lines, however, do not differ markedly from those of the other samples]. The temperature $T_{\min}$ is a quantitative measure for the crossover frequency between the two-level states and higher-frequency excitations.

The temperature dependence of Raman and neutron scattering discussed in Sect. 9.2.4 shows that these higher-frequency excitations are more or less harmonic vibrational modes. If one measures the specific heat to still higher temperatures, one finds a peak in C_p/T^3 at a temperature $T_{\max}$ (see Fig. 9.15). In terms of the vibrational density of states, that corresponds to a maximum in the vibrational density of states *divided* by frequency squared, i.e. a peak in the classical scattering function. That peak, called "boson peak" to stress its harmonic nature, has indeed been observed in many neutron and Raman measurements.

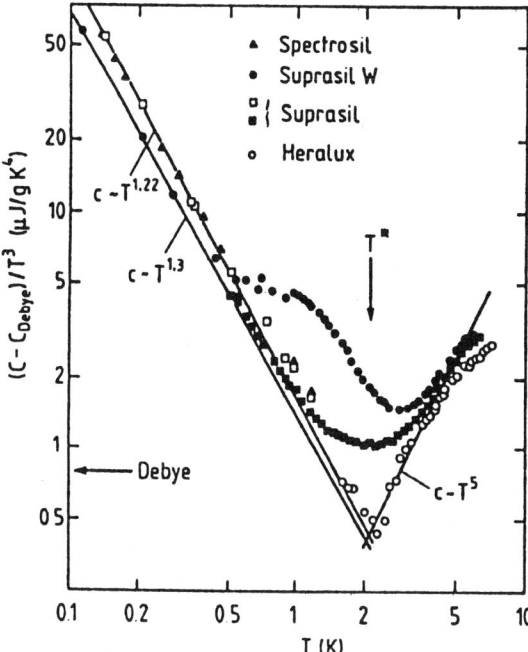

Fig. 9.1. Specific heat C_p of different SiO_2 glasses shown as C_p/T^3 vs. temperature T in a double logarithmic scale. The Debye contribution has been subtracted. T^* corresponds to $T_{\min}$ (see Buchenau et al. (1991)).

9.2.2 Thermal Conductivity

The thermal conductivity of glasses shows a change of behavior at approximately the same temperature as the specific heat, namely a crossover from an increase with T^2 (or better, $T^{2-\delta}$, where $\delta \approx 0.2$) to a more or less pronounced plateau. The tunneling model provides a widely accepted explanation for the initial T^2 rise, attributing the scattering below 1 K to the interaction between the sound waves and tunneling states. On the contrary, there are many conflicting explanations for the thermal conductivity in the plateau region (see Ramos and Buchenau (1997)). This behavior contrasts markedly with crystalline behavior (see Fig. 9.2) showing a scattering mechanism for the sound waves in glasses setting in at about 100 to 200 GHz which does not exist in perfect crystals. That scattering mechanism is strong enough to suppress the thermal conductivity contribution from the whole high frequency part of the sound wave spectrum. Though many people believe that it is due to the scattering of the sound waves from the static disorder in the glass (that mechanism was already proposed in Żeller and Pohl's classical paper in 1971), subsequent estimates by Zaitlin and Anderson (1975) and by Jäckle (1976) threw severe doubts on that explanation. A more plausible explanation was proposed by Karpov and Parshin (1985) and by Yu and Freeman (1987),

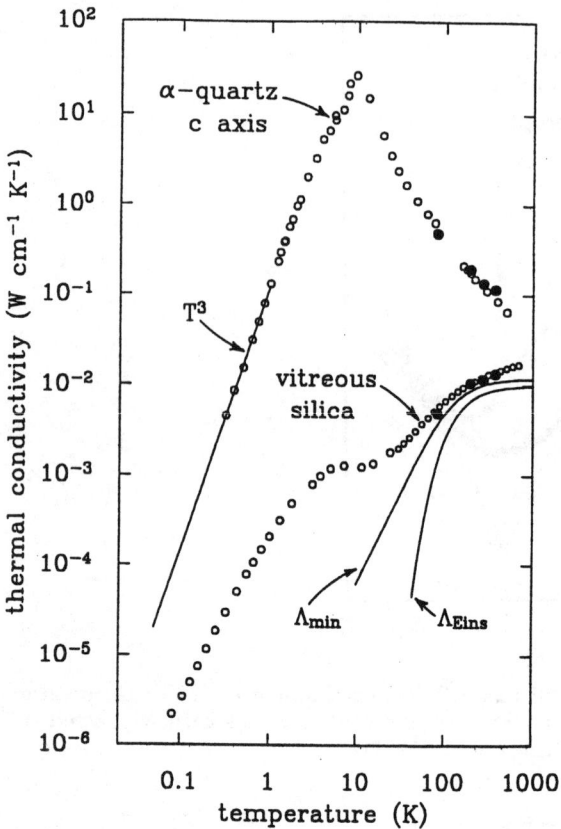

Fig. 9.2. Thermal conductivity of crystalline and amorphous SiO_2 over a wide temperature range in a double logarithmic plot (Cahill (1989)). The two continuous lines [Λ_{min}, Cahill and Pohl (1989), and Λ_{Eins}, Einstein (1911)] correspond to estimates about the lowest thermal conductivity possible in a solid.

linking the plateau in the thermal conductivity to the higher-frequency excitations which cause the rise of C_p/T^3. In these approaches, the plateau is due to the coupling between those higher-frequency excitations and the sound waves, which is assumed to be of the same strength as the one between tunneling states and sound waves.

9.2.3 Coherent Neutron Scattering

The nature of these higher-frequency excitations can be investigated by spectroscopic methods, in particular Raman (Winterling (1975)) and neutron (Buchenau et al. (1986), and (1988)) scattering. Neutron scattering offers the additional advantage of being able to measure the angular dependence of the inelastic scattering intensity (the dynamic structure factor). The dynamic structure factor contains information on the correlation of the atomic

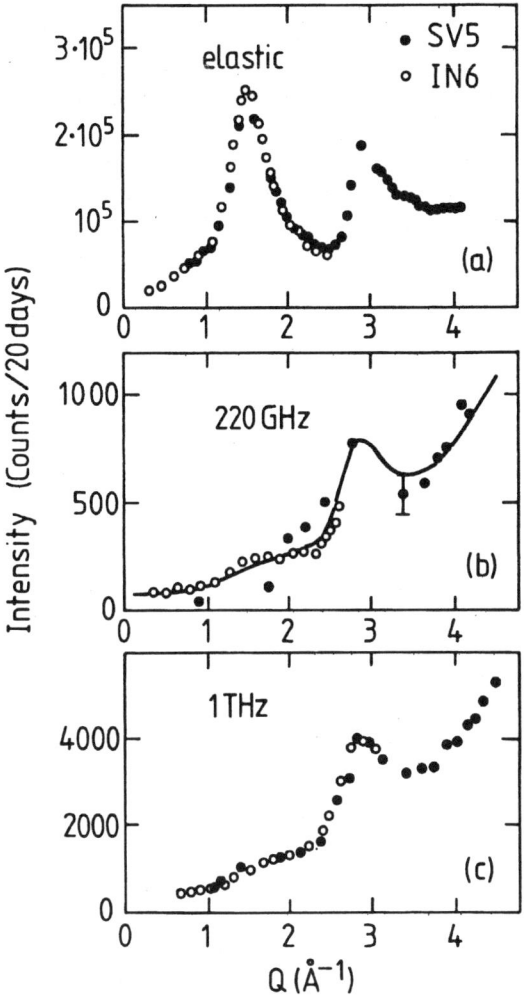

Fig. 9.3. Momentum-transfer dependence of neutron scattering intensities in vitreous silica at 300 K, measured on spectrometers SV5 (DIDO, Jülich) and IN6 (ILL, Grenoble) : (a) elastic; (b) inelastic at 220 GHz; (c) inelastic at 1 THz. The line in (b) is a fit to the points in (c).

motion. If the atoms move essentially in phase, the interference is dominated by the static correlations. In this case, one finds the same maxima and minima in the inelastic scattering as in the elastic one (Buchenau (1985b)). Such a behavior has been found in metallic glasses (Suck and Rudin (1983)).

Vitreous silica is a different case, as shown in Fig. 9.3. Here the first sharp diffraction peak of the elastic scattering at 1.6 Å^{-1} is not reproduced in the inelastic measurements, while the second one at 3.0 Å^{-1} is clearly seen. Even without a detailed analysis, one immediately concludes that not all neighboring atoms move in phase. In particular, those atomic neighbors which give rise to the first sharp diffraction peak have to move out of phase in order to explain the observed inelastic dynamic structure factor of Fig. 9.3b and c. In silica, the main part of the first sharp diffraction peak is due to the second oxygen neighbors on two different corner-connected tetrahedra (see Fig. 9.4). Obviously, the motion of the atoms in the mode must be such that the oxygens at opposite corners do not move in phase with each other.

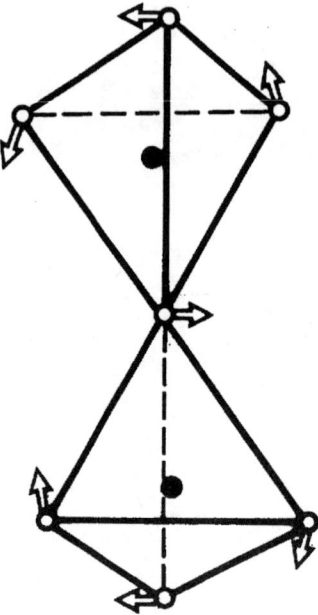

Fig. 9.4. Coupled rotation of two corner-connected SiO$_4$ tetrahedra.

The SiO$_4$ tetrahedra are stable units with strong bonds. Therefore the most likely candidate for a low-frequency motion is a coupled rotation of more or less rigid corner-connected tetrahedra as shown in Fig. 9.4. One can in fact explain the dynamic structure factor in Fig. 9.3b and c quantitatively in terms of coupled librational motions of the SiO$_4$ tetrahedra (Buchenau et al. (1986)).

Coupled tetrahedra librations do also show relatively low vibrational frequencies in crystalline quartz. In particular, the soft mode which gets a very low frequency at the structural phase transformation from α-quartz to β-quartz is of this type (Grimm and Dorner (1975)).

This experimental result is significant not only for the identification of the characteristic modes of a specific system. If these modes in vitreous silica are indeed the specific realization of universal glass properties, one can exclude quite generally all those models of glassy anomalies which explain the low frequencies by large wavelengths. Naturally, for a sound wave, the frequency is inversely proportional to the wavelength. That relation has become so deeply rooted in our thinking that sometimes we tend to identify low frequencies with large wavelengths (large compared to interatomic distances). However, in glasses one should not forget the other alternative, namely a localized mode which by chance happens to have a very small restoring force constant. In that case, the Fourier decomposition of the mode into plane waves can contain a large amount of plane waves with a short wavelength. To put it differently, nearby atoms can move opposite to each other like in the example of Fig. 9.4.

There have been several attempts to explain the excess peak in C_p/T^3 in terms of long wavelength modes, with wavelengths much larger than the interatomic spacing. Examples are the assumption of an anomalous dispersion of the sound waves (Jones et al. (1981)) or the fracton models (Orbach (1985)). The neutron results of Fig. 9.3 allow one to discard that possibility, at least for the prototype glass silica. If one believes in the universality of the low-temperature anomalies, one must discard it also for all other glasses.

Another experiment contradicting these models is the observation of standing waves in thin films of vitreous silica up to 400 GHz with frequencies fitting the sound velocities (Rothenfusser et al. (1983)). The result shows that the sound waves alone cannot account for the high vibrational density of states at 400 GHz because there is no anomalous dispersion. One has to postulate additional modes which are not sound waves in order to explain the excess specific heat above 2 K.

9.2.4 Temperature Dependence of Raman and Neutron Scattering

A further important information from the spectroscopic methods is the temperature dependence of the inelastic signals. That allows one to judge whether one deals with harmonic vibrations or with anharmonic modes (like tunneling or relaxational motion). If one has harmonic modes, the inelastic scattering intensity in energy gain of the scattering particle is proportional to the Bose–Einstein thermal population factor

$$f_B = \frac{1}{e^{\hbar\omega/k_B T} - 1} \ . \tag{9.1}$$

If the thermal energy exceeds $\hbar\omega$, that Bose factor begins to increase linearly with temperature.

Figure 9.5 shows results of Raman (Winterling (1975)) and neutron (Buchenau et al. (1988)) scattering for vitreous silica. Above 500 GHz, one finds the Bose factor dependence expected for harmonic vibrations. At 220 GHz (see Fig. 9.3b), the curve clearly rises more strongly than linear with increasing temperature. The same is true for Raman data at 300 GHz (Winterling (1975)). Obviously, there is a crossover from reasonably harmonic motion at higher frequencies to anharmonic modes at lower frequencies.

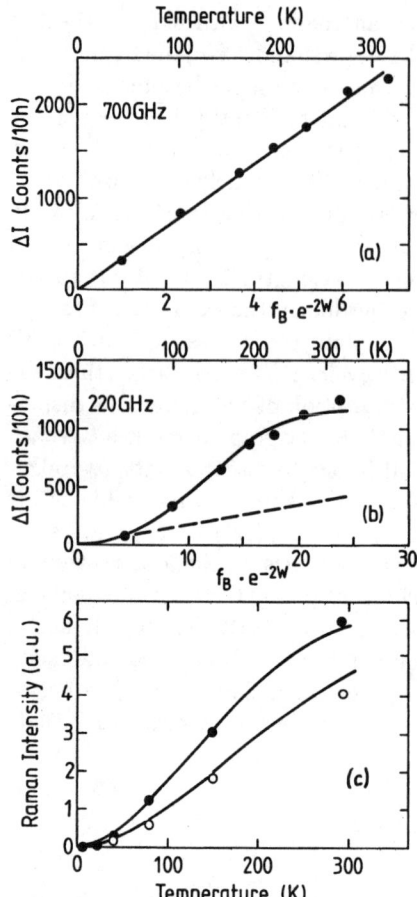

Fig. 9.5. Temperature dependence of inelastic neutron scattering intensities in vitreous silica (f_B, Bose factor; e^{-2W}, Debye–Waller factor) at (a) 700 GHz; (b) 220 GHz. The dashed line in (b) denotes the harmonic contribution. (c) Temperature dependence of Raman intensities at 5 cm^{-1} (*upper curve*) and at 10 cm^{-1} (300 GHz). The curves in (b) and (c) show the fit in terms of relaxations (from Buchenau et al. (1988)).

It is relatively easy to convince oneself that these anharmonic modes are not the tunneling states observed below 1 K in the specific heat. The tunnel splitting corresponds to still lower frequencies, 20 GHz and lower. Also, the temperature behavior of tunneling states is expected to be quite different. The observed one in Fig. 9.3b and c rather corresponds to the one expected for low-barrier relaxational modes, with barrier heights of the order of a few hundred Kelvin. In fact, mechanical and dielectric relaxation data (Hunklinger and Arnold (1976), von Schickfus and Hunklinger (1981)) suggest the existence of such low-barrier relaxational modes in silica.

One can explain the curves in Fig. 9.3b and c, assuming a small harmonic part and a larger part due to thermally activated relaxational jumps over energy barriers (Buchenau et al. (1988)). The barrier distribution giving a good fit agrees with the one fitted to mechanical and dielectric relaxation measurements (Buchenau et al. (1988)) between 10 and 100 K. The simple classical interpretation of the neutron and Raman scattering data implies a crossover from vibrations in single-well potentials at higher frequencies to thermally activated jumps in double-well potentials at lower frequencies. The dynamic structure factor remains the same in both cases (see Fig. 9.3). This shows that soft vibrations and structural relaxations have the same kind of eigenvector, implying a common origin of both. It seems reasonable to include the two-level states in the same picture because their explanation in the tunneling model requires double-well potentials similar to those of the relaxational modes. That concept, a broad distribution of local modes, extending from low-frequency vibrational modes to low-barrier relaxational modes, is the physical basis of the phenomenological soft-potential model.

9.2.5 Comparison Between Neutron and Specific-Heat Data

Figure 9.6 shows the vibrational density of states at low frequencies determined from the harmonic part of the neutron scattering intensities. It is compared with a density of states fitted to the specific heat between 2 and 50 K. The comparison contains no adaptable parameter because both the specific heat and the neutron data give the density of states in an absolute scale. In the neutron case, this is achieved by a quantitative comparison of the inelastic scattering intensities with the elastic scattering, making use of a structural model of the glass and assuming a specific eigenvector of the atomic motion (coupled rotations of corner-connected SiO_4 tetrahedra) in that structural model. The detailed procedure is described by Buchenau et al. (1986). The good agreement between the vibrational densities of states obtained from neutron scattering and the specific heat measurement is an additional proof of the underlying assumption of harmonic vibrations. In C_p/T^3, a mode contributes most strongly at that temperature where its energy splitting is about four times higher than the thermal energy. Thus one samples essentially the first excited level of each vibration. In contrast, the neutron data at temperatures from 100 to 300 K sample the vibrations around 1 THz in states containing two to ten vibrational quanta. If both measurements agree, there must be a reasonably equispaced ladder of vibrational levels in these modes. That in turn means that one deals with reasonably harmonic vibrations.

Note that the broad peak at 3 THz in the density of states (Fig. 9.6) does not correspond to a maximum in the measured spectra. In the measurement, one observes the density of states divided by frequency and multiplied by the Bose factor. At higher temperatures, this is essentially the density of states multiplied by temperature and divided by frequency squared. That measured intensity directly parallels the quantity C_p/T^3 and shows a maximum at

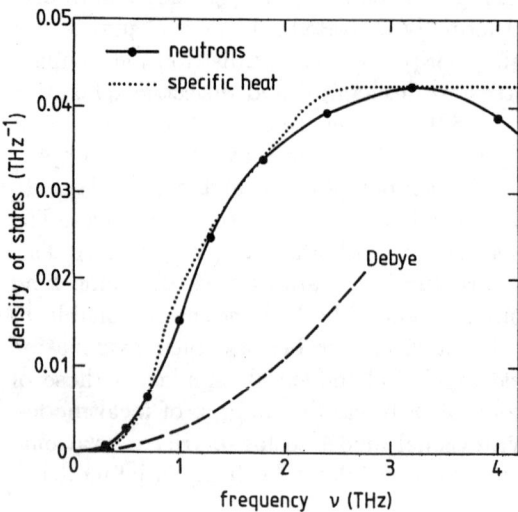

Fig. 9.6. Vibrational density of states of vitreous silica from neutron scattering and from the specific heat (Buchenau et al. (1986)).

about 0.9 THz, the so-called *boson peak*, compatible with the maximum in C_p/T^3 at 10 K (see Fig. 9.15). The boson-peak frequency marks the point where the vibrational density of states has the largest excess ratio to the Debye expectation.

9.2.6 More Recent Neutron Data

So far, our discussion has been limited to vitreous silica. Somehow this is a special case, because the additional excitations at the boson peak heavily outnumber the sound waves (by about a factor of seven). In most other glasses, that ratio is lower. Judging from the maximum in the specific heat over temperature cubed, one has a ratio 3:2 in B_2O_3 at the boson peak (White et al. (1984), Pérez-Enciso et al. (1997)). Therefore one should be able to observe the sound waves in the inelastic coherent neutron scattering, i.e. the first peak in the elastic $S(Q)$ should also show up in the inelastic scattering near the elastic line. This is indeed the case, as first shown by Hannon et al. (1993). Their results are displayed in Fig. 9.7. Interestingly, one sees the first sharp diffraction peak at 1.5 Å^{-1} only at the lowest energies. For higher energies, the data in Fig. 9.7 show that the peak gradually disappears, consistent with the soft-potential picture of an increasing number of soft local modes with increasing frequency. A similar case is polybutadiene, where again one sees the sound-wave contribution in the inelastic neutron scattering signals. Figure 9.8 shows elastic and inelastic data from deuterated polybutadiene at 60 K (Buchenau et al. (1996b)).

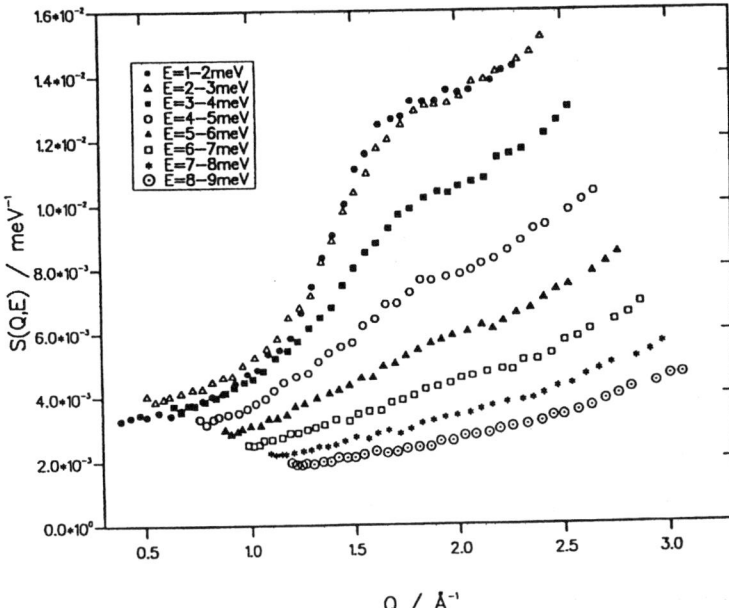

Fig. 9.7. The Q dependence of the scattering function $S(Q, E)$ of vitreous B_2O_3 at low energy (Hannon et al. (1993)).

The elastic data in Fig. 9.8(a) show some small angle scattering, which is of no concern in our context, and a pronounced peak at 1.5 Å^{-1}, which is ascribed to nearest-neighbor polymer chains lying side by side. The inelastic data in Fig. 9.8b also show the same peak, but much more marked at the lower frequency of 1 meV than at the higher frequency of 6 meV. Since the sound waves must reproduce the peak, while the additional soft modes need not necessarily do so, one can use these dynamic structure factors to distinguish between sound waves and additional excitations. That analysis does indeed confirm the soft-potential picture of a mixture of sound waves with other vibrational modes (Buchenau et al. (1996b)).

An implication of the soft-potential model which can be tested by neutron scattering is the so-called "non-Gaussianity" resulting from the existence of quasi-localized soft modes. Since some of these modes have unusually large vibrational or relaxational amplitudes, the mean square displacements of those atoms which participate in that motion are larger than those of the average atom, at least in the direction of motion of the soft mode. Consequently, one has to reckon with a whole distribution of different atomic mean square displacements in the glass. The effect can be most easily seen in the elastic incoherent scattering. In the simplest case where all atomic mean square dis-

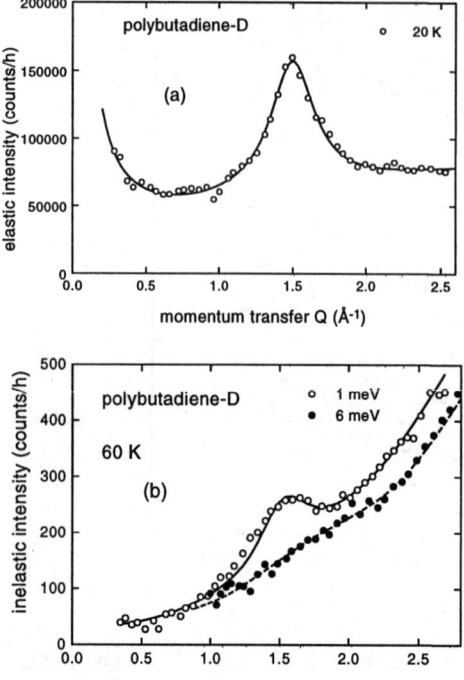

Fig. 9.8. Momentum tranfer Q dependence of the scattering from polybutadiene (a) elastic at 20 K, (b) inelastic at two different frequencies at 60 K (Buchenau et al. (1996b)).

placements are equal, the Q dependence of the incoherent elastic scattering intensity $I_{el}(Q)$ is given by the Debye–Waller factor

$$I_{el}(Q)/I_{el}(0) = e^{-2W} = e^{-\alpha Q^2}, \quad (9.2)$$

where α is the mean square displacement of the atoms in the direction of the momentum transfer vector $\mathbf{Q}$.

Next, consider that each atom i has a different mean square displacement α_i in the direction of $\mathbf{Q}$. Then the Q dependence of the elastic scattering from the N atoms reads as

$$I_{el}(Q)/I_{el}(0) = \frac{1}{N}\sum_{i=1}^{N} e^{-\alpha_i Q^2} \approx e^{(-\overline{\alpha}Q^2 + \frac{1}{2}(\overline{\alpha^2}-\overline{\alpha}^2)Q^4)}, \quad (9.3)$$

where $\overline{\alpha}$ and $\overline{\alpha^2}$ are the averages of α_i and α_i^2, respectively. The dimensionless coefficient $(\overline{\alpha^2} - \overline{\alpha}^2)/\overline{\alpha}^2$ is denoted as the *non-Gaussianity* A_0.

Indications of non-Gaussian behavior were first found in myoglobin (Doster et al. (1989)) and in glycerol (Fujara et al. (1991)). In those papers, however, it was assumed that the non-Gaussianity stemmed from the relaxational part of the scattering alone.

Recently, the deviations from Gaussianity have also been observed in the inelastic incoherent scattering (Buchenau et al. (1996a)) for two polymers. The observations could be consistently explained in terms of quasi-localized vibrational modes in the boson-peak region, with a localization to a few monomers of the polymer chain.

The existence of quasi-localized vibrational modes at low frequencies localized to about twenty atoms has also been established in a number of numerical simulations on model glasses (Schober and Laird (1991), Oligschleger and Schober (1993), Cho et al. (1994), Hafner and Krajci (1994), Bembenek and Laird (1995)). That topic is discussed in more detail in Chap. 8.

9.3 The Soft-Potential Model and its Parameters

9.3.1 The Anharmonic Quartic Potential

Let us consider a localized mode with a very weak or even negative restoring force. The atomic motion in a localized mode can be described by its eigenvector $\mathbf{e}_i$, $i = 1,..N$ for the N atoms of the glass, which fulfils the normalization condition

$$\sum_{i=1}^{N} \mathbf{e}_i^2 = 1. \tag{9.4}$$

The displacement $\mathbf{u}_i$ of the atom i in the mode is given by the mode amplitude $\mathcal{A}$ according to

$$\mathbf{u}_i = \sum_{i=1}^{N} \frac{\mathbf{e}_i}{M_i^{1/2}} \mathcal{A} \tag{9.5}$$

(M_i is the mass of atom i). The mode being both soft and localized, the atomic displacements will be large and the anharmonic terms of the potential have to be taken into account. In the case of a negative restoring force, the anharmonic terms are even needed to stabilize the mode. We first consider the purely quartic potential (see Fig. 9.9)

$$V(\mathcal{A}) = \frac{v_4}{4} \mathcal{A}^4. \tag{9.6}$$

The zero-point energy in the purely quartic potential can be estimated by equating a kinetic confinement energy and the potential energy at a normal coordinate displacement $\mathcal{A}_0$. That relation defines the energy W, the first parameter of the soft-potential model, which will be seen to determine the crossover frequency between tunneling and vibrational motion:

$$\frac{\hbar^2}{2\mathcal{A}_0^2} = \frac{v_4}{4} \mathcal{A}_0^4 \equiv W. \tag{9.7}$$

In the soft-potential model, it is usual to define a dimensionless displacement coordinate

$$x = \frac{\mathcal{A}}{\mathcal{A}_0}. \tag{9.8}$$

The purely quartic potential is a special case. In general, one has to reckon with a finite restoring force and a finite asymmetry of the potential. Therefore, neglecting terms higher than the fourth order, the potential of the soft mode in terms of the dimensionless displacement x reads as

$$V(x) = W\left(D_1 x + D_2 x^2 + x^4\right), \tag{9.9}$$

with a kinetic energy

$$E_{\text{kin}} = \frac{1}{2}\frac{\hbar^2}{2W}\dot{x}^2. \tag{9.10}$$

Here, the origin has been choosen such that the third-order term of the potential vanishes. D_2 characterizes the small positive or negative restoring force and D_1 describes the asymmetry of the potential. The two coefficients determine whether one deals with a tunneling state, a soft vibration or even a relaxational state at more elevated temperatures. Figure 9.10 shows the single- and double-well regions of the potential of (9.9) in the $D_1 - D_2$ plane. The inserts show the potential and the levels of a typical tunneling state (left)

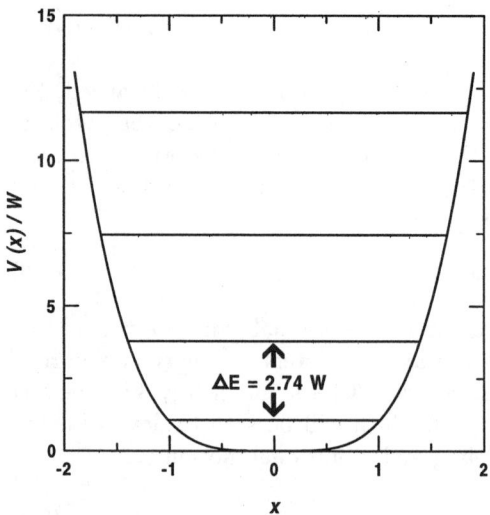

Fig. 9.9. Pure quartic potential $V(x) = W x^4$ at the border between double-well and single-well potentials, and its corresponding energy levels.

9.3 The Soft-Potential Model and its Parameters

and of a typical vibrational state (right). Assuming modes with different D_1 and D_2 values at different places in the glass, one obtains a broad distribution of soft modes ranging from tunneling to vibrational states.

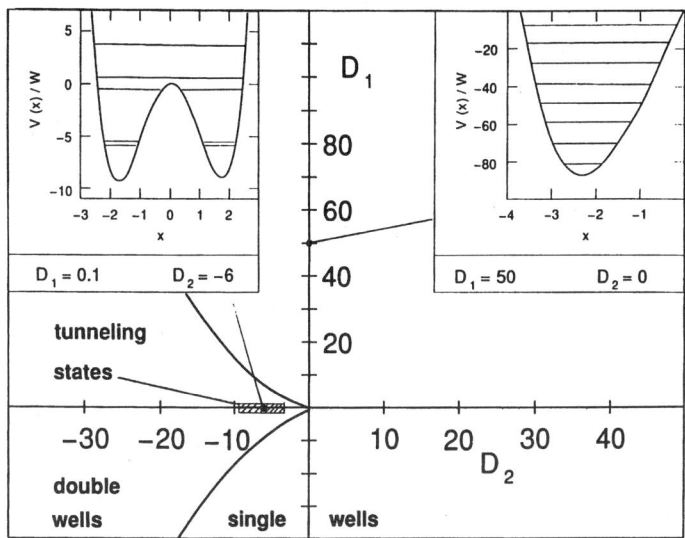

Fig. 9.10. Single- and double-well regions in the $D_1 - D_2$ plane of the soft-potential model. Inserts: Potentials and levels of a typical tunneling state (*left*) and of a typical vibrational state (*right*).

Although W would be expected to vary from mode to mode, it seems often possible in practice to describe the many different low-frequency modes of a single glass by a single W [an exception from this rule has been found in heat-release experiments (Parshin and Sahling (1993), see Chap. 2, which seem to reveal two groups of excitations with different W]. This fundamental parameter W is found to be of the order of the thermal energy at a few Kelvin. From crude estimates of v_4, Buchenau et al. (1991) inferred an effective mass corresponding to about twenty to hundred atoms participating in the mode. This high number of participating atoms has been verified recently by numerical work on the low-frequency vibrations of a model glass (Laird and Schober (1991), Schober and Laird (1991)).

9.3.2 Assumptions

The formulation and assumptions of the soft-potential model (SPM) have been changing as the comparison to experimental data proceeded. The following three assumptions, however, seem to be acknowledged by most scientists working with the model:[1]

(i) The soft modes can on average be characterized by a single energy W.

(ii) D_1 and D_2 are randomly distributed around the origin of the $D_1 - D_2$ plane with a density $P(D_1, D_2) = P(0,0) \equiv P_s$.

(iii) The interaction between the soft modes and the sound waves is bilinear in the displacement of the soft mode and in the strain field of the sound wave. On the average, the coupling can be described by the two coupling constants Λ_l and Λ_t with

$$\delta V_l = \Lambda_l x \epsilon_l \quad ; \quad \delta V_t = \Lambda_t x \epsilon_t. \tag{9.11}$$

Here ϵ_l and ϵ_t denote the longitudinal strain field of a longitudinal sound wave and the shear strain of a transverse sound wave, respectively, and δV is the change in energy.

9.3.3 Level Splittings and Matrix Elements

The analytical description of the tunneling states in the soft-potential model is based on the concepts of the standard tunneling model. The tunneling model postulates localized tunneling states in double-well potentials. The potentials are characterized by an energy difference Δ between the minima (the asymmetry of the potential) and a tunnel splitting $\Delta_0 = \hbar\omega_0 \exp(-\lambda)$. Here, ω_0 is some crystal-like vibrational frequency and λ is a tunnel integral. The parameters Δ and λ are random; their distribution function is assumed to be constant: $P(\Delta, \lambda) = P_0$. The level splitting E is calculated via $E = \sqrt{\Delta_0^2 + \Delta^2}$.

If one wants to calculate the distribution function $P(\Delta, \lambda)$ in the soft-potential model, one needs expressions for Δ and λ as a function of D_1 and D_2. In earlier works with the model, a simplified Wentzel–Kramers–Brillouin (WKB) expression was used:

$$\Delta_0 = W \exp\left(-\frac{\sqrt{2}}{3} \mid D_2 \mid^{3/2}\right). \tag{9.12}$$

---

[1] The notation used here is not the same one employed in some earlier works. It is, however, easily transformed to the original notation used by Karpov et al. (1982), Il'in et al. (1987), Galperin et al. (1989a), Buchenau et al. (1991), Buchenau et al. (1992), Parshin and Sahling (1993), Parshin (1994b), Parshin (1994a), through the relations $x \equiv x/(a\eta_L^{1/2})$ and $P_s = P_0 \eta_L^{5/2}$ [for Buchenau et al. (1991) and Buchenau et al. (1992), $P_s = 2P_0 \eta_L^{5/2}$]. Please note that this P_0 should not be confused with the constant density of states P_0 of the standard tunneling model. In addition, we have changed the definitions of Λ_l and Λ_t as compared to those in Buchenau et al. (1992) by a factor $\eta_L^{1/2}$. The new notation avoids the use of η_L.

For a symmetric double-well potential, the two minima lie at $\pm\sqrt{|D_2|/2}$. For small D_1, the energy difference Δ between the two minima is approximated by

$$\Delta = WD_1\sqrt{|2D_2|}. \tag{9.13}$$

From these equations, one derives

$$P(\Delta, \lambda) = \frac{2P_s}{W|D_2|} \tag{9.14}$$

via the Jacobian of the transformation, as shown by Il'in et al. (1987) and Buchenau et al. (1992). Note that this $P(\Delta, \lambda)$ is no longer constant, but depends on the value of D_2, i.e. on the barrier height.

As discussed in detail in Chap. 4, in the tunneling model one has a coupling to the sound waves given by

$$\gamma_j = \frac{1}{2}\frac{\partial \Delta}{\partial \epsilon_j}, \tag{9.15}$$

where $j = l, t$ for longitudinal and transverse sound waves. With (9.11) and (9.15), the relation between the coupling constant of the two models reads as $\gamma_j = \Lambda_j\sqrt{|D_2|/2}$, a relation which again depends on D_2.

Both the tunneling states and the sound waves are strongly influenced by the coupling. For the tunneling states, the coupling determines their relaxation time τ. Rewriting the expression for the relaxation time of the tunneling model in terms of the parameters of the soft-potential model, one gets

$$\tau^{-1} = A_t\frac{|D_2|}{2}\Delta_0^2 E \coth\left(\frac{E}{2k_\mathrm{B}T}\right) \tag{9.16}$$

with

$$A_t = \frac{1}{2\pi\varrho\hbar^4}\left(\frac{\Lambda_l^2}{v_l^5} + \frac{2\Lambda_t^2}{v_t^5}\right). \tag{9.17}$$

For the sound waves, the influence of the tunneling states on the acoustic properties is determined (see Chaps. 4 and 5) by the parameter combination $C_j = P_0\gamma_j^2/\varrho v_j^2$, where ϱ is the density and v_j is the longitudinal and transverse sound velocity for $j = l$ and $j = t$, respectively. Here, one finds (Buchenau et al. (1992)) from (9.14) and the relation between the coupling constants

$$P_0\gamma_j^2 = \frac{P_s\Lambda_j^2}{W}, \tag{9.18}$$

so this parameter combination turns out to be a constant in both models, at least within the approximations used.

With (9.14), (9.16) and (9.18), one can take over the expressions for the specific heat, the thermal conductivity, the temperature dependence of the sound velocity, the acoustic absorption and many other physical properties at very low temperatures from the tunneling model and thus get a complete description for the soft modes of the soft-potential model in the tunneling region. That has been done by Parshin (1994b) in his review of the model, an excellent paper with a much wider theoretical scope than our more experimentally oriented contribution.

We thought it convenient to improve some of the above expressions in order to obtain more accurate results for comparison with experiment. By making use of numerical calculations which will be described below, Ramos et al. (1993) have shown that (9.12) should be replaced by the more accurate expression

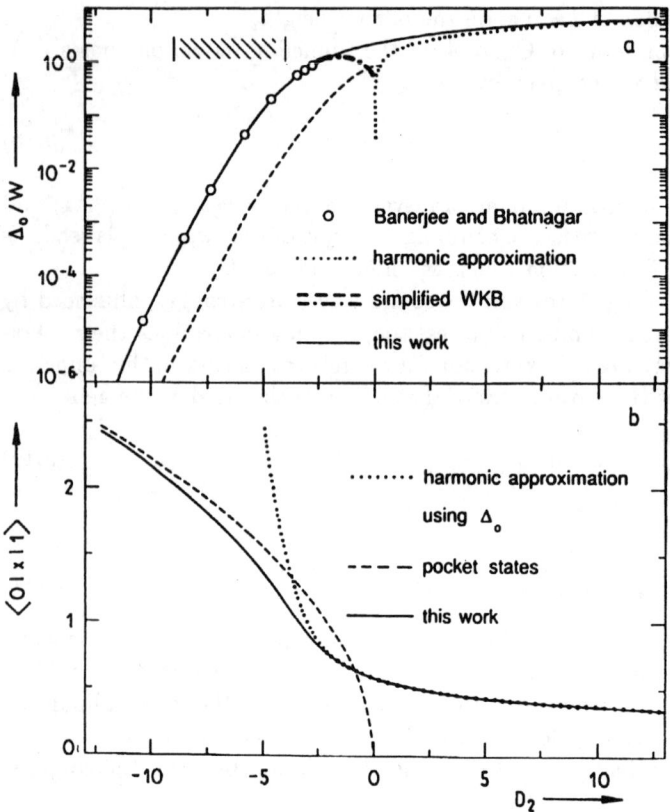

Fig. 9.11. (a) Energy splitting Δ_0 and (b) matrix element $\langle 0 \mid x \mid 1 \rangle$ of the two lowest levels for symmetric potentials of the form of (9.9) as a function of the harmonic coefficient D_2. The continuous lines were calculated numerically. The dashed and dot-dashed lines are analytical approximations in the tunneling region. The dotted lines are approximations in the vibrational region. The open circles in (a) are numerical results from Banerjee and Bhatnagar (1978). The dashed region indicates the range of double-wells giving tunneling states which could contribute significantly to physical properties.

9.3 The Soft-Potential Model and its Parameters

$$\Delta_0 = W \mid D_2 \mid^{3/2} \exp\left(1 - \frac{\sqrt{2}}{3} \mid D_2 \mid^{3/2}\right) \tag{9.19}$$

and (9.13) by

$$\Delta = W D_1 \sqrt{2(\mid D_2 \mid -1)}. \tag{9.20}$$

Similarly, the interaction between sound waves and tunneling states is better described by $\gamma_j = \Lambda_j \sqrt{(\mid D_2 \mid -1)/2}$. Combining these equations as before, one now finds

$$P_0 = \frac{2P_s}{W(\mid D_2 \mid -1)} f(D_2) \tag{9.21}$$

and

$$P_0 \gamma_j^2 = \frac{P_s \Lambda_j^2}{W} f(D_2), \tag{9.22}$$

where the function $f(D_2)$ is close to unity and is given by

$$f(D_2) = \sqrt{\mid D_2 \mid -1} / \left(\sqrt{\mid D_2 \mid} - \frac{3}{\sqrt{2} \mid D_2 \mid}\right). \tag{9.23}$$

Equations (9.21)–(9.23) give the direct relation between the parameters of the two models within an accuracy which should be satisfactory on an experimental accuracy level.

Energy levels and wave functions for potentials of the form of (9.9) have been calculated by Gil et al. (1993) and Ramos et al. (1993), through a numerical search for the solutions of the Schrödinger equation. Figure 9.11a shows the energy splitting Δ_0 of the two lowest levels for the symmetric case $D_1 = 0$ as a function of D_2 on a logarithmic scale. Within the numerical accuracy of about 1 %, they agree with earlier, more accurate calculations by Banerjee and Bhatnagar (1978). The comparison to the analytical approximations shows that the expression (9.19) proposed for the tunnel splitting (the dot-dashed line in Fig. 9.11a) is to be preferred to the one used in earlier work, (9.12) (the dashed line in Fig. 9.11a). However, all three approximations, including the harmonic approximation (9.24) (the dotted line in Fig. 9.11a), break down as one approaches the origin of the $D_1 - D_2$ plane – the center of the distribution. This is understandable, since the well-defined cases tunneling and harmonic vibration reside far away from the center, at large negative and positive values of D_2, respectively. So one cannot expect an all-embracing approximation centered at the origin. This clearly demonstrates the usefulness of numerical work, in particular in the central crossover region where the approximations fail.

It is to be stressed that the failure of (9.12) to relate correctly the splitting energy with the parameters of the double-well potential (see Fig. 9.11a) is not specific of the soft-potential model. It corresponds to a general failure of the WKB approximation for potential barriers of such a low energy as those relevant for the low-temperature properties of glasses. Therefore, the standard tunneling model is also in trouble when one wishes to obtain *quantitative* relationships between splitting energies and barrier heights, for instance.

The situation is similar with respect to the matrix elements $\langle 0 \mid x \mid 1 \rangle$ shown in Fig. 9.11b, again for the symmetric case. Here, however, one can make use of the fact that in the neighborhood of the pure quartic potential the wave functions are fairly alike to those of a harmonic potential with the same level splitting. The breakdown of the quasi-harmonic approximation, i.e.

$$\langle 0 \mid x \mid 1 \rangle^2 = \frac{W}{\Delta_0}, \tag{9.24}$$

the dotted line in Fig. 9.11b, marks the gradual crossover to the tunneling case, where the wave functions are not alike to those of any harmonic oscillator. In the tunneling case, the wave functions are better described by a combination of two pocket states within the two wells of the potential. These pocket states, however, are not exactly centered in the well minima, but are shifted inside by a small amount. The reason for this shift can be seen from the left insert of Fig. 9.10, which shows that the rise of the potential towards the outside is steeper than that towards the barrier. This is the reason for the smaller matrix elements as compared to the analytical approximation $\sqrt{\mid D_2 \mid}/2$ in the tunneling range. The same effect leads to somewhat smaller values of Δ in the asymmetric case, as compared to the analytical approximation given by (9.13). This effect has been taken into account in (9.20)–(9.22), as discussed above.

9.3.4 The Distribution-Limiting Thermal Strain "Ansatz"

With the simplest random distribution around the origin of the $D_1 - D_2$ plane of a constant density $P(D_1, D_2) = P(0,0) \equiv P_s$, the second basic assumption mentioned above, it is not possible to understand the whole of the low-temperature behavior of glasses. As will be described in Sect. 9.4, this unlimited distribution of soft modes predicts a vibrational density of states $g(\nu) \propto \nu^4$, which is able to explain the minimum and subsequent increase in C_p/T^3 around 1–3 K (Il'in et al. (1987), Buchenau et al. (1991)), the plateau in the thermal conductivity and the rapid rise of the acoustic absorption around 5–10 K (Buchenau et al. (1992), Ramos and Buchenau (1997)), as well as a number of relaxation and heat-release phenomena (see Chap. 2). Nevertheless, an explanation of the glassy anomalies at still higher temperatures and frequencies, namely, the ubiquitous maximum in C_p/T^3 and the second rise of the thermal conductivity at a few K, and the related broad maximum in $g(\nu)/\nu^2$ (the *boson peak*) would still be missing.

As pointed out in the introduction, there are two divergent views on the correct explanation of these phenomena at higher frequencies and temperatures. Il'in et al. (1987) and Parshin (1994b) invoke the interaction between the modes as the central mechanism. They argue that the interaction of the modes with the elastic continuum is already strong enough to destroy the picture of isolated quasi-local modes as the number of modes increases towards the boson peak.

However, Chap. 5 shows that the same argument may already be held against the tunneling states: If one assumes the tunneling entities distributed at random over the sample, then their interaction via the distortion of the elastic medium is strong enough to change the tunnel splittings themselves dramatically.

We will therefore adopt a different point of view, namely the one that the tunneling states and the quasi-local modes are already the results of mode interaction, and that the soft-potential model is the correct phenomenological model to deal with that result, as shown by its good agreement with experiment. With respect to the boson-peak region, Gil et al. (1993) and Ramos et al. (1993) have shown that the anomalous features of the thermal properties of glasses could be explained over the entire low-temperature range within the soft-potential model, by adding a fourth assumption to those three mentioned above. This additional assumption postulates that the distribution function $P(D_1, D_2)$ is a Gaussian in the asymmetry D_1 centered around zero, and it is independent of D_2:

$$P(D_1, D_2) = P_s \exp(-AD_1^2). \tag{9.25}$$

It is straightforward to see that some kind of limitation in the $P(D_1, D_2)$ distribution is needed in order to account for the "boson peak" or the maximum in C_p/T^3. Otherwise, the density of soft vibrations would increase as $g(\nu) \propto \nu^4$ without end. Furthermore, a limitation in the assymmetry D_1 has in fact been anticipated, since Il'in et al. (1987) introduced it as being due to small perturbations.

In order to estimate the width of that Gaussian distribution, Gil et al. (1993) used a thermal strain *ansatz*, based on an idea of Ferrari et al. (1987). These authors had postulated a static thermal strain in each degree of freedom, freezing in at the glass transition T_g, which accounted for the additional specific heat of undercooled liquids. Gil et al. (1993) assumed that the symmetric state with $D_1 = 0$ corresponds to an unstrained soft mode, the state of minimum free energy at T_g, and that D_1 results from a static thermal strain, exerted on the soft mode by its surroundings. From the free energy $F = -k_B T \ln Z$ with $Z = \int_{-\infty}^{\infty} dx \exp(-\beta W(D_1 x + D_2 x^2 + x^4))$, one derives the relation

$$\frac{\partial \langle x \rangle}{\partial D_1} = -\frac{W}{k_B T}(\langle x^2 \rangle - \langle x \rangle^2). \tag{9.26}$$

At small D_1, $\langle x \rangle$ can be neglected. For $D_1 = D_2 = 0$, one has (Buchenau (1992)) the mean square displacement $\langle x^2 \rangle = 0.338(k_B T/W)^{1/2}$. Taking (9.26) as the expression for the restoring force against the buildup of the displacement connected with D_1, one calculates that the energy required to generate a small D_1 at $D_2 = 0$ is $0.169 D_1^2 W^{3/2} (k_B T)^{-1/2}$. Weighting each D_1 with the Boltzmann factor corresponding to that energy at T_g, one gets the Gaussian distribution

$$P(D_1, D_2) = P_s \exp(-0.169 D_1^2 (W/k_B T_g)^{3/2}), \qquad (9.27)$$

assuming its validity for all values of D_2. This distribution function, which does not introduce any additional parameter, was used in numerical calculations (Gil et al. (1993), Ramos et al. (1993)) to demonstrate the capability of the soft-potential model to describe the anomalous thermal and vibrational properties of glasses over the whole low-temperature range. These results will be presented in Sect. 9.4.

9.3.5 Other Approaches

The thermal strain *ansatz* used above to estimate quantitatively the range of single-well potentials is appealing, especially since it is able to describe rather accurately several physical properties of glasses over the entire low-temperature region without adding any adjustable parameter. Nevertheless, there is a debate as to whether the approximation of independent soft modes is still valid, neglecting the interactions among soft modes and/or between soft modes and acoustic phonons at those relatively high frequencies. Precisely, around the "boson-peak" frequencies the Debye scheme of sound waves begins to fail. Moreover, as first put forward by Il'in et al. (1987) and later developed by Buchenau et al. (1992) and Gurevich et al. (1993), one can show that the strong resonant scattering of phonons by the quasi-harmonic soft modes causes the phonon mean free path to reach the Ioffe–Regel limit (i.e. the phonon wavelength) at frequencies of that order, hence casting doubts about the applicability of the concept of delocalized plane-wave phonons carrying sound and heat in the usual way. As a matter of fact, Gurevich et al. (1993) have proposed a different approach to account for the necessary limitation of soft modes arising from single-well potentials and hence for the "boson peak". In brief, they estimate that the simple picture of independent quasi-localized harmonic vibrations is lost very close to the Ioffe–Regel limit, roughly at energies of 10–15 W. Therefore, the interaction between soft modes becomes important, leading to the reconstruction of the vibrational density of states. These modes are now much more delocalized, the density of states depending linearly in energy. The corresponding crossover from non-interacting soft modes with $g(\nu) \propto \nu^4$ to these new delocalized soft vibrations with $g(\nu) \propto \nu$ would explain the "boson peak".

Nonetheless, the former approach of the Gaussian distribution $P(D_1, D_2)$ has at least the merit of providing a simple, plausible distribution function cutoff, without adjustable parameters, which is able to account quantitatively for several physical properties in the whole range, as will be shown below. We will include and discuss this additional assumption of the soft-potential model when necessary in Sect. 9.4.

9.4 Predictions of the Soft-Potential Model

9.4.1 Tunneling Density of States in Double-Well Potentials

As already noted in Sect. 9.3.3, the analytical description of the tunneling states within the SPM is directly taken from the standard tunneling model. As a consequence, similar though not identical results are obtained with the soft-potential model in the energy regions dominated by tunneling states. Let us find the corresponding density of tunneling states within the soft-potential model. Following Parshin (1994a), we can choose as variables the total energy separation between the levels E

$$E = \sqrt{\Delta_0^2 + \Delta^2}, \tag{9.28}$$

and the dimensionless parameter

$$u = \frac{\Delta_0}{E}. \tag{9.29}$$

Making use of (9.19) and (9.20), the density of tunneling states is found approximately to be

$$P(E, u) \simeq \left(\frac{2}{9}\right)^{1/3} \frac{P_s}{W} \frac{1}{u\sqrt{1-u^2}} \ln^{-2/3}(40W/Eu), \tag{9.30}$$

also nearly constant in energy. The numerical coefficient 40 in the logarithmic factor arises from averaging the numerical prefactor of (9.19) in the appropriate range. The error introduced by such an average in a logarithmic term will be negligible.

Within the soft-potential model, the double-well potentials region of the $D_1 - D_2$ plane are not only the origin of tunneling states at low temperatures, but also of classical relaxation processes at higher temperatures via thermal activation *above* the barrier, instead of tunneling *through* the barrier. The barrier height of the double-well potentials is given approximately by

$$V_B \simeq W \frac{|D_2|^2}{4}. \tag{9.31}$$

Choosing the energy difference between the minima, $E = |\Delta|$, and the barrier height V_B as independent variables, it is straightforward to find the distribution function of double-well potentials:

$$P(E, V_B) = \frac{P_s}{W^{5/4} V_B^{3/4}}. \tag{9.32}$$

It is important to stress that, in this particular case, the distribution function found following the soft-potential model postulates differs significantly from the corresponding one of the standard tunneling model.

9.4.2 Vibrational Density of States

Nevertheless, the main contribution of the soft-potential model to our understanding of the low-temperature properties of glasses is the complementary right-side region in the $D_1 - D_2$ plane of Fig. 9.10, namely the more or less harmonic vibrational *soft* modes occurring in single-well potentials, which are in fact the great majority.

The density of states $g_s(h\nu)$ of the soft quasi-harmonic vibrations can be found to be

$$g_s(h\nu) = \frac{1}{8} \frac{P_s}{W} \left(\frac{h\nu}{W}\right)^4 . \qquad (9.33)$$

The numerical prefactor differs slightly from earlier expressions (Buchenau et al. (1992)) because the vibrational modes within double-well potentials have also been included (Ramos et al. (1993)).

Let us consider now how to calculate the vibrational density of states $g_s(h\nu)$, when we take into account the limiting distribution function given by (9.25) or (9.27).

For each soft potential given by (9.9), the second derivative at the (absolute) mimimum of the potential at the coordinate $x_{\min}$, together with the effective mass, determines a harmonic frequency ν. For the following derivations, it is useful to define a reduced frequency

$$\nu_r = \frac{h\nu}{W} . \qquad (9.34)$$

It is easy to find

$$\nu_r = 2\sqrt{D_2 + 6x_{\min}^2} \qquad (9.35)$$

and

$$D_1 + 2D_2 x_{\min} + 4x_{\min}^3 = 0 . \qquad (9.36)$$

Combining the two equations, one finds the curve of constant ν_r in the D_1-D_2 plane

$$D_1 = \frac{1}{12\sqrt{6}}(\nu_r^2 - 4D_2)^{1/2}(\nu_r^2 + 8D_2) , \qquad (9.37)$$

which is shown in Fig. 9.12. The most important part is from $D_2 = -\nu_r^2/8$ to $D_2 = \nu_r^2/4$; below $-\nu_r^2/8$, one has only the solutions of the upper well in the double-well case.

The harmonic frequency is zero at the origin of the $D_1 - D_2$ plane and increases as one goes away from that origin in any direction, including those going into the double-well region. As it grows, the splitting of the two lowest levels will gradually approach $h\nu$, reaching the harmonic limit for $h\nu \gg W$. In the following, we will calculate the density of states in this harmonic limit. In order to facilitate the calculation, we consider only the lower of the two wells

9.4 Predictions of the Soft-Potential Model

in the double-well case. The density of states $g_s(\nu)$ of the soft quasi-harmonic vibrations is given by

$$g_s(\nu_r) = \int_{-\nu_r^2/8}^{\nu_r^2/4} dD_2 \int_{-\infty}^{\infty} dD_1 P_s \exp(-AD_1^2) \delta(\nu_r'(D_1, D_2) - \nu_r)$$

$$= 2 \int_{-\nu_r^2/8}^{\nu_r^2/4} dD_2 \frac{\partial D_1}{\partial \nu_r} P_s \exp(-AD_1^2). \qquad (9.38)$$

Differentiating (9.37) with respect to ν_r yields

$$\frac{\partial D_1}{\partial \nu_r} = \frac{1}{4\sqrt{6}} \frac{\nu_r^3}{(\nu_r^2 - 4D_2)^{1/2}}. \qquad (9.39)$$

The integral (9.38) can be simplified considerably by expressing both D_1 and D_2 by t with $3t^2/2 = 1 - 4D_2/\nu_r^2$. Replacing again ν_r by ν, the vibrational density of states of quasi-harmonic vibrations can be expressed as

$$g_s(h\nu) = \frac{1}{8} \frac{P_s}{W} \left(\frac{h\nu}{W}\right)^4 \int_0^1 dt \, \exp[-A(h\nu/2W)^6 t^2 (1-t^2)^2]. \qquad (9.40)$$

Comparing (9.40) and (9.33), it is observed that the effect of including the thermal strain *ansatz* given by (9.27) [or any other similar Gaussian distribution of potential asymmetries as in (9.25)] is to multiply the quartic density of states (9.33) of the most basic soft-potential model by the integral factor of (9.40), a function which is easy to evaluate numerically and decreases

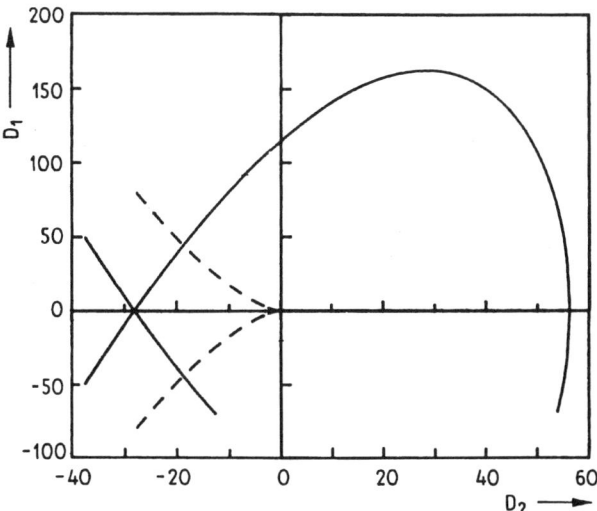

Fig. 9.12. Curve of constant harmonic frequency $h\nu = 15W$ in the $D_1 - D_2$ plane of the soft-potential model (continuous line). The dashed line is the boundary between single- and double-well regions.

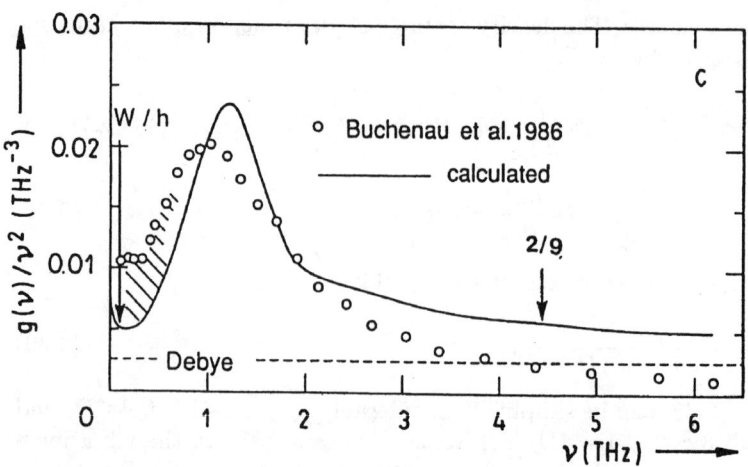

Fig. 9.13. Comparison of the calculated vibrational density of states to neutron data (Buchenau et al. (1986)) of vitreous silica, taken from Ramos et al. (1993).

smoothly from 1 at very low frequencies down to 0 at high frequencies. A broad maximum in $g_s(\nu)/\nu^2$ obviously appears now, which turns out to be located at a frequency $\nu_{\max}$ given by

$$h\nu_{\max} \simeq 2.65 W/A^{1/6}. \tag{9.41}$$

Therefore, equation (9.40) is able to describe the above-mentioned maximum, which is observed in neutron (Buchenau et al. (1986), Phillips et al. (1989)) and Raman (Shuker and Gammon (1970), Winterling (1975), Malinovsky et al. (1990)) data, known as "boson peak". At the peak, it is found

$$g_s(h\nu_{\max})/(h\nu_{\max})^2 \simeq 0.086 \frac{P_s}{W^3}\left(\frac{h\nu_{\max}}{W}\right)^2. \tag{9.42}$$

The vibrational density of states at low frequencies obtained by solving numerically (9.40) are compared to neutron scattering data in Figs. 9.13 and 9.14. Above the peak, the density of states seen in neutron experiments (Buchenau et al. (1986), Phillips et al. (1989)) gradually goes over to that of the crystalline counterparts. Taking the frequency $\nu_{\max}$, somewhat arbitrarily, as the upper limit for what we call soft modes, Ramos et al. (1993) get a total number N_s of soft modes

$$N_s \simeq 0.021 P_s \left(\frac{h\nu_{\max}}{W}\right)^5. \tag{9.43}$$

N_s appears then to be still small, less than 1% of the total number of the three vibrational modes per atom, though it is large compared to the total number of tunneling states of about 10^{-6} states/atom (obtained by taking W as the upper limit of the tunnel splitting).

9.4 Predictions of the Soft-Potential Model

The reconstructed density of quasi-harmonic vibrational states mentioned in Sect. 9.3.5 (Gurevich et al. (1993)) should give a similar picture of $g_s(\nu)/\nu^2$. Hence it may be argued that the usefulness of the Gaussian distribution given by (9.25) or (9.27) is, at least, to provide a functional tool to carry out quantitative calculations of several different physical magnitudes, studying the capabilities of the model to account for the whole of low-energy physical behavior of glasses.

9.4.3 Specific Heat

As described above, the soft-potential model predictions below 1 K deviate little from those of the standard tunneling model. The specific heat is then almost linearly dependent on temperature. Following the arguments of Parshin (1994a) (see also Parshin (1994b)), but with the new notation introduced above and some minor modifications, it can be found approximately that

$$C_{p,\text{TLS}}(T) \simeq \frac{\pi^2}{6} \left(\frac{1}{9}\right)^{1/3} \frac{P_s}{W} k_B^2 T \ln^{1/3}\left(\frac{t_{\text{exp}}}{\tau_{\min}(k_B T)}\right). \tag{9.44}$$

Nevertheless, the most interesting issue is the new predictions of the soft-potential model at not-very-low temperatures. In order to make calculations of physical properties such as the low-temperature specific heat, it seems more convenient to solve numerically the corresponding equations for the whole distribution of soft potentials. The first such calculation was done already by Il'in et al. (1987). Here, we present specific heat curves calculated by a numerical integration over the $D_1 - D_2$ plane (Gil et al. (1993)). The specific heat was calculated from the lowest six levels of each single-well

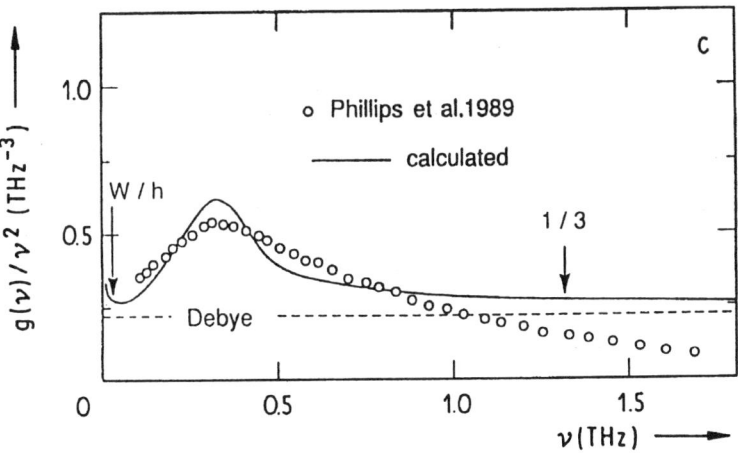

Fig. 9.14. Comparison of the calculated vibrational density of states to neutron data (Phillips et al. (1989)) of amorphous selenium, taken from Ramos et al. (1993).

potential. In the double-well case, the tunneling states were only included if their relaxation time, calculated via (9.16), did not exceed an assumed experimental time t_{exp} of 10 s. In any case, the two lowest levels were used to calculate a density of states. The sound-wave contributions, calculated within the Debye model from the density and the sound velocities, were directly added to that specific heat calculated from soft modes.

Numerical calculations with the two well-studied cases of vitreous silica and amorphous selenium showed that one always gets the minimum in the calculated C_p/T^3 at a temperature

$$k_B T_{\min} \simeq W/1.8, \qquad (9.45)$$

practically independent of the width of the Gaussian distribution $P(D_1, D_2)$, Λ_l and Λ_t, within a physically reasonable range of the other parameters.

Similarly, the calculated specific heat in the tunneling region well below $T_{\min}$ depends only weakly on Λ_l, Λ_t and $P(D_1, D_2)$, and is well described by

$$C_{p,\text{TLS}} \approx 10\, P_s\, k_B \left(\frac{k_B T}{W}\right)^{1.2}, \qquad (9.46)$$

which can be found to be in reasonable agreement with (9.44).

Changing the experimental time by a factor of ten changes the prefactor of this numerical result by less than ten percent. The validity extends up in temperature to about $T_{\min}/5$; above that temperature the specific heat increases more rapidly with temperature.

Above the minimum in C_p/T^3 given by (9.45), which is the hallmark of the crossover from tunneling to vibrational states represented by the pure quartic potential at $D_1 = D_2 = 0$, the excess specific heat over the Debye contribution abruptly rises as $(C_p - C_{\text{Debye}}) \propto T^5$ [Buchenau et al. (1991), see Fig. 9.1], at least at the beginning.

These numerical results of the specific heat of glasses above the well-known tunneling range can be rationalized as follows. The excitations giving rise to this excess specific heat above the *minimum* in C_p/T^3 [see (9.45)] are soft vibrational modes, more and more quasi-harmonic as temperature increases. Therefore, with the exception of temperatures very close to that minimum, where the anharmonic character of the energy potentials is important, the quasi-harmonic approximations (9.38)–(9.42) should be reasonable.

The heat capacity C_h of a single harmonic oscillator is given by

$$C_h(x) = k_B \frac{x^2 e^{-x}}{(1 - e^{-x})^2}, \qquad (9.47)$$

where

$$x = \frac{\hbar \omega}{k_B T} \qquad (9.48)$$

and k_B is the Boltzmann constant. Therefore, the contribution of quasi-

9.4 Predictions of the Soft-Potential Model   557

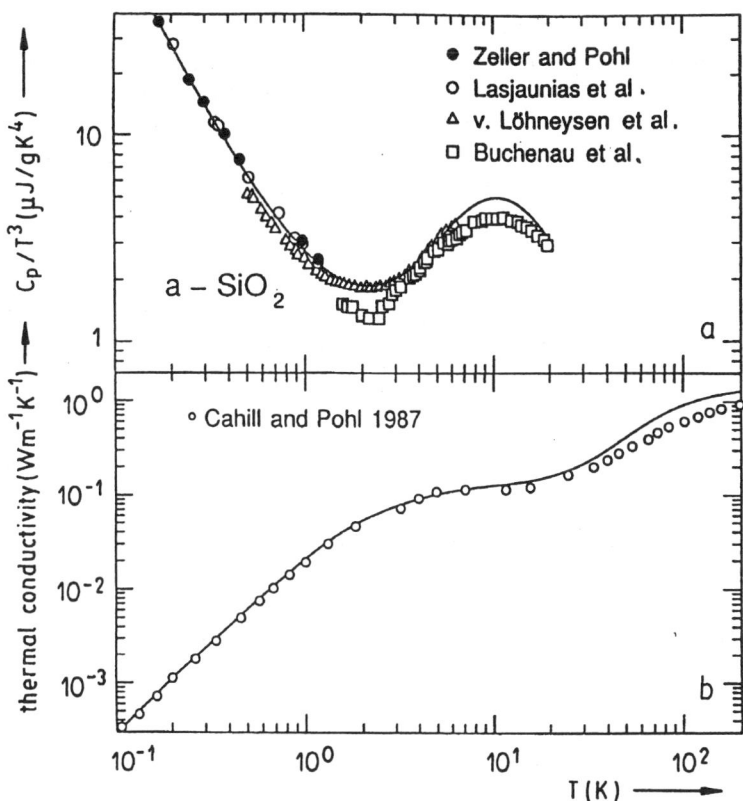

Fig. 9.15. Comparison of the calculated (**a**) specific heat plotted as C_p/T^3, and (**b**) thermal conductivity (continuous lines) within the soft-potential model to experimental data in vitreous silica: Zeller and Pohl (1971), Lasjaunias et al. (1975), von Löhneysen et al. (1985), Buchenau et al. (1986), Cahill and Pohl (1987).

harmonic soft vibrations to the specific heat can be evaluated simply by

$$C_{p,s} = \int_0^\infty d\nu \, g(\nu) \, C_h(x) \,, \tag{9.49}$$

where the soft-modes density of states is given by (9.33) and the contribution from each individual mode is given by (9.47).

With these simple equations, it is straightforward to find that the initial rise for $T > T_{\min}$ (corresponding to the constant distribution P_s) can be expressed by

$$C_p = \frac{2\pi^6}{21} P_s k_B \left(\frac{k_B T}{W}\right)^5 + C_{\text{Debye}} \,, \tag{9.50}$$

where C_{Debye} is the Debye contribution to the specific heat.

At still higher temperatures, the specific heat increases less and less strongly, exhibiting the famous "bump" in C_p/T^3. This is explained by the natural limitation of the density of soft vibrational modes, which has been discussed above in terms of distribution functions such as those from (9.25) or (9.27). Replacing the simple density of soft harmonic vibrations (9.33) by the one obtained with the additional limiting assumption leading to (9.40), the broad maximum in C_p/T^3 (e.g. at $\sim$ 10K in SiO_2, see Fig. 9.15a) arises in a natural way.

The results of the aforementioned numerical calculations are compared to specific heat data in Fig. 9.15a and Fig. 9.16a. Obviously, the assumed Gaussian distribution describes both the peak position and the peak height in C_p/T^3 reasonably well. The calculated peak in C_p/T^3 using (9.27) appears at a temperature T_{max} with approximately

$$T_{\text{max}} \simeq 1.07 \, T_{\text{min}}^{3/4} \, T_g^{1/4}. \tag{9.51}$$

Table 9.1 shows that this relation is not only fulfilled for vitreous silica and amorphous selenium, but also for a number of other glasses, with an average prefactor around 1 and typical deviations within the error resulting from the experimental determination of the three temperatures. It is interesting to note that simultaneously the ratio T_g/T_{min} varies by a factor 16. This demonstrates convincingly that one has to reckon with two different energy scales in the problem: the energy W, which is determined by the fourth-order anharmonic term in the potential together with the number of atoms participating in a single mode, and the thermal energy $k_B T_g$ at the glass transition temperature.

Table 9.1. Test of the relation $T_{\text{max}} = 1.07 \, T_{\text{min}}^{3/4} \, T_g^{1/4}$ for different glasses. T_g is the glass transition temperature, T_{max} and T_{min} stand for the maximum and minimum in C_p/T^3, and $R \equiv T_{\text{max}}/(T_{\text{min}}^{3/4} \, T_g^{1/4})$ should be approximately 1.07, according to (9.51). For references of experimental data, see Ramos et al. (1993), and Ramos and Buchenau (1997).

GLASS	T_g (K)	T_{max}(K)	T_{min}(K)	R
SiO_2	1473	10.0	2.1	0.93
$(SiO_2)_{0.65}(Na_2O)_{0.35}$	717	13.5	4.0	0.92
GeO_2	830	8.1	2.1	0.87
B_2O_3	570	5.2	1.1	0.99
$(B_2O_3)_{0.94}(Na_2O)_{0.06}$	610	7.5	1.9	0.93
$(B_2O_3)_{0.84}(Na_2O)_{0.16}$	720	10.0	2.5	0.97
Se	305	3.1	0.6	1.09
Polybutadiene	186	5.1	1.4	1.07
Polyethylene	240	5.0	1.3	1.04
glycerol	185	8.0	2.2	1.20
$LiCl \cdot 7 H_2O$	144	10.7	3.3	1.26

9.4 Predictions of the Soft-Potential Model   559

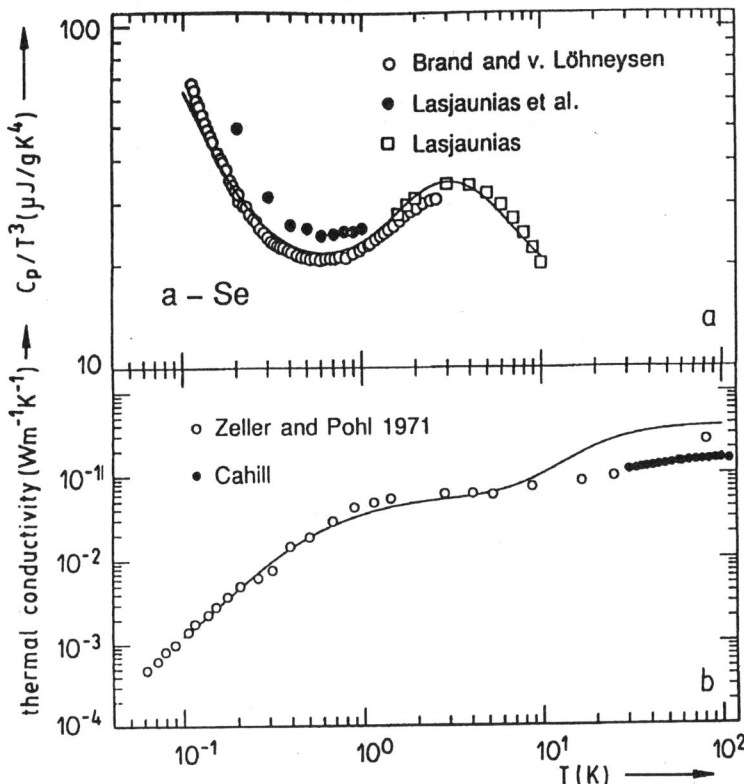

Fig. 9.16. Comparison of the calculated (**a**) specific heat plotted as C_p/T^3, and (**b**) thermal conductivity (continuous lines) within the soft-potential model to experimental data in amorphous selenium: Brand and von Löhneysen (1991), Lasjaunias et al. (1972), Lasjaunias (1969), Zeller and Pohl (1971), Cahill (1989).

For the Gaussian distribution function of (9.25), the position $T_{\max}$ of the maximum in C_p/T^3 is found to be located approximately at

$$k_B T_{\max} \simeq 0.583 W/A^{1/6}, \tag{9.52}$$

which in conjunction with $A = 0.169 \, (W/k_B T_g)^{3/2}$ [see (9.25) and (9.27)] gives a relation rather close to (9.51).

9.4.4 Thermal Conductivity

The low-temperature thermal conductivity κ is given by the well-known expression (Freeman and Anderson (1986))

$$\kappa = \frac{1}{3} \int_0^\infty d\omega \sum_j C_{\text{Debye},j}(\omega, T) v_j l_j, \tag{9.53}$$

where $C_{\text{Debye},j}(\omega,T)d\omega$ is the specific heat per unit volume contributed by longitudinal $(j=l)$ and transverse $(j=t)$ sound waves within the frequency interval $d\omega$ and l_j is the mean free path for the corresponding sound waves at the frequency ω. Taking the upper limit of the integral as infinity is justified by a strong decrease of the mean free path with increasing frequency. The number density of sound waves per unit volume and frequency interval $d\omega$ is given by

$$n_l(\omega) = \frac{\omega^2}{2\pi^2 v_l^3} \quad , \quad n_t(\omega) = \frac{\omega^2}{\pi^2 v_t^3}, \tag{9.54}$$

respectively. Inserting these expressions together with (9.47) into the standard formula for the thermal conductivity (9.53), one obtains

$$\kappa = \frac{1}{6\pi^2} \int_0^\infty d\omega\, \omega^2\, C_h(x) \left(\frac{l_l}{v_l^2} + \frac{2 l_t}{v_t^2} \right). \tag{9.55}$$

According to the soft-potential model treatment given by Buchenau et al. (1992), the inverse mean free path is a sum of three contributions

$$l_j^{-1} = l_{\text{res,tunn}}^{-1} + l_{\text{rel,class}}^{-1} + l_{\text{res,vib}}^{-1}, \tag{9.56}$$

that stem from the scattering of the sound waves either by the tunneling states or by low-barrier structural relaxations or by localized vibrations, respectively. Under certain circumstances, one has to take a fourth contribution into account, which stems from the relaxation of the soft vibrations (Parshin (1994b)), which we neglect here.

For the relatively high-frequency sound waves responsible for the thermal conductivity, the scattering by the tunneling states is given by (Phillips (1981a))

$$l_{\text{res,tunn}}^{-1} = \frac{\pi \omega C_j^{\text{tunn}}}{v_j} \tanh\left[\frac{\hbar \omega}{2 k_\text{B} T} \right], \tag{9.57}$$

where again j stands for l or t in the longitudinal or transverse case, and $C_j^{\text{tunn}} = P_0 \gamma_j^2 / \varrho v_j^2$.

The mean free path of the sound waves under the influence of classical relaxational processes in asymmetric double-well potentials is given by (Buchenau et al. (1992))

$$l_{\text{rel,class}}^{-1} = \frac{\pi \omega C_j}{v_j} \left(\frac{k_\text{B} T}{W} \right)^{3/4} \ln^{-1/4}(1/\omega \tau_0), \tag{9.58}$$

with

$$C_l = \frac{P_s \Lambda_l^2}{W \varrho v_l^2} \quad , \quad C_t = \frac{P_s \Lambda_t^2}{W \varrho v_t^2}. \tag{9.59}$$

C_l and C_t are defined in such a way that they are very nearly equal to the corresponding dimensionless parameter combinations of the tunneling model which are usually denoted by the same letters. The latter are approximately 1.1 times the former [see (9.71) below, Ramos et al. (1993)].

Here, τ_0 is the inverse of an attempt frequency and is of the order of 10^{-13} seconds. Since the frequency for the relevant sound waves is of the order of 10 to 100 GHz, the logarithmic factor $\ln^{-1/4}(1/\omega\tau_0) \approx 0.7$ in the whole relevant range.

Finally, the mean free path for the resonant scattering from the localized vibrations follows the relation

$$l^{-1}_{\text{res,vib}} = \frac{\pi\omega C_j}{v_j} \frac{1}{8} \left(\frac{\hbar\omega}{W}\right)^3. \tag{9.60}$$

Inserting these three equations into (9.55), replacing the variable ω by x and the temperature T by the dimensionless variable

$$z = \frac{k_B T}{W}, \tag{9.61}$$

one obtains the thermal conductivity

$$\kappa = \frac{2k_B}{3\pi} \left(\frac{W}{h}\right)^2 \left(\frac{1}{C_l v_l} + \frac{2}{C_t v_t}\right) F(z), \tag{9.62}$$

where

$$F(z) = \int_0^\infty dx \, \frac{x^3 e^{-x}}{(1-e^{-x})^2} \frac{z^2}{1.1 \tanh(x/2) + 0.7 z^{3/4} + x^3 z^3/8}. \tag{9.63}$$

It is easy to evaluate $F(z)$ numerically. In the relevant range $k_B T \leq 4W$, the function is well approximated by

$$F(z) = \frac{9z^2}{1.1 + 0.7z + 3z^2}. \tag{9.64}$$

Defining an average $\overline{C}$ over C_l and C_t with

$$\overline{C} = \frac{1/v_l + 2/v_t}{1/C_l v_l + 2/C_t v_t}, \tag{9.65}$$

one arrives at the final equation

$$\kappa = \frac{6k_B}{\pi \overline{C}} \left(\frac{W}{h}\right)^2 \left(\frac{1}{v_l} + \frac{2}{v_t}\right) \frac{z^2}{1.1 + 0.7z + 3z^2}. \tag{9.66}$$

The plateau in the thermal conductivity is thus reached roughly for $k_B T > 2W$ (i.e. $z > 2$).

One of the ways to determine the soft-potential parameter W from thermal conductivity data is to plot κ/T versus temperature and look for the maximum at $T_{\max,\kappa}$, since

$$W \simeq 1.6 \, k_B T_{\max,\kappa}. \tag{9.67}$$

Therefore, $T_{\max,\kappa}$ in κ/T is predicted to occur slightly above the minimum $T_{\min}$ in the specific heat C_p/T^3, see (9.45).

Having done that, one can determine the average value $\overline{C}$ for the two constants C_l and C_t via the equation

$$\overline{C} \simeq 0.27 \frac{k_B}{\kappa(T_{\max,\kappa})} \left(\frac{W}{h}\right)^2 \left(\frac{1}{v_l} + \frac{2}{v_t}\right). \tag{9.68}$$

In Fig. 9.17, thermal conductivity data are compared to fits in terms of

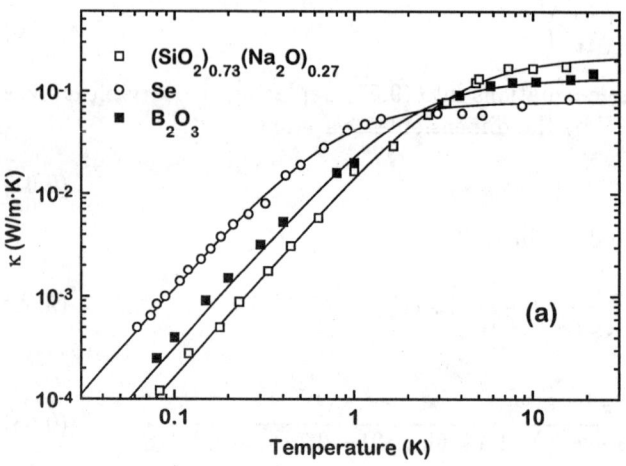

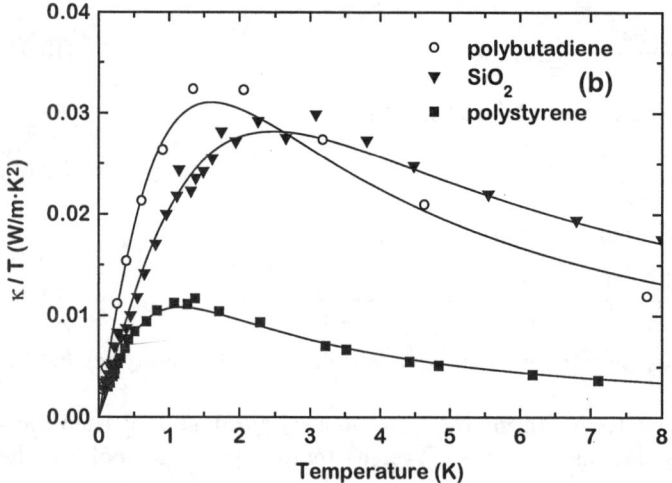

Fig. 9.17. (a) Thermal conductivity data (for references, see Ramos and Buchenau (1997)) compared to fits in terms of (9.66) for several glasses. (b) Plot of κ/T versus temperature for other experimental data with the corresponding fits of the maximum in κ/T at $T_{\max,\kappa}$ which the soft-potential model attributes to the crossover from tunneling states to localized vibrations.

Table 9.2. Set of soft-potential model parameters (W, P_s and $\overline{C}$) for different glasses. W_C and W_κ are obtained from specific heat and thermal conductivity data, respectively. P_s are obtained from the same low-temperature specific heat data using (9.46). $\overline{C}^{\text{SPM}}$ and $\overline{C}^{\text{tunn}}$ are weighted average $\overline{C}$ [defined in (9.65)], obtained either from soft-potential model analysis of thermal conductivity measurements via (9.67)–(9.68) or from tunneling-model fits of the internal friction plateau found in the literature, using (9.71). Sources of experimental data are mostly indicated in Ramos and Buchenau (1997).

GLASS	W_C/k_B (K)	W_κ/k_B (K)	P_s (kg^{-1})	$\overline{C}^{\text{SPM}}$ ($\times 10^{-4}$)	$\overline{C}^{\text{tunn}}$ ($\times 10^{-4}$)
SiO$_2$	3.8	4.0	6.3×10^{19}	2.6	2.7
(SiO$_2$)$_{0.73}$(Na$_2$O)$_{0.27}$	5.7	6.1	2.7×10^{20}	5.2	
GeO$_2$	3.8	3.8	3.9×10^{19}	2.2	2.2
B$_2$O$_3$	2.1	3.2	2.3×10^{19}	3.1	3.3
(B$_2$O$_3$)$_{0.99}$(Na$_2$O)$_{0.01}$	2.7	2.1	9.3×10^{19}	1.7	
(B$_2$O$_3$)$_{0.90}$(Na$_2$O)$_{0.10}$	4.2	4.0	1.6×10^{20}	3.6	
Se	1.1	1.3	8.4×10^{18}	1.9	1.7
Polybutadiene	2.5	2.6	1.2×10^{20}	3.4	
Polystyrene	1.8	1.9	1.1×10^{20}	7.1	
PMMA	2.5	2.0	1.4×10^{20}	3.5	2.9

(9.66). Figure 9.17a shows several experimental data of glasses in the usual form, κ versus temperature in a double logarithmic scale. Figure 9.17b plots κ/T versus temperature for other experimental data (for references, see Table 9.2). In both cases, (9.66) is seen to provide a good fit. Figure 9.17b shows the maximum in κ/T at $T_{\max,\kappa}$, which allows the determination of W via (9.67). Hence the soft-potential model attributes the maximum in κ/T to the crossover from tunneling states to localized vibrations.

Table 9.2 compiles the parameters obtained from such fits for a number of different glasses, together with the SPM parameters obtained from specific heat data. Further details are given by Ramos and Buchenau (1997).

All these calculations with the simple constant distribution $P(D_1, D_2) = P_s$ assumed therefore a density of soft vibrations increasing with ν^4, without any limitation. Let us consider again the Gaussian distribution function of potentials leading to the density of soft modes given by (9.40). Having fixed W by its relation to $T_{\min}$ and P_s by the experimental specific heat data at very low temperatures, and determining Λ_l and Λ_t via (9.22) from tunneling-model fits of Berret and Meißner (1988), the thermal conductivity can be calculated numerically by inserting the previously evaluated density of soft vibrations (see Figs. 9.13 and 9.14). The result so obtained (Gil et al. (1993), Ramos et al. (1993)) is shown as the continuous line in Fig. 9.15b for vitreous silica and, similarly, as the continuous line in Fig. 9.16b for amorphous selenium.

We may also try to correct the analytical expressions shown above in order to include again the distribution-limiting Gaussian function of (9.25) or (9.27). The only changes occurring affect the resonant scattering from quasi-harmonic vibrations (9.60), which should be multiplied by the same integral factor of (9.40), and correspondingly, the last term in the denominator of (9.63) is multiplied by the same factor [notice that $(h\nu/2W) \equiv (xz/2)$].

Though these calculations (see Fig. 9.15b and Fig. 9.16b) overestimate the thermal conductivity at higher temperatures, in particular in the case of amorphous selenium, it shows the onset of the second rise at the right temperature, thus linking the form of the thermal conductivity curve to that of the specific heat in a consistent way [this connection between the two curves was pointed out earlier by Yu and Freeman (1987)]. As shown by Freeman and Anderson (1986), the onset of the second rise in the thermal conductivity occurs universally as the mean free path l of the sound waves reaches the condition $l \approx \lambda$ (λ wavelength). For temperatures above the onset, the thermal conductivity is well described by assuming that the condition continues to hold towards higher frequencies. According to our calculations, this assumption is reasonably well fulfilled, starting from the frequency ν_{max} of the maximum in $g(\nu)/\nu^2$.

9.4.5 Acoustic Attenuation

The acoustic absorption data are usually evaluated in terms of the tunneling model (Hunklinger and Arnold (1976), Hunklinger and Raychaudhuri (1986), Phillips (1987)), as can be seen in detail in Chap. 4. The evaluation provides the parameters C_l^{tunn}, C_t^{tunn} and P_0, the density of tunneling states per unit volume and unit energy. At low frequencies, the resonant absorption of sound discussed in Sect. 9.4.4 is not so important as the relaxational absorption of sound by the tunneling states. Taking over the equations derived within the tunneling model, the soft-potential model naturally reproduces again (Buchenau et al. (1992), Parshin (1994a)) the results of the former. At very low temperatures and/or high frequencies ($\omega\tau_{min} \gg 1$), the inverse relaxational mean free path $l^{-1}_{rel,TLS}$ rises approximately with T^3 according to

$$l^{-1}_{rel,TLS} = (36)^{1/3} \frac{\pi^4}{48} A_t \frac{C_j^{tunn}}{v_j} \ln^{2/3}(20W/k_B T) (k_B T)^3 . \qquad (9.69)$$

At slightly higher temperatures or lower frequencies ($\omega\tau_{min} \ll 1$), the phonon mean free path becomes independent of temperature:

$$l^{-1}_{rel,TLS} = \frac{\pi}{2} C_j^{tunn} \left(\frac{\omega}{v_j}\right) . \qquad (9.70)$$

This plateau in the acoustic attenuation around 1 K is often measured via the internal friction of mechanical resonance experiments. The dimensionless internal friction Q^{-1} is related to the mean free path l^{-1} by $Q^{-1} = v_j l^{-1}/\omega$.

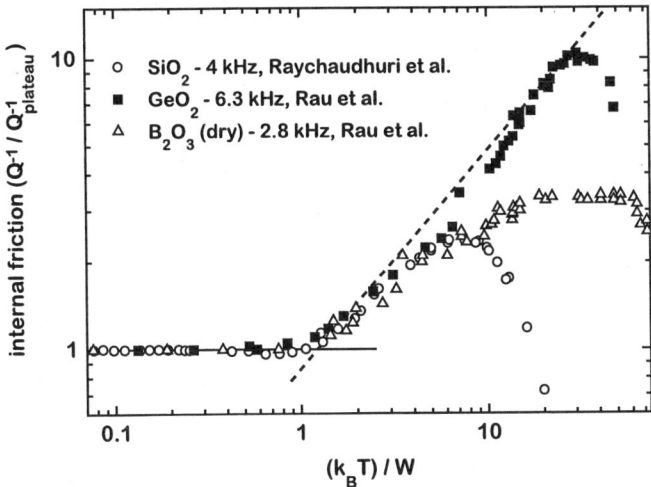

Fig. 9.18. Acoustic attenuation data scaled to the plateau value for SiO$_2$ (Raychaudhuri and Hunklinger (1984)), GeO$_2$ and B$_2$O$_3$ (Rau et al. (1995a)), compared to the soft-potential model predictions below and above the crossover. The dashed line is from (9.71), with W obtained from thermal conductivity data evaluation (see text), hence without any adjustable parameter.

Therefore, the plateau in the internal friction stems from the relaxation of all tunneling states activated at the given temperature in the time-dependent strain field and can be rewritten as

$$Q^{-1}_{\text{plateau}} = \frac{\pi}{2} \overline{C}_j^{\text{tunn}} \simeq 1.1 \frac{\pi}{2} \overline{C}_j^{\text{SPM}}, \tag{9.71}$$

where the superscripts 'tunn' and 'SPM' are used here to emphasize their slightly different values in the two models. Weighted average values $\overline{C}^{\text{tunn}}$ obtained from acoustic absorption data in the literature Berret and Meißner (1988) are compared in Table 9.2 to $\overline{C}^{\text{SPM}}$ values obtained from the soft-potential model (SPM) evaluation of thermal conductivity data in the crossover region as described above.

Again, the soft-potential model predicts a crossover to classical relaxation at higher temperatures, given by (9.58). For a frequency of some kHz, the logarithmic factor in that expression is about 0.47, leading to

$$Q^{-1}_{\text{rel,class}} \simeq 0.86\, Q^{-1}_{\text{plateau}} \left(\frac{k_\text{B} T}{W} \right)^{3/4}. \tag{9.72}$$

As first discussed by Galperin et al. (1989b), this classical relaxation reaches the level of the plateau at a temperature of about $1.2 W/k_\text{B}$ (i.e. at about twice $T_{\text{max},\kappa}$ in κ/T or T_{min} in C_p/T^3). This is a third possibility to determine W.

In Fig. 9.18, acoustic attenuation data are compared to the prediction of the soft-potential model for three substances, SiO$_2$ (Raychaudhuri and Hunklinger (1984)), GeO$_2$ and B$_2$O$_3$ (Rau et al. (1995a)). The internal friction was scaled to the plateau value and the temperature was scaled to W/k_B, taking W from the more accurate thermal conductivity data in Table 9.2 (with the exception of B_2O_3, where we took the average of the two values in Table 9.2, which is close to the values found for B_2O_3 with 1 % of Na_2O, given the uncertainty of experimental data in the plateau region).

Figure 9.18 shows that the onset of the rise of the internal friction does indeed occur at the predicted temperature within experimental error (note that the dashed line in Fig. 9.18 has no adjustable parameter). Thus, if one accepts this explanation for the onset, one has a third independent way of determining W from the acoustic attenuation.

It should be mentioned that the initial rise over the plateau has found an alternative explanation (see Chap. 3) in terms of incoherent tunneling which does not require the additional parameter W. Nevertheless, it does require the existence of tunneling states with splittings larger than W, which, according to the soft-potential model, should not exist (at least not as tunneling states in symmetric double-well potentials; naturally, there could always be levels splittings in asymmetric double-well potentials larger than W, but these should then have such large relaxation times that they should not contribute). Which of the two explanations is the correct one cannot be decided here and now. There is agreement, however, that at higher temperatures, the classical-relaxation concept applies. Therefore, the case of GeO_2, where even the high-temperature data fall on the dashed line, seems to support the soft-potential interpretation.

A second important point to be noted from Fig. 9.18 is that in all three examples, the data fall below the soft-potential expectation at higher temperatures, indicating a decrease of the number of double-well potentials as the barrier height approaches thermal energies corresponding to temperatures of 500 to 1000 K. Obviously, the energy landscape of these glasses is richer in low barriers than in higher ones. From earlier data collections on glasses, this seems to be a general rule (von Schickfuss and Hunklinger (1981)).

9.5 Conclusion and Outlook

In the preceding sections, we described the soft-potential model in terms of an extension of the standard tunneling model for glasses towards a more general picture of soft modes in glasses, encompassing soft quasi-localized vibrational states and local structural rearrangements with low-energy barriers together with the tunneling states. In the view of the authors, the model is an important step towards a fundamental understanding of the glassy anomalies at low temperatures and their relation to another unsolved puzzle, the dynamics of the glass transition.[2]

---

[2] See, for example, the Proceedings of the Conference "Dynamics of Disordered Materials II" (ILL Grenoble 1992) in volume 201 of Physica A (1993).

9.5 Conclusion and Outlook

However, the reader should be aware that the soft-potential model, while sharing the weakness of a phenomenological approach with the standard tunneling model, does not share its unanimous recognition. In particular, the attribution of the plateau in the thermal conductivity to the resonant scattering of the sound waves from localized vibrational modes is by no means a matter of general agreement (Elliott (1992), Sokolov et al. (1997)). Another objection bases on the strong interaction one predicts for these modes via their coupling to elastic distortions. That issue is discussed in more detail in Chap. 5.

Experimentally, the soft-potential picture is strongly supported by a number of neutron scattering results that have been reviewed in Sect. 9.2. The dynamic structure factors of the coherent inelastic scattering at the boson-peak frequency clearly demonstrate that one deals with a mixture of sound waves and other modes corresponding to much shorter wavelengths. The temperature dependence at frequencies below the boson peak show an anharmonicity that is compatible with an ensemble of low-barrier relaxators with a jump vector similar to the eigenvector of the vibrational boson-peak modes. Finally, the non-Gaussianity observed in elastic and inelastic incoherent scattering closely corresponds to the expectation for localized low-frequency modes coexisting with the sound waves.

The theoretical formulation of the soft-potential model, first given by Karpov, Klinger and Ignat'ev (1982) and later extended by Il'in, Karpov and Parshin (1987), has been described in detail in Sect. 9.3. It requires several assumptions on the potential distribution of the local modes which coexist with the sound waves at low frequencies. The central postulate is a common fourth order term in all of these local mode potentials, stabilizing those with a negative harmonic term. The first- and second-order term are assumed to have a wide distribution around zero. The number of atoms participating in a given mode is then again assumed to be the same for all modes (to be more specific, a combination of the fourth-order term and the number of participating atoms, that determines the zero-point energy W in the purely quartic potential, is assumed to be more or less the same for all local modes).

It turns out to be possible to extend the standard tunneling model to the soft-potential model by adding a single parameter, the crossover energy W between tunneling states on the one hand and vibrational or classical relaxational states on the other. In terms of the frequency ω, one passes from tunneling to vibrational states as the frequency crosses $W/\hbar$, in terms of temperature T, one passes from tunneling to classical relaxation as the temperature crosses W/k_B.

The interaction with the longitudinal and transverse sound waves, which in the standard tunneling model requires two coupling parameters, translates into bilinear coupling relations between the elastic distortion and the displacement of the soft mode, again requiring two coupling parameters which can be determined from those of the tunneling model in a straightforward way.

Section 9.4 compares the predictions of the soft-potential model to specific heat, thermal conductivity, acoustic absorption and neutron data in a number of different glasses. The main result of that section is the good agreement between the values of the crossover energy W from thermal conductivity, specific heat and acoustic attenuation data. Obviously, the soft-potential model passes this critical test for its central hypothesis of a close correspondence between tunneling states, low-frequency localized vibrations and low-barrier classical relaxation. The good agreement of the values of W obtained from the thermal conductivity with those determined from specific heat and acoustic attenuation data supports the interpretation of the plateau in the thermal conductivity in terms of a resonant scattering of the sound waves from localized vibrations. It is not a proof beyond any reasonable doubt, but still any other explanation will have to deal with the coincidence of the onset of the plateau with the strong rise of the specific heat and the subsequent rise of the acoustic attenuation which the collected data reflect.

A second remarkable result of Sect. 9.4 is the good agreement between the values $\overline{C}$ determined from the thermal conductivity with those determined from the acoustic absorption. This result is not completely new, at least as far as the region below 1 K is concerned, because it has been seen earlier in fits of the tunneling model. Here, however, it is obtained for a temperature region that includes the plateau in the thermal conductivity.

The soft-potential model provides a quantitative description of the low-temperature thermal conductivity of ten different glasses up to and including the plateau at about 5 K. The parameters W and $\overline{C}$ of that description agree within experimental error with those obtained from specific heat and acoustic attenuation data, showing the internal consistency of the model. There is no second description of the low-temperature anomalies of glasses covering a similarly wide range of experimental data with a similar degree of accuracy.

While the energy, frequency and temperature range around W is thus well described within the model, the uncertainty begins as one goes to higher frequencies and temperatures. The simple explanation of the boson peak in terms of a Gaussian distribution of asymmetries described in Sect. 9.4.2 is by no means generally accepted. One critique is that the number of modes at the boson peak does not scale with the number of tunneling states, showing a marked difference for strong and fragile glasses (Sokolov et al. (1997)). That critique, however, supposes a constant distribution over a rather wide range of values of the harmonic potential coefficient D_2, from negative to large positive values, and need not be taken seriously. A more serious objection (Parshin (1994a)) is that the number of modes increases drastically with increasing frequency, leading to a spatial overlap of different local modes with nearly the same frequency, so the description might fail somewhere around the boson peak.

9.5 Conclusion and Outlook

At the present stage, where we have no valid modification of the harmonic approximation for the many-valley energy landscape of the glass, such an objection obviously cannot be met on a solid theoretical ground. We can only hope that the soft-potential approach will help to find such a valid modification.

References

The numbers in brackets "[...]" indicate the page (within an uncertainty of ± one) and/or the table number where the work is cited in the text.

Abens, S., Sahling, S. (1996): unpublished. [Table 2.2]
Abens, S., Topp, K., Sahling, S., Pohl, R.O. (1996): Proceedings of the 21st International Conference on Low Temperature Physics, Czech. J. Phys. **46**(S4), 2259. [Table 2.2]
Abragam, A. (1961): *The Principles of Nuclear Magnetism*, Oxford University Press, London. [409, 410]
Affleck, I. (1981): Phys. Rev. Lett. **46**, 388. [167]
Alefeld, G. (1970): *Vacancies and Interstitials in Metals*, edited by A. Seeger, D. Schumacher, W. Shilling, J. Diehl. North-Holland, Amsterdam. [416]
Alefeld, G., Völkl, J. (1978): *Hydrogen in Metals I*, Springer, Berlin, Heidelberg, New York. [390, 411]
Al'shits, E.I., Kharlamov, B.M., Personov, R.I. (1986): J. Appl. Spect. (JPS) **45**, 559 (in Russian). [344, 345]
Al'shits, E.I., Kharlamov, B.M., Maslov, V.G. (1987): unpublished. [345]
Al'shits, E.I., Krasheninnikov, V.N., Kulikov, S.G., Kharlamov, B.M. (1996): Opt. Spektr. **81**, 956 (in Russian). [346]
Altounian, Z., Guo-hua, Tu, Strom-Olsen (1982): J. Appl. Phys. **53**, 4755. [195]
Ambrose, W.P., Basché, Th., Moerner, W.E. (1991): J. Chem. Phys. **95**, 7150. [345, 352]
Amore Bonapasta, A., Giannozzi, P., Capizzi, M. (1992): Phys. Rev. B **45**, 11744. [454]
Anderson, P.W., Yuval, G. (1969): Phys. Rev. Lett. **23**, 89; Yuval, G., Anderson, P.W. (1970): Phys. Rev. B **1**, 1522. [58, 62]
Anderson, P.W., Halperin, B.I., Varma, C.M. (1972): Phil. Mag. **25**, 1. [2, 9, 9, 15, 279, 315, 325, 325, 459, 527]
Anderson, I.S., Ross, D.K., Bonnet, J.E. (1989): Z. Phys. Chem. NF **164**, 923. [449]
Anderson, I.S., Berk, N.F., Rush, J.J., Udovic, T.J., Barnes, R.G., Magerl, A., Richter, D. (1990): Phys. Rev. Lett. **65**, 1439. [449, 449, 449, 449, 450, Table 7.1]
Ansari, A., Berendzen, J., Bowne, S.F., Frauenfelder, H., Iben, I.E.T., Sauke, T.B., Shyamsunder, E., Young, R.D. (1985): Proc. Natl. Acad. Sci. USA **82**, 5000. [370]
Anthony, P.J., Anderson, A.C. (1979): Phys. Rev. B **20**, 763. [229]
Araki, H., Park, G., Hikata, A., Elbaum, C. (1979): Solid State Commun. **32**, 625. See also (1980): Phys. Rev. B **21**, 4470. [189]
Arnold, W., Hunklinger, S. (1975): Solid State Commun. **17**, 883. [226, 234, 361]
Arnold, W., Hunklinger, S., Stein, S., Dransfeld, K. (1974): J. Non Cryst. Solids **14**, 192. [148]
Arnold, W., Martinon, C., Hunklinger, S. (1978): J. Phys. Paris Lett. **39**, C6-961. [226]

Arnold, W., Billmann, A., Doussineau, P., Levelut, A. (1982a): Physica **109 & 110** B, 2036; J. Phys. (Paris) **43** C, 9. [156]
Arnold, W., Doussineau, P., Levelut, A. (1982b): J. Phys. Lett. (Paris) **43**, L- 695. [189]
Aslangul, C., Pottier, N., Saint-James, D. (1986): J. Phys. (Paris) **47**, 1657. [62, 73]
Astilean, S., Corval, A., Casalegno, R., Trommsdorff, H.P. (1994): J. Luminescence **58**, 275. [347, 348, 348, 349]
Attenberger, T., Bogner, U., Maier, M. (1991): Chem. Phys. Lett. **180**, 207. [345]
Babbit, W.R., Lesama, A., Mossberg, T.W. (1989): Phys. Rev. B **39**, 1987. [321]
Bachellerie, A. (1974): *Proc. Satellite Symp. Int. Congr. Acoust. Microwave Acoust., 8th*, p. 93. [148]
Bagatskii, M.I., Manzhelii, V.G., Ivanov, M.A., Muromtzev, P.I. (1992): Fiz. Nizk. Temp. **18**, 1142. [310]
Bai, Y.S., Fayer, M.D. (1988a): Chem. Phys. **128**, 135. [354]
Bai, Y.S., Fayer, M.D. (1988b): Phys. Rev. B **37**, 10440. [354]
Bai, Y.S., Littau, K.A., Fayer, M.D. (1989): Chem. Phys. Lett. **162**, 449. [372]
Baier, G., v. Schickfus, M. (1988): Phys. Rev. B **38**, 9952. [232, 241, 249]
Baker, C., Birnbaum, H.K. (1973): Acta Metall. **21**, 865. [416]
Ball, K.D., Berry, R.S., Kunz, R.E., Li, Feng-Yin, Proykova, A., Wales, D.J. (1996): Science **271**, 963. [487, 487]
Banerjee, K., Bhatnagar, S.P. (1978): Phys. Rev. D **18**, 4767. [547]
Banville, M., Harris, R. (1980): Phys. Rev. Lett. **44**, 1136. See also Harris, R., Lewis, L.J. (1982): Phys. Rev. B **25**, 4997. [219]
Baranovskii, S.D., Shklovskii, B.I., Efros, A.L. (1980): JETP **51**, 199. [271, 280, 281, 281, 281, 281, 284, 298, 298, 301, 303, 305, 308]
Barbara, P.F., von Borczyskowski, C., Casalegno, R., Corval, A., Kryschi, C., Romanovskii, Yu.V., Trommsdorff, H.P. (1995): Chem. Phys. **199**, 285. [347, 348, 348, 349]
Barnett, R.N., Cleveland, C.L., Landman, U. (1985): Phys. Rev. Lett. **55**, 2035. [467]
Bartmatz, M., Chen, H.S. (1974): Phys. Rev. B **9**, 4073. [194]
Basché, Th., Ambrose, W.P., Moerner, W.E. (1992): J. Opt. Soc. Am. B **9**, 829. [352]
Beck, R., Götze, W., Prelovsek, P. (1979): Phys. Rev. A **20**, 1140. [57]
Becker, K.W., Keller, J. (1986): Z. Phys. B **62**, 477. [57]
Becker, O.M., Karplus, M. (1997): J. Chem. Phys. **106**, 1495. [491]
Bée, M. (1988): *Quasielastic neutron scattering*, Adam Hilger, Bristol. [408]
Bellessa, G. (1977): J. Phys. C **10**, L-285; Phys. Lett. A **62**, 125. Bellessa, G., Doussineau, P., Levelut, A. (1977): J. de Phys. Lett. **38**, L-65. [185]
Bellessa, G. (1978): Phys. Rev. Lett. **40**, 1456. [122, 122, 123, 124, 155, 167, 167, 182]
Bellessa, G. (1980): J. Phys. **41**, C8. [496]
Bellessa, G. (1983): J. Physique **44**, L-387. [422, Table 7.1]
Bellessa, G. (1985): Phys. Rev. B **32**, 5481. [Table 7.1]
Bellessa, G., Bethoux, G. (1977): Phys. Lett. A **62**, 125. [186]
Bembenek, S.D., Laird, B.B. (1995): Phys. Rev. Lett. **74**, 936. [541]
Berg, M., Walsh, C.A., Narasimhan, L.R., Littau, K.A., Fayer, M.D. (1987): Chem. Phys. Lett. **139**, 66. [326, 354]
Berg, M., Walsh, C.A., Narasimhan, L.R., Fayer M.D. (1987): J. Luminescence **38**, 9. [354]
Bernard, L., Piché, L., Schumacher, G.S., Jofrin, J. (1979): J. Low Temp. Phys. **35**, 411. [241, 249]

Berret, J.F., Meißner, M. (1988): Z. Phys. B **70**, 65. [187, 463, 493, 508, 513, 513, 513, 519, 563, 565, Table 8.2]
Berret, J.F., Pelous, J., Vacher, R., Raychaudhuri, A.K., Schmidt, M. (1986): J. Non-Cryst. Solids **87**, 70. [513, Table 8.4]
Berry, R.S. (1993): Science **98**, 6910. [487]
Berry, B.S., Pritchet, W.C. (1975): IBM J. Res. Dev. **19**, 334. [170]
Berry, B.S., Pritchet, W.C. (1981): Phys. Rev. B **24**, 2299. [414]
Binder, K., Stauffer, D. (1987): *Monte Carlo Methods in Statististical Physics*, edited by K. Binder. Topics Curr. Physics, Vol. 36, Springer, Berlin, Heidelberg, New York. [466]
Birge, N.O., Moon, J.S., Hoadley, D. (1996): Proceedings of the 21st International Conference on Low Temperature Physics, Czech. J. Phys. **46**, 2343. [103, 104, 104, 107, 107]
Birnbaum, H.K., Flynn, C.P. (1976): Phys. Rev. Lett. **37**, 25. [419]
Black, J.L. (1978): Phys. Rev. B **17**, 2740. [20, 20]
Black, J.L. (1981): *Glassy Metals I*, edited by H.-J. Güntherodt, H. Beck. Springer, Berlin, Heidelberg, New York. [155, 155, 426]
Black, J.L., Fulde, P. (1979): Phys. Rev. Lett. **43**, 453. [156, 156, 423, 423, 425, 426, 432]
Black, J.L., Halperin, B.I. (1977): Phys. Rev. B **16**, 2879. [226, 227, 232, 239, 241, 326, 355, 355, 361]
Boiron, A.-M., Lounis, B., Orrit, M. (1996): J. Chem. Phys. **105**, 3969. [334, 334, 345]
Brand, O., Löhneysen, H.v. (1991): Europhys. Lett. **16**, 455. [558]
Brandt, E.H., Kronmüller, H. (1987): J. Phys. F **17**, 1291. [219, 471]
Brawer, S.A. (1981): Phys. Rev. Lett. **46**, 778. [470]
Bray, A.J., Moore, M.A. (1982): Phys. Rev. Lett. **49**, 1545. [58]
Breinl, W., Friedrich, J., Haarer, D. (1984a): J. Chem. Phys. **81**, 3915. [326, 354]
Breinl, W., Friedrich, J., Haarer, D. (1984b): Chem. Phys. Lett. **106**, 487. [342]
Broer, M.M. (1986): Phys. Rev. B **33**, 4160. [326, 354]
Broer, M.M., Golding, B. (1986): JOSA B **3**, 523. [326, 354]
Buchenau, U., (1985a): Solid State Commun. **56**, 889. [507]
Buchenau, U. (1985b): Z. Physik B **58**, 181. [532]
Buchenau, U. (1992): Phil. Mag B **65**, 303. [549]
Buchenau, U., Schober, H.R. (1989): *Phonons 89*, edited by S. Hunklinger, W. Ludwig, G. Weiss. World Scientific, Singapore. [503]
Buchenau, U., Nücker, N., Dianoux, A.J. (1984): Phys. Rev. Lett. **53**, 2316. [521]
Buchenau, U., Prager, M., Nücker, N., Dianoux, A.J., Ahmad, N., Phillips, W.A. (1986): Phys. Rev. B **34**, 5665. [529, 530, 532, 534, 537, 554, 554, 554, 558]
Buchenau, U., Nücker, N., Gilroy, K.S., Phillips, W.A. (1984): Phys. Rev. Lett. **60**, 1318. [521, 532, 535, 536, 536]
Buchenau, U., Galperin, Yu.M., Gurevich, V.L., Schober, H.W. (1991): Phys. Rev. B **49**, 5039. [28, 28, 49, 464, 477, 530, 543, 543, 548, 556]
Buchenau, U., Galperin, Yu.M., Gurevich, V.L., Parshin, D.A., Ramos, M.A., Schober, H.R. (1992): Phys. Rev. B **46**, 2798. [29, 110, 121, 464, 493, 543, 545, 545, 548, 550, 552, 560, 560, 564]
Buchenau, U., Pecharroman, C., Zorn, R., Frick, B. (1996a): Phys. Rev. Lett. **77**, 659. [540]
Buchenau, U., Wischnewski, A., Richter, D., Frick, B. (1996b): Phys. Rev. Lett. **77**, 4035. [538, 538]
Buechner, S., Heuer, A. (1997): preprint. [482]
Bunyatova, G.I., Sahling, A., Sahling, S. (1990): Solid State Commun. **75**, 125. [Table 2.2]

Burin, A.L. (1991a): Soviet J. Low Temp. Phys. **17**, 456. [280, 298]
Burin, A.L. (1991b): JETP Letters **54**, 320. [281]
Burin, A.L. (1995a): J. Low Temp. Phys. **100**(3/4), 309. [278, 278, 278, 279, 279, 294]
Burin, A.L. (1995b): (private communication). [286]
Burin, A.L. (1996): Proceedings of the 21st International Conference on Low Temperature Physics, Czech. J. Phys. **46**(S4), 2269. [284, 294]
Burin, A.L. (1997): preprint. [303]
Burin, A. L. (1997b): private communication. [181].
Burin, A.L., Kagan, Yu. (1993): Physica B **91**, 367. [165, 366]
Burin, A.L., Kagan, Yu. (1994a): Physica B **194-196**, 393; JETP **79**, 347 and **80**, 761. [59, 242, 290, 366, 368]
Burin, A.L., Kagan, Yu. (1994b): Zh. Éksp. Teor. Fiz. **106**, 633; JETP **80**, 761 (1995). [165, 165, 165, 165, 166, 180, 181]
Burin, A.L., Kagan, Yu. (1995): JETP **80**, 761. [242, 243, 246, 259, 366, 368]
Burin, A.L., Kagan, Yu. (1996a): Sov. Phys. JETP **109**, 299. [220, 518]
Burin, A.L., Kagan, Yu. (1996b): JETP **82**, 159. [298, 309]
Burin, A.L., Kagan, Yu. (1996c): Phys. Lett. A **215**, 191. [298, 309]
Burin, A.L., Maksimov, L.A., Polishchuk, I.Ya. (1989): JETP Letters **49** 784. [243, 250]
Cahill, D.G. (1989): Ph. D. Thesis, Cornell. [530, 558]
Cahill, D.G., Pohl, R.O. (1987): Phys. Rev. B **35**, 4067. [558]
Cahill, D.G., Pohl, R.O. (1989): Solid State Commun. **70**, 927. [530]
Calvayrac, Y., Chevalier, J.P., Harmelin, M., Quivy, A., Bigot, J. (1983): Phil. Mag. B **48**, 323. [195]
Cannelli, G., Cantelli, R. (1977): *Proc. 2nd Int. Conf. on Hydrogen in Metals*, Pergamon Press, Oxford. [395, 414]
Cannelli, G., Cantelli, R. (1982): Solid State Commun. **43**, 567. [428, 429]
Cannelli, G., Mazzolai, F. M. (1973): Appl. Phys. **1**, 111. [417]
Cannelli, G., Verdini, L. (1966a): Ric. Sci. **36**, 98. [416]
Cannelli, G., Verdini, L. (1966b): Ric. Sci. **36**, 246. [416]
Cannelli, G., Cantelli, R., Vertechi, G. (1981): Appl. Phys. Lett. **39**, 832. [442, 442]
Cannelli, G., Cantelli, R., Cordero, F. (1982): unpublished. [428]
Cannelli, G., Cantelli, R., Cordero, F. (1983): J. Physique **44**, C9-403. [417]
Cannelli, G., Cantelli, R., Cordero, F. (1985): Phys. Rev. B **32**, 3573. [418]
Cannelli, G., Cantelli, R., Cordero, F. (1986): Phys. Rev. B **34**, 7721. [423, 428, 428, 429, 429, Table 7.1]
Cannelli, G., Cantelli, R., Cordero, F. (1987): Phys. Rev. B **35**, 7264. [424]
Cannelli, G., Cantelli, R., Cordero, F. (1989): Z. Phys. Chem. **NF 164**, 943. [440, 442]
Cannelli, G., Cantelli, R., Cordero, F. (1990): presented at the Int. Symp. on Metal-H Systems (Banff, Canada), unpublished. [442]
Cannelli, G., Cantelli, R., Capizzi, M., Coluzza, C., Cordero, F., Frova, A., Lo Presti, A. (1991a): Phys. Rev. B **44**, 11486. [456]
Cannelli, G., Cantelli, R., Cordero, F., Trequattrini, F., Anderson, I.S., Rush, J.J. (1991b): Phys. Rev. Lett. **67**, 2682. [417, 451, 451]
Cannelli, G., Cantelli, R., Cordero, F., Trequattrini, F. (1993): Z. Phys. Chem. **179**, 317. [390, 440, 443]
Cannelli, G., Cantelli, R., Cordero, F., Trequattrini, F. (1994a): Phys. Rev. B **49**, 15040. [434, 444]
Cannelli, G., Cantelli, R., Cordero, F., Trequattrini, F., Schultz, H. (1994b): J. Alloys and Compounds **211/212**, 80. [418, 418, 444]

Cannelli, G., Cantelli, R., Cordero, F., Trequattrini, F., Schultz, H. (1995): J. Alloys and Compounds **231**, 274. [444, 447]
Cannelli, G., Cantelli, R., Cordero, F., Giovine, E., Trequattrini, F., Capizzi, M., Frova, A. (1996): Solid State Commun. **98**, 873. [456]
Cannelli, G., Cantelli, R., Cordero, F., Trequattrini, F. (1997a): Phys. Rev. B **55**, 14865. [451]
Cannelli, G., Cantelli, R., Cordero, F., Giovine, E., Trequattrini, F. (1997b): to be published. [456, 456]
Cantelli, R., Mazzolai, F.M., Nuovo, M. (1969): Phys. Stat. Sol. **34**, 597. [414]
Cantelli, R., Mazzolai, F.M., Nuovo, M. (1977): J. Physique **32**, C2. [414]
Carruzzo, H.M., Grannan, E.R., Yu, C.C. (1994): Phys. Rev. B **50**, 6685. [229, 231, 241, 256, 258, 278, 278, 279, 286, 294, 312]
Casalegno, R., Corval, A., Astilean, S., Trommsdorff, H.P. (1992): J. Lumin. **53**, 211. [348]
Chakravarty, S., Leggett, A.J. (1984): Phys. Rev. Lett. **52**, 5. [62, 93, 94]
Chen, C.G., Birnbaum, H.K. (1976): Phys. Stat. Sol. (a) **36**, 687. [417]
Chen, H.S. (1980): Rep. Prog. Phys. **43**, 353. [194]
Chen, H.S. (1983): *Amorphous Metallic Alloys*, edited by F.E. Luborsky. Butterworth, Washington, D.C. [194]
Cheng, Y.M., Stavola, M. (1994): Phys. Rev. Lett. **73**, 3419. [456, 456]
Cho, M., Fleming, G.R., Saito, S., Ohmine, I., Stratt, R.M. (1994): J. Chem. Phys. **100**, 6672. [541]
Chun, K., Birge, N.O. (1993): Phys. Rev. B **48**, 11500. [57, 62, 103, 104]
Cibuzar, G., Hikata, A., Elbaum, C. (1984): Phys. Rev. Lett. **53**, 326. [194]
Classen, J., Enss, C., Bechinger, C., Weiss, G., Hunklinger, S. (1994): Ann. Physik (Leipzig) **3**, 315. [179, 181, 184]
Classen, J., Hübner, M., Enss, C., Weiss, G., Hunklinger, S. (1997): to be published in Phys. Rev. B. [220]
Clemens, J.M., Hochstrasser, R.M., Trommsdorff, H.P. (1984): J. Chem. Phys. **80**, 1745. [347, 349, 349, 349, 349]
Cohen, M.H., Grest, G.S. (1980): Phys. Rev. Lett. **45**, 1271; (1981): Solid State Commun. **39**, 143. [219]
Continentino, M. (1980): Phys. Rev. B **22**, 6127. [245]
Coppersmith, S.N. (1991): Phys. Rev. Lett. **67**, 2315. [298]
Coppersmith, S.N. (1993): Phys. Rev. B **48**, 142. [161, 196, 220]
Coppersmith, S.N., Golding, B. (1993): Phys. Rev. B **47**, 4922. [160, 185, 185, 193]
Cordero, F. (1993): Phys. Rev. B **47**, 7674. [400]
Cordie, P., Bellessa, G. (1981): Phys. Rev. Lett. **47**, 106. [185, 185, 188, 189]
Cotts, R.M. (1978): edited by G. Alefeld, J. Völkl. Springer, Berlin, Heidelberg, New York. [409]
Crissman, J., Sauer, J., Woodward, E. (1964): J. of Polymer Science A **2**, 5075. [Table 2.4]
Cukier, R.I., Morillo, M., Chun, K., Birge, N.O. (1995): Phys. Rev. B **51**, 13767. [104]
Dab, D., Heuer, A., Silbey, R. (1995): J. Luminescence **64**, 95. [346, 480]
Dattagupta, S., Grabert, H., Jung, R. (1989): J. Phys. C **1**, 1405. [93]
DeBoer, J. (1948): Physica **XIV**, 139. [509]
Den Hartog, F.T.H., Bakker, M.P., Koedijk, J.M.A., Creemers, T.M.N., Völker, S. (1996): J. Luminescence **66&67**, 1. [352]
Deye, M., Esquinazi, P. (1989): Z. Phys. B **76**, 283. [21, 22]
Deye, M., Esquinazi, P. (1990): *Phonons 89*, edited by S. Hunklinger, W. Ludwig, G. Weiss. World Scientific, Singapore, p. 468. [23, 45]
Dicker, A.I.M., Dobkowski, J., Völker, S. (1981): Chem. Phys. Lett. **84**, 415. [321]

Dosch, H., Peisl, J., Dorner, B. (1987): Phys. Rev. B **35**, 3069. [453]
Dosch, H., Schmid, F., Wiethoff, P., Peisl, J. (1992): Phys. Rev. B **46**, 55. [453, 453, 453]
Doster, W., Cusack, S., Petry, W. (1989): Nature **337**, 754. [540]
Doussineau, P. (1989): *Disordered Systems and New Materials*, edited by M. Borissov, N. Kirov, A. Vavrek. World Scientific, Singapore, p. 66. [178]
Doussineau, P., Robin, A. (1980): *Phonon Scattering in Condensed Matter*, edited by H.J. Maris. Plenum, New York, p. 65. [411]
Doussineau, P., Legros, P., Levelut, A., Robin, J. (1978): J. Phys. Lett. (Paris) **39**, L265. [155, 185]
Doussineau, P., Frenois, C., Leisure, R.G., Levelut, A., Prieur, J.Y. (1980): J. Phys. (Paris) **41**, 1193. [22, 66]
Drescher-Krasicka, E., Granato, A.V. (1985): J. Physique **46**, C10. [422, 423, 423, 428, 429, Table 7.1]
Duppen, K., Molenkamp, L.W., Morsink, J.B.W., Wiersma, D.A., Trommsdorff, H.P. (1981): Chem. Phys. Lett. **84**, 421. [321]
Duquesne, J.-Y., Bellesa, G. (1979): J. de Phys. Lett. **40**, L-193. [Table 2.4]
Duquesne, J.Y., Gellessa, G. (1985): Phil. Mag. B **52**, 821 (1985). [519]
Dyadyna, G.A., Karpov, V.G., Solov'ev, V.N., Khrisanov, V.A. (1989): Sov. Phys. Solid State **31**, 629 (1989). [470]
Ediger, M.D., Angell, C.A., Nagel, S.R. (1996): J. Phys. Chem. **100**, 13200. [519]
Efros, A.L., Shklovskii, B.I. (1985): *Electron-Electron Interaction in Disordered Systems*, edited by A.L. Efros, M. Pollac. North-Holland, Amsterdam, p. 409. [281, 281, 298]
Egami, T., Maeda, K., Vitex, V. (1980): Phil. Mag. A **41**, 883. [219]
Egorov, S.A., Skinner, J.L. (1995): J. Chem. Phys. **103**, 1533. [118, 118]
Ehrl, M., Deeg, F.W., Bräuchle, C., Franke, O., Sobbi, A., Schulz-Ekloff, G., Wöhrle, D. (1994): J. Phys. Chem. **98**, 47. [333]
Einstein, A. (1911): Ann. Phys. **35**, 679. [530]
Ellervee, A., Kikas, J.V., Laisaar, A., Suisalu, A. (1993): J. Luminescence **56**, 151. [352]
Elliott, S.R. (1992): Europhys. Lett. **19**, 201. [566]
Elsässer, C., Fähnle, M., Schimmele, L., Chan, C.T., Ho, K.M. (1994): Phys. Rev. B **50**, 5155. [453]
Enss, C. (1996): private communication. [108]
Enss, C., Hunklinger, S. (1997): Phys. Rev. Lett. **79**, 2831. [59, 168, 182, 459]
Enss, C., Bechinger, C., v. Schickfus, M. (1990): *Phonons 89*, volume 2, edited by S. Hunklinger, W. Ludwig, G. Weiss. World Scientific, Singapore. [241, 258]
Enss, C., Schwoerer, H., Arndt, D., v. Schickfus, M. (1995): Phys. Rev. B **51**, 811. [249]
Enss, C., Gaukler, M., Nullmeier, M., Weis, R., Würger, A. (1997): Phys. Rev. Lett. **78**, 370. [58]
Erickson, L.E. (1977): Phys. Rev. B **16**, 4731. [327]
Esquinazi, P. (1996): Physica Scripta **T66**, 196. [187, 218]
Esquinazi, P., Luzuriaga, J. (1988): Phys. Rev. B **37**, 7819. [194, 195, 195, 196, 212, 213, 214, 216, 216]
Esquinazi, P., de la Cruz, M.E., Ridner, A., de la Cruz, F. (1982): Solid State Commun. **44**, 941. [194, 195, 196]
Esquinazi, P., Ritter, H.-M., Neckel, H., Weiss, G., Hunklinger, S. (1986): Z. Phys. B **64**, 81. [191, 193, 195, 216]
Esquinazi, P., König, R., Pobell, F. (1992): Z. Phys. B **87**, 305. [161, 173, 176, 177, 178, 179, 181, 185, 188, 189, 189, 199, 201, 201, 204, 206, 209, 209, 209, 211, 213, 215, 220, 241, 255, 258, 258, 258, 259, 259, 296, 459]

Esquinazi, P., König, R., Pobell, F. (1993): *Phonon Scattering in Condensed Matter VII*, (Springer Series in Solid State Physics). Springer, Berlin, Heidelberg, New York, Vol. 112, p. 317. [204]
Esquinazi, P., König, R., Valentin, D., Pobell, F. (1994): J. Alloys and Compounds **211/212**, 27. [211, 220, 241, 255, 296]
Esquinazi, P., König, R., Pobell, F. (1996a): Physica B **219&220**, 247. [209, 209]
Esquinazi, P., König, R., Neppert, B., Pobell, F. (1996b): Physica B **219** & **220**, 284. [200]
Federle, G., Hunklinger, S. (1982): J. Phys. (Paris) C **9**, C9. [Table 2.4]
Fenimore, P.W., Weissman, M.B. (1994): J. App. Phys. **76**(2), 6192. [276]
Ferrari, L., Phillips, W.A., Russo, G. (1987): Europhys. Lett **3**, 611. [549]
Fisch, R. (1980): Phys. Rev. B **22**, 3459. [298]
Fisher, M.P.A., Dorsey, A.T. (1985): Phys. Rev. Lett. **54**, 1609. [58, 62, 93]
Flynn C.P., Stoneham, A.M. (1970): Phys. Rev. B **1**, 3966. [58, 412, 412, 412, 417]
Foote, M.C., Anderson, A.C. (1987): Rev. Sci. Instr. **58**, 130. [256]
Frauenfelder, H., Wolunes, P.G. (1994): Physics Today **2**, 58. [370]
Freeman, J.J., Anderson, A.C. (1986): Phys. Rev. B **34**, 5684. [220, 238, 238, 297, 297, 463, 559, 564]
Friedrich, J., Haarer, D. (1984): Angew. Chemie **96**, 96; Angew. Chem. Int. Ed. Engl. **23**, 113. [327, 328, 353]
Friedrich, J., Haarer, D. (1986): *Optical spectroscopy of glasses*, edited by I. Zschokke-Granacher. Reidel, Dordrecht, p. 149. [326, 354]
Fritsch, K., Friedrich, J., Kharlamov, B.M. (1996): J. Chem. Phys. **105**, 1798. [337, 357, 369, 372, 372, 375, 375, 378, 380]
Fritsch, K., Eiker, A., Friedrich, J., Kharlamov, B.M., Vanderkoj, J.M. (1997): Europhys. Lett., submitted. [369]
Frossati, G., Gilchrist, J. le G., Lasjaunias, J. C., Meyer, W. (1977): J. Phys. C **10**, L515. [233]
Fujara, F., Petry, W., Diehl, R.M., Schnauss, W., Sillescu, H. (1991): Europhys. Lett. **14**, 563. [540]
Fukai, Y. (1993): *The metal-hydrogen system. Basic Bulk Properties*. Springer Series in Material Science, Springer, Berlin, Heidelberg, New York. [390, 411]
Furusawa, A., Horie, K. (1991): J. Chem. Phys. **94**, 80. [333]
Gaganidze, E., Esquinazi, P. (1996a): J. de Physique IV, Vol. 6, Colloque C8, p. C8-515. [212]
Gaganidze, E., Esquinazi, P. (1996b): unpublished. Gaganidze, E., Ph.D. Thesis, Universität Bayreuth (1998): unpublished. [185, 189]
Gaganidze, E., Esquinazi, P., König, R. (1995): Europhys. Lett. **31**, 13. [122, 186, 187, 218]
Gaganidze, E., König, R., Esquinazi, P., Zimmer, K., Burin, A. (1997): Phys. Rev. Lett. **79**, 5038. [181, 198, 314, 523, 523]
Galperin, Yu.M., Gurevich, V.L., Parshin, D.A. (1984): Sov. Phys. JETP **59**, 1104. [163]
Galperin, Yu.M., Gurevich, V.L., Parshin, D.A. (1985a): Sov. Phys. JETP **59**, 1004. [189]
Galperin, Yu.M., Gurevich, V.L., Parshin, D.A. (1985b): Phys. Rev. B **32**, 6873. [493]
Galperin, Yu.M., Karpov, V.G., Kozub, V.I. (1989a): Advances in Physics **38**, 669. [543]
Galperin, Yu.M., Gurevich, V.L., Kozub, V.I. (1989b): Europhys. Lett. **10**, 753. [565]
Gebhard, M., Vogt, B., Witthuhn, W. (1991): Phys. Rev. Lett. **67**, 847. [456]
Geva, E., Skinner, J.L. (1997a): J. Chem. Phys., accepted. [325, 326]

Geva, E., Skinner, J.L. (1997b): J. Phys. Chem., submitted. [326]
Geva, E., Reily, P.D., Skinner, J.L. (1996): Acc. Chem. Res. **29**, 579. [326]
Gil, L., Ramos, M.A., Bringer, A., Buchenau, U. (1993): Phys. Rev. Lett. **70**, 182. [464, 529, 547, 549, 549, 549, 550, 555, 563]
Gilroy, K.S., Phillips, W.A. (1981): Phil. Mag. B **43**, 735. [183]
Gloos, K., Smeibidl, P., Kennedy, C., Singsaas, A., Sekowski, P., Mueller, R., Pobell, F. (1988): J. Low Temp. Phys. **73**, 101. [176]
Gloos, K., Mitschka, C., Pobell, F., Smeibidl, P. (1990): Cryogenics **30**, 14. [204]
Goldanskii, V.I., Krupyansky, Y.F., Fleurov, V.N. (1989): *Protein structure: molecular and electronic reactivity*, edited by R. Austin et al. Springer, Berlin, Heidelberg, New York. [296]
Golding, B., Graebner, J.E. (1976): Phys. Rev. Lett. **37**, 852. [16, 169, 227, 290]
Golding, B., Graebner, J.E., Halperin, B.I., Schutz, R.J. (1973): Phys. Rev. Lett. **30**, 223. [16, 148, 151, 168]
Golding, B., Graebner, J.E., Haemmerle, W.H. (1977): *Proc 7th Intl Conf Amorph & Liq Semicond*, edited by W.E. Spear. Edinburgh, p. 367. [16, 227]
Golding, B., Graebner, J.E., Kane, A.B., Black, J.L. (1978): Phys. Rev. Lett. **41**, 1487. [155, 160, 185, 185, 413, 425, 432]
Golding, B., v. Schickfus, M., Hunklinger, S., Dransfeld, K. (1979): Phys. Rev. Lett. **43**, 1817. [290, 291, 459, 522]
Golding, B., Zimmermann, N.M., Coppersmith, S.N. (1992): Phys. Rev. Lett. **68**, 998. [57, 62, 103]
Goldstein, M. (1969): J. Chem. Phys. **51**, 3728. [466, 466, 487]
Gorokhov, D.A., Blatter, G. (1996): Phys. Rev. B (to be published). [167]
Gorokhovskii, A.A., Kaarly, R.K., Rebane, L.A. (1974): JETP Letters **20**, 474 (in Russian). [317, 327, 339]
Gorokhovskii, A.A., Rebane, L.A. (1977): Optics Comm. **20**, 144. [321]
Gorsky, W.S. (1935): Phys. Z. SU **8**, 457. [413]
Götze, W., Sjörgen, L. (1992): Rep. Prog. Phys. **55**, 241. [524]
Götze, W., Vujicic, G.M. (1988): Phys. Rev. B **38**, 9398. [62]
Grabert, H. (1992): Phys. Rev. B **46**, 12753. [58, 79, 91, 105]
Grabert, H., Schober, H.R. (1997): *Hydrogen in Metals III*, edited by H. Wipf. Springer, Berlin, Heidelberg, New York. [58, 63, 91, 411]
Grabert, H., Weiss, U. (1985): Phys. Rev. Lett. **54**, 1605. [58, 62, 93, 99]
Grabert, H., Linkwitz, S., Weiss, U., Dattagupta, S. (1986): Europhys. Lett. **2**, 631. [407, 408, 432, 432]
Grace, J.M., Anderson, A.C. (1989): Phys. Rev. B **40**, 1901. [487, 487, 522]
Graebner, J.E., Golding, B. (1979): Phys. Rev. B **19**, 964. [227, 234]
Gradl, G., Orth, K., Friedrich, J. (1992): Europhys. Lett. **19**, 459. [66]
Gradstein, I.S., Ryshik, I.M. (1981): *Table of sums, integrals, and series*, Academic Press, London. [80, 106]
Granato, A.V. (1992): Phys. Rev. Lett. **68**, 974. [219]
Granato, A.V., Hultman, K.L., Huang, K.-F. (1985): J. Physique **46**, C10. [399, 401, 434, 437]
Grannan, E.R., Randeria, M., Sethna, J.P. (1990): Phys. Rev. B **41**, 7784. [235]
Greywall, D.S. (1978): Phys. Rev. B **18**, 2127. [9]
Grimm, H., Dorner, B. (1975): J. Phys. Chem. Solids **36**, 407. [534]
Grondey, S., v. Löhneysen, H., Schink, H.J., Samwwer, K. (1983): Z. Phys. B **51**, 287. [194, 194, 195]
Gronert, H.W., Herlach, D.M., Schröder, A., van der Berg, R., v. Löhneysen, H. (1986): Z. Phys.B **63**, 173. [194, 195]
Gruzdev, N.V., Vainer, Yu.G. (1993): J. Luminescence **56**, 181. [326, 354]

Guénault, A.M., Pickett, G.R. (1990): *Helium Three*, edited by W.P. Halperin, L.P. Pitaevskii. Elsevier Science Publishers, Amsterdam. [170]
Guénault, A.M., Keith, V., Kennedy, C.J., Pickett, G.R. (1983): Phys. Rev. Lett. **51**, 589. [170]
Gurevich, V.L., Parshin, D.A., Pelous, J., Schober, H.R. (1993): Phys. Rev. B **48**, 16318. [529, 550, 550, 554]
Guttman, L., Rahman, S.M. (1985): Phys. Rev. B **33**, 1506. [471]
Hafner, J., Krajci, M. (1994): J. Phys.: Condens. Matter **6**,4631. [541]
Hanada, R. (1981): Scripta Met. **15**, 1121. [417]
Häner, P., Schilling, R. (1989): Europhys. Lett. **8**, 129. [503, 505]
Hannig, G., Maier, H., Haarer, D., Kharlamov, B.M. (1996): Mol. Cryst. Liq. Cryst. **291**, 11. [336, 337, 362, 364]
Hannon, A.C., Sinclair, R.N., Wright, A.C. (1993): Physica A **201**, 375. [538]
Härdle, H. (1985): Ph.D. Thesis, Universität Heidelberg, unpublished. [198]
Harrison, J.P. (1979): J. Low Temp. Phys. **37**, 467. [234]
Harris, R., Lewis, L.J. (1982): Phys. Rev. B **25**, 4997. [470]
Hartmann, O., Karlsson, E., Wäckelgard, E., Wäppling, R., Richter, D., Hempelmann, R., Niinikoski, T.O. (1988): Phys. Rev. B **37**, 4425. [433]
Hauer, B., Hempelmann, R., Richter, D., Udovic, T.J., Rush, J.J., Bennington, S.M., Dianoux, A.J. (1996): Physica B **226**, 210. [449]
Hayes, J.M., Small, G.J. (1978): Chem. Phys. **27**, 151. [344, 345]
Hayes, J.M., Stout, R.P., Small, G.J. (1981): J. Chem. Phys. **74**, 4266. [325]
Hegarty, J., Yen, W.M. (1979): Phys. Rev. Lett. **43**, 1126. [323, 347]
Heuer, A. (1996): Czechoslovak Journal of Physics **46**, 2245. [487]
Heuer, A. (1997a): Phys. Rev. Lett. **78**, 4051. [487, 490]
Heuer, A. (1997b): Phys. Rev. B **56**, 161. [493, 501, 502]
Heuer, A. (1997c): Europhys. Lett. (submitted). [481]
Heuer, A., Neu, P. (1997): J. Chem. Phys. **107**, 8686. [484, 484]
Heuer, A., Silbey, R.J. (1993a): Phys. Rev. Lett. **70**, 3911. [468, 471]
Heuer, A., Silbey, R.J. (1993b): Phys. Rev. B **48**, 9411. [493, 497]
Heuer, A., Silbey, R.J. (1994): Phys. Rev. B **49**, 1441. [512]
Heuer, A., Silbey, R.J. (1996): Phys. Rev. B **53**, 609. [312, 312, 312, 477, 510, 512]
Heuer, A., Spiess, H.W. (1994): J. of Non-Cryst. Solids **176**, 294. [513]
Hochstrasser, R.M., Trommsdorff, H.P. (1987): Chem. Phys. **115**, 1. [349]
Höhler, R., Münzel, J., Kasper, G., Hunklinger, S. (1991): Phys. Rev. B **43**, 9220. [262]
Holstein, T. (1959): Ann. Phys. (N.Y.) **8**, 325 and 343. [58, 68, 81, 81, 82, 96, 106]
Holtom, G.R., Trommsdorff, H.P., Hochstrasser, R.M. (1986): Chem. Phys. Lett. **131**, 44. [347]
Hu, P., Walker, L.R. (1977): Solid State Commun. **24**, 813. [226, 234]
Huang, K.F., Granato, A.V., Birnbaum, H.K. (1985): Phys. Rev. B **32**, 2178. [428]
Hughes, A.E. (1968): J. Phys. Chem. Solids **29**, 1461. [320]
Hunklinger, S. (1984): *Phonon Scattering in Condensed Matter*, Solid State Sciences Vol. 51, edited by W. Eisenmenger, K. Lassmann, S. Dottinger. Springer, Berlin, Heidelberg, New York, p. 378. [154, 154]
Hunklinger, S. (1989): *Disordered Systems and New Materials*, edited by M. Borissov, N. Kirov, A. Vavrek. World Scientific, Singapore, p. 113. [178]
Hunklinger, S., Arnold, W. (1976): *Physical Acoustics* **12**, edited by R.N. Thurston, W.P. Mason. Academic Press, New York, p. 155. [10, 57, 108, 109, 117, 119, 119, 119, 146, 148, 148, 150, 151, 166, 233, 234, 530, 564]
Hunklinger, S., Piché, L. (1975): Solid State Commun. **17**, 1189. [168, 169]

Hunklinger, S., Raychaudhuri, A.K. (1986): *Progress in low temperature physics, Vol. IX*, edited by D.F. Brewer. Elsevier, Amsterdam. [229, 254, 280, 280, 280, 288, 296, 312, 530, 530]
Hunklinger, S., Schmidt, M. (1984): Z. Phys. **54**, 93. [325]
Hunklinger, S., Arnold, W., Stein, S., Nava, R., Dransfeld, K. (1972): Phys. Lett. A **42**, 253. [147, 168, 169]
Hunklinger, S., Arnold, W., Stein, S. (1973): Phys. Lett. A **45**, 311. [16, 168, 226, 234, 234]
Il'in, M.A., Karpov, V.G., Parshin, D.A. (1987): Zh. Eksp. Teor. Fiz. **92**, 291. [28, 49, 543, 545, 548, 549, 550, 555]
Ivanov, M.A. (1985): Fiz. Tverd. Tela. **27**, 1334. [310]
Ivlev, B.I., Mel'nikov, V.I. (1987): Phys. Rev. B **36**, 6889. [167]
Jaaniso, R., Hagemann, H., Bill, H. (1994): J. Chem. Phys. **101**, 10323. [320]
Jäckle, J. (1972): Z. Physik **257**, 212. [14, 15, 57, 63, 65, 67, 90, 121, 139, 148, 223, 229, 405, 411]
Jäckle, J. (1976): *Proceedings of the 4th International Conference of Non-Crystalline Solids, Clausthal-Zellerfeld*, edited by G.H. Frischat. Trans Tech, Aedermannsdorf, p. 586. [530]
Jackson, B., Silbey, R. (1983): Chem. Phys. Lett. **99**, 331. [325, 325]
Jahn, S., Haarer, D., Kharlamov, B.M. (1991): Chem. Phys. Lett. **181**, 31. [327, 334]
Jalmukhambetov, A.U., Osad'ko, I.S. (1983): Chem. Phys. **77**, 247. [335]
Jankowiak, R., Hayes, J.M., Small, G.J. (1993): Chem. Rev. **93**, 1471. [327, 352]
Joffrin, J., Levelut, A. (1975): J. Phys. (Paris) **36**, 811. [58, 59]
Jones, D.P., Jäckle, J., Phillips, W.A. (1981): *Phonon Scattering in Condensed Matter*, edited by H.J. Maris. Springer, Berlin, Heidelberg, New York, p. 65. [535]
Joòs, J., Haller, E.E., Falicov, L.M. (1980): Phys. Rev. B **22**, 832. [455]
Joyeux, M., Prass, B., Borczyskowski, C., Trommsdorff, H.P. (1993): Chem. Phys. **178**, 433. [337]
Jutzler, M., Schröder, B., Gloos, K., Pobell, F. (1986): Z. Phys.B - Cond. Matt. **64**, 115. [176]
Kador, L. (1991): J. Chem. Phys. **95**, 5574. [322, 355]
Kador, L. (1995): Phys. Stat. Sol. (b) **189**, 11. [331]
Kador, L., Haarer, D. (1987): J. Appl. Phys. **62**, 4226. [383]
Kador, L., Schulte, G., Haarer, D. (1986): J. Phys. Chem. **90**, 1264. [329, 335, 341]
Kagan, Yu. (1992): J. Low Temp. Phys. **87**, 525. [411]
Kagan, Yu., Klinger, M.I. (1974): J. Phys. C **7**, 2791. [58]
Kagan, Yu., Klinger, M.I. (1976): Sov. Phys.-JETP **43**, 132. [412]
Kagan, Yu., Maksimov, L.A. (1980): Sov. Phys. JETP **52**, 688. [62]
Kagan, Yu., Prokof'ev, N.V. (1986): Sov. Phys. JETP **63**, 1276 and **66**, 211. [62, 62]
Kagan, Yu., Prokof'ev, N.V. (1988): Solid State Commun. **65**, 1385; Sov. Phys. JETP **70**, 957 (1990). [158, 158, 160, 193, 193, 193]
Kagan, Yu., Prokof'ev, N.V. (1992): *Quantum Tunneling in Condensed Media*, edited by Yu. Kagan, A.J. Leggett. Elsevier, Amsterdam. [58, 62]
Kahn, J.M., R. E. McMurray, Jr., R.E.McMurray,Jr., Haller, E.E., Falicov, L.M. (1987): Phys. Rev. B **36**, 8001. [455]
Kajitani, T., Hamada, S., Hirabayashi, M. (1989): Z. Phys. Chem. NF **163**, 175. [454]
Karpov, V.G., Parshin, D.A. (1985): Zh. Eksp. Teor. Fiz. **88**, 2212 [Sov. Phys. JETP **61**, 1308]. [530]

Karpov, V.G., Klinger, M.I., Ignat'ev, F.N. (1982): Solid State Commun. **44**, 333. [543]
Karpov, V.G., Klinger, M.I., Ignat'ev, F.N. (1983): Sov. Phys. - JETP **57**, 439. [27, 27, 464]
Kassner, K., Silbey, R. (1989): J. Phys. Cond. Matter **1**, 4599. [361, 369]
Kehr, K. W. (1978): in *Hydrogen in Metals I*, edited by G. Alefeld and J. Völkl, Springer, Berlin, Heidelberg, New York. [411]
Kehrein, S.K., Mielke, A. (1997): Ann. Phys. (Leipzig) **6**, 90. [57]
Keppens, V., and Laermans, C. (1996): Phys. Rev. B **53**, 14849. [199].
Kharlamov, B.M., Personov, R.I., Bykovskaja, L.A. (1974): Optics Comm. **12**, 191. [317, 323, 344, 345]
Kharlamov, B M., Personov, R.I., Bykovskaja, L.A. (1975): Opt. Spektr. **39**, 240 (in Russian). [344, 345, 345]
Kharlamov, B.M., Al'shits, E.I., Personov, R.I. (1984): Isv. AN SSSR, ser. phys. **48**, 1313 (in Russian). [341, 341, 344, 345]
Kharlamov, B.M., Haarer, D., Jahn, S. (1994): Optics and Spectroscopy **76**, 302. [335, 372, 379]
Khodykin, O.V., Ulitsky, N. I., Kharlamov, B.M. (1996): Optics and Spektroscopy **80**, 438 (in Russian). [372, 372, 373]
Khodykin, O.V., Müller, J., Kharlamov, B.M., Haarer, D. (1997): Chem. Phys. Lett., to be published. [334, 334, 335, 335]
Kibble, B.P., Rayner, G.H. (1984): *Coaxial AC Bridges*, Adam Hilger, Bristol, [255]
Kikas, J.V., Schellenberg, P., Friedrich, J. (1993): Chem. Phys. Lett. **207**, 143. [321]
Kirkpatrick, T., Thirumalai, D. (1989): J. Phys. A **22**, L149. [505]
Kirkpatrick, S., Gelatt, C.D., Vecchi, M.P. (1983): Science **220**, 671. [468]
Kittel, C. (1963): *Quantum Theory of Solids*, Wiley, New York. [63, 64]
Klafter, J., Silbey, R. (1981): J. Chem. Phys. **75**, 3973. [344, 345]
Klamt, A., Teichler, H. (1986): Phys. Stat. Sol. (b) **134**, 533. [413, 417]
Klauder, J.R., Anderson, P.W. (1962): Phys. Rev. **125**, 912. [326, 355, 355]
Kleiman, R.N., Agnolet, G., Bishop, D.J. (1987): Phys. Rev. Lett. **59**, 2079. [179, 241, 258]
Klein, M.W. (1990): Phys. Rev. Lett. **65**, 3017. [366, 366, 368]
Klein, M.W., Fischer, B., Anderson, A.C., Anthony, P.J. (1978): Phys. Rev. B **18**, 5887. [220, 220, 262, 297, 298, 298, 312]
Kohler, B., Personov, R.I., Woehl, J.C. (1995): *Laser Techniques in Chemistry*, edited by A.B. Myers, T.R. Rizzo. Techniques of Chemistry Series, Vol. XXIII, John Wiley & Sons. [352]
Köhler, W., Friedrich, J. (1987): Phys. Rev. Lett. **59**, 2199. [343, 344, 346]
Köhler, W., Friedrich, J. (1988a): Europhys. Lett. **7**, 517. [372]
Köhler, W., Friedrich, J. (1988b): J. Chem. Phys. **88**, 6655. [346, 346]
Köhler, W., Friedrich, J. (1989): J. Chem. Phys. **90**, 1270. [370, 373]
Köhler, W., Breinl, W., Friedrich, J. (1985): J. Phys. Chem. **89**, 2473. [329, 335]
Köhler, W., Meiler, J., Friedrich, J. (1987): Phys. Rev. B **35**, 4031. [346, 346]
Köhler, W., Friedrich, J., Scheer, H. (1988): Phys. Rev. A **37**, 660. [341]
Köhler, W., Zollfrank, J., Friedrich, J. (1989): Phys. Rev. B **39**, 5414. [346]
Koike, S., Kojima, A., Kano, M., Otake, M., Kojima, H., Suzuki, T. (1990): J. Phys. Soc. Japan **59**, 584. [454]
Koiwa, M. (1974): Acta Metall. **22**, 1259. [395]
Kokkinidis, M. (1977): PhD Thesis, Techn. Univ., München. [414]
Kokshenev, V.B., Nemes, M.C., Kim, J.L. (1996): Solid State Commun. **98**, 421. [300]
Koláč, M., Neganov, B.S., Sahling, A., Sahling, S. (1986): Solid State Commun. **57**, 425. [45, 50, Table 2.2]

Koláč, M., Neganov, B.S., Sahling, A., Sahling, S. (1987): J. Low Temp. Phys. **68**, 285. [46, Table 2.2, Table 2.5]
Kondo, J. (1976): Physica **84 B**, 40 and 207. [61, 94, 406, 427, 433]
Kondo, J. (1984): Physica **125 B**, 279; Physica **126 B**, 377. [62, 90, 94, 99]
Kondo, J. (1986): Physica **141 B**, 305. [432]
König, R., Esquinazi, P., Pobell, F. (1993): J. Low Temp. Phys. **90**, 55. [206, 206]
König, R., Betat, A., Pobell, F. (1994a): J. Low Temp. Phys. **97**, 311. [170]
König, R., Esquinazi, P., Pobell, F. (1994b): Physica B **194-196**, 417. [216]
König, R., Esquinazi, P., Neppert, B. (1995): Phys. Rev. B **51**, 11424. [174, 175, 207, 207, 208, 216, 257]
Korotaev, O.N., Kalitievski, M.Yu. (1980): JETP **52**, 220. [321]
Kozuch, D.M., Stavola, M., Spector, S.J., Pearton, S.J., Lopata, J. (1993): Phys. Rev. B **48**, 8751. [456]
Kramer, E.J., Bauer, C.L. (1967): Phys. Rev. **163**, 407. [428, 428]
Krause, J.T., Kurkjian, C.R. (1968): J. Am. Ceram. Soc. **51**, 226. [Table 8.4]
Kühn, R., Horstmann, U. (1997): to be printed in Phys. Rev. Lett. [505]
Kümmerl, L., Kliesch, H., Wöhrle, D., Haarer, D. (1994): Chem. Phys. Lett. **227**, 337, [333, 342]
Laermans, C., Esteves, V. (1988): Phys. Lett. A **126**, 341. [199, 220]
Laermans, C., Keppens, V. (1995): Phys. Rev. B **51**, 8158. [521]
Laermans C., Arnold, W., Hunklinger, S. (1977): J. Phys. C **10**, L161. [290, 291, 293]
Lagos, M., Cerón, H. (1988): Solid State Commun. **65**, 535. [416]
Laird, B.B., Schober, H.R. (1991): Phys. Rev. Lett. **66**, 636. [469, 543]
Landau, L.D., Lifshitz, E.M. (1980a): *Classical Theory of Fields*, 4th ed., Pergamon, New York. [244]
Landau, L.D., Lifshitz, E.M. (1980b): *Quantum Mechanics (Non-Relativistic Theory)*, 3rd ed., Pergamon, New York. [253]
Lasjaunias, J.C. (1969): C.R. Acad. Sci. B (France) **269**, 763. [558]
Lasjaunias, J.C., Maynard, R., Thoulouze, D. (1972): Solid State Commun. **10**, 215. [558]
Lasjaunias, J.C., Ravex, A., Vandorpe, M., Hunklinger, S. (1975): Solid State Commun. **17**, 1045. [558]
Lasjaunias, J.C., Maynard, R., Vandorpe, M. (1978): J. Phys. **39**, 973. [261, 262]
Lasjaunias, J.C., Zougmore, F., Bethoux, O. (1981): Solid State Commun. **40**, 853. [520]
Lasjaunias, J.C., Zougmore, F., Bethoux, O. (1986): Solid State Commun. **60**, 35. [520]
Lasocka, M., Matyja, H. (1981): *Ultrarapid Quenching of Liquid Alloys*, Vol. 20 of Treatise on Materials Science and Technology, edited by H. Herman. Academic, New York. [194]
Leggett, A.J. (1991): Physica B **169**, 322. [508, 518]
Leggett, A.J., Garg, A. (1985): Phys. Rev. Lett. **54**, 857. [58]
Leggett, A.J., Chakravarty, S., Dorsey, A.T., Fisher, M.P.A., Garg, A., Zwerger, W. (1987): Rev. Mod. Phys. **59**, 1. [57, 60, 61, 62, 62, 63, 66, 67, 68, 69, 71, 72, 73, 79, 85, 87, 88, 90, 90, 90, 90, 92, 92, 93, 94, 95, 99, 100, 125]
Leibfried, G., Breuer, N. (1978): *Point Defects in Metals I*, Springer, Berlin, Heidelberg, New York. [397]
Leisure, R.G., Schwarz, R.B., Migliori, A., Torgeson, D.R., Svare, I. (1993a): Phys. Rev. B **48**, 893. [411, 451, 451, 451]
Leisure, R.G., Schwarz, R.B., Migliori, A., Torgeson, D.R., Svare, I. (1993b): Phys. Rev. B **48**, 887. [451]
Levitov, L.S. (1990): Phys. Rev. Lett. **64**, 547. [307]

Lichtenberg, F., Raad, H., Moor, W., Weiss, G., Hunklinger, S. (1990): *Phonons 89*, edited by S. Hunklinger, W. Ludwig, G. Weiss. World Scientific, Singapore, p. 471. [191]
Lichty, L.R., Han, J-W., Ibanez-Meier, R., Torgeson, D.R., Barnes, R.G., Seymour, E.F.W., Sholl, C.A.(1989): Phys. Rev. B **39**, 2012. [450, 450]
Limbach, H.H., Hennig, J., Kendrick, R., Yannoi, C.S. (1984): J. Am. Chem. Soc. **106**, 4059. [339]
Lin, S., Fünfschilling, J., Zschokke-Gränacher, I. (1992): Chem. Phys. Lett. **190**, 72. [344]
Lindrum, M., Nickel, B. (1990): Chem. Phys. **144**, 129. [327]
Littau, K.A., Bai, Y.S., Fayer, M.D. (1990): J. Chem. Phys. **92**, 4145. [372]
v. Löhneysen, H., Platte, M. (1979): Z. Phys. B **36**, 113. [530, 530]
v. Löhneysen, H., Rüsing, H., Sander, W. (1985): Z. Phys. B **60**, 323. [530, 530, 558]
Lou, L.F. (1976): Solid State Commun. **19**, 335. [215]
Luke, G.M., Brewer, J.H., Kreitzman, S.R., Noakes, D.R., Celio, M., Kadono, R., Ansaldo, E.J. (1991): Phys. Rev. B **43**, 3284. [433]
Ma, S.K. (1987): *Statistical Mechanics*, World Scientific, Singapore. [517]
Macfarlane, R.M., Shelby, R.M. (1979): Phys. Rev. Lett. **42**, 788. [327]
Macfarclane R.M., Shelby R.M. (1981): Opt. Lett. **6**, 96. [327]
Macfarlane, R.M., Shelby, R.M. (1987): J. Luminescence **36**, 179. [323]
Magerl, A., Rush, J.J., Rowe, J.M., Richter, D., Wipf, H. (1983): Phys. Rev. B **27**, 927. [418]
Magerl, A., Rush, J.J., Rowe, J.M. (1986a): Phys. Rev. B **33**, 2093. [453]
Magerl, A., Dianoux, A.J., Wipf, H., Neumaier, K., Anderson, I.S. (1986b): Phys. Rev. Lett. **56**, 159. [408, 418, 430, 432, Table 7.1]
Mahan G.D. (1981): *Many-particle physics*, Plenum Press, New York. [73, 81]
Maier, H., Haarer, D. (1995): J. Luminescence **64**, 87. [354, 357, 371, 371, 375, 375]
Maier, H., Haarer, D. (1997): J. Luminescence **72-74**, 413. [369]
Maier, H., Wunderlich, R., Haarer, D., Kharlamov, B.M., Kulikov, S.G. (1995): Phys. Rev. Lett. **74**, 5252. [382]
Maier, H., Kharlamov, B.M., Haarer, D. (1996): Phys. Rev. Lett. **76**, 2085. [521]
Maier, H., Müller, K.P., Jahn, S., Haarer, D. (1997): *Macromolecular Systems: Microscopic Interactions and Macroscopic Properties*, edited by H. Hoffmann, M. Schwoerer, Th. Vogtmann. VCH, Weinheim. [362]
Maier, M. (1986): Appl. Phys. B **41**, 73. [352]
Maleev S.V. (1981): Sov. Phys. JETP **52**, 1008. [57]
Maleev S.V. (1983): Sov. Phys. JETP **57**, 149. [57]
Maleev S.V. (1988): Sov. Phys. JETP **67**, 157. [58, 59]
Maleev, S.V. (1989): Sov. Phys. JETP **67**, 157. [245]
Malinovsky, V.K., Novikov, V.N., Parshin, P.P., Sokolov, A.P., Zemlyanov, M.G. (1990): Europhys. Lett. **11**, 43. [554]
Manson, N.B. (1982): Optics Comm. **44**, 32. [327]
Maradudin, A.A. (1968): Solid State Physics **18**, 273; **19**, 1 - Academic Press, New York-London, 1968 [319]
Marshall, W., Lovesy, S.W. (1971): *Theory of thermal neutron scattering*, Clarendon Press, Oxford. [406]
Martin, A.J., Brenig, W. (1974): Phys. Status Solidi B **64**, 163. [529]
Maschhoff, K.R., Granato, A.V. (1985): J. Physique **46**, C10. [428]
Matey, J.R., Anderson, A.C. (1978): Phys. Rev. B **17**, 5029. [194]
Matsumoto, T., Sasaki, Y., Hihara, M. (1975): J. Phys. Chem. Solids **36**, 215. [395]
Mattausch, G., Felsner, T., Hegenbarth, E., Kluge, B., Sahling, S. (1996): Phase Trans. **59**, 189. [Table 2.2]

Matusiewicz, G., Booker, R., Keiser, J., Birnbaum, H.K. (1974): Scripta Met. **8**, 1419. [415]
McCumber, D.E., Sturge, M.D. (1963): J. Appl. Phys. **34**, 1682. [321]
Mebert, J., Maile, B., Eisenmenger, W. (1990): *Phonons 89*, edited by S. Hunklinger, W. Ludwig, G. Weiss. World Scientific, Singapore, p. 495. [198]
Meixner, A.J., Renn, A., Bucher, S.E., Wild, U.P. (1986): J. Phys. Chem. **90**, 6777. [383]
Meixner, A.J., Renn, A., Wild, U.P. (1989): J. Chem. Phys. **91**, 6728. [332, 332]
Meixner, A.J., Renn, A., Wild, U.P. (1990): J. Luminescence **45**, 320. [332]
Messer, R., Blessing, A., Dais, S., Höpfel, D., Majer, G., Schmidt, C., Seeger, A., Zag, W., Lässer, R. (1986): Z. Phys. Chem. **NF 2**, 61. [409, 434]
Miedema, A.R. (1973): J. Less-Common Met. **22**, 117. [395]
Moerner W.E. (1988): *Persistent spectral holeburning: science and applications*, "Topics in current physics", **44**, Springer, Berlin, Heidelberg, New York. [327]
Moerner, W.E., Basché, Th. (1993): Angew. Chem. Int. Ed. Engl. **32**, 457. [331]
Moerner, W.E., Gertz, M., Huston, A.L. (1984): J. Phys. Chem. **88**, 6459. [345]
Moerner, W.E., Plakhotnik, T.V., Irngartinger, Th., Croci, M., Palm, V., Wild, U.P. (1994): J. Phys. Chem. **98**, 7382. [345, 352, 352]
Mogilyanskii, A. A., Raich, M. E. (1989): JETP **68**, 1081. [298]
Molenkamp, L.W., Wiersma, D.A. (1984): J. Chem. Phys. **80**, 3054. [321]
Molenkamp, L.W., Wiersma, D.A. (1985): J. Chem. Phys. **83**, 1. [326, 354]
Mon, K.K., Ashcroft, N.W. (1978): Solid State Commun. **27**, 609. [219]
Mori, H. (1965): Progr. Theor. Phys. **33**, 127; ibid. **34**, 399. [133]
Morishita, M., Kuroda, T., Sawada, A., Satoh, T. (1989): J. Low Temp. Phys. **76**, 387. [170]
Morkel, C., Wipf, H., Neumaier, K. (1978): Phys. Rev. Lett. **40**, 947. [419, 422, 431]
Morr, W., Müller, A., Weiss, G., Wipf, H., Golding, B. (1989): Phys. Rev. Lett. **63**, 2084. [199, 213, 220, 422, 423, 423, 424, 447, 447, Table 7.1, Table 7.2]
Müller, K.P., Haarer, D. (1991): Phys. Rev. Lett. **66**, 2344. [333, 362]
Muro, K., Sievers, A.J. (1986): Phys. Rev. Lett. **57**, 897. [455]
Muromtzev, P.I., Bagatsky, M.I., Manzhelii, V.G., Minchina, I.Y. (1994): Fiz. Nizk. Temp. **20**, 247. [296, 310]
Narasimhan, L.R., Littau, K.A., Pack, D.W., Elschner, A., Bai, Y.S., Fayer, M.D. (1990): Chem. Rev. **90**, 439. [372]
Narasimhan, L. R., Bai, Y. S., Dugan, M. A., Fayer, M. D. (1991): Chem. Phys. Lett. **176**, 335. [326, 354]
Narayanamurti, V., Pohl, R.O. (1970): Rev. Mod. Phys. **42**, 201. [57, 199, 220, 389]
Natelson, D., Rosenberg D., Osheroff, D. D. (1997): to be published. [258, 290]
Nava, R. (1994): Phys. Rev. B **49**, 4295. [187]
Neckel, H., Esquinazi, P., Weiss, G., Hunklinger, S. (1986): Solid State Commun. **57**, 151. [191]
Neu, P., Heuer, A. (1997) J. Chem. Phys. **106**, 1749. [462]
Neu, P., Würger, A. (1994a): Z. Phys. B **95**, 385. [57, 122, 133, 134, 135, 135]
Neu, P., Würger, A. (1994b): Europhys. Lett. **27**, 457. [167, 167, 183, 184, 184, 186, 186, 187, 187, 187, 217, 218]
Neu, P., Reichman, D.R., Silbey, R.J. (1997): Phys. Rev. B **56**, 5250. [369]
Neu, P., Silbey, R., Zilker, S., Haarer, D. (1997): Phys. Rev. B, submitted. [335]
Neumaier, K., Wipf, H., Cannelli, G., Cantelli, R. (1982): Phys. Rev. Lett. **49**, 1423. [440, 441]
Neumaier, K., Steinbinder, D., Wipf, H., Blank, H., Kearley, G. (1989): Z. Phys. B **76**, 359. [Table 7.1]

Neumann, M., Johnson, M.R., von Laue, L., Trommsdorff, H.P. (1996): J. Luminescence 66& 67, 146. [347]
Nieuwenhuizen, T.M. (1993): Europhys. Lett. 24, 191. [300]
Nishiyama, H., Akimoto, H., Okuda, Y., Ishimoto, H. (1992): J. Low Temp. Phys. 89(3/4), 727. [261]
Nittke, A., Esquinazi, P. (1996): Proceedings of the 21st International Conference on Low Temperature Physics, Czech. J. Phys. 46(S4), 2239. [43, Table 2.2]
Nittke, A., Scherl, M., Esquinazi, P., Lorenz, W., Junyun Li, Pobell, F. (1995): J. Low Temp. Phys. 98, 517. [26, 40, 183, 183, 183, 184, Table 2.2, Table 2.4, Table 2.5]
Niu, Q. (1991): J. Stat. Phys. 65, 317. [57, 82, 90]
Nowick, A.S., Berry, B.S. (1972): *Anelastic Relaxation in Crystalline Solids*, Academic Press, New York. [397, 404, 416]
Nozières, P., de Dominicis, C.T. (1969): Phys. Rev. 178, 1097. [58, 62]
Oligschleger, C., Schober, H.R. (1993): Physica A 201, 391. [541]
Oligschleger, C., Schober, H.R. (1995): Solid State Commun. 93, 1031. [469, 481, 483]
Olson, R.W., Lee, H.W.H., Patterson, F., Fayer, M.D., Shelby, R.M., Burum, D.P., Macfarlane, R.M. (1982): J. Chem. Phys. 77, 2283. [321, 347, 348, 348]
Oppenländer, A., Rambaud, Ch., Trommsdorff, H.P., Vial, J.-C. (1989): Phys. Rev. Lett. 63, 1432 [347, 349, 349]
Orbach, R. (1961): Proc. R. Soc. London A 264, 458. [429]
Orbach, R. (1985): Science 231, 814. [535]
Orrit, M., Bernard, J., Personov, R.I. (1993): J. Phys. Chem. 97, 10256. [331]
Orth, D.L., Malsh, R.J., Skinner, J.L. (1993): J. Phys. -Cond. Matter 5, 2533. [320]
Osad'ko, I.S. (1979): Sov. Phys. Usp. 22, 311. [319, 321]
Osad'ko, I.S. (1983): *Modern problems in condensed matter sciences*, edited by V.M. Agranovich, R.M. Hochstrasser. North-Holland, Amsterdam. [319, 321]
Osad'ko, I.S. (1991): Phys. Rep. 206, 43. [319, 321, 321, 325]
Osheroff, D.D., Rogge, S., Natelson, D. (1996): Proceedings of the 21st International Conference on Low Temperature Physics, Czech. J. Phys. 46(S6), 3295. [265, 290]
Pal-Val, P.P., Natsik, V.D., Kaufmann, H.J., Sologubenko, A.S. (1993): Mat. Sci. Forum 119/121, 117. [430]
Park, G., Hikata, A., Elbaum, C. (1981): Phys. Rev. B 23, 5597. [185]
Parker, C.A. (1968): *Photoluminescence of Solutions*, Elsevier Publishing Co., Amsterdam. [331]
Parshin, D.A. (1993): Z. Phys. B 91, 367. [163, 163]
Parshin, D.A. (1994a): Phys. Solid State 36, 991. [296, 529, 543, 550, 555, 564, 568]
Parshin, D.A. (1994b): Phys. Rev. B 49, 9400. [508, 510, 511, 511, 512, 517, 522, 529, 543, 560]
Parshin, D.A., Sahling, S. (1993): Phys. Rev. B 47, 5677. [29, 477, 543]
Parshin, D.A., Würger, A. (1992): Phys. Rev. B 46, 762. [32]
Parshin, D.A., Liu, X., Brand, O., v. Löhneysen, H. (1993): Z. Phys. B 93, 57. [517]
Patterson, F., Lee, H.W.H., Olson, R.W., Fayer, M.D. (1981): Chem. Phys. Lett. 84, 59. [348]
Pearton, S.J., Corbett, J.W., Stavola, M. (1992): *Hydrogen in crystalline semiconductors*, Springer Series in Materials Science. Vol. 16. Springer, Berlin, Heidelberg, New York. [454]
Peeters, E., Laermans, C., Parshin, D., Coeck, M. (1997): REI-9 Conference, Kuoxville (TN),Sept. 1997, to be published in NIMB. [199].
Pérez-Enciso, E., Ramos, M.A., Vieira, S. (1997): Phys. Rev. B 56, 32. [538]

Personov, R.I. (1983): *Modern problems in condensed matter sciences*, edited by V.M. Agronovich, A.A. Maradudin. North-Holland, Amsterdam, N.Y., Oxford. [323, 327]
Personov, R.I. (1992): J. Photochem. Photobiol. A **62**, 321. [352]
Personov, R.I., Kharlamov, B.M. (1986): Laser Chemistry **6**, 181. [327, 328, 352]
Personov, R.I., Al'shits, E.I., Bykovskaja, L.A. (1972): Optics Comm. **6**, 169. [323, 344]
Pfiz, T., Messer, R., Seeger, A. (1989): Z. Phys. Chem. NF **164**, 969. [434]
Phillips, W.A. (1972): J. Low Temp. Physics **7**, 351. [2, 9, 9, 15, 279, 315, 325, 325, 459, 527]
Phillips, W.A. (1981a): *Amorphous Solids – Low-Temperature Properties*, Springer, Berlin, Heidelberg, New York. [1, 121, 530, 560]
Phillips, W.A. (1981b): Phil. Mag. B **43**, 747. [522]
Phillips, W.A. (1987): Rep. Prog. Phys. **50**, 1657. [325, 459, 513, 564]
Phillips, W.A. (1990): *Phonons 89*, edited by S. Hunklinger, W. Ludwig, G. Weiss. World Scientific, Singapore, p. 367. [166]
Phillips, W.A., Buchenau, U., Nücker, N., Dianoux, A.J., Petry, W. (1989): Phys. Rev. Lett. **63**, 2381. [554, 554, 554]
Piché, L., Maynard, R., Hunklinger, S., Jäckle, J. (1974): Phys. Rev. Lett. **32**, 1426. [168, 178]
Pietralla, M., Mayr, P., Weishaupt, K. (1996): J. Non-Cryst. Solids **195**, 199. [513]
Pirc, R., Gosar, P. (1969): Phys. Kond. Mat. **9**, 377. [57, 90, 118]
Plakhotnik, T.V., Moerner, W.E., Palm, V., Wild, U.P. (1995): Optics Comm. **114**, 83. [334]
Plakhotnik, T.V., Donley, E.A., Wild, U.P. (1997): Ann. Rev. Phys. Chem. **48**, 175. [331]
Pobell, F. (1992): *Matter and Methods at Low Temperatures*, Springer, Berlin, Heidelberg, New York. [176, 335, 362]
Poker, D.B., Setser, G.G., Granato, A.V., Birnbaum, H.K. (1984): Phys. Rev. B **29**, 622. [422, 422, 423, 428]
Powell, R.C., Xi, L., Gang, X., Quarles, G.J., Walling, J.C. (1985): Phys. Rev. B **32**, 2788. [321]
Pschierer, H., Schellenberg, P., Friedrich, J. (1994): Mol. Cryst. Liq. Cryst. **253**, 113. [352]
Qi, Zh., Völkl, J., Lässer, R., Wenzl, H. (1983): J. Phys. F: Met. Phys. **13**, 2053. [414]
Ralph, D.C., Buhrman, R.A. (1992): Phys. Rev. Lett. **69**, 2118. [158]
Ramos, M.A., Buchenau, U. (1997): Phys. Rev. B **55**, 5749. [530, 548, 558, 562, 563]
Ramos, M.A., Gil, L., Bringer, A., Buchenau, U. (1993): phys. status solidi A **135**, 477. [529, 546, 547, 549, 550, 552, 554, 554, 558, 560, 563]
Rau, S., Enss, C., Hunklinger, S., Neu, P., Würger, A. (1995a): Phys. Rev. B **52**, 7179. [122, 122, 133, 134, 516, 522, 565]
Rau, S., Baessler, S., Kasper, G., Weiss, G., Hunklinger, S. (1995): Annalen d. Phys. **4**, 91. [522]
Ravex, A., Lasjaunias, J.C., Béthoux (1981): Solid State Commun. **40**, 853. [194]
Ravex, A., Lasjaunias, J.C., Béthoux (1984): J. Phys. F **14**, 329. [194]
Raychaudhuri, A.K., Hunklinger, S. (1984): Z. Phys. B **57**, 113. [169, 170, 178, 185, 186, 186, 189, 191, 565]
Rebane, K.K. (1968): *Elementary theory of vibrational structure of impurity spectra in crystals*, Nauka, Moscow (in Russian). [319]
Rebane, L.A., Gorokhovskii, A.A., Kikas, J.V. (1982): Appl. Phys. B **29**, 235. [335]
Regelmann, T., Schimmele, L., Seeger, A. (1994): Z. Phys. B **95**, 441. [66]

Reichert, P., Schilling, R. (1985): Phys. Rev. B **32**, 5731. [503]
Reinecke, T.L. (1979): Solid State Commun. **32**, 1103. [326, 355, 363]
Reineker, P., Kassner, K. (1986): *Optical spectroscopy of glasses*, edited by I. Zschokke-Granacher. Reidel, Dordrecht. [325]
Renner, T., Deeg, F.W., Bräuchle, C. (1995): in *Spektroskopie amorpher und kristalliner Festkörper*, edited by D. Haarer and W. Spiess, Steinkopff, Darmstadt, 1995. [332]
Richter, D., Alefeld, G., Heidemann, H., Wakabayashi, N. (1977): J. Phys. F: Metal Phys. **7**, 569. [415]
Richter, W., Schulte, G., Haarer, D. (1984): Optics Comm. **51**, 413. [352]
Rivier, N. (1979): Phil. Mag. A **40**, 859. [219]
Rogge, S. (1994): (private communication). [289]
Rogge, S. (1996): PhD thesis, Stanford University. [256, 256, 292]
Rogge, S., Salvino, D.J., Tigner, B., Osheroff, D.D. (1994): Physica B **194-196**, 407. [289]
Rogge, S., Natelson, D., Osheroff, D.D. (1996a): Proceedings of the 21st International Conference on Low Temperature Physics, Czech. J. Phys. **46**(S4), 2263. [290]
Rogge, S., Natelson, D., Osheroff, D.D. (1996b): Phys. Rev. Lett. **76**, 3136. [241, 255, 258, 258, 260, 260, 263, 263, 271, 271, 274, 286, 288, 288, 288, 312, 459]
Rogge, S., Natelson, D., Osheroff, D.D. (1997a): Rev. Sci. Instr. **68**, 1831. [264]
Rogge, S., Natelson, D., Osheroff, D.D. (1997b): J. Low Temp. Phys. **106**, 717. [263, 264, 272, 292, 292, 312]
Rogge, S., Natelson, D., Tigner B., Osheroff, D.D. (1997c): Phys. Rev. B **55**, 11256. [261, 261, 314]
Rothenfusser, M., Dietsche, W., Kinder, H. (1983): Phys. Rev. B **27**, 5196. [198, 535]
Rush, J.J., Magerl, A., Rowe, J.M., Harris, J.M., Provo, J.L. (1981): Phys. Rev .B **24**, 4903. [453]
Sahling, S. (1989): Solid State Commun. **72**, 497. [38, 43, Table 2.5]
Sahling, S. (1990): Solid State Commun. **75**, 125. [Table 2.2]
Sahling, S. (1992): Proc. of the 7th Int. Conf. on Phonon Scattering in Condensed Matter, p. 289. [Table 2.2]
Sahling, S., Sahling, A. (1988): Mod. Phys. Lett. B**2**, 1327. [Table 2.2]
Sahling, A., Sahling, S. (1989): J. Low Temp. Phys. **77**, 450. [Table 2.2]
Sahling, S., Sievert, J. (1990): Solid State Commun. **75**, 237. [Table 2.2]
Sahling, S., Sahling, A., Neganov, B.S., Kol'ač, M. (1986a): J. Low Temp. Phys. **65**, 289. [46, Table 2.2]
Sahling, S., Sahling, A., Neganov, B.S., Kol'ač, M. (1986b): Solid State Commun. **59**, 643. [42, Table 2.2]
Sahling, S., Kol'ač, M., Sahling, A. (1988): J. Low Temp. Phys. **73**, 450. [Table 2.2]
Salvino, D.J. (1993): PhD thesis, Stanford University. [267, 269]
Salvino, D.J., Rogge, S., Tigner, B., Osheroff, D.D. (1994): Phys. Rev. Lett. **73**, 268. [263, 263, 270, 272, 284, 288, 288, 288, 289, 290, 312]
Schaumann, G., Völkl, J., Alefeld, G. (1970): Phys. Stat. Sol. **42**, 401. [414, 414, 414]
Schellenberg, P., Friedrich, J., Kikas, J.V. (1994): J. Chem. Phys. **101**, 9262. [342, 342]
Scheibner, W., Jäckel, M. (1985): Phys. Stat. Sol. (a) **87**, 543. [Table 2.5]
Schickfus, M.v., Hunklinger, S. (1981): *Amorphous Solids – Low-Temperature Properties*, edited by W.A. Phillips. Springer, Berlin, Heidelberg, New York, Chap. 6. [566]
Schiller, P., Schneiders, A. (1975): Phys. Stat. Sol. (a) **29**, 375. [416]

Schmidt, C. (1978): Rev. Sci. Inst. **50**, 454. [204]
Schmidt, Th., Baak, J., van de Straat, D.A., Brom, H.B., Völker, S. (1993): Phys. Rev. Lett. **71**, 3031. [327, 359]
Schmidt, Th., Macfarlane, R.M., Völker, S. (1994): Phys. Rev. B **50**, 15707. [327, 327, 359]
Schober, H.R., Laird, B.B. (1991): Phys. Rev. B **44**, 6746. [541, 543]
Schober, H.R., Oligschleger, C. (1996): Phys. Rev. B **53**, 11469. [469]
Schober, H.R., Stoneham, A.M. (1988): Phys. Rev. Lett. **60**, 2307. [413, 416, 417]
Schober, T., Wenzl, H. (1978): *Hydrogen in Metals II*, edited by G. Alefeld, J. Völkl. Springer, Berlin, Heidelberg, New York. [390, 392, 393, 414, 414]
Schober, H.R., Oligschleger, C., Laird, B.B. (1993): J. of Non-Cryst. Solids **156**, 965. [469, 481]
Schwark, M., Pobell, F., Kubota, M., Mueller, R.M. (1985): J. Low Temp. Phys. **58**, 171. [40, 45, Table 2.2, Table 2.5]
Sellers, G.J., Anderson, A.C., Birnbaum, H.K. (1974): Phys. Rev. B **10**, 2771. [419]
Selzer, P.M., Huber, D.L., Hamilton, D.S., Yen, W.M., Weber, M.J. (1976): Phys. Rev. Lett. **36**, 813. [323]
Semmelhack, H., König, R., Esquinazi, P. (1996): unpublished. [216]
Shelby, R.M., Macfarlane, R.M. (1979): Chem. Phys. Lett. **64**, 545. [327]
Sherington, D., Kirkpatrick, S. (1975): Phys. Rev. Lett. **35**, 1792. [300]
Shimshoni, E., Gefen, M. (1991): Ann. of Phys. **210**, 16. [278]
Shore, H.B., Sanders, L.M. (1975): Phys. Rev. B **12**, 1546. [109]
Shu, L., Small, G.J. (1990): Chem. Phys. **141**, 447. [344]
Shu, L., Small, G.J. (1992): JOSA B **9**, 724. [344]
Shuker, R., Gammon, R.W. (1970): Phys. Rev. Lett. **25**, 222. [529, 554]
Silbey, R., Harris, R.A. (1989): J. Phys. Chem. **93**, 7062. [57]
Silbey, R., Trommsdorff, H.P. (1990): J. Chem. Phys. **89**, 897. [66]
Silbey, R.J., Koedijk, J.M.A., Völker, S. (1996): J. Chem. Phys. **105**, 901. [464]
Skinner, J.L. (1986): Ann. Rev. Phys. Chem. **39**, 4931. [319, 321]
Skinner, J.L., Moerner, W.E. (1996): J. Phys. Chem. **100**, 13251. [318]
Skinner, J.L., Trommsdorff, H.P. (1988a): Chem. Phys. Lett. **165**, 540. [57]
Skinner, J.L., Trommsdorff, H.P. (1988b): J. Chem. Phys. **89**, 897. [347]
Small, G.J. (1983): *Modern problems in condensed matter sciences*, edited by V.M. Agronovich, A.A. Maradudin. North-Holland, Amsterdam. [325, 325, 327]
Smith, D.A. (1978): Phys. Rev. Lett. **42**, 729. [470]
Sokolov, A.P., Rössler, E., Kisliuk, A., Quitmann, D. (1993): Phys. Rev. Lett. **71**, 2062. [519]
Sokolov, A.P., Calemczuk, R., Salce, B., Kisliuk, A., Quitmann, D., Duval, E. (1997): Phys. Rev. Lett. **78**, 2405. [566, 568]
Solovjov, K.N., Zalesskij, I.E., Kotlo, V.N., Shkirman, S.F. (1973): JETP Letters **17**, 463 (in Russian). [339, 341]
Soulen, R.J., Dove, R.B. (1979): SRM 768 *Temperature Reference Standard For Use Below 0.5K* NBS Special Publication. [176]
Steinbinder, D., Wipf, H., Magerl, A., Richter, D., Dianoux, A.-J., Neumaier, K. (1988): Europhys. Lett. **6**, 535. [98, 432, 433, 433, Table 7.1]
Steinbinder, D., Wipf, H., Dianoux, A.-J., Magerl, A., Neumaier, K., Richter, D., Hempelmann, R. (1991): Europhys. Lett. **16**, 211. [408, 430, 432, 432, 434, Table 7.1]
Stephens, R.B. (1973): Phys. Rev. B **8**, 2896. [Table 2.4, Table 2.5]
Stephens, R.B. (1976): Phys. Rev. B **13**, 852. [9]
Stephens, R.B., Cieloszyk, G.S., Salinger, G.L. (1972): Phys. Lett. **28A**, 215. [Table 2.4]
Stillinger, F.H. (1990): Phys. Rev. B **41**, 2409. [491, 491]

Stillinger, F.H. (1995): Science **267**, 1935. [487]
Stillinger, F.H., Weber, T.A. (1983): Phys. Rev. A **28**, 2408. [467, 467, 468, 487, 489, 491]
Stockburger, J., Grifoni, M., Sassetti, M., Weiss U. (1994): Z. Phys. B **94**, 447. [164, 164, 164, 164, 185, 188, 189, 189, 189, 189]
Stockburger, J.T., Grifoni, M., Sassetti, M. (1995): Phys. Rev. B **51**, 2835. [163, 163, 163, 163, 164, 179]
Stoneham, A.M. (1969): Rev. Mod. Phys. **41**, 82. [389]
Stoneham, A.M. (1972): J. Phys. F **2**, 417. [412, 412]
Storm, C.B., Teklu, Y. (1972): J. Am. Chem. Soc. **94**, 1745. [339]
Strehlow, P., Dreyer, W. (1994): Physica B **194-196**, 485. [261]
Suck, J.B., Rudin, H. (1983): *Glassy Metals II*, Springer Topics in Applied Physics **53**, edited by H. Beck, H.-J. Güntherodt. Springer, Berlin, Heidelberg, New York, p. 217. [532]
Sugimoto, H., Fukai, Y. (1980): Phys. Rev. B **22**, 670. [413, 454, 454]
Sussman, J.A. (1967): J. Phys. Chem. Solids **28**, 1643. [405, 411]
Suzuki, T., Namzue, H., Koike, S., Hayakawa, H. (1983): Phys. Rev. Lett. **51**, 798. [454, 454]
Svare, I. (1989): Phys. Rev. B **40**, 11585. [433, 449]
Svare, I., Torgeson, D.R., Borsa, F. (1991): Phys. Rev. B **43**, 7448. [411, 450, 450, 451]
Szabo, A. (1975): Phys. Rev. B **11**, 4512. [327]
Takahashi, Ju.-I., Tsuchiya, Ju., Kawasaki, K. (1994): Chem. Phys. Lett. **222**, 325. [344, 345]
Teichler, H., Seeger, A. (1976): Phys. Lett. A **82**, 91. [58]
Thauer, P., Esquinazi, P., Pobell, F. (1990): Physica B **165/166**, 905. [213, 215]
Thijssen, H.P.H., Dicker, A.I.M., Völker, S. (1982): Chem. Phys. Lett. **92**, 7. [323]
Thomas, N., Arnold, W., Weiss, G., v. Löhneysen, H. (1980): Solid State Commun. **33**, 523. [212]
Thorne, J.R.G., Denning, R.G., Barker, T.J. (1985): J. Luminescence **34**, 147. [321]
Tielbürger, D., Merz, R., Ehrenfels, R., Hunklinger, S. (1992): Phys. Rev. B **45**, 2750. [16, 23, 122, 166, 167, 182, 182, 183, 184, 463, 479, 516, 518]
Tigner, B. (1994): PhD thesis, Stanford University. [292]
Tigner, B., Salvino, D.J., Rogge, S., Osheroff, D.D. (1993): *ICPP Cornell '92*, volume 112, edited by M. Meissner, R.O. Pohl. Springer, Berlin, Heidelberg, New York. [270]
Topp, K.A., Cahill, D.G. (1996): Z. Phys. B **101**, 235. [145, 197]
Tornow, M., Weis, R., Weiss, G., Enss, C., Hunklinger, S. (1994): Physica B **194-196**, 1063. [280]
Trommsdorff, H.P. (1986): *Tunneling*, edited by J. Jortner, B. Pullman, Reidel. Publishing Company, p. 103. [347]
Trommsdorff, H.P., Casalegno, R., Miller, R.J.D., Clemens, J.M., Hochstrasser, R.M. (1984): J. Luminescence **31-32**, 517. [347, 348]
Uhler, W., Schilling, R. (1988): Phys. Rev. B **37**, 5787. [505]
Vainer, Yu.G., Plakhotnik, T.V., Personov, R.I. (1996): Chem. Phys. **209**, 101. [326]
Vajda, P. (1995): *Hydrogen in rare-earth metals, including RH_{2+x} phases*., edited by K.A. Gschneider Jr., L. Eyring. *Handbook on the Physics and Chemistry of Rare Earths*, vol. 20. Elsevier Science, Amsterdam. [390, 393, 414]
Vajda, P., Daou, J.N., Moser, P., Remy, P. (1991): Solid State Commun. **79**, 383. [416]
Van der Zaag, P.J., Galaup, J.P., Völker, S. (1990): Chem. Phys. Lett. **166**, 263. [327]
Vargas, P., Kronmüller, H. (1980): Phil. Mag. A **51**, 59. [395]

Vargas, P., Böhm, M.C., Kronmüller, H. (1985): Z. Phys. Chem. NF **143**, 229. [395]
Vineyard, G.H. (1957): J. Phys. Chem Sol. **3**, 121. [411]
Vladár, K., Zawadowski, A. (1983): Phys. Rev. B **28**, 1564. [157, 158, 191]
Völker, S. (1989): Ann. Rev. Phys. Chem. **40**, 499. [323, 327, 335]
Völker, S., Macfarlane, R.M. (1980): J. Chem. Phys. **73**, 4476. [341, 341]
Völker, S., van der Waals, J.H. (1976): Mol. Phys. **32**, 1703. [339, 341]
Völker, S., Macfarlane, R.M., Genack, A.Z., Trommsdorff, H.P. (1977): J. Chem. Phys. **67**, 1759. [321]
Völker, S., Macfarlane, R.M., van der Waals, J.H. (1978): Chem. Phys. Lett. **53**, 8. [321]
Völkl, J., Wipf, H., Beaudry, B.J., K.A. Gschneider Jr., K.A. Gschneider (1987): Phys. Stat. Sol. (b) **144**, 315. [416]
Vollmayr, K., Kob, W., Binder, K. (1996): Phys. Rev. B **54**, 15808. [466]
Voncken, A.P.J., Naish, J.H., Riese, D., König, R., Pobell, F., Owers-Bradley, J.R. (1997): J. Low Temp. Phys. **100**, 1105. [177]
De Vries, H., Wiersma, D.A. (1980): J. Chem. Phys. **72**, 1851. [334]
Walsh, C.A., Fayer, M.D. (1985): J. Luminescence **34**, 37. [348]
Walsh, C.A., Berg, M., Narasimhan, L.R., Fayer, M.D. (1986): Chem. Phys. Lett. **130**, 6. [323, 326, 354]
Walsh, C.A., Berg, M., Narasimhan, L.R., Fayer, M.D. (1987): J. Chem. Phys. **86**, 77. [354]
Wang, J.L., Weiss, G., Wipf, H., Magerl, A. (1984): *Phonon Scattering in Condensed Matter*, edited by W. Eisenmenger, K. Lassmann, S. Dottinger. Springer, Berlin, Heidelberg, New York, p. 401. [423, 423]
Wang, X., Bridges, F. (1992): Phys. Rev. B **46**, 5122, and references therein. [199]
Wannemacher, R., Smorenburg, H.E., Schmidt, Th., Völker, S. (1992): J. Luminescence **53**, 266. [327]
Wannemacher, R., Koedijk, J.M.A., Völker, S. (1993): Chem. Phys. Lett. **206**, 1. [327]
Wässerbach, W. (1978): Phil. Mag. A **38**, 401. [204]
Wässerbach, W. (1987): Mat. Sci. Eng. **96**, 167. [204]
Watson, S., Pohl, R. (1995): Phys. Rev. B **51**, 8086. [220]
Weber, T.A., Stillinger, T.H. (1985): Phys. Rev. B **32**, 5402. [471]
Weiss, G., Arnold, W., Dransfeld, K., Güntherodt, H.J. (1980): Solid State Commun. **33**, 111. [62]
Weiss, G., Hunklinger, S., v. Löhneysen, H. (1981): Phys. Lett. **85A**, 84. [212]
Weiss, G., Hunklinger, S., v. Löhneysen, H. (1982): Physica B **109 & 110**, 1946. [156]
Weiss, U. (1993): *Quantum Dissipative Systems*, Series in Modern Condensed Matter Physics, Vol. 2, World Scientific, Singapore. [62, 72, 90, 164]
Weiss, U., Wollensak, M. (1989): Phys. Rev. Lett. **62**, 1663. [62, 100, 102]
Westlake, D.G. (1972): Scripta Met. **6**, 887. [415]
White, Jr., B.E., Pohl, R.O. (1995): Phys. Rev. Lett. **75**, 4437. [198, 313, 522]
White, B.E., Pohl, R.O. (1996): Z. Phys. B **100**, 401. [145, 197]
White, G.K., Collocott, S.J., Cook, J.S. (1984): Phys. Rev. B **29**, 4778. [538]
Winterling, G. (1975): Phys. Rev. B **12**, 2432. [529, 532, 535, 535, 554]
Wipf, H. (1997): *Hydrogen in Metals III*, edited by H. Wipf. Springer, Berlin, Heidelberg, New York. [98, 107, 390]
Wipf, H., Alefeld, G. (1974): Phys. Stat. Sol. (a) **23**, 175. [415]
Wipf, H., Neumaier, K. (1984): Phys. Rev. Lett. **52**, 1308. [396, 418, 421, Table 7.1]
Wipf, H., Magerl, A., Shapiro, S.M., Satija, S.K., Thomlinson, W. (1981): Phys. Rev. Lett. **46**, 947. [430, 431]

Wipf, H., Steinbinder, D., Neumaier, K., Gutsmiedel, P., Magerl, A., Dianoux, A.-J. (1987): Europhys. Lett. **4**, 1379. [62, 98, 430, 432, 432, 432, Table 7.1]
Wunderlich, R., Maier, H., Haarer, D., Kharlamov, B.M. (1997): Phys. Rev. B, submitted. [336, 383]
Würger, A. (1990): Z. Phys. B **81**, 273. [66]
Würger, A. (1997a): *From Coherent Tunneling to Relaxation*, Springer Tracts in Modern Physics Vol. 135, Springer, Berlin, Heidelberg, New York. [63, 66, 82, 106, 109, 123, 310, 434]
Würger, A. (1997b): Phys. Rev. Lett. **78**, 1759. [73, 74, 74, 76]
Würger, A. (1997c): J. Phys. Cond. Matt. **9**, 5543. [125, 138]
Würger, A. (1997d): Phys. Lett. A **236**, 571. [62]
Würger, A. (1998a): Phys. Rev. B **57**, 347. [73, 77, 89, 89, 112, 115, 115, 118, 126, 128, 129, 141]
Würger, A. (1998b): Solid State Commun. **106**, 63. [434]
Yagi, E., Kobayashi, T., Nakamura, S., Kano, F., Watanabe, K., Fukai, Y., Koike, S. (1986): Phys. Rev. B **33**, 5121. [454]
Yamada, K., Sakurai, A., Miyazima, S. (1985): Prog. Theoret. Phys. **73**, 1342. [61]
Yen, W.M., Scott, W.C., Schawlow, A.L. (1964): Phys. Rev. A **136**, 271. [321]
de Yoreo, J.J., Knaak, W., Meissner, M., Pohl, R.O. (1986): Phys. Rev. B **34**, 1888. [249]
Yoshinari, O., Tanaka, K., Matsui, H. (1996): Phil. Mag. A **74**, 495. [418]
Yu, C.C. (1985): Phys. Rev. B **32**, 4220. [245]
Yu, C.C., Freeman, J.J. (1987): Phys. Rev. B **36**, 7620. [530, 564]
Yu, C.C., Granato, A.V. (1985): Phys. Rev. B **32**, 4793. [432]
Yu, C.C., Leggett, A.J. (1988): Comm. Cond. Matt. Phys. **14**, 231. [238, 262, 280, 288, 297, 298, 312, 518]
Zaitlin, M.P., Anderson, A.C. (1975): Phys. Rev. B **12**, 4475. [530]
Zapp, P.E., Birnbaum, H.K. (1980): Acta Metall. **28**, 1523. [416]
Zeller, R.C., Pohl, R.O. (1971): Phys. Rev. B **4**, 2029. [1, 9, 9, 57, 199, 242, 527, 528, 530, 558, 558]
Zilker, S.J., Haarer D. (1997): Chem. Phys. **200**, 167. [334, 335]
Zimmermann, J., Weber, G. (1981a): Phys. Rev. Lett. **46**, 661. [9, 35, Table 2.2]
Zimmermann, J., Weber, G. (1981b): Phys. Lett. A **86**, 32. [20, 22, Table 2.5]
Zimmermann, J. (1984): Cryogenics **24**, 27. [Table 2.2, Table 2.4, Table 2.5]
Zimmermann N.M., Golding, B., Haemmerle, W.H. (1991): Phys. Rev. Lett. **67**, 1332. [103]
Zollfrank, J., Friedrich, J. (1990): J. Chem. Phys. **93**, 8586. [341, 342]
Zollfrank, J., Friedrich, J., Tani, T. (1989): Polymer **30**, 231. [344]
Zwerger, W. (1983): Z. Phys. B **53**, 53; ibid **54**, 87. [62]

Index

C
- definition, 152
- from internal friction, 154, 563
- from sound velocity, 154
- from thermal conductivity, 562, 563
- weighted average, 561–563

W, 28, 543
- determination, 561

$\mathcal{Z}$-factor
- blip expansion, 77, 143
- for weak coupling, 130, 131
- mode-coupling approximation, 136

η_L, definition of, 28

γ, coupling constant, 14, 151, 398
- microscopic origin, 494

τ_*, 248

τ_1, 227, 230, 231, 234, 241, 404
- minimum, 229
- single TS
- – phonon, 15, 229
- thermal, 240, 242, 243

τ_2, 148, 185, 227, 230, 234, 239, 241, 248, 253

τ_p, 15

τ_0, 251, 255

τ_{12}, 232

τ_{1T}, 240, 254

τ_c, 278

τ_{max}, 284, 285

τ_{min}, 15, 16, 20, 284

τ_p, 283

tau_1, 405

Acoustic attenuation, 147, 149, 404, 564
- SiO_2, 565

Acoustic intensity
- influence on internal friction, 163
- influence on sound dispersion, 151, 161
- influence on the acoustic properties in metallic systems, 164
- influence on the tunneling splitting, 162
- influence on ultrasonic attenuation, 148

Acoustic properties, 145, 223, 396, 413, 422
- above 1 K in amorphous, 166
- amorphous superconductors, 191
- amorphous thin films, 197
- experiments at very low temperatures, 176
- H trapped by interstitial O,N,C in Nb and Ta, 422–430
- in NbTiH polycrystals, 442–444
- in NbZrH single crystals, 444–449
- influence of incoherent tunneling, 166
- influence of thermal activation, 166
- influence of thermal treatment, 215
- isotope mass dependence, 447
- nonlinear behavior, 161
- nonuniversal behavior at $T > 1$ K, 183
- normal-conducting amorphous metals, 185–190
- normal-conducting polycrystalline metals, 207
- of amorphous dielectrics, 178
- PdSiCu, 122, 186
- polycrystalline metals, 199
- – above 1 K, 217
- – strain-dependent effects, 205
- polycrystalline superconductors, 200
- Polymethylmethacrylate (PMMA), 183
- Polystyrene (PS), 183
- SiO_2, 178–183, 224
- thermally activated relaxation, 184

Acoustic relaxation

- after application of a dc field, 275–277
Adiabatic approximation, 236, 277, 294
Ag
- sound velocity, 211
Aging, 375
Al
- sound velocity, 209
Aluminium
- heat release, 43
Amorphous
- dielectrics, 38, 178, 224
- metals, 185–197
Anelastic response, 396, 398, 413
Arrhenius law, 411, 417
Asymmetry
- change with dc electric field, 278
Asymmetry Δ, 10
- within the SPM, 545

Bare tunneling energy Δ_b, 61
Barrier height V, 11
- within the soft-potential model, 551
Basis
- energy eigenstates, 11, 426
- left-right, 10
Bath spectral function, 61
Benzoic acid, 347
Bi wire, 103, 109
Binding energy of atoms $\mathcal{E}$, 28
BK7
- resonant absorption, 226
- sound velocity, 224
Blip expansion, 72
Blip–blip interactions, 74
Bloch equations, 229
Boron oxide, 522
- internal friction, 565
- neutron scattering in, 538
Bose commutation relations, 60
Bose factor, 128, 535
Boson peak, 529, 530, 548, 554, 567
Bosonic heat bath, 61
Bridge techniques, 256

Chromophore, 318
Clamping
- influence of, 174
Coherent coupling, 165, 243
Coherent motion, 84
Coherent neutron scattering, 407, 532
Coherent tunneling, 93, 407, 430, 449
Collective excitations, 242

Composite operators, 126
Computer simulations
- algorithms to locate nearby minima, 467
- basic problems, 466
- determination of low-temperature parameters, 476
- distribution of DWP's, 472–476
- earlier results, 470
- microscopic properties of DWP's, 478
- pair potential, 471
- scope, 465
- search for DWP's, 472
- search strategies, 481
Cooling process, 25
Cooling rate, 466, 520
Correlation time, 410
Corresponding states, 508
Coulomb gap, 281
Coupled librational motion, 534
Coupled phonon spectrum, 134
Coupled rotations, 507
Coupling constant γ
- table, 420, 514
Coupling constant γ, 12, 14, 151, 463, 545
Coupling parameter, 64, 67
Covalent forces, 518
Crossing diagrams, 137
Crossover temperature T', 252
Crossover temperature T^*, 86, 117
Crossover temperature T_0, 65, 108
- table, 109
Crossover temperature T_c, 31
Crossover temperature T_{el}, 94
Crossover temperature T_{min}, 530
Crossover temperature T_m, 233
Crossover to incoherent tunneling, 86, 94, 117, 408
Crossover to relaxation, 117
Cu
- sound velocity, 211
$Cu_{30}Zr_{70}$
- acoustic properties, 195–197
Cubic spectral density function, 64–66
Cutoff $\Delta_{0,min}$, 261
Cutoff u_{min}, 17, 23

D
- specific heat in NbO_x, 421
Damping
- by conduction electrons, 61
- by phonons, 63

– of TLS, 407
Damping rate
– activated, 82, 96
– beyond NIBA, 83
– by conduction electrons, 95, 102
– in NIBA, 78
DBATT, 334
Dc bias application technique, 264
Debye
– approximation, 64
– frequency, 65
– temperature, 108, 244, 320, 511
– – beyond, 96
– unit, 226
Debye–Waller factor, 69, 70, 319, 540
Decoupling approximation, 134
Defect elastic energy, 397
Defects, 521
Deformation potential, 398, 401, 463, 477, 493, *see* also Coupling constant
– determination, 496
– microscopic origin, 494
– numerical evaluation, 497
Density of states P, 15, 16
– within the SPM, 29, 30, 545, 551
Dephasing rate, 248
Dephasing time, 239, 321
Diaelastic polarization, 399, 401, 402, 422
Dielectric hysteresis, 292
Dielectric properties, 223
– SiO_2, 258–263
– SiO_x, 267–269
Dielectric relaxation, 273
– frequency dependence, 289
Dielectric saturation temperature, 262
Diffusion coefficient, 414
– of H, D and T in V, Nb and Ta, 415
Diffusion of interstitial H, 395, 413
Dilution cryostat for optical spectroscopy, 362
Dimethyl-Siloxan
– heat release, 41
Dipole gap
– density of states
– – single-particle excitations, 281, 282
– – time-dependent, 283, 293
– nonequilibrium response
– – acoustic experiments, 287
– – adiabatic perturbation, 294
– – nonadiabatic perturbation, 282, 284–286
Dipole gap model fit, 291

Dislocation lines, 220
Distribution
– Gaussian, 184, 441
– Lorentzian, 422, 441, 447
Distribution function
– time dependent, 20
Double force tensor, 397, 438
Dressed tunnel energy
– by conduction electrons, 94
Dressed tunnel frequency, 108
Dressed tunnel matrix element, 61, 69, 82
Dressing effect, 69, 86, 108
Dye laser, 328
Dynamic structure factor, 407, 532
Dynamic susceptibility, 402

Echo experiments
– phonon, 227, 234, 249
Effective mass, 481, 518, 543
Effective temperature, 374, 375, 377
Elastic deformation potential, 108
Elastic dipole, 397, 437, 438
Elastic energy, 397
Elastic strain tensor, 14, 63
Electron mean free path, 161, 213
Electron-assisted tunneling, 24, 425
Electron-hole excitations, 61
Electron-polaron effect, 158
Electron-TS's coupling strength, 62, 420
Energy diagram of an organic molecule, 319
Energy dissipated
– by a mechanical oscillator, 176
Energy distribution function, 390
Energy landscape, 466, 487
– attraction basin, 467
– metabasins, 491
– one-dimensional representation, 490
Energy relaxation, 131
Energy splitting
– failure of the WKB approximation, 547
– of a pair, 243
– within the SPM, 544, 547
Equilibrium data
– capacitive response
– – low-temperature saturation, 261–263, 314
– dissipation, 255, 258, 311
– nonlinear effects, 256, 258, 259
Excitation

Index   595

- delocalized, 245–247, 249, 250
- many-center, 299, 304, 307, 308, 312
-- density of states, 300, 311
- pair, 243, 298, 299, 309
-- cutoff radius, 243–245, 247, 252, 306, 308, 309, 313
-- density of states, 246, 305–307, 310
-- effective asymmetry, 243–245, 310
-- effective tunneling amplitude, 243–248, 310
-- energy, 304
-- interaction energy, 247
-- interactions between, 246
-- resonant cluster, 247, 248, 250–252, 261
-- resonant condition, 244
- single-particle, 299
-- density of states, 301, 304, 305, 307
-- energy, 302, 304
- triplet, 250, 263, 307, 314
Extrinsic DWP, 484
- distribution of barrier heights, 486
Extrinsic TLS, 338, 344, 347

$Fe_{80}B_{14}Si_6$
- heat release, 40
Films, 182, 197, 523
Fluorescence excitation, 331, 332
Force matrix, 469
Four-level system, 434
- neutron spectroscopy evidence, 449
Fourier transformation, 140
Fragile glasses, 519
Freezing temperature T^*, 32
Frenkel–Kontorova model, 505

Gaussian distribution ansatz, 550
GeO_2
- internal friction, 565
Glass transition, 487, 513, 519
- fragility, 519
Glass transition temperature
- amorphous $Cu_{30}Zr_{70}$, 195
- table, 514
Gorsky effect, 395, 413

H
- acoustic properties in hcp rare earths, 451
- delocalization in the 4T configuration, 454
- diffusion coefficient in Nb, 394
- in Nb unit cell, 419
- local vibration modes, 452
- octahedral sites, 391
- relaxation rates in bcc and hcp metals, 391
- second tunnel system in Nb, 428
- specific heat in NbO_x, 421
- specific heat of H trapped by substitutionals, 440–442
- tetrahedral sites, 391
Hamiltonian
- interaction, 238, 306
- Ising model in random field, 300
- pair, 235, 237, 242, 243, 304, 305
- phonons, 235
- single TS, 10
-- dc field, 278
-- measuring field, 228
Heat bath, 60
Heat capacity, 239, 242, 249, 298, 313
- nonequilibrium, 263, 314
- phonon, 242, 297
Heat release, 13
- approximation for the, 18
- of amorphous metals, 40
- of crystalline materials, 42
- of dielectric anorganic materials, 38
- of organic materials, 40
- temperature dependence, 44
- time dependence, 18, 38
Heat switch, 37
Hole burning
- acoustic, 226, 227
- dielectric, 270
- optical, 317, 327–338
-- electric field effects, 382
-- time dependence, 343
Hole narrowing, 380
Hole width, 364
Holographic detection, 332
Homogeneous broadening, 321, 323, 324, 326
Hydride phases, 392
Hydrogen Bond, 342

Impurity, 484
Impurity atoms, 108
Incoherent scattering of neutrons, 407
Incoherent tunneling, 25, 85, 94, 122, 135, 166, 408, 432, 449, 566
- at $T < 0.1$ K, 59, 168, 182
Inhomogeneous broadening, 320, 322
Interacting nonadiabatic regime, 284
Interaction
- energy, between two defects, 237, 238, 281

- hole in density of states, 271, 303
-- dipole gap, 271, 281, 298
- mechanism, 225, 235, 242, 297, 299
- strength, dimensionless, 238, 245, 251, 259, 275, 280, 288, 298

Interaction with conduction electrons, 154, 163, 413, 443
- adiabatic, 425
- nonadiabatic, 432, 450
- strong-coupling theory, 157
- superconducting state, 423, 427, 429
- the electron-polaron effect, 158

Interaction with phonons
- first-order Raman process, 22
- multiphonon processes, 406, 412, 415
- one-phonon, 15, 405
- two-phonon, 24

Internal friction, see also Acoustic properties
- definition, 147
- dependence on the acoustic intensity, 190
- PdSiCu, 190
- SiO_2, 565, see also Acoustic properties of

Interstitial hydrogen, 98, 389
Intersystem crossing, 319
Intrinsic DWP, 484
- distribution of barrier heights, 486
Intrinsic TLS, 338
Ioffe–Regel limit, 550
Ising model, 238, 300

KBr–KCN
- rotational echoes in, 249
KCl:Li, 108
Kondo parameter, 62
Kondo temperature, 157
Korringa-like relaxation, 155, 425
Kramers–Kronig relation, 141, 229

Landau–Zener criterion, 278
Laplace transformation, 140
$LiCl \cdot 7H_2O$
- heat release, 38
Line broadening
- origin of the inhomogeneous, 322
Liouville operator, 71, 101, 113, 125
Local modes, 527
Longitudinal rate
- for strong coupling, 116
- for weak coupling, 131
Lorentzian diffusion kernel, 355

Low-frequency vibrations, 464
Low-temperature anomalies, 9, 528

Mass density, 108
Maximum cutoff radius, 308
Maximum tunnel energy, 121
Measurement techniques
- acoustic
-- equilibrium, 170, 257–258
-- magnetic drive, 257
- calorimetric measurements, 37
- dielectric
-- equilibrium, 255–257
- heat release, 35
- hole burning, 328
- nonequilibrium
-- acoustic, 265
-- acoustic hole burning, 226
-- dielectric, 263, 265
-- jumps, 266
-- optical, 370
-- optical hole burning, 371
-- strain perturbation, 264
-- sweeps, 266, 267, 269
- quasi-static measurements, 35
Melt-spun samples, 520
Memory matrix, 113
Mesoscopic wires, 57, 103
Metal-H interaction, 413, 454
Metallic glass, 470, see Amorphous metals
Metallic materials, 91
Mode-coupling approximation, 133
- validity for the spin-phonon model, 138
Model glass
- LJ glass, 471
- metallic glass, 469
- one-dimensional, 504
- selenium, 469
- soft spheres, 469
Molecular dynamics, 467, 482
Mori's reduction method, 111
Mylar
- dielectric response, 270

Neutron scattering, 98, 521, 535
Neutron spectroscopy, 406
- inelastic neutron scattering, 407, 409, 430, 449
- quasi-elastic neutron scattering, 408, 432, 449
NMR experiments on hydrogenated Nb and Ta, 434

Noncrossing approximation, 137
Nonequilibrium data
- acoustic response, 275, 290
-- bias dependence, 276
-- dissipation, 276
- dielectric response
-- T-dependence of the frequency dependence, 271, 274, 289
-- bias dependence, 268, 270
-- dissipation, 272
-- field cooling, 270
-- frequency dependence, 267, 286
-- hysteresis, 292, 293
-- strain perturbation, 272, 274, 290
-- temperature dependence, 267, 286, 290
Nonequilibrium dielectric response, 285
Noninteracting blip approximation, 73, 77, 105
Nonlinear effects, 161
Nonlogarithmic diffusional hole broadening, 365

Ohmic damping, 61, 99
Ohmic dissipation, 62
Optical density, 329
Optical linewidths, 464
Overdamped motion, 83, 85, 94, 118

Pairs
- coupling between resonant, 247
- excitation density, 246
- of TS's, 243
Paraelastic polarization, 399, 413, 437
Pauli matrices, 60
$Pb_{0.915}La_{0.085} \cdot Zr_{0.65}Ti_{0.35}$ (PLZT)
- heat release, 43
$Pd_{30}Zr_{70}$
- acoustic properties, 191–194
PdSiCu, 122
- acoustic properties, 185–190
Pentacene, 349
Persistent hole, 324, 328
- nonphotochemical hole burning, 338, 345–347, 349
- photochemical hole burning, 338
Perturbation series
- for weak coupling, 126
Phase relaxation, 129
Phase shift, 85
Phonon band, 319
Phonon coupling parameter, 64
Phonon dressing, 69, 72, 86, 108, 132

Phonon heat bath, 63
Phonon mean free path, 297, 560, 564
Phonon scattering, 463
Phonon-assisted tunneling, 86, 107
Phonons, 463
PMMA
- electric dipole moment of TS's in, 386
- hole burning on, 363
- internal friction, 184
- nonuniversal behavior, 369
- soft-potential model parameters, 563
- spectral hole broadening, 371
Polaron effect, 86
Polaron transformation, 68
Pole approximation, 77, 93, 115, 130
Polybutadiene
- thermal conductivity, 562
Polycrystalline superconductors, 200
- strain-dependent effects, 205
Polymers, 518
- acoustic properties, 183
- heat release, 41
Polymethylmethacrylate (PMMA)
- heat release, 42
Polystyrene
- electric dipole moment of TS's in, 386
- heat release, 42
- internal friction, 184
- thermal conductivity, 562
Population difference $\hat{N}$, 12
Porphyrins, 333, 339
Pressure, 505, 522
Proton tautomerization, 339, 340
Pseudospins, 60
Pt
- internal friction, 175, 217
- sound velocity, 210, 216

Quantum diffusion
- of H in Nb, 98
- theories of, 411–413
Quantum efficiency, 331
Quartic potential, 541
Quartz
- neutron-irradiated, 199, 521
Quasi-harmonic approximation
- breakdown, 548
Quasi-homogeneous width, 325, 327

Rainbow diagrams, 137
Raman process, 24, 66, 381

Raman scattering, 535
Reduced propagator
- for weak coupling, 125
Relaxation
- crossover from phonons to interaction-dominated, 252–253
- interaction-driven, 242, 249, 250, 258, 295
-- effect on dissipation, 254
-- nonresonant TS, 252, 253
- phonon-driven, 229, 241, 252, 253, 258
Relaxation mechanism
- absorption, 149
- derivation of the dynamic susceptibility, 402
- sound dispersion, 149
Relaxation pole, 119
Relaxation process, 147
- anelastic, 402
- electron-assisted, 24
- influence on attenuation, 149, 402
- influence on internal friction, 151, 402
- influence on sound dispersion, 149, 151, 402
- thermally activated, 25
Relaxation rate, 402, 405, 411, 432
- direct process, 14, 15
- electron-assisted, 25, 425
- Gorsky effect, 413
- H in metals, 390
- H in semiconductors, 456
- minimum $\tau_{p,\min}$, 15
- one-phonon, 405, 447
- Raman process, 24
- spin-lattice (NMR), 410
- thermally activated, 25
Relaxation strength, 398, 405
- of a TLS, 405
Relaxation to conduction electrons
- electron-polaron effect, 158
- relaxation rate, 155
- relaxation rate in the superconducting state, 156, 427
- standard theory, 155
- strong-coupling theory, 157, 432
Renormalization group, 299
Renormalized tunnel frequency, 130, 136
Resonant mechanism
- absorption, 148
Resonant pair of TS's, 245

Resonant process, 147
- diaelastic response, 402
- influence on ultrasonic attenuation, 148
- internal friction, 152
- sound dispersion, 152

Saddle-Point Integration, 80
Sample thermalization, 176
Saturation, 327, 333–335
Screening effect, 70, 94
Selenium, 484
- specific heat, 559
- thermal conductivity, 562
- vibrational density of states, 555
Self-energy
- blip expansion, 72, 92
- weak-phonon coupling, 127
Self-heating effects, 176
Self-trapping, 412, 454
Semiconductors
- doped, 455
Shpol'skii systems, 320, 341
Silica, 10, 21, 120, 466, 471, 493, 501, 507, 513, 516, 521, 522, see also Acoustic properties of SiO_2
- dielectric loss, 259
- dielectric response to a dc electric bias, 267
- dielectric susceptibility, 224
- heat release, 40
- internal friction, 181, 565
- neutron scattering in, 535
- Raman scattering in, 535
- sound velocity, 179
- specific heat, 531, 557
- thermal conductivity, 532
- two-pulse echoes, 119
- vibrational density of states, 538, 554
Single-molecule spectroscopy, 332
Soft modes, 465, 481, 552, 556
Soft sphere glass, 484
Soft vibrational modes, 530
Soft-potential model, 27, 296, 464, 493, 503, 527, 541, 566
- assumptions, 544
- density of states $P(E, u)$, 551
- density of states $P(E, V_B)$, 551
- fourth assumption, 549
- quasi-harmonic modes
-- contribution of, 552
-- density of states, 553
- table of parameters for different glasses, 563

Solid solutions of H, 390
Sound absorption
- at $T \gg 1$ K, 516
Sound attenuation, 404, see also Acoustic properties
- definition, 147
Sound attenuation maximum at T_s
- table, 517
Sound propagation, see also Acoustic properties
- above 5 K, 122
Sound velocity, 108, 120, 495, see also Acoustic properties
- table, 514
Specific heat, 195, 418, 419, 463, 464, 477, 519, 530, 555
- from quasi-harmonic soft vibrations, 557
- linear term, 19
- maximum in C_p/T^3 at $T_{\max}$
-- table, 558
- maximum in C_p/T^3 at $T_{\max}$, 559
- of a two-level system, 18, 395
- of TLS's within the SPM, 556
- Suprasil I, 21
- Suprasil W, 21, 23
- time dependence, 20, 30
- within the renormalization group theory, 314
Spectral diffusion, 225–228, 234, 239–241, 253, 295, 326, 352–354
- electric field-induced, 382
- equilibrium TS dynamics, 355, 360, 362
- nonequilibrium TS dynamics, 355, 370, 372, 374
- thermal cycling, 378, 381
- thermal relaxation, 370, 373
Spectral function $J(\omega)$, 61
Spin glasses, 271, 277
Spin-boson model, 60
Spin-phonon coupling, 111
Sputtered samples, 520
Standard tunneling model, 14
- equilibrium response, 224, 233, 234
- nonequilibrium response, 278
-- adiabatic perturbation, 279, 286
-- dissipation, 279
-- nonadiabatic perturbation, 279
- postulates, 15, 279, 296
- predictions for the acoustic properties, 154, 156

- relaxation to conduction electrons, 155
- relaxation to phonons, 150
Strain u, 14
Strain application technique, 265
Strain field, 463, 494
Stycast
- heat release, 41
Sudden approximation, 278

Temperature slopes
- definition, 233
Thermal conductivity, 195, 196, 239, 298, 463, 464, 478, 531, 559, 562
- plateau, 531, 561, 564
Thermal cycling, 341, 378
Thermal resistance R_{th}, 35
Thermal treatment
- influence on amorphous metals, 194
- influence on polycrystalline metals, 215
Thermal TS's, 240, 244, 251
Thermally activated process, 31, 45, 166
Thermally activated rate, 96
Three-pulse photon echoes, 335
Transmission detection, 330, 332
Transverse rate
- for strong coupling, 84, 116
- for weak coupling, 130
- from NIBA, 83
- mode-coupling approximation, 135
Triplet state, 318, 327
TS electric dipole moment, 382, 385, see also Dielectric properties
TS–TS interaction, 367, 369, 370, see also Interaction
TS-phonon coupling, 14, 228–229, 235
TS-TS interaction
- influence on the acoustic properties of thin films, 197
Tunnel splitting Δ_0, 11
Tunneling model, 150, 461, 544
- introduction of the, 10
- standard predictions, 154, 156, 228, see also Standard tunneling model
Tunneling model (beyond), 527, 566
Tunneling parameter λ, 11
Tunneling parameters, 463, 476, 509
Tunneling system, 10, 15
- effective mass, 461, 478
Tunneling systems
- length scale, 523

- nature of, 503
- origin, 219
- within the SPM, 551

Two-level system, 15
- Hamiltonian, 10

Two-phonon process, 66, 118
Two-pulse echoes, 119
Two-state dynamics
- for weak coupling, 125

Two-state polaron, 58
Two-state pseudospin system, 60
Two-state system, 60

Underdamped motion, 84, 93, 117
Universal $1/f$-noise, 296
Universality, 506, 517
- of the low-temperature properties, 295, 508
- quantitative, 517
- significance of the $1/R^3$-interactions, 297

Vespel
- heat release, 41

Vibrating reed technique, 170
Vibrating wire technique, 170
Virtual phonon exchange, 298
Vitreous silica, *see* Silica, also Acoustic properties of SiO_2

WKB-approximation, 11, 462, 518
Wormlike soft modes, 507

Zero-phonon line, 319–321, 325, 326
Zero-point energy, 10, 166, 182
Zn-TBP, 334